A COURSE IN
ORDINARY DIFFERENTIAL EQUATIONS

SECOND EDITION

STEPHEN A. WIRKUS
ARIZONA STATE UNIVERSITY
GLENDALE, USA

RANDALL J. SWIFT
CALIFORNIA STATE POLYTECHNIC UNIVERSITY
POMONA, USA

CRC Press
Taylor & Francis Group
Boca Raton London New York

CRC Press is an imprint of the
Taylor & Francis Group, an **informa** business

A CHAPMAN & HALL BOOK

CRC Press
Taylor & Francis Group
6000 Broken Sound Parkway NW, Suite 300
Boca Raton, FL 33487-2742

First issued in hardback 2019

© 2015 by Taylor & Francis Group, LLC
CRC Press is an imprint of Taylor & Francis Group, an Informa business

No claim to original U.S. Government works

ISBN-13: 978-1-4665-0908-5 (hbk)

To our families

Erika Tatiana, Alan, Abdi, and Avani

and

Kelly, Kaelin, Robyn, Erin, and Ryley

for bringing us more joy than math and showing us the true concept

and meaning of

infinity

with their tireless

patience, love, and understanding.

Contents

Stephen A. Wirkus completed his Ph.D. at Cornell University under the direction of Richard Rand. He began guiding undergraduate research projects while in graduate school and came to Cal Poly Pomona in 2000 after being a Visiting Professor at Cornell for a year. He co-founded the Applied Mathematical Sciences Summer Institute (AMSSI), an undergraduate research program jointly hosted by Loyola Marymount University, that ran from 2005 through 2007. He came to Arizona State University in 2007 as a tenured Associate Professor and won the 2013 Professor of the Year Award at ASU as well as the 2011 NSF AGEP Mentor of the Year award. He was a Visiting MLK Professor at the Massachusetts Institute of Technology in 2013-2014. He has guided over 80 undergraduate students in research and has served as Chair for 4 M.S. students, and 2 Ph.D. students. He has over 30 publications and technical reports with over 40 students and has received grants from the NSF and NSA for guiding undergraduate research.

Randall J. Swift completed his Ph.D. at the University of California, Riverside under the direction of M. M. Rao. He began his career at Western Kentucky University and taught there for nearly a decade before moving to Cal Poly Pomona in 2001 as a tenured Associate Professor. He is active in research and teaching, having authored more than 80 journal articles, three research monographs and three textbooks in addition to serving as Chair for 25 M.S. students. Now a Professor, he received the 2011-12 Ralph W. Ames Distinguished Research Award from the College of Science at Cal Poly Pomona. The award honors Swift for his outstanding research in both pure and applied mathematics, and his contributions to the mathematics field as a speaker, journal editor, and principal investigator on numerous grants. He was also a Visiting Professor in 2007-2008 at The Australian National University in Canberra Australia as well as having taught at the Claremont Colleges.

This book is based on lectures given by the first author at Cal Poly Pomona, Arizona State University (ASU), and the Massachusetts Institute of Technology (MIT), and by the second author at Western Kentucky University (WKU) and California State Polytechnic University–Pomona (Cal Poly Pomona). The text can be used for a traditional one-semester sophomore-level course in ordinary differential equations (such as WKU's MATH 331). However, there is ample material for a two-quarter sequence (such as Cal Poly Pomona's MAT 216-431), as well as sufficient linear algebra in the text so that it can be used for a one-quarter course that combines differential equations and linear algebra (such as Cal Poly Pomona Math 224), or a one-semester course in differential equations that brings in linear algebra in a significant way (such as ASU's MAT 275 or MIT's 18.03 without the PDEs). Most significantly, computer labs are given in MATLAB®,[1] Maple™, and Mathematica at the end of each chapter so the book may be used for a course to introduce and equip the student with a knowledge of the given software (such as ASU's MAT 275). Near the end of this Preface, we give some sample course outlines that will help show the independence of various sections and chapters. The focus of the text is on applications and methods of solution, both analytical and numerical, with emphasis on methods used in the typical engineering, physics, or mathematics student's field of study. We have tried to provide sufficient problems of a mathematical nature at the end of each section so that even the pure math major will be sufficiently challenged.

Key Features

This second edition of the book keeps many of the key features from the first edition:

- MATLAB, Maple, and Mathematica are incorporated at the end of each chapter, helping students with pages of tedious algebra and many of the differential equations topics; the goal of the software is still to show

---

[1]MATLAB is a registered trademark of The MathWorks, Inc. For product information, please contact:
The Mathworks, Inc.
3 Apple Hill Drive
Natick, MA, 01760-2098 USA
Tel: 508-647-7000
Fax: 508-647-7001
E-mail: info@mathworks.com
Web: www.mathworks.com

students how to make informed use of the relevant software in the field; all three software packages have parallel code and exercises;

- There are numerous problems of varying difficulty for both the applied and pure math major, as well as problems for the nonmathematician (engineers, etc.);
- An appendix that gives the reader a "crash course" in the three software packages; no prior knowledge is assumed;
- Answers to most of the odd problems in the back of the book;
- Chapter reviews at the end of each chapter to help the students review;
- Projects at the end of each chapter that go into detail about certain topics and sometimes introduce new topics that the students are now ready to see;
- An appendix on linear algebra to supplement the treatment within the text, should it be appropriate for the reader/course;
- A full solutions manual for the qualified instructor.

It also incorporates new features, many of which have been suggested by professors and students who have taught/learned from the first edition:

- The computer codes are moved to the end of each chapter as **Computer Labs** to facilitate reading of the book by students and professors who either choose not to use the technology or who do not have access to it immediately;
- The latest software versions are used; significant changes have occurred in certain aspects of MATLAB, Maple, and Mathematica since the first edition in 2006, and the relevant changes are incorporated;
- Much of the linear algebra discussion has been moved to Chapter 5 (from Chapter 3), which deals with linear systems;
- Sections have been added on complex variables (Chapter 3), the exponential response formula for solving nonhomogeneous equations (Chapter 4), forced vibrations (Chapter 4) as well as a subsection on nondimensionalization (Chapter 2), and a combining of the sections on Euler and Runge-Kutta methods (Chapter 2);
- Many rewritten sections highlight applications and modeling within many fields;
- Exercises flow from easiest to hardest;
- Color graphs to help the reader better understand crucial concepts in ordinary differential equations;
- Updated and extended projects at the end of each chapter to reflect changes within the chapters.

Approaches to Teaching Ordinary Differential Equations

The second edition of this book has evolved with our understanding of how to teach the material in the best possible way. Some notable examples from the above list:

1. The structure of the course in covering linear systems in their entirety before covering applications to nonlinear systems (phase plane, etc.) was a direct result of numerous conversations with MIT's Professor Haynes Miller (who frequently teaches MIT's 18.03) as was the incorporation of the new sections on essential topics from complex variables, exponential response and complex replacement (developed by Haynes) for solving nonhomogeneous differential equations, and the s-domain and poles as an important use of Laplace transforms by engineers.

2. Combining the computer codes into Computer Labs at the end of each section rather than having snippets of code embedded throughout the text was a direct result of a switch in ASU's method of teaching this course. Setting aside six class periods for such labs is the way differential equations is now taught at ASU.

3. The presentation of essential linear algebra topics to aid in the understanding of differential equations was helped by discussions with MIT's Professor Gil Strang as well as seeing some of his lectures firsthand.

Most differential equations we have encountered in practice have needed analytical approximations or numerical approximations to gain insight into their behavior. We don't feel that students use technology wisely if they simply ask the computer to solve a given problem. We thus focus on what we consider to be the basics necessary for adequately preparing a student for study in her or his respective fields, including mathematics. We present the syntax from MATLAB, Maple, and Mathematica in Computer Labs at the end of each chapter. We feel that this provides the readers a better understanding of the theory and allows them to gain more insight into real-world problems they are likely to encounter. The vast majority of our students also have *no previous experience with* MATLAB, Maple, or Mathematica and we start from the basics and teach informed use of the relevant mathematical software. The student whom we "typically encounter" has had one year of calculus and is usually a major in a field other than pure mathematics.

Our book is traditional in its approach and coverage of basic topics in ordinary differential equations. However, we cover a number of "modern" topics such as direction fields, phase lines, the Runge-Kutta method, and nondimensionalization in Chapter 2 and epidemiological and ecological models in Chapter 6. As mentioned earlier, we also bring elements from linear algebra, such as eigenvectors, bases, and transformations, in order to best equip the reader of the book for a solid understanding of the material. Besides the Computer Labs there are also Projects at the end of each chapter that give useful insight into past and future topics covered in the book. The topics covered in these projects include a mix of traditional, modeling, numerical, and linear algebra aspects of ordinary differential equations. It is the intent that students who study this book and work *most* of the problems contained in these pages will be very prepared to continue their studies in engineering and mathematics.

While we could not begin to prescribe how this book may best be used for each school, we include some *possible* sections covered for various course outlines. We stress that if you intend to incorporate MATLAB, Maple, or Mathematica into your course, it is crucial to assign Exercises 1-4 (plus a few others) from Appendix A and the Chapter 1 Computer Lab early in the course. Appendix A only requires a knowledge of college algebra and some calculus (Taylor series) while Chapter 1 Computer Lab requires a knowledge of calculus as it is applied to differential equations. Thus both can be assigned within the first 2 weeks of the course (and likely together).

Traditional semester ODE course:

Chap. 1	Chap. 2	Chap. 3	Chap. 4	Chap. 5	Chap. 7	Chap. 8
1.1-1.6	2.1-2.2	3.1-3.3	4.1, 4.3	5.1	7.1-7.4	8.1-8.5
		3.5-3.6	4.5-4.6	5.4-5.8		

Semester ODE course with modeling or application emphasis:

Chap. 1	Chap. 2	Chap. 3	Chap. 4	Chap. 5	Chap. 6	Chap. 7
1.1-1.4	2.1-2.6	3.1-3.2	4.1-4.2	5.1, 5.4	6.1-6.5	7.1-7.5
		3.4-3.7	4.4-4.7	5.5, 5.7		

Semester ODE course with linear algebra emphasis and no separate computer labs:

Ch. 1	Ch. 2	Ch. 3	Ch. 4	Ch. 5	Ch. 6	Ch. 7	App. B
1.1-1.4	2.1-2.2	3.1-3.2	4.1-4.2	5.1-5.5	6.1	7.1-7.7	B.1-B.4
	2.5	3.4-3.7	4.4, 4.7	5.7-5.8			

Semester ODE course with linear algebra emphasis and 6 computer labs:

Ch. 1	Ch. 2	Ch. 3	Ch. 4	Ch. 5	Ch. 7	Comp. Labs
1.1-1.4	2.1-2.2	3.1-3.2	4.1-4.2	5.1-5.5	7.1-7.6	A& 1, 2,
	2.5	3.4-3.7	4.4, 4.7			3, 4, 5&B, 7

Quarter ODE course with linear algebra emphasis:

Ch. 1	Ch. 2	Ch. 3	Ch. 4	Ch. 5	App. B
1.1-1.4	2.1-2.2	3.1-3.2	4.1-4.2	5.1-5.5	B.1-B.4
	2.5	3.4-3.7	4.7		

Acknowledgments

Students, with their questions both in-class and during office hours, helped shaped this second edition as did those professors who used the first edition and/or provided constructive feedback to us, including Erika Camacho, Andrew Knyazev, Luis Melara, Jenny Switkes, Steven Weintraub, and many others. Various chapters were read by Alexandra Churikova, Maytee Cruz-Aponte, Clay Goggil, and Christine Sowa, and their feedback has been of great help. Mike Pappas, in particular, was a big help in proofreading near-final drafts of several chapters. Valerie Cheathon provided a valuable check of all the codes as did Joshua Grosso (MATLAB) and Alan Wirkus-Camacho (Maple and Mathematica). Scott Wilde, again, provided invaluable help in revising the solutions manual.

As texts based upon lecture notes seemingly develop, many of the examples, exercises, and projects have been collected over many years for various courses taught by both authors. Some were taken from others' textbooks and papers. We have tried to give proper credit throughout this text; however, it was not always possible to properly acknowledge the original sources. It is our hope that we repay this explicit debt to earlier writers by contributing our (and their) ideas to further student understanding of differential equations.

We particularly wish to thank our production coordinator, Jessica Vakili, as well as Michele Dimont, Amy Blalock, Hayley Ruggieri, and Sherry Thomas. Bob Stern and Bob Ross, our editors at Chapman & Hall/CRC Press, both deserve a big thanks for believing in this project and for helpful guidance, advice and patience. We sincerely thank all these individuals; without their assistance this text would not have succeeded.

URL for typos and errata:
        http://www.public.asu.edu/~swirkus/ACourseInODEs

Finally, we would appreciate any comments that you might have regarding this book.

Stephen A. Wirkus (e-mail: swirkus@asu.edu*)*
Randall J. Swift (e-mail: rjswift@csupomona.edu*)*

From the first edition:

We owe a very special thanks to Erika Camacho (Arizona State University) for her help in writing the MATLAB and Maple code for this book and for detailed suggestions on numerous sections. John Fay and Gary Etgen reviewed earlier drafts of this text and provided helpful feedback. Scott Wilde provided valuable assistance in writing and preparing the solutions manual for the book. We owe a big thanks to our former students David Monarres, for help in preparing portions of this book, and Walter Sosa and Moore Chung, for their help in preparing solutions. We would also like to acknowledge our Cal Poly Pomona colleagues Michael Green, Jack Hofer, Tracy McDonald, Jim McKinney, Siew-Ching Pye, Dick Robertson, Paul Salomaa, Jenny Switkes, Karen Vaughn, and Mason Porter (Caltech/Oxford) for their willingness to use draft versions of this text in their courses and their important suggestions, which improved the overall readability of the text. The faculty and students of AMSSI and MTBI also deserve a special thanks for comments on early drafts of the computer code. Mary Jane Hill assisted us with certain aspects of the text and helped in typesetting some of the chapters of the initial drafts of the book; her effort is greatly appreciated. The production and support staff at Chapman & Hall/CRC Press have been very helpful. We particularly wish to thank our project coordinator Theresa Del Forn and project editor Prudence Board. Our editor Bob Stern deserves a special thanks for believing in this project and for his guidance, advice, and patience. We sincerely thank all these individuals; without their assistance this text would not have succeeded.

A few remarks for students and professors:

This book will succeed if any fears and reservations about learning one of the three computer algebra systems used in this book are put aside. Computers are not here to supplant us, but rather they are here to help illustrate and illuminate concepts and insights that we have. Nothing is foolproof and we stress the importance of *informed use of the relevant mathematical software.* Numerical answers, although quite accurate most of the time, should always be examined carefully because computers are as smart as the programmer allows them to be. There should never be a blind trust in an answer.

It is essential that the technology that you choose—MATLAB, Maple, or Mathematica—be introduced early in the class, just as it is introduced early in the book. While certain mathematical software packages may be better suited for studying differential equations, none have the versatility that the above three programs have to give insight into other areas of mathematics. The two keys to learning the programs are (1) learning the syntax and (2) learning to use the help menus to figure out some of the commands. Setting aside one class, for example, to give a brief tutorial on one of these software packages in the computer lab is a very worthwhile investment. It is by no means necessary and the typical student will be able to learn the material on his/her own by carefully following Appendix A. For reinforcement, it is crucial to include at least one or two technology problems with each homework assignment. The conscientious student will be well prepared to use the same software package in any upper division course in *any* branch of the mathematical sciences and its applications.

It is not necessary to bring computer demonstrations into the classroom. Both authors have taught their courses successfully without classroom demonstrations; handouts sometimes are useful, especially from the appendices. The students, for better or worse, are generally far less afraid of technology than one might expect. If students are sent to the computer lab with an assignment to do and aided with Appendix A, the vast majority will come back with satisfactory answers. Yes, you may bang your head against your desk in frustration at times, but just ask the person next to you for help and also seek the help menus and you will be able to learn MATLAB, Maple, and Mathematica quite well.

Chapter 1

Traditional First-Order Differential
Equations

A Very Brief History

The study of Differential Equations began very soon after the invention of Differential and Integral Calculus, to which it forms a natural sequel. In 1676 Newton solved a differential equation by the use of an infinite series, only 11 years after his discovery of the *fluxional* form of differential calculus in 1665. These results were not published until 1693, the same year in which a differential equation occurred for the first time in the work of Leibniz (whose account of the differential calculus was published in 1684).

In the next few years progress was rapid. In 1694–1697 John Bernoulli explained the method of "Separating the Variables," and he showed how to reduce a homogeneous differential equation of the first order to one in which the variables were separable. He applied these methods to problems on orthogonal trajectories. He and his brother Jacob (after whom the "Bernoulli Equation" is named; see Section 1.6.1) succeeded in reducing a large number of differential equations to forms they could solve. Integrating Factors were probably discovered by Euler (1734) and (independently of him) by Fontaine and Clairaut, though some attribute them to Leibniz. Singular Solutions, noticed by Leibniz (1694) and Brook Taylor (1715), are generally associated with the name of Clairaut (1734). The geometrical interpretation was given by Lagrange in 1774, but the theory in its present form was not given until much later by Cayley (1872) and M.J.M. Hill (1888).

Today, differential equations are used in many different fields. They can often accurately capture the behavior of continuous models or a large number of discrete objects where the current state of the system determines the future behavior of the system. Such models are called **deterministic** (as opposed to **stochastic** or **random**). The study of **nonlinear** differential equations is still a very active area of research. Although this text will consider some nonlinear differential equations, here the focus will be on the linear case. We will begin with some basic terminology.

1.1 Introduction to First-Order Equations

Order, Linear, Nonlinear

We begin our study of differential equations by explaining what a differential equation is. From our experience in calculus, we are familiar with some differential equations. For example, suppose that the acceleration of a falling object is $a(t) = -32$, measured in ft/sec^2. Using the fact that the derivative of the velocity function $v(t)$ (measured in ft/sec) is the acceleration function $a(t)$, we can solve the equation

$$v'(t) = a(t) \quad \text{or} \quad \frac{dv}{dt} = a(t) = -32.$$

Many different types of differential equations can arise in the study of familiar phenomena in subjects ranging from physics to biology to economics to chemistry. We give examples from various fields throughout the text and engage the reader with many such applications.

It is clearly necessary (and expedient) to study, independently, more restricted classes of these equations. The most obvious classification is based on the nature of the derivative(s) in the equation. A differential equation involving derivatives of a function of one variable (ordinary derivatives) is called an **ordinary differential equation**, whereas one containing partial derivatives of a function of more than one independent variable is called a **partial differential equation**. In this text, we will focus on ordinary differential equations.

The **order** of a differential equation is defined as the order of the highest derivative appearing in the equation.

Example 1 The following are examples of differential equations with indicated orders:
 (a) $dy/dx = ay$ (First-Order)
 (b) $x''(t) - 3x'(t) + x(t) = \cos t$ (second order)
 (c) $(y^{(4)})^{3/5} - 2y'' = \cos x$ (fourth order)
where the superscript $^{(4)}$ in (c) represents the fourth derivative.

Our focus will be on **linear** differential equations, which are those equations that have an unknown function, say y, and each of its higher derivatives appearing in linear functions. That is, we do *not* see them as $y^2, yy', \sin y$, or $(y^{(4)})^{3/5}$.[1] More precisely, a linear differential equation is one in which the dependent variable and its derivatives appear in additive combinations of their first powers. Equations where one or more of y and its derivatives appear in nonlinear functions are called **nonlinear** differential equations. In the above example, only (c) is a nonlinear differential equation.

Example 2 Classify the equations as linear or nonlinear.

---

[1] Most of the equations we consider will involve an unknown function y that depends on x. Two other common variables used are (i) the unknown function y that depends on t and (ii) the unknown function x that depends on t, the latter being used in Example 1b.

(a) $y'' + 3y' - x^2 y = \cos x$
(b) $y'' - 3y' + y^2 = 0$
(c) $y^{(3)} + yy' + \sin y = x^2$

Solution

The first of these equations is linear as it consists of an additive combination of y, y', and y'', each of which is raised to the first power. In contrast to this, the second equation is nonlinear because of the y^2 term. The last equation is nonlinear both because of the yy' term and the $\sin y$ term—either of these terms by itself would have made the equation nonlinear. Our study of nonlinear differential equations will focus on techniques for specific equations or on understanding the qualitative behavior of a nonlinear differential equation, since general techniques of solution are rarely applicable.

Much of this book is concerned with the solutions of linear differential equations. Thus we need to explain what we mean by a solution. First we note that any nth-order differential equation can be written in the form

$$F(x, y, y', ..., y^{(n)}) = 0, \tag{1.1}$$

where n is a positive integer. For example, $y' = x^2 + y^2$ can be written as

$$y' - x^2 - y^2 = 0.$$

Here $F(x, y, y') = y' - x^2 - y^2$. The second-order equation $y'' - 3x^2 y' + 5y = \sin x$ can be written as

$$y'' - 3x^2 y' + 5y - \sin x = 0$$

and we see that $F(x, y, y', y'') = y'' - 3x^2 y' + 5y - \sin x$.

Definition 1.1.1

A *solution* to an nth-order differential equation is a function that is n times differentiable and that satisfies the differential equation. Symbolically, this means that a solution of differential equation (1.1) is a function $y(x)$ whose derivatives $y'(x), y''(x), ..., y^{(n)}(x)$ exist and that satisfies the equation

$$F(x, y(x), y'(x), ..., y^{(n)}(x)) = 0$$

for all values of the independent variable x in some interval (a, b) where

$$F(x, y(x), y'(x), ..., y^{(n)}(x))$$

is defined. (Note that the solution to a differential equation does not contain any derivatives, although the derivatives of this solution exist.) The interval (a, b) may be infinite; that is, $a = -\infty$, or $b = \infty$, or both.

Example 3 The function $y(x) = 2e^{3x}$ is a solution of the differential equation

$$\frac{dy}{dx} = 3y,$$

for $x \in (-\infty, \infty)$ because it satisfies the differential equation by giving an identity:

$$\frac{dy}{dx} = 2\frac{de^{3x}}{dx} = 6e^{3x} = 3y.$$

Initial-Value vs. Boundary-Value Problems

We will soon see that solving a general differential equation gives rise to a solution that has constants. These constants can be eliminated by specifying the initial state of the system or conditions that the solution must satisfy on its domain of definition or "boundary." An example of the first situation is specifying the position and velocity of a mass on a spring. An example of the second is a rope hanging from two supports, given the location of these two supports.

Consider a first-order differential equation

$$\frac{dy}{dx} = f(x, y)$$

and suppose that the solution $y(x)$ was subject to the condition that $y(x_0) = y_0$. This is an example of an **initial-value problem**. The condition $y(x_0) = y_0$ is called an **initial condition** and x_0 is called the **initial point**. More generally, we have the following:

Definition 1.1.2

An *initial-value problem* consists of an nth-order differential equation together with n initial conditions of the form

$$y(x_0) = a_0, \quad y'(x_0) = a_1, ..., \quad y^{(n-1)}(x_0) = a_{n-1}$$

that must be satisfied by the solution of the differential equation and its derivatives at the initial point x_0.

Example 4 The following are examples of initial-value problems:
(a) $dy/dx = 2y - 3x$, $y(0) = 2$ (here $x = 0$ is the initial point)
(b) $x''(t) + 5x'(t) + \sin(tx(t)) = 0$, $x(1) = 0$, $x'(1) = 7$ (here $t = 1$ is the initial point).
(Note that the differential equation in (a) is linear, whereas the equation in (b) is nonlinear.) We define a *solution* to an nth-order initial-value problem as a function that is n times differentiable on an interval (a, b); this satisfies the given differential equation on that interval, and satisfies the n, given initial conditions with the requirement that $x_0 \in (a, b)$. As before, the interval (a, b) might be infinite.

In contrast to an initial-value problem, a **boundary-value problem** consists of a differential equation and a set of conditions *at different x-values* that the solution $y(x)$ must satisfy. Although any number of conditions (≥ 2) may be specified, usually only two are given. Rather than specifying the initial state of the system, we can think of a boundary-value problem as specifying the state of the system at two different physical locations, say $x_0 = a, x_1 = b, a \neq b$.

Example 5 The following are examples of boundary-value problems:
(a) $d^2y/dx^2 + 5xy = \cos x$, $y(0) = 0$, $y'(\pi) = 2$
(b) $dy/dx + 5xy = 0$, $y(0) = y(1) = 2$

Although a boundary-value problem may not seem too different from an initial-value problem, methods of solution are quite varied. We will focus on initial-value problems. We ask whether an initial-value problem has a unique solution. Essentially this is two questions:
1. Is there a solution to the problem?
2. If there is a solution, is it the only one?
As we see in the next two examples, the answer may be "no" to each question.

Example 6 *An initial-value problem with no solution.*
The initial-value problem

$$\left(\frac{dy}{dx}\right)^2 + y^2 + 1 = 0$$

with $y(0) = 1$ has no real-valued solutions, since the left-hand side is always positive for real-valued functions.

Example 7 *An initial-value problem with more than one solution.*

The initial-value problem

$$\frac{dy}{dx} = xy^{1/3}$$

with $y(0) = 0$ has at least two solutions in the interval $-\infty < x < \infty$. Note that the functions

$$y = 0 \text{ and } y = \frac{x^3}{3\sqrt{3}}$$

both satisfy the initial condition and the differential equation.

Two Important Models

One of the most fundamental models in biology deals with population growth and one of the most fundamental models in physics deals with a mass on a spring. In the next two examples, we examine how differential equations describe the behavior of these two phenomena.

Example 8 The change in the population of bacteria with sufficient nutrients and space to grow is known to be proportional to its current population. The differential equation can be written as

$$\frac{dP}{dt} = kP \tag{1.2}$$

where $P(t)$ is the current population of bacteria and k is a constant determined by its growth rate. We can verify that

$$P(t) = P(0)e^{kt} \tag{1.3}$$

is a solution to this differential equation. Because of the presence of the constant $P(0)$, we say that Equation (1.3) is a family of solutions parameterized by the constant $P(0)$. To verify this is indeed a solution we take the derivative to get $P(0)ke^{kt}$. Substituting this into the left side of the differential equation and the supposed solution into the right side:

$$\underbrace{\frac{dP}{dt}}_{P(0)ke^{kt}} = k \cdot \underbrace{P}_{P(0)e^{kt}}$$

we see that with a slight rearrangement of the expressions underneath, we have equality for all t. Thus (1.3) is a solution to differential equation (1.2) for all t and we see the solution describes the *exponential growth* of the population.

Example 9 In a later chapter we will learn that a mass on a spring moving along a slippery[2] surface can be described by the differential equation

$$mx'' + kx = 0$$

where $x(t)$ is the distance the spring has stretched from its resting length, k is the spring constant, and m is the mass, as shown in Figure 1.1. We can verify that $x = \cos\left(\sqrt{\frac{k}{m}}\, t\right)$ is a solution. To do so we take the second derivative to get $x'' = -\frac{k}{m}\cos\left(\sqrt{\frac{k}{m}}\, t\right)$ and substitute it into the equation along with the assumed form of x:

$$m \cdot \left[-\frac{k}{m}\cos\left(\sqrt{\frac{k}{m}}\, t\right) \right] + k \cdot \cos\left(\sqrt{\frac{k}{m}}\, t\right) = 0.$$

Simplification shows that it is indeed a solution and it holds for all t.

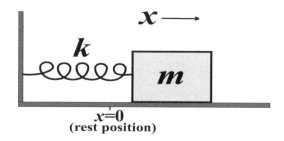

FIGURE 1.1: Model of spring system for Example 9. $x = 0$ marks the position if the spring were at its natural (unstretched) length and we will take x to the right as positive.

In the next several sections we will develop methods for finding solutions to first-order differential equations. We will then discuss existence and uniqueness of solutions in Chapter 2.

Problems

In Problems **1–12**, classify the differential equations by specifying (i) the order, (ii) whether it is linear or nonlinear, and (iii) whether it is an initial-value or boundary-value problem (where appropriate).

---

[2]Physicists use the word "slippery" to mean "ignore frictional forces."

1. $3y'' + y = \sin x$ 2. $y'' = \sin x$
3. $y'' + y' - y = 0$ 4. $y'' + 3y' = 0, \quad y(0) = 1, y'(1) = 0$
5. $y^{(3)} + (\sin x)y^{(2)} + y = x, \quad y(0) = 1, y'(0) = 0, y''(0) = 2$
6. $y'' = 0, \quad y(1) = 1, y'(1) = 2$ 7. $y' + e^x y = y^4, \quad y(0) = 0$
8. $y'' - 3yy' = x$ 9. $y'' + \sin y = 0$
10. $y'' - 4y' + 4y = 0, \quad y(0) = 1, y'(0) = 1$
11. $y'' + e^x y' + y^2 = 0, \quad y(0) = 1, y(\pi) = 0$
12. $x^2 y'' + y' + (\ln x)y = 0$

In Problems **13–24**, verify that the given function is a solution to the differential equation by substituting it into the differential equation and showing that the equation holds true. Assume the interval is $(-\infty, \infty)$ unless otherwise stated. Do NOT attempt to solve the differential equation.

13. $y(x) = 2x^3, \quad x\dfrac{dy}{dx} = 3y$ 14. $y(x) = x, \quad y'' + y = x$

15. $y = 2, \quad \dfrac{dy}{dx} = x^3(y - 2)^2$ 16. $y(x) = x^3, \quad \dfrac{dy}{dx} = 3y^{2/3}$

17. $y(x) = e^x - x, \quad \dfrac{dy}{dx} + y^2 = e^{2x} + (1 - 2x)e^x + x^2 - 1$

18. $y(x) = \sin x + 2\cos x, \quad y'' + y = 0$ 19. $y(x) = x^2 - x^{-1}, \quad x^2 y'' = 2y, \quad x \neq 0$

20. $y(x) = x + C\sin x, \quad y'' + y = x, \quad C = \text{constant}$

21. $y(x) = \dfrac{-1}{x - 3}, \quad \dfrac{dy}{dx} = y^2, \quad (-\infty, 3)$ 22. $y(x) = \dfrac{-1}{5x + 4}, \quad \dfrac{dy}{dx} = 5y^2, \quad (-4/5, \infty)$

23. $y_1(x) = e^x$ and $y_2(x) = e^{2x}, \quad y'' - 3y' + 2y = 0$
24. $y_1(x) = e^x$ and $y_2(x) = xe^x, \quad y'' - 2y' + y = 0$

In Problems **25–28**, detetermine which of the functions solve the given differential equation.

25. $y'' + 6y' + 9y = 0$: (a) e^x, (b) e^{-3x}, (c) xe^{-3x}, (d) $4e^{3x}$, (e) $2e^{-3x} + xe^{-3x}$
26. $y'' + 9y = 0$: (a) $\sin 3x$, (b) $\sin x$, (c) $\cos 3x$, (d) e^{3x}, (e) x^3
27. $y'' - 7y' + 12y = 0$: (a) e^{2x}, (b) e^{3x}, (c) e^{4x}, (d) e^{5x}, (e) $e^{3x} + 2e^{4x}$
28. $y'' + 4y' + 5y = 0$: (a) e^{-2x}, (b) $e^{-2x}\sin 2x$, (c) $e^{-2x}\cos 2x$, (d) $\cos 2x$

In Problems **29–32**, find values of r for which the given function solves the differential equation on $(-\infty, \infty)$.

29. $y(x) = e^{rx}, \quad y'' + 3y' + 2y = 0$ 30. $y(x) = e^{rx}, \quad y'' + 3y' - 4y = 0$
31. $y(x) = xe^{rx}, \quad y'' + 6y' + 9y = 0$ 32. $y(x) = xe^{rx}, \quad y'' + 4y' + 4y = 0$

1.2   Separable Differential Equations

We will now introduce the simplest first-order differential equations. Although these are the simplest class of differential equations we will encounter, they appear in numerous applications and aspects of subsequent theory. We make the following definition:

Definition 1.2.1

A first-order differential equation that can be written in the form

$$g(y)\, y' = f(x) \quad \text{or} \quad g(y)\, dy = f(x)\, dx,$$

where $y = y(x)$, is called a separable differential equation.

Separable differential equations are solved by collecting all the terms involving the dependent variable y on one side of the equation and all the terms involving the independent variable x on the other side. Once this is completed (it may require some algebra), both sides of the resulting equations are integrated. That is, the equation

$$g(y)\, y' = f(x)$$

can be written in "differential form"

$$g(y)\, \frac{dy}{dx} = f(x)$$

so that treating dy/dx as a fraction, we have

$$g(y)\, dy = f(x)\, dx.$$

Here the variables are separated, so that integrating both sides gives

$$\int g(y)\, dy = \int f(x)\, dx. \tag{1.4}$$

The Method of Separation of Variables, which we just applied to (1.4), is the name given to the method we use to solve Separable Equations—it is one of the simplest and most useful methods for solving differential equations. (Incidentally, it is an important technique for solving certain classes of partial differential equations, too.)

Sometimes we will be able to solve (1.4) for y. When we can do so, we will say we can express the **explicit solution** and will write $y = h(x)$. Other times, we will not be able to solve (1.4) or it will not be worth our time and efforts to do so. In these situations, we say that we are giving the **implicit solution** with (1.4). When our solution can be written explicitly, it will be easy to plot solutions in the x-y plane, by hand or with the computer; however, when the solution is implicit, plotting solutions by hand is challenging at best. The various computer programs, discussed in Appendix A and the end of each chapter, will allow us to view plots in the x-y plane without much additional work. We now consider a number of examples.

Example 1 Solve $y' = ky$ where k is a constant.

Solution
Writing y' as $\frac{dy}{dx}$ gives

$$\frac{dy}{dx} = ky.$$

Treating $\frac{dy}{dx}$ as a "fraction" and rearranging terms gives

$$\frac{dy}{y} = k\,dx.$$

This step will only be valid if $y \neq 0$. We note that $y = 0$ is also a solution to the original differential equation. Integrating gives

$$\int \frac{dy}{y} = \int k\,dx,$$

which is

$$\ln|y| = kx + C_1, \Longrightarrow |y| = e^{kx+C_1}.$$

This gives

$$y = \pm e^{kx}e^{C_1}.$$

Now e^{C_1} is a positive constant, so that we may let $C = \pm e^{C_1}$. In the above process, we encountered the constant solution $y = 0$, which also gives us the possibility that $C = 0$. Thus, we have

$$y = Ce^{kx} \tag{1.5}$$

as our solution, where $x \in (-\infty, \infty)$ and C is any real constant. We say that (1.5) defines a **one-parameter family of solutions** of $y' = ky$. It is also important to remember the "trick" used above for getting rid of the absolute values—it will come up quite often in practice! We will consider a few more examples with similar standard "tricks."

Example 2 Solve $\dfrac{dx}{dt} = e^{t-x}$, $x(0) = \ln 2$, for $x(t)$.

Solution
Separating the variables gives

$$\frac{dx}{dt} = e^t e^{-x}$$

and thus

$$e^x\,dx = e^t\,dt.$$

Integrating both sides of this equation gives

$$e^x = e^t + C.$$

Solving for x, we have

$$x = \ln|e^t + C|.$$

Applying the initial condition $x(0) = \ln 2$ yields

$$\ln 2 = \ln|1 + C|, \quad \text{so that} \quad C = 1.$$

Thus

$$x = \ln(e^t + 1),$$

which is defined for all t. Note that $e^t + 1$ is always positive so that we can drop the absolute value signs. *We should also note that after integrating, we could have applied the initial condition to determine C and then proceeded to solve for x instead of first solving for x and then applying the initial condition to determine C. Both methods will result in the same final answer.* See Figure 1.2 for a plot of the solution.

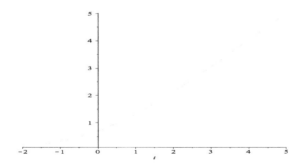

FIGURE 1.2: Plot of solution for Example 2.

Example 3 Solve $(x - 4)\, y^4 - x^3\, (y^2 - 3) \dfrac{dy}{dx} = 0.$

Solution

To separate variables, we divide by $x^3 y^4$, which implicitly assumes that $x \neq 0$ and $y \neq 0$. Doing so gives

$$\frac{x - 4}{x^3}\, dx = \frac{y^2 - 3}{y^4}\, dy.$$

This simplifies to $(x^{-2} - 4x^{-3})\, dx = (y^{-2} - 3y^{-4})\, dy$. Integrating gives

$$\frac{-1}{x} + \frac{2}{x^2} = \frac{-1}{y} + \frac{1}{y^3} + C$$

as the general solution, which is valid when $x \neq 0$ and $y \neq 0$. This is definitely a case where giving the solution in an implicit representation is acceptable! See Figure 1.3 for a plot of the implicit solution. We refer the reader to the end of this chapter for the computer code used to plot these types of solutions with one of the software packages. There is, however, a more important idea that is illustrated by this example. If we assume $x \neq 0$ and $y^2 - 3 \neq 0$, we can rewrite the original differential equation as

$$\frac{dy}{dx} = \frac{(x-4)y^4}{x^3(y^2-3)},$$

and then one can clearly see that $y = 0$ is a solution. (That is, when $y = 0$ is substituted into both sides of the equation we get an identity for all x.) This problem shows that the separation process can lose solutions.

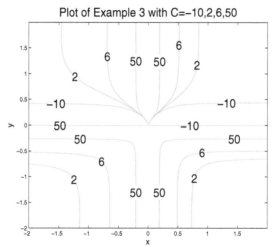

Plot of Example 3 with C=−10,2,6,50

FIGURE 1.3:   Implicit plot for Example 3. The curves plotted here satisfy the implicit solution. We note here that the C-values superimposed on the curves were good for this problem, but it often takes ingenuity, experience, trial and error, or some combination of these to get a "nice" picture.

How can we verify that

$$\frac{-1}{x} + \frac{2}{x^2} = \frac{-1}{y} + \frac{1}{y^3} + C$$

is a solution? We need to substitute it into the differential equation as before. This will require us to find y' and we will do so with implicit differentiation. Taking the derivative of both sides of the equation gives

$$\frac{1}{x^2} - \frac{4}{x^3} = \frac{1}{y^2}y' - \frac{-3}{y^4}y'.$$

We solve for y' and then simplify the complex fraction to obtain

$$y' = \frac{y^4(x-4)}{x^3(y^2-3)},$$

which is an equivalent form of our original differential equation.

Although the separation process will work on any differential equation in the form of Definition 1.2.1, evaluating the integrals in (1.4) can sometimes be a daunting, if not impossible, task. As discussed in calculus, certain indefinite integrals such as

$$\int e^{x^2}\, dx$$

cannot be expressed in finite terms using elementary functions. When such an integral is encountered while solving a differential equation, it is often helpful to use definite integration by assuming an initial condition $y(x_0) = y_0$.

Example 4 Solve the initial-value problem

$$\frac{dy}{dx} = e^{x^2}y^2, \quad y(2) = 1$$

and use the solution to give an approximate answer for $y(3)$.

Solution We would like to divide both sides by y^2 and we note that $y = 0$ is a solution. We set this solution aside and now assume $y \neq 0$, divide by y^2, and integrate from $x = 2$ to $x = x_1$ to obtain

$$\int_2^{x_1} [y(x)]^{-2}\frac{dy}{dx}\, dx = -[y(x)]^{-1}|_2^{x_1}$$

$$= \frac{-1}{y(x_1)} + \frac{1}{y(2)}$$

$$= \int_2^{x_1} e^{x^2}\, dx.$$

If we let t be the variable of integration and replace x_1 by x and $y(2)$ by 1, then we can express the solution to the initial-value problem by

$$y(x) = \frac{1}{1 - \displaystyle\int_2^x e^{t^2}\, dt}.$$

With an explicit solution, we often want to be able to find the corresponding y-value given any x. The right-hand side still cannot be solved exactly but can be approximated if x is given. For example, $y(3) \approx -0.0007007$. We note that we will have a point $x > 2$ that will make the denominator zero (and thus is not in the domain of our solution) and our function will become unbounded.

It is sometimes the case that a substitution or other "trick" will convert the given differential equation into a form that we can solve. A differential equation of the form

$$\frac{dy}{dx} = f(ax + by + k),$$

where $a, b,$ and k are constants, is separable if $b = 0$; however, if $b \neq 0$ the substitution

$$u(x) = ax + by + k$$

makes it a separable equation.

Example 5 Solve

$$\frac{dy}{dx} = (x + y - 4)^2$$

by first making an appropriate substitution.

Solution

We let $u = x + y - 4$ and thus $\frac{dy}{dx} = u^2$. We need to calculate $\frac{du}{dx}$. For this example, taking the derivative with respect to x gives

$$\frac{du}{dx} = 1 + \frac{dy}{dx}.$$

Substitution into the original differential equation gives

$$\frac{du}{dx} - 1 = u^2.$$

This equation is separable. Dividing by $1 + u^2$, we obtain

$$\frac{du}{1 + u^2} = dx$$

and integrating gives

$$\arctan(u) = x + c.$$

Thus $u = \tan(x + c)$. Since $u = x + y - 4$, we then have

$$y = -x + 4 + \tan(x + c),$$

which is defined wherever $\tan(x + c)$ is defined.

Problems

In Problems, **1–20**, solve each of the following differential equations. Explicitly solve for $y(x)$ or $x(t)$ when possible.

1. $\frac{dy}{dx} = \cos x$

2. $x\frac{dy}{dx} = (1+y)^2$

3. $x\frac{dx}{dt} + t = 1$

4. $(1+x)\frac{dy}{dx} = 4y$

5. $\tan x\, dy + 2y\, dx = 0$

6. $\frac{dy}{dx} = 2\sqrt{xy}$

7. $4xy\,dx + (x^2 + 1)dy = 0$

8. $\frac{dy}{dx} = \frac{x^2}{1+y^2}$

9. $y' = 10^{x+y}$

10. $xy' = \sqrt{1 - y^2}$

11. $y' = xye^{x^2}$, $y(0) = 1$. Explain why this differential equation guarantees that its solution is symmetric about $x = 0$.

12. $y' = 2x^2(y^2 + 1)$, $y(0) = 1$

13. $(e^x + 1)\cos y\, dy + e^x(\sin y + 1)\, dx = 0$, $y(0) = 3$

14. $(\tan x)y' = y$, $y\left(\frac{\pi}{2}\right) = \frac{\pi}{2}$

15. $2x(y^2 + 1)\, dx + (x^4 + 1)\, dy = 0$, $y(1) = 1$

16. $(x^2 - 1)y' + 2xy^2 = 0$, $y(\sqrt{2}) = 1$

17. $(y + 2)\, dx + y(x + 4)\, dy = 0$, $y(-3) = -1$

18. $8\cos^2 y\, dx + \csc^2 x\, dy = 0$, $y(\pi/12) = \pi/4$

19. $y' = e^{x^2}$, $y(0) = 0$

20. $\frac{dy}{dx} = \frac{y^3 + 2y}{x^2 + 3x}$, $y(1) = 1$

21. Find the solution of the following equation that satisfies the given conditions for $x \to +\infty$: $x^2y' - \cos 2y = 1$, $y(+\infty) = \frac{9\pi}{4}$.

22. Find the solution of the following equation that satisfies the given conditions for $x \to +\infty$: $3y^2y' + 16x = 2xy^3$, $y(x)$ is bounded for $x \to +\infty$.

In Problems **23–27** make an appropriate substitution to solve each of the following differential equations. Explicitly solve for $y(x)$ or $x(t)$ when possible.

23. $xy\,dx + (x + 1)dy = 0$

24. $y' - y = 2x - 3$

25. $(x + 2y)y' = 1$ $y(0) = -2$

26. $y' = \cos(y - x)$

27. $y' = \sqrt{4x + 2y - 1}$

28. Suppose that the population $N(t)$ of a given species (bacteria, elves, Toolie birds, college students, etc.) is not always zero and varies at a rate proportional to its current value. That is,

$$\frac{dN}{dt} = rN,$$

where $r \in \mathbb{R}$ is some measured constant proportionality factor. If the initial population is assumed to be $N(0) = N_0 > 0$, solve this exponential differential equation and discuss the behavior of the solution as $t \to \infty$ for different values of r.

29. An equivalent way of thinking of the exponential growth problem 28 is to assume the per capita growth rate, $\frac{1}{N}\frac{dN}{dt}$, is constant. That is, we assume $\frac{1}{N}\frac{dN}{dt} = r$. It is more realistic to assume that the per capita growth rate decreases as the population grows. If we assume this decrease is linear and agrees with the exponential growth model for small populations, we can write the equation

$$\frac{1}{N}\frac{dN}{dt} = r\left(1 - \frac{N}{K}\right)$$

where the left-hand side is the per capita growth rate and the right-hand side is a linearly decreasing function in N that has y-intercept r and x-intercept K. Multiplying both sides by N gives

$$\frac{dN}{dt} = r\left(1 - \frac{N}{K}\right)N,$$

which is the well-known *logistic* differential equation. If the initial population is given as $N(0) = N_0 > 0$, solve this differential equation and discuss the behavior of the solution as $t \to \infty$. From this behavior, why is K called a *carrying capacity*?

1.3   Linear Equations

Linear first-order differential equations are perhaps the most commonly arising class of differential equations in applications. A linear differential equation is defined as follows:

Definition 1.3.1

A first-order ordinary differential equation is linear in the dependent variable y and the independent variable x if it can be written as

$$\frac{dy}{dx} + P(x)y = Q(x). \tag{1.6}$$

More generally, we often see equations of the form

$$a_1(x)y' + a_0(x)y = b(x)$$

but, provided $a_1(x) \neq 0$ for all x, we can always divide by $a_1(x)$ and define $P(x) = a_0/a_1$ and $Q(x) = b/a_1$ to obtain an equation of the form of (1.6).

In our work to follow, specifically in Chapters 3 and 4 we will refer to an equation of this form as a "linear nonhomogeneous equation." In the case when $Q(x) = 0$, we refer to the equation as "homogeneous," but we caution the reader to be careful with the word "homogeneous" as it can also have other meanings; see Section 1.6. While it is an unfortunate fact that mathematicians often use the same term for different mathematical notions, our use of it should be clear by context.

In the following pages, we present two techniques for solving linear differential equations. It is likely the case that only one of these methods will be presented in class depending on the emphasis of your course. The first is variation of parameters while the second is the integrating factor technique.

Variation of Parameters

The first method of solving linear equations that we consider has a nice generalization for higher order equations. If we consider (1.6) with $Q(x) = 0$:

$$\frac{dy}{dx} + P(x)\, y = 0$$

we can solve this linear homogeneous equation by using separation of variables. We obtain y_c, the **complementary solution**.[3] We know that we can multiply y_c by any constant and it will still be a solution; however, we instead consider $u y_c$ where u is a function of x and try to find a function u that will make this work. In order for $u(x) y_c$ to be a solution, it needs to satisfy the differential equation. Substituting the assumed solution into (1.6) we obtain

$$(u'(x) y_c + u(x) y_c') + P(x) u(x) y_c = Q(x), \tag{1.7}$$

which we can regroup and then simplify:

$$u'(x) y_c + u(x)[y_c' + P(x) y_c] = Q(x)$$

$$\implies u'(x) y_c = Q(x)$$

since y_c is a solution to the homogeneous equation. We then solve for $u'(x)$ and integrate to obtain:

$$u(x) = \int \frac{Q(x)}{y_c}\, dx. \tag{1.8}$$

As we only care about finding one function $u(x)$ that will work, we don't introduce the typical $+C$ upon integration. Thus we have found a function

---

[3]This solution is sometimes called the **homogeneous solution** and is denoted y_h. The terms are used interchangeably.

$u(x)$ that makes uy_c a solution—we call this a particular solution and denote it y_p. Our general solution to (1.6), with $y_p = u(x)y_c$, is then

$$y = Cy_c + y_p, \tag{1.9}$$

where C is a constant that is determined by the initial condition.

Example 1 Solve $\dfrac{dy}{dx} + 2xy = 3x$ using variation of parameters.

Solution

This equation is linear with $P(x) = 2x$ and $Q(x) = 3x$. We solve the homogeneous equation first:

$$\frac{dy}{dx} = -2xy \Rightarrow \int \frac{dy}{y} = \int -2x\,dx$$

$$\Rightarrow \ln|y| = -x^2 + C_1$$

$$\Rightarrow y = Ce^{-x^2}. \tag{1.10}$$

We now assume that a particular solution can be written as

$$y_p = u(x)y_c = u(x)e^{-x^2}.$$

The function $u(x)$ that will allow this to be a solution of the original linear equation is

$$u(x) = \int \frac{Q(x)}{y_c}\,dx$$

$$= \int \frac{3x}{e^{-x^2}}\,dx$$

$$= 3\int xe^{x^2}\,dx$$

$$= \frac{3}{2}e^{x^2}. \tag{1.11}$$

Recalling that $y_p = u(x)y_c$, our solution is then given by

$$y = Cy_c + y_p$$

$$= \underbrace{Ce^{-x^2}}_{y_c} + \underbrace{\frac{3}{2}e^{x^2}}_{u(x)}\underbrace{e^{-x^2}}_{y_c} \tag{1.12}$$

which simplifies to

$$y = \frac{3}{2} + Ce^{-x^2}.$$

We can easily check that this is a solution of the original differential equation.

Example 2 Solve

$$\frac{dy}{dx} + \left(\frac{2x+1}{x}\right) y = e^{-2x},$$

using variation of parameters.

Solution

This is clearly linear and we first solve the homogeneous equation

$$\frac{dy}{dx} + \left(\frac{2x+1}{x}\right) y = 0.$$

Separation of variables gives us

$$y_c = C\frac{e^{-2x}}{x}.$$

We now assume a particular solution of the form $y_p = u(x)\frac{e^{-2x}}{x}$. From the derivation, we know that things will cancel out so that we need to solve for u in (1.8):

$$u(x) = \int \frac{e^{-2x}}{e^{-2x}/x} = \int x\,dx$$

so that

$$u(x) = \frac{x^2}{2} \implies y_p = u(x)y_c = \frac{x^2}{2}\frac{e^{-2x}}{x} = \frac{1}{2}xe^{-2x}.$$

Our general solution is $y_c + y_p$:

$$y = \frac{C}{x}e^{-2x} + \frac{1}{2}xe^{-2x}.$$

Superposition

A key idea in the study of linear differential equations is that of **superposition**. We have been studying the basic linear equation (1.6)

$$\frac{dy}{dx} + P(x)y = Q(x).$$

We can state a very useful theorem that will serve as an important tool in our further study.

THEOREM 1.3.1   Superposition

Suppose that y_1 is a solution to $y' + P(x)y = Q_1(x)$ and y_2 is a solution to $y' + P(x)y = Q_2(x)$. Then

$$c_1 y_1 \qquad \text{is a solution to} \qquad y' + P(x)y = c_1 Q_1(x)$$

for any constant c_1. For any constants c_1, c_2, we also have that

$$c_1 y_1 + c_2 y_2 \qquad \text{is a solution to} \qquad y' + P(x)y = c_1 Q_1(x) + c_2 Q_2(x).$$

Example 3 Verify that e^{2x} is a solution to $\dfrac{dy}{dx} + y = 3e^{2x}$ and $5x - 5$ is a

solution to $\dfrac{dy}{dx} + y = 5x$. Then find a solution to

$$\frac{dy}{dx} + y = e^{2x} + 4x.$$

Solution
We can easily verify that $y_1 = e^{2x}$ and $y_2 = 5x - 5$ are the solutions of the respective differential equations. Let $Q_1(x) = 3e^{2x}$, $Q_2(x) = 5x$, and $Q(x) = e^{2x} + 4x$ denote the right-hand sides of the three differential equations. We observe that

$$Q(x) = \frac{1}{3}Q_1(x) + \frac{4}{5}Q_2(x).$$

By superposition, it follows that

$$y = \frac{1}{3}y_1 + \frac{4}{5}y_2 = \frac{1}{3}(e^{2x}) + \frac{4}{5}(5x - 5)$$

is a solution of $y' + y = Q(x)$.

Integrating Factor Technique

In studying separable equations, we put all the terms of one variable on the left side of the equation and the terms of the other variable on the right side of the equation. This allowed us to integrate functions of just one variable. Another trick that we will use is to rewrite the left side so that it looks like the result of the product rule (from Calculus). To remind ourselves, for y, μ that are both functions of the same variable, the product rule states that

$$(y\mu)' = y'\mu + \mu'y.$$

We know how to integrate the left hand side so the goal is to somehow rewrite part of our equation so that it looks like the right-hand side. Looking at $y'\mu + \mu'y$ and recalling our basic linear equation $y' + Py = Q$, we want to multiply the left-hand side by a function μ that satisfies

$$\mu' = \mu P. \tag{1.13}$$

In this equation, $P = P(x)$ is known, whereas $\mu = \mu(x)$ (called the **integrating factor**) is unknown. We can find $\mu(x)$ because Equation (1.13) is separable. Thus

$$\frac{d\mu}{\mu} = P(x)\,dx.$$

Integrating gives

$$\mu(x) = e^{\int P(x)\,dx}. \tag{1.14}$$

Since (1.14) is an integrating factor, we have

$$e^{\int P(x)\,dx}\,\frac{dy}{dx} + e^{\int P(x)\,dx}P(x)y = Q(x)e^{\int P(x)\,dx},$$

which is the same as

$$\frac{d}{dx}\left(e^{\int P(x)\,dx}\,y\right) = Q(x)e^{\int P(x)\,dx}.$$

So

$$e^{\int P(x)\,dx}\,y = \int Q(x)e^{\int P(x)\,dx}\,dx + C,$$

which gives

$$y = e^{-\int P(x)\,dx}\left(\int Q(x)e^{\int P(x)\,dx}\,dx + C\right) \tag{1.15}$$

as the solution of the differential equation (1.6). Note that we have explicitly written the constant of integration even though the integral has not yet been evaluated. Depending upon your situation, one can memorize the formula (1.15) for the solution of a first-order linear equation; however, it is just as easy (if not out right preferable) to simply apply the method of solution each time.

Summary: Solving linear equations via an integrating factor
1. Write the linear equation in the form of Equation (1.6).
2. Calculate the integrating factor $e^{\int P(x)\,dx}$.
3. Evaluate the integral $\int Q(x)e^{\int P(x)\,dx}\,dx$ and then multiply this result by $e^{-\int P(x)\,dx}$.
4. The general solution to (1.6) is

$$y = Ce^{-\int P(x)\,dx} + e^{-\int P(x)\,dx}\int Q(x)e^{\int P(x)\,dx}\,dx.$$

In the event that we are given an initial condition $y(x_0) = y_0$, we would apply it at the time of integration, going from x_0 to a final (general) value x. If we let $\bar{p}(x) = \int P(x)dx$, then the general formula becomes

$$y = Ce^{-\bar{p}(x)} + e^{-\bar{p}(x)} \int Q(x)e^{\bar{p}(x)} \, dx,$$

and applying the initial condition gives us the solution

$$y = y_0 e^{\bar{p}(x_0) - \bar{p}(x)} + e^{-\bar{p}(x)} \int_{x_0}^{x} Q(t)e^{\bar{p}(t)} \, dt, \qquad (1.16)$$

where the variable of integration has changed to a dummy variable t.

Example 4 Solve

$$\frac{dy}{dx} + \left(\frac{2x+1}{x}\right) y = e^{-2x}.$$

This is linear with

$$P(x) = \frac{2x+1}{x} \quad \text{and} \quad Q(x) = e^{-2x}$$

so that an integrating factor is

$$e^{\int P(x)dx} = e^{\int \frac{2x+1}{x} \, dx}$$

$$= e^{(2x + \ln|x|)}$$

$$= |x|e^{2x}.$$

We note that integrating factors are not unique. For instance, dropping the absolute value to obtain xe^{2x} gives another integrating factor of the differential equation. Thus, multiplying the original equation by this expression gives

$$xe^{2x}\frac{dy}{dx} + e^{2x}(2x+1)y = x.$$

If we had multiplied by $-xe^{2x}$, we would have obtained the *same* equation. This equation can be simplified to give

$$\frac{d}{dx}(xe^{2x}y) = x.$$

Integrating this equation gives

$$xe^{2x}y = \frac{1}{2}x^2 + C,$$

which becomes
$$y = \frac{1}{2}xe^{-2x} + \frac{C}{x}e^{-2x}.$$

These last few steps could have been avoided by using (1.15).

Example Solve $(x^2+1)\dfrac{dy}{dx} + 4xy = x$ with the initial condition $y(0) = 10$.

Solution

Rewriting this equation gives
$$\frac{dy}{dx} + \left(\frac{4x}{x^2+1}\right)y = \frac{x}{x^2+1},$$

hence
$$P(x) = \frac{4x}{x^2+1} \quad \text{and} \quad Q(x) = \frac{x}{x^2+1}$$

so that an integrating factor is
$$e^{\int P(x)dx} = e^{\int \frac{4x}{x^2+1} dx}$$
$$= e^{\ln(x^2+1)^2}$$
$$= (x^2+1)^2.$$

Once we have our integrating factor, we can use the solution as given in (1.15), first noting that
$$e^{-\int P(x)dx} = e^{-\ln(x^2+1)^2}$$
$$= (x^2+1)^{-2}. \tag{1.17}$$

Then
$$y = \frac{1}{(x^2+1)^2}\left(\int x(x^2+1)\,dx\right)$$
$$= \frac{1}{(x^2+1)^2}\left(\frac{1}{4}x^4 + \frac{1}{2}x^2 + C\right).$$

Now the initial condition, $y(0) = 10$, gives $C = 10$ and thus
$$y = \frac{\frac{1}{4}x^4 + \frac{1}{2}x^2 + 10}{(x^2+1)^2}$$

is the solution we seek.

Now that we know the techniques of solving linear equations, we consider some applications. In Section 1.4, we will consider **Newton's law of cooling** that describes how the temperature of an object changes due to the constant temperature of the medium surrounding it. This is not always realistic, as in some settings the temperature of the surroundings varies. For example, determining the temperature inside a building over a span of a 24-hour day is complicated because the outside temperature varies. If we assume that the building has no heating or air conditioning, the differential equation that needs to be solved to find the temperature $u(t)$ at time t inside the building is

$$\frac{du}{dt} = k(C(t) - u(t)), \tag{1.18}$$

where $C(t)$ is a function that describes the outside temperature and $k > 0$ is a constant that depends on the insulation of the building. Note that (1.18) is a linear equation. According to this equation, if $C(t) > u(t)$, then

$$\frac{du}{dt} > 0,$$

which implies that $u(t)$ increases, and if $C(t) < u(t)$, then

$$\frac{du}{dt} < 0,$$

so that $u(t)$ decreases.

Example 6 Suppose that on a given day during the month of April in Pomona, California, the outside temperature in degrees Fahrenheit is given by

$$C(t) = 70 - 10\cos\left(\frac{\pi t}{12}\right)$$

for $0 \le t \le 24$. Determine the temperature in a building that has an initial temperature of 60°F if $k = 1/4$. See Figure 1.4.

Solution

We see that the average temperature (i.e., the average of $C(t)$) is 70°F because

$$\int_0^{24} \cos\left(\frac{\pi t}{12}\right) dt = 0.$$

The initial-value problem that we must solve is

$$\frac{du}{dt} = k\left(70 - 10\cos\left(\frac{\pi t}{12}\right) - u\right)$$

with initial condition $u(0) = 60$. The differential equation can be rewritten as

$$\frac{du}{dt} + ku = k\left(70 - 10\cos\left(\frac{\pi t}{12}\right)\right),$$

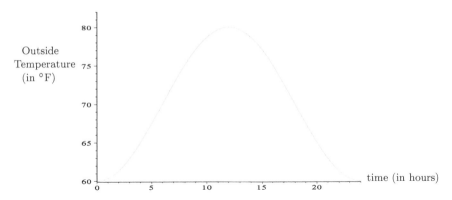

FIGURE 1.4: Outside temperature over 24 hours for Example 6.

which is a linear equation and is thus solvable. This gives (check it!)

$$u(t) = \frac{10}{9 + \pi^2}\left(63 + 7\pi^2 - 9\cos\left(\frac{\pi t}{12}\right) - 3\pi\sin\left(\frac{\pi t}{12}\right)\right) + C_1 e^{-t/4}.$$

We then apply the initial condition $u(0) = 60$ to determine the arbitrary constant C_1 and obtain the solution

$$u(t) = \frac{10}{9 + \pi^2}\left(63 + 7\pi^2 - 9\cos\left(\frac{\pi t}{12}\right) - 3\pi\sin\left(\frac{\pi t}{12}\right)\right) - \frac{10\pi^2}{9 + \pi^2}e^{-t/4}.$$

A graph of this solution is shown in Figure 1.5. The graph shows that the temperature reaches its maximum of about 77°F near $t = 15.5$, which is about 3:30 p.m.

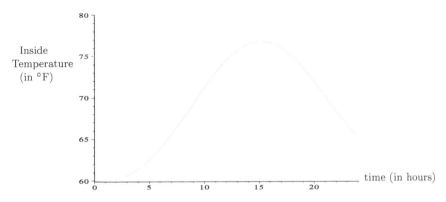

FIGURE 1.5: Inside temperature over 24 hours for Example 6.

Sometimes an equation may not immediately appear to be linear.

Example 7 Consider the differential equation

$$y^2 \, dx + (3xy - 1) \, dy = 0.$$

This equation is <u>not</u> linear in y. What do we do? Look harder. If we consider y as the independent variable and x as the dependent variable, we can write

$$\frac{dx}{dy} = \frac{1 - 3xy}{y^2},$$

which is

$$\frac{dx}{dy} + \frac{3x}{y} = \frac{1}{y^2},$$

and we see that it is in the form

$$\frac{dx}{dy} + P(y) x = Q(y),$$

which <u>is</u> linear in x, so that this equation can be solved using the theory we have just developed.

Hence, an integrating factor is

$$e^{\int P(y) dy} = e^{\int \frac{3}{y} dy} = e^{\ln |y|^3} = y^3.$$

We also have $\exp\left(-\int P(y) dy\right) = 1/y^3$. Then our solution is

$$x = \frac{1}{y^3} \left(\int \frac{1}{y^2} (y^3) dy \right) + \frac{C}{y^3}$$

$$= \frac{1}{y^3} \left(\frac{y^2}{2} \right) + \frac{C}{y^3}.$$

This becomes

$$x = \frac{1}{2y} + \frac{C}{y^3}$$

which is defined for all $y \neq 0$.

● ● ● ● ● ● ● ● ● ● ● ●

Problems

Solve the linear equations in Problems **1–18** by considering y as a function of x, that is, $y = y(x)$.

1. $y' + y = e^x$
2. $y' + 2y = 4$
3. $y' + 2y = -3x$
4. $y' - 2xy = e^{x^2}$
5. $y' - 3x^2 y = x^2$
6. $3xy' + y = 12x$
7. $\dfrac{dy}{dx} + \dfrac{1}{x} y = x$
8. $y' + \dfrac{1}{x} y = e^x$
9. $\dfrac{dy}{dx} - \dfrac{2x}{1 + x^2} y = x^2$
10. $xy' + (1 + x)y = e^{-x} \sin 2x$
11. $\dfrac{dy}{dx} + y = \cos x$
12. $(2x + 1)y' = 4x + 2y$
13. $\dfrac{dy}{dx} - y = 4e^x$, $y(0) = 4$
14. $y' + 2y = xe^{-2x}$, $y(1) = 0$
15. $y' + y \tan x = \sec x$, $y(\pi) = 1$
16. $y' = (1 - y)\cos x$, $y(\pi) = 2$
17. $\frac{dy}{dx} + \frac{y}{x} = \frac{\cos x}{x}$, $y(\frac{\pi}{2}) = \frac{4}{\pi}$, $x > 0$
18. $xy' + 2y = \sin x$, $y(\frac{\pi}{2}) = 1$, $x > 0$

Solve the linear equations in Problems **19–21** by considering x as a function of y, that is, $x = x(y)$.

19. $(x + y^2)dy = y\,dx$
20. $(2e^y - x)y' = 1$
21. $(\sin 2y + x \cot y)y' = 1$

Problems **22–23** address aspects of superposition.

22. Recall that a linear equation is called **homogeneous** if $Q(x) = 0$, i.e., if it can be written as
$$\frac{dy}{dx} + P(x)\,y = 0.$$
(a) Show that $y = 0$ is a solution (called the *trivial* solution).
(b) Show that if $y = y_1(x)$ is a solution and k is a constant, then $y = ky_1(x)$ is also a solution.
(c) Show that if $y = y_1(x)$ and $y = y_2(x)$ are solutions, then $y = y_1(x) + y_2(x)$ is a solution.

23. (a) If $y = y_1(x)$ satisfies the homogeneous linear equation $\dfrac{dy}{dx} + P(x)\,y = 0$ and $y = y_2(x)$ satisfies the nonhomogeneous linear equation $\dfrac{dy}{dx} + P(x)\,y = r(x)$, show that $y = y_1(x) + y_2(x)$ is a solution to the nonhomogeneous linear equation
$$\frac{dy}{dx} + P(x)\,y = r(x).$$

(b) Show that if $y = y_1(x)$ is a solution of $\dfrac{dy}{dx} + P(x)\,y = r(x)$, and $y = y_2(x)$ is a solution of $\dfrac{dy}{dx} + P(x)\,y = q(x)$, then $y = y_1(x) + y_2(x)$ is a solution of
$$\frac{dy}{dx} + P(x)\,y = q(x) + r(x).$$

(c) Use the results obtained in parts (a) and (b) to solve

$$\frac{dy}{dx} + 2y = e^{-x} + \cos x.$$

24. A pond that initially contains $500,000$ gal of unpolluted water has an outlet that releases $10,000$ gal of water per day. A stream flows into the pond at $12,000$ gal/day containing water with a concentration of 2 g/gal of a pollutant. Find a differential equation that models this process and determine what the concentration of pollutant will be after 10 days.

25. When wading in a river or stream, you may notice that microorganisms like algae are frequently found on rocks. Similarly, if you have a swimming pool, you may notice that in the absence of maintaining appropriate levels of chlorine and algaecides, small patches of algae take over the pool surface, sometimes overnight. Underwater surfaces are attractive environments for microorganisms because water removes waste and provides a continuous supply of nutrients. On the other hand, the organisms must spread over the surface without being washed away. If conditions become unfavorable, they must be able to free themselves from the surface and recolonize on a new surface.

The rate at which cells accumulate on a surface is proportional to the rate of growth of the cells and the rate at which the cells attach to the surface. An equation describing this situation is given by

$$\frac{dN(t)}{dt} = r(N(t) + A),$$

where $N(t)$ represents the cell density, r the growth rate, A the attachment rate, and t time.

(a) If the attachment rate, A, is constant, solve

$$\frac{dN(t)}{dt} = r(N(t) + A)$$

with the initial condition $N(0) = 0$.

(b) If $A = 3$ in a particular colony of cells, use the following table to find the growth rate at the end of each hour:

t	1	2	3	4
$N(t)$	3	9	21	45

Using this growth rate, estimate the algae population size at the end of 24 hours and 36 hours.

26. In Section 3.7, you will learn about electric circuits as an application of a second order differential equation. However, consider the circuit with an inductor and resistor only, whose differential equation is first-order and linear and is given by

$$LI' + RI = V,$$

where I is the to-be-determined current in the circuit, L measures the inductance, R measures the resistance, and V is the constant applied voltage. Find an equation describing the current in the circuit.

27. Suppose $a(t) > 0$, and $f(t) \to 0$ for $t \to \infty$. Show that every solution of the equation

$$\frac{dx}{dt} + a(t)x = f(t)$$

approaches 0 for $t \to \infty$.

28. In the same equation suppose that $a(t) > 0$, and let $x_0(t)$ be the solution for which the initial condition $x(0) = b$ is satisfied. Show that for every positive $\varepsilon > 0$ there is a $\delta > 0$, such that if we perturb the function $f(t)$ and the number b by a quantity less than δ, then the solution $x(t), t > 0$, is perturbed by less than ε. The word **perturbed** is understood in the following sense: $f(t)$ is replaced by $f_1(t)$ and b is replaced by b_1 where

$$|f_1(t) - f(t)| < \varepsilon, \quad |b_1 - b| < \delta.$$

This property of the solution $x(t)$ is called **stability for persistent disturbances**.

1.4 Some Physical Models Arising as Separable Equations

Now that we have studied separable equations in detail, we consider some applications. The wide variety of application problems that we will consider all lead to equations in which variables can be separated.

Free Fall, Neglecting Air Resistance

We will begin this application section with an easy problem from elementary physics. This application should be very familiar.

If $x(t)$ represents the position of a particle at time t, then the velocity of the particle is given by

$$v(t) = \frac{dx}{dt}.$$

Similarly, the acceleration of the particle is

$$a(t) = \frac{dv}{dt} = \frac{d^2x}{dt^2}.$$

Thus, if we consider a particle that is in free fall, where the acceleration of the particle is due to gravity alone, we have

$$a(t) = -g.$$

Here g is assumed to be a constant and we use $-g$ as gravity acts downward. For the moment, we ignore the effects of air resistance. Thus,

$$\frac{dv}{dt} = -g,$$

which is a simple separable equation, so that

$$v(t) = -gt + c.$$

If we assume that the particle has an initial velocity v_0, so that $v(0) = v_0$, then $v(t) = -gt + v_0$. Now this gives the separable equation

$$\frac{dx}{dt} = -gt + v_0$$

which has solution

$$x(t) = \frac{-g}{2}t^2 + v_0 t + C_1.$$

If the particle has initial position x_0, then

$$x(0) = x_0$$

which gives

$$x(t) = \frac{-g}{2}t^2 + v_0 t + x_0 \qquad (1.19)$$

as the position $x(t)$ of the particle in free fall, at time t.

Example 1 A man standing on a cliff 60 m high hurls a stone upward at a rate of 20 m/sec. How long does the stone remain in the air and with what speed does it hit the ground below the cliff?

Solution

Here $x_0 = 60$ and $v_0 = 20$. We take $g = 9.8$ m/sec^2. Thus,

$$x(t) = -\frac{9.8}{2}t^2 + 20t + 60$$

and

$$v(t) = -9.8t + 20.$$

The stone is in the air while $x(t) > 0$, so to find the time t that the stone is in the air, we set $x(t) = 0$ and solve for t. Using the quadratic equation,

$$t = \frac{-20 \pm \sqrt{(20)^2 - 4(-4.9)(60)}}{2(-4.9)} = -2.01, \ 6.09.$$

The stone is thus in the air for about 6.1 sec. We use this time to find the velocity upon impact:

$$v(6.1) = -9.8(6.1) + 20 = -39.78 \ \text{m/sec}.$$

Air Resistance

We will now consider the effects of **air resistance**. The amount of air resistance (sometimes called the **drag force**) depends upon the size and velocity of the object, but there is no general law expressing this dependence. Experimental evidence shows that at very low velocities for small objects it is best to approximate the resistance R as proportional to the velocity, while for larger objects and higher velocities it is better to consider it as proportional to the square of the velocity [38].

By Newton's second law $F = ma$, so that if $v(t)$ is the velocity of the object, we have

$$m\frac{dv}{dt} = F_1 + F_2$$

where F_1 is the weight of the object,

$$F_1 = mg,$$

and F_2 is the force of the air resistance on the object as it falls, so

$$F_2 = k_1 v \quad \text{or} \quad F_2 = k_2 v^2$$

where k_1, k_2 are proportionality constants. Note that $k_i < 0$ because air resistance is always opposite the velocity; see examples 2 and 3 below. We also point out that the units of k_1 and k_2 are different. In SI units, force has units of Newtons $= N = $ kg· m/sec^2. Thus k_1 must have the units of kg/sec. On the other hand, k_2 can be written as

$$k_2 = -\frac{1}{2}C\rho A$$

where ρ is the air density (SI units of kg/m^3), A is the cross-sectional area of the object (SI units of m^2), and C is the drag coefficient (unitless) [38].

Example 2   An object weighing 8 pounds falls from rest toward earth from a great height. Assume that air resistance acts on it with a force equal to $2v$. Calculate the velocity $v(t)$ and position $x(t)$ at any time. Find and interpret $\lim_{t\to\infty} v(t)$.

Solution

Remembering that pounds is a **force** (not a mass), we see that we need to calculate the mass of the object in order to apply Newton's second law. Using $g = 32$ ft/sec^2 gives $m = w/g = 8/32 = 1/4$. Thus by Newton's second law

$$m\frac{dv}{dt} = F_1 + F_2,$$

that is

$$\frac{1}{4}\frac{dv}{dt} = 8 - 2v.$$

This is a separable equation and can be written as

$$\frac{dv}{8 - 2v} = 4dt$$

so that upon integrating both sides we have

$$-\frac{1}{2} \ln |8 - 2v| = 4t + c.$$

Using the condition that the object fell from rest, so that $v(0) = 0$, we can determine the constant c and solve for $v(t)$. We have

$$v(t) = 4 - 4e^{-8t}$$

as the velocity of the object at any time. A graph of this velocity is shown in Figure 1.6. Analytically, we see that $v(t)$ approaches 4 as $t \to \infty$. This value is known as the **limiting** or **terminal velocity** of the object.

Now since $\frac{dx}{dt} = v(t)$, we have

$$\frac{dx}{dt} = 4 - 4e^{-8t}.$$

This is easily integrated to obtain $x(t) = 4t + \frac{1}{2}e^{-8t} + c$. If we take the initial position of the object as zero, so that $x(0) = 0$, then

$$x(t) = 4t + \frac{1}{2}e^{-8t} - \frac{1}{2}.$$

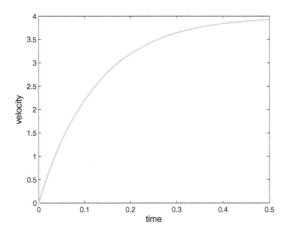

FIGURE 1.6: Approach to terminal velocity of free-falling object of Example 2.

A Cool Problem

In addition to free-fall problems, separable equations arise in some simple thermodynamics applications. One such application is the following example.

Suppose that a pie is removed from a hot oven and placed in a cool room. After a given period of time the pie has a temperature of 150° F. We want to determine the time required to cool the pie to a temperature of 80° F, when we can finally enjoy eating it.

This example is an application of **Newton's law of cooling**, which states

> *the rate at which the temperature $T(t)$ changes in a cooling body is proportional to the difference between the temperature of the body and the constant temperature T_s of the surrounding medium.*

Symbolically we know the rate of change is the derivative and the statement is expressed as

$$\frac{dT}{dt} = k(T - T_s), \qquad (1.20)$$

with the initial temperature of the body $T(0) = T_0$ and k a constant of proportionality. We observe that if the initial temperature T_0 is larger than the temperature of the surrounding T_s, then $T(t)$ will be a decreasing function of t (as the body is cooling), so $dT/dt < 0$, but $T_0 - T_s > 0$ so that the proportionality constant k must be negative. A similar analysis with $T_0 < T_s$ also gives $k < 0$. This condition on k also follows by noting that the temperature of the body will approach that of the surrounding medium as time gets large.

To solve (1.20), we seek a function $T(t)$ that describes the temperature at time t. For this equation, separating the variables we have

$$\frac{dT}{T - T_s} = k\,dt.$$

Integrating both sides of this equation gives

$$\int \frac{dT}{T - T_s} = \int k\,dt.$$

Evaluating both integrals, we obtain

$$\ln |T - T_s| = kt + C,$$

where C is the constant of integration. Exponentiating both sides and simplifying gives

$$|T - T_s| = e^{kt}e^C \implies T - T_s = \pm e^C e^{kt}.$$

Solving for the temperature, we see that

$$T(t) = C_1 e^{kt} + T_s$$

where $C_1 = \pm e^C$. We can then apply the initial condition $T(0) = T_0$, which implies $T_0 = C_1 + T_s$, so that $C_1 = T_0 - T_s$ and the solution is then

$$T(t) = (T_0 - T_s)e^{kt} + T_s. \tag{1.21}$$

We know that the temperature of the body approaches that of its surroundings and this can be seen mathematically as

$$\lim_{t \to \infty} T(t) = T_s,$$

which is true because $k < 0$.

Let's now consider a specific pie-cooling example.

Example 3   Suppose that a pie is removed from a $350°$F oven and placed in a room with a temperature of $75°$F. In 15 min the pie has a temperature of $150°$F. We want to determine the time required to cool the pie to a temperature of $80°$F, when we can finally enjoy eating it.

Solution

Comparing with the above derivation, we see that $T_0 = 350$ and $T_s = 75$. Substituting these values in (1.21) gives

$$T(t) = 275 e^{kt} + 75.$$

We still need to find k or equivalently e^k, which quantifies how fast the cooling of the pie occurs. We were given the temperature after 15 min, i.e., $T(15) = 150$. Thus

$$275 e^{15k} + 75 = 150,$$

and solving for e^k gives

$$e^k = \left(\frac{3}{11}\right)^{1/15},$$

or $k = -0.08662$. Thus

$$T(t) = 275 \left(\frac{3}{11}\right)^{t/15} + 75,$$

and this can be used to find the temperature of the pie at any given time. We can also calculate the time it takes to cool to any given temperature. We want to know when $T(t) = 80°$F. Thus we solve

$$275 \left(\frac{3}{11}\right)^{t/15} + 75 = 80$$

for t to obtain

$$t = \frac{-15 \ln 55}{\ln 3 - \ln 11} \approx 46.264.$$

Thus, the pie will reach a temperature of 80°F after approximately 46 min.

It is interesting to note that the first term in our equation for the pie temperature satisfies

$$275 \left(\frac{3}{11}\right)^{t/15} > 0$$

for all $t > 0$. Thus

$$T(t) = 275 \left(\frac{3}{11}\right)^{t/15} + 75 > 75.$$

The pie never actually reaches room temperature! This is an artifact of our model; we do note, however, that

$$\lim_{t \to \infty} 275 \left(\frac{3}{11}\right)^{t/15} + 75 = 75,$$

which can also be seen in Figure 1.7.

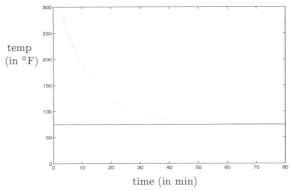

FIGURE 1.7: Graph of pie temperature vs. time of Example 3.

We present another example of Newton's law of cooling from forensic science.

Example 4 In the investigation of a homicide, the time of death is important. The normal body temperature of most healthy people is 98.6°F. Suppose that when a body is discovered at noon, its temperature is 82°F. Two hours later it is 72°F. If the temperature of the surroundings is 65°F, what was the approximate time of death?

Solution

This problem is solved as the last example. Here $T(0)$ represents the temperature when the body was discovered and $T(2)$ is the temperature of the body 2 hours later.

Thus, $T_0 = 82$ and $T_s = 65$ so that (1.21) becomes

$$T(t) = 17e^{kt} + 65.$$

Using $T(2) = 72$, we solve $17e^{2k} + 65 = 72$ for e^k to find

$$e^k = \left(\frac{7}{17}\right)^{1/2}$$

so that

$$T(t) = 17\left(\frac{7}{17}\right)^{t/2} + 65.$$

This equation gives us the temperature of the body at any given time. To find the time of death, we use the fact that the body temperature was at $98.6°$F at this time. Thus we solve

$$17\left(\frac{7}{17}\right)^{t/2} + 65 = 98.6$$

for t and find that

$$t = \frac{2\ln(1.97647)}{\ln 7 - \ln 17} \approx -1.53569.$$

This means that the time of death occurred approximately 1.53 hours before being discovered. Therefore, the time of death was approximately 10:30 a.m. because the body was found at noon.

Mixture Problems

Problems involving mixing typically give rise to separable differential equations. A typical mixture problem is given in the following example.

Example 5 A bucket contains 10 L of water and to it is being added a salt solution that contains 0.3 kg of salt per liter. This salt solution is being poured in at the rate of 2 L/min. The solution is being thoroughly mixed and drained off. The mixture is drained off at the same rate so that the bucket contains 10 L at all times. How much salt is in the bucket after 5 min?

Solution

Let $y(t)$ be the number of kilograms of salt in the bucket at the end of t minutes. We need to derive a differential equation for this problem and we do so by considering change in this system over a small time interval. We first find the amount of salt added to the bucket between time t and time $t + \Delta t$. Each minute, 2 L of solution is added so that in Δt minutes, $2\Delta t$ liters is added.

In these $2\Delta t$ liters the amount of salt is

$$0.3 \text{ kg/L} \times (2\Delta t) \text{ L} = (0.6\Delta t) \text{ kg}.$$

On the other hand, $2\Delta t$ liters of solution is withdrawn from the bucket in an interval Δt. Now at time t the 10 L in the flask contains $y(t)$ kilograms of salt. Then $2\Delta t$ of these liters contains approximately $(0.2\Delta t)(y(t))$ kilograms of salt if we suppose that the change in the amount of salt $y(t)$ is small in the short period of time Δt.

We have computed the amount of salt added in the interval $(t, t + \Delta t)$, as well as the amount subtracted in the same interval. But the difference between the amounts of salt present at times $t + \Delta t$ and t is $y(t + \Delta t) - y(t)$, so that we have obtained the equation

$$y(t + \Delta t) - y(t) = 0.6\Delta t - (0.2\Delta t)(y(t)).$$

We now divide by Δt and let $\Delta t \to 0$. The left side approaches the derivative $y'(t)$, and the right side is $0.6 - 0.2y(t)$. The differential equation is thus

$$y'(t) = 0.6 - 0.2y(t), \tag{1.22}$$

which can be thought of as *the rate of change in the number of kilograms of salt in the bucket $y'(t)$ being equal to the rate of salt (in kg) flowing into the bucket 0.6 ($= 0.3$ kg/L $\times 2$ L) minus the rate of salt flowing out of the bucket $0.2y(t)$.*

Equation (1.22) is a separable equation and can be written as

$$\frac{dy}{0.6 - 0.2y} = dt.$$

Integrating both sides gives

$$\ln|0.6 - 0.2y| = -0.2t + c$$

so that solving for $y(t)$ we obtain

$$y(t) = 3 - Ce^{-0.2t}. \tag{1.23}$$

When t is zero, the amount of salt in the bucket is zero, that is, $y(0) = 0$. Equation (1.23) shows that when $t = 0$, we have

$$y(0) = 3 - C;$$

or $C = 3$. The value of C is now known, so that Equation (1.23) becomes

$$y(t) = 3 - 3e^{-0.2t}.$$

To find y at the end of 5 min, we simply substitute $t = 5$ so that the amount of salt in the bucket is $y(5) \approx 1.9$ kg.

Problems

In Problems **1–7** it will be convenient to take the velocity to be the unknown function.

1. A ball dropped from a building falls for 4.00 sec before it hits the ground. If air resistance is neglected, answer the following questions:
   (a) What was its final velocity just as it hit the ground?
   (b) What was the average velocity during the fall?
   (c) How high was the building?

2. You drop a rock from a cliff, and 5.00 sec later you see it hit the ground. Neglecting air resistance, how high is the cliff?

3. A ball thrown straight up climbs for 3.0 sec before falling. Neglecting air resistance, with what velocity was the ball thrown?

4. Iron Man is flying at treetop level near Paris when he sees the Eiffel Tower elevator start to fall (the cable snapped). He knows Pepper Potts is inside. If Iron Man is 2 km away from the tower, and the elevator falls from a height of 350 m, how long does he have to save Pepper, and what must be his average velocity? Solve this problem assuming no air resistance. (Of course, Tony Stark instantly does the calculations required, as he is an expert in differential equations!)

5. The mass of a football is 0.4 kg. Air resists passage of the ball, the resistive force being proportional to the square of the velocity, and being equal to 0.004 N when the velocity is 1 m/sec. Find the height to which the ball will rise, and the time to reach that height if it is thrown upward with a velocity of 20 m/sec. How is the answer altered if air resistance is neglected?

6. The football of the preceding exercise is released (from rest) at an altitude of 17.1 m. Find its final velocity and time of fall.

7. Assume that air resistance is proportional to the square of velocity. The terminal velocity of a 75-kg human in air of standard density is 60 m/sec [38]. Neglecting the variation of air density with altitude and assuming that the 75-kg parachutist falls from an altitude of 1.8 km, find the velocity. Hint: use the terminal velocity to find the coefficient of v^2.

Problems **8–12** concern Newton's law of cooling.

8. At the request of their children, Randy and Stephen make homemade popsicles. At 2:00 p.m., Kaelin asks if the popsicles are frozen ($0°C$), at which time they test the temperature of a popsicle and find it to be $5°C$. If

they put the popsicles with a temperature of 15°C in the freezer at 12:00 noon and the temperature of the freezer is −2°C, when will Erin, Kaelin, Robyn, Ryley, Alan, Abdi, and Avani be able to enjoy the popsicles?

9. An object cools in 10 min from 100°C to 60°C. The surroundings are at a temperature of 20°C. When will the object cool to 25°C?

10. Determine the time of death if a corpse is 79°F when discovered at 3:00 p.m. and 68°F 3 hours later. Assume that the temperature of the surroundings is 60°F and that normal body temperature is 98.6°F.

11. A thermometer is taken from an inside room to the outside, where the air temperature is 5°F. After 1 minute the thermometer reads 55°F, and after 5 minutes it reads 30°F. Determine the initial temperature of the inside room.

12. A slug of metal at a temperature of 800°F is put in an oven, the temperature of which is gradually increased during an hour from $a°$ to $b°$. Find the temperature of the metal at the end of an hour, assuming that the metal warms kT degrees per minute when it finds itself in an oven that is T degrees warmer.

In Problems **13–17** it is supposed that the amount of gas (or liquid) contained in any fixed volume is constant. Also, thorough mixing is assumed.

13. A 20-L vessel contains air (assumed to be 80% nitrogen and 20% oxygen). Suppose 0.1 L of nitrogen is added to the container per second. If continual mixing takes place and material is withdrawn at the rate at which it is added, how long will it be before the container holds 99% nitrogen?

14. A 100-L beaker contains 10 kg of salt. Water is added at the constant rate of 5 L/min with complete mixing, and drawn off at the same rate. How much salt is in the beaker after 1 hour?

15. A tank contains 25 lb of salt dissolved in 50 gal of water. Brine containing 4 lb/gal is allowed to enter at a rate of 2 gal/min. If the solution is drained at the same rate find the amount of salt as a function $S(t)$ of time t. Find the concentration of salt at time. Suppose the rate of draining is modified to be 3 gal/min. Find the amount of salt and the concentration at time t.

16. Consider a pond that has an initial volume of 10,000 m³. Suppose that at time $t = 0$, the water in the pond is clean and that the pond has two streams flowing into it, stream A and stream B, and one stream flowing out, stream C. Suppose 500 m³/day of water flows into the pond from stream A, 750 m³/day flows into the pond from stream B, and 1250 m³ flows out of the pond via stream C. At $t = 0$, the water flowing into the pond from stream A becomes contaminated with road salt at a concentration of 5 kg/1000 m³. Suppose the water in the pond is well mixed so the concentration of salt at any given time is constant. To make matters worse, suppose also that at time $t = 0$ someone begins dumping

trash into the pond at a rate of 50 m^3/day. The trash settles to the bottom of the pond, reducing the volume by 50 m^3/day. To adjust for the incoming trash, the rate that water flows out via stream C increases to 1300 m^3/day and the banks of the pond do not overflow. Determine how the amount of salt in the pond changes over time. Does the amount of salt in the pond reach 0 after some time has passed?

17. A large chamber contains 200 m^3 of gas, 0.15% of which is carbon dioxide (CO_2). A ventilator exchanges 20 m^3/min of this gas with new gas containing only 0.04% CO_2. How long will it be before the concentration of CO_2 is reduced to half its original value?

Problems **18–20** concern radioactive decay. The decay law states that the amount of radioactive substance that decays is proportional at each instant to the amount of substance present.

18. The strength of a radioactive substance decreases 50% in a 30-day period. How long will it take for the radioactivity to decrease to 1% of its initial value?

19. It is experimentally determined that every gram of radium loses 0.44 mg in 1 year. What length of time elapses before the radioactivity decreases to half its original value?

20. A tin organ pipe decays with age as a result of a chemical reaction that is catalyzed by the decayed tin. As a result, the rate at which the tin decays is proportional to the product of the amount of tin left and the amount that has already decayed. Let M be the total amount of tin before any has decayed. Find the amount of decayed tin $p(t)$.

Problems **21–22** deal with geometric situations where the derivative arises and yields a separable equation.

21. Find a curve for which the area of the triangle determined by the tangent, the ordinate to the point of tangency, and the x-axis has a constant value equal to a^2.

22. Find a curve for which the sum of the sides of a triangle constructed as in the previous problem has a constant value equal to b.

23. On an early Monday morning in February in rural Kentucky (not far from Western Kentucky University) it started to snow. There had been no snow on the ground before. It was snowing at a steady, constant rate so that the thickness of the snow on the ground was increasing at a constant rate. A snowplow began clearing the snow from the streets at noon. The speed of the snowplow in clearing the snow is inversely proportional to the thickness of the snow. The snowplow traveled two miles during the first hour after noon and traveled one mile during the second hour after noon. At what time did it begin snowing?

1.5 Exact Equations

We will now introduce another type of differential equation. Exact equations are not separable equations nor are they necessarily linear. They come up in higher level math in fields such as potential theory and harmonic analysis.

Consider the first-order differential equation $\frac{dy}{dx} = f(x, y)$. We observe that it can always be expressed in the differential form

$$M(x, y)\, dx + N(x, y)\, dy = 0$$

or equivalently as

$$M(x, y) + N(x, y)\frac{dy}{dx} = 0$$

and vice versa. We will now consider a type of differential equation that is not separable, but, nevertheless, has a solution. We need a definition from multivariable calculus to proceed:

Definition 1.5.1

Let $F(x, y)$ be a function of two real variables such that F has continuous first partial derivatives in a domain D. The total differential dF of F is defined by

$$dF(x, y) = \frac{\partial F(x, y)}{\partial x}\, dx + \frac{\partial F(x, y)}{\partial y}\, dy$$

for all $(x, y) \in D$.

Example 1 Suppose $F(x, y) = xy^2 + 2x^3y$; then

$$\frac{\partial F}{\partial x} = y^2 + 6x^2y \quad \text{and} \quad \frac{\partial F}{\partial y} = 2xy + 2x^3$$

so that the total differential dF is given by

$$dF(x, y) = \frac{\partial F(x, y)}{\partial x}\, dx + \frac{\partial F(x, y)}{\partial y}\, dy$$

$$= (y^2 + 6x^2y)\, dx + (2xy + 2x^3)\, dy.$$

Definition 1.5.2

The expression

$$M(x, y)\, dx + N(x, y)\, dy \qquad (1.24)$$

is called an exact differential in a domain D if there exists a function F of two real variables such that this expression equals the total differential $dF(x, y)$ for all $(x, y) \in D$. That is, (1.24) is an exact differential in D if there exists a function F such that

$$\frac{\partial F}{\partial x} = M(x, y) \quad \text{and} \quad \frac{\partial F}{\partial y} = N(x, y)$$

for all $(x, y) \in D$.

If $M(x, y)\, dx + N(x, y)\, dy$ is an exact differential, then the differential equation

$$M(x, y)\, dx + N(x, y)\, dy = 0 \qquad (1.25)$$

is called an *exact differential equation*. As long as $x = C$ (a constant) is not a solution, we consider the equivalent form

$$M(x, y) + N(x, y)\frac{dy}{dx} = 0 \qquad (1.26)$$

as the standard form for an exact equation.

Example 2 The differential equation

$$y^2 + 2xy\frac{dy}{dx} = 0$$

is exact, since if $F(x, y) = xy^2$ then

$$\frac{\partial F}{\partial x} = y^2 \text{ and } \frac{\partial F}{\partial y} = 2xy.$$

Not all differential equations, however, are exact. Consider

$$y + 2x\frac{dy}{dx} = 0.$$

We cannot find an $F(x, y)$ so that

$$\frac{\partial F}{\partial x} = y \text{ and } \frac{\partial F}{\partial y} = 2x.$$

Numerous trials and errors may be enough to convince us that this is the case. What we really need is a method for testing a differential equation for exactness and for constructing the corresponding function $F(x, y)$. Both are contained in the following theorem and its proof.

Consider the differential equation

$$M(x, y) + N(x, y) \frac{dy}{dx} = 0 \tag{1.27}$$

where M and N have continuous first partial derivatives at all points (x, y) in a rectangular domain D. Then the differential equation (1.27) is exact in D, **if and only if**

$$\frac{\partial M(x, y)}{\partial y} = \frac{\partial N(x, y)}{\partial x} \tag{1.28}$$

for all (x, y) in D.

Remark: The proof of this theorem is rather important, as it not only provides a test for exactness, but also a method of solution for exact differential equations.

Proof: To prove one direction of the theorem, we first suppose the differential equation (1.27) is exact in D and show that (1.28) must hold as a result. If (1.27) is exact, then there is a function F such that

$$\frac{\partial F}{\partial x} = M(x, y) \quad \text{and} \quad \frac{\partial F}{\partial y} = N(x, y).$$

So

$$\frac{\partial^2 F}{\partial y \partial x} = \frac{\partial M}{\partial y} \quad \text{and} \quad \frac{\partial^2 F}{\partial x \partial y} = \frac{\partial N}{\partial x}$$

by differentiation. Now we have assumed the continuity of the first partials of M and N in D, so that

$$\frac{\partial^2 F}{\partial y \partial x} = \frac{\partial^2 F}{\partial x \partial y}.$$

This means that

$$\frac{\partial M}{\partial y} = \frac{\partial N}{\partial x},$$

which is the same as (1.28).

To prove the other direction, we assume (1.28) and show that (1.27) must be exact. (Proving this direction will also show us how to construct the solution for a given exact equation.) Thus, we assume

$$\frac{\partial M}{\partial y} = \frac{\partial N}{\partial x}$$

and find an F so that

$$\frac{\partial F}{\partial x} = M(x, y) \quad \text{and} \quad \frac{\partial F}{\partial y} = N(x, y). \tag{1.29}$$

It is clear that we can find an F that satisfies either of these equations, but can we find an F that satisfies both? Let's proceed and see what happens. Suppose that F satisfies

$$\frac{\partial F}{\partial x} = M(x, y).$$

We can integrate both sides of this equation to get

$$F(x, y) = \int M(x, y)\, dx + \phi(y) \tag{1.30}$$

where $\int M(x, y)\, dx$ is the partial integration with respect to x holding y constant. Note that our "constant" of integration, $\phi(y)$, is a function but is a function of y only (it might also include an additive constant, but definitely no x). This is because the expression $\partial F/\partial x$ would result in the loss of any "only y functions." Now we need to find an $F(x, y)$ that satisfies both equations in (1.29). We thus need to make sure the $F(x, y)$ in (1.30) also satisfies $\frac{\partial F}{\partial y} = N(x, y)$. We calculate $\partial F/\partial y$ by differentiating (1.30) with respect to y:

$$\frac{\partial F}{\partial y} = \frac{\partial}{\partial y} \int M(x, y)\, dx + \frac{d\phi(y)}{dy}.$$

Equating with $N(x, y)$ gives

$$N(x, y) = \left(\frac{\partial}{\partial y} \int M(x, y)\, dx \right) + \phi'(y),$$

where $\phi'(y) = d\phi(y)/dy$. Solving for $\phi'(y)$ gives

$$\phi'(y) = N(x, y) - \frac{\partial}{\partial y} \int M(x, y)\, dx.$$

Since $\phi(y)$ is a function of only y, it must also be the case that $\phi'(y)$ is a function of only y. We can see this by showing

$$\frac{\partial}{\partial x} \left(N(x, y) - \frac{\partial}{\partial y} \int M(x, y)\, dx \right) = 0.$$

Evaluating the left-hand side and simplifying give

$$\frac{\partial}{\partial x}\left(N(x,y) - \frac{\partial}{\partial y}\int M(x,y)\,dx\right) = \frac{\partial N}{\partial x} - \frac{\partial^2}{\partial x \partial y}\int M(x,y)\,dx$$

$$= \frac{\partial N}{\partial x} - \frac{\partial^2 F}{\partial x \partial y} \quad \text{(by noting what } F \text{ is)}$$

$$= \frac{\partial N}{\partial x} - \frac{\partial^2 F}{\partial y \partial x} \quad \text{(by continuity)}$$

$$= \frac{\partial N}{\partial x} - \frac{\partial^2}{\partial y \partial x}\int M(x,y)\,dx$$

$$= \frac{\partial N}{\partial x} - \frac{\partial M}{\partial y}$$

$$= 0,$$

where the last equality holds since we have assumed that

$$\frac{\partial N}{\partial x} = \frac{\partial M}{\partial y}.$$

What this means is that

$$N(x,y) - \frac{\partial}{\partial y}\int M(x,y)\,dx$$

cannot depend on x since its derivative with respect to x is zero. Hence,

$$\phi(y) = \int\left(N(x,y) - \frac{\partial}{\partial y}\int M(x,y)\,dx\right)dy$$

and thus

$$F(x,y) = \int M(x,y)\,dx + \phi(y)$$

is a function that satisfies both

$$\frac{\partial F}{\partial x} = M(x,y) \quad \text{and} \quad \frac{\partial F}{\partial y} = N(x,y).$$

Thus,

$$M(x,y) + N(x,y)\frac{dy}{dx} = 0$$

is exact in D. ∎

In short, the criterion for exactness is (1.28):

$$\frac{\partial N}{\partial x} = \frac{\partial M}{\partial y}.$$

If this equation holds, then the differential equation is exact. If this is not true, the differential equation is not exact.

Example 3 We considered the differential equation

$$y^2 + 2xy \frac{dy}{dx} = 0 \qquad (1.31)$$

earlier. We see that

$$M(x,y) = y^2 \quad \text{and} \quad N(x,y) = 2xy.$$

Thus,

$$\frac{\partial M}{\partial y} = 2y = \frac{\partial N}{\partial x},$$

so that the differential equation is exact. On the other hand,

$$y + 2x \frac{dy}{dx} = 0 \qquad (1.32)$$

gives $M(x,y) = y$ and $N(x,y) = 2x$ so that

$$\frac{\partial M}{\partial y} = 1 \neq 2 = \frac{\partial N}{\partial x}.$$

Hence $y + 2x \frac{dy}{dx} = 0$ is not exact.

Example 4 Consider the differential equation

$$(2x \sin y + y^3 e^x) + (x^2 \cos y + 3y^2 e^x) \frac{dy}{dx} = 0.$$

Here

$$M(x,y) = 2x \sin y + y^3 e^x \quad \text{and} \quad N(x,y) = x^2 \cos y + 3y^2 e^x;$$

hence

$$\frac{\partial M}{\partial y} = 2x \cos y + 3y^2 e^x = \frac{\partial N}{\partial x}.$$

Thus the differential equation is exact.

Remark: The test for exactness applies to equations in the form

$$M(x,y) + N(x,y) \frac{dy}{dx} = 0. \qquad (1.33)$$

If the left-hand side is an exact differential, then we can solve the exact differential equation (1.33) by finding a function $F(x,y)$ so that

$$\frac{\partial F(x,y)}{\partial x} dx + \frac{\partial F(x,y)}{\partial y} dy = 0.$$

More simply, using the total differential, we obtain $dF(x, y) = 0$. Thus,

$$F(x, y) = C$$

is a solution to (1.33).

THEOREM 1.5.2

Suppose the differential equation

$$M(x, y) + N(x, y) \frac{dy}{dx} = 0$$

is exact. Then the general solution of this differential equation is given implicitly by

$$F(x, y) = C,$$

where $F(x, y)$ is a function such that

$$\frac{\partial F}{\partial x} = M(x, y) \quad \text{and} \quad \frac{\partial F}{\partial y} = N(x, y).$$

Remark 1: As with separable and homogeneous equations, the constant in Theorem 1.5.2 is determined by an initial condition.

Remark 2: We have an explicit form for $F(x, y)$, namely,

$$F(x, y) = \int M(x, y) \, dx + \phi(y),$$

where $\phi(y) = \int \left(N(x, y) - \frac{\partial}{\partial y} \int M(x, y) \, dx \right) dy$. This form, however, is not always useful. We will see by example how to solve exact differential equations.

Remark 3: We integrated $\partial F/\partial x = M$ and substituted this into $\partial F/\partial y = N$. We instead could have solved $\partial F/\partial y = N$ first (by integrating with respect to y and obtaining a "constant" $\psi(x)$) and then substituted into $\partial F/\partial x = M$. The resulting F is the same but would be written

$$F(x, y) = \int N(x, y) \, dy + \psi(x), \tag{1.34}$$

where $\psi(x) = \int \left(M(x, y) - \frac{\partial}{\partial x} \int N(x, y) \, dy \right) dx$. See Problem 19 at the end of this section.

Example: Show that

$$(3x^2 + 4xy) + (2x^2 + 2y) \frac{dy}{dx} = 0$$

is exact and then solve it by the methods discussed in this section.

We have

$$M(x, y) = 3x^2 + 4xy \quad \text{and} \quad N(x, y) = 2x^2 + 2y$$

so that the equation is exact, since

$$\frac{\partial M}{\partial y} = 4x = \frac{\partial N}{\partial x}.$$

Our goal is to find an $F(x, y)$ that simultaneously satisfies the equations

$$\frac{\partial F}{\partial x} = M(x, y) \quad \text{and} \quad \frac{\partial F}{\partial y} = N(x, y).$$

That is, F must satisfy

$$\frac{\partial F}{\partial x} = 3x^2 + 4xy \quad \text{and} \quad \frac{\partial F}{\partial y} = 2x^2 + 2y.$$

Integrating $\partial F/\partial x$ with respect to x gives

$$F(x, y) = \int (3x^2 + 4xy)\, dx$$

$$= x^3 + 2x^2 y + \phi(y).$$

This same F must also satisfy $\partial F/\partial y = N$ and we then have

$$2x^2 + \phi'(y) = \frac{\partial F}{\partial y} = 2x^2 + 2y.$$

Thus, $\phi'(y) = 2y$. Integrating with respect to y gives

$$\phi(y) = y^2 + C_0$$

so that

$$F(x, y) = x^3 + 2x^2 y + y^2 + C_0.$$

Thus, a one-parameter family of solutions is given by

$$x^3 + 2x^2 y + y^2 = C.$$

We now solve an exact equation by first integrating with respect to y; see Remark 3 above.

Example 6 Show that

$$(2x \cos y + 3x^2 y) + (x^3 - x^2 \sin y - y) \frac{dy}{dx} = 0$$

is exact and solve it subject to the initial condition $y(0) = 2$. Plot the solution.

Solution

We have $M(x, y) = 2x \cos y + 3x^2 y$ and $N(x, y) = x^3 - x^2 \sin y - y$. The equation is exact because

$$\frac{\partial M}{\partial y} = 3x^2 - 2x \sin y = \frac{\partial N}{\partial x}.$$

Now we find an $F(x, y)$ so that

$$\frac{\partial F}{\partial x} = M(x, y) \quad \text{and} \quad \frac{\partial F}{\partial y} = N(x, y).$$

This time we will integrate $\partial F/\partial y = N$ with respect to y. Thus

$$
\begin{aligned}
F(x, y) &= \int N(x, y) \, dy \\
&= \int (x^3 - x^2 \sin y - y) \, dy \\
&= x^3 y + x^2 \cos y - \frac{y^2}{2} + \psi(x).
\end{aligned}
$$

This must also satisfy $\partial F/\partial x = M$. Calculating $\partial F/\partial x$ gives

$$\frac{\partial F}{\partial x} = 3x^2 y + 2x \cos y + \psi'(x).$$

Substituting into

$$\frac{\partial F}{\partial x} = M(x, y)$$

gives $\psi'(x) = 0$, which is easily integrated to obtain $\psi(x) = C_1$. Thus,

$$F(x, y) = x^3 y + x^2 \cos y - \frac{1}{2} y^2 + C_1,$$

and a one-parameter family of solutions is

$$x^3 y + x^2 \cos y - \frac{1}{2} y^2 = C.$$

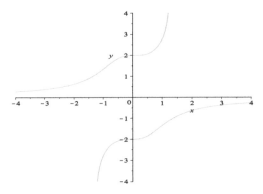

FIGURE 1.8: Implicit plot for Example 6. The upper curve is the solution curve because it passes through the initial condition.

The initial condition $y(0) = 2$ gives $C = -2$. Hence

$$x^2 \cos y + x^3 y - \frac{1}{2}y^2 = -2$$

is the implicit solution that satisfies the given initial condition. The solution curves can be plotted as shown in Figure 1.8.

Note that although both curves in Figure 1.8 satisfy the implicit equation, only one of these curves passes through the given initial condition and thus is the correct solution.

Solution by Grouping

There is a much slicker method for solving exact differential equations and it is known as the **method of grouping.** For better or worse, it requires a "working knowledge" of differentials and a certain amount of ingenuity. We again consider Example 5, this time in its differential form:

$$(3x^2 + 4xy)\, dx + (2x^2 + 2y)\, dy = 0.$$

We rewrite it in the form

$$3x^2\, dx + (4xy\, dx + 2x^2\, dy) + 2y\, dy = 0$$

which is

$$d(x^3) + d(2x^2 y) + d(y^2) = d(C).$$

That is,

$$d(x^3 + 2x^2 y + y^2) = d(C)$$

so that

$$x^3 + 2x^2 y + y^2 = C.$$

Clearly, this procedure is much quicker if we can find the appropriate grouping. Let's try this method one more time by again considering Example 6. We group the terms as

$$(2x \cos y \, dx - x^2 \sin y \, dy) + (3x^2 y \, dx + x^3 \, dy) - y \, dy = 0.$$

Thus, we have

$$d(x^2 \cos y) + d(x^3 y) - d\left(\frac{y^2}{2}\right) = d(C)$$

and so

$$x^2 \cos y + x^3 y - \frac{1}{2}y^2 = C$$

is a one-parameter family of solutions.

Important Note: If we use the method of grouping, we still need to check that the equation is exact for our first step.

In Problems **1–13**, check to see if the equation is exact. If it is, solve it by the methods of this section. If an initial condition is given, graph the solution.

1. $(1 + xy^2) + (1 + x^3 y)\frac{dy}{dx} = 0$
2. $(1 + y^2 \sin 2x) - 2y \cos 2x \frac{dy}{dx} = 0$
3. $2xy + (x^2 - y^2)\frac{dy}{dx} = 0$
4. $(1 + y^2 \sin 2x) - y \cos 2x \frac{dy}{dx} = 0$
5. $2xy^3 + (1 + 3x^2 y^2)\frac{dy}{dx} = 0$
6. $(2 + \frac{y}{x^2})\, dx + (y - \frac{1}{x})\, dy = 0$
7. $3x^2(1 + \ln y) = (2y - \frac{x^3}{y})\frac{dy}{dx}$
8. $(\frac{x}{\sin y} + 2) + \frac{(x^2+1)\cos y}{\cos 2y - 1}\frac{dy}{dx} = 0$
9. $(2xy + 1) + (x^2 + 4y)\frac{dy}{dx} = 0, \; y(0) = 1$
10. $(2y \sin x \cos x + y^2 \sin x) + (\sin^2 x - 2y \cos x)\frac{dy}{dx} = 0, \; y(0) = 3$
11. $(2 - 9xy^2)x + (4y^2 - 6x^3)y \frac{dy}{dx} = 0, \; y(1) = 1$
12. $(y \sec^2 x + \sec x \tan x) + (\tan x + 2y)\frac{dy}{dx} = 0, \; y(0) = 1$
13. $e^{-y} - (2y + xe^{-y})\frac{dy}{dx} = 0, \; y(1) = 3$

In Problems **14–15**, determine the constant A such that the equation is exact. Then solve the resulting exact equation.

14. $(x^2 + 3xy) + (Ax^2 + 4y)\frac{dy}{dx} = 0$
15. $\left(\frac{Ay}{x^3} + \frac{y}{x^2}\right) + \left(\frac{1}{x^2} - \frac{1}{x}\right)\frac{dy}{dx} = 0$

In Problems **16–17**, determine the most general function $(N(x,y)$ or $M(x,y))$ that makes the equation exact.

16. $M(x,y) + (2ye^x + y^2 e^{3x})\frac{dy}{dx} = 0$
17. $(x^3 + xy^2) + N(x,y)\frac{dy}{dx} = 0$

18. Let x represent the units of labor and y represent the units of capital. If $f(x,y)$ measures the number of units produced, a differential equation satisfied by a level curve of it is

$$ax^{a-1}y^{1-a} + (1-a)x^a y^{-a}\frac{dy}{dx} = 0.$$

Solve this equation as (i) a separable equation and (ii) an exact equation. In doing (ii), we obtain the well-known **Cobb-Douglas production function** $f(x,y) = Cx^a y^{1-a}$.

19. By following the proof of Theorem 1.5.1, show that an equivalent formulation of $F(x, y)$ is given by

$$F(x,y) = \int N(x,y)\, \frac{dy}{dx} + \int \left(M(x,y) - \frac{\partial}{\partial x} \int N(x,y)\, \frac{dy}{dx} \right).$$

Although this could easily be obtained by rearranging the previously obtained expression for F (1.30), do *not* simply rearrange terms.

20. By using the substitution $y = vx$, show that the homogeneous equation

$$(Ax + By) + (Cx + Ey)\, \frac{dy}{dx} = 0,$$

where A, B, C, and E are constants, is exact if and only if $B = C$.

21. By using the substitution $y = vx$, show that the homogeneous equation

$$(Ax^2 + Bxy + Cy^2) + (Ex^2 + Fxy + Gy^2)\, \frac{dy}{dx} = 0,$$

where A, B, C, E, F, and G are constants, is exact if and only if $B = 2E$ and $F = 2C$.

1.6　Special Integrating Factors and Substitution Methods

Special Integrating Factors

In solving linear equations, we learned that we could multiply by an appropriate **integrating factor**, thus transforming the equation into a form we can solve. Besides the one we learned, there are other integrating factors that we will now consider.

Definition 1.6.1

If the differential equation

$$M(x, y)\, dx + N(x, y)\, dy = 0 \tag{1.35}$$

is not exact in a domain D but the differential equation

$$\mu(x, y) M(x, y)\, dx + \mu(x, y) N(x, y)\, dy = 0 \tag{1.36}$$

is exact in D, then $\mu(x, y)$ is called an integrating factor of the differential equation (1.35).

Example 1 The differential equation

$$(3y + 4xy^2)\, dx + (2x + 3x^2y)\, dy = 0 \qquad (1.37)$$

is not exact since

$$\frac{\partial M}{\partial y} = 3 + 8xy \neq 2 + 6xy = \frac{\partial N}{\partial x}.$$

If we let $\mu(x, y) = x^2y$, we can use (1.36) to rewrite (1.37) as

$$(x^2y)(3y + 4xy^2)\, dx + (x^2y)(2x + 3x^2y)\, dy = 0.$$

Expanding gives

$$M = 3x^2y^2 + 4x^3y^3 \quad \text{and} \quad N = 2x^3y + 3x^4y^2.$$

Then

$$\frac{\partial}{\partial y}(3x^2y^2 + 4x^3y^3) = 6x^2y + 12x^3y^2 = \frac{\partial}{\partial x}(2x^3y + 3x^4y^2).$$

Thus the new equation is exact and hence $\mu(x, y) = x^2y$ is an integrating factor.

We saw above how multiplying by an appropriate integrating factor converted a linear equation into an exact equation, which we could then solve. Multiplying by an appropriate integrating factor is a technique that will work in other situations as well.

We have seen that if the equation

$$M(x, y)\, dx + N(x, y)\, dy = 0$$

is not exact and if $\mu(x, y)$ is an integrating factor, then the differential equation

$$\mu(x, y)M(x, y)\, dx + \mu(x, y)N(x, y)\, dy = 0$$

is exact. Using the criterion for exactness, we must have

$$\frac{\partial}{\partial y}(\mu(x, y)M(x, y)) = \frac{\partial}{\partial x}(\mu(x, y)N(x, y)).$$

To simplify notation, we will write M, N instead of $M(x, y), N(x, y)$ when taking the partial derivatives, even though both M and N are functions of x and y. The criterion for exactness can then be written

$$\frac{\partial \mu}{\partial y}M(x, y) + \mu(x, y)\frac{\partial M}{\partial y} = \frac{\partial \mu}{\partial x}N(x, y) + \mu(x, y)\frac{\partial N}{\partial x}.$$

Rearranging gives

$$\frac{\partial \mu}{\partial y} M(x, y) - \frac{\partial \mu}{\partial x} N(x, y) = \mu(x, y) \frac{\partial N}{\partial x} - \mu(x, y) \frac{\partial M}{\partial y}. \tag{1.38}$$

Thus $\mu(x, y)$ is an integrating factor if and only if it is a solution of the partial differential equation (1.38). We will not consider the solution of this partial differential equation. We will instead consider (1.38) in the case where μ only depends on x, i.e., $\mu(x, y) = \mu(x)$. (We can also consider the case when $\mu(x, y) = \mu(y)$ and the analogous formulation is left as one of the exercises.) In this situation, (1.38) reduces to

$$-\mu'(x) N(x, y) = \mu(x) \frac{\partial N}{\partial x} - \mu(x) \frac{\partial M}{\partial y}.$$

That is,

$$\frac{1}{\mu} \frac{d\mu}{dx} = \frac{1}{N(x, y)} \left(\frac{\partial M}{\partial y} - \frac{\partial N}{\partial x} \right). \tag{1.39}$$

If the right-hand side of (1.39) involves two dependent variables, we run into trouble. If, however, it depends only upon x, then Equation (1.39) is separable, in which case we obtain

$$\mu(x) = \exp\left[\int \frac{1}{N(x, y)} \left(\frac{\partial M}{\partial y} - \frac{\partial N}{\partial x} \right) dx \right]$$

as an integrating factor.

Example 2 Solve the differential equation

$$(2x^2 + y)\, dx + (x^2 y - x)\, dy = 0.$$

Solution
In this equation,

$$M(x, y) = 2x^2 + y \quad \text{and} \quad N(x, y) = x^2 y - x$$

so that

$$\frac{\partial M}{\partial y} = 1 \neq 2xy - 1 = \frac{\partial N}{\partial x}$$

and the equation is not exact. It can also be shown (try it!) that the differential equation is not separable, homogeneous, or linear. Now

$$\frac{1}{N(x, y)} \left(\frac{\partial M}{\partial y} - \frac{\partial N}{\partial x} \right) = \frac{1}{x^2 y - x}(1 - (2xy - 1))$$

$$= \frac{-2}{x}$$

depends only upon x. Thus,

$$\mu(x) = \exp\left(-\int \frac{2}{x}\, dx\right) = e^{-2\ln|x|} = \frac{1}{x^2}$$

is an integrating factor. If we multiply the equation through by this factor we have

$$\left(2 + \frac{y}{x^2}\right) dx + \left(y - \frac{1}{x}\right) dy = 0.$$

Now this equation is exact since

$$\frac{\partial M}{\partial y} = \frac{1}{x^2} = \frac{\partial N}{\partial x}.$$

We can thus solve this differential equation using the exact method to obtain

$$2x + \frac{y^2}{2} - \frac{y}{x} = C.$$

1.6.1   Bernoulli Equation

We will now consider a class of differential equations that can be reduced to linear equations by an appropriate transformation. These equations are called *Bernoulli* equations and often arise in applications.

Definition 1.6.2

A first-order differential equation of the form

$$\frac{dy}{dx} + P(x)\, y = Q(x)\, y^n \quad n \in \mathbb{R} \tag{1.40}$$

is called a Bernoulli differential equation.

Note that when $n = 0$ or $n = 1$, the Bernoulli equation is actually a linear equation and can be solved as such. When $n \neq 0$ or 1, then we must consider an additional method.

THEOREM 1.6.1

Suppose $n \neq 0$ or 1, then the transformation

$$v = y^{1-n}$$

reduces the Bernoulli equation (1.40) to

$$\frac{dv}{dx} + (1-n)P(x)\,v = (1-n)Q(x), \qquad (1.41)$$

which is a linear equation in v.

Proof: Multiply the Bernoulli equation by y^{-n} and thus obtain

$$y^{-n}\frac{dy}{dx} + P(x)\,y^{1-n} = Q(x). \qquad (1.42)$$

Now let $v = y^{1-n}$ so that

$$\frac{dv}{dx} = (1-n)y^{-n}\frac{dy}{dx}.$$

Hence, Equation (1.42) becomes

$$\frac{1}{1-n}\frac{dv}{dx} + P(x)\,v = Q(x),$$

that is,

$$\frac{dv}{dx} + (1-n)P(x)\,v = (1-n)Q(x).$$

Letting

$$P_1(x) = (1-n)P(x) \quad \text{and} \quad Q_1(x) = (1-n)Q(x)$$

gives

$$\frac{dv}{dx} + P_1(x)\,v = Q_1(x),$$

a linear differential equation in v. ∎

Example 3 Solve the differential equation

$$\frac{dy}{dx} + y = xy^3.$$

Solution
This is a Bernoulli equation with $n = 3$. We thus let $v = y^{1-3} = y^{-2}$, so that

$$\frac{dv}{dx} = -2y^{-3}\frac{dy}{dx}.$$

Using (1.41) we obtain

$$\frac{dv}{dx} - 2v = -2x. \qquad (1.43)$$

This is a linear differential equation with integrating factor

$$\exp\left(\int P(x)\,dx\right) = \exp\left(\int -2\,dx\right) = e^{-2x}.$$

We also calculate $\exp\left(-\int P(x)\,dx\right) = e^{2x}$. Thus the solution of (1.43) can be written

$$v = e^{2x}\left(\int -2xe^{-2x}\,dx\right).$$

Integrating by parts gives

$$v = e^{2x}\left(xe^{-2x} + \frac{1}{2}e^{-2x}\right) + Ce^{2x}.$$

Simplifying gives

$$v = x + \frac{1}{2} + Ce^{2x}.$$

But our original problem was in the variable y. We know $v = y^{-2}$ and thus the solution is

$$\frac{1}{y^2} = x + \frac{1}{2} + Ce^{2x}$$

which can be written as

$$y = \pm\left(\frac{1}{x + \frac{1}{2} + Ce^{2x}}\right)^{1/2}.$$

This solution is defined as long as the denominator is not equal to zero.

1.6.2 Homogeneous Equations of the Form $g(y/x)$

We have now been introduced to separable differential equations and their relative ease of solution. We will now consider a class of differential equations that can be reduced to separable equations by a change of variables.

Remark: Before proceeding, we alert the reader that the use of the word *homogeneous* in this section must not be confused with its use as the type of linear ordinary differential equation whose right-hand side is zero (as in Chapters 3 and 4). Its use in the latter chapters is more common but both have their place.

Example 4   Consider the differential equation

$$\frac{dy}{dx} = \frac{x - y}{x + y}.$$

Solution
After a minute or so of reflection, we see that this is not a separable equation. We can, however, rewrite the equation as

$$\frac{dy}{dx} = \frac{1 - \frac{y}{x}}{1 + \frac{y}{x}} \tag{1.44}$$

so that we can isolate the fraction y/x. This suggests we consider the change of variable

$$v = \frac{y}{x}$$

or equivalently

$$y = vx.$$

Our original problem has dy/dx and thus we take the derivative of both sides of the above equation with respect to x to get

$$\frac{dy}{dx} = v + x\frac{dv}{dx}.$$

Substitution of this and $y = vx$ into (1.44) gives

$$v + x\frac{dv}{dx} = \frac{1 - v}{1 + v}.$$

Simplifying results in the separable equation

$$x\frac{dv}{dx} = \frac{1 - 2v - v^2}{1 + v},$$

and we separate its variables as

$$\frac{1 + v}{1 - 2v - v^2}dv = \frac{dx}{x},$$

and integrate to give

$$\ln|1 - 2v - v^2| = -2\ln x + C_1.$$

Exponentiation of both sides yields

$$|1 - 2v - v^2| = e^{C_1}x^{-2} = C_2 x^{-2}.$$

But, $v = y/x$ so that substitution gives

$$1 - \frac{2y}{x} - \left(\frac{y}{x}\right)^2 = \pm C_2 x^{-2} = C x^{-2}.$$

Multiplying by x^2 to clear the fraction gives

$$x^2 - 2xy - y^2 = C$$

as the implicit solution to the differential equation.

This is an example of a general method of reducing a class of differential equations to that of a separable equation. We need some terminology.

Definition 1.6.3

The first-order differential equation

$$M(x,y) + N(x,y)\frac{dy}{dx} = 0$$

is said to be of homogeneous type (or homogeneous) if, when written in the derivative form

$$\frac{dy}{dx} = f(x,y),$$

there exists a function g such that $f(x,y)$ can be expressed in the form $g(y/x)$.

By classifying the equation as homogeneous, we will be able to apply the above technique in order to reduce the differential equation to one that is separable. It is sometimes not obvious that a given equation can be rewritten as a homogeneous equation. We present two examples now to help clarify this concept.

Example 3 The differential equation

$$(x^2 - 3y^2) + 2xy\frac{dy}{dx} = 0$$

is homogeneous, since the equation can be written in derivative form as

$$\frac{dy}{dx} = \frac{3y^2 - x^2}{2xy},$$

and we can rearrange this as

$$\frac{3y^2 - x^2}{2xy} = \frac{3}{2}\left(\frac{y}{x}\right) - \frac{1}{2}\left(\frac{1}{y/x}\right)$$

so that

$$\frac{dy}{dx} = \frac{3}{2}\left(\frac{y}{x}\right) - \frac{1}{2}\left(\frac{1}{y/x}\right).$$

The right-hand side is of the form $g(y/x)$ for the function

$$g(z) = \frac{3z}{2} - \frac{1}{2z},$$

and so the differential equation is homogeneous.

Example 4 The differential equation

$$\left(y + \sqrt{x^2 + y^2}\right) dx - x\,dy = 0$$

can be written as

$$\frac{dy}{dx} = \frac{y + \sqrt{x^2 + y^2}}{x}.$$

For $x > 0$, we have

$$\frac{y + \sqrt{x^2 + y^2}}{x} = \frac{y}{x} + \sqrt{1 + \left(\frac{y}{x}\right)^2},$$

so

$$\frac{dy}{dx} = \frac{y}{x} + \sqrt{1 + \left(\frac{y}{x}\right)^2}$$

which is of the form $g(y/x)$ for a function of the form

$$g(z) = z + \sqrt{1 + z^2}.$$

If we had considered $x < 0$, we would have obtained

$$g(z) = z - \sqrt{1 + z^2}.$$

In either case, we see that the differential equation is homogeneous.

As we mentioned, we have introduced homogeneous differential equations because they are related to separable equations; in fact, we have the following theorem which formalizes the method used in Example 4.

THEOREM 1.6.2

If

$$M(x, y) + N(x, y) \frac{dy}{dx} = 0 \qquad (1.45)$$

is a homogeneous equation, then the change of variables

$$y = vx$$

transforms (1.45) into a separable equation in the variables v and x.

Note that this change of variables implies that

$$y' = v + xv'$$

by the product rule.

Example 7 Solve

$$y + (x - 2y) \frac{dy}{dx} = 0.$$

Solution

We first observe that this can be rewritten as

$$\frac{dy}{dx} = \frac{y}{2y - x}.$$

Dividing numerator and denominator by x gives

$$\frac{dy}{dx} = \frac{y/x}{2y/x - 1}.$$

The right-hand side is then of the form $g(y/x)$ and making the change of variables $y = vx$ gives

$$v + x\frac{dv}{dx} = \frac{v}{2v - 1},$$

which becomes

$$x\frac{dv}{dx} = \frac{2(v - v^2)}{2v - 1}.$$

This equation is separable! Rearranging gives

$$\frac{2v - 1}{2(v - v^2)}\,dv = \frac{1}{x}\,dx,$$

and integrating both sides yields

$$-\frac{1}{2}\ln|v - v^2| = \ln|x| + C_1.$$

We then use $v = y/x$ to reintroduce the y-variable. Thus

$$-\frac{1}{2}\ln\left|\frac{y}{x} - \left(\frac{y}{x}\right)^2\right| = \ln|x| + C_1,$$

but we can let $C_1 = \ln C$ for an arbitrary constant C, so that

$$-\frac{1}{2}\ln\left|\frac{y}{x} - \left(\frac{y}{x}\right)^2\right| = \ln|Cx|$$

is the implicit solution. We could obtain an explicit solution with a bit more work but choose not to. Again, we could plot these solutions for various C-values with our favorite software package.

Problems

Solve Problems **1–8** by first finding an integrating factor of suitable form.
1. $ydx + (e^x - 1)dy = 0$ 2. $(x^2 + y^2 + x)dx + ydy = 0$
3. $y(x + y)dx + (xy + 1)dy = 0$ 4. $(x^2 - y^2 + y)dx + x(2y - 1)dy = 0$
5. $ydx - xdy = 2x^3 \sin x\, dx$ 6. $(3x^2 + y)\, dx + (x^2y - x)\, dy = 0$
7. $(3x^2y - x^2)dx + dy = 0$ 8. $(x^2 + 2x + y)dx = (x - 3x^2y)dy$

9. Show that if $(\partial N/\partial x - \partial M/\partial y)/(xM - yN)$ depends only on the product xy, that is,

$$\frac{\frac{\partial N}{\partial x} - \frac{\partial M}{\partial y}}{xM - yN} = H(xy),$$

then the equation

$$M(x, y)\,dx + N(x, y)\,dy = 0$$

has an integrating factor of the form $\mu(xy)$. Find the general formula for $\mu(xy)$.

10. We derived a formula for an integrating factor if $\mu(x, y) = \mu(x)$. If $\mu(x, y) = \mu(y)$, derive the integrating factor formula

$$\mu(y) = \exp\left[\int \frac{1}{M(x, y)} \left(\frac{\partial N}{\partial x} - \frac{\partial M}{\partial y}\right) dy\right]. \qquad (1.46)$$

Solve the Bernoulli equations given in Problems **11–21**.

11. $y' + y = xy^2$ **12.** $y' + 3y = y^4$

13. $y' + 2xy = 4y$ **14.** $y' - xy = xy^3$

15. $xy\,dy = (y^2 + x)dx$ **16.** $xy' + 2y + x^5 y^3 e^x = 0$

17. $xy' - 2x^2\sqrt{y} = 4y$ **18.** $y' = y^4 \cos x + y\tan x$

19. $xy^2 y' = x^2 + y^3$ **20.** $(x + 1)(y' + y^2) = -y$

21. Solve the logistic equation $\frac{dN}{dt} = rN\left(1 - \frac{N}{K}\right)$.

For Problems **22–34**, solve the homogeneous differential equation analytically.

22. $(x + y)\,dx - x\,dy = 0$ **23.** $(x + 2y)dx - xdy = 0$

24. $(y^2 - 2xy)dx + x^2 dy = 0$ **25.** $2x^3 y' = y(2x^2 - y^2)$

26. $2x^2 \frac{dy}{dx} = x^2 + y^2$ **27.** $\frac{dy}{dx} = \frac{y^2 + 2xy}{x^2}$

28. $xy' - y = x\tan(\frac{y}{x})$ **29.** $(x^2 + y^2)y' = 2xy$

30. $ydx = (2x + y)dy$ **31.** $(x - y)dx + (x + y)dy = 0$

32. $y' = 2(\frac{y}{x+y})^2$ **33.** $y^2 + x^2 y' = xyy'$

34. $(x + 4y)y' = 2x + 3y$

35. A function F is called **homogeneous of degree** n if

$$F(tx, ty) = t^n F(x, y) \text{ for all } x \text{ and } y.$$

That is, if tx and ty are substituted for x and y in $F(x, y)$ and if t^n is then factored out, we are left with $F(x, y)$. For instance, if $F(x, y) = x^2 + y^2$, we note that

$$F(tx, ty) = (tx)^2 + (ty)^2 = t^2 F(x, y)$$

so that F is homogeneous of degree 2. Homogeneous differential equations and functions that are homogeneous of degree n are related in the following manner. Suppose the functions M and N in the differential equation

$$M(x,y)\,dx + N(x,y)\,dy = 0$$

are both homogeneous of the same degree n.

(a) Show, using the change of variables $t = 1/x$, that

$$M\left(1, \frac{y}{x}\right) = \left(\frac{1}{x}\right)^n M(x,y),$$

which implies that

$$M(x,y) = \left(\frac{1}{x}\right)^{-n} M\left(1, \frac{y}{x}\right).$$

(b) Show, using a similar calculation, that

$$N(x,y) = \left(\frac{1}{x}\right)^{-n} N\left(1, \frac{y}{x}\right),$$

so that the differential equation

$$M(x,y)\,dx + N(x,y)\,dy = 0$$

becomes

$$\frac{dy}{dx} = \frac{-M(x,y)}{N(x,y)} = -\frac{\left(\frac{1}{x}\right)^{-n} M(1, \frac{y}{x})}{\left(\frac{1}{x}\right)^{-n} N(1, \frac{y}{x})}.$$

Simplifying gives

$$\frac{dy}{dx} = -\frac{M(1, \frac{y}{x})}{N(1, \frac{y}{x})}.$$

(c) Show that both numerator and denominator of the right-hand side of

$$\frac{dy}{dx} = -\frac{M(1, \frac{y}{x})}{N(1, \frac{y}{x})}$$

are in the form $g(y/x)$ and conclude that if M and N are both homogeneous functions of the <u>same</u> degree n, then the differential equation

$$M(x,y)\,dx + N(x,y)\,dy = 0$$

is a homogeneous differential equation.

36. Using the idea presented in Problem **35**, show that each of the equations in Problems **22–34** are homogeneous.

37. Suppose that the equation $M(x,y)\,dx + N(x,y)\,dy = 0$ is homogeneous. Show that the transformation $x = r\cos t$, $y = r\sin t$ reduces this equation to a separable equation in the variables r and t.

Use the method of Problem **37** to solve Problems **38–39**.

38. $(x-y)dx + (x+y)dy = 0$ **39.** $(x+y)\,dx - x\,dy = 0$

40. (a) Solve

$$\frac{dy}{dx} = \frac{y-x}{y+x}.$$

(b) Now consider

$$\frac{dy}{dx} = \frac{y-x+1}{y+x+5}. \tag{1.47}$$

(i) Show that this equation is NOT homogeneous. How can we solve this? Consider the equations $y - x = 0$ and $y + x = 0$. They represent two straight lines through the origin. The intersection of $y - x + 1 = 0$ and $y + x + 5 = 0$ is $(-2, -3)$. Check it! Let $x = X - 2$ and $y = Y - 3$. This amounts to taking new axes parallel to the old with an origin at $(-2, -3)$.

(ii) Use this transformation to obtain the differential equation

$$\frac{dY}{dX} = \frac{Y-X}{Y+X}.$$

(iii) Using the solution from part (a), obtain the solution to (1.47).

Use the technique of Problem **40** to solve Problems **41–45**.

41. $(2x + y + 1)dx - (4x + 2y - 3)dy = 0$
42. $x - y - 1 + (y - x + 2)y' = 0$
43. $(x + 4y)y' = 2x + 3y - 5$
44. $(y + 2)dx = (2x + y - 4)dy$
45. $y' = 2\left(\frac{y+2}{x+y-1}\right)^2$

Chapter 1 Review

In Problems **1–7**, determine whether the statement is true or false. If it is true, give reasons for your answer. If it is false, give a counterexample or other explanation of why it is false.

1. The equation $y'' + xy' - y = x^2$ is a linear ordinary differential equation that is considered an initial-value problem.

2. The equation $y^{(4)} - y^2 = \sin x$ is a nonlinear ordinary differential equation.

3. An implicit solution to $y' = f(x, y)$ can be written in the form $y(x) = f(x)$.

4. With an appropriate substitution, any exact equation can be put in the form of a linear equation.

5. A solution to the differential equation $y' = f(x, y)$ must be defined for all x.

6. An equation of the form $M(x, y) + N(x, y)y' = 0$ is considered exact if

$$\frac{\partial M(x, y)}{\partial y} = \frac{\partial N(x, y)}{\partial x}.$$

7. Every equation of the form $y' = f(y)$ is separable.

In Problems **8–11**, verify that the given function is a solution (possibly implicit) to the given differential equation. State the interval on which it is a solution and the C-values for which the solution is valid.

8. $y(x) = 1 + Ce^{x^4}$, $y' = 4x^3(y - 1)$ **9.** $y(x) = Ce^x - x^2 - 2x - 2$, $y' = x^2 + y$

10. $y^2(x) = \dfrac{1}{C - 2x}$, $y' = y^3$ **11.** $y(x) = \dfrac{3x - 1 + 3C}{x + C}$, $y' = (y - 3)^2$

In Problems **12–16**, solve each of the following separable differential equations.

12. $y' - xy^2 = 2xy$ **13.** $2x^2yy' + y^2 = 2$

14. $e^{-x}(1 + \frac{dx}{dt}) = 1$ **15.** $xy' + y = y^2$ $y(1) = 0.5$

16. $y' = 3\sqrt[3]{y^2}$ $y(2) = 0$

In Problems **17–22**, solve each of the following homogeneous differential equations.

17. $xy' = y - xe^{y/x}$ **18.** $xy' - y = (x + y)\ln\frac{x+y}{x}$

19. $xy' = y\cos(\ln(\frac{y}{x}))$ **20.** $(y + \sqrt{xy})dx = xdy$

21. $xy' = \sqrt{x^2 - y^2} + y$ **22.** $(2x - 4y)dx + (x + y)dy = 0$

In Problems **23–25**, solve each of the following exact differential equations.

23. $\dfrac{y}{x} + (y^3 + \ln x)\dfrac{dy}{dx} = 0$

24. $2x(1 + \sqrt{x^2 - y}) - \sqrt{x^2 - y}\dfrac{dy}{dx} = 0$

25. $\dfrac{3x^2 + y^2}{y^2} - \dfrac{2x^3 + 5y}{y^3}\dfrac{dy}{dx} = 0$

In Problems **26–28**, solve each of the following differential equations by first finding an integrating factor of a suitable form.

26. $xy^2(xy' + y) = 1$ **27.** $y^2 dx - (xy + x^3)dy = 0$

28. $(y - \frac{1}{x})dx + \frac{dy}{y} = 0$

In Problems **29–34**, solve the following linear or Bernoulli equations.

29. $xy' + y = x^4$ **30.** $y' - y = \sin x$

31. $y' + 4xy = 5x^3$ **32.** $xy' + y = xy^4$

33. $y' - y = xy^3$ **34.** $2y' + xy = 4xy^3$

Solve the remaining problems using one of the methods of this chapter.

35. *One Big Ol' Pot of Soup* Mike, a professor of mathematics and a part-time evening cook at a local diner, prepares a big pot of soup late at night, just before closing time. He does this so that there would be plenty of soup to feed customers the next day. Being food safety cautious, he knows that refrigeration is essential to preserve the soup overnight; however, the soup is too hot to be put directly into the fridge when it is ready. (The soup had just boiled at 100°C, and the fridge is not powerful enough to accommodate a big pot of soup if it is any warmer than 20°C.) Mike is resourceful (as he is a student of M.M. Rao) and discovered that by cooling the pot in a sink full of cold water (kept running, so that its temperature was roughly constant at 5°C) and stirring occasionally, he could bring the temperature of the soup to 60°C in 10 min. How long before closing time should the soup be ready so that Mike could put it in the fridge?

36. A cup of very hot coffee from a fast food restaurant is at 120°F when it is served. If left on the table in a room with ambient temperature of 70°F, it is found to have cooled to 115°F in 1 min. If burns can result from spilling coffee at a temperature greater than 100°F, how many minutes will the coffee have to sit before the management is safe from lawsuits?

37. A 30-gal tank initially has 15 gal of saltwater containing 6 lb of salt. Saltwater containing 1 lb of salt per gallon is pumped into the top of the tank at the rate of 2 gal/min, while a well-mixed solution leaves the bottom of the tank at a rate of 1 gal/min. Determine the amount of salt in the tank at time t. How long does it take for the tank to fill? What is the amount of salt in the tank when it is full?

38. A 100-gal tank initially contains 100 gal of sugar-water at a concentration of 0.25 lb of sugar per gallon. Suppose sugar is added to the tank at a rate of p lb/min, sugar-water is removed at a rate of 1 gal/min, and the water in the tank is kept well mixed. Determine the concentration of sugar at time t. What value of p should be chosen so that, when 5 gal of sugar solution is left in the tank, the concentration is 0.5 lb of sugar per gallon?

39. Find a curve such that the point of intersection of an arbitrary tangent with the x-axis has an abscissa half as great as the abscissa of the point of tangency.

Chapter 1 Computer Labs

The reader should only complete the computer lab(s) after reading Appendix A, where the introductory details and some basic commands and examples are given. Although every attempt has been made to make the syntax given in this book applicable to a range of versions/releases, certain syntax may be different for an earlier or later version of a given package. We assume the reader has *no* familiarity with any package but has access to at least one of them. From this point, we will assume the reader is familiar with the relevant section of Appendix A.

Chapter 1 Computer Lab: MATLAB

MATLAB Example 1: Enter the following code that demonstrates how to differentiate and integrate functions using the *Symbolic Math Toolbox*. It considers $f(x) = \cos x + e^{2x} + \ln x$ and calculates $f'(x), f''(x), \frac{df(x)}{dt}$, $\int f(x)dx$, $\int_0^\pi f(x)dx$, and $\int_0^x f(t)dt$.

```
>>  clear all
>>  syms x t
>>  f(x)=cos(x)+exp(2*x)+log(x)
>>  diff(f(x),x)
>>  diff(f(x),x,2)
>>  diff(f(x),t)
>>  int(f(x),x)
>>  int(f(x),x,0,pi)
>>  quad('cos(x)+exp(2*x)+log(x)',0,pi) %definite int w/o syms
>>  int(f(x),t)
>>  f(t)=subs(f(x),x,t)
>>  g(x)=int(f(t),0,x)
>>  g(pi)
```

MATLAB Example 2: Enter the following code that demonstrates how to verify that a given function is a solution (possibly implicit) to the differential equation. The *Symbolic Math Toolbox* is again needed. It considers $y'' + 2y' + y = 0$ with explicit solutions $y_1 = e^{-x}$ and $y_2 = xe^{-x}$, followed by $(x-4)y^4 - x^3(y^2 - 3)y' = 0$ with implicit solution $\frac{-1}{x} + \frac{2}{x^2} = \frac{-1}{y} + \frac{1}{y^3} + C$.

```
>>  clear all
```

```
>> syms x y(x) C
>> eq1=diff(y(x),x,2)+2*diff(y(x),x)+y(x)
>> subs(eq1,y(x),exp(-x))
>> subs(eq1,y(x),x*exp(-x))
>> subs(eq1,y(x),exp(x))
>> eq2=(x-4)*y(x)^4-x^3*(y(x)^2-3)*diff(y(x),x)
>> sol2=-1/x+2/x^2+1/y(x)-1/y(x)^3-C
>> dsol2=diff(sol2,x)
>> collect(expand(y(x)^4*x^3*dsol2),diff(y(x),x))
```

MATLAB Example 3: Enter the following code that demonstrates how to plot one implicit function (in this case a solution to a differential equation) and then multiple plots with the same function but different constants. It considers the implicit solution from MATLAB Example 2, $\frac{-1}{x}+\frac{2}{x^2} = \frac{-1}{y}+\frac{1}{y^3}+C$.

```
>> clear all
>> [X,Y]=meshgrid(-2:.01:2,-2:.01:2);
>> Z=-1./X+2./X.^2+1./Y-1./Y.^3;
>> contour(X,Y,Z,[-5 -5])
>> [C,h]=contour(X,Y,Z,[-5 -5]);
>> clabel(C,h)
>> axis([-2 2 0 0.6])
>> xlabel('x'), ylabel('y')
>> title('Implicit Plot with C=-5')
>> [C,h]=contour(X,Y,Z,[-10,2,6,50]);
>> clabel(C,h)
>> xlabel('x'), ylabel('y')
>> title('Implicit Plot with C=-10,2,6,50')
```

MATLAB Example 4: We will now numerically solve

$$(x^2 + 1)y' + 4xy = x, \quad y(0) = 10.$$

In the code below, you will first create a MATLAB function (separate file) called Example4.m. Then you will use MATLAB's **ode45** to plot the approximate numerical solution.

```
%%% Create a function, Example4.m, and save it.
function dydx=Example4(xn,yn)
%
%Rewrite the original ode in the form y'=f(x,y)
%to obtain y'=(x-4xy)/(x^2+1)
%
dydx=(xn-4*xn*yn)/(xn^2+1);
```

```
% Now type the following in the command window.
>>  clear all
>>  [x,y]=ode45(@Example4,[0,4],10);
>>  plot(x,y)
>>  subplot(2,1,1),plot(x,y)
>>  title('Numerical Approximation of Soln of (x^2+1)y{\prime}
    +4xy=x, y(0)=10')
>>  xlabel('x'), ylabel('y')
>>  x1=0:.1:4;
>>  y1=((1/4)*x1.^4+(1/2)*x1.^2+10)./(x1.^2+1).^2;
>>  subplot(2,1,2),plot(x1,y1,'k')
>>  title('Closed Form Soln of (x^2+1)y{\prime}+4xy=x,y(0)=10')
>>  xlabel('x'), ylabel('y')
```

MATLAB Exercises

Turn in both the commands that you enter for the exercises below as well as the output/figures. These should all be in one document. Please highlight or clearly indicate all requested answers. Some of the questions will require you to modify the above MATLAB code to answer them.

1. Enter the commands given in MATLAB Example 1 and submit both your input and output.

2. Enter the commands given in MATLAB Example 2 and submit both your input and output.

3. Enter the commands given in MATLAB Example 3 and submit both your input and output.

4. Enter the commands given in MATLAB Example 4 and submit both your input and output.

5. Find the derivative of $t^2 + \cos t$ with respect to t.

6. Find the derivative of $x^2 + \cos x$ with respect to x.

7. Find the second derivative of $t^2 + \cos t$ with respect to t.

8. Find the second derivative of $x^2 + \cos x$ with respect to x.

9. Find the antiderivative of $t^2 + \cos t$ with respect to t.

10. Find the antiderivative of $x^2 + \cos x$ with respect to x.

11. Find the derivative of $\ln(tx) + \cos^2 x + \sin t^2$ with respect to t.

12. Find the derivative of $e^{tx} + \cos^2 x + \sin t^2$ with respect to x.

13. Find the antiderivative of $\ln(tx) + \cos^2 x + \sin t^2$ with respect to t.

14. Find the antiderivative of $e^{tx} + \cos^2 x + \sin t^2$ with respect to x.

15. Find the definite integral of $x^2 + \cos x$ from $x = 0$ to $x = 3$.

16. Find the definite integral of $\ln x$ from $x = 1$ to $x = 5$.

17. Find the definite integral of e^{-x^2} from $x = -1$ to $x = 3$. Give the answer as a decimal. You may find it easier to use **quad** rather than the *Symbolic Math Toolbox*.

18. Find the definite integral of $\dfrac{\sin x}{x}$ from $x = 1$ to $x = 10$. Give the answer as a decimal. You may find it easier to use **quad** rather than the *Symbolic Math Toolbox*.

19. Numerically verify that $y(x) = \sin x + 2 \cos x$ is a solution on $(-\infty, \infty)$ to $\dfrac{d^2 y}{dx^2} + y = 0$.

20. Numerically verify that $y_1(x) = e^x, y_2(x) = xe^x$ are both solutions on $(-\infty, \infty)$ to $y'' - 2y' + y = 0$.

21. Solve $y' = \sin(x^2)$, $y(0) = 0$ by hand. Then use MATLAB to find $y(1)$, $y(3)$. (Hint: remember, from Calculus, that $y(x) = \int_0^x f(t)dt$ is a function of x and thus $y(1)$ is a number that can be evaluated with the above code.) Use MATLAB to plot the solution from $x = 0$ to $x = 4$.

22. Use MATLAB to numerically solve $y' = \sin(x^2)$, $y(0) = 0$. Plot this numerical solution from $x = 0$ to $x = 4$.

23. Solve $y' = \dfrac{\sin(x)}{x}$, $y(1) = 0$ by hand. Then use MATLAB to find $y(2)$, $y(3)$. (Hint: remember, from Calculus, that $y(x) = \int_1^x f(t)dt$ is a function of x and thus $y(2)$ is a number that can be evaluated with the above code.) Use MATLAB to plot the solution from $x = 1$ to $x = 6$.

24. Use MATLAB to numerically solve $y' = \dfrac{\sin(x)}{x}$, $y(1) = 0$. Plot this numerical solution from $x = 1$ to $x = 6$.

25. Numerically verify that $y - y^6 + \cos x + x \sin x = C$ is an implicit solution to the ODE $(1 - 6y^5)y' = -x \cos x$.

26. Solve $y + xy' = \dfrac{y}{2y - 1}$ by hand. This solution is implicit. Choose 3 different C-values and plot the implicit solutions.

27. Consider the equation

$$\frac{2}{\sqrt{x^2 + (y - 2)^2}} + \frac{1}{\sqrt{(x + 1)^2 + y^2}} = C$$

in which y is an implicit function of x. Give an implicit plot of the equation for $C = 2, 4, 10$ for $x \in [-3, 3], y \in [-3, 4]$.

28. Consider the equation

$$1 - e^{-x^2 - 3y^2 + x} = C$$

in which y is an implicit function of x. Give an implicit plot of the equation for $C = .1, .5, .9$ with $x \in [-3, 3], y \in [-1, 1]$.

29. We will learn in a later chapter that the solution of a differential equation is sometimes written in the form of a series. There is a code in Appendix A that plots the Taylor series of $\sin x$ about $x = 0$ for a given number of terms and over a given interval. Modify this code to show numerically that

$$y = 1 + x + \frac{x^2}{2!} + \frac{3x^3}{3!} + \frac{3x^4}{4!} + \frac{3x^5}{5!} + \frac{3x^6}{6!} + \cdots \qquad (1.48)$$

is the solution of $y' = y + x^2$, $y(0) = 1$. Find the analytical solution of this ODE (either by hand or using `dsolve` [in R2012 or later]) and compare it to the 5-, 6-, and 7-term series expansion (1.48) over the interval $[-4, 4]$.

Chapter 1 Computer Lab: Maple

For Maple, many of the commands are entered via the basic input palette. For example, the integral sign, the exponential function, and fractions are all entered with the palette to make it aesthetically pleasing; alternatively, there are commands that could be entered directly. See Appendix A for more details.

Maple Example 1: Enter the following code that demonstrates how to differentiate and integrate functions. It considers $f(x) = \cos x + e^{2x} + \ln x$ and calculates $f'(x), f''(x), \frac{df(x)}{dt}, \int f(x)dx, \int_0^\pi f(x)dx$, and $\int_0^x f(t)dt$. Please note that the derivative symbol that appears below is always entered from the Expression palette $\frac{d}{dx}f$.

restart

$f := x \mapsto \cos(x) + e^{2 \cdot x} + \ln(x)$

diff(f(x), x)

$\frac{d}{dx}f(x)$ #*the derivative symbol was entered from the Expression palette*

diff(f(x), x\$2)

$\frac{d^2}{dx^2}f(x)$ #*the second derivative symbol was modified from the*
          Expression palette entry by inserting the power of 2

diff(f(x), t)

$\frac{d}{dt}f(x)$

int(f(x), x)

$\int f(x)\,dx$ #*entered from Expression; if from Common Symbols, palette needs to use* d *from there too*

$int(f(x), x = 0..\text{Pi})$

$$\int_0^\pi f(x)\, dx$$

$int(f(x), t)$

$$\int f(x)\, dt$$

$$g := x \mapsto \int_0^x f(t)\, dt$$

$g(\pi)$

$evalf(g(\pi))$

Maple Example 2: Enter the following code that demonstrates how to verify that a given function is a solution (possibly implicit) to the differential equation. It considers $y'' + 2y' + y = 0$ with explicit solutions $y_1 = e^{-x}$ and $y_2 = xe^{-x}$, followed by $(x-4)\, y^4 - x^3\, (y^2 - 3)y' = 0$ with implicit solution $\frac{-1}{x} + \frac{2}{x^2} = \frac{-1}{y} + \frac{1}{y^3} + C$.

restart #checking a solution, plotting implicit solutions

$eq1 := \dfrac{d^2}{dx^2} y(x) + 2 \cdot \dfrac{d}{dx} y(x) + y(x) = 0$

$eval\left(subs\left(y(x) = e^{-x}, eq1\right)\right)$ *#Note that e is entered from the palette*

$eval\left(subs\left(y(x) = x \cdot e^{-x}, eq1\right)\right)$

$eval\left(subs\left(y(x) = e^x, eq1\right)\right)$ *#this is NOT a solution*

$eq2 := (x-4) \cdot y(x)^4 - x^3 \cdot (y(x)^2 - 3) \cdot \dfrac{d}{dx} y(x) = 0$

$sol2 := -\dfrac{1}{x} + \dfrac{2}{x^2} + \dfrac{1}{y} - \dfrac{1}{y^3} = C$

$dsol2 := implicitdiff(sol2, y, x)$ *#this is y'*

$solve\left(eq2, \dfrac{d}{dx} y(x)\right)$ *#this is the rhs of the ODE, written as y'=f(x,y(x))*

Maple Example 3: Enter the following code that demonstrates how to plot one implicit function (in this case a solution to a differential equation) and then multiple plots with the same function but different constants. It considers the implicit solution from Maple Example 2, $\frac{-1}{x} + \frac{2}{x^2} = \frac{-1}{y} + \frac{1}{y^3} + C$.

restart

with(plots) : #loads package for implicit plotting

$sol2 := -\dfrac{1}{x} + \dfrac{2}{x^2} + \dfrac{1}{y} - \dfrac{1}{y^3} = C$

$implicitplot(subs(C = -5, sol2), x = -2..2, y = -2..2,$
    $gridrefine = 4, title = "\text{Implicit Plot with C=-5}")$

$implicitplot([subs(C = -10, sol2), subs(C = 2, sol2), subs(C = 6, sol2),$

$subs\ (C = 50, sol2)]$, $x = -2..2, y = -2..2, gridrefine = 4, title =$ "Implicit Plot with C=-10,2,6,50", $legend = ["C=-10","C=2","C=6",$ "C=50"])

Maple Example 4: We will now numerically solve

$$(x^2 + 1)y' + 4xy = x, \quad y(0) = 10.$$

In the code below, you will use the Maple package `DEtools` to do this.

$restart$
$with(DEtools):$
$with(plots):$
$eq3 := (x^2 + 1) \cdot \dfrac{d}{dx} y(x) + 4 \cdot x \cdot y(x) = x$
$DEplot\ (eq3, y(x), x = 0..4, [y(0) = 10], linecolor = blue, arrows = none,$
   $title = "Numerical\ Approximation\ of\ Solution\ of(x^2 + 1)y\'(x) +$
   $4xy(x) = x, y(0) = 10")$
$sol3 := dsolve\ (\{eq3, y(0) = 10\})$
$plot\ (rhs\ (sol3), x = 0..5, color = "Gray", title = "Closed\ Form\ Solution\ of$
$(x^2 + 1)y\'(x) + 4xy(x) = x, y(0) = 10")$

Maple Exercises
Turn in both the commands that you enter for the exercises below as well as the output/figures. These should all be in one document. Please highlight or clearly indicate all requested answers. Some of the questions will require you to modify the above Maple code to answer them.

1. Enter the commands given in Maple Example 1 and submit both your input and output.
2. Enter the commands given in Maple Example 2 and submit both your input and output.
3. Enter the commands given in Maple Example 3 and submit both your input and output.
4. Enter the commands given in Maple Example 4 and submit both your input and output.
5. Find the derivative of $t^2 + \cos t$ with respect to t.
6. Find the derivative of $x^2 + \cos x$ with respect to x.
7. Find the second derivative of $t^2 + \cos t$ with respect to t.
8. Find the second derivative of $x^2 + \cos x$ with respect to x.
9. Find the antiderivative of $t^2 + \cos t$ with respect to t.
10. Find the antiderivative of $x^2 + \cos x$ with respect to x.

11. Find the derivative of $\ln(tx) + \cos^2 x + \sin t^2$ with respect to t.

12. Find the derivative of $e^{tx} + \cos^2 x + \sin t^2$ with respect to x.

13. Find the antiderivative of $\ln(tx) + \cos^2 x + \sin t^2$ with respect to t.

14. Find the antiderivative of $e^{tx} + \cos^2 x + \sin t^2$ with respect to x.

15. Find the definite integral of $x^2 + \cos x$ from $x = 0$ to $x = 3$.

16. Find the definite integral of $\ln x$ from $x = 1$ to $x = 5$.

17. Find the definite integral of e^{-x^2} from $x = -1$ to $x = 3$. Give the answer as a decimal. You will likely need to use `evalf`.

18. Find the definite integral of $\dfrac{\sin x}{x}$ from $x = 1$ to $x = 10$. Give the answer as a decimal. You will likely need to use `evalf`.

19. Numerically verify that $y(x) = \sin x + 2\cos x$ is a solution on $(-\infty, \infty)$ to $\dfrac{d^2 y}{dx^2} + y = 0$.

20. Numerically verify that $y_1(x) = e^x, y_2(x) = xe^x$ are both solutions on $(-\infty, \infty)$ to $y'' - 2y' + y = 0$.

21. Solve $y' = \sin(x^2)$, $y(0) = 0$ by hand. Then use Maple to find $y(1)$, $y(3)$. (Hint: remember, from Calculus, that $y(x) = \int_0^x f(t)dt$ is a function of x and thus $y(1)$ is a number that can be evaluated with the above code.) Use Maple to plot the solution from $x = 0$ to $x = 4$.

22. Use Maple to numerically solve $y' = \sin(x^2)$, $y(0) = 0$. Plot this numerical solution from $x = 0$ to $x = 4$.

23. Solve $y' = \dfrac{\sin(x)}{x}$, $y(1) = 0$ by hand. Then use Maple to find $y(2)$, $y(3)$. (Hint: remember, from Calculus, that $y(x) = \int_1^x f(t)dt$ is a function of x and thus $y(2)$ is a number that can be evaluated with the above code.) Use Maple to plot the solution from $x = 1$ to $x = 6$.

24. Use Maple to numerically solve $y' = \dfrac{\sin(x)}{x}$, $y(1) = 0$. Plot this numerical solution from $x = 1$ to $x = 6$.

25. Numerically verify that $y - y^6 + \cos x + x \sin x = C$ is an implicit solution to the ODE $(1 - 6y^5)y' = -x \cos x$.

26. Solve $y + xy' = \dfrac{y}{2y - 1}$ by hand. This solution is implicit. Choose 3 different C-values and plot the implicit solutions.

27. Consider the equation

$$\frac{2}{\sqrt{x^2 + (y-2)^2}} + \frac{1}{\sqrt{(x+1)^2 + y^2}} = C$$

in which y is an implicit function of x. Give an implicit plot of the equation for $C = 2, 4, 10$ for $x \in [-3, 3]$, $y \in [-3, 4]$.

28. Consider the equation

$$1 - e^{-x^2 - 3y^2 + x} = C$$

in which y is an implicit function of x. Give an implicit plot of the equation for $C = .1, .5, .9$ with $x \in [-3, 3]$, $y \in [-1, 1]$.

29. We will learn in a later chapter that the solution of a differential equation is sometimes written in the form of a series. There is a Maple procedure in Appendix A that plots the Taylor series of $\sin x$ about $x = 0$ for a given number of terms and over a given interval. Modify this code to show numerically that

$$y = 1 + x + \frac{x^2}{2!} + \frac{3x^3}{3!} + \frac{3x^4}{4!} + \frac{3x^5}{5!} + \frac{3x^6}{6!} + \cdots \qquad (1.49)$$

is the solution of $y' = y + x^2$, $y(0) = 1$. Find the analytical solution of this ODE (either by hand or using `dsolve`) and compare it to the 5-, 6-, and 7-term expansion (1.49) over the interval $[-4, 4]$.

Chapter 1 Computer Lab: Mathematica

For Mathematica, many of the commands are entered via the basic input palette. For example, the integral sign, the exponential function, and fractions are all entered with the palette to make it aesthetically pleasing; alternatively, there are commands that could be entered directly. See Appendix A for more details.

<u>Mathematica Example 1:</u> Enter the following code that demonstrates how to differentiate and integrate functions. It considers $f(x) = \cos x + e^{2x} + \ln x$ and calculates $f'(x), f''(x), \frac{df(x)}{dt}, \int f(x)dx, \int_0^\pi f(x)dx$, and $\int_0^x f(t)dt$.

```
Quit[]
f[x_] = Cos[x] + e^2x + Log[x]  (*all entered from palette*)
D[f[x],x]
f'[x]
D[f[x],{x,2}]
f''[x]
D[f[x],t]
∫ f[x]d x
```

```
Integrate[f[x],x]
```
$\int_0^\pi f[x] dx$
$\int_0^\pi f[x] dt$
$g[x\_] = \int_0^x f[t] dt$
$g[\pi]$
```
N[g[π]]
```

Mathematica Example 2: Enter the following code that demonstrates how to verify that a given function is a solution (possibly implicit) to the differential equation. It considers $y'' + 2y' + y = 0$ with explicit solutions $y_1 = e^{-x}$ and $y_2 = xe^{-x}$, followed by $(x - 4)y^4 - x^3(y^2 - 3)y' = 0$ with implicit solution $\frac{-1}{x} + \frac{2}{x^2} = \frac{-1}{y} + \frac{1}{y^3} + C$.

```
Quit[]
deq1[x_] = y''[x] + 2 y'[x] + y[x]
   (*note the space for multiplication in above line*)
ReplaceAll[deq1[x], {y[x] → e⁻ˣ, y'[x] → D[e⁻ˣ, x],
  y''[x] → D[e⁻ˣ, {x,2}]}]
ReplaceAll[deq1[x], {y[x] → xe⁻ˣ, y'[x] → D[xe⁻ˣ, x],
  y''[x] → D[xe⁻ˣ, {x,2}]}]
FullSimplify[%]
ReplaceAll[deq1[x], {y[x] → eˣ, y'[x] → D[eˣ, x],
  y''[x] → D[eˣ, {x,2}]}] (*This is NOT a solution*)
deq2[x_] = (x - 4) y[x]⁴ - x³ (y[x]² - 3) y'[x]
sol2[x_] = -1/x + 2/x² + 1/y[x] - 1/y[x]³
dsol2 = D[sol2[x], x]
Solve[dsol2 == 0, y'[x]]
Solve[deq2[x] == 0, y'[x]]
```

Mathematica Example 3: Enter the following code that demonstrates how to plot one implicit function (in this case a solution to a differential equation) and then multiple plots with the same function but different constants. It considers the implicit solution from Mathematica Example 2, $\frac{-1}{x} + \frac{2}{x^2} = \frac{-1}{y} + \frac{1}{y^3} + C$.

```
Quit[]
sol2[x_,y_] = -1/x + 2/x² + 1/y - 1/y³
ContourPlot[sol2[x, y]==-5, {x, -2, 2}, {y, -2, 2},
   PlotPoints → 20, PlotLabel → "Implicit Plot with C=-5"]
ContourPlot[{sol2[x, y]==-10, sol2[x, y]==2, sol2[x, y]==6,
  sol2[x, y]==50}, {x, -2, 2}, {y, -2, 2}, PlotPoints →20,
```

```
PlotLabel→ "Implicit Plot with C=-10,2,6,50"]
```

Mathematica Exercises
Turn in both the commands that you enter for the exercises below as well as
the output/figures. These should all be in one document. Please highlight or
clearly indicate all requested answers. Some of the questions will require you
to modify the above Mathematica code to answer them.

1. Enter the commands given in Mathematica Example 1 and submit both
   your input and output.

2. Enter the commands given in Mathematica Example 2 and submit both
   your input and output.

3. Enter the commands given in Mathematica Example 3 and submit both
   your input and output.

4. Enter the commands given in Mathematica Example 4 and submit both
   your input and output.

5. Find the derivative of $t^2 + \cos t$ with respect to t.

6. Find the derivative of $x^2 + \cos x$ with respect to x.

7. Find the second derivative of $t^2 + \cos t$ with respect to t.

8. Find the second derivative of $x^2 + \cos x$ with respect to x.

9. Find the antiderivative of $t^2 + \cos t$ with respect to t.

10. Find the antiderivative of $x^2 + \cos x$ with respect to x.

11. Find the derivative of $\ln(tx) + \cos^2 x + \sin t^2$ with respect to t.

12. Find the derivative of $e^{tx} + \cos^2 x + \sin t^2$ with respect to x.

13. Find the antiderivative of $\ln(tx) + \cos^2 x + \sin t^2$ with respect to t.

14. Find the antiderivative of $e^{tx} + \cos^2 x + \sin t^2$ with respect to x.

15. Find the definite integral of $x^2 + \cos x$ from $x = 0$ to $x = 3$.

16. Find the definite integral of $\ln x$ from $x = 1$ to $x = 5$.

17. Find the definite integral of e^{-x^2} from $x = -1$ to $x = 3$. Give the answer
    as a decimal. You will likely need to use N.

18. Find the definite integral of $\dfrac{\sin x}{x}$ from $x = 1$ to $x = 10$. Give the answer
    as a decimal. You will likely need to use N.

19. Numerically verify that $y(x) = \sin x + 2 \cos x$ is a solution on $(-\infty, \infty)$
    to $\dfrac{d^2 y}{dx^2} + y = 0$.

20. Numerically verify that $y_1(x) = e^x, y_2(x) = xe^x$ are both solutions on
    $(-\infty, \infty)$ to $y'' - 2y' + y = 0$.

21. Solve $y' = \sin(x^2)$, $y(0) = 0$ by hand. Then use Mathematica to find $y(1)$, $y(3)$. (Hint: remember, from Calculus, that $y(x) = \int_0^x f(t)dt$ is a function of x and thus $y(1)$ is a number that can be evaluated with the above code.) Use Mathematica to plot the solution from $x = 0$ to $x = 4$.

22. Use Mathematica to numerically solve $y' = \sin(x^2)$, $y(0) = 0$. Plot this numerical solution from $x = 0$ to $x = 4$.

23. Solve $y' = \dfrac{\sin(x)}{x}$, $y(1) = 0$ by hand. Then use Mathematica to find $y(2)$, $y(3)$. (Hint: remember, from Calculus, that $y(x) = \int_1^x f(t)dt$ is a function of x and thus $y(2)$ is a number that can be evaluated with the above code.) Use Mathematica to plot the solution from $x = 1$ to $x = 6$.

24. Use Mathematica to numerically solve $y' = \dfrac{\sin(x)}{x}$, $y(1) = 0$. Plot this numerical solution from $x = 1$ to $x = 6$.

25. Numerically verify that $y - y^6 + \cos x + x \sin x = C$ is an implicit solution to the ODE $(1 - 6y^5)y' = -x \cos x$.

26. Solve $y + xy' = \dfrac{y}{2y - 1}$ by hand. This solution is implicit. Choose 3 different C-values and plot the implicit solutions.

27. Consider the equation

$$\frac{2}{\sqrt{x^2 + (y - 2)^2}} + \frac{1}{\sqrt{(x + 1)^2 + y^2}} = C$$

in which y is an implicit function of x. Give an implicit plot of the equation for $C = 2, 4, 10$ for $x \in [-3, 3]$, $y \in [-3, 4]$.

28. Consider the equation

$$1 - e^{-x^2 - 3y^2 + x} = C$$

in which y is an implicit function of x. Give an implicit plot of the equation for $C = .1, .5, .9$ with $x \in [-3, 3]$, $y \in [-1, 1]$.

29. We will learn in a later chapter that the solution of a differential equation is sometimes written in the form of a series. There is a Mathematica procedure in Appendix A that plots the Taylor series of $\sin x$ about $x = 0$ for a given number of terms and over a given interval. Modify this code to show numerically that

$$y = 1 + x + \frac{x^2}{2!} + \frac{3x^3}{3!} + \frac{3x^4}{4!} + \frac{3x^5}{5!} + \frac{3x^6}{6!} + \cdots \qquad (1.50)$$

is the solution of $y' = y + x^2$, $y(0) = 1$. Find the analytical solution of this ODE (either by hand or using NDSolve) and compare it to the 5-, 6-, and 7-term expansion (1.50) over the interval $[-4, 4]$.

Chapter 1 Projects

Project 1A: Particles in the Atmosphere

Under normal atmospheric conditions, the density of soot particles, $N(t)$, satisfies the differential equation

$$\frac{dN}{dt} = -k_c N^2(t) + k_d N(t) \tag{1.51}$$

where $k_c = a_c K_c$ and $k_d = a_d K_d$. K_c is the *coagulation constant* (measured in units of cm^3/sec) and relates how well particles stick together; K_d is the *dissociation constant* (measured in molar units) and relates how well particles fall apart; $a_c > 0$ and $a_d > 0$ are the relevant conversion factors that permit the units to work out. Both K_c and K_d depend on temperature, pressure, particle size, and other external forces.

(a) Solve (1.51) using separation of variables to obtain the solution

$$N(t) = \frac{e^{k_d t}}{\left(\frac{k_c}{k_d}\right) e^{k_d t} + C},$$

where C is an arbitrary constant.

(b) Find the value of the constant C that makes $N(t_0) = N_0$. The following table lists typical values of k_c and k_d.

k_c	k_d
163	5
125	26
95	57
49	85
300	26

For *each* pair of values in the table, sketch the graph (use a graphing calculator or computer) of $N(t)$ if $N(0) = 0.01$, 0.05, 0.1, 0.5, 0.75, 1, 1.5, and 2. You will have five graphs with eight solution curves on each graph (corresponding to the eight initial values). Regardless of the initial condition, what do you notice in each case? Do pollution levels seem more sensitive to k_c or k_d?

(c) Show that if $k_d > 0$, then

$$\lim_{t \to \infty} N(t) = \frac{k_d}{k_c}.$$

Why is the assumption $k_d > 0$ reasonable? For each pair in the table, calculate k_d/k_c. Which situation results in the highest pollution levels? How could the situation be changed?

Project 1B: Insights into Graphing

In describing the solution of a differential equation, we use the term *general solution* for a family of solutions of the differential equation. These general solutions contain arbitrary constants. The term *particular solution* is used to discuss solutions that are free of arbitrary constants, usually as a result of requiring that the solution satisfy some initial condition. Sometimes a differential equation has a particular solution that cannot be obtained by selecting a specific value for the arbitrary constant in the general solution.

(a) Using calculus show that the line

$$y = mx + \frac{a}{m}$$

is tangent to the parabola $y^2 = 4ax$ for all values of m.

(b) Show that the slope of the tangent to the parabola is given by

$$\frac{dy}{dx} = \frac{2a}{\sqrt{4ax}}$$

so that at $x = \frac{a}{m^2}$, the slope is m.

(c) We know that at the point of tangency P the tangent line and the parabola have the same direction. Thus, they have a common value of dy/dx, as well as of x and y. Show that at the tangency point P,

$$m = \frac{dy}{dx}$$

and that the tangent satisfies the equation

$$y = \frac{dy}{dx}x + \frac{a}{\frac{dy}{dx}}.$$

(d) Show that the differential equation

$$y = \frac{dy}{dx}x + \frac{a}{\frac{dy}{dx}}$$

also holds for the parabola at P, where x, y, and dy/dx are the same as for the tangent. But, since P may be any point on the parabola, show that the equation of the parabola $y^2 = 4ax$ must be a solution to the differential equation. This solution is a *singular solution* to the differential equation as it contains no arbitrary constants.

(e) Solve the differential equation

$$y = \frac{dy}{dx}x + \frac{a}{\frac{dy}{dx}}.$$

Why is the solution different than that obtained in part d? Does this violate the uniqueness theorem?

Chapter 2

Geometrical and Numerical Methods for First-Order Equations

We have studied methods for solving particular types of differential equations, but as we have seen, there are many equations for which these methods do not apply. In this chapter we will investigate some useful geometrical and computer-assisted methods for analyzing the solution of a differential equation.

2.1 Direction Fields—the Geometry of Differential Equations

It should come as no surprise that most differential equations that one encounters cannot be solved. For example, even when an equation is separable, we have seen that some of the resulting integrals may not have antiderivatives in terms of elementary functions. Even when we exhaust all possible methods for analytically solving ordinary differential equations, we will encounter just as many equations where those methods *don't* apply.

All is not lost, however, as graphical and numerical methods have become commonplace in analyzing the behavior of differential equations. We will now turn to graphical and numerical methods for solving the first-order differential equation

$$\frac{dy}{dx} = f(x, y). \tag{2.1}$$

The numerical methods of solution that we cover in this section apply in an analogous way to higher-order equations and systems of equations.

We will begin our study of the geometry of differential equations by again considering (2.1) and assuming for now that solutions exist and are unique in some rectangle $\{(x, y) | a < x < b,\ c < y < d\}$. Suppose that $y_1(x)$ is a known solution to (2.1) on this rectangle. What does this mean? Recall that a solution gives us an identity when substituted into the differential equation. Thus,

$$\frac{dy_1(x)}{dx} = f(x, y_1(x)) \tag{2.2}$$

on this rectangle. Let us take a closer look at (2.2). The left-hand side of this equation is simply the derivative of the solution and this derivative is given by the right-hand side. But it really didn't take much work to evaluate the right-hand side. In fact, given any pair of values in the rectangle, say (x_0, y_0), we could find the slope of the solution that passes through that point in the x-y plane by simply evaluating $f(x_0, y_0)$. And we could do this for *each* point in this rectangle. Thus, we can calculate the slope of any solution curve without actually knowing an explicit solution.

The implications of this last statement are profound: given an initial condition, we can trace out solution curves without having to know the analytical solution. We can choose a grid of points in the x-y plane and evaluate the slope of the solution at each of these points. We draw a short line segment in the plane at that point with the calculated slope. The collection of these line segments is called the **slope field** or **direction field**. In many examples, the slope of the solution may be the same along a given curve. If we let (x, y) be the points on this curve, then

$$f(x, y) = k.$$

Any member of the family of curves that satisfies this equation is called an *isocline*, which means a curve along which the inclination of the tangents is the same. The use of isoclines is extremely useful if we are sketching a direction field by hand. We will illustrate their use in the first example and leave it to the reader to check their use in the others. Let us consider a simple example.

Example 1 Draw the direction field for the equation

$$\frac{dy}{dx} = y$$

and sketch the solution curve in this direction field that passes through the point $(0, 1)$.

Solution
We draw a representation of the direction field by selecting a grid of points and drawing a short tangent line at each point. The slope of each tangent line is determined by evaluating the right-hand side of our differential equation at the given point. This is very time consuming but is easily done on the computer. Equipped with the computer-generated direction field, we should be able to check that each point (x, y) in the plane has the slope on the graph, as determined by substituting particular values into the right-hand side of the given differential equation. In particular, we observe that the isoclines are given by $y = k$, i.e., by horizontal lines at the y-value of k. For example, all the line segments along $y = 1$ have a slope of 1 and all the line segments along $y = -2$ have a slope of -2. A direction field that only depends on y and not x is called *autonomous* and will be discussed in detail in later sections. The direction field for $\frac{dy}{dx} = y$ is shown in Figure 2.1a.

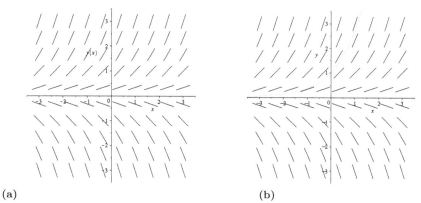

(a)   (b)

FIGURE 2.1: The direction field for $\frac{dy}{dx} = y$ is shown in **(a)**. The graph of $y(x)$ through the point $(0,1)$ on the direction field for $\frac{dy}{dx} = y$ is shown in **(b)**.

Using the condition that $y(x)$ passes through the point $(0,1)$, we can sketch the solution of $y(x)$ on the direction field by following the line segments in the direction field of Figure 2.1a. The sketch of the graph of $y(x)$ is shown in Figure 2.1b.

Notice how the solution passes tangentially through the direction field. That is, each line segment is tangent to the solution in the direction field. Now, of course, we could have easily calculated the solution to

$$\frac{dy}{dx} = y, \quad y(0) = 1$$

as $y(x) = e^x$ and we should observe that this *is* the curve plotted. Let's try this method again for an equation for which we didn't know the solution.

Draw the direction field for the equation

$$\frac{dy}{dx} = x^2 + y^2$$

and sketch several solution curves that pass through the direction field.

Solution

The direction field for $\frac{dy}{dx} = x^2 + y^2$ is shown in Figure 2.2a. The reader should check a few pairs of points on the graph to make sure the direction field is correct, e.g., check the pairs $(3,0)$ and $(0, \frac{1}{2})$. One could also observe that the isoclines are given by $x^2 + y^2 = k$ which are circles of radius $\sqrt{k}$, which indicate that the line segments have the same slope along any circle in the x-y plane. On this direction field, we can plot several graphs of $y(x)$, shown in Figure 2.2b.

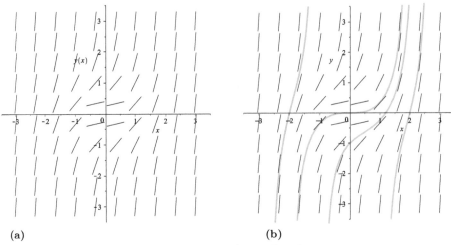

(a)   (b)

FIGURE 2.2: The direction field for $\frac{dy}{dx} = x^2 + y^2$ is shown in (a). Graphs of several solutions $y(x)$ on the direction field for $\frac{dy}{dx} = x^2 + y^2$ are shown in (b).

We note two points that should be obvious:

1. The finer the mesh of the grid for the representation of the slope field, the better the approximate solution curve which we are able to draw. This is the same as saying the tangent line is a good approximation to a curve close to the point of tangency. And the more points we have the better we can sketch this approximation.

2. Drawing a direction field by hand is tedious—it is best to use a computer. It is, however, *essential* that we check a few points of the direction field (generated by the computer) by hand. As much as we would like computers to always give us the answers we want, this will never be the case.

Problems

Without solving the equations and without using the computer, match each differential equation in Problems 1–4 and 5–8 with the graph of its direction field shown here.

1. $\dfrac{dy}{dx} = \dfrac{x^3 + 1}{y^3 + 1}$

2. $\dfrac{dy}{dx} = (x^3 + 1)(y^3 + 1)$

3. $\dfrac{dy}{dx} = \dfrac{y^3 + 1}{x^3 + 1}$

4. $\dfrac{dy}{dx} = \dfrac{1}{(x^3 + 1)(y^3 + 1)}$

5. $\dfrac{dy}{dx} = y(x^2 - y)$

6. $\dfrac{dy}{dx} = x(x^2 - y)$

7. $\dfrac{dy}{dx} = y^2(y - x^2)$

8. $\dfrac{dy}{dx} = \dfrac{x}{y - x^2}$

(a) (b)

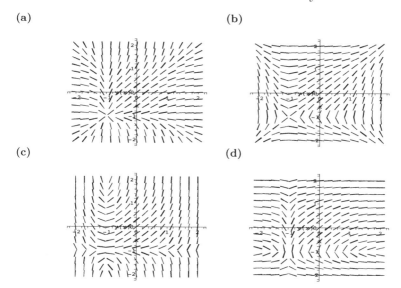

FIGURE 2.3: Direction fields for Problems 1–4.

(a) (b)

(c) (d)

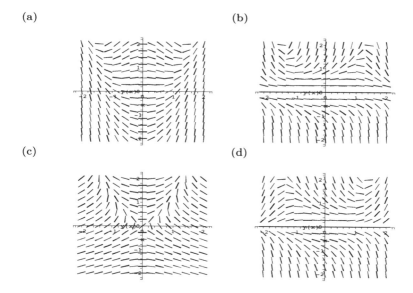

FIGURE 2.4: Direction fields for Problems 5–8.

Sketch the direction field for each of the first-order differential equations given in Problems **9–25**. (Get clarification from your instructor as to whether you should obtain these direction fields by hand or from the computer.) Verify that the direction fields you obtain in problems **10–12** are autonomous by observing that the slopes of the line segments are constant along fixed y-values. For each problem, use your sketch to draw various solution curves. Then draw the solution curve that passes through the given initial condition.

9. $\dfrac{dy}{dx} = \cos y$, $(0,0)$ 10. $\dfrac{dy}{dx} = y^4$, $(1,1)$ 11. $\dfrac{dy}{dx} = \sin y$, $(0,0.1)$

12. $\dfrac{dy}{dx} = e^{-y}$, $(0,1)$ 13. $\dfrac{dy}{dx} = \cos x$, $(0,0)$ 14. $\dfrac{dy}{dx} = x^4$, $(1,1)$

15. $\dfrac{dy}{dx} = \sin x$, $(0,-1)$ 16. $\dfrac{dy}{dx} = e^{-x}$, $(0,1)$ 17. $\dfrac{dy}{dx} = x+y$, $(0,0)$

18. $\dfrac{dy}{dx} = xy$, $(1,1)$ 19. $\dfrac{dy}{dx} = e^{x^2}$, $(0,1)$ 20. $\dfrac{dy}{dx} = x^2(y+1)$, $(0,1)$

21. $y' = \dfrac{x-1}{y-1}$, $(0,0)$ 22. $y' = y(y^2-2)$, $(0,1)$ 23. $y' = \dfrac{x^2-1}{y^2+1}$, $(0,0)$

24. $y' = \dfrac{x^3(y^2+1)}{y^2+x^2}$, $(0,0)$ 25. $y' = xy(x+2)$, $(0,1)$

2.2   Existence and Uniqueness for First-Order Equations

We have now seen some approaches for gaining insight into the solution of an ordinary differential equation. Before going further we address the issue of when we can expect solutions to even exist or, if they exist, to be unique. This will become essential as we further develop numerical solution methods.

We begin with an example based on two direction fields that look very similar.

Example 1   Consider the direction fields for the differential equations

$$y' = 3y^{4/3} \text{ and } y' = 2y^{2/3}$$

(given in Figure 2.5), together with the initial condition $(\tfrac{-3}{2}, -1)$.

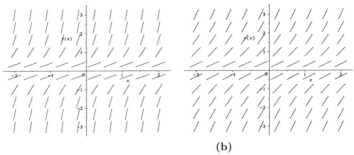

(a)   (b)

FIGURE 2.5:   Direction fields for (a) $y' = 3y^{4/3}$ and (b) $y' = 2y^{2/3}$.

The reader should attempt to sketch the corresponding solution curve. It is not obvious that one of the solution curves passes through $y = 0$ while the other does not. The problem with passing through $y = 0$ is that $y = 0$ is itself a constant solution for both of the differential equations. The point where the two solutions intersect is a point where the solutions are *not unique*. In other words, if we start at the initial condition (henceforth IC) where they intersect, we cannot be sure which solution curve we should follow. We can see this more explicitly by solving the two equations, which easily can be done since both are separable equations. The differential equation

$$y' = 3y^{4/3} \text{ with IC } y\left(\frac{-3}{2}\right) = -1 \text{ has solution } y = \frac{-1}{\left(x + \frac{5}{2}\right)^3}$$

while

$$y' = 2y^{2/3} \text{ with IC } y\left(\frac{-3}{2}\right) = -1 \text{ has solution } y = \left(\frac{2}{3}x\right)^3.$$

The first of these approaches zero as x gets large and the second passes through $(0,0)$; see Figure 2.6.

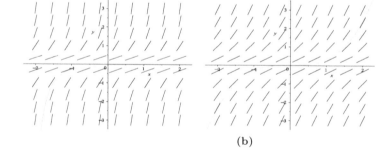

(a)   (b)

FIGURE 2.6: Direction fields for (a) $y' = 3y^{4/3}$ with IC $y\left(\frac{-3}{2}\right) = -1$, which stays in bottom half plane and (b) $y' = 2y^{2/3}$ with IC $y\left(\frac{-3}{2}\right) = -1$, which passes through upper half plane.

Thus, it appears that we do not have any problems with uniqueness of solutions in $y' = 3y^{4/3}$, but we do have a problem with uniqueness of solutions in $y' = 2y^{2/3}$ or at least we do at the point $(0,0)$.

There are two things we need to be aware of at this point. The first is the need for a way of checking whether we can expect solutions to exist and be unique at a given initial condition. The other involves how to best calculate a numerical approximation to the solution of a differential equation—we trusted the computer to accurately sketch a solution curve in the previous section but how can we be sure that it will not overlook the slight differences between two direction fields? We will save our discussion of the latter until the next chapter and will only address the former for now. In order to understand the types of initial value problems that yield a *unique* solution the following result is required.

THEOREM 2.2.1   Existence and Uniqueness

Consider the initial-value problem

$$y' = f(x, y) \quad \text{with} \quad y(x_0) = y_0.$$

If f and $\partial f / \partial y$ are continuous functions on the rectangular region

$$R : a < x < b, \quad c < y < d$$

containing the point (x_0, y_0), then there exists an interval

$$|x - x_0| < h$$

centered at x_0 on which there exists one and only one solution to the differential equation that satisfies the initial condition.

The proof of this theorem will be omitted, as it will take us off course. If you are interested, it can be found in C. Corduneanu's *Principles of Differential and Integral Equations* [13].

Remark: If the condition that $\partial f / \partial y$ is continuous on the rectangular region R containing the point (x_0, y_0) is not included in the assumptions of the theorem, then we say that *at least* one solution exists. We call this more "relaxed" result an **existence theorem** because the uniqueness of the solution is not guaranteed.

Example 2   Solve the differential equation

$$\frac{dy}{dx} = \frac{x}{y}$$

with the initial condition $y(0) = 0$. Does this result contradict the Existence and Uniqueness theorem?

Solution

The equation is separable and thus easily can be solved. Rewriting the equation

$$y \, dy = x \, dx,$$

integration yields the family of solutions

$$y^2 - x^2 = C.$$

Application of the initial condition gives us $C = 0$, so that

$$y^2 - x^2 = 0.$$

Solving for y gives two solutions

$$y = x \quad \text{and} \quad y = -x.$$

Does this result contradict the Existence and Uniqueness theorem? Although more than one solution satisfies this initial-value problem, the Existence and Uniqueness theorem is not contradicted because the function x/y is not continuous at the point $(0,0)$. The requirements of the theorem are not met at the place where uniqueness is violated.

Example 3 Verify that the initial-value problem

$$\frac{dy}{dx} = y$$

with the initial condition $y(0) = 1$ has a unique solution.

Solution
In this problem,

$$f(x, y) = y, \; x_0 = 0, \quad \text{and} \quad y_0 = 1.$$

Hence, both f and $\partial f/\partial y$ are continuous on all rectangular regions containing the point

$$(x_0, y_0) = (0, 1).$$

By the Existence and Uniqueness Theorem, there exists a unique solution to the differential equation that satisfies the initial condition $y(0) = 1$. We verify this by solving the initial-value problem.
    The equation is separable and equivalent to

$$\frac{dy}{y} = dx.$$

A general solution is given by $y = ce^x$ and with the initial condition $y(0) = 1$ gives

$$y = e^x.$$

The Existence and Uniqueness theorem gives sufficient, but not necessary, conditions for the existence of a unique solution of an initial-value problem. That is, there may be differential equations that do not satisfy the theorem but have solutions that are all unique.

Example 4 Consider the initial-value problem

$$\frac{dy}{dx} = \frac{x}{y^{2/3}}$$

with $y(0) = 0$. The Existence and Uniqueness theorem does not guarantee the existence of a solution because $xy^{-2/3}$ is discontinuous at $(0,0)$. The equation is separable and has solution

$$\frac{3}{5}y^{\frac{5}{3}} = \frac{1}{2}x^2 + c$$

so that

$$y = \left(\frac{5}{6}x^2 + c\right)^{3/5}.$$

Using the initial condition gives the *unique* solution

$$y = \left(\frac{5}{6}\right)^{\frac{3}{5}} x^{\frac{6}{5}}.$$

The theorem only provides sufficient conditions for a unique solution to exist. It does <u>not</u> say a unique solution won't exist if the conditions are <u>not</u> satisfied.

Example 5 Consider the initial-value problem

$$x\frac{dy}{dx} = y$$

with $y(0) = 0$. We can rewrite this as

$$\frac{dy}{dx} = \frac{y}{x}.$$

We see that $f(x,y) = y/x$ and thus $\partial f/\partial y = 1/x$. The Existence and Uniqueness theorem does not guarantee the existence of a solution because $f(x,y)$ is not continuous at the point $(0,0)$.

The equation is separable and we can calculate the solution as

$$y = Cx.$$

Now, the initial condition $y(0) = 0$ gives the identity $0 = 0$, which means that $y = Cx$ is a solution to the differential equation for <u>any</u> value of C. Thus, there are infinitely many solutions to the initial-value problem; see Figure 2.7. Hence, the solution of the initial-value problem is *not unique*.

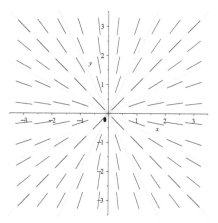

FIGURE 2.7: Direction fields for $\frac{dy}{dx} = \frac{y}{x}$. Note that all the solutions shown here satisfy the initial condition $y(0) = 0$.

Problems

In Problems **1–6** determine whether the Existence and Uniqueness theorem guarantees a unique solution to the following initial-value problems on some interval. Explain.

1. $\dfrac{dy}{dx} + x^2 = y^2$, $y(0) = 0$

2. $y' = x^2 - y^2$, $y(0) = 1$

3. $\dfrac{dy}{dx} + x^2 = y^{-2}$, $y(0) = 0$

4. $y' = x \ln y$, $y(1) = 1$

5. $\dfrac{dy}{dx} = y + \dfrac{1}{1-x}$, $y(1) = 0$

6. $\dfrac{dy}{dx} + \csc x = e^y$, $y(0) = 0$

In Problems **7–14** determine the region(s) of the x-y plane where solutions (i) exist and also where solutions (ii) exist and are unique, according to the Existence and Uniqueness theorem.

7. $\dfrac{dy}{dx} = 3x(y + 2)^{2/3}$

8. $\dfrac{dy}{dx} = 2xy^{2/3}$

9. $\dfrac{dy}{dx} = (x - y)^{1/5}$

10. $\dfrac{dy}{dx} = 2x^{2/3}y$

11. $\dfrac{dy}{dx} = x^2 y^{-1}$

12. $\dfrac{dy}{dx} = xy^{-2/3}$

13. $\dfrac{dy}{dx} = (x + y)^{-2}$

14. $\dfrac{dy}{dx} = xy^{2/3}$

In Problems **15–20**, consider the given initial-value problem. (i) Use the Existence and Uniqueness theorem to determine if solutions will exist and be unique. (ii) Solve the initial-value problem to obtain an analytic solution.

Use this analytic solution to determine whether solution(s) passing through the given IC will be unique.

15. $\frac{dy}{dx} = 5(y-2)^{3/5}$, $y(0) = 2$ **16.** $\frac{dy}{dx} = \frac{x}{y^2}$, $y(1) = 0$

17. $\frac{dy}{dx} = 2y^{1/2}$, $y(1) = 3, y \geq 0$ **18.** $\frac{dy}{dx} = x^{1/3}(1-y^2)$, $y(0) = 1$

19. $\frac{dy}{dx} = 3(y-1)^{1/3}$, $y(0) = 1$ **20.** $\frac{dy}{dx} = \frac{\sin x}{(y-1)^2}$, $y(0) = 1$

21. According to the Existence and Uniqueness theorem, the initial-value problem

$$\frac{dy}{dx} = |y|$$

with $y(1) = 0$ has a solution. Must this solution be unique? Solve this equation by considering two cases, $y \geq 0$ and $y < 0$, and using separation of variables. Is the solution unique?

2.3   First-Order Autonomous Equations—Geometrical Insight

We will now consider a method of geometrical analysis of differential equations to solve **autonomous** first-order equations.

> **Definition 2.3.1**
>
> A first-order differential equation of the form
>
> $$\frac{dy}{dx} = f(y) \tag{2.3}$$
>
> is called autonomous when f does not depend explicitly on the independent variable x.

We develop an important method of analyzing a differential equation to determine its behavior without solving it. *It is recommended that the reader be very familiar with the review material in Section 2.3.1 on quickly graphing factored polynomials.*

We first note that any first-order autonomous equation is separable and can thus be rewritten as

$$\int \frac{dy}{f(y)} = \int dx.$$

The integral on the left, however, is often not a simple integral. Our reason for needing to evaluate the integral was our desire to find the explicit (or implicit) solution, thereby allowing us to plot solutions in the x-y plane. For autonomous equations, however, we will gain insight into our solution in a qualitative manner. We will still be able to sketch solutions in the x-y plane but will lose the ability to answer a question such as "given the initial condition of $y(0) = 1$, what is the value of the solution at $x = 3$?" Our "qualitative description" means that we will be able to describe the long-term behavior of the solutions even though we will not write an explicit or implicit formula for the solution; that is, we will be able to answer the question "given the initial condition of $y(0) = 1$, what is the value of the solution as $x \to \infty$?"

The Phase Line

Our goal is to be able to predict the solution curve of the differential equation given in (2.3). We will gain an understanding of the behavior of the solution by observing that the expression $\frac{dy}{dx}$ is the rate of change of y with respect to x.

Definition 2.3.2

An equilibrium solution of the autonomous equation $y'(x) = f(y)$ is any constant, $y = C$, that is a solution of the differential equation.

By inspection, we see that $\frac{d(C)}{dx} = 0$ for any constant so that an equilibrium solution,[1] y^*, satisfies $f(y^*) = 0$. When $\frac{dy}{dx} > 0$, y is increasing; and when $\frac{dy}{dx} < 0$, y is decreasing. We can say these are *qualitatively different* behaviors of the solution. Thus for (2.3) we can say whether the solution y is increasing or decreasing simply based on the sign of $f(y)$.

Example 1   Consider the exponential differential equation

$$\frac{dy}{dx} = ry. \tag{2.4}$$

The equation is separable with solution $y = Ce^{rx}$, where C is determined by the initial condition. We could plot these solutions but instead we want to use our "qualitative approach." To do so, we observe that the right-hand side is simply a straight line whose slope depends on r. There are two qualitatively different cases: $r > 0$ (exponential growth) and $r < 0$ (exponential decay). The horizontal axis is the y-axis and the vertical axis represents $\frac{dy}{dx}$; see Figure 2.8. When $r > 0$, the derivative is always positive for $y > 0$, implying that the solution y is increasing. This is indicated by an arrow pointing to the right on the y-axis, with right being the direction of increase in y. In the other case, the arrow points to the left because the derivative is always negative. The y-axis with the equilibria and corresponding arrows drawn on

---

[1] We note that some books refer to equilibrium solutions as **critical points** because they satisfy the calculus definition $\frac{dy}{dx} = 0$. We will plot solutions in the y-$\frac{dy}{dx}$ plane and a "critical point" of the graph in this plane is *not* the same as the equilibrium solutions we are focusing on here. Thus, we will refer to y-values that satisfy $\frac{dy}{dx} = 0$ as equilibria.

it is called the **phase line** and contains the essential information necessary to describe the long-term behavior of the solution, i.e., what is $\lim_{x\to\infty} y(x)$ for a given initial condition $y(x_0) = y_0$. More can be said although it does not show in the arrows. In the first diagram, the derivative function is growing, which indicates not only that y is growing, but it is growing more and more rapidly.

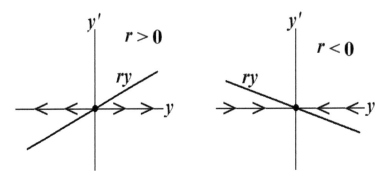

FIGURE 2.8:   Phase line diagrams for Equation (2.4).

We said that an equilibrium solution, y^*, corresponds to $\frac{dy}{dx} = 0$ and we thus have one equilibrium at $y^* = 0$. If our initial condition is *on* the equilibrium solution, e.g., $y(0) = 0$, then we will always remain at that equilibrium solution. For a general equilibrium, y^*, we state this as

$$\lim_{x\to\infty} y(x) = y^* \text{ for the initial condition } y(x_0) = y^*.$$

Note that y still depends on x (even though there is no x that is seen in the picture!) and the arrows indicate what happens to y as x increases.

It is natural for us to ask what happens to the solution of a differential equation as it proceeds from an initial condition close to an equilibrium solution. We will still consider the differential equation of Example 1 to gain insight into this. We will first examine the direction field for the two distinct cases, $r > 0$ and $r < 0$; see Figure 2.9. The unique feature about the direction field of an autonomous equation is that the equilibrium solution is represented by a horizontal line at the value y^*. Solutions starting near this horizontal line will either approach it or diverge away from it.

If we consider (a) corresponding to $r > 0$, we see that if we start with an initial value $y_0 = 0$, we will always stay at $y = 0$. If we start with an initial y-value $y_0 > 0$, our solutions will go to ∞, while if we start with an initial y-value $y_0 < 0$, our solutions will go to $-\infty$. Thus, starting anywhere other than the equilibrium solution will yield a solution that diverges away from $y^* = 0$. We summarize the behavior near the equilibrium solution with a convenient definition. For convenience of notation, we will denote the derivative as y' instead of as $\frac{dy}{dx}$.

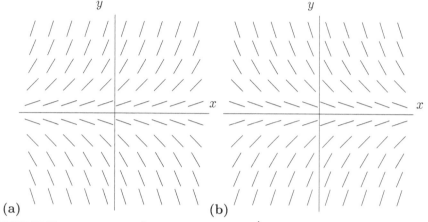

FIGURE 2.9: The y-y' direction field for $\frac{dy}{dx} = ry$. In **(a)**, we have $r > 0$ and in **(b)** we have $r < 0$.

Definition 2.3.3

Consider the autonomous differential equation $y' = f(y)$, where $y = y(x)$. Assume that y^* is an equilibrium solution. Then

1. y^* is said to be **asymptotically stable** if all solutions with an open set of initial conditions close by it approach y^* as $x \to \infty$.

2. y^* is said to be **stable** if all solutions with an open set of initial conditions close by it always remain close by (but do not necessarily approach it) as $x \to \infty$.

3. y^* is said to be **unstable** if it is not stable.

Note that every equilibrium solution that is asymptotically stable is also stable. We should also note that the equilibrium solution always corresponds to horizontal line segments in the direction field. Thus, a quick inspection of a direction field can tell us whether or not it corresponds to an autonomous equation.

Relating this definition to our previous example, we see that the equilibrium solution $y^* = 0$ is unstable if $r > 0$ because solutions with an initial condition close by y^* do not remain close by. When $r < 0$, however, solutions starting close by y^* approach it and thus $y^* = 0$ is asymptotically stable.

Our definition of stability can be put in terms of the traditional ϵ-δ definitions used in calculus. Specifically, y^* is stable if, for each $\epsilon > 0$, there exists a $\delta > 0$ such that

$$|y(x) - y^*| < \epsilon \text{ whenever } |y_0 - y^*| < \delta \text{ for all } x > 0,$$

where y_0 is the initial y-value.

Remark: In the case of a first-order autonomous equation, we can have an equilibrium solution that is **half-stable**. In this case, it is stable from one side but unstable from the other. A simple example is $y' = y^2$ where initial conditions to the left of $y = 0$ approach $y = 0$, whereas initial conditions to the right of $y = 0$ go to ∞.

What was the purpose of using the phase line if we ultimately plotted the direction field and concluded information about the stability of y^*? We actually did not need the direction field to reach the conclusions that we did— everything was already contained in the phase line. Specifically, the arrows on the line were drawn because they described how the solution was changing. In the case of $r > 0$, any solution starting with an initial condition to the right of $y^* = 0$ increased (as x increased) and thus moved away from y^*, whereas any solution starting with an initial condition to the left of $y^* = 0$ decreased (as x increased) and thus moved away from y^*. We can think of the arrows as pushing the solution in a given direction. On both sides of $y^* = 0$, the arrows are pushing the solution away and thus we classify y^* as unstable. For $r < 0$, it is the opposite situation and we see that the arrows are pointing toward $y^* = 0$ from both sides, thus indicating that solutions approach y^* as x increases. In this case, we consider $y^* = 0$ an asymptotically stable equilibrium solution. We can also sketch the solutions in the x-y plane without knowledge of the direction field. The information contained in the phase line is sufficient to sketch qualitatively accurate solutions.

By examining the phase line, we have the ability to explain what happens to the solution as $x \longrightarrow \infty$. As mentioned earlier, this description is a *qualitative* one—we cannot answer the question "given that $y(1) = -2$, what is $y(8)$?" Giving a *quantitative* answer to this question would require having the explicit or implicit solution, which may or may not be an easy solution to obtain. Using the phase line, we can also still give a plot of the solutions in the x-y plane but we again lose our ability to describe the value of the solution at a particular x-value. The usefulness of the phase line is in its ability to *quickly* give us qualitative information about *any* solution as $x \to \infty$.

Example 2   Consider the autonomous equation

$$y' = y^2(1 - y^2). \tag{2.5}$$

Find the equilibria. Draw the corresponding phase line diagram and use it to determine the stability of the equilibrium solutions. Use the information contained in the phase line to sketch solutions in the x-y plane. Then conclude the long-term behavior for *any* initial condition.

Solution

To determine the equilibria, we set $y' = 0$ and solve for y. This gives us equilibria of $y = 0, -1, 1$, where $y = 0$ is a double root. The highest power and corresponding coefficient is $-y^4$ and thus we can use the information from Appendix B to quickly sketch the curve in the y-y' plane (see Figure 2.10). To draw the arrows on the phase line, we simply need to observe where the derivative y' is positive vs. negative. Between the equilibria $y^* = -1$ and $y^* = 0$ as well as between $y^* = 0$ and $y^* = 1$, we see that $y' > 0$ because the curve is above the horizontal axis. Thus we draw arrows going to the right to indicate an increase in the solution value. To the left of $y^* = -1$ and to the right of $y^* = 1$, we see that $y' < 0$ because the curve is below the horizontal axis and thus we draw arrows going to the left.

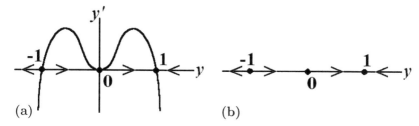

FIGURE 2.10: View of (2.5) in the (a) y-y' plane and (b) phase line.

Examining the phase line from the y-y' plane (see Figure 2.10), we see that the arrows on both sides of $y^* = -1$ are pointing away from it and it is thus unstable. The arrows on both sides of $y^* = 1$ are pointing toward it and it is thus asymptotically stable. For $y^* = 0$, we observe that arrows to the left of it point toward it but arrows to the right of it point away from it. Thus $y^* = 0$ is half-stable.

To sketch solutions in the x-y plane (see Figure 2.11), we first draw the equilibrium solutions as horizontal lines. We next note that solutions will always be unique and thus *we will never be able to cross these equilibrium solutions*. If we begin with the initial condition $y_0 < -1$, the arrows on the phase line indicate that solutions will go to $-\infty$ as $x \to \infty$. We thus draw curves in the x-y plane that go away from the line $y = -1$. If the initial condition satisfies $-1 < y_0 < 0$, the arrows on the phase line indicate that we will approach the equilibrium solution $y^* = 0$. We thus draw curves in the x-y plane that go away from $y = -1$ and toward $y = 0$. Because the value of y' is small near the equilibria, the slope of the corresponding solution curve in the x-y plane will be small. Thus as $x \to -\infty$, the solution will have $y = -1$ as a horizontal asymptote, whereas when $x \to \infty$ the solution will have $y = 0$ as its horizontal asymptote. We can follow similar reasoning to draw solutions that begin with initial condition $0 < y_0 < 1$ and $y > 1$.

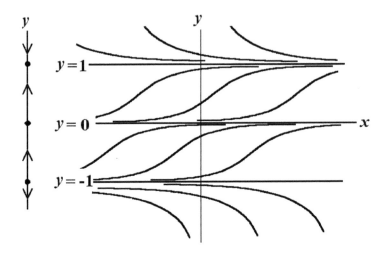

FIGURE 2.11: Sketch of some solutions of (2.5), based exclusively on phase line information; phase line (on left) from Figure 2.10 is now drawn vertically for easier comparison.

Equipped with the above information, we can now answer the final question of this example. Mathematically, we want to know $\lim_{x \to \infty} y(x)$ for any given $y(x_0) = y_0$. This would be a very challenging task if we had the explicit solution in front of us. With the phase line, however, we simply need to describe what happens to solutions with initial conditions between equilibrium solutions.

(i) if $y_0 < -1$, then $y(x) \to -\infty$ as $x \to \infty$
(ii) if $-1 < y_0 < 0$, then $y(x) \to 0$ as $x \to \infty$
(iii) if $0 < y_0 < 1$, then $y(x) \to 1$ as $x \to \infty$
(iv) if $y_0 > 1$, then $y(x) \to 1$ as $x \to \infty$

It is this type of "qualitative description" and knowledge of the "long-term behavior" that we are able to gain from considering the phase line instead of the explicit solution. Depending on the situation you may be considering, one of these approaches may be more desirable than the other.

Stability of Equilibrium via Taylor Series

Up to this point, our description of the equilibrium solutions has been based on insight gained from the phase line. Although this has been useful, it is often desirable to have an analytical method for determining stability of equilibria. This will be completely independent of the graphical interpretation we just learned but we will see that a combination of both methods will often be useful in analyzing a given problem.

Recall that the Taylor series expansion of $f(y)$ near y^* is given by

$$f(y) = f(y^*) + f'(y^*)(y - y^*) + \frac{1}{2!}f''(y^*)(y - y^*)^2 + \frac{1}{3!}f^{(3)}(y^*)(y - y^*)^3 + \cdots .$$

(2.6)

Taylor series are often useful in approximating the value of a function near a point y^* by only considering a finite number of terms of the Taylor series. In some situations, we are able to gain very good approximations with only the first two or three terms in the expansion.

We now expand the right-hand side of our autonomous equation in a Taylor series about the equilibrium solution, y^*:

$$y' = f(y^*) + f'(y^*)(y - y^*) + \cdots .$$

How has this helped us? We first observe that $f(y^*) = 0$ because y^* is an equilibrium solution. Then we observe that for y-values very close to y^*, the terms $(y - y^*)^2, (y - y^*)^3$, etc. are small compared to $f'(y^*)(y - y^*)$. Thus, close by the equilibrium y^*, we see that

$$y' \approx f'(y^*)(y - y^*)$$

(2.7)

is a good approximation. But this differential equation is separable and we know the solution is an exponential function. Moreover, the behavior of the solution will depend only on the sign of $f'(y)$. We can summarize this result in a useful theorem.

THEOREM 2.3.1

Consider the autonomous differential equation $y' = f(y)$, with equilibrium solution y^*. If f has a Taylor expansion about y^*, then
(a) y^* is stable when $f'(y^*) < 0$ and
(b) y^* is unstable when $f'(y^*) > 0$.

Note 1: In (a), the condition $f'(y^*) < 0$ says that the graph in the y-y' plane *decreases* through the horizontal axis, whereas in (b), the condition $f'(y^*) > 0$ says that the graph in the y-y' plane *increases* through the horizontal axis.

Note 2: The stability of y^* cannot be determined by this theorem in the case when $f'(y^*) = 0$. In this situation, the stability of y^* is determined by the higher-order terms of the Taylor expansion. We would need to use other means, e.g., the phase line, in order to conclude anything about the stability of the equilibrium.

Example 3: Find the equilibrium solutions of the autonomous equation $y' = y^2(1 - y^2)$. Use Theorem 2.3.1 to determine their stability.

Solution

We see the equilibria are $y = -1, 0, 1$. Our function is $f(y) = y^2(1 - y^2) = y^2 - y^4$ and the derivative is $f'(y) = 2y - 4y^3$. Applying the previous theorem shows that

(i) $f'(-1) = 2(-1) - 4(-1)^3 = 2 > 0 \implies y^* = -1$ is unstable

(ii) $f'(1) = 2(1) - 4(1)^3 = -2 < 0 \implies y^* = 1$ is stable

(iii) $f'(0) = 2(0) - 4(0)^3 = 0 \implies$ stability of $y^* = 0$ is inconclusive. In order to determine the stability of $y^* = 0$, we could use the graphical method of Example 2 to find that it is half-stable.

2.3.1   Graphing Factored Polynomials

We give a brief reminder on how to graph factored polynomials quickly. We will not be concerned with maxima, minima, inflection points, etc. Instead we care only about whether a graph is positive or negative between roots, how it passes through a root, and its behavior for large x-values. Knowing how to graph these quickly will be especially helpful in Sections 2.3-2.4, which deal with first-order autonomous equations and their applications.

We will assume that we begin with a polynomial $P(x)$ in the form

$$a_0 x^{k_0} (a_1 x - b_1)^{k_1} (a_2 x - b_2)^{k_2} \cdots (a_n x - b_n)^{k_n} (c_m x^m + \cdots c_1 x + c_0)^{k_q}, \quad (2.8)$$

where the a_i, b_i, c_i are real numbers and where the last factor $c_m x^m + \cdots c_1 x + c_0$ is assumed to be a polynomial with no real roots. That is, it does not factor any further over the field of real numbers. Note that $c_m x^m + \cdots c_1 x + c_0$ might be able to be factored as a product of quadratic factors but since our concern will be with the term with the highest power and it leading coefficient, we will leave it in this expanded form. For example, it could be written as

$$(x^2 + 1)(x^2 + 6)^2$$

or it could be written out as

$$x^6 + 13x^4 + 48x^2 + 36.$$

The latter form matches the final factor in (2.8). To be a bit more explicit, we are requiring

$$a_i \neq 0, \quad i = 0, 1, 2, \ldots n, \qquad c_m \neq 0, \quad c_0 \neq 0.$$

The roots of (2.8) are given by

$$r_i = \frac{b_i}{a_i} \quad \text{with } r_i \neq r_j \text{ if } i \neq j, \tag{2.9}$$

and each root r_i occurs k_i times where $k_i \geq 0$ is a nonnegative integer. For the $a_0 x^{k_0}$ factor, we should observe that $b_0 = 0$.

In order to sketch the curve, we need to know how many times each root occurs (its multiplicity), the highest power in the polynomial $P(x)$ given in (2.8), and the sign of the coefficient of this highest power. The roots are given by (2.9). For the last two needed items, we multiply the terms with highest power of x from each factor. That is, we consider

$$a_0 x^{k_0} (a_1 x)^{k_1} (a_2 x)^{k_2} \cdots (a_n x)^{k_n} (c_m x^m)^{k_q} \tag{2.10}$$

which can be simplified slightly as

$$a_0 (a_1)^{k_1} (a_2)^{k_2} \cdots (a_n)^{k_n} (c_m)^{k_q} x^{k_0 + k_1 + k_2 + \cdots + k_n + m k_q} \equiv A x^K, \tag{2.11}$$

where

$$A = a_0 (a_1)^{k_1} (a_2)^{k_2} \cdots (a_n)^{k_n} (c_m)^{k_q} \quad \text{and} \quad K = k_0 + k_1 + k_2 + \cdots + k_n + m k_q. \tag{2.12}$$

With the knowledge of A, K, and the roots (and their multiplicity), we can sketch the curve with the two charts seen in Figure 2.12.

We finish this brief review with two examples.

Example 4 Sketch the polynomial

$$P(x) = (x - 1)^2 (2 + 3x)(4 - x)^3 (x + 2).$$

Solution

We see that the polynomial has the following roots and multiplicities:

root	x=1	x=-2/3	x=4	x=-2
multiplicity	2	1	3	1

Calculating the highest power and its coefficient gives

$$(x)^2 (3x)(-x)^3 (x) = -3x^7.$$

We are in the case K odd and $A < 0$ and we can identify the long term behavior from Figure 2.12. To find the behavior near the roots, we go from left to right. The first root we encounter is $x = -2$ and it occurs once. Since we are coming from above, we must pass through the root as in 1a of Figure 2.12(b). The second root we encounter is $x = -2/3$ and it occurs once. Since we are coming from below, we must pass through it as in 1b of Figure 2.12(b). The next root we encounter is $x = 1$ and it occurs twice. Since we are coming from above, we must pass through it as in 2b. The final root is $x = 4$ and it occurs 3 times. Since we are now coming from above, we must pass through it as in 3a of Figure 2.12(b). Note that we can match up with the long term behavior for large positive x. A sketch is given in Figure 2.13. Note that for these rough sketches, we do *not* care about the relative maxima, etc. Our goal is to know when the graph is positive versus negative and how it passes through the roots. Thus, we will not draw a scale on the y-axis.

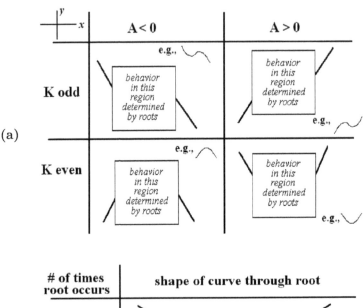

(a)

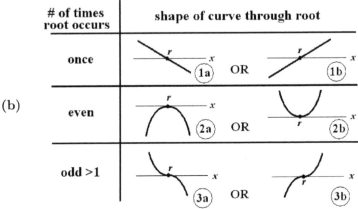

(b)

FIGURE 2.12: (a) Behavior of factored polynomial (2.8) for large $|x|$. The curves obey $\lim_{x\to-\infty} P(x)$ and $\lim_{x\to\infty} P(x)$. (b) Behavior near the roots of factored polynomial (2.8).

Example 5 Sketch the polynomial

$$P(x) = -x^3(x^2 - 4)(x - 1)(5 - x)(3 + 3x)(4 - 2x)^2(x^2 + 1).$$

Solution

Noting that $(x^2 - 4)$ factors into $(x + 2)(x - 2)$, we see that this polynomial has the following roots and multiplicities:

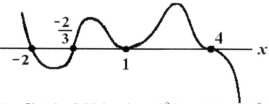

FIGURE 2.13: Sketch of $P(x) = (x-1)^2(2+3x)(4-x)^3(x+2)$.

root	x=0	x=2	x=-2	x=1	x=5	x=-1
multiplicity	3	3	1	1	1	1

Calculating the highest power and its coefficient gives

$$-x^3(x^2)(x)(-x)(3x)(-2x)^2(x^2) = +12x^{12}.$$

We are in the case K even and $A > 0$ and we can identify the long term behavior from Figure 2.12. To find the behavior near the roots, we go from left to right. The first root we encounter is $x = -2$. It occurs only once and since we are coming from above, we must pass through it as in 1a of Figure 2.12. The next root we encounter is $x = -1$. We are now coming from below and because it occurs once, we pass through it as in 1b. The next root is $x = 0$ and it occurs three times. Thus we pass through it as in 3a of Figure 2.12. The next root is $x = 1$ and it occurs once. Since we are now coming from below we pass through it as in 1b. The next root is $x = 2$ which occurs three times. We are now coming from above and thus pass through it as in 3a. The final root is $x = 5$ and occurs only once. Thus we pass through as in 1b. Note that we can match up with the long term behavior for large positive x. A sketch is given in Figure 2.14.

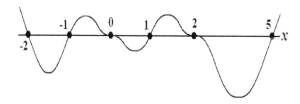

FIGURE 2.14: Sketch of $P(x) = -x^3(x^2-4)(x-1)(5-x)(3+3x)(4-2x)^2(x^2+1)$.

2.3.2   Bifurcations of Equilibria

Often we have a situation where the stability of a given equilibrium solution depends on the value of a *parameter*. We saw this in the first example when we considered $y' = ry$. In this case, r was the parameter and we saw that $y^* = 0$ was stable when $r < 0$, whereas $y^* = 0$ was unstable when $r > 0$. Thus changing the r-value changes the stability of the equilibrium solution. We could also have a change in the number of equilibria and we state the following definition:

Definition 2.3.4

Consider the equation $y' = f(r, y)$, where f is an infinitely differentiable function and r is a parameter. A **bifurcation** is a qualitative change in the number or stability of equilibrium solutions for this equation.

There are three bifurcations that we will consider: transcritical, saddle-node, and pitchfork. We will give examples of the first two and will leave the third as an exercise.

Example 6 (Transcritical Bifurcation)

Consider the equation

$$y' = ry - y^2, \tag{2.13}$$

where $r \in \mathbb{R}$. The equilibria are $y^* = 0, r$ and, since r may change, we consider the three cases: $r < 0$, $r = 0$, and $r > 0$. We could check the stability analytically using the Taylor series method previously discussed. However, it is more instructive to see the change in the stability of the solutions with the phase line; see Figure 2.15.

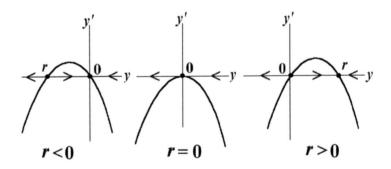

FIGURE 2.15: Phase line view of a transcritical bifurcation. The bifurcation happens **at** the value $r = 0$ but we determine the type of bifurcation by observing what happens before (for $r < 0$) and after (for $r > 0$).

We first note that $y^* = 0$ is always an equilibrium solution. When $r < 0$, there is another equilibrium solution at $y^* = r$, lying to the left of $y^* = 0$. Drawing arrows on the phase line, we see that $y^* = r$ is unstable and $y^* = 0$

is stable. As r increases, the non-zero equilibrium approaches $y^* = 0$. At the bifurcation value of $r = 0$, the equilibrium solution $y^* = 0$ is a double root and is half-stable. As r becomes positive, the non-zero equilibrium solution is now to the right of $y^* = 0$. Drawing arrows on the phase line, we see that $y^* = 0$ is now unstable and $y^* = r$ is stable.

This bifurcation is known as a **transcritical bifurcation**. It is usually easiest to think of in terms of the parameter r. When r is negative, we have two equilibrium solutions; as r continues to increase, the solutions approach each other and "crash" into each other when $r = 0$, switching stability as they do so; for $r > 0$, the solution $y^* = 0$ is now the unstable one and $y^* = r$ is the stable one. We stated early in this example that there were three cases to examine (before, at, and after the bifurcation) and we finish by noting that sometimes we may not be able to figure the qualitative situations without first doing a bit of work.

Example 7 (Saddlenode Bifurcation)
Consider the equation

$$y' = r - y^2, \tag{2.14}$$

where $r \in \mathbb{R}$. The equilibria are $y^* = \pm\sqrt{r}$, which exist when $r \geq 0$. We again have three cases to consider: $r < 0$, $r = 0$, and $r > 0$. When $r < 0$, there are no equilibrium solutions. Drawing arrows on the phase line, we see that any initial condition gives a solution that diverges to $-\infty$. When $r = 0$, an equilibrium solution appears at $y^* = 0$ and is half-stable. Solutions beginning with an initial condition $y_0 < 0$ go to $-\infty$, whereas an initial condition $y_0 > 0$ gives a solution that approaches 0. When $r > 0$, there are now two equilibria. Drawing arrows on the phase line shows that the equilibria at $y^* = -\sqrt{r}$ is unstable and $y^* = \sqrt{r}$ is stable; see Figure 2.16.

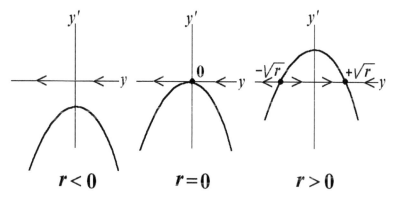

FIGURE 2.16: Phase line view of a saddlenode bifurcation. The bifurcation happens **at** the value $r = 0$ but we determine the type of bifurcation by observing what happens before (for $r < 0$) and after (for $r > 0$).

This bifurcation is known as a **saddlenode bifurcation** and we again think of it in terms of the parameter r. When r is negative, we have no equilibrium solutions; as r continues to increase, an equilibrium solution appears on the phase line, seemingly out of nowhere and is half-stable. For $r > 0$, this equilibrium then breaks into two parts, an unstable and a stable equilibrium.

It is sometimes useful to plot the equilibrium solutions versus the parameter. We call this a **bifurcation diagram** and these are shown in Figure 2.17.

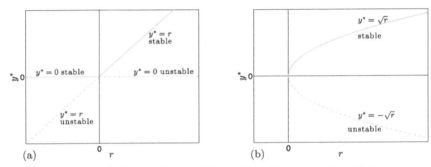

FIGURE 2.17:   (a) Transcritical bifurcation diagram.   Equilibria curves obtained from Figure 2.15. (b) Saddlenode bifurcation diagram. Equilibria curves obtained from Figure 2.16.

In both of these examples, we were able to vary the one parameter r in order to observe the bifurcation. We thus say that there was a **one parameter family** associated with this bifurcation. We can also have two parameters present and both can change to give these bifurcations. In these cases, we would say that we have a **two parameter family** that gives us the bifurcations. However, we could always perform various substitutions to rewrite the equation so that only one parameter is really needed in each of the bifurcations. Because of this, we refer to the transcritical and saddlenode bifurcations as being of **co-dimension 1**. The two types of pitchfork bifurcations that are explored in the Problems are also co-dimension 1 bifurcations.

•   •   •   •   •   •   •   •   •   •   •   •

Problems

For Problems **1–20**, (i) sketch the differential equation in the y-y' plane by hand. (ii) Draw the phase line diagram, clearly stating the location and stability of the equilibria. (iii) State the long-term behavior *for all* initial conditions. (iv) Assuming $y = y(x)$, sketch the corresponding graph in the traditional x-y plane.

1. $y' = 2y + 3$
2. $y' = -3y + 2$
3. $y' = y^2 - y - 6$
4. $y' = y(y+2)(y-3)$
5. $y' = y^2 + 4y + 4$
6. $y' = -y^2$
7 $y' = y^2(2-y)$
8. $y' = (y-1)^2(y-2)^3(1+y)$
9. $y' = (y-2)^3(y^2-9)$
10. $y' = \cos y + 1, \quad -2\pi < y < 2\pi$
11. $y' = \sin y, \quad -2\pi < y < 2\pi$

For the next two problems, assume that $v = v(t)$. Continue with parts (i)-(iv) from the previous problems but note that (i) will be in the v-v' and (iv) will be in the t-v plane. The equations were discussed in detail in Section 1.4 as they are used to model the motion of a free-falling object. If $v(t)$ represents velocity, interpret the stability of the resulting equilibria.

12. $v' = g - \frac{k}{m}v$
13. $v' = g - \frac{k}{m}v^2$

For the next seven problems, continue with (i)-(iv) but assume that $x = x(t)$. Note that (i) will be sketched in the x-x' plane and (iv) will be sketched in the t-x plane.

14. $x' = (2-x)^3(x^2+4)$
15. $x' = (2-x)^3(x^2+4)^2$
16. $x' = -x^2(4-x)(9-x^2)$
17. $x' = x^5(1-x)(1-x^3)$
18. $x' = x(x-3)(1+x^3)(1-x^2)^2$
19. $x' = x^2(1-2x)^3(x^2-1)$
20. $x' = x^3(x^2+5)(x-4)^2(x+5)$

Use the analytic approach of Theorem 2.3.1 to determine the stability of the equilibria in Problems **21–24**.

21. $y' = y^2 - 1$
22. $y' = 1 - y^2$
23. $y' = y^3 + 1$
24. $y' = -y^3$

In Problems **25–27**, let the parameter r vary and determine the bifurcation that occurs by drawing the two qualitatively different phase line portraits. State the stability of the equilibria for both pictures. Draw the bifurcation diagram as well.

25. $y' = r + y^2$
26. $y' = 1 - r + y^2$
27. $y' = ry + y^2$

28. A pitchfork bifurcation occurs when two equilibria appear out of one and change the stability of the existing equilibria as they appear; three equilibria are present after the bifurcation has occurred. Draw the phase line view of the following two bifurcations (before, at, and after the bifurcation). Then draw the bifurcation diagram.
(a) $y' = ry - y^3$ (**supercritical pitchfork bifurcation**)
(b) $y' = ry + y^3$ (**subcritical pitchfork bifurcation**)
For both types of pitchfork bifurcations, you will need to consider the three different cases $r < 0$, $r = 0$, and $r > 0$.

29. Consider the autonomous equation $y' = f(y)$ and suppose it is known that $f'(y^*) = 0$, $f''(y^*) = 0$, $f^{(3)}(y^*) < 0$ for the equilibrium solution y^*. Can we conclude anything about the stability of this solution?

2.4   Modeling in Population Biology

The autonomous equations of the previous section have application in the area of population modeling. We can derive some important equations by thinking of the *per capita* rate of change of a population. We will let $x(t)$ denote the given population. The rate of change is simply the derivative with respect to time and the per capita rate of change is then given by

$$\frac{1}{x}\frac{dx}{dt}.$$

We have considered the simplest model of population growth, where we assume the per capita rate of change is constant and is equal to some parameter r:

$$\frac{1}{x}\frac{dx}{dt} = r.$$

Rearranging this gives the familiar exponential growth differential equation that we have seen before: $x' = rx$. The assumption that the rate of change is constant is not usually realistic for large populations. Why? Think of bacteria growing in a petri dish with finite space and nutrients with which to grow. When the population is small, the exponential growth model is acceptable, but when limited space and resources become an issue, the assumptions of this model break down. A more accurate model arises if we assume a linear decrease in the per capita rate of change of the population. This basically says that the more bacteria there are, the less quickly the overall population will grow due to limited resources. If there are too many bacteria, the overall population level may not change or may even decrease. The equation can be written

$$\frac{1}{x}\frac{dx}{dt} = r - k_1 x,$$

which we will rewrite as

$$\frac{dx}{dt} = rx\left(1 - \frac{x}{K}\right),$$

where $K = r/k_1$. This is known as the **logistic** differential equation. For small populations, we note that the logistic equation is approximated well by the exponential model. To see this, observe that if x is much smaller than K, then $\frac{x}{K}$ is very small and the $rx(1 - \frac{x}{K})$ term is approximately rx. Thus for small x, the differential equation is approximately $\frac{dx}{dt} = rx$. For the logistic equation, we can conclude that if $x(t)$ is small relative to K, then $x(t)$ exhibits nearly exponential growth. When $x(t)$ is near K, it levels out to be nearly constant. The behavior in-between is harder to predict, but it would seem reasonable that there is a smooth transition from exponential growth to no growth. If x is greater than K, the results are similar except the derivative is negative, so $x(t)$ starts off decaying exponentially until it nears K where it

levels out. These behaviors are illustrated in Figure 2.18(a) and we sketched
these solution curves without explicitly calculating the solution!

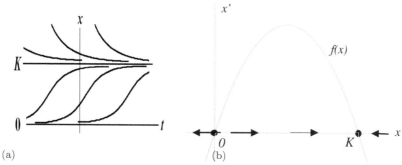

(a)   (b)

FIGURE 2.18:   (a) Qualitative behavior of the solutions to the logistic
equation. (b) Phase line for the logistic equation.

The constant r gives the intrinsic rate of growth of the population and the
constant K is called the **carrying capacity** since, if x represents a population,
the population rises or decays until it reaches this level. This can easily be
seen with the corresponding phase line given in Figure 2.18(b). If we start
with an initial condition x_0 in the interval $(0, K)$, then the solution $x(t)$ will
approach K as $t \to \infty$; if we start with an initial condition $x_0 > K$, then
$x(t)$ will again approach K as $t \to \infty$; if the initial condition satisfies $x_0 < 0$
(biologically meaningless but okay mathematically), then $x(t)$ will approach
$-\infty$ as $t \to \infty$. As mentioned earlier, we can also observe that the larger the
magnitude of $f(x)$, the larger the change in the solution. Thus, the solution
changes most rapidly at the value $K/2$. The change is very small near the
equilibria.

Since linear and exponential models are solved readily, one does not need to
resort to the phase line to analyze the behavior of their solutions. However,
the use of phase line analysis in these cases gives a quick illustration of the
behavior of solutions in the long run. With the quadratic or logistic model,
the value of the phase line analysis method becomes clear.

Again considering the logistic model, we could have solved it as a separable
equation using the technique of partial fractions in an intermediate step (or
as a Bernoulli equation) . The solution can be understood by looking at it in
the right way. But the qualitative analysis provided here gives us a way to
quickly describe the long-term behavior of the solution for any initial x-value
(i.e., any initial condition). In particular, we can readily see that *any* initial
population between 0 and K will grow. In terms of modeling, this is actually
a weakness in the logistic model. It suggests that a non-zero population, no
matter how small, will result in growth. There are several resolutions to this
problem. One is to put restrictions on the range of model validity. One might

say that this model holds only for populations above the size of, say, 100. On the other hand, one might want a model that tends toward zero if the population gets too small. One such model is an **Allee effect model**.

An Allee effect model is an improvement over the logistic model in that if a population gets too small the growth rate becomes negative. For small population or large population, we thus have the per capita growth rate being negative and we can write

$$\frac{1}{x}\frac{dx}{dt} = r(a - x)(x - b),$$

where $b > a > 0$. This is used to handle a number of complex processes that occur with small populations, such as difficulties finding mates or inbreeding effects, without complicating an otherwise simple model. The Allee effect model is also the next natural step in our progression for it involves $\frac{dx}{dt}$ with a cubic equation.

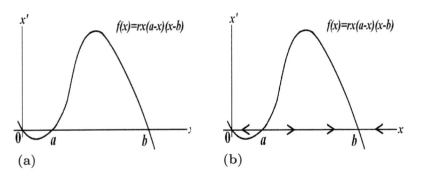

(a)   (b)

FIGURE 2.19:   (a) Cubic from Allee effect model.   (b) Phase line diagram for Allee effect model.

An Allee effect model is a cubic whose graph is given in Figure 2.19. The most useful forms to see this cubic equation are in the form $\frac{dx}{dt} = rx(a-x)(x-b)$, $r > 0$, $b > a > 0$, which is useful if the equilibria $x = a$, $x = b$ are known, and in the form $\frac{dx}{dt} = x(r - a(x - b)^2)$ (where the constants r, a, b would need to be chosen appropriately in either case to give the desired cubic shape). A typical Allee effect growth curve and phase line are presented in Figure 2.19. There are three equilibria: one corresponding to the carrying capacity, one corresponding to zero population, and one at the threshold between "large" and "small" population sizes. With this model, both the zero population and the carrying capacity are stable equilibria, which generally corresponds to intuition. The "large-small" threshold point is an unstable equilibrium. Populations that start out above this value tend toward the carrying capacity;

populations below this point tend to extinction.

We again note that the differential equation for the Allee effect model is separable and the resulting integral would involve partial fractions. Our qualitative phase line analysis was much quicker. We have now worked through analysis of three specific cases: linear, quadratic, and cubic. From this foundation, generalizations to other functions are obvious. A very nice discussion of population models with polynomial growth rates can be found in the technical report by Bewernick et al. [7].

Example 1 Consider a population that is governed by the Allee model with a carrying capacity of 20, a minimum threshold population of 5, and an intrinsic growth rate of 2. Write this differential equation and analyze its stability.

Solution

If we let N be the population, then comparing with the Allee model we see that $r = 2$, $a = 5$, $b = 20$ so that

$$\frac{dN}{dt} = 2N(5 - N)(N - 20).$$

The equilibrium solutions are $N^* = 0, 5, 20$. Analyzing stability either on the phase line or by examining $f'(N^*)$, we see that $N^* = 0$ is stable, $N^* = 5$ is unstable, and $N^* = 20$ is stable.

2.4.1   Nondimensionalization

In applications, it is often the case that parameters will arise in your model. As was seen in the previous section, changes in these parameters can often result in bifurcations in the system. It is quite often the case that it is not simply one parameter that can be changed but multiple parameters that can be changed. One way to simplify analytical and numerical computation that may be involved is to **nondimensionalize** the equations. In doing so, we rewrite the system in a form without units and often can group parameters together in such a way that it is easier to see the ones that are really affecting the system.

Example 2 Consider the logistic equation with constant effort harvesting, which will be derived in the next subsection:

$$\frac{dN}{dt} = rN\left(1 - \frac{N}{K}\right) - HN,$$

where N is the population, r its intrinsic growth rate, K the carrying capacity, and H the coefficient of harvesting.

We define a new population variable x and a new time variable τ:

$$x = \frac{N}{A}, \qquad \tau = \frac{t}{T}. \tag{2.15}$$

We note that x will be a dimensionless (i.e., without units) population if A has units of population and τ will be dimensionless time if T has units of time. We substitute for N and t in the ODE. On the left-hand side, we obtain

$$\frac{dN}{dt} = \frac{d(xA)}{d\tau}\frac{d\tau}{dt} = A\frac{dx}{d\tau}\frac{1}{T} = \frac{A}{T}x', \tag{2.16}$$

where $'$ denotes the derivative with respect to τ. On the right-hand side, we obtain

$$r(xA)\left(1 - \frac{xA}{K}\right) - H(xA) = A\left[rx\left(1 - \frac{x}{K/A}\right) - Hx\right]. \tag{2.17}$$

Substituting both into the ODE then yields

$$\frac{A}{T}x' = A\left[rx\left(1 - \frac{x}{K/A}\right) - Hx\right], \tag{2.18}$$

from which we immediately see that the A cancels from out front of both sides. Multiplying by T gives

$$x' = rxT\left(1 - \frac{x}{K/A}\right) - HTx. \tag{2.19}$$

At this point, we pause to see if we can make convenient choices for A and T that will simplify things for us. The rule of thumb is usually to start with the innermost functions, work your way out, and then lump remaining parameters together. Proceeding in this fashion, we choose $A = K$ as that will simplify the parentheses. The two choices we have remaining are to set $T = \frac{1}{r}$ or $T = \frac{1}{H}$. Both are equally correct but in finding bifurcations by hand, it's probably best to choose the former since graphing a line will be easier. Thus, we set $T = \frac{1}{r}$ and $h = HT = \frac{H}{r}$. We note that indeed the units of A are population and the units of T are time so that x and τ are dimensionless. Our equivalent ODE is then

$$x' = x(1 - x) - hx. \tag{2.20}$$

Mathematically, we see that this equation, with its lone parameter h, will be much easier to analyze than the original formulation with three parameters. We note that x is a parameter that is now scaled to the carrying capacity and τ is scaled to the intrinsic growth rate of the population.

For the general situation, let n be the order of the ODE or the number of first-order equations of a system of ODEs. In nondimensionalizing either of these, we introduce n new dimensionless variables and one new dimensionless time variable. We proceed as above and will be able to at least eliminate $n+1$ parameters. Sometimes the problem is structured so that we can eliminate more than that. If we start out with less than $n+1$ parameters, it will depend on the structure of the equation as to how "nice" we can make things look.

Harvesting

The example we considered in our nondimensionalization was a modification of the logistic equation under what we called **constant effort harvesting**. Let's go back to our formulation of the logistic equation in terms of per capita growth:

$$\frac{1}{N}\frac{dN}{dt} = r\left(1 - \frac{N}{K}\right),$$

where $N \geq 0$, $r, K > 0$. We consider a model of a fishery in which fish are harvested (i.e., caught) at a rate that is proportional to their population size, which is also referred to as constant effort harvesting. To model this, we substract off a constant H from this equation to denote the per capita removal rate:

$$\frac{1}{N}\frac{dN}{dt} = r\left(1 - \frac{N}{K}\right) - H.$$

Multiplying by N gives the familiar equation

$$\frac{dN}{dt} = rN\left(1 - \frac{N}{K}\right) - HN. \tag{2.21}$$

We want to be able to understand the long-term behavior of the solutions. However, in its present form (2.21) has 3 parameters and this can be difficult to analyze. Luckily, as we saw previously, we can nondimensionalize this and obtain an ODE with equivalent dynamics

$$x' = x(1 - x) - hx,$$

where $x \geq 0$, $h > 0$. To find equilibrium solutions, we set $x' = 0$ and obtain

$$x(1 - x) - hx = 0, \qquad \text{which factors as} \qquad x(1 - h - x) = 0.$$

We thus see that the equilibria are $x = 0, 1 - h$. Because the biological application requires $x \geq 0$, we observe that for increasing values of h, our system goes from having 2 equilibria to 1 equilibria. If we ignored the biological restriction of $x \geq 0$ and considered the mathematical version $x \in \mathbb{R}$, then we see that we go from 2 equilibria to 1 equilibrium (when $h = 1$) to 2 equilibria. Thus, our system has likely undergone a transcritical bifurcation. To check the stability, we set $f(x) = x(1 - x) - hx$ and evaluate $f'(x^*)$:

$$f'(x) = 1 - h - 2x \Longrightarrow f'(0) = 1 - h \qquad and \qquad f'(1 - h) = h - 1.$$

For $h < 1$ we see that $x^* = 0$ is unstable and $x^* = 1 - h$ is stable, while for $h > 1$ we see that $x^* = 0$ is stable and $x^* = 1 - h$ is unstable (and not biologically relevant). Our system did indeed undergo a transcritical bifurcation.

● ● ● ● ● ● ● ● ● ● ● ●

Problems

1. Consider the equation
$$x' = x^2(1 - x),$$

   where $x \geq 0$ represents a population.
   (a) Determine the equilibria and their stability. Do the stability results differ from the logistic equation?
   (b) Compare the growth rate of the two models for small x.

2. Consider the Allee model
$$\frac{1}{x}\frac{dx}{dt} = r(a - x)(x - b),$$

   with $r = 2, a = 1, b = 6$. Determine the equilibria and their stability.

3. Consider the Allee model
$$x' = rx(a - x)(x - b),$$

   with $r = 1, a = 2, b = 10$. Determine the equilibria and their stability.

4. Consider the equation
$$x' = x^2(2 - x)(x - 7),$$

   where $x \geq 0$ represents a population.
   (a) Determine the equilibria and their stability. Do the stability results differ from an equation with the Allee effect?
   (b) Compare the growth rate of the two models for small x.

5. Consider the equation
$$x' = x^2(1 - x)(x - 4),$$

   where $x \geq 0$ represents a population.
   (a) Determine the equilibria and their stability. Do the stability results differ from an equation with the Allee effect?
   (b) Compare the growth rate of the two models for small x.

6. When is the exponential population model appropriate? When is the logistic population model appropriate? When is an Allee model appropriate? Discuss the benefits of each of these models and their drawbacks.

7. Suppose that a certain harmful bacteria, once introduced into the body, always persists. The body's immune system is usually able to keep the levels of the bacteria low unless the level of bacteria introduced is initially too large. We thus consider the equation

$$x' = x(x - a)(x - 5),$$

where $x \geq 0$ represents the population of this harmful bacteria and $0 < a < 5$ is a parameter.
(a) Determine the equilibria and their stability.
(b) Describe what happens to the bacteria for any initial condition. (You will have two cases to consider.)
(c) Give a biological interpretation of the parameter a. In particular, determine whether a healthy person is likely to have a larger or smaller a-value than an unhealthy person. Are there other factors that may change the a-value?

8. Suppose a certain harmful bacteria is governed by the equation

$$x' = x(10 - x)[(x - 5)^2 - a],$$

where $x \geq 0$ represents the population of this harmful bacteria and $a < 25$ is a parameter (a can also be negative). Suppose that a bacteria level $x > 8$ represents a lethal level (and the person thus dies).
(a) Determine the equilibria and their stability when $a > 0$.
(b) Determine the equilibria and their stability when $a = 0$.
(c) Determine the equilibria and their stability when $a < 0$.
(d) What type of bifurcation did the system undergo?
(e) What is the biological significance of going from $a > 0$ to $a < 0$? Does the patient survive this change in the parameter?
(f) Give possible biological factors that may influence the value of the parameter a.

9. Write a differential equation describing a population of bacteria that has each of the following characteristics:
(i) Below a level $x = 1$, the bacteria will die off.
(ii) If the initial level of bacteria satisfies $1 < x_0 < 10$, the body is able to keep the bacteria at the "safe" level of $x = 6$.
(iii) If the initial level of bacteria satisfies $x > 10$, then the number of bacteria grows without bound.

10. Write a differential equation describing a population of bacteria that has each of the following characteristics:
(i) Below a level $x = 1/2$, the bacteria will die off.
(ii) If the initial level of bacteria satisfies $1 < x_0 < 8$, the body is able to keep the bacteria at the "safe" level of $x = 2$.
(iii) If the initial level of bacteria satisfies $x > 8$, then the number of bacteria grows without bound.

11. Suppose that the per capita growth rate (x'/x) of a certain population is described by the polynomial

$$x(2-x)^2(x-4).$$

(a) Write the corresponding differential equation.
(b) Determine the equilibria and their stability.

12. Suppose that the per capita growth rate (x'/x) of a certain population is described by the polynomial

$$r(x-a)(x-1), \quad r, a > 0.$$

(a) Write the corresponding differential equation.
(b) Determine the equilibria and their stability. You will have three cases to consider $(a < 1, a = 1, a > 1)$.
(c) Give a biological interpretation for what happens to the population levels for each of your answers in part b.

13. Show that the logistic equation

$$\frac{dN}{dt} = rN\left(1 - \frac{N}{K}\right)$$

can be written in dimensionless form as

$$\frac{dx}{d\tau} = x(1-x)$$

for suitable choices of x and τ.

14. Another harvesting model that we could consider takes the form

$$\frac{dN}{dt} = RN\left(1 - \frac{N}{K}\right) - \frac{HN}{B+N}$$

for $N \geq 0$, $R, H, B > 0$. Show that this can be written in dimensionless form as

$$\frac{dx}{d\tau} = x(1-x) - \frac{hx}{b+x}.$$

15. A well-known model of a biochemical switch (see [48] and references therein) is

$$\frac{dg}{dt} = k_1 s_0 - k_2 g + \frac{k_3 g^2}{k_4^2 + g^2},$$

where g is the concentration of gene product (e.g., mRNA) and $k_i > 0$ are constant. Show that this can be written in dimensionless form as

$$\frac{dx}{d\tau} = s - rx + \frac{x^2}{1+x^2}.$$

We note that this is one of the "lucky" situations where we began with 5 parameters and our dimensionless equation only had 2 parameters.

16. Consider the harvesting model in Problem 14:

$$\frac{dN}{dt} = RN\left(1 - \frac{N}{K}\right) - \frac{HN}{B+N}$$

for $N \geq 0$, $R, H, B > 0$.

(a) Give a biological interpretation of this type of harvesting. In particular, explain what happens for small and large N.

(b) Now consider the dimensionless form

$$\frac{dx}{d\tau} = x(1-x) - \frac{hx}{b+x}.$$

Show that the system has 1, or 2 biologically relevant equilibrium solutions depending on h, b. Classify their stability.

2.5 Numerical Approximation: Euler and Runge-Kutta Methods

In this section we will consider two methods for numerically approximating solutions to ordinary differential equations. While Euler's Method is straightforward and easy to understand, it suffers from the fact that errors accumulate quickly! A modification of it is the 4th order Runge-Kutta method and we will see that errors accumulate much slower for a comparable number of computations.

Euler's Method

We will now consider Euler's method, the first of two methods of numerically solving a differential equation of the form

$$\frac{dy}{dx} = f(x, y).$$

In examining direction fields, we observed that solutions must pass through the field tangentially. A natural goal is to want the computer to somehow sketch the solution curves. There are numerous methods for solving differential equations, some explicit and some implicit. We can choose methods that will give us as much accuracy as we want; however, the greater the accuracy desired, the more the computational work involved. As mentioned, we highlight two explicit methods here and leave an implicit method for the exercises. As we will soon see, Euler's method is a simple-to-understand method and turns out to be very unreliable. It does, however, give us insight into the second method we will consider—the Runge-Kutta method, which is widely used in practical applications due to its balance of computational work vs. accuracy.

Extremely important note: At this point we caution the reader about the potential pitfalls of computing numerical solutions. There is not a single method that will always work. Some methods have a wider applicability than others; some are easier to work with than others; all of the methods will run into difficulties for specialized problems. The reader should always view a graphical and/or numerical output with a bit of skepticism. Even if no other approach seems to work, the reader should attempt some type of calculation or analysis to make sure the solution behaves as it is expected. The best advice we can offer is *don't blindly trust your computer output.* You should always be able to justify why the computer output is a believable answer.

The Method

The first method is essentially an algorithm that formalizes the method of solution we used to draw a direction field. The method is an old one, essentially due to Euler.

Here is how it works. Pick an initial value, say $y(0) = a$. Compute the equation of the tangent line at this point, namely,

$$y_0(x) = a + f(0, a)x.$$

This approximate solution is good for very small changes. Euler's method is sensitive and tends to diverge away from a good approximation. We specify the **step size,** h, which tells the algorithm how far to go before recomputing a tangent line. In essence we are choosing a mesh size that is uniform on the x-axis but not the y-axis. The next equation of a tangent line is

$$y_1(x) = y_0(h) + f(h, y_0(h))h.$$

The process is continued until the last desired x-value is reached. The smaller the h-value, the smaller we expect the error to be *and* the more calculations we require to reach the specified value of x. Usually when Euler's method is used the equations of the lines are not written; rather the series of points that the lines connect is given. The coordinates of each point $x_{i+1} = x_i + h = x_0 + ih$ are calculated by the Euler method formula:

$$y_{i+1} = y_i + hf(x_i, y_i), \quad i = 0, 1, 2, \ldots. \tag{2.22}$$

Note that the value y_{i+1} is an approximation to the true solution. We obviously want the two to be as close together as possible. See Figure 2.20 for a graphical interpretation of (2.22).

To generate these approximate solutions, we apply the following algorithm:

1. Specify (x_0, y_0), h

2. Divide the interval along the x-axis by the step size h to obtain the total number of steps n and a sequence of x-values: $x_i = x_0 + ih$, with $i = 0, 1, 2, 3, \ldots, n$

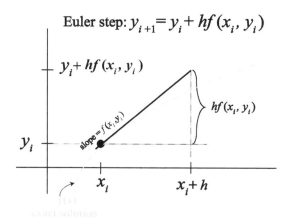

FIGURE 2.20: One step of Euler's method. Given the point (x_i, y_i), the next y-value is calculated by $y_{i+1} = y_i + h$. Note that no information regarding the function on the interval $(x_i, x_i + h]$ is used to calculate the approximation at $x_i + h$.

3. For each x_i, calculate the next approximate solution value $y_{i+1} = y_i + hf(x_i, y_i)$

Euler's method is easy to implement on a spreadsheet, calculator, or with any programming language geared toward mathematical computations.

Use Euler's method with $h = 0.1$ to approximate the solution of

$$\frac{dy}{dx} = xy$$

with $y(0) = 1$ on $0 \le x \le 1$. Determine the exact solution and compare the results.

Since $\frac{dy}{dx} = xy$ we have $f(x, y) = xy$. The initial condition $y(0) = 1$ says that $x_0 = 0$ and $y_0 = 1$ so with $h = 0.1$ we have

$$x_1 = x_0 + h = 0.1, \quad x_2 = x_0 + 2h = 0.2, \quad \cdots, \quad x_{10} = x_0 + 10h = 1.0.$$

Euler's formula (2.22) applied to this example becomes

$$y_{i+1} = y_i + (0.1)(x_i)(y_i).$$

We calculate y_1 as

$$y_1 = y_0 + (.1)(x_0)(y_0) = 1 + (.1)(0)(1) = 1.$$

We then use x_1 and the newly found y_1 to calculate y_2:

$$y_2 = y_1 + (.1)(x_1)(y_1) = 1 + (.1)(.1)(1) = 1.01.$$

Since we are solving the equation on the interval $0 \le x \le 1$, we need to carry out these calculations until $x_i = 1$, that is, eight additional times. The results are summarized in the table in Figure 2.21.

x_i	y_i	x_i	y_i
0.0	1.0	0.6	1.15873
0.1	1.0	0.7	1.22825
0.2	1.01	0.8	1.31423
0.3	1.0302	0.9	1.41937
0.4	1.06111	1.0	1.54711
0.5	1.10355		

FIGURE 2.21:   Euler's method with $h = 0.1$ for the equation $\frac{dy}{dx} = xy$.

The closed form solution to $\dfrac{dy}{dx} = xy$ with the condition $y(0) = 1$ is found, using separation of variables, to be

$$y(x) = e^{x^2/2}.$$

(Give it a try!) The table in Figure 2.22 compares the values using Euler's method and the exact solution. We see that initially the approximation is good, but its accuracy declines as x increases. A smaller step size h would increase the accuracy of the approximation but, as mentioned earlier, requires more steps.

Figure 2.23 shows the graph of $y(x) = e^{x^2/2}$ plotted together with the ten points from Euler's method from Figure 2.22.

There are many methods that are considered superior for accuracy to Euler's method, but Euler's method has the advantage of simplicity, intuition, and clarity. Regarding the accuracy, the table in Figure 2.22 compared the approximate answer via Euler's method vs. the exact method. In general, we can place a bound on this difference, known as the **local truncation error**, by saying that

$$|y_i - y(x_i)| \le Ch,$$

where C is a constant that depends on the function and the specified interval. See, for example, Burden and Faires [10]. It is also an important method to understand because when $h = 1$, it treats differential equations as difference

x_i	Euler y_i	Exact $y(x_i)$	Error $|y_i - y(x_i)|$
0.0	1.0	1.0	0
0.1	1.0	1.0050	0.005
0.2	1.01	1.0202	0.0102
0.3	1.0302	1.0460	0.0158
0.4	1.06111	1.08329	0.02218
0.5	1.10355	1.13315	0.02960
0.6	1.15873	1.19722	0.03849
0.7	1.22825	1.27762	0.04937
0.8	1.31423	1.37713	0.06290
0.9	1.41937	1.49930	0.07993
1.0	1.54711	1.64872	0.10161

FIGURE 2.22: Euler's method with $h = 0.1$ compared with the exact solution for the equation $\frac{dy}{dx} = xy$.

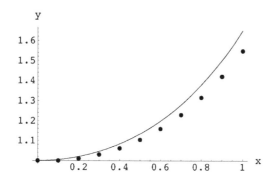

FIGURE 2.23: Plot of $y(x) = e^{x^2/2}$ with points obtained via Euler's method.

equations. (Differential equations depend on a continuous independent variable, whereas difference equations depend on a discrete independent variable. As an example, difference equations can be used to model an insect population in which the adults lay eggs and die before the young hatch—the independent variable is the generation of the population.) In particular, we can use a numerical differential equation solver to give numerical solutions to difference equations provided we specify that Euler's method be used and that a step size $h = 1$ is always used. See, for example, Martelli [34].

Runge-Kutta Method

We have just employed Euler's method for solving a first-order differential equation. There are many other methods. Two that are frequently encountered are second-order and fourth-order Runge-Kutta methods. The method was named after the German mathematicians Carl Runge (1856–1927) and

Wilhelm Kutta (1867–1944), who developed the theory long before the advent of modern computers. These work in manners similar to Euler's method except that instead of using first-order approximations (lines), they use second-order (parabolas) and fourth-order (quartics) curves, although they do it in ways that avoid the computation of higher-order derivatives. The Runge-Kutta methods can also be interpreted as multiple applications of Euler's method at *intermediate* values, that is, between x_i and $x_i + h$. It is this interpretation that we will use for graphical purposes.

The algorithms are more complicated than Euler's but, nevertheless, are easily enough programmed. We will consider the Runge-Kutta fourth-order algorithm which is by far the method most commonly used by scientists and engineers in their respective fields. The idea is to use a weighted average of the slopes of field segments at four points "near" the current point (x_i, y_i). The location of each slope evaluation is determined by an application of Euler's method.

In order to figure out a good formula for giving weights to the function value at the four points we assume we have computed y_i, which is the approximation to $y(x_i)$. In order to compute the next approximate solution value y_{i+1}, we observe that the fundamental theorem of calculus gives

$$y(x_{i+1}) - y(x_i) = \int_{x_i}^{x_{i+1}} y'(x)dx = \int_{x_i}^{x_i+h} y'(x)dx. \qquad (2.23)$$

We then use Simpson's rule to obtain an approximation to this integral:

$$y(x_{i+1}) - y(x_i) \approx \frac{h}{6}\left[y'(x_i) + 4y'\left(x_i + \frac{h}{2}\right) + y'(x_{i+1})\right].$$

We said that we want to use a weighted average of the slopes at four points "near" the current point. We obtain a fourth point by performing *two* evaluations at the middle point $x_i + \frac{h}{2}$. Thus our approximation is rewritten

$$y(x_{i+1}) - y(x_i) \approx \frac{h}{6}\left[y'(x_i) + 2y'\left(x_i + \frac{h}{2}\right) + 2y'\left(x_i + \frac{h}{2}\right) + y'(x_{i+1})\right]$$

$$(2.24)$$

and we will update the slope at the successive evaluations.

In order to see the evaluations at the four points, we give a graphical interpretation of the Runge-Kutta fourth-order method and then state the formula for calculating the next approximate value, y_{i+1}. See Figure 2.24 for the graphical interpretation. We begin at the point (x_i, y_i) and calculate the coordinates of the next point of the approximate solution (x_{i+1}, y_{i+1}).

1. First evaluation $k_1 = f(x_i, y_i)$

We calculate the slope of the solution that passes through our current value (x_i, y_i) by evaluating $f(x_i, y_i)$. Call this slope k_1.

Preparation for 2nd evaluation: Euler's method tells us to continue at this slope for one step size and the Euler approximation of the next value is

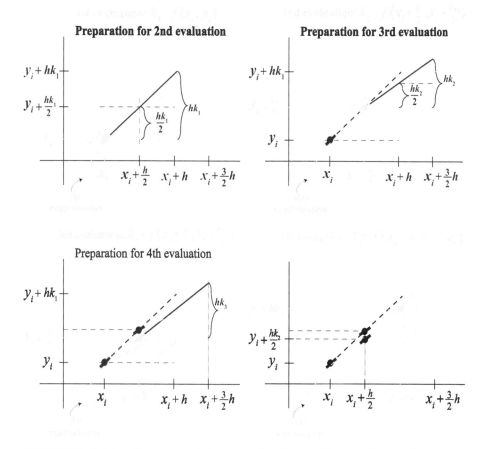

FIGURE 2.24: One step of fourth-order Runge-Kutta. Given the point (x_i, y_i), the next y-value is calculated by $y_{i+1} = y_i + h(\quad +2\quad +2\quad +\quad)/6$. The graph here is shown for a solution that is increasing and concave down on the interval $(x_i, x_i + h)$.

$(x_i + h, y_i + hk_1)$. The Runge-Kutta method takes *half* of this increase in the y-coordinate, which is easily seen to be $hk_1/2$, and goes to the x-value *halfway* between x_i and the next step $x_i + h$.

We now perform our first of two "mid-value" calculations. Our "mid-value" x-value is the halfway point $x_i + \frac{h}{2}$ and the y-value is $y_i + \frac{hk_1}{2}$; we evaluate the slope of the solution passing through this point, which is $f(x_i + \frac{h}{2}, y_i + \frac{hk_1}{2})$. For simplicity, call this value k_2.

Preparation for 3rd evaluation: If we followed this slope for one step size (to the x-value $x_i + \frac{3h}{2}$) we would be at the y-value $y_i + \frac{hk_1}{2} + hk_2$. We again take *half* of this increase, which is seen to be $hk_2/2$. We go back to the "mid-value" x-value and a different y-value, $y_i + \frac{hk_2}{2}$.

3. Third evaluation $k_3 = f(x_i + \frac{h}{2}, y_i + \frac{hk_2}{2})$
We now perform our second of two "mid-value" calculations. Our "mid-value" x-value is still the halfway point $x_i + \frac{h}{2}$ and the y-value is now $y_i + \frac{hk_2}{2}$; we again evaluate the slope of the solution passing through this point, which is $f(x_i + \frac{h}{2}, y_i + \frac{hk_2}{2})$. For simplicity, call this value k_3.

Preparation for 4th evaluation: If we followed this slope for one step size (to the x-value $x_i + \frac{3h}{2}$) we would be at the y-value $y_i + \frac{hk_2}{2} + hk_3$. We note the increase, which is seen to be hk_3. We now go to the x-value $x_i + h$ and the y-value $y_i + hk_3$, which takes into account this last estimated increase.

4. Fourth evaluation $k_4 = f(x_i + h, y_i + hk_3)$
We now do our final evaluation at this point by calculating the slope of the solution which passes through the value $(x_i + h, y_i + hk_3)$. Call this slope k_4 and note that the x- and y-coordinates in this evaluation were *not* the original one predicted by Euler's method.

We thus have four slopes close by our current (x_i, y_i) pair: one at x_i, one at $x_i + h$, and two at $x_i + \frac{h}{2}$. The Runge-Kutta fourth-order method takes a weighted average of these slopes to calculate the next point:

$$y_{i+1} = y_i + \frac{h}{6}(k_1 + 2k_2 + 2k_3 + k_4). \tag{2.25}$$

As with the Euler method, we can put a bound on the error between the exact solution value $y(x_i)$ and the approximate value calculated by the Runge-Kutta method. The local truncation error, which is simply this difference, satisfies

$$|y_i - y(x_i)| \leq Mh^4$$

where M is a constant that depends on the function and the specified interval. For a more detailed explanation, see Burden and Faires [10].

We summarize the above in the following algorithm:

1. Specify (x_0, y_0), h
2. Divide the interval along the x-axis by the step size h to obtain the total number of steps n and a sequence of x-values: $x_i = x_0 + ih$, $i = 0, 1, 2, 3, \ldots, n$
3. For each x_i, calculate $k_{j,i}$, $j = 1, 2, 3, 4$ (where the additional subscript will be used to denote the calculation from the ith value), defined by
   a. $k_{1,i} = f(x_i, y_i)$
   b. $k_{2,i} = f(x_i + h/2, y_i + (hk_{1,i})/2)$
   c. $k_{3,i} = f(x_i + h/2, y_i + (hk_{2,i})/2)$
   d. $k_{4,i} = f(x_i + h, y_i + hk_{3,i})$
4. Calculate $y_{i+1} = y_i + h(k_{1,i} + 2k_{2,i} + 2k_{3,i} + k_{4,i})/6$, which is the next y-value of the approximate solution, as calculated by the fourth-order Runge-Kutta method.

Example 2 We will now use the Runge-Kutta algorithm with $h = 0.1$ to approximate the solution of

$$\frac{dy}{dx} = xy$$

with $y(0) = 1$ on $0 \le x \le 1$. This is the same equation we approximated using Euler's method.

Solution

The initial condition $y(0) = 1$ means $x_0 = 0$, $y_0 = 1$, so with $h = 0.1$ we have

$$k_{1,0} = f(0,1) = 0,$$

$$k_{2,0} = f\left(0 + 0.05, 1 + (0.1)\left(\frac{0}{2}\right)\right) = 0.05,$$

$$k_{3,0} = f\left(0 + 0.05, 1 + (0.1)\left(\frac{0.05}{2}\right)\right) = 0.0501,$$

$$k_{4,0} = f(0 + 0.1, 1 + (0.1)(0.0501)) = 0.1005,$$

and thus the next value is

$$y_1 = 1 + \left(\frac{0.1}{6}\right)(0 + 2(0.05) + 2(0.0501) + 0.1005) = 1.005012.$$

The remaining calculations are performed similarly and are summarized in the table in Figure 2.25 along with the true solution. Note the close agreement of the Runge-Kutta approximate solution with the true value of y_n.

x_i	Runge-Kutta y_i	True $y(x_i)$	Error $|y_i - y(x_i)|$
0.0	1.0	1.0	0.0
0.1	1.0050125	1.0050125	0.0000000
0.2	1.0202013	1.0202013	0.0000000
0.3	1.0460279	1.0460279	0.0000000
0.4	1.0832871	1.0832872	0.0000000
0.5	1.1331485	1.1331484	0.0000001
0.6	1.1972174	1.1972173	0.0000001
0.7	1.2776213	1.2776213	0.0000000
0.8	1.3771278	1.3771277	0.0000001
0.9	1.4993025	1.4993024	0.0000001
1.0	1.6487213	1.6487210	0.0000003

FIGURE 2.25: Runge-Kutta's method with $h = 0.1$ compared with the analytic solution $(e^{x^2/2})$ for the equation $\frac{dy}{dx} = xy$.

It is often better to have closed-form solutions as they may provide theoretical insights, and there are no problems with errors associated with approximate methods. Unfortunately most of the differential equations one encounters cannot be solved. Numerical solvers allow us to proceed with our analysis, but are limited in that one must explicitly express all parameters as numbers and give all necessary initial conditions.

• • • • • • • • • • • •

Problems

In Problems 1–6, let $h = 0.1$ and approximate the solutions to $y(x_1)$ and $y(x_2)$ by hand using (a) Euler's Method and (b) the fourth-order Runge-Kutta method. Then compare the results with the given explicit solution at x_1 and x_2. It will be easiest if you display the results as we did in the table of Figure 2.25.

1. $y' = x^3$, $y(1) = 1$; explicit solution: $y = \frac{1}{4}(x^4 + 3)$

2. $y' = -y^2$, $y(0) = 1$; explicit solution: $y = \frac{1}{x+1}$

3. $y' = x^4 y$, $y(1) = 1$; explicit solution: $y = e^{(x^5 - 1)/5}$

4. $y' = -y^2 \cos x$, $y(0) = 1$; explicit solution: $y = \dfrac{1}{1 + \sin x}$

5. $y' = \frac{\sin x}{y^3}$, $y(\pi) = 2$; explicit solution: $y = (12 - 4\cos x)^{1/4}$

6. $y' = ye^{-x}$, $y(0) = 1$; explicit solution: $y = \exp(1 - e^{-x})$

In Problems 7–12, solve with your computer software package using (a) Euler's Method and (b) the fourth-order Runge-Kutta method with $h = 0.1$ to find approximate values for $y(x_1), y(x_2), y(x_3), y(x_4),\ y(x_5), y(x_6), y(x_7),$ and $y(x_8)$. Then compare the results with the given explicit solution at $x_1, x_2, x_3, x_4, x_5, x_6, x_7,$ and x_8.

7. $y' = e^{-y}$, $y(0) = 2$; explicit solution: $y = \ln(x + e^2)$

8. $y' = -xy^2$, $y(0) = 1$; explicit solution: $y = \dfrac{2}{2 + x^2}$

9. $y' = y + \cos x$, $y(0) = 0$; explicit solution: $y = \frac{1}{2}(\sin x - \cos x + e^x)$

10. $y' = y + \sin x$, $y(0) = 2$; explicit solution: $y = \frac{-1}{2}(\cos x + \sin x - 5e^x)$

11. $y' = x + y$, $y(0) = 0$; explicit solution: $y = -x - 1 + e^x$

12. $y' = (x+1)(y^2 + 1)$, $y(0) = 0$; explicit solution: $y = \tan(\frac{1}{2}x^2 + x)$

In Problems 13–17, use MATLAB, Maple, or Mathematica and the fourth-order Runge-Kutta method with $h = 0.01$ to plot the numerical solution. Choose a viewing window that will allow you to see the behavior of the solution. If your solution appears to diverge rapidly or illustrate strange behavior, try reducing your step size to $h = .001$.

13. $y' = e^{-x^2}$, $y(0) = 1$

14. $y' = x^3 e^y + 3x^2 \sin y$, $y(.5) = 1$

15. $y' = x^3 y - x^2 y^2$, $y(-1) = 1$

16. $y' = |1 - x^2|y + x^3$, $y(1) = 1$

17. $y' = y\sqrt{x^2 + y^2 + 1} + \cos(xy)$, $y(0) = 1$

18. Use the fourth-order Runge-Kutta method to numerically solve the following two logistic differential equations for time t between 0 and 1. Assume that $x(0) = 100$. Solve both using the following h values: 1, 0.5, 0.25, and 0.1. Compare the solutions of the first equation for the different h-values. Explain what is happening as h gets small. Repeat the comparison for the second equation. Compare the solutions of the two equations to each other for the different h-values. Explain why two differential equations that are identical except for the parameter values exhibit different behaviors while h is in the process of "getting small."

(a) $\dfrac{dx}{dt} = 1.5x \left(1 - \dfrac{x}{1000}\right)$,   (b) $\dfrac{dx}{dt} = 2.5x \left(1 - \dfrac{x}{1000}\right)$

19. Consider the initial value problem

$$\frac{dy}{dx} = f(x, y), \ a < x < b, \ y(x_0) = y_0,$$

with true solution $y(x)$. Derive the formula used in Euler's method at the ith step in the following manner: assume the approximate and exact solutions agree at $(x_i, y(x_i))$ and Taylor expand the exact solution at the next calculated x-value $y(x_i + h)$. Ignore terms of $O(h^2)$.

2.6   An Introduction to Autonomous Second-Order Equations

We will now consider a special class of second-order differential equations, useful in applications, that can be solved by methods we know. With first-order equations, we considered autonomous equations to be those that did not have the independent variable appearing explicitly in the problem. The definition here is similar.

Definition 2.6.1

A second-order differential equation of the form

$$y''(t) = h(y, y') \tag{2.26}$$

is called autonomous when h does not depend explicitly on the independent variable t.

Equation (2.26) can be interpreted as a differential equation whose solutions provide the position $y(t)$ of a body that moves according to Newton's law for a special kind of forcing function h.

One, perhaps unexpected, way to solve (2.26) is to consider the velocity as a function of the position. That is, we let

$$v(y) = \frac{dy}{dt}$$

so that v is a function of y, rather than t. We also note that this determines the acceleration as

$$
\begin{aligned}
\frac{d^2 y}{dt^2} &= \frac{d}{dt}\frac{dy}{dt} \\
&= \frac{d}{dt}v(y) \\
&= \frac{dv}{dy}\frac{dy}{dt} \\
&= v\frac{dv}{dy}.
\end{aligned}
$$

Thus, the task of solving the second-order equation (2.26) is reduced to solving the first-order equation

$$v\frac{dv}{dy} = h(y, v).$$

Solutions can be plotted in the y-v plane and are called **orbits**. If we can solve this transformed equation for $v(y)$, the remaining separable differential equation

$$v(y) = \frac{dy}{dt}$$

can be solved to obtain $y(t)$.

Example 1 A simple application of the preceding method is given by the motion of a frictionless spring whose deflection y satisfies the initial value problem

$$m\frac{d^2 y}{dt^2} + ky = 0$$

for a positive mass m and a positive spring constant k, with prescribed initial values $y(0) = y_0$ and $y'(0) = v_0$ for the position and velocity. (We will give a formal study of harmonic motion later.) Letting

$$v(y) = \frac{dy}{dt},$$

the equation becomes

$$mv\frac{dv}{dy} = -ky$$

which is separable and exact (check it!), so that

$$mv \, dv + ky \, dy = 0.$$

Integrating gives

$$\frac{m}{2} v^2 + \frac{k}{2} y^2 = c.$$

The initial conditions $y(0) = y_0$ and $y'(0) = v_0$ give

$$c = \frac{m}{2} v_0^2 + \frac{k}{2} y_0^2.$$

This gives

$$\frac{m}{2} v^2 + \frac{k}{2} y^2 = \frac{m}{2} v_0^2 + \frac{k}{2} y_0^2,$$

which can be interpreted as saying that the sum of the kinetic and potential energies must remain constant, i.e., the total energy must remain constant. Solving this expression for $v(y) = dy/dt$ gives

$$\frac{dy}{dt} = v(y) = \pm \sqrt{\frac{k}{m}} \sqrt{\alpha^2 - y^2}, \qquad (2.27)$$

where

$$\alpha^2 = \frac{m}{k} v_0^2 + y_0^2$$

is a constant determined by the initial conditions. The sign of v is uniquely determined by the prescribed initial velocity v_0. It now remains for us to solve (2.27) for $y(t)$. But (2.27) is separable and thus

$$\frac{dy}{\sqrt{\alpha^2 - y^2}} = \pm \sqrt{\frac{k}{m}} \, dt,$$

so that

$$\int \frac{dy}{\sqrt{\alpha^2 - y^2}} = \int \pm \sqrt{\frac{k}{m}} \, dt$$

which is

$$\sin^{-1} \left(\frac{y}{\alpha} \right) = \pm \sqrt{\frac{k}{m}} \, t + \phi.$$

Here ϕ is some constant. Thus, the deflection $y(t)$ of the spring is given by

$$y(t) = \alpha \sin \left(\pm \sqrt{\frac{k}{m}} \, t + \phi \right)$$

$$= (\pm \alpha \cos \phi) \sin \left(\sqrt{\frac{k}{m}} \, t \right) + (\alpha \sin \phi) \cos \left(\sqrt{\frac{k}{m}} \, t \right)$$

where this last expression follows from the trigonometric identity $\sin(u\pm v) = \sin(u)\cos(v) \pm \cos(u)\sin(v)$. Letting

$$C_1 = \pm a\cos\phi \quad \text{and} \quad C_2 = a\sin\phi$$

gives

$$y(t) = C_1 \sin\left(\sqrt{\frac{k}{m}}\,t\right) + C_2 \cos\left(\sqrt{\frac{k}{m}}\,t\right).$$

Now using the initial conditions $y(0) = y_0$ and $y'(0) = v_0$ we have

$$y(t) = y_0 \sin\left(\sqrt{\frac{k}{m}}\,t\right) + v_0 \cos\left(\sqrt{\frac{k}{m}}\,t\right),$$

which is the familiar equation for **simple harmonic motion.**

● ● ● ● ● ● ● ● ● ● ● ●

Problems

1. Show that the autonomous equation

$$yy'' = (y')^2$$

has solution

$$y = ce^{kx}$$

for constants c and k.

2. In a first physics course, students derive the equation of motion for a frictionless simple pendulum as

$$mL\theta'' + mg\sin\theta = 0 \tag{2.28}$$

where θ is the angle that the pendulum makes with the vertical; however, the next step is to assume the angle is small and use the small angle approximation $(\sin\theta \approx \theta + \cdots)$ to rewrite this equation as

$$\theta'' + \omega^2\theta = 0, \qquad \omega^2 = \frac{g}{L},$$

which is conveniently the equation for simple harmonic motion. This approximation obviously fails if θ becomes too large. Let's revisit (2.28) and consider this as an autonomous equation. Let $v = d\theta/dt$ and use the methods of this section to derive a total energy formula. Use this expression to verify the plot of some of the orbits in Figure 2.27. Alternatively, plot this total energy formula to obtain the orbits in Figure 2.27. Interpret the three qualitatively different orbits, keeping in mind that the pendulum is allowed to whirl over the top.

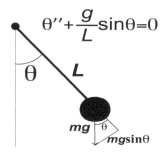

FIGURE 2.26:   Free body diagram for simple pendulum without friction.

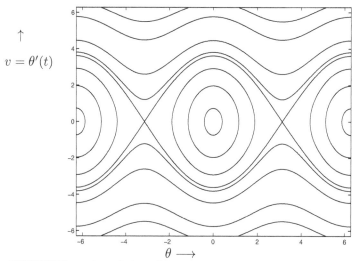

FIGURE 2.27:   Orbits for simple pendulum with $g = 9.8, L = 1$.

3. Suppose the position of a body of mass m that is shot up from the surface of the Earth satisfies the inverse square law

$$m\,x''(t) = -mg\frac{R^2}{x^2(t)},$$

where $x(t)$ is the distance from the Earth's center at time t, R is the Earth's radius, and the initial conditions are

$$x(0) = R > 0 \quad \text{and} \quad x'(0) = v_0 > 0.$$

If $v_0^2 \geq 2gR$, show that the body will never return to Earth. The solution defines the **escape velocity** for the body. (Hint: Determine the expression for $v(x)$ and its sign.)

4. The nonlinear equation

$$\frac{d^2y}{dx^2} = \sqrt{1 + \left(\frac{dy}{dx}\right)^2}$$

describes the position y of a suspension cable, which either supports a bridge or hangs under its own weight. The equation is of the form (2.26), but it is also of the form

$$y''(x) = f(x, y'),$$

where $y = y(x)$ and the right-hand side f is independent of y.
(a) Let $u(x) = dy/dx$ and derive a first-order equation in terms of only u, x, that is of the form:

$$\frac{du}{dx} = f(x, u).$$

(b) Solve this equation to obtain

$$x - x_0 = \ln(u + \sqrt{1 + u^2})$$

for some constant x_0.
(c) Exponentiate the above equation and the negative of it and combine appropriately to obtain the expression

$$u = \frac{1}{2}(e^{x-x_0} + e^{-(x-x_0)}).$$

(d) Using the original substitution $u(x) = dy/dx$, obtain the equation

$$y(x) = \cosh(x - x_0) + c,$$

which is the (familiar) equation for the *catenary*, where x_0 is the location of the zero of y' (i.e., the minimum of y) and c is 0 if that minimum value is 1.

Chapter 2 Review

In Problems **1–6**, determine whether the statement is true or false. If it is true, give reasons for your answer. If it is false, give a counterexample or other explanation of why it is false.

1. The Existence and Uniqueness theorem allows us to conclude that solutions to $y' = \tan y$ will always exist and be unique.

2. The Existence and Uniqueness theorem allows us to state that the solution to $y' = x^2 y^{1/3}$ passing through the initial condition $y(1) = 0$ will not be unique.

3. The Runge-Kutta method for approximating the solution to an ODE is superior to Euler's method because it uses 1/2 the step size of Euler's method to calculate the approximate solution.

4. The Euler and Runge-Kutta methods are both used for numerically approximating the solution to a differential equation. Euler's method is superior to the Runge-Kutta method because the step size is smaller and thus the approximations are better at each step.

5. The Allee model of population growth is often considered superior to the Logistic model because it models the need for a critical population level in order for the species to survive.

6. Autonomous equations can be written as $\frac{dx}{dt} = f(x)$ and are useful because we are able to obtain long-term behavior of the solutions quickly.

In Problems **7–13**, determine the region(s) of the x-y plane where solutions (i) exist and also where solutions (ii) exist and are unique, according to the Existence and Uniqueness theorem.

7. $\dfrac{dy}{dx} = 2x^{2/3} - y$

8. $\dfrac{dy}{dx} = (xy - 1)^{1/3}$

9. $\dfrac{dy}{dx} = 2x - y^{2/3}$

10. $\dfrac{dy}{dx} = y \tan x$

11. $\dfrac{dy}{dx} = \sin x - \tan y$

12. $\dfrac{dy}{dx} = \sec y$

13. $\dfrac{dy}{dx} = e^{y/(x+1)}$

14. In the direction field given in Figure 2.28(a), sketch the solution to $y' = x^2 - y$ that passes through the initial condition $y(-2) = 0$.

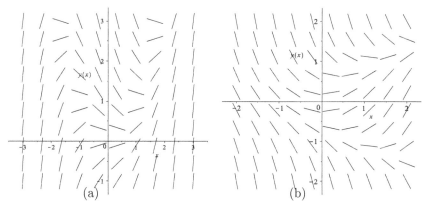

FIGURE 2.28: (a) Graph for Problem **14**. (b) Graph for Problem **15**.

15. In the direction field given in Figure 2.28(b), sketch the solution to $y' = x - y^2$ that passes through the initial condition $y(0) = 0$.

16. Consider the autonomous equation $y' = (3 + y)(2 - y)$.
    (a) Find the equilibria.
    (b) Draw a graph in the y-y' plane and indicate the stability of these points by drawing the appropriate arrows on the phase line.
    (c) Explicitly state the stability of the equilibrium solutions.
    (d) Sketch solutions in the x-y plane.

17. The following equation has been proposed to model the level of a certain harmful bacteria in the body:

$$x' = x(x - a)(x - 4),$$

where $0 < a < 4$, $x = x(t), x \geq 0$.
    (a) Draw the phase line diagram and determine the stability of the equilibrium solutions.
    (b) Describe the long-term behavior of the solution for any initial condition.
    (c) Sketch solutions in the t-x plane.

18. Consider the autonomous ODE $y' = (1 - y)(2 + y)$.
    (a) Find the equilibria.
    (b) Draw a graph in the y-y' plane and indicate the stability of these points by drawing the appropriate arrows on the phase line.
    (c) Explicitly state the stability of the equilibrium solutions.
    (d) Sketch solutions in the x-y plane.

19. Match the two given direction fields in Figure 2.29 with their respective equations.
    (a) $y' = y^2(xy - 1)$, (b) $y' = x^2(xy^2 - 1)$, (c) $y' = xy(1 - x^2)$,
    (d) $y' = y(xy - 1)$

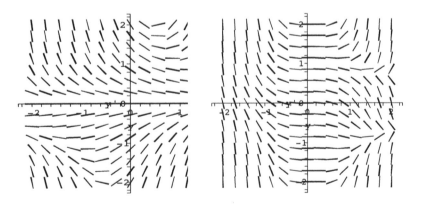

FIGURE 2.29:   Direction fields for Problem 19.

20. Match the two given direction fields in Figure 2.30 with their respective equations.

(a) $y' = y(x^2 - 1)$, (b) $y' = y \sin \pi x$, (c) $y' = y(1 - x^2)$, (d) $y' = y(y^2 - 1)$

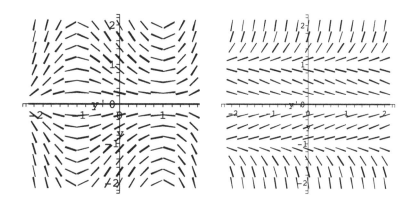

FIGURE 2.30: Direction fields for Problem **20**.

In Problems **21–26**, graph the equation in the x-x' plane and then use phase line analysis to determine the stability of the equilibria.

21. $x' = x^2(x - 2)(4 - 3x)$

22. $x' = (x^2 - 9)^2(x - 1)$

23. $x' = (1 + x)(3 + 2x)(5 - x)^7$

24. $x' = x^4(2x-3)^3(x^2+4)(x^2+2x+1)$

25. $x' = (16 - 3x^2)(x^2 - 4)^3$

26. $x' = x(4 - 3x)^3(2x + 1)^4(x^3 - 1)$

In Problems **27–29**, let the parameter r vary and determine the bifurcation that occurs by drawing the two qualitatively different phase line portraits. State the stability of the equilibria for both pictures. Draw the bifurcation diagram as well.

27. $y' = 1 - r - y^2$ **28.** $y' = r - y - e^{-y}$ **29.** $y' = y(1 - y) - ry$

30. Show that the **constant yield harvesting** equation

$$\frac{dN}{dt} = rN\left(1 - \frac{N}{K}\right) - H$$

can be non-dimensionalized to give $x' = x(1 - x) - h$. What type of bifurcation occurs? Discuss why this is not a good model for harvesting.

Use MATLAB, Maple, or Mathematics to solve each of Problems **31–38**. Find the numerical solution using the fourth-order Runge-Kutta method with $h = 0.1$. Compare this to the solution obtained using Euler's method with $h = 0.1$ (either with a table or by plotting).

31. $2xy' + y^2 = 1$, $y(1) = \pi$ **32.** $xy' + x^2 + xy - y = 0$, $y(1) = 1$

33. $y - y' = y^2 + xy'$, $y(2) = 0$ **34.** $(x + 2y^3)y' = y$, $y(-1) = 1$

35. $x^2 y' = y(x + y)$, $y(\sqrt{2}) = 1$ **36.** $(1 - x^2)y' + xy = 0$, $y(0) = 5$

37. $(2xy^2 - y)^2 + xy' = 0$, $y(\pi) = 1$ **38.** $y^2 + 2(x-1)y' - 2y = 0$, $y(0) = -3$

Chapter 2 Computer Labs

Chapter 2 Computer Lab: MATLAB

MATLAB Example 1: Enter the following code that demonstrates how to plot direction fields in MATLAB for $y' = \dfrac{1}{x^2 + y^2}$.

```
>> clear all
>> [X,Y]=meshgrid(-3:.55:3,-3:.55:3); %Note the CAPITAL letters
>> DY=1./(X.^2+Y.^2); %DY is rhs of original equation
>> DX=ones(size(DY));
>> DW=sqrt(DX.^2+DY.^2); %the length of each vector
>> quiver(X,Y,DX./DW,DY./DW,.5,'.'); %plots normalized vectors
>> xlabel('x');
>> ylabel('y');
```

MATLAB Example 2: Enter the following code that demonstrates how to plot solution curves for $y' = \dfrac{1}{x^2 + y^2}$. You will first need to create a function Ch2Example.m that contains the ODE that will be solved numerically. Do not close your figure from Example 1 above.

```
%Create a function Ch2Example.m and enter the following commands
function dy= Ch2Example(x,y)
%
%The original ode is dy/dx=1/(x^2+y^2)
%
dy = 1/(x^2 + y^2);
%end of function Ch2Example.m
```

After having created and saved the above function, type the following in the Command Window.

```
>> x0=-3; xf=3; y0=-3;
>> options=odeset('refine',10,'AbsTol',1e-15);
>> [x,y]=ode45(@Ch2Example,[x0,xf],y0,options);
>> x0=-3; xf=3; y0=-2;
>> [x1,y1]=ode45(@Ch2Example,[x0,xf],y0,options);
>> x0=-3; xf=3; y0=-1;
>> [x2,y2]=ode45(@Ch2Example,[x0,xf],y0,options);
>> x0=-3; xf=3; y0=0;
>> [x3,y3]=ode45(@Ch2Example,[x0,xf],y0,options);
>> hold on %keeps direction field plot open
>> plot(x,y) %superimpose solutions
>> axis([-3 3 -3 3])
>> plot(x1,y1)
>> plot(x2,y2)
>> plot(x3,y3)
>> hold off
```

MATLAB Example 3: Enter the following code that creates three functions Euler.m, RK4.m, and Ch2NumExample.m that respectively contain the Euler method, 4th order Runge-Kutta methods, and the ODE that will be solved numerically. Then enter the commands to call these functions in order to compare their output. The final group of commands will superimpose the RK4 solution onto the direction field. The IVP we consider is $y' = xy$, $y(0) = 1$.

```
%Create a function Euler.m and enter the following commands
function [xout,yout]=Euler(fname,xvals,y0,h)
%This code implements the Euler Method for numerically
%solving y'=f(x,y)
%
%fname=the function f for the ODE
%xvals = vector that contains the initial x0 and xf
%y0 = initial y
%h = stepsize

x0=xvals(1); xf=xvals(2);
x=x0; y=y0;
steps=(xf-x0)/h;
xout=zeros(steps+1, 1);% allocates space for xout
yout=zeros(steps+1, 1);% allocates space for yout
xout(1)=x; yout(1)=y;
```

```
for j=1:steps
    f=feval(fname,x,y);
    x=x+h;
    y=y+h*f;
    xout(j+1)=x;
    yout(j+1)=y;
end
%end of function Euler.m


%Create a function RK4.m and enter the following commands
function [xout,yout]=RK4(fname,xvals,y0,h)
%This code implements the 4th order Runge-Kutta method for
%numerically solving the ODE y'=f(x,y)
%
%fname = the function f for the equation.
%x0 =initial x
%n = number of steps to be taken.
%y0 =initial y
%h =stepsize


x0=xvals(1); xf=xvals(2);
x=x0; y=y0;
steps=(xf-x0)/h;
f=feval(fname,x,y);
xout=zeros(steps+1, 1);% allocates space for xout
yout=zeros(steps+1, length(f));% allocates space for yout
y=y'; %The ' is needed to match syntax of ode45 in higher dim
xout(1,1)=x; yout(1,:)=y;

for i=1:steps
    k1 = h*f;
    k2 = h*feval(fname,x+(h/2),y+(k1/2));
    k3 = h*feval(fname,x+(h/2),y+(k2/2));
    k4 = h*feval(fname,x+h,y+k3);
    ynext = y +(k1 + 2*k2 + 2*k3 + k4)/6;
    xnext = x+h;
    f = feval(fname,xnext,ynext);
    xout(i,:)=xnext;
    yout(i,:)=ynext'; %we again need the '
    x=xnext;
    y=ynext;
end
%end of function RK4.m
```

```
%Create a function Ch2NumExample.m and enter the following
commands
function dy= Ch2NumExample(x,y)
%
%The original ode is dy/dx=x*y
%
dy = x*y;
%end of function Ch2NumExample.m
```

After having created and saved the above three functions, type the following in the Command Window.

```
>> clear all
>> close all
>> x0=0; xf=3.1; y0=1; h=.1;
>> [x,y]=Euler(@Ch2NumExample,[x0, xf],y0,h);
>> [x y]
>> [x1,y1]=RK4(@Ch2NumExample,[x0 xf],y0,h);
>> [x1 y1]
>> x2=x0:h:xf;
>> y2= exp(x2.^2/2); %this is the analytical solution
>> [x2' y2']
>> plot(x,y,'b:')  %Now graphically compare Euler with RK4
>> hold on
>> plot(x1,y1,'k--')
>> xlabel('x')
>> ylabel('y')
>> title('Numerical solns of y\{prime} =xy using EM and RK4
   with h=.1, dotted/blue is EM, dashed/black is RK4')
>> hold off
```

The following commands will superimpose the plots of the numerical solution from RK4.m (obtained above) and the direction field.

```
>> plot(x1,y1,'k--','LineWidth',2)
>> hold on
>> [X,Y]=meshgrid(-3:.3:3,0:.3:4);
>> DY=X.*Y; %DY is rhs of original equation
>> DX=ones(size(DY));
>> DW=sqrt(DX.^2+DY.^2);
>> quiver(X,Y,DX./DW,DY./DW,.4,'.');
>> xlabel('x');
>> ylabel('y');
```

```
>> axis([-3  3  0  4])
>> hold off
```

%End of Example 3.

MATLAB Example 4: If you covered the topic of bifurcations in class or are instructed to read this on your own by your instructor, enter the following commands to illustrate how to animate the qualitatively different solutions of the ODE with a supercritical bifurcation, $y' = ry - y^3$. The final set of commands superimposes three different qualitative solutions observed during the animation. Although it can be entered into the Command Window, the following code should be entered into a MATLAB Script instead.

```
%Enter into a Script for easier manipulation
clear all
close all
r=-2:.1:2;
y=-2.5:.1:2.5;
figure
axis([-2.5 2.5 -1.5 1.5])
for j = 1:length(r)
    eq1 = r(j)*y-y.^3; % rhs of supercritical pitchfork bifn
    plot(y,eq1)
    axis([-2.5  2.5  -1.5  1.5])
    title(['r=',num2str(r(j))])
    xlabel('y'); ylabel('y\{prime}');
    grid
    drawnow
    pause(.05);
end
figure
plot(y,-2*y-y.^3,'b') %plot for r=-2
hold on
plot(y,-y.^3,'r') %plot for r=0
plot(y,2*y-y.^3,'k') %plot for r=2
legend('r<0', 'r=0', 'r>0')
axis([-2.5  2.5  -1.5  1.5])
hold off
```

Dfield in MATLAB
For a first-order system, a very user-friendly software supplement

called **dfield**, written by John C. Polking, may be found at
<http://math.rice.edu/~dfield/>, exists for MATLAB and it is freely
available for educational use. There are two or three programs that you will
need to download (depending on your version of MATLAB) and install in your
working directory. The program dfield (like its two-dimensional counterpart
pplane that we will encounter in Chapter 6) is much easier to implement than
either of the above methods for plotting direction fields and superimposing
numerical solutions. We give a brief introduction here.

Once you have placed the relevant programs in your working directory, type
dfield8. A new window should pop up; see the Figure 2.31. (Note that this
is for MATLAB 7.7 or later. Download the appropriate files for other versions
of MATLAB.)

The dependent variable, by default, is $x(t)$ but we can easily change these
in the window so that everything will be in terms of $y(x)$. We consider the
equations given below and enter them in the dfield window as

y'=1/(x^2+y^2)

and we must also specify that the independent variable is x (as opposed to
the default variable t).

For other problems, we may have parameters or expressions to enter. For
this problem, we also need to modify the minimum and maximum values of x
and y. We set the minimum x-value to be -3, the maximum x-value to be 3,
the minimum y-value to be -4, and the maximum y-value to be 4. We click
the *Proceed* button on the window that we have been typing to generate the
direction field. To plot solutions, we simply click on the direction field at a
desired initial condition. Repeat a few times. The reader should experiment
with the *Options* on the display window and see other options for plotting.

MATLAB Exercises
Turn in both the commands that you enter for the exercises below as well as
the output/figures. These should all be in one document. Please highlight or
clearly indicate all requested answers. Some of the questions will require you
to modify the above MATLAB code to answer them.

1. Enter the commands given in MATLAB Example 1 and submit both your
   input and output.

2. Enter the commands given in MATLAB Example 2 and submit both your
   input and output.

3. Enter the commands given in MATLAB Example 3 and submit both your
   input and output.

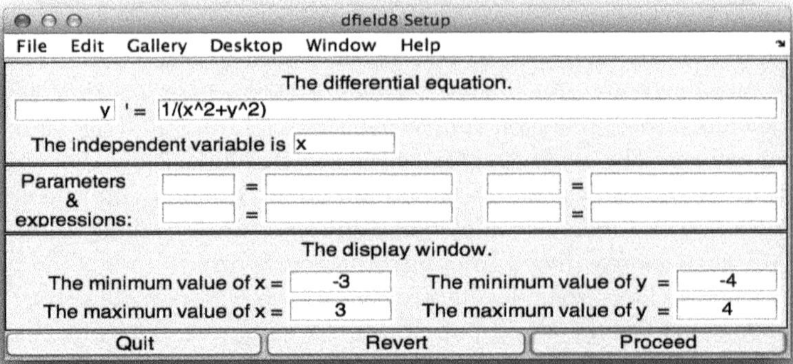

FIGURE 2.31: Pop-up window for `dfield8` program, used to obtain numerical solutions to first-order equations.

4. Enter the commands given in MATLAB Example 4 and submit both your input and output.

5. Plot the direction fields for the following ODEs:

   (a) $y' = y + x^2$ (b) $y' = \sin(x^2)$

   (c) $y' = xy^2$ (d) $y' = 1 - y^2$.

6. Superimpose three solutions onto the direction fields of the ODEs from #5:

   (a) $y' = y + x^2$ (b) $y' = \sin(x^2)$

   (c) $y' = xy^2$ (d) $y' = 1 - y^2$.

7. Plot the direction fields for the following ODEs:

   (a) $y' = y(1 - y)$ (b) $y' = \sin(y^2)$

   (c) $y' = y^2(3 - y^2)$ (d) $y' = x^3(1 - y^2)$.

8. Superimpose three solutions onto the direction fields of the ODEs from #7:

   (a) $y' = y(1 - y)$ (b) $y' = \sin(y^2)$

   (c) $y' = y^2(3 - y^2)$ (d) $y' = x^3(1 - y^2)$.

9. For $y' = y + x^2$ with $y(0) = -1$, use a stepsize of $h = 0.1$ to obtain the solutions from the Euler method and the 4th order Runge-Kutta method. For the same ODE and IC, also obtain the analytical solution (either by hand or by using `dsolve` [in R2012 or later]). Compare the values of the three solutions at $x = 0, 0.1, 1, 5$.

10. Repeat the instructions in #9 for the stepsize $h = 0.01$.

11. For $y' = x(y + 1)^2$ with $y(0) = 2$, use a stepsize of $h = 0.25$ to obtain the solutions from the Euler method and the 4th order Runge-Kutta method.

For the same ODE and IC, also obtain the analytical solution (either by hand or by using `dsolve`). Compare the values of the three solutions at $x = 0, 0.5, 2, 5$.

12. Repeat the instructions in #11 for the stepsize $h = 0.1$.

13. For $y' = (x - 1)\cos^2(y)$ with $y(0) = 0$, use a stepsize of $h = 0.2$ to obtain the solutions from the Euler method and the 4th order Runge-Kutta method. For the same ODE and IC, also obtain the analytical solution (either by hand or by using `dsolve`). Compare the values of the three solutions at $x = 0, 0.4, 2, 5$.

14. Repeat the instructions in #13 for the stepsize $h = 0.05$.

15. If you covered the topic of bifurcations in class or are instructed to read this on your own by your instructor, animate the qualitatively different solutions of the ODE with a transcritical bifurcation, $y' = ry - y^2$. Then superimpose the plots of three qualitatively different solutions, i.e., one for each of $r < 0$, $r = 0$, and $r > 0$.

16. If you covered the topic of bifurcations in class or are instructed to read this on your own by your instructor, animate the qualitatively different solutions of the ODE $y' = ry^3 - y^5$. What type of bifurcation is this? Then superimpose the plots of three qualitatively different solutions, i.e., one for each of $r < 0$, $r = 0$, and $r > 0$.

17. If you covered the topic of bifurcations in class or are instructed to read this on your own by your instructor, animate the qualitatively different solutions of the ODE $y' = r + 3y - 4y^3$. What are the *two* bifurcations that occur? Then superimpose the plots of five qualitatively different solutions, i.e., one for each of $r < -1$, $r = -1$, $-1 < r < 1$, $r = 1$, and $r > 1$.

Chapter 2 Computer Lab: Maple

Maple Example 1: Enter the following code that demonstrates how to plot direction fields in Maple for $y' = \dfrac{1}{x^2 + y^2}$. Please note that the derivative symbol that appears below is always entered from the Expression palette $\dfrac{d}{dx} f$.

restart
with(DEtools): #*Loads the needed diff eqns package*

$eq1 := \dfrac{\mathrm{d}}{\mathrm{d}x} y(x) = \dfrac{1}{x^2 + y(x)^2}$

dfieldplot($eq1, y(x), x = -3..3, y = -3..3, arrows = line, dirgrid = [10, 10]$)

Maple Example 2: Enter the following code that demonstrates how to plot solution curves for $y' = \dfrac{1}{x^2 + y^2}$. Do not type **restart** as you will need the ODE *eq1* from Example 1 above.

with(*plots*):
$IC1 := y(-3) = -3$
$soln1 := dsolve(\{eq1, IC1\}, y(x), numeric, range = -10..10)$:
$IC2 := y(-3) = -2$
$soln2 := dsolve(\{eq1, IC2\}, y(x), numeric, range = -10..10)$:
$IC3 := y(-3) = -1$
$soln3 := dsolve(\{eq1, IC3\}, y(x), numeric, range = -10..10)$:
$IC4 := y(-3) = 0$
$soln4 := dsolve(\{eq1, IC4\}, y(x), numeric, range = -10..10)$:
$eq2 := odeplot(soln1, [x, y(x)], numpoints = 200, thickness = 2)$:
$eq3 := odeplot(soln2, [x, y(x)], numpoints = 200, thickness = 2)$:
$eq4 := odeplot(soln3, [x, y(x)], refine = 2, thickness = 2)$:
$eq5 := odeplot(soln4, [x, y(x)], refine = 2, thickness = 2)$:
$eq6 := dfieldplot(eq1, y(x), x = -3..3, y = -3..3, arrows = line, dirgrid$
    $= [10, 10])$:
indexMaple commands! **dfieldplot** *display*($[eq2, eq3, eq4, eq5, eq6]$)

Maple Example 3: Enter the following code that utilizes Maple's built-in Euler method and 4th order Runge-Kutta method. The first set of commands will obtain these numerical solutions while the final group of commands will superimpose the Runge-Kutta solution onto the direction field. The IVP we consider is $y' = xy$, $y(0) = 1$.

restart
with(*plots*):
$eq1 := \dfrac{\mathrm{d}}{\mathrm{d}x} y(x) = x \cdot y(x)$
$IC := y(0) = 1$
$solnEM := dsolve(\{IC, eq1\}, y(x), numeric, method = classical[foreuler],$
    $stepsize = .1)$:
$solnRK4 := dsolve(\{IC, eq1\}, y(x), numeric, method = classical[rk4],$
    $stepsize = .1)$:
solnEM(.3)
solnRK4(.3)

solnEM(3.1)

solnRK4(3.1)

soln := *dsolve*({*IC, eq1*}, *y*(*x*))

evalf(*subs*(*x* = .3, *soln*))

evalf(*subs*(*x* = 3.1, *soln*))

> *#Note that solnRK4 compares well with the exact answer but solnEM compares horribly*

eq2 := *odeplot*(*solnEM*, [*x, y*(*x*)], *view* = [−3..3, 0..120], *labels* = [*x, y*], *numpoints* = 500, *color* = *blue, linestyle* = 2)

eq3 := *odeplot*(*solnRK4*, [*x, y*(*x*)], *view* = [−3..3, 0..120], *labels* = [*x, y*], *numpoints* = 500, *color* = *black, linestyle* = 3) :

display([*eq2, eq3*], *title* = "Numerical solns of y'=xy using EM and RK4 with h=.1, dotted/blue is EM, dashed/black is RK4")

#Now we superimpose solnRK4 onto the direction field

with(*DEtools*):

eq4 := *odeplot*(*solnRK4*, [*x, y*(*x*)], *view* = [−3..3, 0..4], *labels* = [*x, y*], *numpoints* = 500) :

eq5 := *dfieldplot*(*eq1, y*(*x*), *x* = −3..3, *y* = 0..4, *arrows* = *line*) :

display([*eq4, eq5*])

Maple Example 4: If you covered the topic of bifurcations in class or are instructed to read this on your own by your instructor, enter the following commands to illustrate how to animate the qualitatively different solutions of the ODE with a supercritical bifurcation, $y' = ry - y^3$. The last command superimposes three different qualitative solutions observed during the animation.

restart

with(*plots*):

eq1 := $r \cdot y - y^3$ *#the rhs of y'=f(r,y) for a supercritical pitchfork bifurcation*

animate(*plot*, [*eq1, y* = −2.5..2.5, −1.5..1.5], *r* = −2..2)

> *#When figure appears, right click on it, then go to* Animation → Play. *Alternatively, in the toolbar beneath the button* Animation *and to the right of* Current Frame, *move the vertical bar to animate figure.*

plot([*subs*(*r* = −2, *eq1*), *subs*(*r* = 0, *eq1*), *subs*(*r* = 2, *eq1*)], *y* = −2.5..2.5, −1.5..1.5, *legend* = ["r < 0", "r = 0", "r > 0"])

Maple Exercises

Turn in both the commands that you enter for the exercises below as well as the output/figures. These should all be in one document. Please highlight or clearly indicate all requested answers. Some of the questions will require you to modify the above Maple code to answer them.

1. Enter the commands given in Maple Example 1 and submit both your input and output.

2. Enter the commands given in Maple Example 2 and submit both your input and output.

3. Enter the commands given in Maple Example 3 and submit both your input and output.

4. Enter the commands given in Maple Example 4 and submit both your input and output.

5. Plot the direction fields for the following ODEs:

    (a) $y' = y + x^2$ (b) $y' = \sin(x^2)$
    (c) $y' = xy^2$ (d) $y' = 1 - y^2$.

6. Superimpose three solutions onto the direction fields of the ODEs from #5:

    (a) $y' = y + x^2$ (b) $y' = \sin(x^2)$
    (c) $y' = xy^2$ (d) $y' = 1 - y^2$.

7. Plot the direction fields for the following ODEs:

    (a) $y' = y(1 - y)$ (b) $y' = \sin(y^2)$
    (c) $y' = y^2(3 - y^2)$ (d) $y' = x^3(1 - y^2)$.

8. Superimpose three solutions onto the direction fields of the ODEs from #7:

    (a) $y' = y(1 - y)$ (b) $y' = \sin(y^2)$
    (c) $y' = y^2(3 - y^2)$ (d) $y' = x^3(1 - y^2)$.

9. For $y' = y + x^2$ with $y(0) = -1$, use a stepsize of $h = 0.1$ to obtain the solutions from the Euler method and the 4th order Runge-Kutta method. For the same ODE and IC, also obtain the analytical solution (either by hand or by using `dsolve`). Compare the values of the three solutions at $x = 0, 0.1, 1, 5$.

10. Repeat the instructions in #9 for the stepsize $h = 0.01$.

11. For $y' = x(y + 1)^2$ with $y(0) = 2$, use a stepsize of $h = 0.25$ to obtain the solutions from the Euler method and the 4th order Runge-Kutta method. For the same ODE and IC, also obtain the analytical solution (either by hand or by using `dsolve`). Compare the values of the three solutions at $x = 0, 0.5, 2, 5$.

12. Repeat the instructions in #11 for the stepsize $h = 0.1$.

13. For $y' = (x - 1)\cos^2(y)$ with $y(0) = 0$, use a stepsize of $h = 0.2$ to obtain the solutions from the Euler method and the 4th order Runge-Kutta method. For the same ODE and IC, also obtain the analytical solution (either by hand or by using `dsolve`). Compare the values of the three solutions at $x = 0, 0.4, 2, 5$.

14. Repeat the instructions in #13 for the stepsize $h = 0.05$.

15. If you covered the topic of bifurcations in class or are instructed to read this on your own by your instructor, animate the qualitatively different solutions of the ODE with a transcritical bifurcation, $y' = ry - y^2$. Then superimpose the plots of three qualitatively different solutions, i.e., one for each of $r < 0$, $r = 0$, and $r > 0$.

16. If you covered the topic of bifurcations in class or are instructed to read this on your own by your instructor, animate the qualitatively different solutions of the ODE $y' = ry^3 - y^5$. What type of bifurcation is this? Then superimpose the plots of three qualitatively different solutions, i.e., one for each of $r < 0$, $r = 0$, and $r > 0$.

17. If you covered the topic of bifurcations in class or are instructed to read this on your own by your instructor, animate the qualitatively different solutions of the ODE $y' = r + 3y - 4y^3$. What are the *two* bifurcations that occur? Then superimpose the plots of five qualitatively different solutions, i.e., one for each of $r < -1$, $r = -1$, $-1 < r < 1$, $r = 1$, and $r > 1$.

Chapter 2 Computer Lab: Mathematica

Mathematica Example 1: Enter the following code that demonstrates how to plot direction fields in Maple for $y' = \dfrac{1}{x^2 + y^2}$.

```
Quit[]
eq1[x_,y_]=  1
           ──────
           x² + y²
p0 = VectorPlot[{1, eq1[x, y]}, {x, -3, 3}, {y, -3, 3},
   VectorScale→{.05, .05, None}, VectorPoints→10,
   VectorStyle→ "Segment"]
```

Mathematica Example 2: Enter the following code that demonstrates how to plot solution curves for $y' = \dfrac{1}{x^2 + y^2}$. Do not type `Quit[ ]` as you will need the direction field from Example 1 above.

```
eq2[x_]=   1
         ─────────
         x² + y[x]²
sol1 = NDSolve[{y'[x]==eq2[x], y[-3]==-3}, y, {x, -10, 10}]
   (*Result is an Interpolating Function.*)
```

```
sol2 = NDSolve[{y'[x]==eq2[x], y[-3]==-2}, y, {x, -10, 10}];
sol3 = NDSolve[{y'[x]==eq2[x], y[-3]==-1}, y, {x, -10, 10}];
sol4 = NDSolve[{y'[x]==eq2[x], y[-3]==0}, y, {x, -10, 10}];
 (*y[x]/.sol1 is the solution we seek.*)
p1 = Plot[Evaluate[y[x]/.sol1], {x, -3, 3}, PlotRange→{-3, 3},
  PlotStyle→{Thickness[.01]}];
p2 = Plot[Evaluate[y[x]/.sol2], {x, -3, 3}, PlotRange→{-3, 3},
  PlotStyle→{Thickness[.01]}];
p3 = Plot[Evaluate[y[x]/.sol3], {x, -3, 3}, PlotRange→{-3, 3},
  PlotStyle→{Thickness[.01]}];
p4 = Plot[Evaluate[y[x]/.sol4], {x, -3, 3}, PlotRange→{-3, 3},
  PlotStyle→{Thickness[.01]}];
Show[p0, p1, p2, p3, p4]
```

Mathematica Example 3: Enter the following code that utilizes Mathematica's built-in Euler method and 4th order Runge-Kutta method. The first set of commands will obtain these numerical solutions while the final group of commands will superimpose the Runge-Kutta solution onto the direction field. The IVP we consider is $y' = xy$, $y(0) = 1$.

```
Quit[]
de[x_]=x y[x] (*don't forget <space> for multiplication*)
solutionEM=NDSolve[{y'[x]==de[x], y[0]==1}, y, {x, 0, 4},
  StartingStepSize→.1,
  Method→{FixedStep, Method→ExplicitEuler}]
solutionRK=NDSolve[{y'[x]==de[x], y[0]==1}, y, {x, 0, 4},
  StartingStepSize→.1,
  Method→{FixedStep, Method→ExplicitRungeKutta}]
yEM[x_] = y[x]/.solutionEM[[1]]
yRK[x_] = y[x]/.solutionRK[[1]]
yEM[.3]
yRK[.3]
yEM[3.1]
yRK[3.1]
soln = DSolve[{y'[x]==de[x], y[0]==1}, y[x], x]
ReplaceAll[soln[[1]][[1]], x→.3]
ReplaceAll[soln[[1]][[1]], x→3.1]
p1 = Plot[yEM[x], {x, 0, 3}, PlotRange→{0, 120},
PlotStyle→{Blue, Dotted}];
p2 = Plot[yRK[x], {x, 0, 3}, PlotRange→{0, 120},
PlotStyle→{Black, Dashed}];
Show[p1, p2, PlotLabel→"Numerical Solns of y'=xy using
  EM and RK4 with h=.1, dotted/blue is EM, dashed/black is RK4"]
```

```
de[x_,y_] = x y
p0 = VectorPlot[{1, de[x, y]}, {x, 0, 3}, {y, 0, 4},
  VectorScale→{.05, .05, None}, VectorPoints→10,
  VectorStyle→"Segment"];
Show[p0, p2]
```

Mathematica Example 4: If you covered the topic of bifurcations in class or are instructed to read this on your own by your instructor, enter the following commands to illustrate how to animate the qualitatively different solutions of the ODE with a supercritical bifurcation, $y' = ry - y^3$. The last command superimposes three different qualitative solutions observed during the animation.

```
Quit[]
f[r_,y_] = r y -y^3 (*don't forget <space> for multiplication*)
Manipulate[Plot[f[r, y], {y, -2.5, 2.5}], {r, -2, 2}]
  (*Now move the circle on the r-scale to animate*)
Plot[{ReplaceAll[f[r, y], r→-2], ReplaceAll[f[r, y], r→0],
  ReplaceAll[f[r, y], r→2]}, {y, -2.5, 2.5},
  PlotLegends→{"r<0", "r=0", "r>0"}]
```

Mathematica Exercises

Turn in both the commands that you enter for the exercises below as well as the output/figures. These should all be in one document. Please highlight or clearly indicate all requested answers. Some of the questions will require you to modify the above Mathematica code to answer them.

1. Enter the commands given in Mathematica Example 1 and submit both your input and output.

2. Enter the commands given in Mathematica Example 2 and submit both your input and output.

3. Enter the commands given in Mathematica Example 3 and submit both your input and output.

4. Enter the commands given in Mathematica Example 4 and submit both your input and output.

5. Plot the direction fields for the following ODEs:
   (a) $y' = y + x^2$ (b) $y' = \sin(x^2)$
   (c) $y' = xy^2$ (d) $y' = 1 - y^2$.

6. Superimpose three solutions onto the direction fields of the ODEs from #5:
   (a) $y' = y + x^2$ (b) $y' = \sin(x^2)$
   (c) $y' = xy^2$ (d) $y' = 1 - y^2$.

7. Plot the direction fields for the following ODEs:
   (a) $y' = y(1 - y)$ (b) $y' = \sin(y^2)$
   (c) $y' = y^2(3 - y^2)$ (d) $y' = x^3(1 - y^2)$.

8. Superimpose three solutions onto the direction fields of the ODEs from #7:

   (a) $y' = y(1 - y)$ (b) $y' = \sin(y^2)$
   (c) $y' = y^2(3 - y^2)$ (d) $y' = x^3(1 - y^2)$.

9. For $y' = y + x^2$ with $y(0) = -1$, use a stepsize of $h = 0.1$ to obtain the solutions from the Euler method and the 4th order Runge-Kutta method. For the same ODE and IC, also obtain the analytical solution (either by hand or by using dsolve). Compare the values of the three solutions at $x = 0, 0.1, 1, 5$.

10. Repeat the instructions in #9 for the stepsize $h = 0.01$.

11. For $y' = x(y + 1)^2$ with $y(0) = 2$, use a stepsize of $h = 0.25$ to obtain the solutions from the Euler method and the 4th order Runge-Kutta method. For the same ODE and IC, also obtain the analytical solution (either by hand or by using dsolve). Compare the values of the three solutions at $x = 0, 0.5, 2, 5$.

12. Repeat the instructions in #11 for the stepsize $h = 0.1$.

13. For $y' = (x - 1)\cos^2(y)$ with $y(0) = 0$, use a stepsize of $h = 0.2$ to obtain the solutions from the Euler method and the 4th order Runge-Kutta method. For the same ODE and IC, also obtain the analytical solution (either by hand or by using dsolve). Compare the values of the three solutions at $x = 0, 0.4, 2, 5$.

14. Repeat the instructions in #13 for the stepsize $h = 0.05$.

15. If you covered the topic of bifurcations in class or are instructed to read this on your own by your instructor, animate the qualitatively different solutions of the ODE with a transcritical bifurcation, $y' = ry - y^2$. Then superimpose the plots of three qualitatively different solutions, i.e., one for each of $r < 0$, $r = 0$, and $r > 0$.

16. If you covered the topic of bifurcations in class or are instructed to read this on your own by your instructor, animate the qualitatively different solutions of the ODE $y' = ry^3 - y^5$. What type of bifurcation is this? Then superimpose the plots of three qualitatively different solutions, i.e., one for each of $r < 0$, $r = 0$, and $r > 0$.

17. If you covered the topic of bifurcations in class or are instructed to read this on your own by your instructor, animate the qualitatively different solutions of the ODE $y' = r + 3y - 4y^3$. What are the *two* bifurcations that occur? Then superimpose the plots of five qualitatively different solutions, i.e., one for each of $r < -1$, $r = -1$, $-1 < r < 1$, $r = 1$, and $r > 1$.

Chapter 2 Projects

Project 2A: Spruce Budworm

We consider the famous spruce budworm model which was used to describe the dynamics of this population on a forest in Canada. Periodically, there were large outbreaks of the budworm that devastated the forest. The forest would recover and then a similar outbreak would occur. A simple first-order autonomous ODE was proposed to explain the behavior [33],[37],[48]:

$$\frac{dN}{dt} = \rho N \left(1 - \frac{N}{K}\right) - \frac{\alpha N^2}{\beta^2 + N^2}. \tag{2.29}$$

Here, N represents the population of the budworm, ρ is the growth rate (assumed to be logistic), and K is the carrying capacity. From calculus, we can observe that the second term is 0 when $N = 0$, approaches α as N gets large, and reaches its steepest ascent when $N = \beta$.

- Show that (2.29) can be written in the dimensionless form

$$\frac{dx}{d\tau} = rx \left(1 - \frac{x}{k}\right) - \frac{x^2}{1 + x^2}. \tag{2.30}$$

- Show that the system has between one and three biologically meaningful equilibria. (Hint: if it helps, graph the two terms separately but on the same graph.)

- Determine the stability of the equilibria. Your answer will have a dependence on the values of the parameters r and k.

- Find the equation, in terms of the parameters r and k, that describes where the system changes from one to three equilibria.

- Give a biological interpretation for the system with one equilbrium point vs. three equilibria.

Project 2B: Multistep Methods of Numerical Approximation

Many numerical methods for solving differential equations exist. In this project, we introduce the student to a *multistep* method that takes into account previous (and sometimes future!) values of the solution in order to calculate the next solution. Multitstep methods of all orders exist but we focus on second-order methods for simplicity of formulation [10]. Students will need to have some programming knowledge in one of MATLAB, Maple, or Mathematica in order to make progress on this project. We consider the differential equation

$$\frac{dy}{dx} = xy, \quad y(0) = 1$$

discussed in Section 2.5.

Explicit Method: Consider the formula

$$y_{i+1} = y_i + \frac{h}{2}\left(3f(x_i, y_i) - f(x_{i-1}, y_{i-1})\right), \quad i = 1, 2, \cdots, N-1, \quad (2.31)$$

which is used to approximate the solution of the initial-value problem, $y' = f(x, y)$, $y(x_0) = y_0$. The local truncation error is given by $\frac{5}{12}y^{(3)}(\xi_i)h^2$, for some $\xi_i \in (x_{i-2}, x_{i+1})$.

1. Using MATLAB, Maple, or Mathematica, implement the above formula, known as the *second-order Adams-Bashforth formula*, an explicit method, to approximate the solution at $x = 1$. Use a step size of $h = .1$ and use the approximate value of $y(.1)$ from Runge-Kutta with $h = .1$ for your value of y_1.

2. Compare the error with that of Euler's method (a first-order method) and the Runge-Kutta method (a fourth-order method).

Implicit Method: Now consider the formula

$$y_{i+1} = y_i + \frac{h}{12}\left(5f(x_{i+1}, y_{i+1}) + 8f(x_i, y_i) - f(x_{i-1}, y_{i-1})\right), \quad (2.32)$$

$i = 1, 2, \cdots, N-1$, which is again used to approximate the solution of the initial value problem, $y' = f(x, y)$, $y(x_0) = y_0$. The local truncation error is given by $\frac{-1}{24}y^{(4)}(\xi_i)h^2$, for some $\xi_i \in (x_{i-1}, x_{i+1})$. This is known as the *second-order Adams-Moulton formula*, an implicit method. Note here that in order to approximate the solution at y_{i+1}, we need to know the value of $f(x_{i+1}, y_{i+1})$! How can this be done?

1. Substitute the original differential equation for $f(x_{i+1}, y_{i+1})$ in the Adams-Moulton formula. Solve for y_{i+1}.

2. Using MATLAB, Maple, or Mathematica, implement the resulting formula of part 1 with $h = .1$ again using the Runge-Kutta approximation for y_1.

3. Compare the error with that of Euler's method (a first-order method) and the Runge-Kutta method (a fourth-order method).

Predictor-Corrector Method: This method combines the explicit and implicit methods used above. Instead of solving the Adams-Moulton (AM) formula for y_{i+1}, we use the approximate result for y_{i+1} obtained from the Adams-Bashforth (AB) formula. The result of substituting the AB result into the AM result usually gives a more accurate answer than the original AB answer and is almost always quicker than solving the AM formula by hand, as done above. For the instructions below, again use $h = .1$ and the Runge-Kutta approximation for y_1.

1. Using MATLAB, Maple, or Mathematica, implement as follows:

a. Perform one step of AB.

b. Substitute this value into the AM formula.

c. Perform one step of AM.

d. If desired, resubstitute this last AM result into the AM expression again.

e. Compare the results with the AB, AM, and exact solutions.

This method is called a *predictor-corrector method* because the AB method predicts the function value at x_{i+1} and the AM method corrects this with a higher-order approximation.

Chapter 3

Elements of Higher-Order Linear
Equations

Following our work with first-order equations, it is natural for us to consider higher-order equations. The first methods of solving differential equations of second or higher-order with constant coefficients were due to Euler. D'Alembert dealt with the case when the auxiliary equation had equal roots. Some of the symbolic methods of finding the particular solution were not given until about 100 years later by Lobatto (1837) and Boole (1859).

3.1 Introduction to Higher-Order Equations

While first-order equations are appropriate for certain models, numerous others require either at least one higher derivative or at least one additional variable to better describe the system in question. We begin our study with a definition.

Definition 3.1.1

A linear differential equation of order n in the dependent variable y and the independent variable x is an equation that is in, or can be expressed in, the form

$$a_n(x) \frac{d^n y}{dx^n} + a_{n-1}(x) \frac{d^{n-1} y}{dx^{n-1}} + \ldots + a_1(x) \frac{dy}{dx} + a_0(x) y = F(x) \quad (3.1)$$

where a_n is not identically zero.

We shall assume throughout that $a_n, a_{n-1}, \ldots, a_0$ and F are continuous real functions on a real interval $a \le x \le b$ and that

$$a_n(x) \neq 0 \text{ for any } x \text{ on } a \le x \le b.$$

The right-hand side $F(x)$ is called the *nonhomogeneous term*. If F is identically zero, then

$$a_n(x) \frac{d^n y}{dx^n} + a_{n-1}(x) \frac{d^{n-1} y}{dx^{n-1}} + \ldots + a_1(x) \frac{dy}{dx} + a_0(x) y = 0 \quad (3.2)$$

is called a **homogeneous** equation. It is sometimes called the *reduced equation* of the given nonhomogeneous equation.

Example 1 The equation

$$\frac{d^2y}{dx^2} + 7x\frac{dy}{dx} + x^5y = e^x$$

is a linear nonhomogeneous differential equation of the second order.

Example 2 The equation

$$\frac{d^3y}{dx^3} + x\frac{d^2y}{dx^2} + 3x\frac{dy}{dx} + y = 0$$

is a linear homogeneous differential equation of the third order.

Existence and Uniqueness of Solutions

Our goal is to consider methods of solving these higher-order differential equations. In particular, we will be interested in second-order linear equations as these equations occur often in physics and engineering applications and give rise to the special functions we will consider. Before we pursue methods of solution, it is best to give an existence theorem for initial-value problems associated with a linear ordinary differential equation of order n.

THEOREM 3.1.1   Existence and Uniqueness

Consider the linear ordinary differential equation of order n (3.1) where $a_n, a_{n-1}, \ldots, a_0$ and F are continuous real functions on a real interval $a \leq x \leq b$ and

$$a_n(x) \neq 0 \text{ for any } x \text{ on } a \leq x \leq b.$$

If x_0 is any point of the interval $a \leq x \leq b$ and $c_0, c_1, \ldots, c_{n-1}$ are n arbitrary constants, then there exists a unique solution f of (3.1) such that

$$f(x_0) = c_0, \quad f'(x_0) = c_1, \quad \ldots, \quad f^{(n-1)}(x_0) = c_{n-1},$$

and this solution is defined over the entire interval $a \leq x \leq b$.

What this theorem says is that for the differential equation (3.1) there is precisely one solution that passes through the value c_0 at x_0 and whose k^{th} derivative takes on the value c_k for each $k = 1, 2, \ldots, n-1$ at $x = x_0$. Further, the theorem asserts that this unique solution is defined for $a \leq x \leq b$.

Example 3 For the initial-value problem

$$\frac{d^2y}{dx^2} + 7x\frac{dy}{dx} + x^5y = e^x \quad \text{with} \quad y(2) = 1 \quad \text{and} \quad y'(2) = 5,$$

there exists a unique solution since the coefficients $1, 7x, x^5$ as well as the nonhomogenous term e^x are all continuous on $(-\infty, \infty)$.

Example 4 Determine if $y = c_1 + c_2 x^2$ gives rise to a unique solution for $xy'' - y' = 0$, $y(0) = 0, y'(0) = 1$ on $(-\infty, \infty)$.

Solution

We can easily verify that this is a solution. However, if we simply apply the IC, we obtain the two equations

$$0 = y(0) = c_1 + c_2 \cdot 0^2$$
$$1 = y'(0) = 2c_2 \cdot 0,$$

the second of which gives us the inconsistent statement $1 = 0$. What happened? A look at the differential equation shows that $a_0(x) = 0$ when $x = 0$, which is the x-value of the initial condition. Thus Theorem 3.1.1 does not apply for the initial condition we were given.

The theorem has an important corollary in the homogeneous case.

THEOREM 3.1.2

Let f be the solution of the nth order homogeneous linear differential equation (3.2) such that

$$f(x_0) = 0, \ f'(x_0) = 0, \ \ldots, \ f^{(n-1)}(x_0) = 0$$

where x_0 is a point of the interval $a \leq x \leq b$ in which the coefficients $a_n, a_{n-1}, \ldots, a_0$ are all continuous and $a_n(x) \neq 0$. Then $f(x) = 0$ for all x on $a \leq x \leq b$.

Example 5 The third-order linear homogeneous ordinary differential equation

$$\frac{d^3 y}{dx^3} + x \frac{d^2 y}{dx^2} + 3x \frac{dy}{dx} + y = 0 \ \text{ with } \ f(8) = f'(8) = f''(8) = 0$$

has the unique (trivial) solution $f(x) = 0$ for all x.

We have now talked about existence and uniqueness of solutions. Our goal will be to obtain the most general possible solution. We will come up with various ways to obtain solutions of the given differential equation but we need to know how to find the minimum number of solutions that can be combined to give every possible solution. This is absolutely essential to understand. Finding this minimum possible list of solutions that can generate all other solutions can be thought of (in a very rough sense) as analogous to finding the prime factors of a given integer. This minimum set is known as a **fundamental set of solutions** and satisfies the property that each element is a solution and each "contributes something new" to the set.

The Need for Sufficiently Different Solutions

We are now going to consider a case where we can easily obtain a solution and use a second example as motivation for studying linear independence and other related concepts from linear algebra. Most of the equations we will consider in this chapter have y as the dependent variable and x as the independent variable. The simplest homogeneous linear equation we have considered thus far is first-order, with constant coefficients. It is of the form

$$a_1 \frac{dy}{dx} + a_0 y = 0.$$

What is the form of the solution? In this simple case, we observe that the equation is separable and the solution will be an exponential function $e^{(-a_0/a_1)x}$. If we consider the second-order homogeneous constant coefficient linear equation

$$a_2 \frac{d^2 y}{dx^2} + a_1 \frac{dy}{dx} + a_0 y = 0,$$

we ask if it is still possible to have a solution of the form $y = e^{rx}$, since it worked in the first-order case. To find the value of r that will make it a solution we substitute into the differential equation and find values for r that will satisfy the resulting condition. Let's try this with a specific example.

Example 6 Consider the equation

$$\frac{d^2 y}{dx^2} + 3 \frac{dy}{dx} + 2y = 0. \tag{3.3}$$

If we substitute $y = e^{rx}$, we have

$$r^2 e^{rx} + 3r e^{rx} + 2 e^{rx} = 0, \quad \text{which gives} \quad e^{rx}(r^2 + 3r + 2) = 0.$$

Because e^{rx} is never zero, we consider only the terms in parentheses. This equation then gives the conditions for the r-values that make e^{rx} a solution:

$$r^2 + 3r + 2 = 0,$$

which factors into $(r + 2)(r + 1) = 0$. This holds if $r = -1$ or $r = -2$. Thus, two solutions to the differential equation are e^{-x} and e^{-2x}.

It is not a coincidence that we found *two* different solutions to this *second-order* equation. We observe that for this situation

$$C_1 e^{-x} + C_2 e^{-2x} \tag{3.4}$$

is also a solution, as the reader should check! It turns out that *any* solution to the differential equation (3.3) must be of the form (3.4). Will things always be so nice? Let's try another example to see.

Example Consider the equation

$$\frac{d^2 y}{dx^2} + 2 \frac{dy}{dx} + y = 0. \tag{3.5}$$

If we substitute $y = e^{rx}$, we see that it will be a solution whenever

$$r^2 + 2r + 1 = 0,$$

which factors into $(r + 1)^2 = 0$. This holds if $r = -1$. A solution of (3.5) can thus be written as $y = e^{-x}$. Is this the only solution? The answer is a resounding "no!" and we observe by substitution that xe^{-x} is also a solution. The details of why we might guess xe^{-x} as a second solution are left for discussion in Section 3.3.

What went wrong in this last example? And how do we deal with such situations? To give satisfactory answers to these questions, we first need some basic concepts from linear algebra, including an understanding of linear independence and linear combination.

3.1.1   Operator Notation

We will now introduce some notation to make our subsequent work a bit easier. The nth derivative of a function $y(x)$ is given in **operator notation** by

$$D^n y = \frac{d^n y}{dx^n}.$$

Using this notation, the left-hand side of the nth-order linear homogeneous differential equation (3.2)

$$a_n(x) \frac{d^n y}{dx^n} + a_{n-1}(x) \frac{d^{n-1} y}{dx^{n-1}} + \ldots + a_1(x) \frac{dy}{dx} + a_0(x) y = 0$$

can be expressed as

$$a_n(x) \frac{d^n y}{dx^n} + a_{n-1}(x) \frac{d^{n-1} y}{dx^{n-1}} + \ldots + a_1(x) \frac{dy}{dx} + a_0(x) y$$

$$= a_n(x) D^n y + a_{n-1}(x) D^{n-1} y + \ldots + a_1(x) Dy + a_0(x) y$$

$$= \left(a_n(x) D^n + a_{n-1}(x) D^{n-1} + \ldots + a_1(x) D + a_0(x) \right) y.$$

Thus, we can write (3.2) as

$$\left(a_n(x) D^n + a_{n-1}(x) D^{n-1} + \ldots + a_1(x) D + a_0(x) \right) y = 0,$$

and in fact we can write the nth-order linear nonhomogeneous equation (3.1) as

$$\left(a_n(x) D^n + a_{n-1}(x) D^{n-1} + \ldots + a_1(x) D + a_0(x) \right) y = F(x).$$

With this in mind, the following definition is natural.

Definition 3.1.2

The expression

$$P(D) = a_n(x) D^n + a_{n-1}(x) D^{n-1} + \ldots + a_1(x) D + a_0(x),$$

where $a_n(x), a_{n-1}(x), \ldots, a_0(x)$ are (possibly constant) real-valued functions with $a_n(x) \neq 0$, is called an nth-order linear differential operator.

With this notation, we can write an nth-order linear homogeneous differential equation compactly as

$$P(D)y = 0$$

and an nth-order linear nonhomogeneous equation as

$$P(D)y = F(x).$$

Example 8 Write the differential equation $2y''' - 3y'' + 5y' - y = 2x$ in operator form.

Solution
The operator is given as

$$P(D) = 2D^3 - 3D^2 + 5D - 1$$

so that the equation is

$$P(D)y = 2x.$$

Example 9 Write the differential equation $2xy'' - (3x^2 + 1)y' + e^x y = \cos x$ in operator form.

Solution

The operator is given as

$$P(D) = 2xD^2 - (3x^2 + 1)D + e^x$$

so that the equation is

$$P(D)y = \cos x.$$

Before we proceed with our further study of nonhomogeneous equations, it will be of future benefit if we obtain some results for computing with differential operators. As we almost always will be dealing with differential operators with **constant coefficients**, we state a few important results with these.

Let y_1 and y_2 both be n times continuously differentiable functions and let $P(D)$ be an nth-order linear differential operator with constant coefficients. We have

$$P(D)(y_1 + y_2) = \left(a_n D^n + a_{n-1} D^{n-1} + \ldots + a_1 D + a_0 \right)(y_1 + y_2)$$

$$= a_n D^n(y_1 + y_2) + a_{n-1} D^{n-1}(y_1 + y_2) + \ldots$$

$$+ a_1 D(y_1 + y_2) + a_0 (y_1 + y_2)$$

$$= a_n D^n y_1 + a_{n-1} D^{n-1} y_1 + \ldots + a_1 Dy_1 + a_0 y_1$$

$$+ a_n D^n y_2 + a_{n-1} D^{n-1} y_2 + \ldots + a_1 Dy_2 + a_0 y_2$$

$$= P(D)y_1 + P(D)y_2.$$

Thus, we have the **linearity property**

$$P(D)(y_1 + y_2) = P(D)y_1 + P(D)y_2. \tag{3.6}$$

It can also be shown, but with a bit more effort, that if $P_1(D)$ and $P_2(D)$ are two linear differential operators with constant coefficients, then

$$P_1(D)P_2(D) = P_2(D)P_1(D) \tag{3.7}$$

so that constant-coefficient linear differential operators *commute*. It is essential to realize, however, that applying an operator to a function is not really a multiplication, although it seems like it at times.

Example 10 Consider $P_1(D) = xD, P_2(D) = D, y = 3x$. Then

$$P_2(D)P_1(D)(y) = (D)(xD)(3x) = (D)(xD(3x)) = (D)(x(3)) = (D)(3x) = 3$$

but

$$P_1(D)P_2(D)(y) = (xD)(D)(3x) = (xD)(D(3x)) = (xD)(3) = x(0) = 0.$$

We again stress that *the order can be exchanged only in the case of differential operators with constant coefficients.* To see this in general, we write

$$P_1(D)P_2(D)$$

$$= \left(a_n\, D^n + a_{n-1}\, D^{n-1} + \ldots + a_1\, D + a_0\right)\left(b_n\, D^n + b_{n-1}\, D^{n-1} + \ldots + b_0\right)$$

$$= (a_n b_n)D^{2n} + (a_{n-1}b_n + a_n b_{n-1})D^{2n-1}$$

$$+ (a_{n-2}b_n + a_{n-1}b_{n-1} + a_n b_{n-2})D^{2n-2} + \cdots$$

$$+ (a_n b_0 + a_{n-1}b_1 + a_{n-2}b_2 + \cdots + a_2 b_{n-2} + a_1 b_{n-1} + a_0 b_n)D^n + \cdots$$

$$+ (a_2 b_0 + a_1 b_1 + a_0 b_2)D^2 + (a_1 b_0 + a_0 b_1)D + a_0 b_0$$

$$= (b_n a_n)D^{2n} + (b_{n-1}a_n + b_n a_{n-1})D^{2n-1}$$

$$+ (b_{n-2}a_n + b_{n-1}a_{n-1} + b_n a_{n-2})D^{2n-2} + \cdots$$

$$+ (b_n a_0 + b_{n-1}a_1 + b_{n-2}a_2 + \cdots + b_2 a_{n-2} + b_1 a_{n-1} + b_0 a_n)D^n + \cdots$$

$$+ (b_2 a_0 + b_1 a_1 + b_0 a_2)D^2 + (b_1 a_0 + b_0 a_1)D + b_0 a_0$$

$$= \left(b_n\, D^n + b_{n-1}\, D^{n-1} + \ldots + b_0\right)\left(a_n\, D^n + a_{n-1}\, D^{n-1} + \ldots + a_1\, D + a_0\right)$$

$$= P_2(D)P_1(D). \tag{3.8}$$

It is left as an exercise to use sigma notation to show this property in a more elegant fashion. With the property $P_1(D)P_2(D) = P_2(D)P_1(D)$, we see that we can always factor a constant-coefficient linear differential operator into powers of first-order terms, just as we would a polynomial, *as long as we allow the use of complex numbers.* Allowing for complex roots will be essential in finding the solutions of constant-coefficient homogeneous linear differential equations. The properties of the differential operator will also be essential when studying the Annihilator method for solving certain constant coefficient nonhomogeneous differential equations; see Section 4.3.

Example 11 The linear differential operator

$$P(D) = D^4 + 5D^3 + 6D^2$$

can be written as

$$P(D) = D^2(D^2 + 5D + 6) = D^2(D+2)(D+3).$$

Note that it could also be written as $P(D) = (D+2)(D+3)D^2$, for example, since the factors commute.

Example 12 The linear differential operator

$$P(D) = D^2 + 9$$

can be written as

$$P(D) = (D + 3i)(D - 3i).$$

Example 13 The linear differential operator

$$P(D) = D^2 + 2D + 2$$

has roots $-1 \pm i$ and thus can be written as

$$P(D) = (D - (-1 + i))(D - (-1 - i)) = (D + 1 - i)(D + 1 + i).$$

In practice we will usually not need to factor the operator completely into linear factors but it is conceptually important to realize that we, in fact, are able to do so.

Example 14 Apply $D^2 + 3$ to the function

$$f(x) = x^4 - \cos x.$$

We apply the operator to this function to obtain

$$(D^2 + 3)(f(x)) = D^2 f(x) + 3f(x) = f''(x) + 3f(x).$$

We thus need to calculate the second derivative of this function. Taking two derivatives gives

$$f'(x) = 4x^3 + \sin x, \quad f''(x) = 12x^2 + \cos x.$$

We thus obtain

$$f''(x) + 3f(x) = (12x^2 + \cos x) + 3(x^4 - \cos x)$$
$$= 3x^4 + 12x^2 - 2\cos x. \tag{3.9}$$

The operator notation is useful for writing differential equations in compact form and we will see that it will be especially useful in the next section.

In Problems **1–5**, determine whether the differential equation is guaranteed to have a unique solution passing through the given initial condition. Then try to find constants to make the proposed function y a solution to the IVP.

1. $y'' - 4y = 0$, $y(0) = 4$, $y'(0) = 0$; $y = c_1 e^{2x} + c_2 e^{-2x}$

2. $y''' + y' = 0$, $y(0) = 0$, $y'(0) = 1$, $y''(0) = 0$; $y = c_1 + c_2 \cos x + c_3 \sin x$

3. $xy'' - 2xy' + 2y = 0$, $y(0) = 0$, $y'(0) = 0$; $y = c_1 x + c_2 x^2$

4. $x^2 y'' - xy' - 3y = 0$, $y(0) = 0$, $y'(0) = 0$; $y = c_1 x^3 + \frac{c_2}{x}$

5. $x^2 y'' - 6xy' + 12y = 2x^2$, $y(0) = 0$, $y'(0) = 0$; $y = c_1 x^3 + c_2 x^4 + x^2$

In Problems **6–13**, find all values of (x, y) where Theorem 3.1.1 guarantees the existence of a unique solution.

6. $y'' + 3xy' + y = \sin x$

7. $xy'' + 3x^2 y = \sin x$

8. $y'' + 5y' + 6y = \frac{\cos x}{x}$

9. $y^{(4)} + 4y'' + 4x^2 y = e^x$

10. $y'' + 6y' - 16y = e^{-x^2} - 1$

11. $y'' + 3xy' + xy = \ln x$

12. $(x - 1)^2 y'' + 5(x^2 - 1)y' + 6y = 1$

13. $y''' + 3(\tan x)y' + xy = x$

In Problems **14–19**, apply the given differential operator $P(D)$ to the functions and simplify as much as possible.

14. $P(D) = D - 5$; $f_1(x) = 3x + 7$, $f_2(x) = \cos x$, $f_3(x) = e^{5x}$

15. $P(D) = D + 1$; $f_1(x) = e^{-x}$, $f_2(x) = 4 + \sin x$, $f_3(x) = e^{-2x}$

16. $P(D) = D^2 - 1$; $f_1(x) = 4e^{-x} + e^{2x}$, $f_2(x) = x^3 + 4$, $f_3(x) = e^x - 5e^{-x}$

17. $P(D) = D^2 + 1$; $f_1(x) = x^3 + 2x$, $f_2(x) = \cos x + \sin x$, $f_3(x) = \sin 2x$

18. $P(D) = D^2 - 4D + 5$; $f_1(x) = e^{2x} \cos x$, $f_2(x) = 4x^2 + e^x$, $f_3(x) = x^3 - \cos 2x$

19. $P(D) = D^2 + 2D + 1$; $f_1(x) = e^x$, $f_2(x) = xe^{-x}$, $f_3(x) = \sin x$

In Problems **20–27**, calculate $P(D)Q(D)(y)$ and $Q(D)P(D)(y)$. Compare the results.

20. $P(D) = D$, $Q(D) = D + 3x$, $y = 2x + 1$

21. $P(D) = D$, $Q(D) = D + 3x$, $y = \sin x$

22. $P(D) = D$, $Q(D) = D + x$, $y = e^{-x^2}$

23. $P(D) = D$, $Q(D) = D + x$, $y = 1 + \cos x$

24. $P(D) = D^2 - D$, $Q(D) = D^3 + 2D$, $y = x^6 - 2x^2$

25. $P(D) = D - 1$, $Q(D) = D^2 + D - 3$, $y = x^2 + \sin x$

26. $P(D) = D^2 + xD + 2$, $Q(D) = D + 1$, $y = e^x$

27. $P(D) = D^2 + xD + 2$, $Q(D) = D + 1$, $y = x^3 + 5x$

28. (a) Use the commutative property of the constant coefficient differential operator to quickly evaluate $D^3(D-2)(D+3)(x^2-2x+7)$.
(b) Use the linearity property of the differential operator to quickly evaluate $D^3(x^2-5+e^{2x})$.

In Problems **29–34**, factor the operators into linear factors and/or irreducible quadratic factors.

29. $D^2 - D - 6$ **30.** $D^4 - 2D^2 + 1$ **31.** $D^4 + 2D^2 + 1$
32. $D^3 - D^2 + 4D$ **33.** $D^3 - 1$ **34.** $D^3 + 8$

In Problems **35–40**, rewrite the following differential equations using operator notation and factor completely.

35. $y'' + 10y' + 16y = 0$ **36.** $y'' + 3y' - 4y = 0$ **37.** $y''' + 6y'' + 9y' = 0$
38. $y^{(5)} + 4y''' + 4y' = 0$ **39.** $y''' - 8y = 0$ **40.** $y''' + 27y = 0$

41. Use sigma notation to show that the constant-coefficient differential operator commutes. That is, let

$$P_1(D) = \sum_{i=0}^{n} a_{n-i} D^{n-i} \quad \text{and} \quad P_2(D) = \sum_{i=0}^{n} b_{n-i} D^{n-i}$$

and show that $P_1(D)P_2(D) = P_2(D)P_1(D)$.

3.2 Linear Independence and the Wronskian

Our discussion of topics from linear algebra will deal with both general functions as well as solutions of differential equations. The interplay between linear algebra and differential equations is a rich one and we encourage the reader to embrace it. In the ensuing discussion, we will not assume that the reader has had a course in linear algebra. The reader may find it useful to read Appendix B.1–B.2 for an introduction into the subject of linear algebra.

Now we will borrow some terminology from linear algebra.

Definition 3.2.1

If $f_1, f_2, \ldots, f_m$ are m given functions and $c_1, c_2, \ldots, c_m$ are m constants, then

$$c_1 f_1 + c_2 f_2 + \ldots + c_m f_m$$

is called a linear combination of $f_1, f_2, \ldots, f_m$.

Our goal is to tie this in with our study of differential equations and this is done with the following theorem:

THEOREM 3.2.1

Let $f_1, f_2, \ldots, f_m$ be **any** m **solutions** of the homogeneous linear differential equation (3.2), then

$$c_1 f_1 + c_2 f_2 + \ldots + c_m f_m$$

is also a **solution** of (3.2) where $c_1, c_2, \ldots, c_m$ are arbitrary constants.

Thus, this theorem gives the useful fact that **any linear combination of solutions to (3.2) is a solution to (3.2)**.

Example 1 Consider the differential equation

$$\frac{d^2 y}{dx^2} + y = 0.$$

We note that $y_1(x) = \sin x$ and $y_2(x) = \cos x$ are solutions. (Check it!) Thus,

$$y(x) = c_1 \sin x + c_2 \cos x$$

is also a solution for any constants c_1 and c_2. (Check this, too!)

We have found that solutions exist and are unique. We also know that any linear combination of solutions of a homogeneous linear differential equation is still a solution. The question is: When we have solutions to (3.2), how do we know when we have the most general solution? To answer this, we need to introduce some additional concepts from linear algebra.

Definition 3.2.2

The n functions $f_1, f_2, \ldots, f_n$ are linearly dependent on $a \leq x \leq b$ if there exist constants $c_1, c_2, \ldots, c_n$, not all zero, such that

$$c_1 f_1(x) + c_2 f_2(x) + \ldots + c_n f_n(x) = 0$$

for all x such that $a \leq x \leq b$. We say that the n functions $f_1, f_2, \ldots, f_n$ are *linearly independent* on $a \leq x \leq b$ if they are not linearly dependent there. That is, the functions $f_1, f_2, \ldots, f_n$ are linearly independent on $a \leq x \leq b$ if

$$c_1 f_1(x) + c_2 f_2(x) + \ldots + c_n f_n(x) = 0$$

for all x such that $a \leq x \leq b$ implies $c_1 = c_2 = \ldots = c_n = 0$.

Example 2 The functions x and $-3x$ are linearly dependent on $0 \leq x \leq 1$ since there are constants c_1 and c_2, not both zero, such that

$$c_1 x + c_2(-3x) = 0$$

for all x on $0 \le x \le 1$. Just take $c_1 = 3$ and $c_2 = 1$.

Example The functions e^x and $2e^x$ are linearly dependent, since

$$-2 \cdot (e^x) + 1 \cdot (2e^x) = 0.$$

The constants are $c_1 = -2$ and $c_2 = 1$.

Example Show that x and x^2 are linearly independent on $-1 \le x \le 1$.

Solution

To see this, we show that

$$c_1 x + c_2 x^2 = 0 \tag{3.10}$$

for all x on $0 \le x \le 1$ implies both $c_1 = 0$ and $c_2 = 0$.

Since the equation holds for all x, it must hold when $x = 1$. This gives

$$c_1(1) + c_2(1)^2 = 0.$$

It must also hold when $x = -1$ and this gives

$$c_1(-1) + c_2(-1)^2 = 0.$$

Solving the second equation gives $c_1 = c_2$ and substitution into the first gives

$$2c_2 = 0,$$

which only holds when $c_2 = 0$. Thus we also have $c_1 = 0$. Hence x and x^2 are linearly independent on $0 \le x \le 1$.

In the previous example, we note that we could have chosen any number of x-values *except* $x = 0$. This is because both functions are zero at $x = 0$. Determining whether a set of functions is linearly independent can sometimes become cumbersome by using the definition if there are many functions to consider. Sometimes it may be useful to use our computer software packages to determine the values of the coefficients and we now illustrate this.

Example Determine whether $f_1 = 1 + 3x$, $f_2 = 4 - 2x^2$, $f_3 = 5x - x^2$ are linearly independent on the real line.

Solution

We need to see if

$$c_1 \cdot f_1 + c_2 f_2 + c_3 f_3 = 0 \quad \text{for all} \quad x \tag{3.11}$$

can only happen when $c_1 = c_2 = c_3 = 0$. We can rearrange (3.11) by grouping like powers of x:

$$(c_1 + 4c_2) + x(3c_1 + 5c_3) + x^2(-2c_2 - c3) = 0. \tag{3.12}$$

Since this holds for all x, by assumption, we must have the quantities in parentheses equal to zero:

$$c_1 + 4c_2 = 0$$
$$3c_1 + 5c_3 = 0$$
$$-2c_2 - c_3 = 0.$$

We want to see if there are non-zero c-values that will make this set of equations true. This simply means we need to solve the equations simultaneously and our computer packages can do this quickly (see the end of the chapter). Regardless of the program (or if we did it by hand), we see that the only way for (3.11) to be true is when $c_1 = c_2 = c_3 = 0$. This shows that the functions are linearly independent for all x-values.

With this additional concept of linear independence, we can give the following theorem.

THEOREM 3.2.2

The nth order homogeneous linear differential equation (3.2) always possesses n solutions that are linearly independent. Further, if $f_1, f_2, \ldots, f_n$ are n linearly independent solutions of (3.2), then every solution f of (3.2) can be expressed as a linear combination

$$c_1 f_1(x) + c_2 f_2(x) + \ldots + c_n f_n(x) \qquad (3.13)$$

of these n linearly independent solutions by proper choice of the constants $c_1, c_2, \ldots, c_n$. The expression (3.13) is called the **general solution** of (3.2) and is defined on (a, b), the interval on which solutions exist and are unique.

We thus have that the solutions $f_1, \ldots, f_n$ can be combined to give us any solution we desire. Could another set of n functions also work or is this set unique? Actually, any set of functions that satisfies the following three conditions will work and we call the set a **fundamental set of solutions** of (3.2).

Three Conditions of a Fundamental Set of Solutions
1. The number of functions (elements) in this set must be the same as the order of the ODE.
2. Each function in this set must be a solution to the ODE.
3. The functions must be linearly independent.

We again note that a fundamental set of solutions is *not* unique. In order to obtain a fundamental set, we may need to add or remove functions from a given set. Once we have a fundamental set of solutions, we can construct all possible solutions from it.

What remains now is to determine whether or not the n solutions of (3.2) are linearly independent. This is necessary in order to determine a fundamental set of solutions.

The following definition is for n general functions and does *not* assume the functions are solutions of a differential equation.

Definition 3.2.3

Let $f_1, f_2, \ldots, f_n$ be n real functions each of which has an $(n-1)^{st}$ derivative on the interval $a \leq x \leq b$. The determinant $W(x) =$

$$W(f_1, f_2, \ldots, f_n)(x) = \begin{vmatrix} f_1(x) & f_2(x) & \ldots & f_n(x) \\ f_1'(x) & f_2'(x) & \ldots & f_n'(x) \\ \vdots & \vdots & \ddots & \vdots \\ f_1^{(n-1)}(x) & f_2^{(n-1)}(x) & \ldots & f_n^{(n-1)}(x) \end{vmatrix}, \text{ in which}$$

primes denote derivatives, is called the Wronskian of these n functions.

The determinant is a useful concept in linear algebra. The formulas for calculating it are straightforward but very computationally expensive (i.e., it takes a long time to do it). A general formula for the determinant exists but we will just state it for the "2 × 2" and "3 × 3" cases:

$$\begin{vmatrix} a & b \\ c & d \end{vmatrix} = ad - bc \tag{3.14}$$

and

$$\begin{vmatrix} a_1 & a_2 & a_3 \\ b_1 & b_2 & b_3 \\ c_1 & c_2 & c_3 \end{vmatrix} = a_1(b_2 c_3 - b_3 c_2) - a_2(b_1 c_3 - b_3 c_1) + a_3(b_1 c_2 - b_2 c_1). \tag{3.15}$$

We refer the reader to Appendix B for a more thorough introduction to determinants.

From the formula for determinant, we observe that

$$W(x) = W(f_1, f_2, \ldots, f_n)(x)$$

is a real function defined on $a \leq x \leq b$.

THEOREM 3.2.3

Let $f_1, f_2, \ldots, f_n$ be defined as in Definition 3.2.3.
1. If $W(x_0) \neq 0$ for some $x_0 \in (a, b)$, then it follows that $f_1, f_2, \ldots, f_n$ are linearly independent on (a, b).
2. If $f_1, f_2, \ldots, f_n$ are linearly dependent on (a, b), then $W(x) = 0$, for all $x \in (a, b)$

We note that this theorem *does not* say that "if $W(x) = 0$ for all $x \in (a, b)$, then the $f_i(x)$ are linearly dependent."[1] This last statement will only be true

---

[1]We also note that part 2 of this theorem is just the *contrapositive* of part 1 and hence equivalent to it. But we list both for emphasis.

over some restricted domain; see Problem 19. *To show linear dependence of functions, we must use the definition.* However, if the functions under consideration are solutions to a differential equation, then linear dependence is much easier to determine, as this next theorem shows.

THEOREM 3.2.4

Let $f_1, f_2, \ldots, f_n$ be defined as in Definition 3.2.3. Suppose that the f_i are each a solution of a given linear homogenous differential equation of order n. Then exactly one of the following statements is true:
1. $W(x) \neq 0$, for all $x \in (a, b)$
2. $W(x) = 0$, for all $x \in (a, b)$

Moreover, $W(x) \neq 0$ for all $x \in (a, b)$ if and only if the $\{f_i\}$ are linearly independent on (a, b). Similarly, $W(x) = 0$ for all $x \in (a, b)$ if and only if the $\{f_i\}$ are linearly dependent on (a, b).

Note that this theorem tells us that if we have solutions of the same differential equation, we only need to check the Wronskian at *one* point in order to determine whether the set is linearly dependent. The previous theorem thus gives us an "easy" check for both linear independence and linear dependence in this case. We show its usefulness with an example.

Example 6 We have seen that $\sin x$ and $\cos x$ are solutions of $\dfrac{d^2 y}{dx^2} + y = 0$.

Given our discussion, we should ask if the two functions are linearly independent. To check this, we calculate the Wronskian:

$$W(\sin x, \cos x) = \begin{vmatrix} \sin x & \cos x \\ \cos x & -\sin x \end{vmatrix}$$

$$= -\sin^2 x - \cos^2 x$$

$$= -(\sin^2 x + \cos^2 x)$$

$$= -1 \neq 0 \text{ for all real } x.$$

Thus, $\sin x$ and $\cos x$ are linearly independent and so are a fundamental set of solutions to the differential equation and

$$y(x) = c_1 \sin x + c_2 \cos x$$

is the general solution.

As mentioned previously, $W(x)$ can show the linear independence of functions, even if they are *not* solutions of a differential equation. Table 3.1 gives a summary for checking linear independence (or dependence) of a set of functions.

Table 3.1: Showing Linear Independence of Solutions vs. Functions

	Solutions of Same ODE	General Functions
Linearly Independent	by definition *or* show $W(x) \neq 0$ for *some* x	by definition *or* show $W(x) \neq 0$ for *some* x
Linearly Dependent	by definition *or* show $W(x) = 0$ for *some* x	by definition *only*

Note that the entire column under "Solutions of same ODE" could also read "for *all* x" since the two are equivalent by Theorem 3.2.4.

Example Show the functions e^x, e^{-x}, and e^{2x} are linearly independent.

Solution

If we calculate the Wronskian $W(x)$ we have

$$W(e^x, e^{-x}, e^{2x}) = \begin{vmatrix} e^x & e^{-x} & e^{2x} \\ e^x & -e^{-x} & 2e^{2x} \\ e^x & e^{-x} & 4e^{2x} \end{vmatrix},$$

which is expanded as

$$W(e^x, e^{-x}, e^{2x}) = e^x \begin{vmatrix} -e^{-x} & 2e^{2x} \\ e^{-x} & 4e^{2x} \end{vmatrix} - e^{-x} \begin{vmatrix} e^x & 2e^{2x} \\ e^x & 4e^{2x} \end{vmatrix} + e^{2x} \begin{vmatrix} e^x & -e^{-x} \\ e^x & e^{-x} \end{vmatrix}$$

$$= e^x(-4e^x - 2e^x) - e^{-x}(4e^{3x} - 2e^{3x}) + e^{2x}(1+1)$$

$$= -6e^{2x} \neq 0 \text{ for all real } x.$$

We could have also tried $x = 0$ (for example) to see that the Wronskian was non-zero. The only drawback for "guessing" values if we want to show linear independence in the case when we *do not* have solutions is that an answer of zero at a specific x-value tells us nothing.

It would be simple to try other values of x for which we wanted to substitute. This would be useful as the following example shows the potential problem of guessing values of x that may give a non-zero value for the determinant.

Example Determine whether the functions $\sin 2x$ and $\sin x$ are linearly dependent or independent by using the Wronskian.

Solution

We can calculate the Wronskian as

$$W(x) = \begin{vmatrix} \sin 2x & \sin x \\ 2\cos 2x & \cos x \end{vmatrix}. \tag{3.16}$$

We can easily check that $W(0) = 0$, which doesn't help. $W(\pi) = 0$ as well. We might be tempted to stop guessing and switch to a check by definition, but "wise" guessing shows that $W(\pi/2) = 2$. Thus the two functions are linearly independent.

While it may be easier to substitute an x-value into the matrix and then calculate the determinant, this is not recommended when you have access to one of the programs. As can be seen in this example, it is just as easy to calculate $W(x)$, especially if we use MATLAB, Maple, or Mathematica.

We note that the problem that arose here is that the functions are both zero at the same point. We simply need to evaluate the functions at places where they (i) don't intersect tangentially or (ii) aren't zero at the same point. Neither of these problems occurs with solutions of linear, homogeneous second-order differential equations because the uniqueness theorem disallows any intersections of solutions. And again we point out that it is often useful to have our computer software packages available to aid us in trying to understand the course material.

●   ●   ●   ●   ●   ●   ●   ●   ●   ●   ●   ●

Problems

1. Show that x and $2x$ are linearly dependent on $[0, 1]$.

2. Show that e^{2x} and xe^{2x} are linearly independent for all x.

3. Show that e^x and $x + 1$ are linearly independent for all x.

4. Show that $\sin x$ and x are linearly independent for all x.

5. Show that if y_1 and y_2 are solutions of the first-order differential equation $y' + p(x)y = 0$ then y_1 and y_2 are linearly dependent.

In Problems **6–18**, determine whether the following sets of functions are linearly independent for all x where the functions are defined.

6. $\{e^x, e^{5+x}\}$ **7.** $\{\sin 2x, \cos 2x\}$ **8.** $\{x^3 - 4, x, 3x\}$
9. $\{x^3 - 4x, x, 2x^3\}$ **10.** $\{\sqrt{x} - 4x, x, 2\sqrt{x}\}$ **11.** $\{x, 2x - 2, x + 3\}$
12. $\{x^3 - 3, x^3 - 3x, x\}$ **13.** $\{e^x, e^{-x}, 1\}$ **14.** $\{x^3, 1 - x\}$
15. $\{1 + x^2, x, x^2\}$ **16.** $\{x^2, x + 2, x^2 - \frac{x}{2} - 1\}$
17. $\{x, e^{2x+3}, e^{2x-1}\}$ **18.** $\{x, e^{2x+2}, e^{2x-5}\}$

19. Consider the functions $f_1 = x, f_2 = |x|$.
   (a) Show that $\{f_1, f_2\}$ is linearly dependent on $[0, 1]$.
   (b) Show that $\{f_1, f_2\}$ is linearly dependent on $[-1, 0]$.
   (c) Show that $\{f_1, f_2\}$ is linearly *independent* on $[-1, 1]$.
   (d) Show that $W(x, |x|) = 0$ for all x.
   Thus we have found a set of functions, $\{x, |x|\}$, whose Wronskian is always zero even though the set is linearly independent.

20. Repeat (a)–(d) in Problem **19** for the functions $f_1 = x^3$, $f_2 = |x|^3$.

In Problems **21**–**23**, verify that the given function is a **two-parameter family of solutions** to the given second-order ODE. Determine whether this solution can give rise to a unique solution that passes through the given ICs. Does this violate the existence and uniqueness theorem?

21. $y = c_1 x + c_2 x \ln x$, $0 < x < \infty$, $x^2 y'' - xy' + y = 0$; $y(1) = 3$, $y'(1) = -1$

22. $y = c_1 + c_2 x^2$, $xy'' - y' = 0$ on $(0, \infty)$; $y(0) = 0$, $y'(0) = 1$

23. $y = c_1 x^2 + c_2 x$, $x^2 y'' - 2xy' + 2y = 0$ on $(-\infty, \infty)$; $y(0) = 3$, $y'(0) = 1$

In Problems **24**–**33**, show that the given set of functions forms a fundamental set of solutions to the differential equation on $(-\infty, \infty)$.

24. $\{3 + x, 2 - 5x\}$, $y'' = 0$ **25.** $\{4, x - 1\}$, $y'' = 0$

26. $\{\sin 2x, \cos 2x\}$, $y'' + 4y = 0$ **27.** $\{e^{-2x}, e^{3x}\}$, $y'' - y' - 6y = 0$

28. $\{e^{2x}, e^{5x}\}$, $y'' - 7y' + 10y = 0$ **29.** $\{e^{-3x}, e^{x/2}\}$, $2y'' + 5y' - 3y = 0$

30. $\{e^x, xe^x\}$, $y'' - 2y' + y = 0$ **31.** $\{1, \cos x, \sin x\}$, $y''' + y' = 0$

32. $\{1, e^{-x}, e^{-2x}\}$, $y''' + 3y'' + 2y' = 0$

33. $\{1, \cos x, \sin x, x \cos x, x \sin x\}$, $y^{(5)} + 2y''' + y' = 0$

In Problems **34**–**51**, determine whether the given set forms a fundamental set of solutions to the differential equation. Take the interval to be $(-\infty, \infty)$ unless otherwise stated. Clearly state your reasons for your answers.

34. $\{x, e^x\}$, $y'' + y = 0$ **35.** $\{x^2 + 4, 5 - x\}$, $y'' + y = 0$

36. $\{\sin 2x, \cos 3x\}$, $y'' + 4y = 0$ **37.** $\{\sin x, e^x\}$, $y'' - y = 0$

38. $\{1, e^{3x}\}$, $y'' - 3y' = 0$ **39.** $\{\sin x - 5 \cos x, 3 \sin x\}$, $y'' + y = 0$

40. $\{\sin 5x + \cos 5x, \cos 5x - \sin 5x\}$, $y'' + 25y = 0$

41. $\{x^2, x, e^{-x}\}$, $y^{(4)}(x) - y'''(x) = 0$ **42.** $\{e^{4x}, e^{2x}\}$, $y'' - 6y' + 8y = 0$

43. $\{e^{-x}, xe^{-x}\}$, $y'' + 2y' + y = 0$ **44.** $\{e^{-3x}, xe^{-3x}\}$, $y'' - 6y' + 9y = 0$

45. $\{x, 3, e^x\}$, $y''' - y'' = 0$ **46.** $\{3, e^{2x}, e^{2x} + 2\}$, $y''' - 4y' = 0$

47. $\{e^{-2x} \sin x, e^{-2x} \cos x\}$, $y'' + 4y' + 5y = 0$

48. $\{x^3, x^{-1}\}$, $x^2 y'' + 6xy' + 12y = 0$ on $(0, \infty)$

49. $\{x^3, x^4\}$, $x^2 y'' - 6xy' + 12y = 0$ on $(0, \infty)$

50. $\left\{ \dfrac{1}{\sqrt{x}}, \dfrac{1}{x}, \dfrac{\sqrt{x} + 2}{x} \right\}$, $2x^2 y'' + 5xy' + y = 0$ on $(0, \infty)$

51. $\{e^{-x}, e^x, xe^{-x}, xe^x\}$, $y^{(4)} - 2y'' + y = 0$

3.3   Reduction of Order—the Case $n = 2$

In this section we develop a method of simplifying an nth-order differential equation if we already know a solution. The following theorem is useful.

Definition 3.3.1

Let y_1 be a solution of the nth-order homogeneous linear differential equation (3.2), with $y_1 \neq 0$ for all $x \in (a, b)$. Then the transformation

$$y_2 = y_1(x)v$$

reduces (3.2) to an $(n-1)$st order homogeneous linear differential equation in the dependent variable

$$w = \frac{dv}{dx}.$$

This result is very useful, for if we know one nontrivial solution then we can reduce the order of the differential equation.

We will consider this theorem for the case with $n = 2$. Suppose y_1 is a known, nontrivial solution of the second-order homogeneous linear equation

$$a_2(x)\frac{d^2y}{dx^2} + a_1(x)\frac{dy}{dx} + a_0(x)y = 0. \tag{3.17}$$

Let $y_2 = y_1 v$ where y_1 is the known solution and v is a function of x to be determined.

Taking derivatives and simplifying gives

$$\frac{dy_2}{dx} = y_1\frac{dv}{dx} + y_1'v$$

and

$$\frac{d^2y_2}{dx^2} = y_1\frac{d^2v}{dx^2} + 2y_1'\frac{dv}{dx} + y_1''v.$$

Substituting the derivatives into the differential equation (3.17) gives

$$a_2(x)\underbrace{\left(y_1\frac{d^2v}{dx^2} + 2y_1'\frac{dv}{dx} + y_1''v\right)}_{y_2''} + a_1(x)\underbrace{\left(y_1\frac{dv}{dx} + y_1'v\right)}_{y_2'} + a_0(x)\underbrace{y_1 v}_{y_2} = 0.$$

Rearranging the terms then yields

$$a_2(x)y_1\frac{d^2v}{dx^2} + (2a_2(x)y_1' + a_1(x)y_1)\frac{dv}{dx}$$
$$+ \underbrace{(a_2(x)y_1'' + a_1(x)y_1' + a_0(x)y_1)}_{=\,0}v = 0, \tag{3.18}$$

where the coefficient of v is zero because we assumed that y_1 is a solution of

$$a_2(x)\frac{d^2y}{dx^2} + a_1(x)\frac{dy}{dx} + a_0(x)y = 0.$$

So the Equation (3.18) becomes

$$a_2(x)y_1\frac{d^2v}{dx^2} + (2a_2(x)y_1' + a_1(x)y_1)\frac{dv}{dx} = 0.$$

We now make the substitution

$$w = \frac{dv}{dx} \tag{3.19}$$

and obtain the first-order homogeneous linear differential equation

$$a_2(x)y_1w' + (2a_2(x)y_1' + a_1(x)y_1)\,w = 0.$$

This equation is separable and you can show that

$$w(x) = \frac{c\exp\left[-\int\frac{a_1(x)}{a_2(x)}dx\right]}{y_1^2(x)}$$

is the solution. By (3.19), we have $\frac{dv}{dx} = w$ and thus integrating gives us

$$v = c\int\frac{\exp\left[-\int\frac{a_1(x)}{a_2(x)}dx\right]}{y_1^2(x)}\,dx$$

and we only need the constant c to satisfy $c \neq 0$. We also have $y_2(x) = y_1(x)v(x)$ and thus

$$y_2 = y_1 c\int\frac{\exp\left[-\int\frac{a_1(x)}{a_2(x)}dx\right]}{y_1^2(x)}\,dx \tag{3.20}$$

is a solution of the differential equation (3.17).

Recall that our goal is to find two linearly independent solutions to (3.17). We assumed that y_1 is a solution and we showed that y_2 defined in (3.20) is a solution. We can show that the solutions y_1 and y_2 are linearly independent by considering their Wronskian:

$$\begin{vmatrix} y_1 & y_2 \\ y_1' & y_2' \end{vmatrix} = y_1y_2' - y_2y_1',$$

where

$$y_2'(x) = y_1\,c\frac{\exp\left[-\int\frac{a_1(x)}{a_2(x)}dx\right]}{y_1^2(x)} + y_1'\,c\int\frac{\exp\left[-\int\frac{a_1(x)}{a_2(x)}dx\right]}{y_1^2(x)}\,dx.$$

Substituting in for y_2 and y_2' in the Wronskian gives

$$y_1 y_2' - y_2 y_1' = y_1 \left(y_1 c \frac{\exp\left[-\int \frac{a_1(x)}{a_2(x)} dx\right]}{y_1^2(x)} + y_1' c \int \frac{\exp\left[-\int \frac{a_1(x)}{a_2(x)} dx\right]}{y_1^2(x)} dx \right)$$

$$\underbrace{\qquad\qquad\qquad\qquad\qquad\qquad}_{y_2'}$$

$$- \underbrace{\left(y_1 c \int \frac{\exp\left[-\int \frac{a_1(x)}{a_2(x)} dx\right]}{y_1^2(x)} dx \right) y_1'}_{y_2}$$

$$= y_1^2(x) c \int \frac{\exp\left[-\int \frac{a_1(x)}{a_2(x)} dx\right]}{y_1^2(x)} dx \qquad\qquad (3.21)$$

$$\neq 0. \qquad\qquad (3.22)$$

Thus y_1 and y_2 are linearly independent and the linear combination

$$c_1 y_1 + c_2 y_2 = c_1 y_1 + c_2 y_1 c \int \frac{\exp\left[-\int \frac{a_1(x)}{a_2(x)} dx\right]}{y_1^2(x)} dx \qquad\qquad (3.23)$$

$$= c_1 y_1 + \tilde{c}_2 y_1 \int \frac{\exp\left[-\int \frac{a_1(x)}{a_2(x)} dx\right]}{y_1^2(x)} dx \qquad\qquad (3.24)$$

is the general solution of (3.17) where $\tilde{c}_2 = c_2 \cdot c$.

While equation (3.24) gives you the general solution, please check with your instructor as to whether you should apply it or go through the derivation of it. In the next two examples, we take the latter approach.

Example 1 Given that $y_1 = e^{2x}$ is a solution of

$$\frac{d^2 y}{dx^2} + \frac{dy}{dx} - 6y = 0,$$

find a solution that is linearly independent of $y_1 = e^{2x}$ by reducing the order.

Solution
Let $y_2 = v e^{2x}$; this gives

$$\frac{dy_2}{dx} = e^{2x} \frac{dv}{dx} + 2e^{2x} v \quad \text{and} \quad \frac{d^2 y_2}{dx^2} = e^{2x} \frac{d^2 v}{dx^2} + 4e^{2x} \frac{dv}{dx} + 4e^{2x} v. \qquad (3.25)$$

Substituting these derivatives gives

$$\left(e^{2x} \frac{d^2 v}{dx^2} + 4e^{2x} \frac{dv}{dx} + 4e^{2x} v \right) + \left(e^{2x} \frac{dv}{dx} + 2e^{2x} v \right) - 6\left(e^{2x} v \right) = 0.$$

This simplifies to

$$e^{2x}\frac{d^2v}{dx^2} + 5e^{2x}\frac{dv}{dx} = 0.$$

Now let $w = \frac{dv}{dx}$ so that

$$e^{2x}\frac{dw}{dx} + 5e^{2x}w = 0.$$

This last equation is separable with solution $w = ce^{-5x}$.

We now choose a specific c-value to obtain the particular solution. For example, choosing $c = -5$ gives

$$\frac{dv}{dx} = -5e^{-5x}.$$

Integrating gives $v(x) = e^{-5x}$. Thus,

$$y_2 = e^{2x}v = e^{2x}e^{-5x} = e^{-3x}$$

and the general solution is

$$y = c_1e^{2x} + c_2e^{-3x}.$$

We note that choosing $c = -5$ in the previous example only made our integral "nice." If we chose, for example, $c = 1$, we still would have obtained an equally valid solution since we ultimately multiplied the second solution by a constant c_2. Reducing the order of the equation works on any linear homogeneous equation, even if the coefficients are not constant as we will see in the next example.

Example 2 Given that $y_1 = x$ is a solution of

$$(x^2 + 1)\frac{d^2y}{dx^2} - 2x\frac{dy}{dx} + 2y = 0,$$

find a solution that is linearly independent of $y_1 = x$ by reducing the order.

Solution

We see that $y_1 = x$ is indeed a solution, since $y_1' = 1$ and $y_1'' = 0$ so that substitution gives

$$(x^2 + 1)(0) - 2x(1) + 2x = 0.$$

We now let $y_2 = vx$ so that

$$\frac{dy_2}{dx} = x\frac{dv}{dx} + v \quad \text{and} \quad \frac{d^2y_2}{dx^2} = x\frac{d^2v}{dx^2} + 2\frac{dv}{dx}. \tag{3.26}$$

Substituting these derivatives gives

$$(x^2 + 1)\left(x\frac{d^2v}{dx^2} + 2\frac{dv}{dx}\right) - 2x\left(x\frac{dv}{dx} + v\right) + 2xv = 0,$$

which implies

$$x(x^2 + 1)\frac{d^2v}{dx^2} + 2\frac{dv}{dx} = 0.$$

Now let $w = \frac{dv}{dx}$ so that

$$x(x^2 + 1)\frac{dw}{dx} + 2w = 0.$$

This equation is separable. After we separate the variables we have

$$\frac{1}{w}\,dw = \frac{-2}{x(x^2 + 1)}\,dx = \left(\frac{2x}{x^2 + 1} - \frac{2}{x}\right)dx,$$

where the last equality results from the partial fraction expansion. We then integrate and solve for w to obtain

$$w = \frac{c(x^2 + 1)}{x^2}.$$

We again choose a specific c-value to obtain the particular solution, say $c = 1$, and obtain

$$\frac{dv}{dx} = \frac{(x^2 + 1)}{x^2} = 1 + \frac{1}{x^2}.$$

Integrating gives $v(x) = x - \dfrac{1}{x}$. Thus,

$$y_2 = xv = x\left(x - \frac{1}{x}\right) = x^2 - 1.$$

So the general solution is

$$y = c_1 x + c_2(x^2 - 1).$$

• • • • • • • • • • • •

In Problems **1–18**, use the given solution to reduce the order of the differential equation. Use the methods of Chapters 1, 2, and Section 3.1 to solve the reduced equation.

1. $y'' - 2y' + y = 0$, $y_1 = e^x$
2. $y'' - 8y' + 16y = 0$, $y_1 = e^{4x}$
3. $y'' - 6y' + 9y = 0$, $y_1 = e^{3x}$
4. $4y'' + 4y' + y = 0$, $y_1 = e^{-x/2}$
5. $2y'' + 5y' - 3y = 0$, $y_1 = e^{x/2}$
6. $y'' - 9y = 0$, $y_1 = e^{3x}$
7. $y'' + 25y = 0$, $y_1 = \cos 5x$
8. $xy'' - y' = 0$, $y_1 = x^2$
9. $xy'' + y' = 0$, $y_1 = \ln x$
10. $x^2 y'' - xy' - 3y = 0$, $y_1 = x^3$
11. $x^2 y'' - xy' + y = 0$, $y_1 = x \ln x$
12. $(x + 1)^2 y'' - 3(x + 1)y' + 3y = 0$, $y_1 = x + 1$
13. $(2x + 1)y'' - 4(x + 1)y' + 4y = 0$, $y_1 = e^{2x}$
14. $(x^2 - 1)y'' - 2xy' + 2y = 0$, $y_1 = x^2 + 1$
15. $y''' - 2y'' - y' + 2y = 0$, $y_1 = e^{2x}$
16. $y''' + 5y'' = 0$, $y_1 = x$
17. $y''' - 4y'' + 5y' - 2y = 0$, $y_1 = e^x$
18. $y''' - 5y'' + 8y' - 4y = 0$, $y_1 = xe^{2x}$

19. Show that the separable differential equation

$$a_2(x)y_1w'(x) + (2a_2(x)y_1' + a_1(x)y_1)w(x) = 0$$

has solution

$$w(x) = \frac{c\exp\left[-\int \frac{a_1(x)}{a_2(x)}dx\right]}{y_1^2(x)}.$$

20. Show that if y_1 is a solution to the differential equation

$$a_2(x)\frac{d^2y}{dx^2} + a_1(x)\frac{dy}{dx} + a_0(x)y = 0$$

and y_2 is another solution given by

$$y_2 = y_1 c \int \frac{\exp\left[-\int \frac{a_1(x)}{a_2(x)}dx\right]}{y_1^2(x)}dx,$$

then $\{y_1, y_2\}$ forms a fundamental solution set.

21. Consider the hyperbolic functions

$$\cosh x = \frac{e^x + e^{-x}}{2} \quad \text{and} \quad \sinh x = \frac{e^x - e^{-x}}{2}.$$

Show that
(a) $\dfrac{d}{dx}\cosh x = \sinh x$.
(b) $\dfrac{d}{dx}\sinh x = \cosh x$.
(c) $\cosh^2 x - \sinh^2 x = 1$.
(d) $\cosh x$ and $\sinh x$ are linearly independent functions.
(e) A general solution of $y'' - y = 0$ is

$$y = c_1 \cosh x + c_2 \sinh x.$$

3.4 Numerical Considerations for nth-Order Equations

Although we will learn how to solve nth-order homogeneous equations that have constant coefficients in Section 3.6, we will often encounter situations where the coefficients are not constant. In addition, sometimes it will be difficult to exactly find the roots of the characteristic polynomial that arises from our differential equation. In both of these situations, we still need to know the behavior of the solution and it is often helpful to calculate this solution numerically. The method we learn in this section also applies to *nonlinear* equations and this will be useful in later chapters (and real life!). The methods that we learned in Section 2.5 have analogous formulations in higher dimensions, with the Runge-Kutta method still being the much preferred method. We will not go through the details of this method again, but will refer the interested reader to many of the texts given in the References, e.g., Burden and Faires [10]. Instead, we will show how to use MATLAB, Maple, and Mathematica to calculate solutions of an nth-order equation. For the Maple and Mathematica code, if we want to plot solutions in the x-y plane we only need to change the equation that is input. For the MATLAB code, we will also need to change the m-file containing the equation. (This will be necessary for MATLAB, Maple, and Mathematica when we want to plot solutions in the y-y' **phase plane**. But first we will need to reduce the nth-order equation to a system of n first-order equations.)

For both implementations we will consider the two differential equations:

$$2xy''(x) + x^2y'(x) + 3x^3y(x) = 0, \ y(1) = 2, y'(1) = 0 \qquad (3.27)$$

and

$$y^{(4)}(x) + x^2y'(x) + y(x) = \cos(x), \ y(0) = 1, y'(0) = 0, y''(0) = 0, y'''(0) = 1. \qquad (3.28)$$

In both cases, we will want to plot the solution from the initial x-value until $x = 5$ using a step size of $h = 0.05$. We will also give the approximate solution at the value $x = 5$.

Converting an nth Order Equation to a System of n First-Order Equations

We said that we will skip the details of the Runge-Kutta method for numerically solving higher-order equations. But we will briefly discuss how the method is applied. For a first-order equation, Section 2.5 showed how we considered four different slopes of the direction field in our efforts to calculate one step. For a second-order equation, we will convert our differential equation to the form

$$u_1' = f(x, u_1, u_2)$$
$$u_2' = g(x, u_1, u_2). \qquad (3.29)$$

In doing so we then see that we can carry out the first-order Runge-Kutta calculation in *each* variable. Thus we now have *two* directions to worry about besides the independent x-direction. But the idea is exactly the same as before—we will still consider four different slopes of the direction field (now three-dimensional!) in order to take one step. The direction field, as you might imagine, gets rather ugly and it is hopeless to even attempt to draw it for a third- or higher-order equation.

The obvious question is "how do we go from a second-order equation to something of the form of (3.29)?" Let us consider the equation

$$a_2(x)y'' + a_1(x)y' + a_0(x)y = F(x) \qquad (3.30)$$

as our second-order equation that we would like to convert. We need two variables, u_1 and u_2. We then set $u_1 = y$ and $u_2 = y'$. The left-hand sides of (3.29) are taken care of and we need to relate the right-hand side to (3.30). We have

$$u_1' = y'$$
$$u_2' = y''$$

but need to write the right-hand sides in terms of our new variables. The first equation is simple enough—y' is the same as u_2. What is y''? Well, we can solve the original equation (3.30) for y'' and then substitute. That is, we write

$$y'' = \frac{1}{a_2}\left(-a_1(x)y' - a_0(x)y + F(x)\right).$$

The right-hand side still needs to be put in terms of the new variables and we substitute $u_1 = y$ and $u_2 = y'$ to obtain

$$y'' = \frac{1}{a_2}\left(-a_1(x)u_2 - a_0(x)u_1 + F(x)\right).$$

Our system of two first-order equations is thus

$$u_1' = u_2$$
$$u_2' = \frac{1}{a_2}\left(-a_1(x)u_2 - a_0(x)u_1 + F(x)\right). \tag{3.31}$$

If we were given initial conditions, these could be converted to give $u_1(x_0) = y_0$, $u_2(x_0) = y_1$. The situation is analogous for a higher-order equation.

How to convert an nth-order ODE into a system of n first-order equations

(1) Introduce the same number of variables as the order of the equation. Rename the 0th through $(n-1)$st derivatives of the function.

(2) Solve the original nth-order equation for the highest derivative, $y^{(n)}$.

(3) Write the first derivative of each new variable as the left-hand side of an equation (you will have n equations). Then rewrite the corresponding right-hand sides in terms of the new variables.

We can always convert an nth-order equation to a system of first-order equations but we cannot always easily take a general first-order system and make it into an nth-order differential equation. And although we illustrated this conversion with a linear equation, we can actually convert *any* nth-order equation to a system of n first-order equations, provided we are able to solve for the highest derivative (i.e., perform Step (2)). Before giving two examples of the above method, we mention that u_1 is the solution of the original nth-order equation. This will be useful when we want to plot the solution. As a bonus, we also will have $n-1$ derivatives of this solution. For physical problems, for example, it may be useful to have both position and velocity as functions of the independent variable time.

Example 1 Reduce the equations in (3.27),

$$2xy''(x) + x^2y'(x) + 3x^3y(x) = 0, \ y(1) = 2, y'(1) = 0,$$

to a system of first-order equations.

Solution
We are given a second-order equation and thus we will obtain two first-order equations. We solve for the highest derivative to obtain

$$y'' = \frac{-x}{2}y' - \frac{3x^2}{2}y.$$

We then set $u_1 = y, \ u_2 = y'$. Our system of first-order equations is thus

$$u_1' = u_2$$

$$u_2' = \frac{-x}{2}u_2 - \frac{3x^2}{2}u_1. \tag{3.32}$$

The initial condition then becomes $u_1(1) = 2, u_2(1) = 0$.

Example 2 Reduce the equations in (3.28),

$$y^{(4)}(x) + x^2y'(x) + y(x) = \cos(x), \ y(0) = 1, y'(0) = 0, y''(0) = 0, y'''(0) = 1,$$

to a system of first-order equations.

Solution
We have a fourth-order equation and thus we will have four first-order equations. Although this system is not homogeneous, this will not affect the steps we take. We solve for the highest derivative to obtain

$$y^{(4)} = -x^2y'(x) - y(x) + \cos(x).$$

We then set $u_1 = y, \ u_2 = y', u_3 = y'', \ u_4 = y'''$. Our system of first-order equations is thus

$$u_1' = u_2$$
$$u_2' = u_3$$
$$u_3' = u_4$$
$$u_4' = -x^2 u_2 - u_1 + \cos(x). \tag{3.33}$$

The initial condition then becomes $u_1(0) = 1, u_2(0) = 0, u_3(0) = 0, u_4(0) = 1$.

* * *

Problems

There are three parts to each of these problems: (i) convert the equations in Problems **1–16** to a system of first-order equations (do this part even if your software allows you to type in the equation without having to first convert it); (ii) use your software package and the 4th order Runge-Kutta method with the given initial conditions at x_0 to estimate the numerical solution at the value $x = x_0 + 5.0$ for the step sizes $h = 0.5, 0.01$ (this will require two different runs of the software code).

1. $7y'' + 4y' - 3y = 0, \; y(0) = 0, y'(0) = 1$
2. $y'' + x^2 y' + 12y = 0, \; y(0) = 0, y'(0) = 7$
3. $y'' + 2y' + 10y = \sin x, \; y(\pi) = e^1, y'(\pi) = 1$
4. $y'' + y' + 2xy = x, \; y(-1) = 1, y'(-1) = 0$
5. $(x + 2)y'' + 3y = x, \; y(0) = 0, y'(0) = 4$
6. $y'' + 4y' + 3\sin(y) = 1, \; y(0) = -1, y'(0) = \pi$
7. $(x^2 + 2)y'' + 3y^2 = e^x, \; y(0) = 1, y'(0) = 2$
8. $x^3 y'' + (\sin x)y' + y = 3x, \; y(1) = 1, y'(1) = \frac{3}{2}$
9. $8y''' + y'' = 0, \; y(0) = 1, y'(0) = 0, y''(0) = 2$
10. $y''' + xy'' + y' + y = e^x, \; y(1) = 1, y'(1) = 3, y''(1) = 1$
11. $x^2 y''' + xy''y' + \sin(y) = e^{-x}, \; y(1) = 1, y'(1) = 0, y''(1) = 1$
12. $3y''' + xy'' + xy' + y^3 = \ln x, \; y(1) = 0, y'(1) = 0, y''(1) = 1$
13. $y''' - 8\frac{1}{x^3}y = 0, \; y(1) = 1, y'(1) = 1, y''(1) = 0$
14. $2y''' - (y'')^2 + y' + xy = e^{-x^2}, \; y(0) = 1, y'(0) = 0, y''(0) = 0$
15. $(x+2)y^{(4)} + 3y' - 2y = 0, \; y(-1) = 3, y'(-1) = -1, y''(-1) = 0, y'''(-1) = 1$
16. $y^{(5)} - 9y' = 0, \; y(1) = 2, y'(1) = 0, y''(1) = 0, y'''(1) = 1, y^{(4)}(1) = 1$

* * *

3.5   Essential Topics from Complex Variables

This section gives us a quick overview of some algebra and calculus that we can do with complex numbers. Understanding the material through the discussion of Euler's formula (3.34) will be useful for Sections 3.6, 3.7, and 4.7 while the remaining material of this section will be needed only for Section 4.4.

In one variable, we can only have real numbers for solutions of a linear algebraic equation. Things change when we consider a quadratic or higher-order equation. In applying the quadratic formula to solve $a_2 x^2 + a_1 x + a_0 = 0$ where we assume a_0, a_1, a_2 are real numbers, we can have the new situation where we may be required to take the square root of a negative number; that is, we consider

$$x = \frac{-a_1 \pm \sqrt{a_1^2 - 4a_2 a_0}}{2a_2}, \quad \text{for} \quad a_1^2 < 4a_2 a_0.$$

We define a new number

$$i = \sqrt{-1}$$

so that our answer is made up of a part with i and (potentially) a part without i. We note that $i^2 = -1$. These are called **complex roots** of the equation and are denoted $a \pm ib$, where a, b are both real numbers. Our first important point to realize is that complex numbers come in **conjugate pairs**—we can't ever just have $a + ib$ without also having $a - ib$. Another key point is that we can't combine real numbers together to get a complex number without introducing the **imaginary number** i into the problem. We typically write a complex variable as z and observe that, in general, it is made up of a real part a and an imaginary part b. We use the following to denote this:

$$\text{Re}(z) = a, \quad \text{Im}(z) = b.$$

We point out that the imaginary part of z is not ib but is instead just the coefficient b.

We can add, subtract, multiply, and divide complex numbers as well as multiply by constants (either real or complex). Let $z = a + ib$, $z_1 = a_1 + ib_1$, and $z_2 = a_2 + ib_2$ with $a, b, a_1, b_1, a_2, b_2, c$ real numbers. Then we have the following:

Addition and subtraction of complex numbers:

$$z_1 + z_2 = (a_1 + ib_1) + (a_2 + ib_2) = (a_1 + a_2) + i(b_1 + b_2)$$
$$z_1 - z_2 = (a_1 + ib_1) - (a_2 + ib_2) = (a_1 - a_2) + i(b_1 - b_2)$$

Multiplication of complex numbers:

$$z_1 z_2 = (a_1 + ib_1)(a_2 + ib_2) = a_1 a_2 + i(a_1 b_2 + b_1 a_2) + i^2 b_1 b_2$$
$$= (a_1 a_2 - b_1 b_2) + i(a_1 b_2 + b_1 a_2)$$

Multiplication of a complex number by a real constant: $cz = ca + icb$

In order to show division, we first note that we don't ever leave complex numbers in a denominator. In other words, we always simplify a fraction to the form $a + ib$ where a, b are real numbers. In order to do this, we need the idea of a complex conjugate, denoted with a line (or bar) over the complex number. This means we change the sign of the imaginary part:

$$\bar{z} = \overline{a + ib} = a - ib.$$

We also observe that

$$z\bar{z} = (a + ib)(a - ib) = a^2 + b^2.$$

Thus if we don't want a complex number in the denominator, we will need to multiply by its complex conjugate. Thus we can show:

Division of complex numbers:

$$\frac{z_1}{z_2} = \frac{a_1 + ib_1}{a_2 + ib_2} = \frac{a_1 + ib_1}{a_2 + ib_2} \frac{a_2 - ib_2}{a_2 - ib_2} = \frac{(a_1 a_2 + b_1 b_2) + i(-a_1 b_2 + b_1 a_2)}{a_2^2 + b_2^2}$$

$$= \left(\frac{a_1 a_2 + b_1 b_2}{a_2^2 + b_2^2}\right) + i\left(\frac{-a_1 b_2 + b_1 a_2}{a_2^2 + b_2^2}\right),$$

where we highlight that the final answer is in the form $a + ib$ for real numbers a, b.

Because we can't get a complex number from real numbers alone, this fundamentally means that we need an extra dimension in order to view complex numbers. To do this, we consider the horizontal axis as the **real axis** and the vertical axis as the **imaginary axis** and refer to this as the **complex plane**; see Figure 3.1

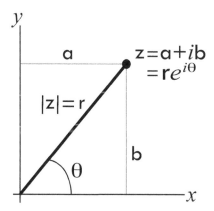

FIGURE 3.1: Cartesian and polar representations of a complex number z.

Any complex number $z = a + ib$ can be represented as a point in the complex plane. Now is where things get interesting because we will additionally view this point in the complex plane with its polar coordinates that correspond to the point (a, b). In doing so, we know that its length is $r = \sqrt{a^2 + b^2}$ while it makes an angle $\theta = \arctan\left(\frac{b}{a}\right)$ with the real axis. We can also write $a = r\cos\theta$ and $b = r\sin\theta$ using polar coordinates. A useful and interesting fact from the theory of complex variables is **Euler's formula**:

$$e^{i\theta} = \cos\theta + i\sin\theta \tag{3.34}$$

which is valid for all θ. This expression can be obtained in many ways and we refer the interested reader to the exercises for two such excursions. We give one remarkable application of Euler's formula; namely, letting $\theta = \pi$ we have

$$e^{i\pi} = \cos\pi + i\sin\pi.$$

That is, $e^{i\pi} = -1$ or

$$e^{i\pi} + 1 = 0.$$

This astounding expression relates the five most famous (and useful) constants $0, 1, e, \pi$, and i. Now, let's get back on track.

With this formula in hand, we can observe that any point in the complex plane can also be written as $re^{i\theta}$ since

$$re^{i\theta} = r\cos\theta + ir\sin\theta = a + ib.$$

Because any point in the plane can be written as a complex number, this means that any number can be written in the form $re^{i\theta}$.

Example 1 Write the following numbers in the form $re^{i\theta}$: 1, $-i$, $1 - i$, and $1 + 3i$.

Solution
(i) 1 is on the real axis so we know $\theta = 0$ or $\theta = 2\pi$. Thus, two possibilities are $1 = e^{i \cdot 0}$ or $1 = e^{i2\pi}$.
(ii) $-i$ is on the imaginary axis so that $\theta = \frac{3\pi}{2}$. Thus, one choice is $i = e^{i3\pi/2}$.
(iii) $1 - i$ is on the -45 degree line; that is, it makes the angle $\frac{-\pi}{4}$ with the real axis. It is easy to calculate $r = \sqrt{1^2 + (-1)^2} = \sqrt{2}$. Thus we can write $1 - i = \sqrt{2}e^{i\pi/4}$.
(iv) For $1 + 3i$, we can calculate the angle it makes with the real axis as $\theta = \arctan(\frac{3}{1})$ and the length as $r = \sqrt{1^2 + 3^2} = \sqrt{10}$. Thus $1 + 3i = \sqrt{10}e^{i\arctan(3)}$.

In the above example, we always had multiple possibilities to choose from for the angle θ. The notation that is often used is $\arg(x + iy)$. Thus, for example,

$$\arg(1 - i) = -\frac{\pi}{4}, -\frac{9\pi}{4}, \cdots, \frac{7\pi}{4}, \frac{15\pi}{4}, \cdots$$

where we have simply added $\pm 2n\pi, n = 0, 1, \cdots$ to the first answer to obtain the others. We sometimes will specify that we want the **principle value**, where we specify that $-\pi \leq \theta < \pi$ and we write $\text{Arg}(x + iy)$ to denote this. Thus

$$\text{Arg}(1 - i) = -\frac{\pi}{4}.$$

Writing complex numbers in this form also allows us to find roots of any complex number.

Example (i) Find all z satisfying $z^3 = 1$.
(ii) Find all z satisfying $z^4 = -1$.

Solution

(i) We again rewrite 1 as $e^{i2\pi}$ and note that *roots must be equally spaced apart*. Taking the cube root of both sides gives $z = e^{i2\pi/3}$. Since roots must be equally spaced and we are looking for 3 of them, the roots must be separated by $2\pi/3$. The other two roots are then

$$e^{i(2\pi/3 + 2\pi/3)}, \qquad e^{i(2\pi/3 + 4\pi/3)}.$$

Thus our roots, written in the form $a + ib$, are

$$e^{i2\pi/3} = \frac{-1}{2} + i\frac{\sqrt{3}}{2}, \quad e^{i4\pi/3} = \frac{-1}{2} - i\frac{\sqrt{3}}{2}, \quad e^{i6\pi/3} = 1.$$

(ii) Since we are looking for 4 roots, they must be spaced apart by $\frac{2\pi}{4} = \frac{\pi}{2}$. The first of these roots is obtained by taking the fourth root of $-1 = e^{i\pi}$, which gives $e^{i\pi/4}$. The other three roots are thus $e^{i(\pi/4 + \pi/2)}$, $e^{i(\pi/4 + 2\pi/2)}$, and $e^{i(\pi/4 + 3\pi/2)}$. Written in the form $a + ib$, we have

$$e^{i\pi/4} = \frac{1}{\sqrt{2}} + i\frac{1}{\sqrt{2}}, \quad e^{i3\pi/4} = \frac{-1}{\sqrt{2}} + i\frac{1}{\sqrt{2}},$$

$$e^{i5\pi/4} = \frac{-1}{\sqrt{2}} - i\frac{1}{\sqrt{2}}, \quad e^{i7\pi/4} = \frac{1}{\sqrt{2}} - i\frac{1}{\sqrt{2}}.$$

Example Let $p(x) = x^2 + 3x - 2$ be a polynomial. Calculate $p(i)$ and $p(-1 + 3i)$.

Solution

We treat $p(z)$ in the natural way that we would think of doing:
(i) $p(i) = (i)^2 + 3(i) - 2 = -1 + 3i - 2 = -3 + 3i$.
(ii) $p(-1 + 3i) = (-1 + 3i)^2 + 3(-1 + 3i) - 2 = (1 - 9 + i(-6)) + (-3 + 9i) - 2 = -13 + 3i$.

Sometimes it will be easier to deal with complex numbers and functions rather than real-valued ones. One of the key facts that make this helpful goes back to our ability to separate out real and imaginary parts. Specifically, we can take a real-valued function, write it as a complex-valued function, manipulate it according to our known rules, and then take the real part. Our answer in the end will be the same as if we hadn't made it complex! We first take a look at an example where we *complexify* our real-valued function.

Example 4 (i) Write $\sin 2x$ as the imaginary part of a complex function.
(ii) Write $e^{-3x}\cos x$ as the real part of a complex function.
(iii) Show that $A\sin\omega t + B\cos\omega t = \sqrt{A^2 + B^2}\cos(\omega t - \phi)$, where $\phi = \arctan\left(\frac{A}{B}\right)$ and A, B, ω, t are real-valued constants.

Solution
(i) From Euler's formula, we know that $e^{ix} = \cos x + i\sin x$, from which it follows that $e^{i2x} = \cos 2x + i\sin 2x$. Thus

$$\sin 2x = \text{Im}(e^{i2x}).$$

(ii) We multiply Euler's formula by e^{-3x} to obtain $e^{-3x+ix} = e^{-3x}\cos x + ie^{-3x}\sin x$. Thus

$$e^{-3x}\cos x = \text{Re}(e^{x(-3+i)}).$$

(iii) Using similar ideas as in (i) and (ii), we write

$$A\sin\omega t + B\cos\omega t = A\cdot\text{Im}(e^{i\omega t}) + B\cdot\text{Re}(e^{i\omega t}) = A\text{Re}(-ie^{i\omega t}) + B\text{Re}(e^{i\omega t})$$

$$= \text{Re}((B - Ai)e^{i\omega t}) = \text{Re}(\underbrace{\sqrt{A^2 + B^2}e^{-i\arctan(A/B)}}_{B - Ai \text{ in polar form}}e^{i\omega t})$$

$$= \text{Re}\left(\sqrt{A^2 + B^2}e^{i(\omega t - \arctan(A/B))}\right)$$

$$= \text{Re}\left(\sqrt{A^2 + B^2}e^{i(\omega t - \phi)}\right), \quad \phi = \arctan\left(\frac{A}{B}\right). \qquad (3.35)$$

The way we would hope to be able to take derivatives and integrals still holds: we just perform the operations on real and imaginary parts separately, treating i as a constant.

Example 5 (i) Evaluate $\dfrac{d}{dx}(\cos 2x + i\sin 2x)$.

(ii) Evaluate $\dfrac{d}{dx}(e^{ix})$.

Solution
(i) $\dfrac{d}{dx}(\cos 2x + i\sin 2x) = -2\sin 2x + i2\cos 2x$.

(ii) $\dfrac{d}{dx}(e^{ix}) = e^{ix}\dfrac{d}{dx}(ix) = ie^{ix}$.

Combining these last few comments will finish our discussion of complex numbers.

Example 6 Evaluate $\dfrac{d}{dx}(e^{-x}\sin 2x)$ in two ways:
(i) Take its derivative as written (don't convert to complex functions).
(ii) First write the function as the imaginary part of a complex function, take the derivative of the complex function, then separate out the imaginary part.

Solution

(i) This is a straight product and chain rule application:

$$\frac{d}{dx}(e^{-x}\sin 2x) = e^{-x}\frac{d}{dx}\sin 2x + \sin 2x\frac{d}{dx}e^{-x} = 2e^{-x}\cos 2x - e^{-x}\sin 2x.$$

(ii) We first write this as $e^{-x}\sin 2x = \text{Im}(e^{-x+i2x})$. Taking this derivative gives

$$\frac{d}{dx}(e^{-x+i2x}) = (-1+2i)e^{-x+i2x} = (-1+2i)(e^{-x}\cos 2x + ie^{-x}\sin 2x)$$

$$= (-e^{-x}\cos 2x - 2e^{-x}\sin 2x) + i(-e^{-x}\sin 2x + 2e^{-x}\cos 2x).$$

Our final answer is the imaginary part of this:

$$\frac{d}{dx}(e^{-x}\sin 2x) = \text{Im}\left(\frac{d}{dx}(e^{-x+i2x})\right) = -e^{-x}\sin 2x + 2e^{-x}\cos 2x.$$

Example 7 Evaluate $\displaystyle\int e^{(3+2i)x}\,dx$. Write the final answer in two equivalent forms:
(a) $zw(x)$ for complex-valued constant z and complex-valued function $w(x)$ and
(b) $f(x) + ig(x)$ for real-valued functions f, g.

Solution

We treat i as the constant:
(a) $\displaystyle\int e^{(3+2i)x}\,dx = \frac{e^{(3+2i)x}}{3+2i} + C = \frac{e^{(3+2i)x}}{3+2i}\cdot\frac{3-2i}{3-2i} + C = \frac{3-2i}{13}e^{(3+2i)x} + C.$
(b) From the last line of (a), we then have

$$\int e^{(3+2i)x}\,dx = \frac{3-2i}{13}e^{(3+2i)x} + C = \frac{(3-2i)e^{3x}(\cos 2x + i\sin 2x)}{13} + C$$

$$= \frac{e^{3x}(3\cos 2x + 2\sin 2x)}{13} + i\frac{e^{3x}(3\sin 2x - 2\cos 2x)}{13} + C.$$

We note that C may be a *complex-valued* constant.

We will finish our discussion of complex numbers with an algebraic example in which our constants may be complex.

Example 8 Solve the system of equations

$$(2+i)A_1 = 3 \tag{3.36}$$
$$A_1 + 3A_2 i = 0, \quad A_1, A_2 \in \mathbb{C}. \tag{3.37}$$

Solution

From (3.36), we have

$$A_1 = \frac{3}{2+i} = \frac{3}{2+i} \cdot \frac{2-i}{2-i} = \frac{3(2-i)}{4+1} = \frac{6}{5} - i\frac{3}{5}.$$

Substituting A_1 into (3.37) and solving for A_2 gives

$$A_2 = \frac{-A_1}{3i} = \frac{-\frac{6}{5} + i\frac{3}{5}}{3i} \cdot \left(\frac{-3i}{-3i}\right) = \frac{\frac{9}{5}}{9} + i\frac{\frac{18}{5}}{9} = \frac{1}{5} + i\frac{2}{5}.$$

Problems

In Problems **1–10**, write the following numbers in the form $re^{i\theta}$.
1. $4i$ **2.** $i\sqrt{2}$ **3.** $-1+i$ **4.** $1+\sqrt{3}i$
5. $-1+\sqrt{3}i$ **6.** $3-\sqrt{3}i$ **7.** $4+3i$ **8.** $3+5i$
9. $2-i$ **10.** $-2+3i$

In Problems **11–16**, find all z satisfying the given equation.
11. $z^2 - 1 = 0$ **12.** $z^2 + 4 = 0$ **13.** $z^3 - i = 0$ **14.** $z^3 + i = 0$
15. $z^4 + 16 = 0$ **16.** $z^4 + 64 = 0$

In Problems **17–22**, evaluate the polynomial at the given complex number.
17. $p(x) = 2x^2 - x + 1$; $1+i$, $2i$ **18.** $p(x) = x^2 + x + 1$, $1-i$, $3i$
19. $p(z) = \frac{1}{z^2-z+1}$; i, $1+i$ **20.** $p(z) = \frac{1}{z^2+2z-1}$; $-i$, i
21. $p(z) = \frac{e^z}{z^2+1}$; $2i$, $1-i$ **22.** $p(z) = \frac{e^z}{z^2+z+1}$; $2i$, $1-i$

In Problems **23–30**, write the given function as the real part or imaginary part of a complex-valued function.

23. $\cos x$ **24.** $\sin 3x$ **25.** $e^x \sin 2x$
26. $e^{-2x} \sin x$ **27.** $\cos x - \sin x$ **28.** $3\cos 2x + \sin 2x$
29. $2e^{-x}\cos 3x + e^{-x}\sin 3x$ **30.** $e^{-x}\cos 2x + 3e^{-x}\sin 2x$

In Problems **31–38**, evaluate the following expressions. Write in the form $f(x) + ig(x)$ for real-valued functions f, g.

31. $\frac{d}{dx}(i\sin(2x))$ **32.** $\frac{d}{dx}(e^{3ix})$ **33.** $\frac{d}{dx}(e^{ix}\sin(x))$

34. $\frac{d}{dx}(e^{2ix}\cos(3x)) + \sin x$ **35.** $\int e^{-ix}dx$

36. $\int e^{(1-2i)x}dx$ **37.** $\int e^{(1+2i)x}dx$ **38.** $\int ie^{2ix}dx$

In Problems **39–42**, solve the system of equations for the possibly complex-valued unknowns.

39. $iA_1 = 1,\ A_1 + iA_2 = 0$ **40.** $iA_2 = 1,\ 2A_1 + iA_2 = 0$

41. $A_1 - iA_2 = 1,\ 2A_1 + iA_2 = i$ **42.** $iA_1 + A_2 = 1,\ A_1 + 2iA_2 = 1$

43. Show that e^{ix} is a solution to $x'' + x = 0$.

44. Show that e^{i2x} is a solution to $x'' + 4x = 0$.

45. Use separation of variables to show that $e^{(a+ib)t}$ is a solution to $x' = (a + ib)x$.

46. Show that, for real-valued constants A and B,

$$A\sin\omega t + B\cos\omega t = \sqrt{A^2 + B^2}\sin(\omega t + \phi), \quad \phi = \arctan\left(\frac{B}{A}\right).$$

(Hint: Follow Example 4(iii) except using Im instead of Re.)

47. Use Taylor series for e^x, $\cos x$, and $\sin x$ to show Euler's formula. (Hint: Expand e^{ix}.)

48. As another method of obtaining Euler's formula, consider

$$z_1 = e^{i\theta} \text{ and } z_2 = \cos\theta + i\sin\theta.$$

Show that these two expressions are both solutions of the complex-valued initial-value problem

$$\frac{dz}{d\theta} = iz \qquad \text{with } z(0) = 1$$

and give reasons why you can conclude that they must be identical.

3.6 Homogeneous Equations with Constant Coefficients

Now we will consider how to actually find linearly independent solutions for

$$a_n\frac{d^n y}{dx^n} + a_{n-1}\frac{d^{n-1}y}{dx^{n-1}} + \ldots + a_1\frac{dy}{dx} + a_0\,y = 0, \tag{3.38}$$

a homogeneous linear equation with constant coefficients $a_n, a_{n-1}, \ldots, a_0 \in \mathbb{R}$.

In the case of a second-order, constant-coefficient equation, we "guessed" a solution of the form e^{rx} (in Section 3.1) based on what happened in the case of the linear first-order homogeneous equation. For the nth-order equation, we are looking for a function with the property

$$\frac{d^k}{dx^k}[f(x)] = cf(x) \text{ for all } x.$$

This is a property of the exponential function:

$$\frac{d^k}{dx^k}e^{rx} = r^k e^{rx}.$$

Thus we substitute $y = e^{rx}$ into (3.38) and evaluate the derivatives to obtain

$$a_n r^n e^{rx} + a_{n-1} r^{n-1} e^{rx} + \ldots + a_1 r e^{rx} + a_0 e^{rx} = 0,$$

which can be rewritten as

$$e^{rx}(a_n r^n + a_{n-1} r^{n-1} + \ldots + a_1 r + a_0) = 0.$$

We again observe that $e^{rx} \neq 0$ so that our solutions are obtained from solving a polynomial in r:

$$\underbrace{a_n r^n + a_{n-1} r^{n-1} + \ldots + a_1 r + a_0}_{= P(r)} = 0.$$

This polynomial, which we denote $P(r)$, is known as the **characteristic equation** and its roots are known as **eigenvalues or characteristic values**. (Sometimes the equation is also referred to as the **auxiliary equation**.) Corresponding to each eigenvalue is the **eigenvector** $y = e^{rx}$, which is a solution to (3.38).

For the equivalent formulation in operator notation, we observe that $D^k e^{rx} = r^k e^{rx}$ so that

$$P(D)e^{rx} = P(r)e^{rx},$$

where $P(D)y = 0$ is our differential equation and $P(r) = 0$ is the resulting characteristic equation.

Now, recalling the **Fundamental Theorem of Algebra**, we know *an nth degree polynomial has n roots if we allow for the possibility of complex roots* (and we count repeated roots, too). There are three cases to consider based on these n roots. We may have roots that are real and distinct, roots that are repeated, or roots that are complex. In general, we may have a combination of all three. We will consider each of these cases.

Case 1: Distinct Real Roots

We begin with a theorem that characterizes this case.

For the nth order homogeneous linear differential equation (3.38) with constant coefficients, if the characteristic equation has n distinct real roots $r_1, r_2, \ldots, r_n$, then $e^{r_1 x}, e^{r_2 x}, \cdots, e^{r_n x}$ are linearly independent solutions of (3.38). The general solution is given by

$$y = c_1 e^{r_1 x} + c_2 e^{r_2 x} + \ldots + c_n e^{r_n x}$$

where $c_1, c_2, \ldots, c_n$ are arbitrary constants.

It should be clear that $e^{r_1 x}, e^{r_2 x}, \cdots, e^{r_n x}$ are solutions; see Figure 3.2. We can see they are linearly independent by calculating $W(0)$, that is, the Wronskian evaluated at $x = 0$. Since they are solutions, recall that $W(x)$ is either always 0 or never 0.

For instance, in the case of a third-order equation, we would have

$$W(0) = \begin{vmatrix} e^{r_1 x} & e^{r_2 x} & e^{r_3 x} \\ r_1 e^{r_1 x} & r_2 e^{r_2 x} & r_3 e^{r_3 x} \\ r_1^2 e^{r_1 x} & r_2^2 e^{r_2 x} & r_3^2 e^{r_3 x} \end{vmatrix}_{x=0} = \begin{vmatrix} 1 & 1 & 1 \\ r_1 & r_2 & r_3 \\ r_1^2 & r_2^2 & r_3^2 \end{vmatrix},$$

which can then be expanded and simplified as

$$
\begin{aligned}
W(0) &= 1 \cdot (r_2 r_3^2 - r_2^2 r_3) - 1 \cdot (r_1 r_3^2 - r_1^2 r_3) + 1 \cdot (r_1 r_2^2 - r_1^2 r_2) \\
&= r_2 (r_3^2 - r_2 r_3 + r_1 r_2) - r_1 (r_3^2 - r_1 r_3 + r_1 r_2) \\
&= r_2 (r_3^2 - r_2 r_3 + r_1 r_2) - r_1 (r_3^2 - r_1 r_3 + r_1 r_2) - r_2 r_1 r_3 + r_1 r_2 r_3 \\
&= r_2 (r_3^2 - r_2 r_3 + r_1 r_2 - r_1 r_3) - r_1 (r_3^2 - r_1 r_3 + r_1 r_2 - r_2 r_3) \\
&= r_2 (r_3 - r_1)(r_3 - r_2) - r_1 (r_3 - r_1)(r_3 - r_2) \\
&= (r_2 - r_1)(r_3 - r_1)(r_3 - r_2) \\
&\neq 0
\end{aligned}
\tag{3.39}
$$

where the last line is true since the roots are distinct.

In general, we have for an nth-order equation that

$$\prod_{i>j} (r_i - r_j)(-1)^{n+1},$$

where the symbol Π (capital Pi) denotes the product of all the factors that follow.

Example 1 Consider the differential equation

$$\frac{d^2 y}{dx^2} - \frac{dy}{dx} - 6y = 0.$$

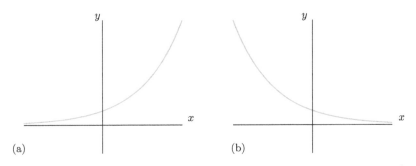

(a)   (b)

FIGURE 3.2:   Basic shapes of solution curves for real, distinct roots: (a) e^{rx} for $r > 0$; (b) e^{rx} for $r < 0$.

The characteristic equation is

$$r^2 - r - 6 = 0$$

which has the roots

$$r = -2 \quad \text{and} \quad r = 3.$$

Thus, e^{-2x} and e^{3x} are linearly independent solutions and the general solution is thus

$$y = c_1 e^{-2x} + c_2 e^{3x}.$$

Example 2 Consider the differential equation

$$\frac{d^3 y}{dx^3} - 4\frac{d^2 y}{dx^2} + \frac{dy}{dx} + 6y = 0.$$

The characteristic equation is

$$r^3 - 4r^2 + r + 6 = 0$$

for which we have some difficulty solving for r. In most cases, the characteristic equation will not factor nicely and numerical methods will perhaps be needed. It is also useful to employ a computer algebra system or recall some college algebra techniques.

In the present case we get lucky, of course, because $r = -1$ is a root. This means that $r + 1$ is a factor. The other factors can be found by division to be $r - 2$ and $r - 3$. That is

$$r^3 - 4r^2 + r + 6 = (r + 1)(r - 2)(r - 3).$$

Hence, e^{-x}, e^{2x}, e^{3x} are solutions. Thus

$$y = c_1 e^{-x} + c_2 e^{2x} + c_3 e^{3x}$$

is the general solution.

Many times we will have a root that is repeated and this requires a modification of the solution from Case 1.

Example 3 Consider the differential equation

$$\frac{d^2y}{dx^2} - 6\frac{dy}{dx} + 9y = 0.$$

The characteristic equation is

$$r^2 - 6r + 9 = 0 \Longrightarrow (r-3)^2 = 0.$$

The roots are

$$r_1 = 3 \quad \text{and} \quad r_2 = 3 \text{ (double root!)}.$$

So we have the solution e^{3x} corresponding to $r_1 = 3$, and the solution e^{3x} corresponding to $r_2 = 3$. Clearly, e^{3x} is not linearly independent of e^{3x}. So we have a small problem.

Since we already know one solution is e^{3x}, we can reduce the order of the equation (see Section 3.3). Let

$$y = e^{3x}v.$$

Thus

$$\frac{dy}{dx} = e^{3x}\frac{dv}{dx} + 3e^{3x}v$$

and

$$\frac{d^2y}{dx^2} = e^{3x}\frac{d^2v}{dx^2} + 6e^{3x}\frac{dv}{dx} + 9e^{3x}v.$$

Substituting into the original differential equation and simplifying gives

$$e^{3x}\frac{d^2v}{dx^2} = 0.$$

Letting $w = \frac{dv}{dx}$, we have

$$e^{3x}\frac{dw}{dx} = 0.$$

This gives that $dw/dx = 0$ and thus $w = c$. We can let $c = 1$ so

$$v = x + c_0$$

and thus

$$v(x)e^{3x} = (x + c_0)e^{3x}$$

is a solution to the second-order equation. Now we know that $(x + c_0)e^{3x}$ and e^{3x} are linearly independent (check the Wronskian!), so taking $c_0 = 0$, we have corresponding to the double root 3, two linearly independent solutions

$$e^{3x} \quad \text{and} \quad xe^{3x}.$$

Thus, the general solution is

$$y = c_1 e^{3x} + c_2 x e^{3x}.$$

Generalizing this example, if a second-order homogeneous linear equation with constant coefficients has r as a double root to its characteristic equation, then e^{rx} and xe^{rx} are the corresponding linearly independent equations; see Figure 3.3. Extending this idea further, we have the following theorem.

THEOREM 3.6.2

Consider the nth order homogeneous linear differential equation (3.38) with constant coefficients.

1. If the characteristic equation has the real root r occurring k times, then the part of the general solution corresponding to this k-fold repeated root is

$$(c_1 + c_2 x + c_3 x^2 + \ldots + c_k x^{k-1})e^{rx}.$$

2. If, further, the remaining roots of the characteristic equation are the distinct real numbers $r_{k+1}, r_{k+2}, \ldots, r_n$, then the general solution is

$$y = (c_1 + c_2 x + c_3 x^2 + \ldots + c_k x^{k-1})e^{rx} + c_{k+1}e^{r_{k+1}x} + \ldots + c_n e^{r_n x}.$$

3. If, however, any of the remaining roots are also repeated, then the parts of the general solution to (3.38) corresponding to each of these other repeated roots are expressions similar to that corresponding to r in part **1.**

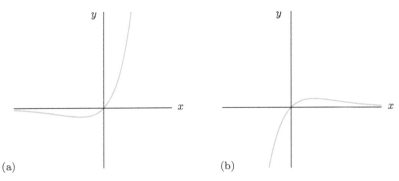

(a)   (b)

FIGURE 3.3: Basic shapes of solution curves for real, repeated roots: (a) xe^{rx} for $r > 0$; (b) xe^{rx} for $r < 0$. The figures here are to be combined with the corresponding part of Figure 3.2. For example, a second-order equation with $r < 0$ as a repeated root will be a linear combination of Figures 3.2(b) and 3.3(b).

Example 4 Solve the equation

$$\frac{d^3y}{dx^3} - 4\frac{d^2y}{dx^2} - 3\frac{dy}{dx} + 18y = 0.$$

Solution

The characteristic equation is

$$r^3 - 4r^2 - 3r + 18 = 0$$

which has roots $3, 3, -2$, (check it!). Theorems 3.6.1 and 3.6.2 tell us that three linearly independent solutions are e^{3x}, xe^{3x}, and e^{-2x}. The general solution is thus

$$y = c_1 e^{3x} + c_2 x e^{3x} + c_3 e^{-2x}.$$

Example 5 Suppose a sixth order homogeneous linear differential equation with constant coefficients had the following roots of the characteristic equation:

$$-1, -1, 2, 3, 3, 3.$$

Then six linearly independent solutions are $e^{-x}, xe^{-x}, e^{2x}, e^{3x}, xe^{3x}, x^2 e^{3x}$ and the general solution to the differential equation is

$$y = c_1 e^{-x} + c_2 x e^{-x} + c_3 e^{2x} + c_4 e^{3x} + c_5 x e^{3x} + c_6 x^2 e^{3x}.$$

Case 3: Complex Roots (Non-Real)

Now suppose that the characteristic equation has $a + bi$, a complex number, as a root. (Here, a and b are real numbers, $b \neq 0$ and $i^2 = -1$.)

Because complex roots of a polynomial with real coefficients always come in conjugate pairs, we know that $a - bi$ is also a root. In obtaining the characteristic equation, we observed that e^{rx} solves the differential equation but *did not* require that r be real. Thus, $e^{(a+ib)x}$ and $e^{(a-ib)x}$ are linearly independent solutions and the corresponding part of the general solution is

$$k_1 e^{(a+bi)x} + k_2 e^{(a-bi)x} \tag{3.40}$$

where k_1 and k_2 are arbitrary (real) constants.

Note that the solutions defined by $e^{(a+bi)x}$ and $e^{(a-bi)x}$ are complex functions of the real variable x. We are interested in real linearly independent solutions.

If we apply Euler's formula to Equation (3.40), we have

$$k_1 e^{(a+bi)x} + k_2 e^{(a-bi)x} = e^{ax}\left(k_1 e^{ibx} + k_2 e^{-ibx}\right)$$
$$= e^{ax}\left(k_1(\cos bx + i\sin bx) + k_2(\cos bx - i\sin bx)\right)$$
$$= e^{ax}\left((k_1 + k_2)\cos bx + i(k_1 - k_2)\sin bx\right)$$
$$= e^{ax}\left(c_1 \sin bx + c_2 \cos bx\right),$$

where $c_1 = k_1 + k_2$ and $c_2 = i(k_1 - k_2)$. Thus, corresponding to the roots $a \pm bi$ is the solution

$$e^{ax}\left(c_1 \sin bx + c_2 \cos bx\right),$$

where $e^{ax}\sin bx$ and $e^{ax}\cos bx$ are two, linearly independent, real-valued solutions; see Figure 3.4. We have the following theorem.

THEOREM 3.6.3
Consider the nth order homogeneous linear differential equation (3.38) with constant coefficients.
1. If the characteristic equation has conjugate complex roots $a + bi$ and $a - bi$, neither repeated, then $e^{ax}\sin bx$ and $e^{ax}\cos bx$ are linearly independent solutions. The corresponding part of the general solution may be written as
$$e^{ax}\left(c_1 \sin bx + c_2 \cos bx\right).$$

2. If, however, $a + bi$ and $a - bi$ are each roots of multiplicity k of the characteristic equation, then the corresponding part of the general solution may be written

$$e^{ax}\left(c_1 + c_2 x + c_3 x^2 + \ldots + c_k x^{k-1}\right)\sin bx$$
$$+e^{ax}\left(c_{k+1} + c_{k+2}x + c_{k+3}x^2 + \ldots + c_{2k}x^{k-1}\right)\cos bx. \quad (3.41)$$

Example 6 Solve the equation $\dfrac{d^2 y}{dx^2} + 9y = 0$.

Solution
The characteristic equation is

$$r^2 + 9 = 0$$

which has solution $r = \pm 3i$, so that $a = 0$ and $b = 3$. Then $\sin 3x$ and $\cos 3x$ are two linearly independent solutions and the general solution is

$$y = c_1 \sin 3x + c_2 \cos 3x.$$

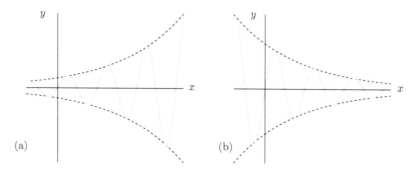

FIGURE 3.4: Basic shapes of solution curves for complex roots: (a) linear combination of $\{e^{ax}\cos(bx), e^{ax}\sin(bx)\}$ for $a > 0$; (b) linear combination of $\{e^{ax}\cos(bx), e^{ax}\sin(bx)\}$ for $a < 0$. Notice the **envelope** $\pm e^{ax}$ that bounds the solution in both parts.

Example Solve the equation

$$(D^2 - 6D + 25)(y) = 0.$$

Solution

Try not to be confused by the use of the operator notation—it is actually easier to write the characteristic equation. The characteristic equation is

$$r^2 - 6r + 25 = 0$$

which has roots $r = 3 \pm 4i$. Two linearly independent solutions are $e^{3x}\sin 4x$ and $e^{3x}\cos 4x$ and thus

$$y = e^{3x}(c_1 \sin 4x + c_2 \cos 4x)$$

is the general solution.

Example Solve the equation

$$\frac{d^4y}{dx^4} - 4\frac{d^3y}{dx^3} + 14\frac{d^2y}{dx^2} - 20\frac{dy}{dx} + 25y = 0.$$

Solution

The characteristic equation is

$$r^4 - 4r^3 + 14r^2 - 20r + 25 = 0.$$

This is nontrivial to solve by hand; however, in our respective computer programs we could easily find the roots as

$$1 + 2i, \ 1 - 2i, \ 1 + 2i, \ \text{and} \ 1 - 2i.$$

These are double complex roots, and linearly independent solutions are $e^x \sin 2x$, $xe^x \sin 2x$, $e^x \cos 2x$, and $xe^x \cos 2x$. The general solution is thus

$$y = e^x \left[(c_1 + c_2 x) \sin 2x + (c_3 + c_4 x) \cos 2x \right].$$

We have yet to consider any initial-value problems, but these are very straightforward as can be seen in this next example.

Example 9 Solve the initial-value problem

$$\frac{d^2 y}{dx^2} - 6 \frac{dy}{dx} + 25 y = 0,$$

with the conditions $y(0) = -3$ and $y'(0) = -1$. Plot the solution on the interval $-1 \le x \le 1$.

Solution

In Example 7 above, we obtained the general solution as

$$y = e^{3x} (c_1 \sin 4x + c_2 \cos 4x).$$

We observe that $y(0) = -3$ implies that $c_2 = -3$ and $y'(0) = -1$ gives

$$4c_1 + 3c_2 = -1$$

so that $c_1 = 2$. Thus the solution is

$$y = e^{3x} (2 \sin 4x - 3 \cos 4x).$$

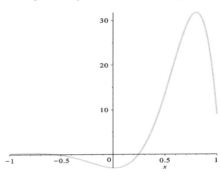

FIGURE 3.5: Plot of solution for Example 9.

Second-Order Linear Homogeneous Equations with Constant Coefficients

The theory we have developed in our method of solving nth-order linear homogeneous equations with constant coefficients relies upon finding the roots of the corresponding nth degree characteristic polynomial. Since polynomials of degree two can be explicitly solved, all solutions of the second-order linear homogeneous equation with constant coefficients

$$a_2 \frac{d^2 y}{dx^2} + a_1 \frac{dy}{dx} + a_0 \, y = 0 \tag{3.42}$$

are obtained from the roots of the quadratic equation

$$a_2 r^2 + a_1 r + a_0 = 0.$$

These roots are easily obtained by the quadratic formula as

$$r_{1,2} = \frac{-a_1 \pm \sqrt{a_1^2 - 4a_2 a_0}}{2a_2} \tag{3.43}$$

so that the general solution to (3.42) is

$$y(x) = \begin{cases} c_1 e^{r_1 x} + c_2 e^{r_2 x} & \text{if } a_1^2 > 4a_2 a_0 \\ (c_1 + c_2 x) e^{-a_1 x / 2a_2} & \text{if } a_1^2 = 4a_2 a_0 \\ e^{-a_1 x / 2a_2} \left[c_1 \cos(wx) + c_2 \sin(wx) \right] & \text{if } a_1^2 < 4a_2 a_0, \end{cases} \tag{3.44}$$

where c_1 and c_2 are arbitrary constants and $w = \dfrac{1}{a_2} \sqrt{a_2 a_0 - \dfrac{a_1^2}{4}}$.

Example 10 Solve the equation

$$\frac{d^2 y}{dx^2} + 8 \frac{dy}{dx} + 15y = 0.$$

Solution

Since the characteristic equation is $r^2 + 8r + 15 = 0$, we see that it has two real roots, which are

$$r_1 = -3 \text{ and } r_2 = -5.$$

Thus the general solution is

$$y(x) = c_1 e^{-3x} + c_2 e^{-5x}.$$

Example 11 Solve the equation

$$\frac{d^2 y}{dx^2} + 2 \frac{dy}{dx} + 2y = 0.$$

Solution

Since the characteristic equation is $r^2 + 2r + 2 = 0$, we see that it has two complex roots. Applying Equation (3.44) gives

$$\frac{1}{a_0} \sqrt{a_0 a_2 - \frac{a_1^2}{4}} = 1.$$

Thus the general solution is

$$y(x) = e^x \left[c_1 \cos x + c_2 \sin x \right].$$

Example 12 Solve the equation

$$\frac{d^2 y}{dx^2} + 6 \frac{dy}{dx} + 9y = 0.$$

Solution

Since the characteristic equation is $r^2 + 6r + 9 = 0$, we see that it has a repeated root of $r = -3$. The solution is thus

$$y(x) = (c_1 + c_2 x) \left(e^{-3x} \right).$$

Problems

In Problems **1–22**, find a general solution of each equation.

1. $y'' + 8y' + 12y = 0$

2. $y'' - 36y = 0$

3. $y'' - 3y' + 4y = 0$

4. $2y'' - 7y' + 3y = 0$

5. $y'' - 2y' = 0$

6. $8y''' + y'' = 0$

7. $4y'' + 4y' + y = 0$

8. $2y'' - 3y' + 4y = 0$

9 $y'' - 4y' + 5y = 0$

10. $y^{(4)} - y = 0$

11. $(D^4 - 16)(y) = 0$

12. $(D^2 + 1)^2(y) = 0$

13. $(D^2 + 4)^2(y) = 0$

14. $y''' + 3y'' - 4y' - 12y = 0$

15. $y''' - 3y'' + 3y' - y = 0$

16. $(D - 4)^2(D^2 - D - 2)(y) = 0$

17. $y^{(4)} - 5y'' + 4y = 0$

18. $y^{(5)} + 8y''' + 16y' = 0$

19. $y^{(5)} - 10y''' + 9y' = 0$

20. $D^3(D^2 - 6D + 9)(y) = 0$

21. $D^2(D - 1)^2(D^2 + 1)(y) = 0$

22. $D(D^2 + 4)(D^2 - 2D + 1)(y) = 0$

For Problems **23–30**, find the solution to each of the IVP.

23. $y'' + y' - 12y = 0$, $y(0) = 0$, $y'(0) = 7$

24. $9y'' + 6y' + 4y = 0$, $y(0) = 3$, $y'(0) = 4$

25. $y'' + 4y' + 4y = 0$, $y(0) = 0$, $y'(0) = 1$

26. $4y'' - 4y' - 3y = 0$, $y(0) = 1$, $y'(0) = 5$

27. $(D^2 + 2D + 10)(y) = 0$, $y(-\pi) = 0$, $y'(-\pi) = 1$

28. $2y'' + 5y' + 2y = 0$, $y(0) = \pi$, $y'(0) = 1$

29. $(D^2 + 4)(y) = 0$, $y(\pi) = 1$, $y'(\pi) = 1$

30. $y''' + 2y'' - 5y' - 6y = 0$, $y(0) = 0$, $y'(0) = 0$, $y''(0) = 1$

For Problems **31–34**, find a differential equation for which the characteristic equation has roots r_i with corresponding multiplicities k_i.

31. $r_1 = -2$, $k_1 = 1$; $r_2 = 0$, $k_2 = 2$

32. $r_1 = 3i$, $k_1 = 2$; $r_{2,3} = 1 \pm i$, $k_2 = 1$

33. $r_1 = 0$, $k_1 = 4$; $r_{2,3} = 2 \pm 3i$, $k_2 = 3$

34. $r_1 = 2 \pm 3i$, $k_1 = 2$; $r_3 = -5$, $k_3 = 1$; $r_4 = 2$, $k_4 = 3$

35. Show that a general solution of the differential equation

$$ay'' + 2by' + cy = 0, \qquad b^2 - ac > 0$$

can be written as

$$y = e^{-bx/a} \left[c_1 \cosh\left(\frac{x\sqrt{b^2 - ac}}{a} \right) + c_2 \sinh\left(\frac{x\sqrt{b^2 - ac}}{a} \right) \right].$$

3.7   Mechanical and Electrical Vibrations

We will now consider the motion of a mass attached to a spring by taking into account the resistance of the medium and the possible external forces acting on the mass. We note that although we have used $y(x)$ as our variable for much of the book, we switch to $x(t)$ for this section in order to stay consistent with the notation used in much of engineering and physics; see Figure 3.6. We need a fact from physics to begin our in-depth examination.

Hooke's law: It is experimentally observed that the magnitude of the force needed to produce a certain elongation of a spring is directly proportional to the amount of the elongation, provided the elongation is not too great. That is,

$$|F| = ks \tag{3.45}$$

where $|F|$ is the magnitude of the force F, s is the amount of elongation, and k is a constant of proportionality, called the spring constant, which depends upon the characteristics of the spring.

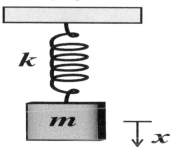

FIGURE 3.6:   Mass on a spring at rest position, $x = 0$.

Example 1   If a 30-lb weight stretches a spring 2 ft, then Hooke's law gives

$$30 = 2k$$

or

$$k = 15 \text{ lb/ft}.$$

When a mass is hung upon a spring that has a spring constant k and produces elongation x, the force F of the mass upon the spring has magnitude kx. At the same time, the spring exerts a force upon the mass called the **restoring force**. This restoring force is equal in magnitude, but opposite in sign to F and hence is $-kx$.

Formulation of the problem: Suppose the spring has natural (unstretched) length L. The mass m is attached to the spring and the spring stretches to its equilibrium position, stretching the spring an amount ℓ. The stretched length is $L + \ell$.

For convenience, we assume the origin of a coordinate system at this equilibrium position. We also assume the positive direction as down. Thus, the value x of this coordinate system is positive, zero, or negative, depending upon whether the mass is below, at, or above equilibrium, respectively.

Forces acting upon the mass: In this coordinate system, forces tending to pull the mass downward are positive, while those tending to pull it upward are negative. The forces are:

1. F_1, the force of gravity. This is given as

$$F_1 = mg,$$

where m is the mass and g is gravity. F_1 is positive since it acts downward.

2. F_2, the restoring force of the spring. Since $x + \ell$ is the total elongation of the spring, by Hooke's law, the magnitude of this force is $k(x + \ell)$. When the mass is <u>below</u> the end of the unstretched spring, the force acts upward and is thus <u>negative</u>. Since $x + \ell$ is positive, we have in this case

$$F_2 = -k(x + \ell).$$

Similarly, if the mass is above the end of the unstretched spring, the spring is acting downward and thus this force is <u>positive</u>. However, $x + \ell$ is <u>negative</u>, so that in this case we again have

$$F_2 = -k(x + \ell).$$

At the equilibrium point, the force of gravity is equal to the restoring force, so that

$$-mg = -k(0 + \ell)$$

or

$$mg = k\ell.$$

Hence, we have the equation

$$F_2 = -kx - mg.$$

3. F_3, the resisting force of the medium. This force is also known as the **damping force**. The magnitude of this force is not known exactly; however, it is known that for small velocities, it is approximately

$$|F_3| = b\left|\frac{dx}{dt}\right|$$

where $b > 0$ a constant, which is known as the **damping constant**. Note that when the mass is moving <u>downward</u>, F_3 acts in the <u>upward</u> direction so that $F_3 < 0$. Here, moving downward implies x increases so that $dx/dt > 0$. Thus, when moving downward

$$F_3 = -b\frac{dx}{dt}.$$

Similarly, when moving upward x decreases, so that $dx/dt < 0$. Thus, in this case, we again have

$$F_3 = -b\frac{dx}{dt}.$$

4. F_4, **any external forces that act upon the mass.** We will let the resultant of all such forces at time t be $F(t)$ and write

$$F_4 = F(t).$$

Now we can apply Newton's second law:

$$F = ma$$

where

$$F = F_1 + F_2 + F_3 + F_4,$$

the sum total of the forces involved. We thus have

$$m\frac{d^2x}{dt^2} = mg - kx - mg - b\frac{dx}{dt} + F(t)$$

or

$$m\frac{d^2x}{dt^2} + b\frac{dx}{dt} + kx = F(t). \tag{3.46}$$

This is the differential equation A for the motion of the mass on a spring. It is a nonhomogeneous second-order linear differential equation with constant coefficients.

If $b = 0$, the motion is called **undamped**; otherwise, it is called *damped*. If $F(t) = 0$ for all t, the motion is called *free*; otherwise, it is called *forced*.

Undamped Oscillations

In the case of free undamped motion, both $b = 0$ and $F(t) = 0$ for all t. This gives the differential equation

$$m\frac{d^2x}{dt^2} + kx = 0,$$

where $m > 0$ is the mass and $k > 0$ is the spring constant. Thus, if we divide through by m, we have

$$\frac{d^2x}{dt^2} + \frac{k}{m}x = 0.$$

Since $m, k > 0$, we let $\omega_n = \sqrt{\frac{k}{m}}$ (where ω_n is known as the **natural angular frequency**) and substitution yields

$$\frac{d^2x}{dt^2} + \omega_n^2 x = 0.$$

The corresponding characteristic equation is

$$r^2 + \omega_n^2 = 0$$

which has roots

$$r = \pm \omega_n i.$$

The general solution is thus

$$x(t) = c_1 \sin \omega_n t + c_2 \cos \omega_n t.$$

Now we suppose that the mass was initially displaced a distance x_0 from the equilibrium with an initial velocity v_0. That is,

$$x(0) = x_0 \text{ and } x'(0) = v_0.$$

Using these initial conditions, we find that

$$c_1 = \frac{v_0}{\omega_n} \text{ and } c_2 = x_0$$

so that

$$x(t) = \frac{v_0}{\omega_n} \sin \omega_n t + x_0 \cos \omega_n t. \tag{3.47}$$

Although Equation (3.47) completely describes the motion of the spring at any given time, it is often easier to picture the solution curves if we write the solution in an alternate form (**amplitude-phase form**), as

$$x(t) = A \cos(\omega_n t - \phi),$$

where A will be an **amplitude** and ϕ the **phase constant**. The quantity $(\omega_n t - \phi)$ is called the **phase of the motion**.

To accomplish this we first observe that the maximum amplitude of (3.47) is given by

$$A = \sqrt{\left(\frac{v_0}{\omega_n}\right)^2 + x_0^2}. \tag{3.48}$$

Each of the two terms contributes to this A. To see this contribution, we consider the fractions

$$\frac{x_0}{A} \text{ and } \frac{v_0/\omega_n}{A}.$$

These quantities are both between 0 and 1 (Why?). They also satisfy the relation

$$\left(\frac{x_0}{A}\right)^2 + \left(\frac{v_0/\omega_n}{A}\right)^2 = 1,$$

which clues us in to set

$$\cos\phi = \frac{x_0}{A} \text{ and } \sin\phi = \frac{v_0/\omega_n}{A}$$

for some ϕ. Rewriting these two expressions and substituting in for x_0 and v_0/ω_n gives

$$x(t) = A\cos\omega_n t \cos\phi + A\sin\omega_n t \sin\phi.$$

Recalling the trig identity

$$\cos(a \pm b) = \cos a \cos b \mp \sin a \sin b$$

gives us that

$$x(t) = A\cos(\omega_n t - \phi)$$
$$= \sqrt{\left(\frac{v_0}{\omega_n}\right)^2 + x_0^2} \cos(\omega_n t - \phi). \tag{3.49}$$

Note that we could have obtained a similar expression

$$x(t) = A\sin(\omega_n t + \phi) \tag{3.50}$$

if we had set

$$\sin\phi = \frac{x_0}{A} \text{ and } \cos\phi = \frac{v_0/\omega_n}{A},$$

and used the trig identity $\sin(a \pm b) = \sin a \cos b \pm \cos a \sin b$. We thus have two identities that can be quite useful:

$$c_1 \sin(\omega t) + c_2 \cos(\omega t) = \begin{cases} A\cos(\omega t - \phi), & A = \sqrt{c_1^2 + c_2^2}, \ \phi = \tan^{-1}\left(\dfrac{c_1}{c_2}\right) \\ \text{or} \\ A\sin(\omega t + \phi), & A = \sqrt{c_1^2 + c_2^2}, \ \phi = \tan^{-1}\left(\dfrac{c_2}{c_1}\right). \end{cases}$$

Back to (3.49), this equation gives the displacement, x, of the mass from the equilibrium as a function of time t. The motion described by $x(t)$ is called **simple harmonic motion**. The constant A given by Equation (3.48) is the amplitude of the motion and gives the maximum (positive) displacement. The motion is periodic, with the mass oscillating between $x = -A$ and $x = A$. We have $x = A$ if and only if

$$\sqrt{\frac{k}{m}}t - \phi = \pm 2n\pi.$$

So the maximum displacement occurs if and only if

$$t = \sqrt{\frac{m}{k}}(2n\pi + \phi) > 0.$$

The time interval between maxima is called the *period*, T. Thus

$$T = \frac{2\pi}{\sqrt{\frac{k}{m}}} = \frac{2\pi}{\omega_n}.$$

Just as the spring constant is inherent to each spring, the quantity

$$\omega_n = \sqrt{\frac{k}{m}} \tag{3.51}$$

is inherent to the mass-spring system, hence the name *natural* angular frequency.

Example 2 An 8-lb weight is placed upon the lower end of a coil spring suspended from the ceiling. The weight comes to rest in its equilibrium position, thereby stretching the spring 6 in. The weight is then pulled down 3 in. below its equilibrium position and released at $t = 0$ with an initial velocity of 1 ft/sec, directed downward. Neglecting the resistance of the medium and assuming that no external forces are present, determine the amplitude, period, and frequency of the resulting motion.

Solution
This is an example of free undamped motion. Since the 8-lb weight stretches the spring 6 in. $= 1/2$ ft, Hooke's law gives $8 = k(1/2)$ or

$$k = 16 \text{ lb/ft}.$$

Further, mass $=$ weight/gravity so

$$m = \frac{8}{32} \text{ slugs}.$$

Thus, the differential equation describing free undamped motion becomes

$$\frac{8}{32}\frac{d^2x}{dt^2} + 16x = 0 \implies \frac{d^2x}{dt^2} + 64x = 0.$$

Now since the weight was released downward with an initial velocity of 1 ft/sec, from a position 3 in. or 1/4 ft below equilibrium we have

$$x(0) = \frac{1}{4} \text{ and } x'(0) = 1.$$

The characteristic polynomial is $r^2 + 64 = 0$ so that $r = \pm 8i$. This gives

$$x(t) = c_1 \sin 8t + c_2 \cos 8t.$$

Using this with the initial conditions gives $c_1 = \frac{1}{8}$ and $c_2 = \frac{1}{4}$. Hence,

$$x(t) = \frac{1}{8} \sin 8t + \frac{1}{4} \cos 8t.$$

A graph of this displacement is shown in Figure 3.7.

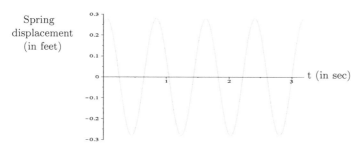

FIGURE 3.7:   Displacement of the spring in Example 2.

To obtain amplitude-phase form we have

$$A = \sqrt{\left(\frac{1}{8}\right)^2 + \left(\frac{1}{4}\right)^2} = \frac{\sqrt{5}}{8}$$

as the amplitude. Thus

$$x(t) = \frac{\sqrt{5}}{8} \cos(8t + \phi),$$

which is of the form $A \cos(\omega t + \phi)$. The period, often denoted T, is $\frac{2\pi}{8} = \frac{\pi}{4}$ sec and the frequency is $4/\pi$. We can find ϕ by solving

$$\cos \phi = \frac{x_0}{A} = \frac{2\sqrt{5}}{5} \text{ and } \sin \phi = \frac{v_0}{\omega_n A} = \frac{\sqrt{5}}{5}$$

for ϕ. From these equations we find

$$\phi \approx 0.46 \text{ rad.}$$

Damped Oscillations

When we allow for friction in our system, some very useful and important concepts arise. In the case of this damped motion, we again consider

$$m\frac{d^2x}{dt^2} + b\frac{dx}{dt} + kx = 0, \tag{3.52}$$

now with $b \neq 0$ but still with $F(t) = 0$. We again have a homogeneous constant coefficient equation and thus we can solve it. Calculating the roots of the characteristic equation gives

$$r = \frac{-b \pm \sqrt{b^2 - 4mk}}{2m} = \frac{-b}{2m} \pm \sqrt{\left(\frac{b}{2m}\right)^2 - \frac{k}{m}}. \tag{3.53}$$

We will have three different cases depending on the sign of the quantity under the square root. The case of critical damping is when $\frac{b}{2m} - \sqrt{\frac{k}{m}} = 0$, which can equivalently be written as

$$\frac{b/(2m)}{\sqrt{k/m}} = \frac{b/(2m)}{\omega_n} = 1.$$

These roots and quantities motivate us to consider a slightly different form of (3.52) where we first divide by m as we did in the undamped case and introduce the **damping ratio** $\zeta = \dfrac{b/(2m)}{\omega_n}$, noting that $\zeta > 0$ and critical damping occurs when $\zeta = 1$. Equation (3.52) is equivalently written as

$$\frac{d^2x}{dt^2} + 2\zeta\omega_n\frac{dx}{dt} + \omega_n^2 x = 0, \tag{3.54}$$

with roots of the characteristic equation

$$r = \omega_n \left(-\zeta \pm \sqrt{\zeta^2 - 1}\right). \tag{3.55}$$

Recalling the three different cases that arose in our study of constant coefficient equations, we see that each case can be realized here depending on the value of $b^2 - 4mk$ or, equivalently, $\zeta^2 - 1$.

Case 1: $b^2 > 4mk$, i.e., $\zeta^2 > 1$

Then the roots are real and distinct. If we denote the roots r_1 and r_2, the general solution can be written as

$$x(t) = c_1 e^{r_1 t} + c_2 e^{r_2 t}. \tag{3.56}$$

Note that $r_1 < 0$ *and* $r_2 < 0$ (check this!). Thus the solution will approach 0 as $t \to \infty$. Depending on the initial condition,[2] it may cross the rest position at most one time but the spring motion will not oscillate as it dies out. This case is called **overdamped** motion; see Figure 3.8.

---

[2]This condition is given by $|v_0| > |x_0 r_2|$, where v_0 (the initial velocity of the mass) and x_0 (the initial displacement) have opposite sign, and $r_2 = \omega_n(-\zeta - \sqrt{\zeta^2 - 1})$ is the smaller root of the characteristic equation; see Problem 22.

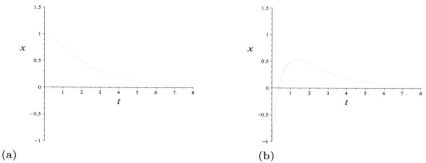

(a)   (b)

FIGURE 3.8: Overdamped motion: $x'' + (5/2)x' + x = 0$. The initial conditions are again $x(0) = 1$, $x'(0) = 0$ in (**a**) and $x(0) = -1$, $x'(0) = 4$ in (**b**). Note that we have overshoot in (**b**) as the initial condition was chosen such that the mass passes through its rest position exactly one time.

Case 2: $b^2 = 4mk$, i.e., $\zeta^2 = 1$

Then the roots are real and equal. If we denote the root r, the general solution can be written as

$$x(t) = c_1 e^{rt} + c_2 t e^{rt}. \tag{3.57}$$

Note that $r < 0$ and the solution again approaches 0 as $t \to \infty$. This last statement is obvious for the first term $c_1 e^{rt}$. The reader can check (for instance, using L'Hospital's rule) that the second term also approaches zero. Depending on the initial condition, the mass may cross the rest position at most one time before dying out without oscillation.[3] This case is called **critically damped** motion; see Figure 3.9. It also represents the motion that will die off to zero fastest.

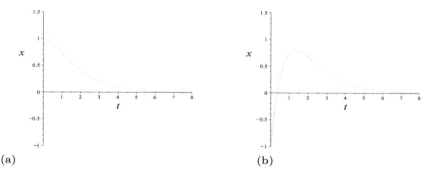

(a)   (b)

FIGURE 3.9: Critically damped motion: $x'' + 2x' + x = 0$. The initial conditions are $x(0) = 1$, $x'(0) = 0$ in (**a**) and are $x(0) = -1$, $x'(0) = 4$ in (**b**). Note that we again have overshoot in (**b**). We also note that the motion dies off faster than in the overdamped case, even though both systems show *no oscillations*.

---

[3]This condition is given by $|v_0| > |bx_0/(2m)|$, where v_0 (the initial velocity of the mass) and x_0 (the initial displacement) have opposite sign; see Problem 22.

Case 3: $b^2 < 4mk$, i.e., $\zeta^2 < 1$

Then the roots are complex conjugates. If we denote the roots as $-\zeta\omega_n \pm i\omega_d$ (that is, we set $\omega_d = \omega_n\sqrt{1-\zeta^2}$ and note that $0 < \zeta < 1$ and $\omega_d \in \mathbb{R}$, where ω_d is the **damped angular frequency** of the system), then the general solution can be written as

$$x(t) = c_1 e^{-\zeta\omega_n t} \sin \omega_d t + c_2 e^{-\zeta\omega_n t} \cos \omega_d t. \tag{3.58}$$

We again state that $\zeta\omega_n > 0$ (and thus $-\zeta\omega_n < 0$) and the solution again approaches 0 as $t \to \infty$. In this case, the spring will oscillate as its amplitude dies off to zero. This case is called **underdamped** motion; see Figure 3.10.

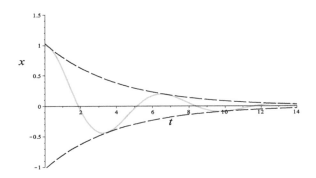

FIGURE 3.10: Underdamped motion: $x'' + (1/2)x' + x = 0$. The initial conditions are $x(0) = 1$, $x'(0) = 0$. The envelope $\pm e^{-t/4}$ is also drawn to show the decaying function that bounds the oscillations.

We can use the earlier trigonometric tricks to rewrite this general solution as[4]

$$x(t) = Ae^{-\zeta\omega_n t} \cos(\omega_d t - \phi), \tag{3.59}$$

where

$$A = \sqrt{c_1^2 + c_2^2}, \quad \cos\phi = \frac{c_2}{A}, \quad \sin\phi = \frac{c_1}{A}. \tag{3.60}$$

The angular frequency of oscillation is now given by

$$\omega_d = \omega_n\sqrt{1-\zeta^2} \tag{3.61}$$

and the amplitude of the oscillations is bounded by the exponentially decreasing functions $\pm Ae^{-\zeta\omega_n t}$, which form an envelope that governs this decay.

---

[4]We again note that if we set $A = \sqrt{c_1^2 + c_2^2}$, $\cos\phi = \frac{c_1}{A}$, $\sin\phi = \frac{c_2}{A}$ we would obtain $x(t) = Ae^{-\zeta\omega_n t} \sin(\omega_d t + \phi)$.

The differential equation of motion of a damped spring is given by

$$3x''(t) + 5x'(t) + x(t) = 0.$$

Classify its motion as overdamped, critically damped, or underdamped.

To classify the motion, we simply need to know whether the roots of the characteristic equation are real and distinct, real and repeated, or complex. In other words, we need to know the sign of $b^2 - 4mk$. In this example, we have

$$b^2 - 4mk = 5^2 - 4 \cdot 3 \cdot 1 = 13.$$

Thus the motion is overdamped. We could have rewritten the equation as in (3.54):

$$x'' + 2\left(\frac{5\sqrt{3}}{6}\right)\left(\frac{1}{\sqrt{3}}\right)x'(t) + \frac{1}{3}x(t) = 0$$

and obtained the same results: $\zeta^2 - 1 = \left(\frac{5\sqrt{3}}{6}\right)^2 - 1 > 0$.

We stress that the case in which we end up always depends on the relationship between the mass, the spring constant, and the damping coefficient. In the case of no damping, we always end up in case 3. We also note that in each of the three cases, the motion of the spring dies out as $t \to \infty$. This motion is called the **transient** motion. We have not yet considered how to solve this problem when $F(t) \neq 0$, and will learn how to solve such problems in Section 4.1. The presence of a forcing function $F(t) \neq 0$ leads to an interference in the motion of the spring and will change the long-term or **steady-state** behavior of the spring. In the event that the frequency of the forcing is close to the natural frequency of the spring, a phenomenon called **resonance** occurs. In this situation the amplitude of the motion is amplified, sometimes dramatically.

In a second semester physics course, one of the topics students study is the flow of electrons. Two common things to consider are the amount of charge of these electrons or the change in the charge of the electrons. In considering the amount of charge, $Q(t)$, we typically measure this in coulombs. For the change in the charge over time, that is $I(t) = \frac{dQ}{dt}$, we call this the current and it is measured in amperes. The quantity that we choose to measure often depends on the components that we have in the circuit. A more common quantity used instead of charge is voltage, which is defined as the potential difference in charge between two points. To write our differential equation, we state the following conservation of energy law:

Kirchhoff's law: the voltage changes through a closed path in a circuit must sum to zero.

Circuits will typically have up to 4 different types of components:

- a supplied voltage of $\mathcal{E}$ volts (denoted $\mathcal{E}$ for electromotive force or emf)

- a resistor of resistance R ohms, where the voltage across it is RI

- a capacitor of capacity C farads, where the voltage across it is $\frac{Q}{C}$

- an inductance of L henrys, where the voltage across it is $L\frac{dI}{dt}$,

and we consider $R, C, L \geq 0$. If all the components are connected in *series*, the resulting differential equation is then

$$L\frac{d^2Q}{dt^2} + R\frac{dQ}{dt} + \frac{1}{C}Q = \mathcal{E}. \tag{3.62}$$

We can equivalently rewrite this in terms of I by taking the derivative of each term:

$$L\frac{d^2I}{dt^2} + R\frac{dI}{dt} + \frac{1}{C}I = \mathcal{E}'. \tag{3.63}$$

Perhaps a bit different from the spring analogy is the possibility of removing L from the circuit and still having a situation that make sense.

Table 3.2: Comparison of Mass-Spring System & Electric Circuit

Mass-Spring System	Electric Circuit
mass	inductance
friction	resistor
spring	capacitor
location of mass	current in circuit

We thus consider the electric circuit in Figure 3.11, which we will consider in an analogous fashion to the mass on the spring; see Table 3.2. The direction of the current, I, traditionally denotes the flow of positive charge and thus flows in the direction opposite that of the moving electrons. The first case that is of interest to us is the one in which the switch is open for a while and then moves to position 2 at time $t = 0$. We can solve this with the methods of Section 3.6. Another situation that we could consider is if the switch is in position 1 for a while so that the supplied voltage (given as $\mathcal{E}$) will supply a forcing current that depends on time. At time $t = 0$, we move the switch to position 2. We will thus have some initial current in the circuit apart from the other components. A final situation is when the switch is open for a while and then moves to position 1 at time $t = 0$. This is a forced differential equation that we will learn to solve in the next chapter. When we consider

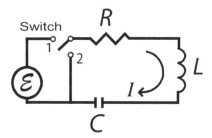

FIGURE 3.11: Basic RLC circuit with differential equations (3.62) or, equivalently, (3.63). The components are connected in series.

the homogeneous situation where there is no applied voltage, we can easily see that the characteristic equation of this circuit is $Lr^2 + Rr + \frac{1}{C}$, with roots

$$r = \frac{-R}{2L} \pm \frac{\sqrt{R^2 - 4\frac{L}{C}}}{2L} = \frac{-R}{2L} \pm \sqrt{\left(\frac{R}{2L}\right)^2 - \omega_0^2},$$

where $\omega_0 = \frac{1}{\sqrt{LC}}$ is called the **resonant frequency** of the circuit. These roots can be real and distinct, real and repeated, or complex. For the case of real and repeated roots, we have

$$R^2 - 4\frac{L}{C} = 0,$$

which can be rewritten as $\frac{R}{2}\sqrt{\frac{C}{L}} = 1$. This motivates us to define a **damping ratio**, analogous to that of the mass-spring system, as

$$\zeta_e = \frac{R}{2}\sqrt{\frac{C}{L}} = \frac{R/(2L)}{\omega_0}, \tag{3.64}$$

where we additionally use a subscript e to denote our application to electric circuits. If we go back to our original equation (3.63), we can rewrite it using ζ_e and ω_0 by first dividing by L and then manipulating to obtain

$$\frac{d^2 I}{dt^2} + 2\zeta_e \omega_0 \frac{dI}{dt} + \omega_0^2 I = 0. \tag{3.65}$$

The roots of the characteristic equation are then easily found as

$$r = \omega_0 \left(-\zeta_e \pm \sqrt{\zeta_e^2 - 1}\right). \tag{3.66}$$

Completely analogous to the mass-spring system, we will again have three different types of current to consider

- overdamped: $\zeta > 1$, i.e., $R^2 > 4\frac{L}{C}$

- critically damped: $\zeta = 1$, i.e., $R^2 = 4\frac{L}{C}$

- underdamped: $\zeta < 1$, i.e., $R^2 < 4\frac{L}{C}$

Thus, when there is no external voltage supplied (that is, $\mathcal{E}' = 0$), the presence of the resistor will cause the current to eventually go to zero.

Example 4 Consider the RLC circuit of Figure 4.3. If the values of the inductor, capacitor, and resistor are .5 henrys, .5 farads, and .2 ohms, respectively, determine whether the transient behavior of the circuit is over-, critically, or underdamped.

Solution
We want the current, which means that (3.63) would be the easier equation to use. Substituting the given values of L, C, R into the characteristic equation show that $R^2 - \frac{4L}{C} = -1 < 0$ so that we are in the underdamped case.

Problems

1. A 12-lb weight is placed upon the lower end of a suspended coil spring. The weight comes to rest in its equilibrium position, stretching the spring 1.5 in. The weight is then pulled down 2 in. below its equilibrium position and released from rest at $t = 0$. Find the displacement of the weight as a function of time; determine the amplitude, period, and the frequency of the resulting motion. Assume there is no resistance of the medium and no external force.

2. A 16-lb weight is placed upon the lower end of a suspended coil spring. The weight comes to rest in its equilibrium position, stretching the spring 6 in. The weight is then pulled down 4 in. below its equilibrium position and released at $t = 0$ with an initial velocity downward of 2 ft/sec. Find the displacement of the weight as a function of time; determine the amplitude, period, and the frequency of the resulting motion. Assume there is no resistance of the medium and no external forces.

3. A 6-lb weight is placed upon the lower end of a suspended coil spring. The weight comes to rest in its equilibrium position, stretching the spring 6 in. At $t = 0$, the weight is set into motion with an initial velocity downward of 2 ft/sec. Find the displacement of the weight as a function of time; determine the amplitude, period, and the frequency of the resulting motion. Assume there is no resistance of the medium and no external force.

4. A 10-lb weight is placed upon the lower end of a suspended coil spring. The weight comes to rest in its equilibrium position, stretching the spring

5 in. At $t = 0$, the weight is set into motion with an initial velocity downward of 3 ft/sec. Find the displacement of the weight as a function of time; determine the amplitude, period, and the frequency of the resulting motion. Assume there is no resistance of the medium and no external force.

In Problems **5–12**, we consider some mass-spring analogs.

5. Two pulleys are attached to a shaft; they have moments of inertia I_1 and I_2. To twist one of the pulleys an angle ϕ with respect to the other requires an elastic shaft-deforming torque of $K\phi$. Find the frequency of the torsional oscillations of the shaft.

6. A weight of mass m is attached to one end of an elastic rod. The other end moves so that its position at time t has coordinate $B \sin \omega t$. The elastic force exerted by the rod is proportional to the difference in displacements of its ends. Neglecting the mass of the rod, and friction, find the amplitude A of the forced vibrations of the mass. Can the relation $A > B$ hold?

7. An electric circuit consists of a voltage source that supplies constant voltage V volts, a resistor of resistance R ohms, and an inductance of L henrys, together with a switch that is closed at time $t = 0$. Find the current as a function of the time. Assume $I(0) = 0$, $I'(0)\frac{V}{L}$.

8. Solve the preceding problem, replacing the inductance L by a capacitor of capacity C farads. The capacitor is uncharged when the switch is closed; assume $I(0) = \frac{V}{R}$.

9. A resistor of resistance R ohms is connected to a capacitor of capacity C farads that has a charge Q_0 coulombs at time $t = 0$. The circuit is closed at $t = 0$. Find the charge as a function of time for $t > 0$.

10. An inductor, resistor, and capacitor are connected in series. At time $t = 0$, the circuit is closed, the capacitor having a charge of Q_0 coulombs at that time and $Q'(0) = 0$. Find the current as a function of time and determine when there will be oscillations.

11. A voltage source supplies a constant voltage in a circuit consisting of the voltage source, and a resistor and capacitor in series. Find the steady-state charge in the circuit.

12. A voltage source supplies a constant voltage and is connected to a resistor, inductor, and capacitor in series. Find the steady-state charge in the circuit.

Problems **13–21** discuss topics from the "Damped Oscillations" subsection. For each problem, classify the motion of the mass on a spring as either underdamped, critically damped, or overdamped. Assume $x = x(t)$, that is, x is a function of t.

13. $x'' + x' + x = 0$ **14.** $4x'' + x' + 3x = 0$ **15.** $x'' + 4x' + 3x = 0$
16. $x'' + 5x' + x = 0$ **17.** $x'' + 2x' + x = 0$ **18.** $x'' + 6x' + 9x = 0$
19. $3x'' + 6x' + x = 0$ **20.** $2x'' + 3x' + 9x = 0$ **21.** $9x'' - 12x' + 4x = 0$

22. For both overdamped and critically damped motion, it is still possible for the mass to pass through its rest position at most *one* time before coming to rest. Use the general solution of the damped mass-spring system to show that the respective conditions are given by $|v_0| > |x_0 r_2|$ and $|v_0| > |bx_0/(2m)|$, where v_0 (the initial velocity of the mass) and x_0 (the initial displacement) have opposite sign, and $r_2 = (-b - \sqrt{b^2 - 4mk})/2m$ is the smaller root of the characteristic equation.

23. In a first physics course, the motion of a damped spring is often given as

$$x(t) \approx c_1 e^{-bt/2m} \cos\left(\sqrt{\frac{k}{m}}\, t\right) + c_2 e^{-bt/2m} \sin\left(\sqrt{\frac{k}{m}}\, t\right), \qquad (3.67)$$

with the qualification that this approximation is valid if the damping constant b is small. Verify this approximation by using a Taylor series expansion about $b = 0$.

In Problems **24–29**, consider the roots from mass on a spring equation, (3.52). Two of the constants m, b, k are given. Determine the range on the remaining one that will result in the described motion.

24. $m = 1, b = 3$, overdamped   **25.** $m = 1, b = 2$, underdamped
26. $m = 2, k = 2$, critically damped   **27.** $m = 1, k = 3$, critically damped
28. $b = 3, k = 2$, underdamped   **29.** $b = 2, k = 2$, overdamped

In Problems **30–34** involve numerical explorations with the computer. They concern the generalized equation of motion for a mass on a spring given by

$$m(t)x''(t) + b(t)x'(t) + k(t)x(t) = F(t).$$

Here, we allow the possibility of changing mass, coefficient of friction, and spring constant, as well as a forcing function. Thus, we will not be able to use $b^2 - 4mk$ to determine the motion of the spring and we will instead use the computer to find the numerical solution.

30. Consider a container attached to a spring. Suppose the container is full of water that is evaporating in the hot Los Angeles sun. Suppose the mass function is given by $m(t) = 2e^{-t/10} + 1$. Let $b(t) = .2$, $k(t) = 1$, and $F(t) = 0$. Plot numerical solutions from $t = 0$ to $t = 50$ from the initial conditions $x(0) = 1$, $x'(0) = 0$.

31. Consider a block of constant mass $m = 2$ sliding back and forth on a sheet of ice. As the block slides, the ice melts and eventually exposes a rougher surface beneath. Assume the damping coefficient is given by $b(t) = \arctan(t - 20) + \pi/2$. Let $k(t) = 4$, $F(t) = 0$. Plot the motion of the spring over a large enough range for t so that you are able to see the results. Assume $x(0) = 1$, $x'(0) = 0$.

32. Springs typically lose some of their stiffness over time. Suppose a spring coefficient is given by $k(t) = 5e^{-t/25}$. Suppose also that $b(t) = 1$, $m(t) = 12$, $F(t) = 0$. Plot the motion of the spring over a large enough range for t so that you are able to see the results. Assume $x(0) = 1$, $x'(0) = 0$.

33. Consider the equation $4x'' + 0.1x' + 4x = \sin(\omega t)$. (i) Let $\omega = 10$. What happens to the motion for large t? Does it decay to zero? (ii) Now let $\omega = 1$. Again, what happens to the motion for large t? (iii) Numerically explore the behavior of the solution for various ω-values. Is there an ω-value that gives the largest oscillations?

34. Consider the equation $4x'' + x' + 4x = \sin(\omega t)$. (i) Let $\omega = 10$. What happens to the motion for large t? Does it decay to zero? (ii) Now let $\omega = 1$. Again, what happens to the motion for large t? (iii) Numerically explore the behavior of the solution for various ω-values. Is there an ω-value that gives the largest oscillations?

Chapter 3 Review

In Problems **1–6**, determine whether the statement is true or false. If it is true, give reasons for your answer. If it is false, give a counterexample or other explanation of why it is false.

1. The Existence and Uniqueness theorem allows us to state that the solution to $xy'' - y = 0$ passing through the initial condition $y(0) = 0, y'(0) = 1$ will not be unique.

2. The Existence and Uniqueness theorem allows us to state that the solution to $y'' - y = \cosh x$ passing through the initial condition $y(0) = 0, y'(0) = 1$ will not be unique.

3. Reducing the order of a differential equation requires the knowledge of a solution.

4. A mass on a spring system can be written as $mx''(t) + bx'(t) + kx(t)$ for constants $m, b, k \in (-\infty, \infty)$.

5. All differential operators commute; that is, if $P(D)Q(D) = Q(D)P(D)$.

6. Only nth-order *linear* differential equations can be rewritten as a system of first-order equations.

In Problems **7–9**, determine whether the differential equation is guaranteed to have a unique solution passing through the given initial condition. Then try to find constants to make the proposed function y a solution to the IVP.

7. $y'' - 4y = 0$, $y(0) = 4, y'(0) = 0$; $y = c_1 e^{2x} + c_2 e^{-2x}$

8. $y'' - 2y' + y = \dfrac{e^x}{x}$, $y(1) = 0, y'(1) = 1$; $y = c_1 e^x + c_2 x e^x + e^x x (\ln x - 1)$

(with $x > 0$)

9. $x^2 y'' - 2y = 0$, $y(0) = 1, y'(0) = 1$; $y = c_1 x^2 + c_2 \frac{1}{x}$

In Problems **10–13**, find all values of (x, y) where Theorem 3.1.1 guarantees the existence of a unique solution.

10. $y'' + 3xy' + y = \frac{\sin x}{x-1}$ 　　　　 **11.** $3y'' + 5x^2 y' + y = \frac{4}{e^x}$

12. $y'' + 7x^2 y = \sin x$ 　　　　　　　 **13.** $y'' - \ln|x|y' + 2y = \frac{\cos x}{x^2+1}$

In Problems **14–22**, determine whether the following sets of functions are linearly independent for all x (where the functions are defined).

14. $\{3x - 2, x + 1\}$ 　　　 **15.** $\{x + 1, x^2 - 1\}$ 　　　 **16.** $\{e^x, x + 1\}$

17. $\{x^2, x^3\}$ 　　　　　 **18.** $\{\sin x, \tan x\}$ 　　　 **19.** $\{\ln x, x - 1\}$

20. $\{\sqrt{x} - 4x, x, 2x\}$ 　 **21.** $\{x^2 - 4x, 3x, x^2\}$ 　 **22.** $\{2, \cos^2 x, \sin^2 x\}$

In Problems **23–29**, determine whether the given set forms a fundamental set of solutions to the differential equation on $(-\infty, \infty)$. Clearly state your reasons for your answers.

23. $\{e^{-3x}, e^{3x}\}$, $y'' - 9y = 0$ 　　　 **24.** $\{e^x, e^{-2x}\}$, $y'' - y' - 2y = 0$

25. $\{e^{-x}, e^{5x}\}$, $(D^2 - 6D + 5)(y) = 0$ **26.** $\{e^x, xe^x\}$, $(D + 1)^2(y) = 0$

27. $\{e^{3x} \sin x, e^{3x} \cos x\}$, $y'' - 6y' + 10y = 0$

28. $\{e^{2x} \sin 2x, e^{2x} \cos 2x\}$, $(D^2 - 4D + 8)(y) = 0$

29. $\{e^{2x} \sin x, e^{2x} \cos x, 3\}$, $y''' - 6y'' + 10y' = 0$

In Problems **30–34**, use the given solution, y_1, to reduce the order of the differential equation. Find a second solution.

30. $y'' + 4y' + 4y = 0$, $y_1 = e^{-2x}$ 　　 **31.** $y'' - 4y' - 5y = 0$, $y_1 = e^{-x}$

32. $y'' + 2y' - 35y = 0$, $y_1 = e^{5x}$ 　　 **33.** $y'' + 10y' + 25y = 0$, $y_1 = e^{-2x}$

34. $y'' + 10y' + 9y = 0$, $y_1 = e^{-9x}$

In Problems **35–40**, rewrite the following differential equations using operator notation and factor completely.

35. $y''' - y' - 12y = 0$ 　　　　　　 **36.** $y''' - 5y'' + 4y' = 0$

37. $y''' - 16y' = 0$ 　　　　　　　 **38.** $y''' + 3y'' - 4y' = 0$

39. $y''' - 3y'' + 3y' - y = 0$ 　　　　 **40.** $y''' + 3y'' + 3y' + 1 = 0$

In Problems **41–46**, (i) convert the equations to a system of first-order equations; (ii) use your software package and the given initial conditions at x_0 to estimate the numerical solution at the value $x = x_0 + 5.0$ for the step sizes $h = 0.1$.

41. $y'' - 4y' + 3y = 0$, $y(0) = 0, y'(0) = 1$

42. $2y''' + y' = 0$, $y(0) = 1, y'(0) = 0, y''(0) = 2$

43. $x^2 y''' - 2y = 0$, $y(1) = 0, y'(1) = 0, y''(1) = 1$

44. $y'' + \sin y = 0$, $y(0) = 1, y'(0) = 1$

45. $y'' + (1 - y^2)y' + 2y = \cos x, \ y(-1) = 1, y'(-1) = 1$

46. $y'' + 2y' + \sin y = 0, \ y(0) = 3, y'(0) = 1$

In Problems **47–50**, write the given function as the real part or imaginary part of a complex-valued function.

47. $3 \sin 2x$ **48.** $\cos \omega t$ **49.** $e^{-2x} \cos x$ **50.** $e^{-x} \sin 2x$

In Problems **51–54**, evaluate the polynomial at the given complex number.

51. $p(z) = \dfrac{1}{z^2 + z + 2}; \ i, 1 + i$ **52.** $p(z) = \dfrac{1}{z^2 + 2z + 1}; \ i, -i$

53. $p(z) = \dfrac{e^z}{z^2 + 4}; \ 3i, i$ **54.** $p(z) = \dfrac{e^z}{z^2 + z + 1}; \ 2i, 1 - i$

Write the general solution of the equations given in **55–62**.

55. $y'' - 5y' + 6y = 0$ **56.** $y'' - 3y' - 10y = 0$

57. $D^2(D + 1)(D - 4)(y) = 0$ **58.** $(D^2 + 1)(D - 2)(y) = 0$

59. $D(D + 1)^3(y) = 0$ **60.** $(D - 1)^2(D + 2)(y) = 0$

61. $D(D + 7)(D^2 - 4)(y) = 0$ **62.** $(D^3 - 27)(y) = 0$

Find the solution of the IVPs given in **63–69**.

63. $y'' + 6y' + 9y = 0, \ y(0) = 1, y'(0) = 0$

64. $y'' + 11y' + 10y = 0, \ y(0) = 1, y'(0) = -1$

65. $y'' + 6y' + 8y = 0, \ y(0) = 1, y'(0) = -1$

66. $y'' + 2y' + 2y = 0, \ y(\pi) = 0, y'(\pi) = 1$

67. $y'' + 4y' + 13y = 0 \ y(\pi/2) = 2, y'(\pi/2) = 1$

68. $(D^3 + 4D)(y) = 0, \ y(0) = 0, y'(0) = 1, y''(0) = 0$

69. $(D^3 + D)(y) = 0, \ y(0) = 2, y'(0) = -1, y''(0) = 0$

For Problems **70–81**, classify the motion of the mass on a spring as either underdamped, critically damped, or overdamped, or find the range of the parameter value that will give the desired result. Assume $x = x(t)$, that is, x is a function of t.

70. $x'' + x' = 0$ **71.** $x'' + 4x' + x = 0$

72. $x'' + \frac{3}{2}x' + x = 0$ **73.** $x'' + 3x' + 2x = 0$

74. $2x'' + 8x' + 8x = 0$ **75.** $x'' + 10x' + 25x = 0$

76. $x'' + bx' + 4x = 0$, underdamped

77. $x'' + bx' + 4x = 0$, overdamped

78. $x'' + 3x' + kx = 0$, underdamped

79. $2x'' + x' + kx = 0$, critically damped

80. $mx'' + 7x' + 3x = 0$, overdamped

81. $mx'' + 2x' + 4x = 0$, critically damped

Chapter 3. Computer Labs

Chapter 3 Computer Lab: MATLAB

MATLAB Example 1: Enter the code below to demonstrate, using *Symbolic Math Toolbox*, the following:
(a) the functions $f_1(x) = 1 + 3x$, $f_2(x) = 4 - 2x^2$, $f_3(x) = 5x - x^2$ are linearly independent;
(b) the functions $g_1(x) = 2 + x$, $f_2(x) = 4 - x + 5x^2$, $g_3(x) = 2 - 5x + 10x^2$ are linearly dependent.

```
>> clear all
>> syms x c1 c2 c3 c4
>> f1=1+3*x
>> f2=4-2*x^2
>> f3=5*x-x^2
>> eq1=c1*f1+c2*f2+c3*f3
>> eq2=coeffs(eq1,x)
>> eq3=solve(eq2(1)==0,eq2(2)==0,eq2(3)==0)
>> c1=eq3.c1
>> c2=eq3.c2
>> c3=eq3.c3
>> %Thus the only c-values that will work are c1=c2=c3=0.
>> %It follows that the three functions are lin. independent.
>> clear all
>> syms x c1 c2 c3 c4
>> g1=2+x
>> g2=4-x+5*x^2
>> g3=2-5*x+10*x^2
>> eq4=c1*g1+c2*g2+c3*g3
>> eq5=coeffs(eq4,x)
>> eq6=solve(eq5(1)==0,eq5(2)==0,eq5(3)==0) %warning appears
>> c1=eq6.c1
>> c2=eq6.c2
>> c3=eq6.c3
>> %z is arbitrary.  With z=1, we have 3*g1-2*g2+1*g3=0.
>> %Thus the three functions are linearly dependent.
```

MATLAB Example 2: Enter the following code that demonstrates how to calculate the Wronskian, $W(x)$, of the functions $f_1(x) = e^x$, $f_2(x) = e^{-x}$, $f_3(x) = e^{2x}$. For linear independence, we need $W(x) \neq 0$ for at least one x-value. It uses the *Symbolic Math Toolbox*.

```
>> clear all
>> syms x
>> f1=exp(x)
>> f2=exp(-x)
>> f3=exp(2*x)
>> diff([f1 f2 f3],x) %2nd row of Wronskian
>> A=[f1,f2,f3; diff([f1,f2,f3],x); diff([f1,f2,f3],x,2)]
>> Wronsk=det(A)
>> subs(Wronsk,x,0) %This calculates W(0).
```

MATLAB Example 3: Enter the following code that demonstrates how to perform calculations with complex numbers using the *Symbolic Math Toolbox*.

```
>> clear all
>> syms r1 r2 z x A1 A2
>> r1=2+3*i
>> real(r1)
>> imag(r1)
>> conj(r1)
>> r2=1/(4+5*i)
>> exp(3+4*i)
>> exp(pi*i)
>> z=1-3*i
>> R=abs(z)
>> theta=angle(z)
>> R.*exp(i*theta)
>> f(x)=x^4
>> solve(f(x)-pi,x)
>> p(x)=x^2+6*x+25
>> solve(p(x),x)
>> subs(p(x),x,i)
>> subs(p(x),x,2+i)
>> [solA1, solA2]=solve((2+i)*A1==3, A1+3*A2*i==0)
```

MATLAB Example 4: Enter the following code that uses RK4 (previously created, or use ode45) to numerically solve the following ODEs and then plots their solutions:
(a) $2xy'' + x^2y' + 3x^3y = 0$, with $y(1) = 2$, $y'(1) = 0$;

(b) $y^{(4)} + x^2 y' + y = \cos(x)$, with $y(0) = 1$, $y'(0) = 0$, $y''(0) = 0$, $y'''(0) = 1$. Give the numerical value of the solution at $x = 5$. You will need to convert each ODE to a system of first-order equations.

```
%Create the function Ch3NumExample1.m:
function dy= Ch3NumExample1(x,y)
%
%The original ode is 2*x*y"+x^2*y'+3*x^3*y=0
%The system of first-order equations is
%u1'=u2
%u2'=(-x/2)*u2-((3*x^2)/2)*u1
%
%We let y(1)=u1, y(2)=u2
%
dy=zeros(2,1); %dy is a column vector!
dy(1) = y(2);
dy(2) = (-1/2)*x*y(2)-(3/2)*x^2*y(1);
%end of function Ch3NumExample1.m
```

```
%Create the function Ch3NumExample2.m:
function dy= Ch3NumExample2(x,y)
%
%The original ode is y^(4)+x^2*y'+y=cos(x)
%The system of first-order equations is
%u1'=u2
%u2'=u3
%u3'=u4
%u4'=-x^2*u2-u1+cos(x)
%
%We let y(1)=u1, y(2)=u2, y(3)=u3, y(4)=u4
%
dy=zeros(4,1); %dy is a column vector!
dy(1) = y(2);
dy(2) = y(3);
dy(3) = y(4);
dy(4) = -x^2*y(2)-y(1)+cos(x);
%end of function Ch3NumExample2.m
```

After having created and saved the above two functions, type the following in the Command Window.

```
>> clear all
>> x0=1; xf=5;
>> y0=[2,0];
```

```
>>  [x1,y1]=RK4(@Ch3NumExample1,[x0,xf],y0,.05);
>>  subplot(2,1,1),plot(x1,y1(:,1))
>>  xlabel('x'); ylabel('y')
>>  subplot(2,1,2),plot(x1,y1(:,2))
>>  xlabel('x'); ylabel('y')
>>  [x1(end), y1(end,:)]  %This shows the last entry of
    vector x1 and matrix y1.  Note that we set xf=5.
>>  figure %calls a new blank figure window
>>  x0=1; xf=5;
>>  y0=[1,0,0,1];
>>  [x2,y2]=RK4(@Ch3NumExample2,[x0,xf],y0,.05);
>>  plot(x2,y2)
>>  legend('y(1)','y(2)','y(3)','y(4)')
>>  xlabel('x'); ylabel('y-values')
>>  [x2(end), y2(end,:)]
>>  figure
>>  [x2,y2]=ode45(@Ch3NumExample2,[x0,xf],y0); %Or use ode45.
>>  plot(x2,y2)
>>  legend('y(1)','y(2)','y(3)','y(4)')
>>  xlabel('x'); ylabel('y-values')
>>  [x2(end), y2(end,:)]
```

MATLAB Example 5: Enter the code below to demonstrate, using *Symbolic Math Toolbox*, how to find the roots of the characteristic equation of the given homogeneous ODEs.
(i) $y'' + 3y' - 4y = 0$
(ii) $y^{(4)} - 4y''' + 6y'' - 4y' - 15y = 0$
(iii) $y^{(4)} + y''' - 2y' - y = 0$
(iv) $y''' + y' - 3y = 0$
(v) $y^{(5)} + 3y'' - y = 0$
You will need to first write the characteristic polynomial in order to find the roots, which are not necessarily real-valued.

```
>>  clear all
>>  %Next three lines do NOT require the Symbolic Math Toolbox
>>  p=[1 0 3 0 1] %these are the coefficients in (v)
>>  %if a coefficient is 0, you must put 0 in its position
>>  roots(p)
>>  %Now we use the Symbolic Math Toolbox:
>>  syms x r
>>  f(r)=r^2+3*r-4
>>  solve(f(r),r)
>>  g(x)=r^4-4*r^3+6*r^2-4*r-15
```

```
>> solve(g(r),r)
>> h(r)=r^4+r^3-2*r-1
>> solve(h(r),r)
>> F(r)=r^3+r-3
>> Fsoln=solve(F(r),r)
>> Fsoln(1) % This is the first root
>> double(Fsoln(1))
>> double(Fsoln)
>> G(r)=r^5+3*r^2-1
>> Gsoln=solve(G(r),r)
```

MATLAB Exercises

Turn in both the commands that you enter for the exercises below as well as the output/figures. These should all be in one document. Please highlight or clearly indicate all requested answers. Some of the questions will require you to modify the above MATLAB code to answer them.

1. Enter the commands given in MATLAB Example 1 and submit both your input and output.
2. Enter the commands given in MATLAB Example 2 and submit both your input and output.
3. Enter the commands given in MATLAB Example 3 and submit both your input and output.
4. Enter the commands given in MATLAB Example 4 and submit both your input and output.
5. Enter the commands given in MATLAB Example 5 and submit both your input and output.
6. Determine whether the following functions are linearly independent:
   (a) $f_1(x) = 2 + 5x - 3x^2$, $f_2(x) = 2 - x + x^2$, $f_3(x) = -2 - 23x + 15x^2$;
   (b) $f_1(x) = 1 + 4x + 2x^2$, $f_2(x) = 3 - x + 2x^2$, $f_3(x) = 5 - 2x - 2x^2$.
7. Determine whether the following sets of functions are linearly independent:
   (a) $\{1+2x-3x^2+x^3, 2-x+x^2+3x^3, -4+x-x^2+2x^3, -6x+8x^2+9x^3\}$;
   (b) $\{3+4x-3x^2+x^3, x-2x^2+4x^3, -4+x^2-x^3, -4-5x-x^2+3x^3\}$.
8. Use the Wronskian to determine whether the following sets of functions are linearly independent.
   (a) $\{\cos x, 3\cos x + \sin 2x, \sin x\}$ (b) $\{\cos x, 3\cos x + \sin x, 4\sin x\}$.
9. Use the Wronskian to determine whether the following sets of functions are linearly independent.
   (a) $\{e^x, xe^x, x^2\}$ (b) $\{\sin(4x) + \cos(4x), \cos(4x) - \sin(4x)\}$.
10. Any number can be written in polar form in the complex plane, $re^{i\theta}$. Find the r and θ values corresponding to the polar form of the following

complex numbers.

(a) $2 - 3i$ (b) $2i$ (c) $-1 - 5i$ (d) $3 + i$

11. Any number can be written in polar form in the complex plane, $re^{i\theta}$. Find the r and θ values corresponding to the polar form of the following complex numbers.

(a) $2 + 3i$ (b) $-3i$ (c) $-2 - i$ (d) $3 + 7i$

12. Consider the polynomial $p(z) = z^2 + 4z + 8$. Evaluate it at the following values:

(a) $z = 2i$ (b) $z = 3 - 2i$ (c) $z = -1 + 4i$ (d) $z = -2 + 2i$

13. Consider the polynomial $p(z) = z^3 - 1$. Evaluate it at the following values:

(a) $z = 2i$ (b) $z = 3 - 2i$ (c) $z = -1 + 4i$ (d) $z = 1$

14. Solve the following systems of equations, where A_1, A_2 may be complex-valued numbers.

(a) $\begin{cases} (2+i)A_1 + A2 = 3 \\ A_1 + 3A_2 i = 0 \end{cases}$ (b) $\begin{cases} 2A_1 + A2 = 3 \\ A_1 - A_2 i = -i \end{cases}$

15. Solve the following systems of equations, where A_1, A_2 may be complex-valued numbers.

(a) $\begin{cases} A_1 - A2 = 1 \\ A_1 - A_2 i = 0 \end{cases}$ (b) $\begin{cases} A_1 i - A2 = 1 \\ A_1 - A_2 i = 5 \end{cases}$

16. For $y'' + y' + y = 0$ with $y(0) = 1, y'(0) = 2$, use a stepsize of $h = 0.1$ to obtain the solution from the 4th order Runge-Kutta method. Plot the solution for $0 \le x \le 20$. Give the numerical value of the solution at $x = 20$.

17. For $y'' + 2y' + 17y = \sin(2x)$ with $y(0) = 0, y'(0) = 0$, use a stepsize of $h = 0.1$ to obtain the solution from the 4th order Runge-Kutta method. Plot the solution for $0 \le x \le 20$. Give the numerical value of the solution at $x = 20$.

18. For $y''' + 4y'' + 21y' + 34y = \cos(3x)$ with $y(0) = 1, y'(0) = 0, y''(0) = .5$, use a stepsize of $h = 0.1$ to obtain the solution from the 4th order Runge-Kutta method. Plot the solution for $0 \le x \le 20$. Give the numerical value of the solution at $x = 20$.

19. For $y''' + x^2 y'' + y = \sin(x)$ with $y(0) = 1, y'(0) = -2, y''(0) = 1$, use a stepsize of $h = 0.1$ to obtain the solution from the 4th order Runge-Kutta method. Plot the solution for $0 \le x \le 20$. Give the numerical value of the solution at $x = 20$.

20. Find the roots of the characteristic equation of the following homogeneous ODEs: (a) $y'' + y' + y = 0$ (b) $y''' + 8y'' + 37y' + 50y = 0$ (c) $y^{(4)} - 4y^{(3)} - 2y'' + 36y' - 63y = 0$

21. Find the roots of the characteristic equation of the following homogeneous ODEs: (a) $y'' + 2y' + 17y$ (b) $y''' - y'' + 11y' - 51y = 0$ (c) $4y^{(4)} - 4y^{(3)} + 13y'' - 4y' - 9y = 0$

22. Find the roots of the characteristic equation of the following homogeneous ODEs: (a) $y''' + 4y'' + 21y' + 34y$ (b) $y^{(5)} + 3y^{(4)} + 9y''' + 27y'' = 0$
(c) $4y^{(5)} + 12y^{(4)} + 25y''' + 12y'' + 21y' = 0$

Chapter 3 Computer Lab: Maple

Maple Example 1: Enter the code below to demonstrate the following:
(a) the functions $f_1(x) = 1 + 3x$, $f_2(x) = 4 - 2x^2$, $f_3(x) = 5x - x^2$ are linearly independent;
(b) the functions $g_1(x) = 2 + x$, $f_2(x) = 4 - x + 5x^2$, $g_3(x) = 2 - 5x + 10x^2$ are linearly dependent.

```
restart
f₁ := x → 1 + 3 · x
f₂ := x → 4 − 2 · x²
f₃ := x → 5 · x − x²
eq1 := c₁ · f₁(x) + c₂ · f₂(x) + c₃ · f₃(x)
eq2 := collect(eq1, [x, x²])
eq3a := coeff(eq2, x, 2)
eq3b := coeff(eq2, x, 1)
eq3c := coeff(eq2, x, 0)
eq4 := solve({eq3a = 0, eq3b = 0, eq3c = 0}, {c₁, c₂, c₃})
# Thus only 0·f1 +0·f2 +0·f3 gives 0.
g₁ := x → 2 + x
g₂ := x → 4 − x + 5 · x²
g₃ := x → 2 − 5 · x + 10 · x²
eq5 := c₁ · g₁(x) + c₂ · g₂(x) + c3 · g₃(x)
eq6a := coeff(eq5, x, 2)
eq6b := coeff(eq5, x, 1)
eq6c := coeff(eq5, x, 0)
eq7 := solve({eq6a = 0, eq6b = 0, eq6c = 0}, {c₁, c₂, c₃})
# c3 is arbitrary. Thus with c3 =1, we have 3·g1 -2·g2 +1·g3=0.
```

Maple Example 2: Enter the following code that demonstrates how to calculate the Wronskian, $W(x)$, of the functions $f_1(x) = e^x$, $f_2(x) = e^{-x}$, $f_3(x) = e^{2x}$. For linear independence, we need $W(x) \neq 0$ for at least one x-value.

restart

with(*LinearAlgebra*)#*Loads the needed linear algebra package*

$f_1 := x \rightarrow e^x$

$f_2 := x \rightarrow e^{-x}$

$f_3 := x \rightarrow e^{2x}$

$A := Matrix\left(\left[[f_1(x), f_2(x), f_3(x)], \dfrac{d}{dx}([f_1(x), f_2(x), f_3(x)]),\right.\right.$

$\left.\left.\dfrac{d^2}{dx^2}([f_1(x), f_2(x), f_3(x)])\right]\right)$

Wronsk := *simplify*(*Determinant*(*A*))

simplify(*subs*(*x* = 0, *Wronsk*))

Maple Example 3: Enter the following code that demonstrates how to perform calculations with complex numbers.

restart

$r_1 := 2 + 3 \cdot I$

Re(r_1)

Im(r_1)

conjugate(r_1)

$r_2 := \dfrac{1}{4 + 5 \cdot I}$

$f_1 := e^{3 + 4 \cdot I}$

evalc(f_1)

$e^{I \cdot \pi}$

$z := 1 - 3 \cdot I$

$R := $ abs(z)

$\theta := $ argument(z)

$R \cdot e^{I \cdot \theta}$

$f := x \rightarrow x^4$

solve($f(x) = \pi, x$)

$p := x \rightarrow x^2 + 6 \cdot x + 25$

solve($p(x), x$)

subs($x = I, p(x)$)

subs($x = 2 + I, p(x)$)

solve($\{(2 + I) \cdot A_1 = 3, A_1 + 3 \cdot A_2 \cdot I = 0\}$)

Maple Example 4: Enter the following code that uses Maple's built-in Runge-Kutta method to numerically solve the following ODEs and then plots their solutions:

(a) $2xy'' + x^2 y' + 3x^3 y = 0$, with $y(1) = 2$, $y'(1) = 0$;

(b) $y^{(4)} + x^2 y' + y = \cos(x)$, with $y(0) = 1$, $y'(0) = 0$, $y''(0) = 0$, $y'''(0) = 1$.

Give the numerical value of the solution at $x = 5$.

```
restart
with(plots):
with(DEtools):
eq1 := 2 · x · y''(x) + x² · y'(x) + 3 · x³ · y(x) = 0
IC1 := y(1) = 2, D(y)(1) = 0
DEplot(eq1, y(x), x = 1..6, [[IC1]], stepsize = .05)
#If we just needed a plot (and not the soln at 5), we could can use DEplot.
#If we need to plot and we need the soln at 5, we use dsolve and odeplot.
soln1 := dsolve ({IC1, eq1}, y(x), numeric, method = classical[rk4],
    stepsize = .05)
soln1(5.0)
odeplot(soln1, [x, y(x)], 1..6, numpoints = 300)
eq2 := y⁽⁴⁾(x) + x² · y'(x) + y(x) = cos(x)
IC2 := y(0) = 1, D(y)(0) = 0, D(D(y))(0) = 0, D(D(D(y)))(0) = 1
DEplot(eq2, y(x), x = 0..6, [[IC2]], stepsize = .05)
soln2 := dsolve ({IC2, eq2}, y(x), numeric, method = classical[rk4],
    stepsize = .05) :
soln2(5.0)
odeplot(soln2, [x, y(x)], 0..6, numpoints = 300)
```

Maple Example 5: Enter the code below to demonstrate how to use Maple to find the roots of the characteristic equation of the given homogeneous ODEs. Although there is a built-in command to solve these, we will only use the computer to simplify our by-hand steps.
(i) $y'' + 3y' - 4y = 0$
(ii) $y^{(4)} - 4y''' + 6y'' - 4y' - 15y = 0$
(iii) $y^{(4)} + y''' - 2y' - y = 0$
(iv) $y''' + y' - 3y = 0$
(v) $y^{(5)} + 3y'' - y = 0$
You will need to first write the characteristic polynomial in order to find the roots, which are not necessarily real-valued.

```
restart
f := r → r² + 3 · r - 4
fsoln := solve(f(r), r)
g := r → r⁴ - 4 · r³ + 6 · r² - 4 · r - 15
gsoln := solve(g(r), r)
h := r → r⁴ + r³ - 2 · r - 1
hsoln := solve(h(r), r)
hsoln1 := allvalues(hsoln[1])
evalf(hsoln1)
hsoln2 := allvalues(hsoln[2])
evalf(hsoln2)
```

$hsoln3 := allvalues(hsoln[3])$
$hsoln4 := allvalues(hsoln[4])$
$F := r \rightarrow r^3 + r - 3$
$Fsoln := solve(F(r), r)$
$Fsoln1 := allvalues(Fsoln[1])$
$evalf(Fsoln1)$
$evalf(Fsoln)$
$G := r \rightarrow r^5 + 3 \cdot r^2 - 1$
$Gsoln := solve(G(r), r)$
$allvalues(Gsoln[1])$
$evalf(Gsoln)$

Maple Exercises

Turn in both the commands that you enter for the exercises below as well as the output/figures. These should all be in one document. Please highlight or clearly indicate all requested answers. Some of the questions will require you to modify the above Maple code to answer them.

1. Enter the commands given in Maple Example 1 and submit both your input and output.

2. Enter the commands given in Maple Example 2 and submit both your input and output.

3. Enter the commands given in Maple Example 3 and submit both your input and output.

4. Enter the commands given in Maple Example 4 and submit both your input and output.

5. Enter the commands given in Maple Example 5 and submit both your input and output.

6. Determine whether the following functions are linearly independent:
   (a) $f_1(x) = 2 + 5x - 3x^2$, $f_2(x) = 2 - x + x^2$, $f_3(x) = -2 - 23x + 15x^2$;
   (b) $f_1(x) = 1 + 4x + 2x^2$, $f_2(x) = 3 - x + 2x^2$, $f_3(x) = 5 - 2x - 2x^2$.

7. Determine whether the following sets of functions are linearly independent:
   (a) $\{1 + 2x - 3x^2 + x^3, 2 - x + x^2 + 3x^3, -4 + x - x^2 + 2x^3, -6x + 8x^2 + 9x^3\}$;
   (b) $\{3 + 4x - 3x^2 + x^3, x - 2x^2 + 4x^3, -4 + x^2 - x^3, -4 - 5x - x^2 + 3x^3\}$.

8. Use the Wronskian to determine whether the following sets of functions are linearly independent.
   (a) $\{\cos x, 3\cos x + \sin 2x, \sin x\}$ (b) $\{\cos x, 3\cos x + \sin x, 4\sin x\}$.

9. Use the Wronskian to determine whether the following sets of functions are linearly independent.
   (a) $\{e^x, xe^x, x^2\}$ (b) $\{\sin(4x) + \cos(4x), \cos(4x) - \sin(4x)\}$.

10. Any number can be written in polar form in the complex plane, $re^{i\theta}$. Find the r and θ values corresponding to the polar form of the following complex numbers.
    (a) $2 - 3i$ (b) $2i$ (c) $-1 - 5i$ (d) $3 + i$

11. Any number can be written in polar form in the complex plane, $re^{i\theta}$. Find the r and θ values corresponding to the polar form of the following complex numbers.
    (a) $2 + 3i$ (b) $-3i$ (c) $-2 - i$ (d) $3 + 7i$

12. Consider the polynomial $p(z) = z^2 + 4z + 8$. Evaluate it at the following values:
    (a) $z = 2i$ (b) $z = 3 - 2i$ (c) $z = -1 + 4i$ (d) $z = -2 + 2i$

13. Consider the polynomial $p(z) = z^3 - 1$. Evaluate it at the following values:
    (a) $z = 2i$ (b) $z = 3 - 2i$ (c) $z = -1 + 4i$ (d) $z = 1$

14. Solve the following systems of equations, where A_1, A_2 may be complex-valued numbers.
    (a) $\begin{cases} (2+i)A_1 + A2 = 3 \\ A_1 + 3A_2 i = 0 \end{cases}$ (b) $\begin{cases} 2A_1 + A2 = 3 \\ A_1 - A_2 i = -i \end{cases}$

15. Solve the following systems of equations, where A_1, A_2 may be complex-valued numbers.
    (a) $\begin{cases} A_1 - A2 = 1 \\ A_1 - A_2 i = 0 \end{cases}$ (b) $\begin{cases} A_1 i - A2 = 1 \\ A_1 - A_2 i = 5 \end{cases}$

16. For $y'' + y' + y = 0$ with $y(0) = 1, y'(0) = 2$, use a stepsize of $h = 0.1$ to obtain the solution from the 4th order Runge-Kutta method. Plot the solution for $0 \le x \le 20$. Give the numerical value of the solution at $x = 20$.

17. For $y'' + 2y' + 17y = \sin(2x)$ with $y(0) = 0, y'(0) = 0$, use a stepsize of $h = 0.1$ to obtain the solution from the 4th order Runge-Kutta method. Plot the solution for $0 \le x \le 20$. Give the numerical value of the solution at $x = 20$.

18. For $y''' + 4y'' + 21y' + 34y = \cos(3x)$ with $y(0) = 1, y'(0) = 0, y''(0) = .5$, use a stepsize of $h = 0.1$ to obtain the solution from the 4th order Runge-Kutta method. Plot the solution for $0 \le x \le 20$. Give the numerical value of the solution at $x = 20$.

19. For $y''' + x^2 y'' + y = \sin(x)$ with $y(0) = 1, y'(0) = -2, y''(0) = 1$, use a stepsize of $h = 0.1$ to obtain the solution from the 4th order Runge-Kutta method. Plot the solution for $0 \le x \le 20$. Give the numerical value of the solution at $x = 20$.

20. Find the roots of the characteristic equation of the following homogeneous ODEs: (a) $y'' + y' + y = 0$ (b) $y''' + 8y'' + 37y' + 50y = 0$ (c) $y^{(4)} - 4y^{(3)} - 2y'' + 36y' - 63y = 0$

21. Find the roots of the characteristic equation of the following homogeneous ODEs: (a) $y'' + 2y' + 17y$ (b) $y''' - y'' + 11y' - 51y = 0$ (c) $4y^{(4)} - 4y^{(3)} + 13y'' - 4y' - 9y = 0$

22. Find the roots of the characteristic equation of the following homogeneous ODEs: (a) $y''' + 4y'' + 21y' + 34y$ (b) $y^{(5)} + 3y^{(4)} + 9y''' + 27y'' = 0$ (c) $4y^{(5)} + 12y^{(4)} + 25y''' + 12y'' + 21y' = 0$

Chapter 3 Computer Lab: Mathematica

Mathematica Example 1: Enter the code below to demonstrate the following:
(a) the functions $f_1(x) = 1 + 3x$, $f_2(x) = 4 - 2x^2$, $f_3(x) = 5x - x^2$ are linearly independent;
(b) the functions $g_1(x) = 2 + x$, $f_2(x) = 4 - x + 5x^2$, $g_3(x) = 2 - 5x + 10x^2$ are linearly dependent.

```
Quit[]
f1[x_] = 1 + 3x (*don't forget <space> for multiplication*)
f2[x_] = 4 - 2x²
f3[x_] = 5x - x²
eq1 = c1 f1[x] + c2 f2[x] + c3 f3[x]
eq2 = Collect[eq1, {x, x²}]
eq3 = CoefficientList[eq2, x]
eq3a = Coefficient[eq2, x, 2]
eq3b = Coefficient[eq2, x, 1]
eq3c = Coefficient[eq2, x, 0]
eq4 = Solve[{eq3a==0, eq3b==0, eq3c==0}]
  (*Thus only 0f1 +0f2 +0f3 gives 0.*)
g1[x_] = 2 + x
g2[x_] = 4 - x + 5x²
g3[x_] = 2 - 5x + 10x²
eq5 = c1 g1[x] + c2 g2[x] + c3 g3[x]
eq6 = CoefficientList[eq5, x]
eq7 = Solve[{eq6[[1]]==0, eq6[[2]]==0, eq6[[3]]==0}]
  (*c1 is arbitrary. Thus c1=3 (e.g.,) gives 3g1-2g2+1g3=0.*)
```

Mathematica Example 2: Enter the following code that demonstrates how to calculate the Wronskian, $W(x)$, of the functions $f_1(x) = e^x$, $f_2(x) = e^{-x}$,

$f_3(x) = e^{2x}$. For linear independence, we need $W(x) \neq 0$ for at least one x-value.

```
Quit[]
f1[x_] = e^x
f2[x_] = e^-x
f3[x_] = e^{2x}
Wronskian[{f1[x], f2[x], f3[x]}, x]
```

Mathematica Example 3: Enter the following code that demonstrates how to perform calculations with complex numbers.

```
Quit[]
  (*Recall that e, i, π are entered from the palette*)
r1 = 2 + 3i
Re[r1]
Im[r1]
Conjugate[r1]
r2 = 1/(4 + 5i)
f1 = e^{3+4i}
ComplexExpand[f1]
e^{iπ}
z = 1 - 3i
R = Abs[z]
θ = Arg[z]
R e^{iθ}
ComplexExpand[%]
f[x_] = x^4
Solve[f[x]==π, x]
p[x_] = x^2 + 6x + 25
Solve[p[x]==0, x]
ReplaceAll[p[x], x→i]
ReplaceAll[p[x], x→2+i]
Solve[{(2 + i) A1==3, A1 + 3 A2 i==0}]
```

Mathematica Example 4: Enter the following code that uses Mathematica's built-in Runge-Kutta method to numerically solve the following ODEs and then plots their solutions:
(a) $2xy'' + x^2y' + 3x^3y = 0$, with $y(1) = 2$, $y'(1) = 0$;
(b) $y^{(4)} + x^2y' + y = \cos(x)$, with $y(0) = 1$, $y'(0) = 0$, $y''(0) = 0$, $y'''(0) = 1$.
Give the numerical value of the solution at $x = 5$.

```
Quit[]
de1[x_] = 2 x y''[x] + x² y'[x] + 3 x³ y[x]
solution1 = NDSolve[{de1[x]==0, y[1]==2, y'[1]==0}, y,
   {x, 0, 6}, StartingStepSize→ .05, Method→{FixedStep,
   Method→ExplicitRungeKutta}]
y1[x_] = y[x]/.solution1[[1]]
y1[5]
Plot[y1[x], {x, 0, 6}, AxesLabel→{"x", "y"}]
de2[x_] = y''''[x] + x² y'[x] + y[x]
solution2 = NDSolve[{de2[x]==Cos[x], y[0]==1, y'[0]==0,
   y''[0]==0, y'''[0]==1}, y, {x, 0, 6}, StartingStepSize→.05,
   Method→{FixedStep, Method→ExplicitRungeKutta}]
y2[x_] = y[x]/.solution2[[1]]
y2[5]
Plot[y2[x], {x, 0, 6},PlotRange→{-6, 25},AxesLabel→{"x", "y"}]
```

Mathematica Example 5: Enter the code below to demonstrate how to use Mathematica to find the roots of the characteristic equation of the given homogeneous ODEs. Although there is a built-in command to solve these, we will only use the computer to simplify our by-hand steps.
(i) $y'' + 3y' - 4y = 0$
(ii) $y^{(4)} - 4y''' + 6y'' - 4y' - 15y = 0$
(iii) $y^{(4)} + y''' - 2y' - y = 0$
(iv) $y''' + y' - 3y = 0$
(v) $y^{(5)} + 3y'' - y = 0$
You will need to first write the characteristic polynomial in order to find the roots, which are not necessarily real-valued.

```
Quit[]
f[r_] = r² + 3 r - 4
Solve[f[r]==0, r]
g[r_] = r⁴ - 4 r³ + 6 r² - 4 r - 15
Solve[g[r]==0, r]
h[r_] = r⁴ + r³ - 2 r - 1
hsol = Solve[h[r]==0, r]
N[hsol]
F[r_] = r³ + r - 3
Fsol = Solve[F[r]==0, r]
N[Fsol]
N[Fsol[[1]]]
G[r_] = r⁵ + 3 r² - 1
Gsol = Solve[G[r]==0, r]
N[Gsol]
```

`N[Gsol[[2]]]`

Mathematica Exercises

Turn in both the commands that you enter for the exercises below as well as the output/figures. These should all be in one document. Please highlight or clearly indicate all requested answers. Some of the questions will require you to modify the above Mathematica code to answer them.

1. Enter the commands given in Mathematica Example 1 and submit both your input and output.

2. Enter the commands given in Mathematica Example 2 and submit both your input and output.

3. Enter the commands given in Mathematica Example 3 and submit both your input and output.

4. Enter the commands given in Mathematica Example 4 and submit both your input and output.

5. Enter the commands given in Mathematica Example 5 and submit both your input and output.

6. Determine whether the following functions are linearly independent:
   (a) $f_1(x) = 2 + 5x - 3x^2$, $f_2(x) = 2 - x + x^2$, $f_3(x) = -2 - 23x + 15x^2$;
   (b) $f_1(x) = 1 + 4x + 2x^2$, $f_2(x) = 3 - x + 2x^2$, $f_3(x) = 5 - 2x - 2x^2$.

7. Determine whether the following sets of functions are linearly independent:
   (a) $\{1+2x-3x^2+x^3, 2-x+x^2+3x^3, -4+x-x^2+2x^3, -6x+8x^2+9x^3\}$;
   (b) $\{3+4x-3x^2+x^3, x-2x^2+4x^3, -4+x^2-x^3, -4-5x-x^2+3x^3\}$.

8. Use the Wronskian to determine whether the following sets of functions are linearly independent.
   (a) $\{\cos x, 3\cos x + \sin 2x, \sin x\}$ (b) $\{\cos x, 3\cos x + \sin x, 4\sin x\}$.

9. Use the Wronskian to determine whether the following sets of functions are linearly independent.
   (a) $\{e^x, xe^x, x^2\}$ (b) $\{\sin(4x) + \cos(4x), \cos(4x) - \sin(4x)\}$.

10. Any number can be written in polar form in the complex plane, $re^{i\theta}$. Find the r and θ values corresponding to the polar form of the following complex numbers.
    (a) $2 - 3i$ (b) $2i$ (c) $-1 - 5i$ (d) $3 + i$

11. Any number can be written in polar form in the complex plane, $re^{i\theta}$. Find the r and θ values corresponding to the polar form of the following complex numbers.
    (a) $2 + 3i$ (b) $-3i$ (c) $-2 - i$ (d) $3 + 7i$

12. Consider the polynomial $p(z) = z^2 + 4z + 8$. Evaluate it at the following values:
    (a) $z = 2i$ (b) $z = 3 - 2i$ (c) $z = -1 + 4i$ (d) $z = -2 + 2i$

13. Consider the polynomial $p(z) = z^3 - 1$. Evaluate it at the following values:

    (a) $z = 2i$ (b) $z = 3 - 2i$ (c) $z = -1 + 4i$ (d) $z = 1$

14. Solve the following systems of equations, where A_1, A_2 may be complex-valued numbers.

    (a) $\begin{cases} (2+i)A_1 + A2 = 3 \\ A_1 + 3A_2 i = 0 \end{cases}$ (b) $\begin{cases} 2A_1 + A2 = 3 \\ A_1 - A_2 i = -i \end{cases}$

15. Solve the following systems of equations, where A_1, A_2 may be complex-valued numbers.

    (a) $\begin{cases} A_1 - A2 = 1 \\ A_1 - A_2 i = 0 \end{cases}$ (b) $\begin{cases} A_1 i - A2 = 1 \\ A_1 - A_2 i = 5 \end{cases}$

16. For $y'' + y' + y = 0$ with $y(0) = 1, y'(0) = 2$, use a stepsize of $h = 0.1$ to obtain the solution from the 4th order Runge-Kutta method. Plot the solution for $0 \leq x \leq 20$. Give the numerical value of the solution at $x = 20$.

17. For $y'' + 2y' + 17y = \sin(2x)$ with $y(0) = 0, y'(0) = 0$, use a stepsize of $h = 0.1$ to obtain the solution from the 4th order Runge-Kutta method. Plot the solution for $0 \leq x \leq 20$. Give the numerical value of the solution at $x = 20$.

18. For $y''' + 4y'' + 21y' + 34y = \cos(3x)$ with $y(0) = 1, y'(0) = 0, y''(0) = .5$, use a stepsize of $h = 0.1$ to obtain the solution from the 4th order Runge-Kutta method. Plot the solution for $0 \leq x \leq 20$. Give the numerical value of the solution at $x = 20$.

19. For $y''' + x^2 y'' + y = \sin(x)$ with $y(0) = 1, y'(0) = -2, y''(0) = 1$, use a stepsize of $h = 0.1$ to obtain the solution from the 4th order Runge-Kutta method. Plot the solution for $0 \leq x \leq 20$. Give the numerical value of the solution at $x = 20$.

20. Find the roots of the characteristic equation of the following homogeneous ODEs: (a) $y'' + y' + y = 0$ (b) $y''' + 8y'' + 37y' + 50y = 0$ (c) $y^{(4)} - 4y^{(3)} - 2y'' + 36y' - 63y = 0$

21. Find the roots of the characteristic equation of the following homogeneous ODEs: (a) $y'' + 2y' + 17y$ (b) $y''' - y'' + 11y' - 51y = 0$ (c) $4y^{(4)} - 4y^{(3)} + 13y'' - 4y' - 9y = 0$

22. Find the roots of the characteristic equation of the following homogeneous ODEs: (a) $y''' + 4y'' + 21y' + 34y$ (b) $y^{(5)} + 3y^{(4)} + 9y''' + 27y'' = 0$ (c) $4y^{(5)} + 12y^{(4)} + 25y''' + 12y'' + 21y' = 0$

Chapter 3 Projects

Project 3A: Runge–Kutta Order 2

In Sections 2.5 and 3.4, we considered the fourth-order Runge-Kutta method for numerically approximating the solution of

$$y' = f(x, y(x)). \tag{3.68}$$

Here we derive the simpler second-order formula. For simplicity, let's consider $y \in \mathbb{R}^1$. Given the solution at $y(x)$, we want to approximate the solution at $y(x + h)$. To do so we will need to recall the single- and multi-variable Taylor expansions:

$$y(x + h) = y(x) + hy'(x) + \frac{h^2}{2}y''(x) + O(h^3) \tag{3.69}$$

$$f(x + h, y + k) = f(x, y) + h\frac{\partial f}{\partial x}(x, y) + \frac{\partial f}{\partial y}(x, y)k + \cdots . \tag{3.70}$$

(1) Differentiate Equation (3.68) and use the Chain Rule to obtain

$$y''(x) = \frac{\partial f}{\partial x}(x, y) + \frac{\partial f}{\partial y}(x, y)f(x, y). \tag{3.71}$$

(2) Use Equation (3.68) and Equation (3.71) to rewrite Equation (3.69) as

$$y(x + h) = y(x) + \frac{h}{2}f(x, y)$$
$$+ \frac{h}{2}\left[f(x, y) + h\frac{\partial f}{\partial x}(x, y) + h\frac{\partial f}{\partial y}(x, y)f(x, y)\right] + O(h^3). \tag{3.72}$$

(3) Rewrite the expression in brackets in Equation (3.72), using Equation (3.70), to obtain

$$y(x + h) = y(x) + \frac{h}{2}f(x, y) + \frac{h}{2}f(x + h, y + hf(x, y)) + O(h^3). \tag{3.73}$$

(4) Use Equation (3.73) to obtain the following second-order Runge-Kutta method:

$$y_{i+1} = y_i + h\left(\frac{1}{2}k_1 + \frac{1}{2}k_2\right) \tag{3.74}$$

$$\text{with } k_1 = f(x, y), k_2 = f(x + h, y + hk_1). \tag{3.75}$$

Project 3B: Stiff Differential Equations

We studied the equation for a damped, unforced mass on a spring:

$$mx'' + bx' + kx = 0,$$

where $m, b, k > 0$ are constant. There are many situations that arise where these constants are not close to each other in magnitude. One specific area of interest is when the spring constant k is large. Such systems are said to be *stiff*. Here we will consider the equation

$$x'' + x' + 1000x = 0. \tag{3.76}$$

(1) Solve this constant coefficient ODE and write its analytical solution. Use the initial condition $x(0) = 1$, $x'(0) = 0$.

(2) Using MATLAB, Maple, or Mathematica solve this equation using the fourth-order Runge-Kutta method introduced in this chapter. Try seven different runs, each with a different step size, using step sizes of $h = 0.2, 0.1, 0.05, 0.02, 0.01, 0.005, 0.002$ and estimating the solution at $t = 20$.

(3) Compare the analytical solution from (1) with the numerical approximation from (2). Comment on the accuracy of the numerical answers.

Besides decreasing the step size, stiff differential equations are often handled with other methods, typically implicit ones. See Project 2B in this book for two methods and numerous others in books such as [10].

Chapter 4

Techniques of Nonhomogeneous
Higher-Order Linear Equations

In this chapter, we examine how to solve linear equations in which there is a forcing function. In Sections 4.2-4.3, we consider two different methods for solving linear, constant coefficient ODEs whose forcing functions are polynomials, exponentials, sinusoidal, or a combination of these. While the reader may read both of these sections, it is our intention that most will choose *one* of these methods for solving this type of equation (and thus skip the other section). In Section 4.4, we present an alternative method that is much quicker for forcing functions that are exponential or sinusoidal (viewed as complex exponentials) and thus might be ideal for engineering students; however, forcing functions that are multiplied by polynomials require a bit more work and one of the methods of Section 4.2 or 4.3 is needed to additionally solve an ODE that has polynomial-only forcing in order to complete the problem. Since most realistic engineering applications have forcing functions that involve decaying exponentials or sinusoidal functions, focusing mainly on the exponential or sinusoidal functions in Section 4.4 will likely be sufficient. These three sections are in contrast to Variation of Parameters that does not require a prescribed type of forcing function nor does the equation need to have constant coefficients. Our reason for presenting all four methods is the greatly varied audience that is learning from this book, and each method has its place. We begin with a little background on nonhomogeneous equations.

4.1   Nonhomogeneous Equations

We will now begin discussing the nonhomogeneous equation

$$a_n(x)\frac{d^n y}{dx^n} + a_{n-1}(x)\frac{d^{n-1}y}{dx^{n-1}} + \ldots + a_1(x)\frac{dy}{dx} + a_0(x)\,y = F(x), \qquad (4.1)$$

originally given at the beginning of Chapter 3. Up to this point we have been focusing on the solution of the homogeneous equation

$$a_n(x)\frac{d^n y}{dx^n} + a_{n-1}(x)\frac{d^{n-1}y}{dx^{n-1}} + \ldots + a_1(x)\frac{dy}{dx} + a_0(x)\,y = 0, \qquad (4.2)$$

and concentrating on the case in which (4.2) has constant coefficients. We will see how our work with homogeneous equations will play a role. In general it is not easy to solve this equation; sometimes we can get lucky, while other times we may only be able to guess a particular solution. Fortunately, many real world phenomena are accurately modeled with sinusoidal forcing functions and we will be able to solve these with the methods of this chapter.

We begin by considering (4.1) written in operator notation:

$$P(D)y = F(x),$$

where $P(D) = a_n(x)D^n + a_{n-1}(x)D^{n-1} + \ldots + a_1(x)D + a_0(x)$. We define a **particular solution**, y_p, to be any solution of (4.1). This means that

$$P(D)y_p = F(x).$$

Example 1 Verify that $y_p = x^2 - 2x$ is a particular solution to $y'' + y' = 2x$.

Solution

We need to substitute this into the differential equation and show that it holds.

$$\underbrace{2}_{y_p''} + \underbrace{2x - 2}_{y_p'} = 2x,$$

which clearly holds for all x. Thus y_p is a particular solution.

THEOREM 4.1.1

Let y_p be any particular solution of the nonhomogeneous nth order linear differential equation (4.1). Let u be any solution of the corresponding homogeneous equation (4.2); then $u + y_p$ is also a solution of the given nonhomogeneous equation.

Example 2 Note that $y = x$ is a solution of the nonhomogeneous equation

$$\frac{d^2y}{dx^2} + y = x$$

and that $y = \sin x$ is a solution of the corresponding homogeneous equation

$$\frac{d^2y}{dx^2} + y = 0.$$

Thus, the sum $y = x + \sin x$ is a solution to

$$\frac{d^2y}{dx^2} + y = x.$$

Example Verify that $y_p = \frac{1}{9}e^{-2x}$ is a particular solution to $y'' + 4y' + 13y = e^{-2x}$. Then use this information to find a particular solution to $y'' + 4y' + 13y = -4e^{-2x}$.

Solution

Subsituting into the differential equation gives

$$\frac{4}{9}e^{-2x} + 4 \cdot \frac{-2}{9}e^{-2x} + 13 \cdot \frac{1}{9}e^{-2x} = e^{-2x}.$$

Simplification on the left gives e^{-2x} and thus y_p is a particular solution. For the second part of the question, we rewrite the left side using operator notation: $P(D)y = e^{-2x}$. Since y_p is a solution, this means that

$$P(D)y_p = e^{-2x}.$$

From the linearity of the operator (see Section 3.1.1), for any constant c we have $cP(D)y_p = P(D)(cy_p)$. Because the homogeneous part of both equations in our current example, $P(D)y$, is the *exact same* we can multiply the first equation by an appropriate constant so that the right-hand sides match:

$$-4 \cdot \left(P(D)y = e^{-2x}\right) \qquad \text{so that} \qquad P(D)(-4y) = -4e^{-2x}.$$

Thus if y_p is a particular solution for the first equation, then

$$P(D)y_p = e^{-2x} \implies P(D)(-4y_p) = -4e^{-2x}$$

and thus $-4y_p = \frac{-4}{9}e^{-2x}$ is a particular solution of the second equation.

The two key items in the previous example are that the operator is linear and that the left-hand sides, $P(D)y$, are the exact same. We have been careful in the discussion so far to not refer to *the* particular solution but rather to *a* particular solution. The reason for this is captured in this next theorem.

THEOREM 4.1.2

Let y_p and Y be two solutions of the nonhomogeneous equation $P(D)y = F(x)$, where $P(D)$ is a linear operator. Then

$$Y = y_p + c_1y_1 + c_2y_2 + \cdots + c_ny_n,$$

where $\{y_1, \cdots, y_n\}$ is a fundamental set of solutions for $P(D)y = 0$.

In other words, once we have one particular solution, y_p, any other particular solution, Y, that we may find is just a linear combination of the particular solution we already had plus the solution from the homogeneous equation. Thus, we only need to find *a* particular solution because any other solution that we may find is simply a combination of this plus the homogeneous solution.

<u>Proof:</u> Since both are particular solutions, we have $P(D)y_p = F(x)$ and $P(D)Y = F(x)$. We consider $Y - y_p$ and apply the operator $P(D)$ to it:

$$P(D)(Y - y_p) = P(D)Y - P(D)y_p \qquad \text{(by linearity)}$$
$$= F(x) - F(x) = 0,$$

from which we conclude that $Y - y_p$ solves the homogeneous equation. It follows that $Y - y_p$ can be written as a linear combination of the solutions in the fundamental set, i.e.,

$$Y - y_p = c_1 y_1 + c_2 y_2 + \cdots + c_n y_n \qquad \text{for some constants } c_1, \cdots, c_n.$$

The conclusion of the theorem follows.∎

THEOREM 4.1.3

Let y_p be a particular solution of the *n*th order nonhomogeneous linear differential equation (4.1). Let

$$y_c = c_1 y_1 + c_2 y_2 + \ldots + c_n y_n$$

be the general solution of the corresponding homogeneous equation (4.2). Then every solution ϕ of the *n*th order nonhomogeneous linear differential equation (4.1) can be expressed in the form

$$\phi = y_c + y_p.$$

The general solution of (4.2), y_c, is called the **complementary function** of (4.1). (Note that the general solution of (4.2) is sometimes called the **homogeneous solution** and is denoted y_h. Thus both y_c and y_h denote the same thing.) We can summarize this theorem as

general solution of nonhomogeneous

$$= \text{general solution of homogeneous}$$
$$+ \text{particular solution of nonhomogeneous}$$

Note that we still have not discussed exactly how to find a particular solution for a given differential equation—we are simply stating that once we do know a particular solution, we can write the general solution to the nonhomogeneous equation.

Example 4 Consider the differential equation

$$\frac{d^2y}{dx^2} + y = x.$$

The complementary function is the general solution

$$y_c = c_1 \sin x + c_2 \cos x$$

of the homogeneous equation. A particular solution is given by $y_p = x$ as mentioned in Example 2, so that the general solution of the nonhomogeneous equation is

$$y = c_1 \sin x + c_2 \cos x + x.$$

As mentioned before this example, we will show how to obtain this y_p in the next section, but for now we pursue further how a particular solution fits into the larger theory of solving nonhomogeneous linear differential equations.

Example 5 Consider the differential equation

$$y'' - 4y = \sin x. \tag{4.3}$$

A particular solution is given by

$$y_p = \frac{-1}{5} \sin x.$$

We can see that y_p is actually a solution since

$$y_p' = \frac{-1}{5} \cos x \quad \text{and} \quad y_p'' = \frac{1}{5} \sin x.$$

Substituting into the original differential equation gives

$$y_p'' - 4y = \frac{1}{5} \sin x - 4\left(\frac{-1}{5} \sin x\right) = \sin x,$$

which shows that $y_p = \frac{-1}{5} \sin x$ is a particular solution.

Now it is straightforward to calculate the homogeneous solution

$$y_c = c_1 e^{-2x} + c_2 e^{2x}.$$

Thus the general solution is

$$y(x) = y_c(x) + y_p(x) = c_1 e^{-2x} + c_2 e^{2x} - \frac{1}{5} \sin x.$$

Now let's consider two differential equations with the same operator:

$$P(D)y = F_1(x), \qquad P(D)y = F_2(x).$$

Let us assume we have a particular solution Y_1 for the first and Y_2 for the second. Can we say anything about the particular solutions of $P(D)y = F_1(x) + F_2(x)$? By linearity of the operator, $P(D)$, we know that

$$P(D)(Y_1 + Y_2) = P(D)Y_1 + P(D)Y_2 = F_1(x) + F_2(x).$$

In other words, $Y_1 + Y_2$ is a particular solution of $P(D)y = F_1(x) + F_2(x)$.

Example 6 Given that $Y_1 = x$ is a solution to $y'' + y = x$ and $Y_2 = \frac{e^x}{2}$ is a solution to $y'' + y = e^x$, find a particular solution to $y'' + y = x + 3e^x$.

Solution

Since $\frac{e^x}{2}$ is a solution to $y'' + y = e^x$, it follows that $\frac{3e^x}{2}$ is a solution to $y'' + y = 3e^x$. Thus a particular solution to $y'' + y = x + 3e^x$ is given by

$$y_p = x + \frac{3e^x}{2}.$$

Forced Mass on a Spring

Perhaps the most studied example of nonhomogeneous equations is that of a forced mass on a spring. We again consider the differential equation describing the motion of a mass on a spring and will let $x(t)$ denote the displacement:

$$m\frac{d^2x}{dt^2} + b\frac{dx}{dt} + kx = F(t). \tag{4.4}$$

We have discussed the case when $F(t) = 0$ in Section 3.7 and will now consider the more general case when $F(t) \neq 0$. The physical interpretation of this type of situation is one in which the support of the spring, i.e., the object to which the spring is attached, is shaken or forced according to some known function. Figure 4.1 gives one manner in which this may be done and Problems **10–12** in Section 4.7 explore alternative formulations for a forced mass on a spring in which the damping is through a **dashpot**. A typical assumption is that the forcing function can be written as

$$F(t) = F_0 \sin(\omega t), \tag{4.5}$$

where F_0 is the magnitude of the forcing function and ω is the frequency of this forcing. We note that this ω is completely independent of the inherent natural frequency, $\omega_0 = \sqrt{\frac{k}{m}}$, of the undamped mass-spring system. Many types of forced vibrations can be well approximated with such an assumption. This is also mathematically convenient because we can "guess" that the form of the particular solution should involve $\sin(\omega t)$ and maybe even $\cos(\omega t)$. What we really want to understand is how changing the frequency of the forcing oscillation, ω, can affect the motion of the attached mass.

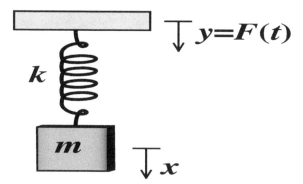

FIGURE 4.1: Mass on a spring forced through spring support.

Example Consider a forced mass on a spring governed by the equation

$$x''(t) + \frac{1}{2}x'(t) + x(t) = \sin(2t). \tag{4.6}$$

Physically, we have that our mass and spring constant are both one but, more importantly, we see that the damping coefficient is small in comparison. Calculating $b^2 - 4mk$ shows that our motion is underdamped. How does the forcing frequency $\omega = 2$ affect things?

Solution
We can calculate the complementary solution as

$$x_c(t) = c_1 e^{-t/4} \sin\left(\frac{\sqrt{15}}{4}t\right) + c_2 e^{-t/4} \cos\left(\frac{\sqrt{15}}{4}t\right) \tag{4.7}$$

because we again have constant coefficients. A particular solution is given by

$$x_p(t) = \frac{-3}{10}\sin(2t) - \frac{1}{10}\cos(2t) \tag{4.8}$$

and the reader should check this. Our general solution is then $x_c(t) + x_p(t)$.

It is very worthwhile at this point to scrutinize these two parts of the solution. The solution of the homogeneous equation, $x_c(t)$, is exactly Case 3 from Section 3.7 describing the possible motions of a damped spring. We know from those results (or could easily calculate if we already forgot) that

$$\lim_{t \to \infty} x_c(t) = \lim_{t \to \infty} c_1 e^{-t/4} \sin\left(\frac{\sqrt{15}}{4}t\right) + c_2 e^{-t/4} \cos\left(\frac{\sqrt{15}}{4}t\right) = 0.$$

(This is another instance where writing the solution in the amplitude-phase form makes the evaluation of this limit easy.) As mentioned earlier, we call this motion **transient** because it dies off as $t \to \infty$; however, it is **not** the case that $x_p(t)$ dies off. Indeed,

$$\lim_{t \to \infty} x_p(t) \text{ does not exist}$$

because the sine and cosine continue to oscillate as $t \to \infty$. What does this mean? It simply (and importantly!) indicates that the motion of the mass does not stop. We can write bounds on the motion of these **steady-state** oscillations, (4.8), using our standard trigonometric identities; see (3.48):

$$\sqrt{\left(\frac{3}{10}\right)^2 + \left(\frac{1}{10}\right)^2} = \frac{1}{\sqrt{10}}.$$

Figure 4.2(a) gives a graph of the solution with the initial condition $x(0) = 1$, $x'(0) = 0$.

It may seem completely obvious that by forcing the spring support, we perpetuate the motion of the spring. What may seem strange is that, for a given mass-spring system, the frequency at which we choose to force the support may have a significant effect on the motion of the mass.

Example 8   Consider the same mass-spring system as in Example 7 but now we will use a different forcing frequency:

$$x''(t) + \frac{1}{2}x'(t) + x(t) = \sin\left(\sqrt{\frac{7}{8}}t\right). \tag{4.9}$$

The homogeneous solution is the same as before and we can check that

$$x_p(t) = \frac{-8\sqrt{14}}{15}\cos\left(\sqrt{\frac{7}{8}}t\right) + \frac{8}{15}\sin\left(\sqrt{\frac{7}{8}}t\right) \tag{4.10}$$

is a particular solution. The homogeneous (transient) solution again dies off and the particular (steady-state) solution determines the long-term behavior. We can again write bounds on the motion of these steady-state oscillations, (4.10), using our standard trigonometric identities:

$$\sqrt{\left(\frac{8\sqrt{14}}{15}\right)^2 + \left(\frac{8}{15}\right)^2} = \frac{8}{\sqrt{15}}.$$

For the *same* initial condition and mass-spring system, changing the frequency of the forcing function dramatically increased the amplitude of the steady-state oscillations; see Figure 4.2b.

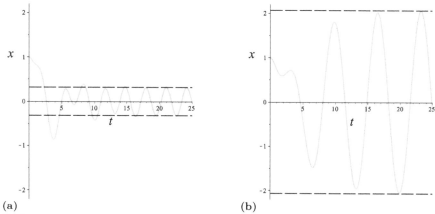

FIGURE 4.2: Forced damped motion with $m = 1$, $b = 1/2$, $k = 1$ for the initial condition $x(0) = 1$, $x'(0) = 0$. In (a), the forcing function is $F(t) = \sin(2t)$; see Example 7. In (b), the forcing function is $F(t) = \sin(\sqrt{7/8}\, t)$; see Example 8. Note the dramatic difference in the amplitude of the steady-state oscillations.

This phenomenon of periodically forcing an oscillating system in a way that excites or amplifies the motion is known as **resonance**. We will discuss resonance in more detail after we learn how to find particular solutions. But we mention two often-used examples. The first is that of a group of marching soldiers that will go out-of-step as they walk across a bridge. They do this because they don't want to introduce a periodic forcing that might put their own lives in danger! The second was a catastrophic event that actually was *not* due to resonance. In 1940, the Tacoma Narrows bridge collapsed due to strong winds that caused the bridge to sway with large oscillations.[1] Resonance was originally used to explain its collapse even though it's clear that a bridge naturally has large damping (and a strong effect due to resonance requires small damping). It wasn't until the 1990s that it became better understood how *nonlinear* effects actually caused the collapse of the bridge [28], [31]. We will not go into any details of the mathematics but encourage the reader to use the numerical techniques of past sections to explore some of this research.

Overview

Examining the forced mass on a spring has given us sufficient motivation as to the importance of the study of nonhomogeneous equations. *Our goal in the next few sections will be to find ways to obtain a particular solution.* The two

---
[1]Try a web search using "Tacoma Narrows bridge" and you should be able to find video footage, pictures, and additional information.

methods on which we will focus are the method of undetermined coefficients and variation of parameters—both are simply ways of accomplishing our goal.

Method of Undetermined Coefficients (Section 4.2 or 4.3)

This method will only work when the right-hand side of the equation, i.e., the forcing function $F(x)$, has the form of one of the solutions of a linear homogeneous equation with constant coefficients and when the left-hand side is linear with constant coefficients. For example, we will be able to solve

$$y^{(5)} + 2y'' + 7y = \sin(3x) + x^2 e^{2x} \cos(x) + e^{-2x}$$

but we will not be able to solve equations such as

$$y'' + xy = \sin(x) \quad \text{or} \quad y'' + y = \tan(x).$$

To use this method, we will assume a certain form of the particular solution, with coefficients to be found, and we will substitute this assumed form into the differential equation and find the coefficients.

We will initially focus on the mechanics of obtaining the particular solution from an assumed form. We will then examine two methods used to obtain the "assumed form" of the particular solution. The first will be via **super-position** (or more simply thought of as tables, memorization, or copying the form of F) and the second will be via **annihilation**.

Exponential Response and Complex Replacement (Section 4.4)

This method will work quickest in the case of exponential or sinusoidal forcing when these terms are not multiplied by a polynomial. In the case of sinusoidal forcing, our approach utilizes complex numbers extensively. In these situations, we will first rewrite the sinusoidal forcing term as a complex exponential. We will then divide by an appropriate complex number and we then have the complex form of our solution. Depending on our original forcing function, $F(x)$, we may have to pull out the real or complex parts of this soluion. When the (possibly complex) exponential forcing is multiplied by a polynomial, we will additionally be required to use the Method of Undetermined Coefficients to solve a polynomial-only forcing function. Thus this particular method works great if only exponential or sinusoidal forcing functions are present, which is the situation that occurs in most engineering applications.

Variation of Parameters (Section 4.5)

This method will work in two places where the method of undetermined coefficients fails—it works when the coefficients are non-constant and for any form of forcing function $F(x)$. For example, it will work on all three of the examples given above:

$$y^{(5)} + 2y'' + 7y = \sin(3x) + x^2 e^{2x} \cos(x) + e^{-2x},$$
$$y'' + xy = \sin(x), \quad \text{and} \quad y'' + y = \tan(x).$$

You may wonder why we should even learn the method of undetermined co-efficients, but it turns out that if we have a simple equation such as

$$y'' + y = x$$

the method of undetermined coefficients will be far less work than variation of parameters.

The main drawback of the variation of parameters, as we will see, is that it requires us to know a fundamental set of solutions before beginning. If our equation has constant coefficients, this will not be a problem. But if we have a non-constant coefficient equation, we must obtain the solution by whatever means necessary.

In Problems **1–8**, consider the given differential equation with forcing $P(D)y = F_1(x)$ and corresponding particular solution y_p. Find a particular solution for the new forcing function F_2.

1. $y'' - 5y' - 6y = 3e^x$ with $y_p = \dfrac{-3e^x}{10}$; $F_2(x) = -5e^x$

2. $y'' + 4y' = 8x - 3$ with $y_p = x^2 - \dfrac{5}{4}x$; $F_2(x) = 4x - \dfrac{3}{2}$

3. $y'' + y' - 6y = \sin x$ with $y_p = \dfrac{-1}{50}\cos x - \dfrac{7}{50}\sin x$; $F_2(x) = 3\sin x$

4. $y'' - 4y = 17\cos x$ with $y_p = -4\sin x - \cos x$; $F_2(x) = -3\cos x$

5. $y'' + 9y = \cos x$ with $y_p = \dfrac{1}{8}\cos x$; $F_2(x) = 12\cos x$

6. $y'' + y' = 2x$ with $y_p = x^2 - 2x$; $F_2(x) = Ax$ for any constant $A \neq 0$

7. $y'' - 5y' + 6y = 1$ with $y_p = \dfrac{1}{6}$, $y'' - 5y' + 6y = x$ with $y_p = \dfrac{x}{6} + \dfrac{5}{36}$, and $y'' - 5y' + 6y = e^x$ with $y_p = \dfrac{e^x}{2}$; $F_2(x) = 2 - 12x + 6e^x$

8. $y'' + 4y' + 4y = x$ with $y_p = \dfrac{x - 1}{4}$ and $y'' + 4y' + 4y = e^{-2x}$ with $y_p = \dfrac{x^2}{2}e^{-2x}$; $F_2(x) = 8x - 3e^{-2x}$

For Problems **9–13**, use MATLAB, Maple, or Mathematica to numerically solve the differential equation subject to the given forcing frequencies. In all cases, use the initial condition of $x(0) = 1$, $x'(0) = 0$ and numerically plot the solution from $t = 0$ to $t = 30$. From your graphs, (i) determine which of the three frequencies is the resonant frequency; (ii) approximate the steady-state amplitude of the solution.

9. $x'' + 2x' + 6x = \sin(\omega t)$, with $\omega = 1, 2, 3$

10. $x'' + 2x' + 3x = \sin(\omega t)$, with $\omega = 1, 2, 3$

11. $3x'' + x' + x = \sin(\omega t)$, with $\omega = \sqrt{1/18}, \sqrt{5/18}, \sqrt{1/2}$

12. $3x'' + 2x' + 4x = \sin(\omega t)$, with $\omega = \sqrt{5/9}, \sqrt{7/9}, \sqrt{10/9}$

13. $x'' + 2x' + 5x = \sin(\omega t)$, with $\omega = 1, \sqrt{2}, \sqrt{3}$

4.2   Method of Undetermined Coefficients via Superposition

In this section, we give the first of two approaches for using the method of undetermined coefficients. Our overall goal, as mentioned earlier, is to find a particular solution given some function $F(x)$. We will discuss how to obtain the general form of the particular solution for certain functions $F(x)$. The method of undetermined coefficients then takes this general form of the particular solution and uses the original differential to determine the coefficients of the assumed form of the particular solution.

We again consider the nonhomogeneous differential equation

$$a_n \frac{d^n y}{dx^n} + a_{n-1} \frac{d^{n-1} y}{dx^{n-1}} + \ldots + a_1 \frac{dy}{dx} + a_0 \, y = F(x), \qquad (4.11)$$

which is simply (4.1) with the specification that $a_0, a_1, \ldots, a_n$ are constants and $F(x)$, the nonhomogeneous term, is a general nonzero function of x.

Recall that the general solution may be written

$$y = y_c + y_p$$

where y_c is the complementary function, that is, the solution to the corresponding homogeneous equation; y_p is a particular solution. We will now consider the first of the two methods that fall under the category of the method of undetermined coefficients.

Example 1 Consider

$$\frac{d^2 y}{dx^2} - 2\frac{dy}{dx} - 3y = 2e^{4x}.$$

We seek a particular solution, y_p, of this nonhomogeneous equation. The right side is the exponential $2e^{4x}$. We know that we need to find a function y_p that will be a solution. That is, when we substitute y_p into the left-hand side of the equation it should equal the right-hand side. Looking for a function that when added to its first and second derivatives gives $2e^{4x}$ suggests that the particular solution should also be an exponential of the form

$$y_p = Ae^{4x}.$$

Here A is a constant, which we call the "undetermined coefficient."

Thus

$$y_p' = 4Ae^{4x} \text{ and } y_p'' = 16Ae^{4x}$$

which gives

$$16Ae^{4x} - 2(4Ae^{4x}) - 3Ae^{4x} = 2e^{4x}.$$

This simplifies to

$$5Ae^{4x} = 2e^{4x}$$

which must satisfy the differential equation for all x, thus

$$5A = 2$$

so that $A = 2/5$. Thus, a particular solution is

$$y_p = \frac{2}{5}e^{4x}.$$

Example 2 Now consider the differential equation

$$\frac{d^2y}{dx^2} - 2\frac{dy}{dx} - 3y = 2e^{3x}.$$

This is the same equation we just considered, except that the right-hand side is now $2e^{3x}$. Let's proceed similarly and consider

$$y_p = Ae^{3x}.$$

Thus

$$y_p' = 3Ae^{3x} \text{ and } y_p'' = 9Ae^{3x}.$$

Substitution gives

$$9Ae^{3x} - 2(3Ae^{3x}) - 3Ae^{3x} = 2e^{3x}$$

which simplifies to

$$0 = 2e^{3x}.$$

But, this last equation <u>does not</u> hold for any real x, so that there is <u>no</u> particular solution of the form

$$y_p = Ae^{3x}.$$

What went wrong? Consider the reduced equation of the nonhomogeneous equation:

$$\frac{d^2y}{dx^2} - 2\frac{dy}{dx} - 3y = 0.$$

The corresponding characteristic equation

$$r^2 - 2r - 3 = 0$$

has roots 3 and -1, so that e^{3x} and e^{-x} are linearly independent solutions. This is why the method failed; the particular solution

$$y_p = Ae^{3x}$$

is already a solution to the homogeneous equation! That is, since Ae^{3x} satisfies the homogeneous equation, it reduces the left side to zero, not $2e^{3x}$.

How do you find a particular solution in this case?

Recall we showed (in the double root case) that e^{rx} and xe^{rx} are linearly independent, so we try a solution of the form

$$y_p = Axe^{3x}.$$

Thus,

$$y_p' = 3Axe^{3x} + Ae^{3x} \text{ and } y_p'' = 9Axe^{3x} + 6Ae^{3x}$$

so that upon substitution we have

$$9Axe^{3x} + 6Ae^{3x} - 2(3Axe^{3x} + Ae^{3x}) - 3Axe^{3x} = 2e^{3x}.$$

This simplifies to

$$4Ae^{3x} = 2e^{3x},$$

which must be valid for all x, so that $4A = 2$, which implies $A = 1/2$.
A particular solution is

$$y_p = \frac{1}{2}xe^{3x}.$$

We need some terminology to help formalize this method of undetermined coefficients.

Definition 4.2.1
We shall call a function f a *UC function* if it is either
1. A function defined by one of the following:
  i) x^n, where n is a nonnegative integer
  ii) e^{ax}, where a is a nonzero constant
  iii) $\sin(bx + c)$ and/or $\cos(bx + c)$, where b and c are constants, $b \neq 0$
2. A function defined as a finite product of two or more functions of these types.

The method of undetermined coefficients applies when the nonhomogeneous function F in the differential equation is a finite linear combination of UC functions. So you can think of the UC functions as "building blocks" of F.

Note here that for a given UC function f, each successive derivative of f is either itself a constant multiple of a UC function or else a linear combination of UC functions.

For a UC function f, we will define the *UC set* of f as the set of functions consisting of (i) f itself and (ii) all successive derivatives of f that are linearly independent of f.

The elements arising from (i) and (ii) above must be nonzero. They may be multiplied by arbitrary nonzero constants without changing the essential "building blocks" contained in the UC set. Thus a UC set of the form $\{3x, 5\}$ is equivalent to the UC set $\{x, 1\}$ and we will choose the latter formulation for convenience. Table 4.1 gives specific examples of UC functions and their corresponding UC sets.

Table 4.1: Examples of UC Functions and Corresponding UC Set

UC Function	UC Set
1. 7	$\{1\}$
2. $x^2 + 3$	$\{x^2, x, 1\}$
3. $2e^{3x}$	$\{e^{3x}\}$
4. $\sin(2x)$	$\{\sin(2x), \cos(2x)\}$
5. $5\cos(2x)$	$\{\sin(2x), \cos(2x)\}$
6. $\sin(2x) - 32\cos(2x)$	$\{\sin(2x), \cos(2x)\}$
7. $x^3 e^{-x}$	$\{x^3 e^{-x}, x^2 e^{-x}, xe^{-x}, e^{-x}\}$
8. $3e^{6x}\cos(x-1)$	$\{e^{6x}\sin(x-1), e^{6x}\cos(x-1)\}$
9. $x\sin(2x)$	$\{x\sin(2x), x\cos(2x), \sin(2x), \cos(2x)\}$
10. $x\sin(2x) + 3\cos(2x)$	$\{x\sin(2x), x\cos(2x), \sin(2x), \cos(2x)\}$
11. $x^2 e^{5x}\sin(3x)$	$\{x^2 e^{5x}\sin(3x), x^2 e^{5x}\cos(3x), xe^{5x}\sin(3x),$ $xe^{5x}\cos(3x), e^{5x}\sin(3x), e^{5x}\cos(3x)\}$

Example 3 The function $f(x) = x^3$ is a UC function. Now

$$f'(x) = 3x^2, \quad f''(x) = 6x, \quad f'''(x) = 6, \quad \text{and} \quad f^{(n)}(x) = 0, \text{ for all } n \geq 4,$$

so the linearly independent UC functions of which successive derivatives of f are either constant multiples or linear combinations are those given by x^2, x, and 1. Thus the UC set of x^3 is

$$\{x^3, x^2, x, 1\}.$$

Example 4 For the UC function $f(x) = \sin 2x$,

$$f'(x) = 2\cos 2x, \quad \text{and} \quad f''(x) = -4\sin 2x,$$

so the UC set of $\sin 2x$ is

$$\{\sin 2x, \cos 2x\}.$$

Table 4.2 gives the general situation as it summarizes our results concerning UC functions.

Table 4.2: UC Functions and Their Corresponding UC Set	
UC Function	**UC Set**
1. x^n	$\{x^n, x^{n-1}, \ldots, x, 1\}$
2. e^{ax}	$\{e^{ax}\}$
3. $\sin(bx + c)$ or $\cos(bx + c)$	$\{\sin(bx + c), \cos(bx + c)\}$
4. $x^n e^{ax}$	$\{x^n e^{ax}, x^{n-1} e^{ax}, \ldots, x e^{ax}, e^{ax}\}$
5. $x^n \sin(bx + c)$ or $x^n \cos(bx + c)$	$\{x^n \sin(bx + c), x^n \cos(bx + c),$ $x^{n-1} \sin(bx + c), x^{n-1} \cos(bx + c), \ldots,$ $x \sin(bx + c), x \cos(bx + c),$ $\sin(bx + c), \cos(bx + c)\}$
6. $e^{ax} \sin(bx + c)$ or $e^{ax} \cos(bx + c)$	$\{e^{ax} \sin(bx + c), e^{ax} \cos(bx + c)\}$
7. $x^n e^{ax} \sin(bx + c)$ or $x^n e^{ax} \cos(bx + c)$	$\{x^n e^{ax} \sin(bx + c), x^n e^{ax} \cos(bx + c),$ $x^{n-1} e^{ax} \sin(bx + c), x^{n-1} e^{ax} \cos(bx + c),$ $\ldots, x e^{ax} \sin(bx + c), x e^{ax} \cos(bx + c),$ $e^{ax} \sin(bx + c), e^{ax} \cos(bx + c)\}$

The Method

Equipped with the table in Table 4.2, we are able to give an outline of the method of undetermined coefficients for finding a particular solution y_p of (4.11) for which the nonhomogeneous term F is a linear combination of UC functions.

There are 5 steps to the method:

1. For each of the UC functions $u_1, u_2, \ldots, u_m$ of which the nonhomogeneous term F is a linear combination, form the corresponding UC set, that is
$$S_1, S_2, \ldots, S_m.$$

2. Suppose that one of the UC sets formed, say S_j, is identical to (or contained in) another, say S_k. Then omit the set S_j from your list.

3. For the UC sets remaining after step **2**, examine the list to see if the complementary solution y_c is listed, say in S_ℓ. If this is the case, multiply <u>each</u> member of this UC set S_ℓ by the lowest positive

integer power of x so that the resulting revised set will contain no members that are solutions of the corresponding homogeneous equation.

4. In general, there remain (i) certain elements of the original UC sets and (ii) revised UC sets. From these UC sets, form a linear combination of <u>all</u> the elements of these sets in each of these categories. The linear combination is formed with unknown constant coefficients.

5. Determine the unknown coefficients by substituting the linear combination formed in step **4** into the differential equation and demanding it identically satisfy the differential equation.

The first 4 steps above are the easiest and create the correct general form of a particular solution. That is, after the first 4 steps of the UC method we end up with a linear combination of terms with unknown constant coefficients. Step 5 requires us to find these unknown (i.e., undetermined) coefficients so that the function will be a solution of the differential equation in question.

Example 5 Solve the differential equation

$$\frac{d^2y}{dx^2} - 2\frac{dy}{dx} - 3y = 2e^x - 10\sin x.$$

Solution
The corresponding homogeneous equation is

$$\frac{d^2y}{dx^2} - 2\frac{dy}{dx} - 3y = 0,$$

which has characteristic equation

$$r^2 - 2r - 3 = 0.$$

This equation has roots $r = 3$ and $r = -1$. The complementary solution is

$$y_c = c_1 e^{3x} + c_2 e^{-x}.$$

The nonhomogeneous term is the linear combination $2e^x - 10\sin x$ of the UC functions e^x and $\sin x$. We now apply the method just outlined to first obtain the correct general form of the particular solution (steps 1–4) and then to find the specific particular solution for this problem.

1. Form the UC set for each of these; we have

$$S_1 = \{e^x\}$$

and

$$S_2 = \{\sin x, \cos x\}.$$

2. Neither of these are identical or contained in each other, so both are retained.

3. The terms e^{3x} and e^{-x} from the complementary solution are not in S_1 and S_2, so there are no revisions required.

4. Form the linear combination

$$y_p = \underbrace{Ae^x}_{\text{from } S_1} + \underbrace{B_1 \sin x + B_2 \cos x}_{\text{from } S_2}$$

as a particular solution.

5. Find A, B_1, and B_2 such that y_p is a solution of the differential equation. We differentiate to obtain

$$y_p' = Ae^x + B_1 \cos x - B_2 \sin x \text{ and } y_p'' = Ae^x - B_1 \sin x - B_2 \cos x.$$

Substitution gives

$$Ae^x - B_1 \sin x - B_2 \cos x - 2(Ae^x + B_1 \cos x - B_2 \sin x)$$
$$-3(Ae^x + B_1 \sin x + B_2 \cos x) = 2e^x - 10 \sin x,$$

which simplifies as

$$-4Ae^x + (-4B_1 + 2B_2)\sin x + (-4B_2 - 2B_1)\cos x = 2e^x - 10 \sin x.$$

This equation is true for all x, so that

$$-4A = 2, \quad -4B_1 + 2B_2 = -10, \quad \text{and} \; -4B_2 - 2B_1 = 0.$$

The system solves to give

$$A = -\frac{1}{2}, \; B_1 = -2, \quad \text{and} \quad B_2 = 1.$$

Thus, a particular solution is

$$y_p = -\frac{1}{2}e^x - 2\sin x + \cos x.$$

Hence, the general solution is

$$y = y_c + y_p = c_1 e^{3x} + c_2 e^{-x} - \frac{1}{2}e^x - 2\sin x + \cos x.$$

Example 6   Solve the differential equation

$$\frac{d^2y}{dx^2} - 3\frac{dy}{dx} + 2y = 2x^2 + e^x + 2xe^x + 4e^{3x}.$$

The corresponding homogeneous equation has the characteristic equation

$$r^2 - 3r + 2 = 0$$

which has roots $r = 2$ and $r = 1$. Thus, the corresponding solutions are e^{2x} and e^x, which gives the complementary solution as

$$y_c = c_1 e^x + c_2 e^{2x}.$$

The nonhomogeneous term is $2x^2 + e^x + 2xe^x + 4e^{3x}$, which corresponds to the UC functions x^2, e^x, xe^x, and e^{3x}.

1. For each of these functions, we form the corresponding UC set:

$$S_1 = \{x^2, x, 1\},$$

$$S_2 = \{e^x\},$$

$$S_3 = \{xe^x, e^x\},$$

and

$$S_4 = \{e^{3x}\}.$$

2. Note that S_2 is completely contained in S_3, so that S_2 is omitted from further consideration.

3. Further note that

$$S_3 = \{xe^x, e^x\}$$

includes e^x which is part of the complementary function; thus multiply each member of S_3 by x to obtain the revised set

$$S_3' = \{x^2 e^x, xe^x\}$$

which does not contain solutions of the corresponding homogeneous equation.

4. From the revised and remaining UC sets, we have the six UC functions $x^2, x, 1, e^{3x}, x^2 e^x$, and xe^x, from which we form the linear combination

$$A_1 x^2 + A_2 x + A_3 + \quad B e^{3x} \quad + C_1 x^2 e^x + C_2 xe^x.$$

5. A particular solution is thus of the form

$$y_p = A_1 x^2 + A_2 x + A_3 + B e^{3x} + C_1 x^2 e^x + C_2 xe^x.$$

In order to find these unknown coefficients, we differentiate y_p to obtain

$$y_p' = 2A_1 x + A_2 + 3B e^{3x} + 2C_1 xe^x + C_1 x^2 e^x + C_2 e^x + C_2 xe^x$$

and

$$y_p'' = 2A_1 + 9B e^{3x} + C_1 x^2 e^x + 4C_1 xe^x + 2C_1 e^x + C_2 xe^x + 2C_2 e^x.$$

Substituting for y'', y' and y gives

$$2A_1 + 9Be^{3x} + C_1x^2e^x + 4C_1xe^x + 2C_1e^x + C_2xe^x + 2C_2e^x$$

$$- 3(2A_1x + A_2 + 3Be^{3x} + 2C_1xe^x + C_1x^2e^x + C_2e^x + C_2xe^x)$$

$$+ 2(A_1x^2 + A_2x + A_3 + Be^{3x} + C_1x^2e^x + C_2xe^x)$$

$$= 2x^2 + e^x + 2xe^x + 4e^{3x}.$$

This simplifies as

$$(2A_1 - 3A_2 + 2A_3) + (2A_2 - 6A_1)x + 2A_1x^2 + 2Be^{3x}$$
$$-2C_1xe^x + (2C_1 - C_2)e^x$$
$$= 2x^2 + e^x + 2xe^x + 4e^{3x}. \tag{4.12}$$

Comparing coefficients gives

$$2A_1 - 3A_2 + 2A_3 = 0, \quad 2A_2 - 6A_1 = 0, \quad 2A_1 = 2,$$
$$2B = 4, \quad -2C_1 = 2, \quad 2C_1 - C_2 = 1. \tag{4.13}$$

Solving these six equations for the six unknown constants gives $A_1 = 1$, $A_2 = 3$, $A_3 = 7/2$, $B = 2$, $C_1 = -1$, and $C_2 = -3$. A particular solution is

$$y_p = x^2 + 3x + \frac{7}{2} + 2e^{3x} - x^2e^x - 3xe^x$$

and the general solution is

$$y = y_c + y_p = c_1e^x + c_2e^{2x} + x^2 + 3x + \frac{7}{2} + 2e^{3x} - x^2e^x - 3xe^x.$$

Example 7 Solve the differential equation

$$\frac{d^4y}{dx^4} + \frac{d^2y}{dx^2} = 3x^2 + 4\sin x - 2\cos x.$$

Solution
The corresponding homogeneous equation is

$$\frac{d^4y}{dx^4} + \frac{dy^2}{dx^2} = 0$$

which has characteristic equation

$$r^4 + r^2 = 0.$$

This polynomial factors as

$$r^2(r^2 + 1) = 0$$

so that $r = 0$ is a double root and $r = \pm i$. The complementary solution is thus

$$y_c = c_1 + c_2 x + c_3 \sin x + c_4 \cos x.$$

The nonhomogeneous term is the linear combination $3x^2 + 4 \sin x - 2 \cos x$ of the UC functions x^2, $\sin x$, and $\cos x$.

1. The UC sets are

$$S_1 = \{x^2, x, 1\},$$
$$S_2 = \{\sin x, \cos x\},$$
$$S_3 = \{\cos x, \sin x\}.$$

2. S_2 and S_3 are identical; we only retain one. Thus, the UC sets are

$$S_1 = \{x^2, x, 1\},$$
$$S_2 = \{\sin x, \cos x\}.$$

3. Notice

$$S_1 = \{x^2, x, 1\}$$

contains 1 and x, which <u>are</u> solutions of the corresponding homogeneous differential equation. Thus, we multiply <u>each</u> element by x^2 to obtain

$$S_1' = \{x^4, x^3, x^2\},$$

none of which are solutions. Similarly, $\sin x$ and $\cos x$ are solutions of the corresponding homogeneous differential equation. Thus, we multiply each by an x, so that

$$S_2' = \{x \sin x, x \cos x\}.$$

4. We form the linear combination of these five UC functions, i.e.,

$$\underbrace{A_1 x^4 + A_2 x^3 + A_3 x^2}_{\text{from } S_1'} + \underbrace{B_1 x \sin x + B_2 x \cos x}_{\text{from } S_2'},$$

where A_1, A_2, A_3, B_1, and B_2 are undetermined.

5. Substitute the particular solution

$$y_p = A_1 x^4 + A_2 x^3 + A_3 x^2 + B_1 x \sin x + B_2 x \cos x$$

into the differential equation in order to determine the unknown coefficients.

Differentiating gives

$$y_p' = 4A_1 x^3 + 3A_2 x^2 + 2A_3 x + B_1 x \cos x + B_1 \sin x - B_2 x \sin x + B_2 \cos x,$$

$$y_p'' = 12A_1 x^2 + 6A_2 x + 2A_3 - B_1 x \sin x + 2B_1 \cos x - B_2 x \cos x - 2B_2 \sin x,$$

$$y_p''' = 24A_1 x + 6A_2 - B_1 x \cos x - 3B_1 \sin x + B_2 x \sin x - 3B_2 \cos x,$$

and

$$y_p^{(4)} = 24A_1 + B_1 x \sin x - 4B_1 \cos x + B_2 x \cos x + 4B_2 \sin x.$$

Substitution gives

$$24A_1 + B_1 x \sin x - 4B_1 \cos x + B_2 x \cos x + 4B_2 \sin x$$

$$+ 12A_1 x^2 + 6A_2 x + 2A_3 - B_1 x \sin x + 2B_1 \cos x - B_2 x \cos x - 2B_2 \sin x$$

$$= 3x^2 + 4 \sin x - 2 \cos x.$$

Equating coefficients gives

$$24A_1 + 2A_3 = 0, \quad 6A_2 = 0, \quad 12A_1 = 3,$$
$$-2B_1 = -2, \quad \text{and } 2B_2 = 4. \tag{4.14}$$

Solving this system of equations gives

$$A_1 = \frac{1}{4}, \quad A_2 = 0, \quad A_3 = -3, \quad B_1 = 1, \quad B_2 = 2,$$

so that a particular solution is

$$y_p = \frac{1}{4}x^4 - 3x^2 + x \sin x + 2x \cos x.$$

This gives the general solution

$$y = c_1 + c_2 x + c_3 \sin x + c_4 \cos x + \frac{1}{4}x^4 - 3x^2 + x \sin x + 2x \cos x.$$

It may often be easier to do steps **4** and **5** in the above method with the help of our computer programs and examples are seen at the end of the chapter.

● ● ● ● ● ● ● ● ● ● ●

Problems

In Problems **1–8**, write the form of the particular solution but do not solve the differential equation. Make sure to solve for y_c first!

1. $y'' + y = e^{-x} + x^2$
2. $y'' + 3y' = x^2 e^{2x} + \cos x$
3. $y'' + y = 4 \sin x + e^x \cos x$
4. $y'' + 4y' + y = \sin(2x) + \cos x$
5. $y''' + y = xe^{2x} \cos x + \sin x$
6. $y'' - y = e^x + xe^{-x}$
7. $y''' + 8y = 4 \sin x + \cos x$
8. $y^{(5)} + y'' = x^3 + xe^x \cos x$

In Problems **9–15**, use the method of undetermined coefficients (step **5** only) to find the coefficients of the particular solution for the given equations.

9. $y'' + 4y' + 13y = e^{-2x}, \quad y_p = Ae^{-2x}$
10. $y'' + 4y' = x^2 - 3, \quad y_p = Ax^3 + Bx^2 + Cx$
11. $y'' - 4y' = \cos x + 3 \sin x, \quad y_p = A \cos x + B \sin x$
12. $9y'' + y = \cos x + \sin(2x), \quad y_p = A \cos x + B \sin x + C \cos(2x) + E \sin(2x)$
13. $4y'' + 25y = x \cos x, \quad y_p = A \cos x + B \sin x + Cx \cos x + Ex \sin x$
14. $y'' + 2y' + 17y = e^x + 2, \quad y_p = Ae^x + B$
15. $y'' + 3y' - 10y = xe^x + 2x, \quad y_p = Ae^x + Bxe^x + Cx + E$

For Problems **16–28**, find the general solution of the given differential equations.

16. $y'' + y' - 6y = 2x$

17. $y'' - 2y' - 8y = 4e^{2x} - 21e^{-3x}$

18. $4y'' - 4y' - 3y = \cos 2x$

19. $y'' - 2y' + 2y = x\cos x$

20. $y'' - 16y = 2e^{4x}$

21. $y'' - 2y' + 5y = e^x \sin 2x$

22. $4y'' + 4y' + y = 3xe^x$

23. $y'' - 2y' + 5y = 2xe^x + e^x \sin 2x$

24. $y''' + y' = 2 - \sin x$

25. $y'' + 4y' + 3y = \cosh x$

26. $y'' + 2y' + y = 2e^{-x}$

27. $y''' + 10y'' + 34y' + 40y = xe^{-4x}$

28. $y^{(4)} - 18y'' + 81y = e^{3x}$

For Problems **29–34**, find the general solution of the given differential equations. Then find the solution that passes through the given initial condition.

29. $y'' - 4y = 32x$, $y(0) = 0, y'(0) = 6$

30. $y'' + y = \cos x$, $y(0) = 1$, $y'(0) = -1$

31. $y'' - 3y' + 2y = e^x$, $y(0) = 1, y'(0) = 0$

32. $y'' + 4y = x^2 + 3e^x$, $y(0) = 0, y'(0) = 2$

33. $y'' + 6y' + 10y = 3xe^{-3x} - 2e^{3x}\cos x, y(0) = 1, y'(0) = -2$

34. $y'' - 2y' - 3y = 3xe^{2x}$, $y(0) = 1$, $y'(0) = 0$

35. The method of undetermined coefficients can be used to solve first-order constant coefficient nonhomogeneous equations. Use this method to solve the following problems:

(a) $y' - 4y = x^2$

(b) $y' + y = \cos 2x$

(c) $y' - y = e^{4x}$

36. Suppose that $y_1(x)$ and $y_2(x)$ are solutions of

$$a\frac{d^2y}{dx^2} + b\frac{dy}{dx} + cy = f(x),$$

where a, b, and c are positive constants.

(a) Show that

$$\lim_{x \to \infty} [y_2(x) - y_1(x)] = 0.$$

(b) Is the result of (a) true if $b = 0$?

(c) Suppose that $f(x) = k$, where k is a constant. Show that

$$\lim_{x \to \infty} y(x) = \frac{k}{c}$$

for every solution $y(x)$ of $a\dfrac{d^2y}{dx^2} + b\dfrac{dy}{dx} + cy = k.$

(d) Determine the solution $y(x)$ of $a\dfrac{d^2y}{dx^2} + b\dfrac{dy}{dx} = k$. Find $\lim_{x \to \infty} y(x)$.

(e) Determine the solution $y(x)$ of $a\dfrac{d^2y}{dx^2} = k$. Find $\lim_{x \to \infty} y(x)$.

37. (a) Let $f(x)$ be a polynomial of degree n. Show that, if $b \neq 0$, there is always a solution that is a polynomial of degree n for the equation $y'' + ay' + by = f(x)$.

(b) Find a particular solution of $y'' + 3y' + 2y = 9 + 2x - 2x^2$.

38. In many physical applications, the nonhomogeneous term $F(x)$ is specified by different formulas in different intervals of x.

(a) Find a general solution of the equation

$$y'' + y = \begin{cases} x, & 0 \leq x \leq 1, \\ 1, & 1 \leq x. \end{cases}$$

Note that the solution is not differentiable at $x = 1$.

(b) Find a particular solution of

$$y'' + y = \begin{cases} x, & 0 \leq x \leq 1, \\ 1, & 1 \leq x \end{cases}$$

that satisfies the initial conditions $y(0) = 0$ and $y'(0) = 1$.

4.3   Method of Undetermined Coefficients via Annihilation

We saw in the previous section how to obtain the particular solution by first assuming the correct general form (based on $F(x)$) and then substituting into the differential equation to obtain the specific particular solution that satisfies the equation. Here, we present an alternative method to obtaining the correct general form based on $F(x)$. The operator notation gives us a useful idea, as can be seen in the following example.

Example 1   Consider the nonhomogeneous constant coefficient differential equation

$$\frac{d^2y}{dx^2} - y = x,$$

which can be written as

$$(D^2 - 1)y = x \tag{4.15}$$

using operator notation. The general solution to this nonhomogeneous equation, $y = y_c + y_p$, must satisfy this equation upon substitution into it. If we differentiate the given equation twice, we obtain the fourth-order linear homogeneous equation with constant coefficients

$$\frac{d^4y}{dx^4} - \frac{d^2y}{dx^2} = 0, \tag{4.16}$$

which can also be written as

$$D^2(D^2 - 1)y = D^2 x = 0. \tag{4.17}$$

The general solution to (4.15) must also satisfy the differential equation (4.17). We note that (4.17), or equivalently, (4.16), is *homogeneous* with constant coefficients. This equation has the characteristic polynomial

$$r^4 - r^2 = 0,$$

which can be further factored as

$$r^2(r + 1)(r - 1) = 0.$$

Thus the general solution of (4.17) must be of the form

$$y(x) = A_1 + A_2 x + c_1 e^{-x} + c_2 e^{x}$$

for constants A_1, A_2, c_1, c_2. Substituting this form of the solution into the original equation gives

$$(c_1 e^{-x} + c_2 e^{x}) - (c_1 e^{-x} + c_2 e^{x} + k_1 + k_2 x) = x$$

so that

$$A_1 = 0 \text{ and } A_2 = -1$$

by comparing coefficients. Thus,

$$y(x) = -x + c_1 e^{-x} + c_2 e^{x}.$$

This example motivates the idea of an **annihilator**. Notice we were able to obtain a homogeneous equation by "annihilating" the nonhomogeneous term x.

Definition 4.3.1

The linear differential operator $Q(D)$ annihilates a function $y(x)$ if

$$Q(D)y(x) = 0$$

for all x. In this case, $Q(D)$ is called an annihilator of $y(x)$.

Example 2 The linear differential operator $Q_1(D) = D^3$ annihilates the function $y(x) = x^2$ because $D^3 x^2 = 0$. Similarly, the linear differential operator $Q_2(D) = D - 2$ annihilates the function $y(x) = e^{2x}$ as

$$Q_2(D)(e^{2x}) = De^{2x} - 2e^{2x} = 0.$$

Example 3 Using the functions of the previous example, we can also observe that

$$Q_1(D)Q_2(D)(x^2 + e^{2x}) = D^3(D - 2)(x^2 + e^{2x}) = 0.$$

This property holds in general. That is, if $Q_1(D)$ and $Q_2(D)$ are linear differential operators having constant coefficients with Q_1 annihilating $y_1(x)$ and Q_2 annihilating $y_2(x)$, then Q_1Q_2 annihilates $c_1y_1(x) + c_2y_2(x)$, where c_1 and c_2 are constants. This is easy to see, as

$$Q_1Q_2(c_1y_1(x) + c_2y_2(x)) = Q_1Q_2(c_1y_1(x)) + Q_1Q_2(c_2y_2(x))$$

$$= c_1Q_1Q_2(y_1(x)) + c_2Q_1Q_2(y_2(x))$$

$$= c_1Q_2Q_1(y_1(x)) + c_2Q_1Q_2(y_2(x))$$

$$= c_1Q_2(0) + c_2Q_1(0)$$

$$= 0.$$

It is important to note that Q_2Q_1 also annihilates $c_1y_1(x) + c_2y_2(x)$ because the operator notation is commutative (in the constant coefficient case); see (3.8).

Annihilators of Familiar Functions

It will be useful for us to know annihilators for our familiar functions. The differential equation

$$\frac{d^ny}{dx^n} = 0$$

in operator form is

$$D^ny = 0,$$

and has solution

$$y = c_1 + c_2x + c_3x^2 + \ldots + c_{n-1}x^{n-1}.$$

Thus, D^n annihilates the functions

$$1, x, x^2, \ldots, x^{n-1}$$

as well as any linear combination of these functions. This result means that any nth degree polynomial in x will be annihilated by an $(n+1)$st differential operator.

A general solution of the differential equation

$$(D - r)^ny = 0$$

is

$$y = c_1 e^{rx} + c_2 x e^{rx} + c_3 x^2 e^{rx} + \ldots + c_{n-1} x^{n-1} e^{rx}$$

since we see that r is a root of the characteristic equation and the root occurs n times. Thus, the differential operator $(D - r)^n$ annihilates the functions

$$e^{rx}, x e^{rx}, x^2 e^{rx}, \ldots, x^{n-1} e^{rx}.$$

We show this for the case $n = 2$:

$$
\begin{aligned}
(D - r)^2 &(c_1 e^{rx} + c_2 x e^{rx}) \\
&= (D - r)^2 (c_1 e^{rx}) + (D - r)^2 (c_2 x e^{rx}) \\
&= c_1 (D - r)^2 (e^{rx}) + c_2 (D - r)^2 (x e^{rx}) \\
&= c_1 (D - r)(r e^{rx} - r e^{rx}) + c_2 (D - r)(r x e^{rx} + e^{rx} - r x e^{rx}) \\
&= c_2 (D - r)(e^{rx}) \\
&= 0.
\end{aligned}
\tag{4.18}
$$

If the characteristic equation of a differential equation has complex roots

$$r_{1,2} = a \pm bi$$

with $b \neq 0$, we can write the characteristic equation as

$$(r - (a + bi))(r - (a - bi)) = r^2 - 2ar + (a^2 + b^2) = 0$$

which corresponds to the differential equation

$$y'' - 2ay' + (a^2 + b^2)y = [D^2 - 2aD + (a^2 + b^2)]y = 0.$$

Thus, the differential operator

$$D^2 - 2aD + (a^2 + b^2)$$

annihilates the functions $e^{ax} \cos bx$ and $e^{ax} \sin bx$.

If the complex roots are repeated with multiplicity n, they correspond to the characteristic equation

$$((r - (a + bi))(r - (a - bi)))^n = (r^2 - 2ar + (a^2 + b^2))^n = 0,$$

but this corresponds to an order $2n$ differential equation, which in operator form is

$$(D^2 - 2aD + (a^2 + b^2))^n y = 0.$$

Thus, the differential operator

$$(D^2 - 2aD + (a^2 + b^2))^n$$

annihilates the functions

$$e^{ax} \cos bx, e^{ax} \sin bx, x e^{ax} \cos bx, x e^{ax} \sin bx, \ldots, x^{n-1} e^{ax} \cos bx, x^{n-1} e^{ax} \sin bx.$$

These results are summarized in Table 4.3.

Table 4.3: Annihilators of Familiar Functions

Function	Annihilator
$1, x, x^2, \ldots, x^{n-1}$	D^n
$e^{rx}, xe^{rx}, \ldots, x^{n-1}e^{rx}$	$(D-r)^n$
$e^{ax}\cos bx, xe^{ax}\cos bx, \ldots x^{n-1}e^{ax}\cos bx$	$(D^2 - 2aD + (a^2 + b^2))^n$
$e^{ax}\sin bx, xe^{ax}\sin bx, \ldots x^{n-1}e^{ax}\sin bx$	$(D^2 - 2aD + (a^2 + b^2))^n$

With the ideas of annihilators in place, we return to the problem of solving nonhomogeneous equations with constant coefficients. If the nth-order linear nonhomogeneous equation with constant coefficients can be written as

$$P(D)y = F(x)$$

where $F(x)$ is one of the functions (or combinations of functions) listed in Table 4.3, then we can find another differential operator $Q(D)$ that annihilates $F(x)$.

If the differential operator $Q(D)$ annihilates $F(x)$, applying $Q(D)$ to the nonhomogeneous equation gives

$$Q(D)P(D)y = Q(D)F(x) = 0,$$

which is a homogeneous equation.

Let's see how to apply these ideas.

Example 4 Solve the nonhomogeneous equation

$$y'' + y = x^2.$$

Solution

If we write this equation in operator notation we have

$$(D^2 + 1)y = x^2.$$

Now the annihilator of x^2 is D^3. We thus can rewrite our original equation as the homogeneous equation

$$D^3(D^2 + 1)y = 0,$$

which has characteristic equation

$$r^3(r^2 + 1) = 0.$$

This gives the eigenvalues $0, 0, 0, i, -i$ so that the solution to the homogeneous equation is

$$y(x) = \underbrace{A_1 + A_2x + A_3x^2}_{\text{from } r = 0, 0, 0} + \underbrace{B_1 \sin x + B_2 \cos x}_{\text{from } r = \pm i}. \tag{4.19}$$

Now if we solve the corresponding homogeneous equation

$$y'' + y = 0$$

we have

$$y_c(x) = c_1 \sin x + c_2 \cos x.$$

Eliminating these terms from solution (4.19) indicates the particular solution has the form

$$y_p(x) = A_1 + A_2 x + A_3 x^2.$$

Notice that we would have obtained the same form for y_p if we used Steps 1–4 from the method introduced in the previous section. The name "method of undetermined coefficients" really applies to the methods once the correct general form of y_p has been assumed. At this point, we thus proceed with Step 5 as before; see Section 4.1. In order to use the original differential equation, we will need to obtain the second derivative to y_p in order to substitute. Calculating the derivatives gives

$$y_p' = A_2 + 2A_3 x \quad \text{and} \quad y_p'' = 2A_3.$$

Substitution gives

$$y_p'' + y_p = 2A_3 + A_1 + A_2 x + A_3 x^2.$$

We know that $y_p'' + y_p = x^2$ and thus

$$x^2 = 2A_3 + A_1 + A_2 x + A_3 x^2.$$

Equating coefficients of like terms gives

$$A_1 = -2, A_2 = 0 \text{ and } A_3 = 1.$$

Thus a particular solution is

$$y_p(x) = x^2 - 2,$$

and the general solution to the nonhomogeneous equation is

$$y(x) = y_c(x) + y_p(x) = c_1 \sin x + c_2 \cos x + x^2 - 2.$$

Example 5 Solve the nonhomogeneous equation

$$y''' + y' = \cos x + x.$$

Solution We first will find the annihilator of the nonhomogeneous term $\cos x + x$. Since $D^2 + 1$ annihilates $\cos x$ and D^2 annihilates x, the annihilator of $\cos x + x$ is

$$Q(D) = D^2(D^2 + 1).$$

If we apply this annihilator to both sides of the nonhomogeneous equation, we have
$$D^2(D^2+1)(D^3+D)y = D^2(D^2+1)(\cos x + x).$$

The right-hand side is thus zero and our equation simplifies to the homogeneous equation
$$D^2(D^2+1)(D^3+D)y = 0.$$

The corresponding characteristic equation is
$$r^2(r^2+1)(r^3+r) = 0,$$

which can be factored as
$$r^3(r^2+1)^2 = 0.$$

The eigenvalues are thus
$$0,0,0,-i,-i,i,i$$

so that the solution of this homogeneous equation is
$$y(x) = \underbrace{A_1 + A_2x + A_3x^2}_{\text{from } r = 0,0,0} + \underbrace{B_1\cos x + B_2\sin x + B_3x\cos x + B_4x\sin x}_{\text{from } r = \pm i, \pm i}.$$

As we want to assume the simplest form for the particular solution that will satisfy the nonhomogeneous equation, we ignore any factors that also occur in the homogeneous equation. Solving the corresponding homogeneous equation $y''' + y' = 0$, we find the characteristic polynomial is $r(r^2+1)$ and the complementary solution is thus
$$y_c(x) = c_1 + c_2\cos x + c_3\sin x.$$

Comparing $y(x)$ and $y_c(x)$ we determine that the particular solution must be of the form
$$y_p(x) = A_2x + A_3x^2 + B_3x\cos x + B_4x\sin x.$$

This is again the same form we could have obtained with Steps 1–4 of Section 4.1. We again use the original nonhomogeneous equation to determine the constants A_2, A_3, B_3, and B_4. Differentiating y_p three times and substituting into the left-hand side of the nonhomogeneous equation gives
$$y_p''' + y_p' = -3B_3\cos x + B_3x\sin x - 3B_4\sin x - B_4x\cos x$$

$$+A_2 + 2A_3x + B_3\cos x - B_3x\sin x + B_4\sin x + B_4x\cos x$$

$$= -2B_3\cos x - 2B_4\sin x + 2A_3x + A_2.$$

This must equal the right-hand side of the nonhomogeneous equation and so we have
$$-2B_3\cos x - 2B_4\sin x + 2A_3x + A_2 = \cos x + x.$$

Comparing coefficients gives

$$A_2 = 0, \ A_3 = \frac{1}{2}, \ B_3 = -\frac{1}{2}, \ \text{and} \ B_4 = 0.$$

This gives the particular solution of the nonhomogeneous equation as

$$y_p(x) = \frac{1}{2}x^2 - \frac{1}{2}x \cos x.$$

Thus, the general solution of the nonhomogeneous equation is

$$y(x) = c_1 + c_2 \cos x + c_3 \sin x + \frac{1}{2}x^2 - \frac{1}{2}x \cos x.$$

Example Solve $y'' + 2y' - 3y = 4e^x - \sin x$ with the conditions that $y(0) = 0$ and $y'(0) = 1$.

Solution

We first note that this problem differs from previous ones because of the presence of initial conditions, but these do not come into the problem until the last step. Proceeding as in previous problems, since $D - 1$ annihilates e^x and $D^2 + 1$ annihilates $\sin x$, the nonhomogeneous term $4e^x - \sin x$ is annihilated by $Q(D) = (D - 1)(D^2 + 1)$. Thus if we apply this annihilator to both sides of the differential equation we have

$$(D - 1)(D^2 + 1)(D^2 + 2D - 3)y = (D - 1)(D^2 + 1)(4e^x - \sin x)$$

and our problem becomes that of solving the homogeneous equation

$$(D - 1)^2(D^2 + 1)(D + 3)y = 0.$$

The corresponding characteristic equation is

$$(r - 1)^2(r^2 + 1)(r + 3) = 0$$

which has roots $-3, 1, 1, -i, i$. The solution is then

$$y(x) = \quad Ae^{-3x} \quad + B_1 e^x + B_2 x e^x + C_1 \cos x + C_2 \sin x. \tag{4.20}$$

The solution to the corresponding homogeneous equation $y'' + 2y' - 3y = 0$ with characteristic equation $r^2 + 2r - 3 = (r + 3)(r - 1) = 0$ is

$$y_c(x) = c_1 e^{-3x} + c_2 e^x.$$

We eliminate these terms from (4.20) to obtain the form for the particular solution as

$$y_p(x) = B_2 x e^x + C_1 \cos x + C_2 \sin x.$$

Differentiating $y_p(x)$ gives

$$y_p'(x) = B_2 e^x + B_2 x e^x - C_1 \sin x + C_2 \cos x$$
$$y_p''(x) = 2B_2 e^x + B_2 x e^x - C_1 \cos x - C_2 \sin x.$$

Substitution of these three equations into the original nonhomogeneous equation gives

$$4B_2 e^x + (2C_2 - 4C_1) \cos x + (-4C_2 - 2C_1) \sin x = 4e^x - \sin x,$$

so that

$$4B_2 = 4, \ \ 2C_2 - 4C_1 = 0, \ \ \text{and} \ -4C_2 - 2C_1 = -1.$$

Thus, $B_2 = 1, C_1 = \frac{1}{10}$, and $C_2 = \frac{1}{5}$ and the particular solution y_p is

$$y_p(x) = xe^x + \frac{1}{10} \cos x + \frac{1}{5} \sin x.$$

This gives the general solution as

$$y(x) = c_1 e^{-3x} + c_2 e^x + xe^x + \frac{1}{10} \cos x + \frac{1}{5} \sin x.$$

We now apply the initial conditions to determine the constants c_1 and c_2. Since

$$0 = y(0) = c_1 + c_2 + \frac{1}{10} \ \ \text{and} \ 1 = y'(0) = c_2 - 3c_1 + \frac{6}{5}$$

we can solve the two equations simultaneously to obtain

$$c_1 = \frac{1}{40} \ \ \text{and} \ c_2 = -\frac{1}{8}.$$

Thus the solution to the initial-value problem is

$$y(x) = \frac{1}{40} e^{-3x} - \frac{1}{8} e^x + xe^x + \frac{1}{10} \cos x + \frac{1}{5} \sin x.$$

Summary of the Annihilator Method

We can summarize the steps of the annihilator method of solving the nth-order linear nonhomogeneous equation with constant coefficients

$$P(D)y = F(x)$$

as follows:

1. Determine an annihilator of $F(x)$. That is, find $Q(D)$ such that

$$Q(D)F(x) = 0.$$

2. Apply the annihilator to both sides of the nonhomogeneous differential equation

$$Q(D)P(D)y = Q(D)F(x) = 0.$$

3. Solve the homogeneous equation

$$Q(D)P(D)y = 0.$$

4. Find the corresponding solution $y_c(x)$ to the homogeneous equation

$$P(D)y = 0$$

which corresponds to the original nonhomogeneous equation.

5. Eliminate the terms of the homogeneous equation $y_c(x)$ from the general solution obtained in Step **3.** The expression that remains is the correct form of a particular solution.

6. Solve for the unknown coefficients in the particular solution $y_p(x)$ by substituting it into the original nonhomogeneous equation

$$P(D)y = F(x).$$

7. A general solution of the nonhomogeneous equation is

$$y(x) = y_c(x) + y_p(x).$$

Problems

In Problems 1–8, use the annihilator method to write the form of the particular solution but do not solve the differential equation. You will need to do Steps 1–5 of the summary.

1. $y'' + y = e^{-x} + x^2$
2. $y'' + 3y' = x^2 e^{2x} + \cos x$
3. $y'' + y = 4\sin x + e^x \cos x$
4. $y'' + 4y' + y = \sin(2x) + \cos x$
5. $y''' + y = xe^{2x} \cos x + \sin x$
6. $y'' - y = e^x + xe^{-x}$
7. $y''' + 8y = 4\sin x + \cos x$
8. $y^{(5)} + y'' = x^3 + xe^x \cos x$

In Problems **9–15**, use the method of undetermined coefficients (only Step 6 of the summary) to find the coefficients of the particular solution for the given equations.

9. $y'' + 4y' + 13y = e^{-2x}$, $\quad y_p = Ae^{-2x}$
10. $y'' + 4y' = x^2 - 3$, $\quad y_p = Ax^3 + Bx^2 + Cx$

11. $y'' - 4y' = \cos x + 3 \sin x$, $y_p = A_1 \cos x + A_2 \sin x$

12. $9y'' + y = \cos x + \sin(2x)$, $y_p = A_1 \cos x + A_2 \sin x + B_1 \cos(2x) + B_2 \sin(2x)$

13. $4y'' + 25y = x \cos x$, $y_p = A_1 \cos x + A_2 \sin x + A_3 x \cos x + A_4 x \sin x$

14. $y'' + 2y' + 17y = e^x + 2$, $y_p = Ae^x + B$

15. $y'' + 3y' - 10y = xe^x + 2x$, $y_p = A_1 e^x + A_2 xe^x + B_1 x + B_2$

For Problems **16–28**, find the general solution of the given differential equations using the method of undetermined coefficients.

16. $y'' + y' - 6y = 2x$

17. $y'' - 2y' - 8y = 4e^{2x} - 21e^{-3x}$

18. $4y'' - 4y' - 3y = \cos 2x$

19. $y'' - 2y' + 2y = x \cos x$

20. $y'' - 16y = 2e^{4x}$

21. $y'' - 2y' + 5y = e^x \sin 2x$

22. $4y'' + 4y' + y = 3xe^x$

23. $y'' - 2y' + 5y = 2xe^x + e^x \sin 2x$

24. $y''' + y' = 2 - \sin x$

25. $y'' + 4y' + 3y = \cosh x$

26. $y'' + 2y' + y = 2e^{-x}$

27. $y''' + 10y'' + 34y' + 40y = xe^{-4x}$

28. $y^{(4)} - 18y'' + 81y = e^{3x}$

For Problems **29–34**, find the general solution of the given differential equations. Then find the solution that passes through the given initial condition.

29. $y'' - 4y = 32x$, $y(0) = 0, y'(0) = 6$

30. $y'' + y = \cos x$, $y(0) = 1$, $y'(0) = -1$

31. $y'' - 3y' + 2y = e^x$, $y(0) = 1, y'(0) = 0$

32. $y'' + 4y = x^2 + 3e^x$, $y(0) = 0, y'(0) = 2$

33. $y'' + 6y' + 10y = 3xe^{-3x} - 2e^{3x} \cos x$, $y(0) = 1, y'(0) = -2$

34. $y'' - 2y' - 3y = 3xe^{2x}$, $y(0) = 1$, $y'(0) = 0$

35. The annihilator method can be used to solve first-order constant coefficient nonhomogeneous equations. Use this method to solve the following problems:

(a) $y' - 4y = x^2$

(b) $y' + y = \cos 2x$

(c) $y' - y = e^{4x}$

36. Suppose that $y_1(x)$ and $y_2(x)$ are solutions of

$$a\frac{d^2 y}{dx^2} + b\frac{dy}{dx} + cy = f(x),$$

where a, b, and c are positive constants.

(a) Show that

$$\lim_{x \to \infty} [y_2(x) - y_1(x)] = 0.$$

(b) Is the result of (a) true if $b = 0$?

(c) Suppose that $f(x) = k$, where k is a constant. Show that

$$\lim_{x \to \infty} y(x) = \frac{k}{c}$$

for every solution $y(x)$ of $a\dfrac{d^2y}{dx^2} + b\dfrac{dy}{dx} + cy = k$.

(d) Determine the solution $y(x)$ of $a\dfrac{d^2y}{dx^2} + b\dfrac{dy}{dx} = k$. Find $\lim\limits_{x\to\infty} y(x)$.

(e) Determine the solution $y(x)$ of $a\dfrac{d^2y}{dx^2} = k$. Find $\lim\limits_{x\to\infty} y(x)$.

37. (a) Let $f(x)$ be a polynomial of degree n. Show that, if $b \neq 0$, there is always a solution that is a polynomial of degree n for the equation $y'' + ay' + by = f(x)$.
(b) Find a particular solution of $y'' + 3y' + 2y = 9 + 2x - 2x^2$.

38. In many physical applications, the nonhomogeneous term $F(x)$ is specified by different formulas in different intervals of x.
(a) Find a general solution of the equation

$$y'' + y = \begin{cases} x, & 0 \le x \le 1, \\ 1, & 1 \le x. \end{cases}$$

Note that the solution is not differentiable at $x = 1$.
(b) Find a particular solution of

$$y'' + y = \begin{cases} x, & 0 \le x \le 1, \\ 1, & 1 \le x \end{cases}$$

that satisfies the initial conditions $y(0) = 0$ and $y'(0) = 1$.

4.4 Exponential Response and Complex Replacement

This is the third method we will consider for finding a particular solution when the forcing function, $F(x)$, is a polynomial, sinusoidal, exponential, or a product or combination of the three. The method is elegant and very quick when the forcing function is an exponential and will require an additional method (e.g., undetermined coefficients for a polynomial-only forcing function) if either is multiplied by a polynomial. It utilizes the operator notation together with complex variables. We begin with our general nonhomogeneous with constant coefficients equation

$$P(D)y = F(x), \tag{4.21}$$

where $P(D) = a_n D^n + a_{n-1} D^{n-1} + \ldots + a_1 D + a_0$. From Section 3.1.1 and 3.6, we know that $D^k e^{rx} = r^k e^{rx}$ for any $k = 0, 1, \cdots$. Thus, our linear differential operator, $P(D)$, satisfies

$$P(D)e^{rx} = P(r)e^{rx}, \tag{4.22}$$

where $P(r)$ is the characteristic polynomial.

If we consider (4.21) with $F(x) = ae^{rx}$, then rearranging (4.22) gives

$$\frac{a}{P(r)} \cdot P(D)e^{rx} = ae^{rx}. \tag{4.23}$$

However, two key elements we learned in Section 4.1 were that (i) we always look for *a* particular solution and (ii) for a linear $P(D)$ we can pull constants through it. Thus (4.23) can be rewritten as

$$P(D)\left(\frac{a}{P(r)}e^{rx}\right) = ae^{rx}. \tag{4.24}$$

Since the ODE we are currently considering is $P(D)y = ae^{rx}$, we can conclude that

$$y_p = \frac{a}{P(r)}e^{rx} \tag{4.25}$$

is a particular solution, provided $P(r) \neq 0$. Since we just want to find any particular solution, we are finished (at least for this simple $F(x)$). $P(r) = 0$ will be a problem we will need to address but that will come later in this section.

Example 1 Find a particular solution for

$$y'' + 5y' + 10y = 3e^{-2x}.$$

Solution
Our operator is $P(D) = D^2 + 5D + 10$ so that $P(r) = r^2 + 5r + 10$. A particular solution is given for this forcing function by (4.25):

$$y_p = \frac{3}{P(-2)} \cdot e^{-2x} = \frac{3}{4}e^{-2x}.$$

This is a very efficient way to get a particular solution but it will only work quickly for forcing functions that are exponential as in our previous example. For most engineering applications, a system is forced with a sinusoidal term (via some type of motor) and the present method will again work because we know that sines and cosines can be written as the real and imaginary part of a complex exponential. The only extra amount of work will be to first rewrite the cosine or sine as the real or complex part of a complex exponential and then, after using the above method, keep only the part that is needed. This will work because the real and imaginary parts do not affect each other; see Section 3.5 for more discussion. Let's consider an example.

Example 2 Find a particular solution for

$$y'' + 3y' + 7y = 4\sin 2x.$$

Solution

Our first task is to rewrite $F(x) = 2\sin x$ as the imaginary part of a complex exponential. Euler's formula states $e^{ix} = \cos x + i\sin x$ so that

$$4\sin 2x = \mathrm{Im}(4e^{i2x}). \qquad (4.26)$$

Thus, we consider the complex version of our differential equation:

$$z'' + 3z' + 7z = 4e^{i2x},$$

where we have changed our variable from the real-valued y to the complex-valued z. A careful look back at our derivation of (4.25) should convince us that everything holds for the case of a complex variable. In other words,

$$P(D)z = ae^{irx} \implies z_p = \frac{a}{P(ir)}e^{irx} \qquad (4.27)$$

is a particular solution. Then, in our current example, a particular solution is given by

$$
\begin{aligned}
z_p &= \frac{4}{P(2i)}e^{i2x} = \frac{4e^{i2x}}{(2i)^2 + 3(2i) + 7} = \frac{4e^{i2x}}{-4 + 6i + 7} \\
&= \frac{4e^{i2x}}{3 + 6i} = \frac{4e^{i2x}}{3 + 6i} \cdot \frac{3 - 6i}{3 - 6i} = \frac{(12 - 24i)e^{i2x}}{9 + 36} \\
&= \frac{(12 - 24i)(\cos 2x + i\sin 2x)}{45} \\
&= \frac{12\cos 2x + 24\sin 2x}{45} + i\frac{12\sin 2x - 24\cos 2x}{45}. \qquad (4.28)
\end{aligned}
$$

Since we began with only the imaginary part in (4.26), we keep the imaginary part of z_p:

$$
\begin{aligned}
y_p &= \mathrm{Im}(z_p) = \mathrm{Im}\left(\frac{12\cos 2x + 24\sin 2x}{45} + i\frac{12\sin 2x - 24\cos 2x}{45}\right) \\
&= \frac{1}{45}(12\cos 2x + 24\sin 2x).
\end{aligned}
$$

Obviously, material from Section 3.5 will come in useful here. We present a few different situations of how to *complexify* a forcing function that is sinusoidal or a sum of sinusoidal functions and then address the situation of when they are multiplied by polynomials.

Example 3 Rewrite

$$2 \cos x + 3 \sin x$$

as a complex exponential.

Solution

If we had either term separately or if the coefficients were the same we could proceed as in Example 2. However, we instead rewrite this with a trigonometric identity as we did in Section 3.7:

$$c_1 \cos(\omega t) + c_2 \sin(\omega t) = A \cos(\omega t - \phi),$$

where $A = \sqrt{c_1^2 + c_2^2}$ and $\tan \phi = \frac{c_2}{c_1}$. Thus we have

$$2 \cos x + 3 \sin x = \sqrt{13} \cos(x - \phi), \qquad \text{where } \phi = \tan^{-1}\left(\frac{3}{2}\right)$$

from which it follows that

$$2 \cos x + 3 \sin x = \sqrt{13} \cos(x - \phi) = \text{Re}\left(\sqrt{13}e^{i(x-\phi)}\right).$$

In the case when we have different arguments of the sinusoidal terms, e.g., $3 \cos x + \sin 2x$, we remember one of our theorems from Section 4.1 that stated that if Y_1 is a particular solution to $P(D)y = F_1(x)$ and Y_2 is a particular solution to $P(D)y = F_2(x)$, then $Y_1 + Y_2$ is a particular solution to $P(D)y = F_1(x) + F_2(x)$. Thus we can find a particular solution by the methods we have learned in this section. If we multiply a sinusoidal term by an exponential we can again proceed as we have thus far.

Example 4 Find a particular solution to

$$y'' + 2y' + 4y = e^{-x} \cos 3x.$$

Solution

In order to use the methods of this section, we first rewrite the forcing function as $e^{-x} \cos 3x = \text{Re}(e^{x(-1+3i)})$ and then complexify the equation as

$$z'' + 2z' + 4z = e^{x(-1+3i)}.$$

We can find a particular solution as we did before:

$$z_p = \frac{1}{P(-1+3i)}e^{x(-1+3i)} = \frac{e^{x(-1+3i)}}{(-1+3i)^2 + 2(-1+3i) + 4}$$

$$= \frac{e^{x(-1+3i)}}{1 - 9 - 6i - 2 + 6i + 4} = \frac{e^{x(-1+3i)}}{-6}$$

$$= \frac{-e^{-x} \cos 3x}{6} + i\frac{-e^{-x} \sin 3x}{6}, \tag{4.29}$$

so that

$$y_p = \operatorname{Re}(z_p) = \frac{-e^{-x} \cos 3x}{6}$$

is a particular solution.

We have two other situations to consider. The first is when our operator $P(r) = 0$ because we would be dividing by zero. The second is when $F(x)$ is a polynomial of degree 1 or higher (degree 0, i.e., constants can be treated as $e^{0 \cdot x}$). When $P(r) = 0$, this says that the r from the e^r or e^{irx} forcing function (either real or complexified) is also a root of the characteristic equation. Thus, our method from this section is not going to give a linearly independent solution. In order to obtain a linearly independent solution in the case when r has multiplicity one in $P(r) = 0$, we consider xe^{rx} as a possibility and observe what results when we apply our operator, $P(D)$:

$$D(xe^{rx}) = (1 + rx)e^{rx}, \quad D^2(xe^{rx}) = r(2 + rx)e^{rx},$$
$$\cdots, \quad D^k(xe^{rx}) = r^{k-1}(k + rx)e^{rx}, \tag{4.30}$$

so that

$$
\begin{aligned}
P(D)(xe^{rx}) &= (a_0 D^n + a_1 D^{n-1} + \ldots + a_{n-1}D + a_n)(xe^{rx})\\
&= [a_0 r^{n-1}(n + rx) + \ldots + a_{n-1}(1 + rx) + a_n x]e^{rx}\\
&= [a_0 r^{n-1}n + a_1 r^{n-2}(n-1) + \ldots + a_{n-1}]e^{rx}\\
&\quad + [a_0 r^{n-1}rx + a_1 r^{n-2}rx + \ldots + a_{n-1}rx + a_n x]e^{rx}\\
&= [a_0 r^{n-1}n + a_1 r^{n-2}(n-1) + \ldots + a_{n-1}]e^{rx}\\
&\quad + [a_0 r^n + a_1 r^{n-1} + \ldots + a_{n-1}r + a_n]xe^{rx}\\
&= P'(r)e^{rx}. \tag{4.31}
\end{aligned}
$$

Thus if we consider $P(D)y = ae^{rx}$ with $P(r) = 0$, we have

$$P(D)(xe^{rx}) = P'(r)e^{rx} \implies P(D)\left(\frac{axe^{rx}}{P'(r)}\right) = ae^{rx}, \tag{4.32}$$

which tells us that $\dfrac{axe^{rx}}{P'(r)}$ is a particular solution of $P(D)y = ae^{rx}$ when r is a root of multiplicity one in $P(r) = 0$. By a similar argument and calculation, we can find that particular solutions of $P(D)y = ae^{rx}$ for repeated roots r of

$P(r) = 0$ are

$$y_p = \frac{ax^2 e^{rx}}{P''(r)} \quad \text{for } P(r) = P'(r) = 0, P''(r) \neq 0, \tag{4.33}$$

$$y_p = \frac{ax^3 e^{rx}}{P'''(r)} \quad \text{for } P(r) = P'(r) = P''(r) = 0, P'''(r) \neq 0, \tag{4.34}$$

$$\vdots$$

$$y_p = \frac{ax^k e^{rx}}{P^{(k)}(r)} \quad \text{for } P(r) = \cdots = P^{(k-1)}(r) = 0, P^{(k)}(r) \neq 0. \tag{4.35}$$

Example 5 Find a particular solution of

$$y''' - 6y'' + 9y' = 4e^{3x}.$$

Solution
We can easily see that the characteristic equation factors as $r(r - 3)^2 = 0$ so that $r = 3$ from the forcing function $4e^{3x}$ is a root that already occurs twice in $P(r) = 0$. From our previous discussion, (4.33) tells us that a particular solution is given by

$$y_p = \frac{4x^2 e^{3x}}{P''(3)} = \frac{4x^2 e^{3x}}{6(3) - 12} = \frac{2}{3}x^2 e^{3x}.$$

If our forcing function $P(r)$ had repeated roots due to sines and cosines, we would first complexify our differential equation, proceed as above, and then isolate the real or imaginary part as appropriate.

Exponential Shift
The other case to consider for $F(x)$ is when it is a polynomial in x, written as $F(x) = g(x)e^{rx}$. For reasons we will see shortly, we will rewrite this as $F(x) = ve^{rx}$ for some polynomial $v(x)$ (where we could always set $r = 0$). Applying the product rule, we observe that

$$D(ve^{rx}) = e^{rx}Dv + vre^{rx} = e^{rx}(D + r)v.$$

Taking a second derivative gives

$$D^2(ve^{rx}) = D(e^{rx}(D + r)v)$$
$$= e^{rx}D((D + r)v) + ((D + r)v)re^{rx}$$
$$= e^{rx}(D + r)^2 v.$$

In general, we have

$$D^k(ve^{rx}) = e^{rx}(D + r)^k v.$$

Our linear operator with constant coefficients can be written as a combination of these:

$$P(D)(ve^{rx}) = e^{rx}P(D+r)v. \tag{4.36}$$

We had begun our recent discussion by considering $P(D)y = g(x)e^{rx}$ with the goal of finding a particular solution. If we let $y = ve^{rx}$, then the ODE together with (4.36) gives

$$g(x)e^{rx} = P(D)y = P(D)(ve^{rx}) = e^{rx}P(D+r)v$$

$$\implies P(D+r)v = g(x) \tag{4.37}$$

after canceling the e^{rx}. If we solve equation (4.37) to find a particular solution v_p, then we have a particular solution $y_p = v_p e^{rx}$ for our original problem. With this, we are able to handle the same types of equations as in the previous two sections.

Example 6 Find a particular solution to $y'' + 3y' - 10y = 4xe^{-2x}$.

Solution

The homogeneous equation has characteristic equation $(r+5)(r-2) = 0$ and these roots cannot give rise to solutions that are the current forcing function. With no repeated roots we can proceed as in (4.37) with $g(x) = 4x$ and $r = -2$ to solve:

$$P(D-2)v = 4x \implies \left((D-2)^2 + 3(D-2) - 10\right)v = 4x$$
$$\implies (D^2 - D - 12)v = 4x. \tag{4.38}$$

This can be solved using the method of undetermined coefficients where we guess a solution of the form $v_p = A_1 x + A_2$. Substituting into (4.38) gives

$$-A_1 - 12(A_1 x + A_2) = 4x.$$

We then equate coefficients and solve to obtain $A_1 = \frac{-1}{3}$ and $A_2 = \frac{1}{36}$. Our particular solution is then

$$v_p = \frac{-x}{3} + \frac{1}{36} \implies y_p = \left(\frac{-x}{3} + \frac{1}{36}\right)e^{-2x}. \tag{4.39}$$

In the case that r from the forcing function $F(x) = g(x)e^{rx}$ is a k-fold root of $P(D)$, we rewrite our operator as

$$P(D) = Q(D)(D-r)^k \tag{4.40}$$

for polynomial operator $Q(D)$, which is equivalent to

$$P(D+r) = Q(D+r)D^k \qquad (4.41)$$

as can be seen by a simple change of variables. Equation (4.41) will be useful given our previous formula for the exponential shift (4.36). In order to find a particular solution, we can't guess ve^{rx} because this would not give us any new solutions using our previous exponential shift derivation. If we think back to previous situations when we had repeated roots, we could multiply by x^k and thus examine

$$P(D)(x^k v e^{rx}) = e^{rx} P(D+r)(x^k v)$$
$$= e^{rx} Q(D+r)D^k(x^k v).$$

We now rename $x^k v = \tilde{v}$ and note that we initially needed to have the x^k explicit in order to obtain a linearly independent solution. Using the fact that we need a particular solution, we then have

$$g(x)e^{rx} = P(D)(\tilde{v}e^{rx}) = e^{rx} Q(D+r)D^k \tilde{v}$$
$$= e^{rx} Q(D+r)u, \quad \text{where } u = D^k \tilde{v} \qquad (4.42)$$

so that $Q(D+r)u = g(x)$ is the equation whose particular solution we first need to find. This requires the method of undetermined coefficients to determine u_p. Because $u_p = D^k \tilde{v}_p$, we integrate k times to obtain $\tilde{v}_p$ and note that we can ignore any constants of integration because terms that arise from them can always be grouped into the homogeneous solution. Finally, we create the particular solution of $P(D)y = Q(D)(D-r)^k y = g(x)e^{rx}$ as

$$y_p = \tilde{v}_p e^{rx}.$$

Example 7 Find a particular solution to $(D-3)^2(D^2+1)y = 5x^4 e^{3x}$.

Solution
The root $r = 3$ from the forcing function occurs twice in the homogeneous solution (because of $(D-3)^2$) and we observe that $Q(D) = D^2 + 1$ so that $Q(D+3) = (D+3)^2 + 1 = D^2 + 6D + 10$. We then need to set $u = D^2 \tilde{v}$ and then find a particular solution of $Q(D+3)u = 5x^4$:

$$u'' + 6u' + 10u = 5x^4$$

where we assume
$$u_p = A_1 x^4 + A_2 x^3 + A_3 x^2 + A_4 x + A_5.$$

Substituting and equating coefficients gives us

$$u_p = \frac{237}{625} - \frac{144}{125}x + \frac{39}{25}x^2 - \frac{6}{5}x^3 + \frac{1}{2}x^4.$$

We then find $\tilde{v}_p$ by integrating u_p twice and then multiplying by e^{3x} to obtain our particular solution

$$y_p = e^{3x}\left(\frac{237}{1250}x^2 - \frac{24}{125}x^3 + \frac{13}{100}x^4 - \frac{3}{50}x^5 + \frac{1}{60}x^6\right).$$

Summary of the Exponential Response Method

We can summarize the steps of the exponential response method of solving the nth-order linear nonhomogeneous equation with constant coefficients $P(D)y = F(x)$ in which $F(x)$ is exponential, sinusoidal, a polynomial, or a combination of these as follows:

---

1. Solve the homogeneous equation $P(D)y = 0$ to obtain the fundamental set of solutions.
2. Write the forcing function $F(x)$ as an exponential, complexifying the problem if necessary.
3. Consider $F(x)$ from **2**.
  a. If the roots r from $F(x)$ do not occur in the fundamental set of solutions from **1**, use (4.25) or (4.27) to write the particular solution. If (4.27) was used, be sure to give the particular solution, y_p, by separating the real or imaginary part as shown earlier in this section.
  b. If $F(x) = g(x)e^{rx}$ and the roots r do not occur in the homogeneous solution, then apply the exponential shift operator and consider the new equation

$$P(D+r)v = g(x).$$

A particular solution, v_p, to this equation can be obtained via the method of undetermined coefficients and then used to create the particular solution, $y_p = v_p e^{rx}$, for the original system.
  c. If the roots r from $F(x) = g(x)e^{rx}$ occur k times, factor the operator as $P(D) = Q(D)(D - r)^k$. Then find a particular solution to

$$Q(D+r)u = g(x) \text{ with } u_p = D^k(v_p).$$

Finally, integrate k times to obtain v_p (ignoring constants of integration) and set $y_p = v_p e^{rx}$.

We finish with an example of a sinusoidal function with repeated roots.

Example 8 Find the general solution to $(D - 3)(D^2 + 1)y = 4x\cos x$.

Solution

Unlike the other problems in the section, we are now asked to find the general solution. To find a particular solution, we first complexify the problem as

$$(D - 3)(D^2 + 1)y = 4x\text{Re}(e^{ix}) \to (D - 3)(D + i)(D - i)z = 4xe^{ix},$$

where we have written $D^2 + 1$ as a product of its (complex) linear factors. This allows us to see that, in the complexified form, $r = i$ is a root that occurs once in the homogeneous equation. We thus need to consider $Q(D) = (D-3)(D+i)$. Applying the exponential shift operator and canceling the e^{ix} terms gives

$$Q(D+i)u = (D+i-3)(D+i+i)u = 4x,$$

where $u_p = Dv_p$ and $z_p = e^{ix}v_p$. Expanding this gives us

$$(D^2 + (-3+3i)D + (-2-6i))u = 4x. \tag{4.43}$$

This is a differential equation with complex coefficients but we can proceed as usual with the method of undetermined coefficients. We assume $u_p = A_1x + A_2$, where A_1, A_2 are complex numbers, and substitute into Equation (4.43):

$$\underbrace{(-3+3i)\,A_1}_{u_p'} + (-2-6i)\underbrace{(A_1x+A_2)}_{u_p} = 4x. \tag{4.44}$$

We equate coefficients of x to find

$$A_1 = \frac{2}{-1-3i} = \frac{-1}{5} + \frac{3}{5}i, \quad A_2 = \frac{(3-3i)A_1}{-2-6i} = \frac{-21}{50} + \frac{3}{50}i.$$

We integrate u_p to find v_p:

$$v_p = \left(\frac{-1}{10} + \frac{3}{10}i\right)x^2 + \left(\frac{-21}{50} + \frac{3}{50}i\right)x.$$

Finally, the particular solution to the complexified equation is $z_p = e^{ix}v_p$ and we want the real part:

$$y_p = \left(\frac{-1}{10}x^2 - \frac{21}{50}x\right)\cos x + \left(\frac{-3}{10}x^2 - \frac{3}{50}x\right)\sin x. \tag{4.45}$$

For the solution of the homogeneous equation, we see the roots of the characteristic equation are $r = 3, \pm i$ so that

$$y_c = C_1e^{3x} + C_2\sin x + C_3\cos x.$$

Thus our general solution is

$$y = C_1e^{3x} + C_2\sin x + C_3\cos x + \left(\frac{-1}{10}x^2 - \frac{21}{50}x\right)\cos x + \left(\frac{-3}{10}x^2 - \frac{3}{50}x\right)\sin x.$$

In Problems **1–16**, use the exponential response formula to find a particular solution. The exponential shift is not needed for any of these.

1. $y'' + y = e^{-x}$
2. $y'' + 4y = 3e^{2x}$
3. $y'' - y = 4e^{3x}$
4. $y'' + 2y' = e^x$
5. $y'' + 4y = 3$
6. $y'' - 4y = 5$
7. $y'' + 2y' + 2y = 3\cos 2x$
8. $y'' + 4y' + 5y = 4\sin 3x$
9. $y'' - 4y = -e^{2x}$
10. $y'' + y' = 5e^{-x}$
11. $y'' - 9y = e^{-3x}$
12. $y'' + 2y' = 3e^{-2x}$
13. $y'' + 2y' + y = 2e^{-x}$
14. $y'' + 4y' + 4y = e^{-2x}$
15. $y'' + 2y' + 2y = -2e^{-x}\cos x$
16. $y'' + 4y' + 5y = e^{-2x}\sin x$

In Problems **17–24**, use the exponential shift formula followed by the exponential response formula to find a particular solution.

17. $y'' + 4y = xe^x$
18. $y'' + y = 4xe^{3x}$
19. $y'' + 9y = x^2 + 3x - 1$
20. $y'' + y' = x\sin x$
21. $y'' - 9y = xe^{-3x}$
22. $y'' + 2y' = 2x^2e^{-2x}$
23. $y'' + 2y' + 2y = x^2e^{-x}\cos x$
24. $y'' + 4y' + 5y = xe^{-2x}\sin x$

In Problems **25–34**, find the general solution. If an initial condition is given, find the solution that passes through it.

25. $y'' - 2y' - 8y = 4e^{2x} - 21e^{-3x}$
26. $y'' + y' - 6y = 2x$
27. $y'' - 2y' + 2y = x\cos x$
28. $4y'' - 4y' - 3y = \cos 2x$
29. $y''' + 10y'' + 34y' + 40y = xe^{-4x}$
30. $y^{(4)} - 18y'' + 81y = e^{3x}$
31. $y'' - 4y = 32x, y(0) = 0, y'(0) = 6$
32. $y'' + y = \cos x, y(0) = 1, y'(0) = -1$
33. $y'' - 3y' + 2y = e^x, \ y(0) = 1, y'(0) = 0$
34. $y'' - 2y' - 3y = 3xe^{2x}, \ y(0) = 1, \ y'(0) = 0$

35. The exponential response method can be used to solve first-order constant coefficient nonhomogeneous equations. Use this method to solve the following problems:
(a) $y' - 4y = x^2$
(b) $y' + y = \cos 2x$
(c) $y' - y = e^{4x}$

36. Suppose that $y_1(x)$ and $y_2(x)$ are solutions of

$$a\frac{d^2y}{dx^2} + b\frac{dy}{dx} + cy = f(x),$$

where $a, b,$ and c are positive constants.
(a) Show that

$$\lim_{x \to \infty} [y_2(x) - y_1(x)] = 0.$$

(b) Is the result of (a) true if $b = 0$?

(c) Suppose that $f(x) = k$, where k is a constant. Show that

$$\lim_{x \to \infty} y(x) = \frac{k}{c}$$

for every solution $y(x)$ of $a\dfrac{d^2y}{dx^2} + b\dfrac{dy}{dx} + cy = k$.

(d) Determine the solution $y(x)$ of $a\dfrac{d^2y}{dx^2} + b\dfrac{dy}{dx} = k$. Find $\lim\limits_{x \to \infty} y(x)$.

(e) Determine the solution $y(x)$ of $a\dfrac{d^2y}{dx^2} = k$. Find $\lim\limits_{x \to \infty} y(x)$.

37. (a) Let $f(x)$ be a polynomial of degree n. Show that, if $b \neq 0$, there is always a solution that is a polynomial of degree n for the equation $y'' + ay' + by = f(x)$.

(b) Find a particular solution of $y'' + 3y' + 2y = 9 + 2x - 2x^2$.

38. In many physical applications, the nonhomogeneous term $F(x)$ is specified by different formulas in different intervals of x.

(a) Find a general solution of the equation

$$y'' + y = \begin{cases} x, & 0 \le x \le 1, \\ 1, & 1 \le x. \end{cases}$$

Note that the solution is not differentiable at $x = 1$.

(b) Find a particular solution of

$$y'' + y = \begin{cases} x, & 0 \le x \le 1, \\ 1, & 1 \le x \end{cases}$$

that satisfies the initial conditions $y(0) = 0$ and $y'(0) = 1$.

4.5　Variation of Parameters

The method of undetermined coefficients (either via tables or annihilation) *does not* apply to functions $F(x)$ that are *not* already solutions to some linear constant-coefficient homogeneous differential equation, nor to equations with variable coefficients. We will now develop a general method that applies to all nonhomogeneous linear equations. The only catch is that it requires knowledge of the homogeneous solution yet does not give us a method for finding this solution. If the homogeneous part of the equation has constant coefficients, we can use the previous methods to find the complementary solution. If the homogeneous equation has *non-constant* coefficients, we must be given the solutions in order to apply this method.

Consider the nonhomogeneous equation

$$y'' + y = \tan x. \tag{4.46}$$

The nonhomogeneous term

$$F(x) = \tan x$$

is not of the form

$$x^j, x^j e^{rx}, x^j e^{ax} \cos bx \text{ or } x^j e^{ax} \sin bx$$

so our previous methods will not work here. The variation of parameters approach, which was discovered by Lagrange, is a method that finds a particular solution for the differential equation. The idea is to use the general solution of the corresponding homogeneous equation

$$y'' + y = 0$$

to find a particular solution. In this case, the characteristic polynomial is

$$r^2 + 1 = 0,$$

which gives the solution

$$y_c(x) = c_1 \sin x + c_2 \cos x.$$

We will seek a particular solution of the form

$$y_p(x) = u_1(x) \sin x + u_2(x) \cos x \tag{4.47}$$

where we have replaced the constants c_1 and c_2 by unknown functions $u_1(x)$ and $u_2(x)$.

Thus, we see how the name of this method applies. We *vary the parameters* c_1 and c_2 by allowing them to be *functions* of x instead of *constants*. Notice that there should be many choices for $u_1(x)$ and $u_2(x)$ because we have two unknown functions and only one equation, the nonhomogeneous differential equation, to use to find them. Because we know that y_p must be a solution to (4.46), we substitute it into this equation. Differentiating y_p gives

$$y_p'(x) = u_1(x) \cos x + u_1'(x) \sin x - u_2(x) \sin x + u_2'(x) \cos x.$$

Now to simplify the process of finding $u_1(x)$ and $u_2(x)$, we impose our *first restriction*

$$u_1'(x) \sin x + u_2'(x) \cos x = 0 \tag{4.48}$$

on $u_1(x)$ and $u_2(x)$. We can think of this condition as arising from the assumed form of the particular solution in the case where the two functions $u_1(x)$ and $u_2(x)$ are actually constants (and thus we would have $u_1' = u_2' = 0$). Even though, in general, our two functions $u_1(x), u_2(x)$ are actually *not* constant, we are simply imposing a condition that will allow us to find suitable functions that will make y_p a solution.

Because of our restriction in (4.48), we have

$$y_p'(x) = u_1(x) \cos x - u_2(x) \sin x.$$

The second derivative of y_p is thus

$$y_p''(x) = -u_1(x) \sin x + u_1'(x) \cos x - u_2(x) \cos x - u_2'(x) \sin x.$$

Now we substitute y_p and y_p'' into the nonhomogeneous equation to obtain

$$y_p''(x) + y_p(x) = \underbrace{-u_1(x) \sin x + u_1'(x) \cos x - u_2(x) \cos x - u_2'(x) \sin x}_{y_p''}$$

$$\underbrace{+u_1(x) \sin x + u_2(x) \cos x}_{y_p}$$

$$= u_1'(x) \cos x - u_2'(x) \sin x.$$

Since y_p is a particular solution, we know that it must satisfy the nonhomogeneous equation. We thus have our *second restriction* on $u_1(x)$ and $u_2(x)$, namely,

$$u_1'(x) \cos x - u_2'(x) \sin x = \tan x. \tag{4.49}$$

Our two restrictions, Equations (4.48) and (4.49), give us a system of two equations in the two unknown functions $u_1'(x), u_2'(x)$,

$$\begin{cases} u_1'(x) \sin x + u_2'(x) \cos x = 0 \\ -u_1'(x) \cos x + u_2'(x) \sin x = \tan x, \end{cases} \tag{4.50}$$

which we can solve for $u_1'(x)$ and $u_2'(x)$. We could use Cramer's rule (see Appendix B.2.2 for a self-contained review of it) to solve this or simply manipulate the equations to solve.

In the former case, we have that

$$u_1'(x) = \frac{\begin{vmatrix} 0 & \cos x \\ \tan x & -\sin x \end{vmatrix}}{\begin{vmatrix} \sin x & \cos x \\ \cos x & -\sin x \end{vmatrix}}$$

$$= \frac{(0)(-\sin x) - (\cos x)(\tan x)}{(\sin x)(-\sin x) - (\cos x)(\cos x)} = \frac{-\cos x \tan x}{-1}$$

and

$$u_2'(x) = \frac{\begin{vmatrix} \sin x & 0 \\ \cos x & \tan x \end{vmatrix}}{\begin{vmatrix} \sin x & \cos x \\ \cos x & -\sin x \end{vmatrix}}$$

$$= \frac{(\sin x)(\tan x) - (\cos x)(0)}{(\sin x)(-\sin x) - (\cos x)(\cos x)} = \frac{\sin x \tan x}{-1}.$$

To find the two functions, we simply need to integrate. This gives

$$u_1(x) = \int \frac{-\cos x \tan x}{-1} \, dx = \int \sin x \, dx = -\cos x$$

and

$$u_2(x) = \int \frac{\sin x \tan x}{-1} \, dx = \int \frac{-\sin^2 x}{\cos x} \, dx = \sin x - \ln|\sec x + \tan x|.$$

Note that *any* antiderivative of $u_1'(x)$ and $u_2'(x)$ is possible as a choice for $u_1(x)$ and $u_2(x)$, respectively. We have simply taken the one with constant zero. Substitution into (4.47) gives the particular solution

$$y_p(x) = u_1(x)y_1(x) + u_2(x)y_2(x)$$

$$= -\cos x \sin x + (\sin x - \ln|\sec x + \tan x|)\cos x.$$

Thus, the general solution is

$$y(x) = y_c(x) + y_p(x)$$
$$= c_1 \sin x + c_2 \cos x \quad -\cos x \sin x + (\sin x - \ln|\sec x + \tan x|)\cos x.$$

In general, to solve the second-order linear nonhomogeneous differential equation

$$a_2(x)y''(x) + a_1(x)y'(x) + a_0(x)y(x) = F(x),$$

where

$$y_c(x) = c_1 y_1(x) + c_2 y_2(x)$$

is a general solution of the corresponding homogeneous equation

$$a_2(x)y''(x) + a_1(x)y'(x) + a_0(x)y(x) = 0,$$

we divide by the lead coefficient $a_2(x)$ to write

$$y''(x) + p(x)y'(x) + q(x)y(x) = G(x).$$

At this point, we need to be given or be able to calculate a fundamental solution set $\{y_1(x), y_2(x)\}$ for the homogeneous equation. If the homogeneous equation has constant coefficients this is done by our previous methods; if the coefficients are not constant, we must be given the fundamental set as the methods for calculating these are beyond the scope of this text. We next assume that a particular solution has a form similar to the general solution y_c by varying the parameters c_1 and c_2, that is, we let

$$y_p(x) = u_1(x)y_1(x) + u_2(x)y_2(x).$$

We need to determine $u_1(x)$ and $u_2(x)$, so we seek two equations by substituting y_p into the second-order linear nonhomogeneous differential equation

$$y''(x) + p(x)y'(x) + q(x)y(x) = G(x).$$

So, differentiating y_p gives

$$y_p'(x) = u_1(x)y_1'(x) + u_1'(x)y_1(x) + u_2(x)y_2'(x) + u_2'(x)y_2(x)$$

which simplifies to

$$y_p'(x) = u_1(x)y_1'(x) + u_2(x)y_2'(x)$$

by making the assumption

$$u_1'(x)y_1(x) + u_2'(x)y_2(x) = 0.$$

The second derivative is

$$y_p''(x) = u_1(x)y_1''(x) + u_1'(x)y_1'(x) + u_2(x)y_2''(x) + u_2'(x)y_2'(x).$$

Now we substitute these expressions into the second-order linear nonhomogeneous differential equation to obtain

$$
\begin{aligned}
y_p''(x) &+ p(x)y_p'(x) + q(x)y_p(x) \\
&= u_1(x)y_1''(x) + u_1'(x)y_1'(x) + u_2(x)y_2''(x) + u_2'(x)y_2'(x) \\
&\quad + p(x)\left[u_1(x)y_1'(x) + u_2(x)y_2'(x)\right] + q(x)\left[u_1(x)y_1(x) + u_2(x)y_2(x)\right] \\
&= u_1(x)\,\underbrace{\left[y_1''(x) + p(x)y_1'(x) + q(x)y_1(x)\right]}_{=\,0} \\
&\quad + u_2(x)\underbrace{\left[y_2''(x) + p(x)y_2'(x) + q(x)y_2(x)\right]}_{=\,0} + u_1'(x)y_1'(x) + u_2'(x)y_2'(x),
\end{aligned}
$$

where the "$= 0$" terms hold because $y_1(x)$ and $y_2(x)$ are solutions of the corresponding homogeneous equation $y''(x) + p(x)y'(x) + q(x)y(x) = 0$. Since

$$y''(x) + p(x)y'(x) + q(x)y(x) = G(x),$$

it follows that

$$u_1'(x)y_1'(x) + u_2'(x)y_2'(x) = G(x).$$

Hence, we have the second equation of our system needed to determine $u_1(x)$ and $u_2(x)$. Specifically,

$$\begin{cases} u_1'(x)y_1(x) + u_2'(x)y_2(x) = 0 \\ u_1'(x)y_1'(x) + u_2'(x)y_2'(x) = G(x). \end{cases}$$

Using Cramer's rule to solve this system (see Appendix B.2.2 for a brief review), we have the unique solution

$$u_1'(x) = \frac{\begin{vmatrix} 0 & y_2(x) \\ G(x) & y_2'(x) \end{vmatrix}}{W(x)} = \frac{-y_2(x)G(x)}{W(x)}$$

and

$$u_2'(x) = \frac{\begin{vmatrix} y_1(x) & 0 \\ y_1'(x) & G(x) \end{vmatrix}}{W(x)} = \frac{y_1(x)G(x)}{W(x)},$$

where

$$W(x) = W(y_1(x), y_2(x)) = \begin{vmatrix} y_1(x) & y_2(x) \\ y_1'(x) & y_2'(x) \end{vmatrix}.$$

We note that $W(x) \neq 0$ because $\{y_1(x), y_2(x)\}$ is a fundamental solution set and thus we know that $y_1(x)$ and $y_2(x)$ are linearly independent.

Summary of Variation of Parameters Method

Rewrite the given second-order equation so that the leading coefficient is one:

$$y''(x) + p(x)y'(x) + q(x)y(x) = G(x).$$

Then follow the following 5 steps:

1. Find or be given a complementary (homogeneous) solution

$$y_c(x) = c_1 y_1(x) + c_2 y_2(x)$$

and fundamental solutions

$$S = \{y_1(x), y_2(x)\}$$

of the corresponding homogeneous equation

$$y''(x) + p(x)y'(x) + q(x)y(x) = 0.$$

2. Let
$$u_1'(x) = \frac{-y_2(x)G(x)}{W(x)} \text{ and } u_2'(x) = \frac{y_1(x)G(x)}{W(x)}.$$

3. Integrate $u_1'(x)$ and $u_2'(x)$ to obtain $u_1(x)$ and $u_2(x)$.

4. A particular solution of $y''(x) + p(x)y'(x) + q(x)y(x) = G(x)$ is given by
$$y_p(x) = u_1(x)y_1(x) + u_2(x)y_2(x).$$

5. The general solution of $y''(x)+p(x)y'(x)+q(x)y(x) = G(x)$ is given by
$$y(x) = y_c(x) + y_p(x).$$

Example 1 Solve $y'' - 2y' + y = e^x \ln x$ for $x > 0$.

Solution
The corresponding homogeneous equation $y'' - 2y' + y = 0$ has the characteristic polynomial
$$m^2 - 2m + 1 = 0$$

so that the corresponding solution is
$$y_c(x) = c_1 e^x + c_2 x e^x.$$

Thus,
$$y_1(x) = e^x \text{ and } y_2(x) = xe^x.$$

So, $S = \{e^x, xe^x\}$ and
$$W(x) = \begin{vmatrix} e^x & xe^x \\ e^x & e^x + xe^x \end{vmatrix} = (e^x)(e^x + xe^x) - (e^x)(xe^x) = e^{2x}.$$

This gives
$$u_1(x) = \int \frac{-xe^x(e^x \ln x)}{e^{2x}} dx$$

and
$$u_2(x) = \int \frac{e^x(e^x \ln x)}{e^{2x}} dx.$$

Both of these integrals can be obtained by integrating by parts and are
$$u_1(x) = \frac{1}{4}x^2 - \frac{1}{2}x^2 \ln x$$

and
$$u_2(x) = x \ln x - x.$$

The particular solution is

$$y_p(x) = u_1(x)y_1(x) + u_2(x)y_2(x)$$

$$= \left(\frac{1}{4}x^2 - \frac{1}{2}x^2 \ln x\right) e^x + (x \ln x - x) \, xe^x$$

$$= \frac{1}{2}x^2 e^x \ln x - \frac{3}{4}x^2 e^x.$$

This gives the general solution as

$$y(x) = y_c(x) + y_p(x)$$

$$= c_1 e^x + c_2 x e^x + \frac{1}{2}x^2 e^x \ln x - \frac{3}{4}x^2 e^x.$$

It is often convenient to use MATLAB, Maple, or Mathematica to quickly calculate our determinants and evaluate the integrals.

Example 2 Solve the nonhomogeneous equation

$$y'' + \frac{1}{4}y = \sec\frac{x}{2} + \csc\frac{x}{2} \tag{4.51}$$

on the interval $0 < x < \pi$.

Solution
The characteristic equation is

$$r^2 + \frac{1}{4} = 0$$

with corresponding solution

$$y_c(x) = c_1 \cos\frac{x}{2} + c_2 \sin\frac{x}{2}.$$

We thus let

$$y_1(x) = \cos\frac{x}{2} \text{ and } y_2(x) = \sin\frac{x}{2}$$

and calculate

$$W(x) = \begin{vmatrix} \cos\frac{x}{2} & \sin\frac{x}{2} \\ -\frac{1}{2}\sin\frac{x}{2} & \frac{1}{2}\cos\frac{x}{2} \end{vmatrix}$$

$$= \left(\cos\frac{x}{2}\right)\left(\frac{1}{2}\cos\frac{x}{2}\right) - \left(\sin\frac{x}{2}\right)\left(-\frac{1}{2}\sin\frac{x}{2}\right) = \frac{1}{2}.$$

This gives

$$u_1(x) = \int -2\sin\frac{x}{2}\left(\sec\frac{x}{2} + \csc\frac{x}{2}\right) dx$$

$$= -2x + 4\ln\left|\cos\frac{x}{2}\right|,$$

$$u_2(x) = \int 2\cos\frac{x}{2}\left(\sec\frac{x}{2} + \csc\frac{x}{2}\right) dx$$

$$= 2x + 4\ln\left|\sin\frac{x}{2}\right|.$$

Thus, a particular solution is given by

$$y_p(x) = u_1(x)y_1(x) + u_2(x)y_2(x)$$

$$= \underbrace{\left(-2x + 4\ln\left|\cos\frac{x}{2}\right|\right)}_{u_1}\underbrace{\cos\frac{x}{2}}_{y_1} + \underbrace{\left(2x + 4\ln\left|\sin\frac{x}{2}\right|\right)}_{u_2}\underbrace{\sin\frac{x}{2}}_{y_2}$$

and the general solution is

$$y(x) = y_c(x) + y_p(x)$$

$$= c_1\cos\frac{x}{2} + c_2\sin\frac{x}{2}$$

$$+ \left(-2x + 4\ln\left|\cos\frac{x}{2}\right|\right)\cos\frac{x}{2} + \left(2x + 4\ln\left|\sin\frac{x}{2}\right|\right)\sin\frac{x}{2}.$$

Example 3 Verify that $y_c(x) = c_1 x^{-1} + c_2 x$ solves the homogeneous part of

$$x^2\frac{d^2y}{dx^2} + x\frac{dy}{dx} - y = 3e^{2x} \tag{4.52}$$

for $x > 0$. Then use variation of parameters to solve the nonhomogeneous equation.

Solution
We can easily show that y_c is a solution by substituting in y_c, y_c', y_c'' into the homogeneous equation. Letting

$$y_1(x) = x^{-1} \text{ and } y_2(x) = x$$

we have

$$W(x) = \begin{vmatrix} x^{-1} & x \\ -x^{-2} & 1 \end{vmatrix} = 2x^{-1},$$

which we note is not zero for any real x. This gives

$$u_1'(x) = \frac{-3}{2}x^2 e^{2x} \text{ and } u_2'(x) = \frac{-3}{2}e^{2x},$$

so

$$u_1(x) = \int \frac{-3}{2}x^2 e^{2x} \, dx$$

$$= -\frac{3}{8}(2x^2 - 2x + 1)e^{2x}$$

integrating by parts twice and

$$u_2(x) = \int \frac{-3}{2}e^{2x} \, dx = -\frac{3}{4}e^{2x}.$$

So a particular solution is given by

$$y_p(x) = u_1(x)y_1(x) + u_2(x)y_2(x)$$

$$= \left(-\frac{3}{8}(2x^2 - 2x + 1)e^{2x}\right)x^{-1} + \left(-\frac{3}{4}e^{2x}\right)x$$

and the general solution is

$$y(x) = y_c(x) + y_p(x)$$

$$= c_1 x^{-1} + c_2 x - \frac{3}{8}(2x - 2 + x^{-1})e^{2x} - \frac{3}{4}xe^{2x}.$$

Higher-Order Equations

Variation of Parameters extends to higher dimensions. We consider an nth-order equation with leading coefficient equal to one:

$$y^{(n)} + p_1(x)y^{(n-1)} + \ldots + p_{n-1}(x)y' + p_n(x)\, y = G(x). \tag{4.53}$$

If $\{y_1, y_2, \cdots, y_n\}$ is a fundamental set of solutions for the reduced equation, then we assume a particular solution of the form

$$y_p = u_1(x)y_1 + u_2(x)y_2 + \cdots + u_n(x)y_n$$

where the $u_i(x)$ are determined by the system of equations

$$
\begin{array}{ccccccc}
y_1 u_1' & + & y_2 u_2' & + \cdots + & y_n u_n' & = & 0 \\
y_1' u_1' & + & y_2' u_2' & + \cdots + & y_n' u_n' & = & 0 \\
\vdots & & \vdots & \ddots & \vdots & & \vdots \\
y_1^{(n-1)}u_1' & + & y_2^{(n-1)}u_2' & + \cdots + & y_n^{(n-1)}u_n' & = & G(x).
\end{array}
\tag{4.54}
$$

Using Cramer's rule, we can calculate the functions $u_i'(x)$ that we will integrate. In the case of the third-order equation with fundamental set of solutions $\{y_1, y_2, y_3\}$, we write a particular solution in the form

$$y_p = u_1(x)y_1 + u_2(x)y_2 + u_3(x)y_3, \qquad (4.55)$$

where

$$u_1'(x) = \frac{\begin{vmatrix} 0 & y_2 & y_3 \\ 0 & y_2' & y_3' \\ G(x) & y_2'' & y_3'' \end{vmatrix}}{\begin{vmatrix} y_1 & y_2 & y_3 \\ y_1' & y_2' & y_3' \\ y_1'' & y_3' & y_3'' \end{vmatrix}}, \quad u_2'(x) = \frac{\begin{vmatrix} y_1 & 0 & y_3 \\ y_1' & 0 & y_3' \\ y_1'' & G(x) & y_3'' \end{vmatrix}}{\begin{vmatrix} y_1 & y_2 & y_3 \\ y_1' & y_2' & y_3' \\ y_1'' & y_3' & y_3'' \end{vmatrix}}, \quad u_3'(x) = \frac{\begin{vmatrix} y_1 & y_2 & 0 \\ y_1' & y_2' & 0 \\ y_1'' & y_3' & G(x) \end{vmatrix}}{\begin{vmatrix} y_1 & y_2 & y_3 \\ y_1' & y_2' & y_3' \\ y_1'' & y_3' & y_3'' \end{vmatrix}}.$$

We then integrate the u_i' and plug into (4.55) to obtain a particular solution. For higher-order equations, we again solve (4.54) and then integrate to obtain the $u_i(x)$. The particular solution is created and thus we are able to write the general solution.

• • • • • • • • • • • •

Problems

In Problems **1–15**, find the general solution using variation of parameters. If an initial condition is given, find the solution that passes through it.

1. $y'' = \dfrac{1}{1+x^2}$

2. $y'' - 2y' + y = e^x \sqrt{x}$

3. $y'' - 2y' + y = \dfrac{e^x}{x^2}$

4. $y'' + 2y' + y = \dfrac{e^{-x}}{x^4}$

5. $y'' + 4y = \csc 2x$

6. $(D-1)^3(y) = \dfrac{e^x}{x}$

7. $D(D+1)(D-2)(y) = x^3$

8. $y^{(4)} - 16y' = e^{4x}$

9. $y'' - 7y' + 10y = e^{3x}, y(0) = 1, y'(0) = 2$

10. $y'' + 5y' + 6y = e^{-x}, y(0) = 1, y'(0) = 2$

11. $y'' + y = \sec x, y(0) = 1, y'(0) = 2$

12. $y'' + y = 6x, \ y(0) = 1, y'(0) = 1$

13. $y'' + y = \sin^2 x, y(0) = 1, y'(0) = 0$

14. $y'' + y = \tan x, y(0) = -1, y'(0) = 1$

15. $y'' + y = \sec^2 x, y(0) = 0, y'(0) = 1$

16. Solve $y'' + 4y' + 3y = 65 \cos 2x$ by
   (a) the method of undetermined coefficients or exponential response;
   (b) the method of variation of parameters.
   Which method is more easily applied?

In Problems **17–20**, consider the differential equation and fundamental set of solutions. First check that the given fundamental set of solutions is actually a fundamental set of solutions. Then use variation of parameters to solve the equation.

17. $2x^2y''' + 6xy'' = 1$, $x > 0$; $\{1, 1/x, x\}$

18. $y'' - \dfrac{1}{x}y' + \dfrac{1}{x^2}y = \dfrac{1}{x}$, $x > 0$; $\{x, x\ln x\}$

19. $y'' + \dfrac{1}{x}y' + \dfrac{1}{x^2}y = \dfrac{1}{x}$, $x > 0$; $\{\sin(\ln x), \cos(\ln x)\}$

20. $y'' + \dfrac{x-1}{x}y' = \dfrac{e^x}{1+x}$, $x > 0$; $\{1, (1+x)e^{-x}\}$

21. Show that the solution of the initial-value problem

$$y''(x) + a_1(x)y'(x) + a_0(x)y(x) = f(x)$$

with $y(x_0) = y_0$, $y'(x_0) = y_0'$ can be written as $y(x) = u(x) + v(x)$ where u is the solution of

$$u''(x) + a_1(x)u'(x) + a_0(x)u(x) = 0$$

with $u(x_0) = y_0$, $u'(x_0) = y_0'$ and v is the solution of

$$v''(x) + a_1(x)v'(x) + a_0(x)v(x) = f(x)$$

with $v(x_0) = 0$, $v'(x_0) = 0$.

4.6   Cauchy-Euler (Equidimensional) Equation

Up to now, all of the second-order differential equations that we have considered were linear with constant coefficients. As we have seen, these equations occur in applications, but they are not the only type of second-order differential equations for which we can develop a method of obtaining an explicit solution. We will now consider the **Cauchy-Euler equation**[2] which is defined by the second-order differential equation

$$a_0x^2y'' + a_1xy' + a_2y = f(x), \tag{4.56}$$

where the coefficients a_i are constant.

Electric Potential of a Charged Spherical Shell

Following Lomen and Lovelock [32], a differential equation that occurs in describing the electric potential of a charged spherical shell is given by

$$x^2y'' + 2xy' - n(n+1)y = 0 \tag{4.57}$$

where x is the distance from the center of the spherical shell and y is the potential. Here n is a positive constant. This differential equation is of a

---

[2]This equation is also called an equidimensional equation, or Euler-Cauchy equation.

form we have not previously considered: it is second order, but does not have constant coefficients. Comparing it with Equation (4.56), we see that it is a homogeneous Cauchy-Euler equation.

To solve this equation, we will now consider a method of solution of (4.57) analogous to our work in Section 3.1; however, here we replace xy' and x^2y''. To do this, we will change variables from the independent variable x to a new independent variable t, where

$$\frac{dy}{dt} = x\frac{dy}{dx}.$$

Applying the chain rule, we have that

$$\frac{dy}{dt} = \frac{dy}{dx}\frac{dx}{dt}$$

so that we can relate x and t by the differential equation

$$x'(t) = x(t).$$

This the differential equation which has solution

$$x(t) = ce^t$$

and thus suggests that we use the change of variables

$$x = e^t, \text{ if } x > 0 \quad \text{and } x = -e^t, \text{ if } x < 0. \tag{4.58}$$

In this example, $x > 0$ as distance is measured positively, so we take $x = e^t$ which gives $t = \ln x$. Differentiation with respect to x then gives

$$\frac{dt}{dx} = \frac{1}{x}.$$

Thus, using the chain rule,

$$\frac{dy}{dt} = \frac{dy}{dx}\frac{dx}{dt} = \frac{dy}{dx}e^t = x\frac{dy}{dx},$$

that is

$$x\frac{dy}{dx} = \frac{dy}{dt}. \tag{4.59}$$

Using this in (4.57) we can replace the $2xdy/dx$ term by $2dy/dt$. Now similarly, the $x^2 d^2y/dx^2$ term can be replaced; if we differentiate (4.59) with respect to x, we have

$$x\frac{d^2y}{dx^2} + \frac{dy}{dx} = \frac{d}{dx}\frac{dy}{dt} = \frac{d^2y}{dt^2}\frac{dt}{dx}.$$

Multiplying by x gives

$$x^2\frac{d^2y}{dx^2} + x\frac{dy}{dx} = \frac{d^2y}{dt^2}$$

so that

$$x^2 \frac{d^2y}{dx^2} = \frac{d^2y}{dt^2} - \frac{dy}{dt}. \tag{4.60}$$

Substituting (4.59) and (4.60) into (4.57), we have

$$\frac{d^2y}{dt^2} + \frac{dy}{dt} - n(n+1)y = 0. \tag{4.61}$$

This is a homogenous second-order linear differential equation with characteristic equation

$$r^2 + r - n(n+1) = 0$$

which has solution

$$r = n \text{ and } r = -n - 1.$$

Thus, the solution to (4.61) is

$$y(t) = c_1 e^{nt} + c_2 e^{-(n+1)t}.$$

Expressing this equation in the original variable x, we have the general solution to (4.57) as

$$y(x) = c_1 x^n + c_2 x^{-(n+1)}.$$

If we consider the potential inside the spherical shell, from symmetry considerations we expect the potential to be zero at the center $x = 0$, which we have if we choose $c_2 = 0$. If we are dealing with the potential outside the spherical shell, we expect the potential to go to 0 as $x \to \infty$, so we would choose $c_1 = 0$. Thus, for either situation we obtain a bounded solution of the differential equation.

Following the work in this example, we see that the homogeneous Cauchy-Euler equation

$$a_0 x^2 y'' + a_1 x y' + a_2 y = 0, \tag{4.62}$$

where the coefficients a_i are constant, has solutions of the form $y = x^r$. Substituting $y = x^r$ into (4.62) gives

$$a_0 x^2 (r(r-1)) x^{r-2} + a_1 x r x^{r-1} + a_2 x^r = 0,$$

so that simplifying we have

$$x^r (a_0 r^2 + (a_1 - a_0)r + a_2) = 0.$$

Noting that $x = 0$ yields the trivial solution, we have the characteristic equation

$$a_0 r^2 + (a_1 - a_0)r + a_2 = 0. \tag{4.63}$$

We again have three different cases to consider depending on the roots of this equation: if the roots are real and distinct $(r_1 \neq r_2)$, if the roots are real and repeated $(r_1 = r_2)$, and if the roots are complex $(r_1 = a + bi)$. However, we note here that we do not need to reinvent the wheel. We know what the solutions look like in each of these cases; the only difference here is that we need to apply the appropriate change of variables $x = \ln t$ as motivated in (4.58).

In the case that the roots are real and distinct, two linearly independent solutions are x^{r_1} and x^{r_2} and the general solution is

$$y(x) = c_1 x^{r_1} + c_2 x^{r_2}.$$

If the roots are real and repeated $(r_1 = r_2)$, then two linearly independent solutions are x^{r_1} and $x^{r_1} \ln x$ and the general solution is

$$y(x) = c_1 x^{r_1} + c_2 x^{r_1} \ln x.$$

If the roots are complex again, non-real $(r_1 = a + bi)$, then two real-valued linearly independent solutions are $x^a \sin(b \ln x)$ and $x^a \cos(b \ln x)$ so that the general solution is

$$y(x) = c_1 x^a \sin(b \ln x) + c_2 x^a \cos(b \ln x).$$

Example 1 Solve

$$2x^2 \frac{d^2 y}{dx^2} + 3x \frac{dy}{dx} - y = 0.$$

Changing variables from x to t where $x = e^t$ gives

$$2 \frac{d^2 y}{dt^2} + \frac{dy}{dt} - y = 0,$$

which has characteristic equation

$$2r^2 + r - 1 = 0.$$

The roots of this equation are $r_1 = 1/2$ and $r_2 = -1$ so that the general solution is

$$y = c_1 x^{\frac{1}{2}} + c_2 x^{-1}.$$

Example 2 Solve

$$x^2 \frac{d^2 y}{dx^2} + 2x \frac{dy}{dx} + 2y = 0$$

for $x > 0$.

Applying the change of variables $x = e^t$ gives

$$\frac{d^2 y}{dt^2} + \frac{dy}{dt} + 2y = 0,$$

which has characteristic equation

$$r^2 + r + 2 = 0.$$

The roots of this equation are

$$r = \frac{-1 \pm \sqrt{7}i}{2}$$

so that the general solution is

$$y(x) = c_1 \frac{\sin(\sqrt{7}\ln x)}{\sqrt{x}} + c_2 \frac{\cos(\sqrt{7}\ln x)}{\sqrt{x}}.$$

Problems

In Problems **1–16**, solve the homogeneous Cauchy-Euler equation by finding the general solution of the following differential equations, assuming $x > 0$.

1. $x^2 y'' + 4xy' + 2y = 0$
2. $x^2 y'' + 3xy' + y = 0$
3. $x^2 y'' + xy' + 4y = 0$
4. $x^2 y'' + 3xy' + 2y = 0$
5. $x^2 y'' + 11xy' + 21y = 0$
6. $2x^2 y'' + xy' - y = 0$
7. $2x^2 y'' - 3xy' + 3y = 0$
8. $x^2 y'' - 3xy' + 3y = 0$
9. $x^2 y'' - 3xy' + 4y = 0$
10. $x^2 y'' + 3xy' - 2y = 0$
11. $x^2 y'' + 5xy' - 3y = 0$
12. $3x^2 y'' + 3xy' + 9y = 0$
13. $7x^2 y'' + 5xy' + y = 0$
14. $x^2 y'' + 5xy' + 8y = 0$
15. $3x^2 y'' + 13xy' + 11y = 0$
16. $3x^2 y'' + 5xy' + y = 0$

17. Show that if $y_1(t)$ and $y_2(t)$ are linearly independent functions of t and satisfy a second-order linear differential equation with constant coefficients, then $Y_1(x) = y_1(\ln x)$ and $Y_2(x) = y_2(\ln x)$ are linearly independent functions of x for $x \neq 0$.

Solve the nonhomogenous Cauchy-Euler equations in Problems **18–20**.

18. $x^2 y'' + 4xy' + 2y = e^x$
19. $x^2 y'' + 3xy' + y = \ln x$
20. $x^2 y'' + xy' + 4y = 2$

4.7 Forced Vibrations

In Section 4.1, we motivated studying nonhomogeneous equations by using a forced mass on a spring as motivation. Equipped with our knowledge of the method from this chapter, we can take a closer look at the phenomenon of resonance. We will consider the case of undamped motion and then damped motion and then we will investigate the analogous electric circuit equations.

Resonance of a Forced Mass on a Spring with No Damping

We begin with a classic example of a mass on a spring without friction.

$$m\frac{d^2x}{dt^2} + kx = F_0 \sin(\omega t), \tag{4.64}$$

where ω denotes the forcing frequency. The solution of the homogeneous equation is easily seen to be

$$x_c(t) = C_1 \sin(\omega_n t) + C_2 \cos(\omega_n t), \quad \omega_n = \sqrt{\frac{k}{m}},$$

where $\omega_n = \sqrt{\frac{k}{m}}$ as before. Any of the methods in Sections 4.2-4.5 will work in the following derivation and we choose the method of undetermined coefficients. A particular solution has the form

$$x_p(t) = A \sin(\omega t) + B \cos(\omega t)$$

and we can substitute this into the equation to determine A and B. In doing this, we obtain

$$A = \frac{F_0}{m(\omega_n^2 - \omega^2)}, \quad B = 0. \tag{4.65}$$

Our particular solution is then

$$x_p(t) = \frac{F_0}{m(\omega_n^2 - \omega^2)} \sin(\omega t).$$

We immediately see a peculiar feature of our particular solution, x_p: it becomes unbounded as $\omega \to \omega_n$ and this phenomenon is known as **pure resonance**. That is, our oscillations will grow without bound as the forcing frequency approaches the natural frequency of the system. In any real system, we will always have some amount of damping; however, if the damping is small enough, we can still have sufficiently large oscillations that will cause a breakdown of the mechanical system.

Resonance of a Forced Mass on a Spring with Damping

While unbounded oscillations were possible in the undamped case, we will see this cannot happen when damping is present. We first consider the familiar equation

$$m\frac{d^2x}{dt^2} + b\frac{dx}{dt} + kx = F_0 \sin(\omega t) \tag{4.66}$$

and then immediately rewrite it as

$$\frac{d^2x}{dt^2} + 2\zeta\omega_n\frac{dx}{dt} + \omega_n^2 x = \frac{F_0}{m} \sin(\omega t), \tag{4.67}$$

where w denotes the forcing frequency, $\zeta = \frac{b/(2m)}{w_n}$ denotes the damping ratio, and w_n denotes the natural frequency (see Section 3.7). As in the undamped case, any of the methods in Sections 4.2-4.5 will work in the following derivation and we will use the method of undetermined coefficients. We thus assume the particular solution has the form

$$x_p(t) = A\sin(wt) + B\cos(wt).$$

We leave it as an exercise for the reader to show that substitution into the differential equation (4.67) gives

$$A = \frac{F_0(w_n^2 - w^2)}{m[4\zeta^2 w_n^2 w^2 + (w^2 - w_n^2)^2]} \tag{4.68}$$

$$B = \frac{-2w\zeta w_n F_0}{m[4\zeta^2 w_n^2 w^2 + (w^2 - w_n^2)^2]}. \tag{4.69}$$

With the details again left as an exercise, we use our now-favorite trigonometric identities to obtain[3]

$$x_p(t) = \frac{F_0/m}{\sqrt{4\zeta^2 w_n^2 w^2 + (w^2 - w_n^2)^2}}\cos(wt - \phi), \tag{4.70}$$

where

$$\cos\phi = \frac{B}{\sqrt{A^2 + B^2}} \quad \text{and} \quad \sin\phi = \frac{A}{\sqrt{A^2 + B^2}}. \tag{4.71}$$

We know that the homogeneous solution decays exponentially and it is only the steady-state solution that determines the long-term behavior of the system. The amplitude of these steady-state solutions is thus given by the factor

$$g(w) = \frac{F_0/m}{\sqrt{4\zeta^2 w_n^2 w^2 + (w^2 - w_n^2)^2}}. \tag{4.72}$$

We often refer to (4.72) as the **gain** for (4.67) and together with the **phase lag** given in (4.71) we refer to this as the **frequency response formula**. Engineers will often consider **complex gain**, which just means that we begin by writing our forcing function as a complex exponential

$$\frac{d^2 z}{dt^2} + 2\zeta w_n \frac{dz}{dt} + w_n^2 z = \frac{F_0}{m}e^{iwt}, \tag{4.73}$$

and then go through the same derivations to obtain

$$\tilde{g}(iw) = \frac{F_0/m}{P(iw)} = \frac{F_0/m}{w_n^2 - w^2 + i2\zeta w_n w}. \tag{4.74}$$

---

[3]As in Section 3.7, we also could have used $\cos\phi = \frac{A}{\sqrt{A^2+B^2}}$ and $\sin\phi = \frac{B}{\sqrt{A^2+B^2}}$ to obtain $\sin(wt + \phi)$ instead of $\cos(wt - \phi)$ in (4.70).

We can observe that Equation (4.72) is simply the absolute value of (4.74):

$$g(\omega) = |\tilde{g}(i\omega)|.$$

The gain is a way for us to characterize how the forcing frequency affects the amplitude response. For a given forced mass-spring system, we can only change the frequency of the forcing, that is, ω. The frequency that makes Equation (4.72) as large as possible is the resonant frequency. Because we consider ω as the only variable, finding the resonant frequency reduces to the problem of finding the ω that makes this function have its maximum value. We leave it for the reader to show in Problem **9** that the resonant frequency is

$$\omega_{res} = \omega_n \sqrt{1 - 2\zeta^2}, \tag{4.75}$$

where $\zeta = \frac{b/(2m)}{\omega_n}$, $\omega_n = \sqrt{k/m}$. This can be done by taking the derivative, setting it equal to zero, and finding the critical points. When Equation (4.75) is real-valued, it will give the maximum gain; otherwise $\omega_{res} = 0$ will give the maximum gain. Using the original equation containing m, b, k, this formula becomes

$$\omega_{res} = \sqrt{\frac{k}{m} - 2\left(\frac{b}{2m}\right)^2}. \tag{4.76}$$

There is no longer guesswork in determining the resonant frequency for Equation (4.66). For forcing functions that are exclusively in terms of cosine or in terms of both sine and cosine, the derivation is similar.

Example 1 Determine the resonant frequency for $3x'' + 2x' + 2x = F_0 \sin(\omega t)$.

Solution
For this problem, we have $m = 3$, $b = 2$, $k = 2$. Plugging into (4.76) gives

$$\omega_{res} = \frac{2}{3}.$$

In the alternative formulation given by (4.75), we have $\omega_n = \sqrt{2/3}$, $\zeta = 1/\sqrt{6}$, and obtain the same answer.

Forced Resonance in Electric Circuits

We again consider the electric circuit from Section 3.7. We had found the resulting differential equation as

$$L\frac{d^2Q}{dt^2} + R\frac{dQ}{dt} + \frac{1}{C}Q = \mathcal{E}. \tag{4.77}$$

We can equivalently rewrite this in terms of I by taking the derivative of each term:

$$L\frac{d^2I}{dt^2} + R\frac{dI}{dt} + \frac{1}{C}I = \mathcal{E}'. \tag{4.78}$$

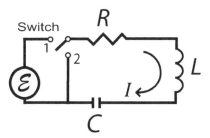

FIGURE 4.3: Basic RLC circuit with differential equations (4.77) or, equivalently, (4.78). We extend the work from Section 3.7 by now considering forcing from the supplied voltage $\mathcal{E} \neq 0$, often known as the electromotive force or emf.

The analogous formulation for no damping in the mass-spring system is to remove the resistor from the electric circuit. In doing so, our differential equation becomes

$$L\frac{d^2 I}{dt^2} + \frac{1}{C}I = \mathcal{E}'. \tag{4.79}$$

Dividing by L, letting $\omega_0 = \frac{1}{\sqrt{LC}}$, and considering $\mathcal{E}' = F_0 \sin(\omega t)$ allow us to rewrite (4.79) as

$$\frac{d^2 I}{dt^2} + \omega_0^2 I = \frac{F_0}{L}\sin(\omega t). \tag{4.80}$$

Analogous to (4.64), solutions to the homogeneous equation are of the form

$$I_c(t) = C_1 \sin(\omega_0 t) + C_2 \cos(\omega_0 t), \quad \omega_0 = \frac{1}{\sqrt{LC}}.$$

A particular solution is found to be

$$x_p(t) = \frac{F_0}{L(\omega_0^2 - \omega^2)}\sin(\omega t). \tag{4.81}$$

As with the mechanical analog, we see that tuning the forcing frequency such that $\omega \to \omega_0$ will cause solutions to grow without bound.

Including resistance in the circuit gives us a differential equation (4.80)

$$\frac{d^2 I}{dt^2} + \frac{R}{L}\frac{dI}{dt} + \frac{1}{LC}I = \frac{F_0}{L}\sin(\omega t). \tag{4.82}$$

We make the substitutions $\omega_0 = \frac{1}{\sqrt{LC}}$ and $\zeta_e = \frac{R}{2}\sqrt{\frac{C}{L}}$ and rewrite this as

$$\frac{d^2 I}{dt^2} + 2\zeta_e\omega_0\frac{dI}{dt} + \omega_0^2 I = \frac{F_0}{L}\sin(\omega t). \tag{4.83}$$

This is the same form as the forced mass on the spring! We note that the parameter ζ_e is again called the **damping ratio** as it has the analogous meaning as it did in the mass-spring system. A particular solution takes the same form as before and we again find its amplitude:

$$g(\omega) = \frac{F_0/L}{\sqrt{4\zeta_e^2 \omega_0^2 \omega^2 + (\omega^2 - \omega_0^2)^2}}. \tag{4.84}$$

The resonant frequency corresponds to the ω that makes this a maximum:

$$\omega_{res} = \omega_0 \sqrt{1 - 2\zeta_e^2} \tag{4.85}$$

or, equivalently,

$$\omega_{res} = \sqrt{\frac{1}{LC} - 2\left(\frac{R}{2L}\right)^2}. \tag{4.86}$$

● ● ● ● ● ● ● ● ● ● ● ●

Problems

In Problems **1–4**, find the gain and resonant frequency for each of the forced oscillators (either a spring-mass system or circuit). Then plot the gain as a function of ω.

1. $x'' + 9x = 5 \sin \omega t$
2. $x'' + 3x' + 4x = 4 \sin \omega t$
3. $x'' + 6x' + 9x = 2 \sin \omega t$
4. $x'' + 6x' + 25x = 3 \sin \omega t$

In Problems **5–6**, find the general solution for the current in the RLC circuit with given parameter values.

5. $R = 300\Omega$, $L = 1.25\text{H}$, $C = .0001\text{F}$, $\mathcal{E} = 100 \sin 10t \text{V}$
6. $R = 100\Omega$, $L = .02\text{H}$, $C = .00001\text{F}$, $\mathcal{E} = 120 \sin 30t \text{V}$

7. Derive Equations (4.68)-(4.69), which give the constants of the particular solution.

8. Derive Equation (4.70), thus obtaining the amplitude-phase form for the particular solution.

9. Derive the resonant frequency equation (4.75) for the forced mass-spring system.

Problems **10–12** focus on alternative formulation for a forced mass on a spring. After deriving the equation of motion, solve the equation for forcing function $y = \sin \omega t$.

10. Consider the spring-mass-dashpot system given in Figure 4.4.

Use Newton's 2nd law to show that the equation of motion is given by

$$mx'' + bx' + kx = ky,$$

where b is the damping from the dashpot.

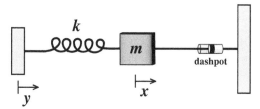

FIGURE 4.4: Spring-mass-dashpot system for Problem 10 with damping coefficient b from the dashpot.

11. Consider the spring-mass-dashpot system given in Figure 4.5.

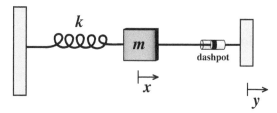

FIGURE 4.5: Spring-mass-dashpot system for Problem 11 with damping coefficient b from the dashpot.

Use Newton's 2nd law to show that the equation of motion is given by

$$mx'' + bx' + kx = by',$$

where b is the damping from the dashpot.

12. Consider the spring-mass-dashpot system given in Figure 4.6.

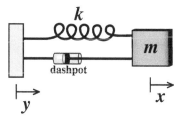

FIGURE 4.6: Spring-mass-dashpot system for Problem 12 with damping coefficient b from the dashpot.

Use Newton's 2nd law to show that the equation of motion is given by

$$mx'' + bx' + kx = by' + ky,$$

where b is the damping from the dashpot.

13. Consider the ODE $mx'' + bx' + kx = by'$, derived in Problem 11. If the forcing is given by $y = F_0 \sin \omega t$, show that the gain, cf. (4.72), is given by

$$g(\omega) = \frac{bF_0\omega}{\sqrt{b^2\omega^2 + m^2(\omega^2 - \omega_n^2)^2}}.$$

14. Consider the ODE $mx'' + bx' + kx = by' + ky$, derived in Problem 12. If the forcing is given by $y = F_0 \sin \omega t$, show that the gain, cf. (4.72), is given by

$$g(\omega) = \frac{F_0\sqrt{b^2\omega^2 + k^2}}{\sqrt{b^2\omega^2 + m^2(\omega^2 - \omega_n^2)^2}}.$$

Chapter 4 Review

In Problems **1–6**, determine whether the statement is true or false. If it is true, give reasons for your answer. If it is false, give a counterexample or other explanation of why it is false.

1. Every linear homogeneous equation with constant coefficients has e^{rx} as a solution, where r is a constant that may be real or complex.

2. For a forced mass on a spring system, the steady-state solution corresponds to the complementary (or homogeneous) solution while the transient solution corresponds to the particular solution.

3. For a forced mass on a spring system, the transient solution dies off as $t \to \infty$ while the steady-state solution persists.

4. If y_1 is a particular solution of $P(D)(y) = \sin x$ and y_2 is a particular solution of $P(D)(y) = x^2$, then $3y_1 - 5y_2$ is a particular solution of $P(D)(y) = 3\sin x - 5x^2$.

5. The method of undetermined coefficients (either via Tables or via the Annihilator Method) applies to any nonhomogeneous term as long as the associated homogeneous equation has constant coefficients.

6. Variation of parameters works for all differential equations.

In Problems **7–11**, consider the particular solutions of their respective forced equations $P(D)(y) = F_1(x)$ and $P(D)(y) = F_2(x)$. Find a particular solution for the new forcing function in $P(D)(y) = F_3$.

7. $y'' + y' - 6y = 50$ with $y_p = -\dfrac{25}{3}$ and $y'' + y' - 6y = 36x$ with $y_p = -6x - 1$; $F_2(x) = 100 - 18x$

8. $y'' + 2y' + 6y = 2$ with $y_p = \dfrac{1}{3}$ and $y'' + 2y' + 6y = e^x$ with $y_p = \dfrac{1}{9}e^x$; $F_2(x) = 6 + 9e^x$

9. $4y'' + 4y' + y = e^{2x}$ with $y_p = \dfrac{1}{25}e^{2x}$ and $4y'' + 4y' + y = \sin 2x + \cos 2x$

with $y_p = -\dfrac{23}{289}\cos 2x - \dfrac{7}{289}\sin 2x$; $F_2(x) = 13e^{2x} + 2\sin 2x + 2\cos 2x$

10. $y'' + 4y' + 5y = \cos x$ with $y_p = \dfrac{1}{8}(\sin x + \cos x)$ and $y'' + 4y' + 5y = 3x$

with $y_p = \dfrac{-12}{25} + \dfrac{3}{5}x$; $F_2(x) = \frac{1}{4}\cos x - x$

11. $y'' + 6y' + 25y = e^{-3x}\sin 4x$ with $y_p = \dfrac{-x}{8}e^{-3x}\cos 4x$ and $y'' + 6y' + 25y =$

e^{4x} with $y_p = \dfrac{1}{65}e^{4x}$; $F_2(x) = 8e^{-3x}\sin 4x + 5e^{4x}$

In Problems **12–19**, find the general solution using
(a) Method of Undetermined Coefficients or Exponential Response
(b) Variation of Parameters

12. $y'' + 4y' = xe^{-x}$ **13.** $y'' - 4y' = \cos x$
14. $y'' + 16y = 2\cos 4x$ **15.** $y'' + 4y' + 4y = e^{-2x}$
16. $9y'' + y = x$ **17.** $4y'' + 25y = 2e^{-x}$
18. $y'' + 2y' + 17y = e^{-x}\cos 4x$ **19.** $y'' + 3y' - 10y = x + e^{-3x}$

In Problems **20–23**, find the general solution of the following Cauchy-Euler problems.

20. $2x^2y'' + 3xy' - y = 3x$ **21.** $2x^2y'' + 5xy' + y = x + 3$
22. $4x^2y'' + 4xy' + y = x$ **23.** $3x^2y'' + 3xy' + y = x^2$

In Problems **24–25**, find the gain and resonant frequency for each of the forced oscillators (either a spring-mass system or circuit). Then plot the gain as a function of ω.

24. $x'' + 6x' + 13x = \sin \omega t$ **25.** $x'' + 2x' + 17x = 3\cos \omega t$

Chapter 4 Computer Labs

Chapter 4 Computer Lab: MATLAB

MATLAB Example 1: Enter the code below to demonstrate, using *Symbolic Math Toolbox*, how you obtain the coefficients of a particular solution using the method of undetermined coefficients for the ODE

$$y'' + 2y' - 3y = 4e^x - \sin x$$

```
>>  clear all
>>  syms x A B C
>>  eqyp=A*x*exp(x)+B*cos(x)+C*sin(x)
>>  eq1=diff(eqyp,x,2)+2*diff(eqyp,x)-3*eqyp -4*exp(x)+sin(x)
>>  %The above eq1 is the original ode written as lhs-rhs=0,
>>  %where the unwritten '=0' at the end is understood by
>>  %matlab.  We now type the relationships of the
>>  %coeffs by inspection of the calculated ode eq1
>>  eq2a=4*A-4 %coefficient of exp(x)
>>  eq2b=2*C-4*B %coefficient of cos(x)
>>  eq2c=-4*C-2*B+1 %coefficient of sin(x)
>>  [A,B,C]=solve(eq2a,eq2b,eq2c)
>>  yp=subs(eqyp)
```

MATLAB Example 2: Enter the code below to demonstrate, using *Symbolic Math Toolbox*, how you obtain the coefficients of a particular solution using the method of undetermined coefficients for the ODE

$$y^{(4)} + y'' = 3x^2 + 4\sin x - 2\cos x.$$

```
>>  clear all
>>  syms x y(x) A1 A2 A3 B1 B2
>>  eqODE=diff(y(x),x,4)+diff(y(x),x,2)-3*x^2-4*sin(x)+2*cos(x)
>>  %The above eqODE is the original ode written as lhs-rhs=0,
>>  %where the unwritten =0 at the end is understood by MATLAB
>>  eqyp=A1*x^4+A2*x^3+A3*x^2+B1*x*sin(x)+B2*x*cos(x)
>>  eq1=subs(eqODE,y(x),eqyp)
>>  [c1,t1]=coeffs(eq1,x)
>>  eq2a=c1(1) %coefficient of x^2
>>  eq2b=c1(2) %coefficient of x
>>  c1(3) %this does NOT give the constant terms
>>  %Now we need to sub sin(x)=0,cos(x)=0 to get constant terms
>>  eq2c=subs(c1(3),{sin(x),cos(x)},{0,0}) %the const terms
>>  [c2,t2]=coeffs(eq1,cos(x))
>>  eq2d=c2(1) %coefficient of cos(x)
>>  [c3,t3]=coeffs(eq1,sin(x))
>>  eq2e=c3(1) %coefficient of sin(x)
>>  [A1,A2,A3,B1,B2]=solve(eq2a,eq2b,eq2c,eq2d,eq2e)
>>  yp=subs(eqyp)
```

MATLAB Example 3: Enter the code below to demonstrate, using *Symbolic Math Toolbox*, how you obtain a particular solution using variation of parameters for the ODE

$$y'' - 2y' + y = e^x \ln x, \quad \text{for } x > 0.$$

```
>> clear all
>> syms x c1 c2
>> y1=exp(x) %The solns y1 and y2 are found by hand
>> y2=x*exp(x) %with yc=c1*y1+c2*y2
>> G=exp(x)*log(x)
>> A=[y1,y2; diff(y1,x),diff(y2,x)]
>> W=simplify(det(A))
>> u1=int(-y2*G/W,x)
>> u2=int(y1*G/W,x)
>> yp=simplify(u1*y1+u2*y2)
>> gensoln=c1*y1+c2*y2+yp
```

MATLAB Example 4: Enter the code below to demonstrate, using the ERF, how to obtain the gain and resonant frequency of Equation (4.67),

$$\frac{d^2x}{dt^2} + 2\zeta\omega_n\frac{dx}{dt} + \omega_n^2 x = \frac{F_0}{m}\sin(\omega t).$$

```
>> clear all
>> syms r zeta wn omega F0 m t
>> zeta=sym('zeta','real')
>> zeta=sym('zeta','positive')
>> wn=sym(wn,'real')
>> wn=sym(wn,'positive')
>> omega=sym('omega','real')
>> omega=sym('omega','positive')
>> F0=sym(F0,'real')
>> F0=sym(F0,'positive')
>> m=sym(m,'real')
>> m=sym(m,'positive')
>> t=sym(t,'real')
>> p(r)=r^2+2*zeta*wn*r+wn^2
>> zpsol=(F0/m)*exp(i*omega*t)/p(i*omega) %complex gain
>> zpsol1=simplify(zpsol*conj(zpsol))/conj(zpsol)
>> g1=abs(zpsol1) %real gain but not nice looking
>> c1=imag(zpsol)
>> c2=real(zpsol)
>> g=simplify(sqrt(c1^2+c2^2)) %also real gain
>> eq1=diff(g,omega)
>> solve(eq1,omega)
```

MATLAB Exercises

Turn in both the commands that you enter for the exercises below as well as the output/figures. These should all be in one document. Please highlight or clearly indicate all requested answers. Some of the questions will require you to modify the above MATLAB code to answer them.

1. Enter the commands given in MATLAB Example 1 and submit both your input and output.

2. Enter the commands given in MATLAB Example 2 and submit both your input and output.

3. Enter the commands given in MATLAB Example 3 and submit both your input and output.

4. Enter the commands given in MATLAB Example 4 and submit both your input and output.

5. Using the ERF or the method of undetermined coefficients, find a particular solution for the ODE $y'' + 2y' - 3y = 4e^x - \sin x$.

6. Using the ERF or the method of undetermined coefficients, find a particular solution for the ODE $y'' - 2y' - 8y = 4e^{2x} - 21e^{-3x}$.

7. Using the ERF or the method of undetermined coefficients, find a particular solution for the ODE $y'' - 2y' + 2y = x \cos x$.

8. Using the ERF or the method of undetermined coefficients, find a particular solution for the ODE $y'' - 2y' + 5y = e^x \sin 2x$.

9. Using the ERF or the method of undetermined coefficients, find a particular solution for the ODE $y'' - 2y' + 5y = 2xe^x + e^x \sin 2x$.

10. Using the ERF or the method of undetermined coefficients, find a particular solution for the ODE $y''' + 10y'' + 34y' + 40y = xe^{-4x}$.

11. Using the ERF or the method of undetermined coefficients, find a particular solution for the ODE $y'' + 6y' + 10y = 3xe^{-3x} - 2e^{3x} \cos x$.

12. Use variation of parameters to find a particular solution for the ODE
$$y'' = \frac{1}{1 + x^2}.$$

13. Use variation of parameters to find a particular solution for the ODE
$$y'' - 2y' + y = \frac{e^x}{x^2}.$$

14. Use variation of parameters to find a particular solution for the ODE
$$y'' + y = \sec x.$$

15. Use variation of parameters to find a particular solution for the ODE
$$y'' + y = \sin^2 x.$$

16. Use variation of parameters to find a particular solution for the ODE $2x^2y''' + 6xy'' = 1$, $x > 0$ with fundamental set of solutions $\{1, 1/x, x\}$.

17. Use variation of parameters to find a particular solution for the ODE
$$y'' + \frac{1}{x} y' + \frac{1}{x^2} y = \frac{1}{x}, \quad x > 0 \text{ with fundamental set of solutions}$$
$\{\sin(\ln x), \cos(\ln x)\}$.

18. Consider the equation $mx'' + bx' + kx = F_0 \cos \omega t$, which models a forced mass on a spring. Use the ERF or method of undetermined coefficients to derive a formula for the gain and the resonant frequency.

19. Consider the equation $mx'' + bx' + kx = by'$ with forcing $y = F_0 \cos \omega t$, which models a forced mass on a spring. Use the ERF or method of undetermined coefficients to derive a formula for the gain and the resonant frequency.

20. Consider the equation $mx'' + bx' + kx = by' + ky$ with forcing $y = F_0 \cos \omega t$, which models a forced mass on a spring. Use the ERF or method of undetermined coefficients to derive a formula for the gain and the resonant frequency.

Chapter 4 Computer Lab: Maple

Maple Example 1: Enter the code below to demonstrate how you obtain the coefficients of a particular solution using the method of undetermined coefficients for the ODE

$$y'' + 2y' - 3y = 4e^x - \sin x$$

restart
$yp := x \to A \cdot x \cdot e^x + B \cdot \cos(x) + C \cdot \sin(x)$
$eqODE := y''(x) + 2 \cdot y'(x) - 3 \cdot y(x) = 4 \cdot e^x - \sin(x)$
$eq1 := expand(subs(y(x) = yp(x), eqODE))$
        #We need to inspect the powers of like terms and
        write down the expressions involving A,B,C.
$eq2a := subs(e^x = 1, \cos(x) = 0, \sin(x) = 0, eq1)$ *#coeff of exp*
$eq2b := subs(e^x = 0, \cos(x) = 1, \sin(x) = 0, eq1)$ *#coeff of cos*
$eq2c := subs(e^x = 0, \cos(x) = 0, \sin(x) = 1, eq1)$ *#coeff of sin*
$eq3 := solve(\{eq2a, eq2b, eq2c\}, \{A, B, C\})$
$eq4 := y_p = subs(eq3, yp(x))$

Maple Example 2: Enter the code below to demonstrate how you obtain the coefficients of a particular solution using the method of undetermined coefficients for the ODE

$$y^{(4)} + y'' = 3x^2 + 4 \sin x - 2 \cos x.$$

restart
$eqODE := y^{(4)}(x) + y''(x) = 3 \cdot x^2 + 4 \cdot \sin(x) - 2 \cdot \cos(x)$

$eqyp := A_1 \cdot x^4 + A_2 \cdot x^3 + A_3 \cdot x^2 + B_1 \cdot x \cdot \sin(x) + B_2 \cdot x \cdot \cos(x)$
$eq1 := expand(subs(y(x) = eqyp, eqODE))$
$eq2a := coeff(lhs(eq1), \cos(x)) = coeff(rhs(eq1), \cos(x))$
$eq2b := coeff(lhs(eq1), \sin(x)) = coeff(rhs(eq1), \sin(x))$
$eq2c := coeff(rhs(eq1), x, 2)$ # *this should give you an error because Maple*
  won't compute a coeff of a power of x with $\cos(x)$ and $\sin(x)$ still
  in the expression. We now get rid of $\cos(x)$ and $\sin(x)$ in eq3:
$eq3 := subs(\cos(x) = 0, \sin(x) = 0, eq1)$
$eq3a := coeff(lhs(eq3), x, 2) = coeff(rhs(eq3), x, 2)$
$eq3b := coeff(lhs(eq3), x, 1) = coeff(rhs(eq3), x, 1)$
$eq3c := coeff(lhs(eq3), x, 0) = coeff(rhs(eq3), x, 0)$
$soln := solve(\{eq2a, eq2b, eq3a, eq3b, eq3c\}, \{A_1, A_2, A_3, B_1, B_2\})$
$eq4 := y_p = subs(soln, eqyp)$

Maple Example 3: Enter the code below to demonstrate how you obtain a particular solution using variation of parameters for the ODE

$$y'' - 2y' + y = e^x \ln x, \quad \text{for } x > 0.$$

restart
$y1 := e^x$
$y2 := x \cdot e^x$
$G := e^x \cdot \ln(x)$
$with(LinearAlgebra) : with(VectorCalculus) :$
$W := Wronskian([y1, y2], x, 'determinant')$
$W[2]$
$numer1 := Matrix([[0, y2], [G, y2']])$
$numer2 := Matrix([[y1, 0], [y1', G]])$
$u1 := \displaystyle\int \frac{Determinant(numer1)}{W[2]} \, dx$
$u2 := \displaystyle\int \frac{Determinant(numer2)}{W[2]} \, dx$
$y_p = simplify(u1 \cdot y1 + u2 \cdot y2)$

Maple Example 4: Enter the code below to demonstrate, using the ERF, how to obtain the gain and resonant frequency of Equation (4.67),

$$\frac{d^2 x}{dt^2} + 2\zeta \omega_n \frac{dx}{dt} + \omega_n^2 x = \frac{F_0}{m} \sin(\omega t).$$

restart
$p := r^2 + 2 \cdot zeta \cdot w_n \cdot r + w_n^2$ # *type zeta, don't use symbol from palette*
$zpsol := \dfrac{\left(\dfrac{F_0}{m}\right) \cdot e^{I \cdot \omega \cdot t}}{subs(r = I \cdot \omega, p)}$ #*complex gain*

$abs(zpsol)$
$evalc(abs(zpsol))$
$g := simplify(evalc(abs(zpsol)))$#real gain
$gprime := diff(g, \omega)$
$w_{res} := solve(gprime, \omega)$

Maple Exercises

Turn in both the commands that you enter for the exercises below as well as the output/figures. These should all be in one document. Please highlight or clearly indicate all requested answers. Some of the questions will require you to modify the above Maple code to answer them.

1. Enter the commands given in Maple Example 1 and submit both your input and output.

2. Enter the commands given in Maple Example 2 and submit both your input and output.

3. Enter the commands given in Maple Example 3 and submit both your input and output.

4. Enter the commands given in Maple Example 4 and submit both your input and output.

5. Using the ERF or the method of undetermined coefficients, find a particular solution for the ODE $y'' + 2y' - 3y = 4e^x - \sin x$.

6. Using the ERF or the method of undetermined coefficients, find a particular solution for the ODE $y'' - 2y' - 8y = 4e^{2x} - 21e^{-3x}$.

7. Using the ERF or the method of undetermined coefficients, find a particular solution for the ODE $y'' - 2y' + 2y = x \cos x$.

8. Using the ERF or the method of undetermined coefficients, find a particular solution for the ODE $y'' - 2y' + 5y = e^x \sin 2x$.

9. Using the ERF or the method of undetermined coefficients, find a particular solution for the ODE $y'' - 2y' + 5y = 2xe^x + e^x \sin 2x$.

10. Using the ERF or the method of undetermined coefficients, find a particular solution for the ODE $y''' + 10y'' + 34y' + 40y = xe^{-4x}$.

11. Using the ERF or the method of undetermined coefficients, find a particular solution for the ODE $y'' + 6y' + 10y = 3xe^{-3x} - 2e^{3x} \cos x$.

12. Use variation of parameters to find a particular solution for the ODE $y'' = \dfrac{1}{1 + x^2}$.

13. Use variation of parameters to find a particular solution for the ODE $y'' - 2y' + y = \dfrac{e^x}{x^2}$.

14. Use variation of parameters to find a particular solution for the ODE $y'' + y = \sec x$.

15. Use variation of parameters to find a particular solution for the ODE $y'' + y = \sin^2 x$.

16. Use variation of parameters to find a particular solution for the ODE $2x^2 y''' + 6xy'' = 1$, $x > 0$ with fundamental set of solutions $\{1, 1/x, x\}$.

17. Use variation of parameters to find a particular solution for the ODE $y'' + \dfrac{1}{x} y' + \dfrac{1}{x^2} y = \dfrac{1}{x}$, $x > 0$ with fundamental set of solutions $\{\sin(\ln x), \cos(\ln x)\}$.

18. Consider the equation $mx'' + bx' + kx = F_0 \cos \omega t$, which models a forced mass on a spring. Use the ERF or method of undetermined coefficients to derive a formula for the gain and the resonant frequency.

19. Consider the equation $mx'' + bx' + kx = by'$ with forcing $y = F_0 \cos \omega t$, which models a forced mass on a spring. Use the ERF or method of undetermined coefficients to derive a formula for the gain and the resonant frequency.

20. Consider the equation $mx'' + bx' + kx = by' + ky$ with forcing $y = F_0 \cos \omega t$, which models a forced mass on a spring. Use the ERF or method of undetermined coefficients to derive a formula for the gain and the resonant frequency.

Chapter 4 Computer Lab: Mathematica

Mathematica Example 1: Enter the code below to demonstrate how you obtain the coefficients of a particular solution using the method of undetermined coefficients for the ODE

$$y'' + 2y' - 3y = 4e^x - \sin x$$

```
Quit[]
yp[x_] = A x e^x + B Cos[x] + C1 Sin[x] (* <space> for mult. *)
  (*Mathematica uses C for naming constants so we use C1*)
dey[x_] = y''[x] + 2 y'[x] - 3 y[x] - 4 e^x + Sin[x]
lhs = Simplify[ReplaceAll[dey[x], y→yp]]
  (*Inspect like terms to get coefficients*)
eq1 = Coefficient[lhs, e^x]
```

```
eq2 = Coefficient[lhs, Cos[x]]
eq3 = Coefficient[lhs, Sin[x]]
sol = Solve[{eq1==0, eq2==0, eq3==0}, {A, B, C1}]
ReplaceAll[yp[x], sol]
```

Mathematica Example 2: Enter the code below to demonstrate how you obtain the coefficients of a particular solution using the method of undetermined coefficients for the ODE

$$y^{(4)} + y'' = 3x^2 + 4\sin x - 2\cos x.$$

```
Quit[]
yp[x_] = A1 x⁴ + A2 x³ + A3 x² + B1 x Sin[x] + B2 x Cos[x]
dey[x_] = y''''[x] + y''[x] - 3 x² - 4 Sin[x] + 2 Cos[x]
lhs = ReplaceAll[dey[x], y→yp]
   (*need to get constant term, not with Cos[x] and Sin[x]*)
eq1 = Coefficient[ReplaceAll[lhs, {Cos[x]→0, Sin[x]→0}], x, 0]
   (*eq1 is the constant term*)
eq2 = Coefficient[lhs, x]
eq3 = Coefficient[lhs, x, 2]
eq4 = Coefficient[lhs, Cos[x]]
eq5 = Coefficient[lhs, Sin[x]]
sol = Solve[{eq1==0, eq2==0, eq3==0, eq4==0 eq5==0},
   {A1, A2, A3, B1, B2}]
ReplaceAll[yp[x], sol]
```

Mathematica Example 3: Enter the code below to demonstrate how you obtain a particular solution using variation of parameters for the ODE

$$y'' - 2y' + y = e^x \ln x, \quad \text{for } x > 0.$$

```
Quit[]
y1[x_] = eˣ
y2[x_] = x eˣ
G[x_] = eˣ Log[x] (*rhs of ODE*)
W = Wronskian[{y1[x], y2[x]}, x]
numer1 = Det[{{0, y2[x]}, {G[x], y2'[x]}}]
numer2 = Det[{{y1[x], 0}, {y1'[x], G[x]}}]
u1 = ∫ (numer1/W) dx
u2 = ∫ (numer2/W) dx
```

```
yp = u1 y1[x] + u2 y2[x]
```

Mathematica Example 4: Enter the code below to demonstrate, using the ERF, how to obtain the gain and resonant frequency of Equation (4.67),

$$\frac{d^2x}{dt^2} + 2\zeta\omega_n\frac{dx}{dt} + \omega_n^2 x = \frac{F_0}{m}\sin(\omega t).$$

```
Quit[]
p[r_] = r² + 2 zeta wn r + wn²
        FO eⁱ ω t
        m
zpsol = ──────────────────────
        ReplaceAll[p[r], r→i ω]
Abs[zpsol]
g[ω_] = ComplexExpand[Abs[zpsol]]
gprime = g'[ω]
wres = Solve[gprime==0, ω]
```

Mathematica Exercises

Turn in both the commands that you enter for the exercises below as well as the output/figures. These should all be in one document. Please highlight or clearly indicate all requested answers. Some of the questions will require you to modify the above Mathematica code to answer them.

1. Enter the commands given in Mathematica Example 1 and submit both your input and output.

2. Enter the commands given in Mathematica Example 2 and submit both your input and output.

3. Enter the commands given in Mathematica Example 3 and submit both your input and output.

4. Enter the commands given in Mathematica Example 4 and submit both your input and output.

5. Using the ERF or the method of undetermined coefficients, find a particular solution for the ODE $y'' + 2y' - 3y = 4e^x - \sin x$.

6. Using the ERF or the method of undetermined coefficients, find a particular solution for the ODE $y'' - 2y' - 8y = 4e^{2x} - 21e^{-3x}$.

7. Using the ERF or the method of undetermined coefficients, find a particular solution for the ODE $y'' - 2y' + 2y = x\cos x$.

8. Using the ERF or the method of undetermined coefficients, find a particular solution for the ODE $y'' - 2y' + 5y = e^x \sin 2x$.

9. Using the ERF or the method of undetermined coefficients, find a particular solution for the ODE $y'' - 2y' + 5y = 2xe^x + e^x \sin 2x$.

10. Using the ERF or the method of undetermined coefficients, find a particular solution for the ODE $y''' + 10y'' + 34y' + 40y = xe^{-4x}$.

11. Using the ERF or the method of undetermined coefficients, find a particular solution for the ODE $y'' + 6y' + 10y = 3xe^{-3x} - 2e^{3x}\cos x$.

12. Use variation of parameters to find a particular solution for the ODE
$$y'' = \frac{1}{1+x^2}.$$

13. Use variation of parameters to find a particular solution for the ODE
$$y'' - 2y' + y = \frac{e^x}{x^2}.$$

14. Use variation of parameters to find a particular solution for the ODE $y'' + y = \sec x$.

15. Use variation of parameters to find a particular solution for the ODE $y'' + y = \sin^2 x$.

16. Use variation of parameters to find a particular solution for the ODE $2x^2y''' + 6xy'' = 1$, $x > 0$ with fundamental set of solutions $\{1, 1/x, x\}$.

17. Use variation of parameters to find a particular solution for the ODE
$$y'' + \frac{1}{x}y' + \frac{1}{x^2}y = \frac{1}{x}, \quad x > 0 \text{ with fundamental set of solutions}$$
$\{\sin(\ln x), \cos(\ln x)\}$.

18. Consider the equation $mx'' + bx' + kx = F_0\cos\omega t$, which models a forced mass on a spring. Use the ERF or method of undetermined coefficients to derive a formula for the gain and the resonant frequency.

19. Consider the equation $mx'' + bx' + kx = by'$ with forcing $y = F_0\cos\omega t$, which models a forced mass on a spring. Use the ERF or method of undetermined coefficients to derive a formula for the gain and the resonant frequency.

20. Consider the equation $mx'' + bx' + kx = by' + ky$ with forcing $y = F_0\cos\omega t$, which models a forced mass on a spring. Use the ERF or method of undetermined coefficients to derive a formula for the gain and the resonant frequency.

Chapter 4 Projects

Project 4A: Forced Duffing Equation [26], [48]

One formulation of the forced Duffing equation is

$$x'' + bx' + kx + \delta x^3 = F_0 \sin(\omega t), \qquad (4.87)$$

where $x = x(t)$. When $\delta = 0$, the equation reduces to that of the forced mass on a spring of Section 4.7. Thus we can think of the left-hand side of Equation (4.87) as describing a *nonlinear spring* with attached mass $m = 1$. Let's take $k = \delta = 1$, for simplicity, and see what this means. For small x-values, i.e., the mass on the spring is not too far from rest position, the x^3 term is probably negligible and we can approximate the solution by dropping this term. However, for moderate or large x-values we cannot neglect this term.

(1) The restoring force, $F = kx + \delta x^3$, is nonlinear. Give a physical interpretation of this term. For example, is the restoring force greater for small x or large x? What physical implications does this have?

(2) Set $k = \delta = 1$, $F_0 = 0$, $b = 0.2$. Using MATLAB, Maple, or Mathematica with $h = 0.01$, let $x'(0) = 1.0$ and begin with the five different initial conditions $x(0) = 0, 0.5, 1.0, 3.0, 4.0$ to plot five different solutions in the t-x plane. What are the transient and long-term behaviors of the solution? Numerically solve the equation to at least $t = 100$.

(3) Now we will observe what happens as we force this nonlinear oscillator by changing F_0 and ω. Still keeping $k = \delta = 1$, $b = 0.2$, set $\omega = 1$ and plot solutions for $F_0 = .1, .5, 2, 10, 15, 20, 25, 30$. Use the initial condition $x(0) = 2$, $x'(0) = 1$. What are the transient and long-term behaviors of each solution? Comment on any difference observed in the plots.

(4) Repeat the above step but now numerically solve Duffing's equation using the initial condition $x(0) = 1$, $x'(0) = 2$. What are the transient and long-term behaviors of each solution? Comment on any difference observed in the plots, both in this part and when compared with those of Part 3.

(5) Now fix $F_0 = 25$ (with $k = \delta = 1$, $b = .2$) and vary ω. Choose the values $\omega = .1, .2, .5, .8, 1, 1.2, 1.5, 2.0, 5.0$, and others, if you so desire. What are the transient and long-term behaviors of each solution? Comment on any difference observed in the pictures.
What effect does the nonlinearity and forcing have on the motion $x(t)$? Is it predictable?

Project 4B: Forced van der Pol Oscillator [27], [48]

One formulation of the forced van der Pol equation is

$$x'' + \epsilon x'(x^2 - 1) + x = F_0 \sin(\omega t), \qquad (4.88)$$

where $x = x(t)$, $\epsilon \in \mathbb{R}$. This was originally considered as an electric circuit with a triode and later as an electric circuit with a tunnel diode. We can think of x as a dimensionless voltage. When $\epsilon = 0$, the equation reduces to that of the forced LC circuit of Section 4.7. Thus we can think of the left-hand side of Equation (4.88) as describing an RLC circuit with *nonlinear resistance*. Let's take $\epsilon = 1$, for simplicity, and see what this means. For large x-values, the coefficient of $x'(t)$ is positive and we would expect damping to occur. However, for small values of x, the coefficient of $x'(t)$ is negative and thus small oscillations are amplified (not damped). (1) The equation of the electric circuit tunnel diode is given as

$$V'' - \frac{\alpha - 3\gamma V^2}{C} V' + \frac{1}{LC} V = 0, \qquad (4.89)$$

where V is the voltage, C the capacitance, L the inductance, and α, γ are parameters [46]. Derive the unforced van der Pol oscillator by nondimensionalizing (4.89). Show that $x = \sqrt{\frac{3\gamma}{\alpha}} V$, $\tau = \frac{1}{\sqrt{LC}} t$, $\epsilon = \sqrt{\frac{1}{LC}} \alpha$ are the required substitutions.
(2) Set $\epsilon = 1$, $F_0 = 0$, $b = 0.2$. Using MATLAB, Maple, or Mathematica with $h = 0.01$, let $x'(0) = 1.0$ and begin with the five different initial conditions $x(0) = 0, 0.5, 1.0, 3.0, 4.0$ to plot five different solutions in the t-x plane. What are the transient and long-term behaviors of the solution? Numerically solve the equation to at least $t = 100$.
(3) Now we will observe what happens as we force this nonlinear oscillator by changing F_0 and ω. Still keeping $\epsilon = 1$, $b = 0.2$, set $\omega = 1$ and plot solutions for $F_0 = .1, .5, 2, 10, 15, 20, 25, 30$. Use the initial condition $x(0) = 2$, $x'(0) = 1$. What are the transient and long-term behaviors of each solution? Comment on any difference observed in the plots.
(4) Repeat the above step but now numerically solve van der Pol's equation using the initial condition $x(0) = 1$, $x'(0) = 2$. What are the transient and long-term behaviors of each solution? Comment on any difference observed in the plots, both in this part and when compared with those of Part 3.
(5) Now fix $F_0 = 25$ (with $\epsilon = 1$, $b = .2$) and vary ω. Choose the values $\omega = .1, .2, .5, .8, 1, 1.2, 1.5, 2.0, 5.0$ and others, if you so desire. What are the transient and long-term behaviors of each solution? Comment on any difference observed in the pictures.
What effect does the nonlinearity and forcing have on the motion $x(t)$? Is it predictable?

Chapter 5

Fundamentals of Systems of
Differential Equations

Earlier we briefly considered systems of equations when converting an nth-order equation to a system of n first-order equations. But systems of differential equations arise in their own right—whenever there is more than one dependent variable for an independent variable. For example, one might consider a system with two or more interacting species with the population sizes changing over time. The most common of these are known as **Lotka-Volterra models** and are discussed in detail in Section 6.5. One of these models is a **predator–prey** system in which the prey population growth depends upon the number of predators that kill the prey. Similarly, the rate at which the predator population grows depends on the size of their food supply, namely, the prey population. In general, these conditions produce nonlinear equations that are very difficult to solve analytically. This is just one scenario we can consider. In this chapter, we discuss methods for solving these types of systems.

The objects of study of the first few sections of this chapter are linear systems of equations, which are differential equations of the form

$$
\begin{aligned}
\frac{dx_1}{dt} &= a_{11}(t)x_1 + a_{12}(t)x_2 + \cdots + a_{1n}(t)x_n + f_1(t) \\
\frac{dx_2}{dt} &= a_{21}(t)x_1 + a_{22}(t)x_2 + \cdots + a_{2n}(t)x_n + f_2(t) \\
&\ \ \vdots \qquad\quad \vdots \qquad\quad \vdots \qquad \ddots \qquad\quad \vdots \\
\frac{dx_n}{dt} &= a_{n1}(t)x_1 + a_{n2}(t)x_2 + \cdots + a_{nn}(t)x_n + f_n(t),
\end{aligned}
\tag{5.1}
$$

where the variables x_i, f_i and coefficients a_{ij} are all functions of t. The situation when an nth-order system is derived from an nth-order equation is simply a special case. In the event that the coefficients a_{ij} are constant and the f_i are zero, we refer to system (5.1) as an **autonomous** system of n first-order equations and there are special techniques that will apply. System (5.1) has aspects of matrix analysis, so that at this point, the reader is strongly encouraged to review the material in Appendix B.1 before proceeding.

5.1   Useful Terminology

Many of the terms that we are about to encounter have arisen at earlier points in this book. The significance here is that they apply to a system of equations as well. We consider (5.1) written in matrix notation:

$$
\begin{bmatrix} \dfrac{dx_1}{dt} \\ \dfrac{dx_2}{dt} \\ \vdots \\ \dfrac{dx_n}{dt} \end{bmatrix} = \begin{bmatrix} a_{11}(t) & a_{12}(t) & \cdots & a_{1n}(t) \\ a_{21}(t) & a_{22}(t) & \cdots & a_{2n}(t) \\ \vdots & \vdots & \ddots & \vdots \\ a_{n1}(t) & a_{n2}(t) & \cdots & a_{nn}(t) \end{bmatrix} \begin{bmatrix} x_1 \\ x_2 \\ \vdots \\ x_n \end{bmatrix} + \begin{bmatrix} f_1(t) \\ f_2(t) \\ \vdots \\ f_n(t) \end{bmatrix}, \qquad (5.2)
$$

or, more compactly, simply as

$$
\frac{d\mathbf{x}}{dt} = \mathbf{A}\mathbf{x} + \mathbf{F}, \qquad (5.3)
$$

where $\mathbf{x}$, $\mathbf{A}$, $\mathbf{F}$ are possibly all functions of t and $\mathbf{A}$ is known as the **coefficient matrix**. If $\mathbf{F}(t) = 0$, we say that the system is homogeneous and write it as

$$
\frac{d\mathbf{x}}{dt} = \mathbf{A}\mathbf{x}. \qquad (5.4)
$$

Before continuing, we formally address the notion of the **derivative of a matrix**.

Definition 5.1.1
The derivative of the $m \times n$ matrix $\mathbf{A}(t) = [a_{ij}(t)]$, whose elements $a_{ij}(t)$ are differentiable on some interval I, is defined by taking the derivative of each component:
$$
\frac{d\mathbf{A}}{dt} = \left[\frac{da_{ij}}{dt} \right].
$$
The integral of the $m \times n$ matrix $\mathbf{A}(t) = [a_{ij}(t)]$, whose elements $a_{ij}(t)$ are continuous on some interval I, is defined by taking the integral of each component:
$$
\int_{t_0}^{t} \mathbf{A} = \left[\int_{t_0}^{t} a_{ij}(s)\, ds \right].
$$

Example 1   Consider the matrix

$$
\mathbf{A} = \begin{bmatrix} 1 & 0 & \sin t \\ e^{2t} & -t & 3t^2 - 4 \end{bmatrix}.
$$

Find the derivative of $\mathbf{A}$. Then find the integral of $\mathbf{A}$ (from 0 to t).

To find the derivative, we differentiate componentwise. Thus

$$\frac{d\mathbf{A}}{dt} = \begin{bmatrix} \dfrac{d}{dt}(1) & \dfrac{d}{dt}(0) & \dfrac{d}{dt}(\sin t) \\ \dfrac{d}{dt}(e^{2t}) & \dfrac{d}{dt}(-t) & \dfrac{d}{dt}(3t^2 - 4) \end{bmatrix} = \begin{bmatrix} 0 & 0 & \cos t \\ 2e^{2t} & -1 & 6t \end{bmatrix}.$$

For the integral, we similarly integrate componentwise. We note that we use the dummy variable s to do this integration:

$$\int_0^t \mathbf{A} = \begin{bmatrix} \displaystyle\int_0^t 1 \, ds & \displaystyle\int_0^t 0 \, ds & \displaystyle\int_0^t \sin s \, ds \\ \displaystyle\int_0^t e^{2s} \, ds & \displaystyle\int_0^t -s \, ds & \displaystyle\int_0^t (3s^2 - 4) \, ds \end{bmatrix}$$

$$= \begin{bmatrix} t & 0 & 1 - \cos t \\ \dfrac{1}{2}e^{2t} - \dfrac{1}{2} & \dfrac{-t}{2} & t^3 - 4t \end{bmatrix}.$$

Using this idea, some of our previous work can be extended. For example, if the components of a vector $\mathbf{x}$ are differentiable on an open interval (a, b) and satisfy the system of differential equations, then we say that $\mathbf{x}$ is a *solution* of (5.2).

Example 2 Verify that the vectors

$$\mathbf{x}_1 = \begin{bmatrix} 0 \\ 0 \\ 1 \end{bmatrix} e^{-2t}, \quad \mathbf{x}_2 = \begin{bmatrix} 1 \\ 0 \\ 1 \end{bmatrix} e^t, \quad \text{and} \quad \mathbf{x}_3 = \begin{bmatrix} 5 \\ 10 \\ 3 \end{bmatrix} e^{3t}$$

are each a solution to the system

$$\frac{d\mathbf{x}}{dt} = \mathbf{A}\mathbf{x}, \quad \text{where} \quad \mathbf{A} = \begin{bmatrix} 1 & 1 & 0 \\ 0 & 3 & 0 \\ 3 & 0 & -2 \end{bmatrix}.$$

This is done by substitution of each vector into the differential equation:

$$\frac{d\mathbf{x}_1}{dt} = \begin{bmatrix} 0 \\ 0 \\ 1 \end{bmatrix} (-2e^{-2t}) \quad \text{and} \quad \mathbf{A}\mathbf{x}_1 = \begin{bmatrix} 1 & 1 & 0 \\ 0 & 3 & 0 \\ 3 & 0 & -2 \end{bmatrix} \begin{bmatrix} 0 \\ 0 \\ 1 \end{bmatrix} e^{-2t} = \begin{bmatrix} 0 \\ 0 \\ -2 \end{bmatrix} e^{-2t};$$

$$\frac{d\mathbf{x}_2}{dt} = \begin{bmatrix} 1 \\ 0 \\ 1 \end{bmatrix} (e^t) \quad \text{and} \quad \mathbf{A}\mathbf{x}_2 = \begin{bmatrix} 1 & 1 & 0 \\ 0 & 3 & 0 \\ 3 & 0 & -2 \end{bmatrix} \begin{bmatrix} 1 \\ 0 \\ 1 \end{bmatrix} e^t = \begin{bmatrix} 1 \\ 0 \\ 1 \end{bmatrix} e^t;$$

$$\frac{d\mathbf{x}_3}{dt} = \begin{bmatrix} 5 \\ 10 \\ 3 \end{bmatrix} (3e^{3t}) \quad \text{and} \quad \mathbf{A}\mathbf{x}_3 = \begin{bmatrix} 1 & 1 & 0 \\ 0 & 3 & 0 \\ 3 & 0 & -2 \end{bmatrix} \begin{bmatrix} 5 \\ 10 \\ 3 \end{bmatrix} e^{3t} = \begin{bmatrix} 15 \\ 30 \\ 9 \end{bmatrix} e^{3t}.$$

In each case, we see that the vector $\mathbf{x}_i$ satisfies the differential equation for all t and thus is a solution for all t.

As we have seen, it is straightforward to check if a given vector is a solution. Just as we have done in previous chapters, it is useful to know *when* we can actually expect to have a solution. Because we are currently considering a linear system of equations, the following theorems are an asset.

THEOREM 5.1.1   Existence and Uniqueness
Consider the initial-value problem for the system

$$\frac{d\mathbf{x}}{dt} = \mathbf{A}\mathbf{x} + \mathbf{F}, \quad \mathbf{x}(t_0) = \mathbf{x}_0, \tag{5.5}$$

where $\mathbf{A}(t)$ is an $m \times n$ matrix and $\mathbf{F}(t)$ is an m-component vector of continuous real functions on the interval $a \le t \le b$. If t_0 is any point in the interval $[a, b]$, then there exists a unique solution to (5.5) and this solution is defined over the entire interval $a \le t \le b$.

For the homogeneous case (5.4), we can consider linear combinations of these solutions.

THEOREM 5.1.2
Let $\mathbf{x}_1, \mathbf{x}_2, \cdots, \mathbf{x}_k$ be **any** k solutions of the homogeneous system $\mathbf{x}' = \mathbf{A}(t)\mathbf{x}$. Then

$$c_1\mathbf{x}_1 + c_2\mathbf{x}_2 + \cdots + c_k\mathbf{x}_k$$

is also a solution of the homogeneous system, where $c_1, c_2, \cdots, c_k$ are arbitrary constants.

Example 3 In Example 2, we showed the vectors

$$\mathbf{x}_1 = \begin{bmatrix} 0 \\ 0 \\ 1 \end{bmatrix} e^{-2t}, \quad \mathbf{x}_2 = \begin{bmatrix} 1 \\ 0 \\ 1 \end{bmatrix} e^t, \quad \text{and} \quad \mathbf{x}_3 = \begin{bmatrix} 5 \\ 10 \\ 3 \end{bmatrix} e^{3t}$$

are each a solution of the system

$$\frac{d\mathbf{x}}{dt} = \mathbf{A}\mathbf{x}, \quad \text{where} \quad \mathbf{A} = \begin{bmatrix} 1 & 1 & 0 \\ 0 & 3 & 0 \\ 3 & 0 & -2 \end{bmatrix}.$$

According to Theorem 5.1.2, we have that

$$\mathbf{x}(t) = c_1\mathbf{x}_1 + c_2\mathbf{x}_2 + c_3\mathbf{x}_3 = \begin{bmatrix} c_2e^t + c_35e^{3t} \\ c_310e^{3t} \\ c_1e^{-2t} + c_2e^t + c_33e^{3t} \end{bmatrix}$$

is also a solution of $\mathbf{x}' = \mathbf{A}\mathbf{x}$. To see this, we first take the derivative of this linear combination to get

$$\mathbf{x}'(t) = \begin{bmatrix} c_2e^t + c_315e^{3t} \\ c_330e^{3t} \\ c_1(-2)e^{-2t} + c_2e^t + c_39e^{3t} \end{bmatrix}.$$

We also evaluate the right-hand side, $\mathbf{A}\mathbf{x}(t)$, to get

$$\begin{bmatrix} 1 & 1 & 0 \\ 0 & 3 & 0 \\ 3 & 0 & -2 \end{bmatrix} \begin{bmatrix} c_2e^t + c_35e^{3t} \\ c_310e^{3t} \\ c_1e^{-2t} + c_2e^t + c_33e^{3t} \end{bmatrix} = \begin{bmatrix} c_2e^t + c_315e^{3t} \\ c_330e^{3t} \\ c_1(-2)e^{-2t} + c_2e^t + c_39e^{3t} \end{bmatrix}.$$

The equality of the results of these two calculations shows that $c_1\mathbf{x}_1 + c_2\mathbf{x}_2 + c_3\mathbf{x}_3$ is indeed a solution.

In Section 3.2, we considered the above theorems for the case of functions and not vectors. We wanted to write the general solution and we needed to know when functions were linearly independent. We can formulate a similar definition for vectors.

Definition 5.1.2

The k functions $\mathbf{x}_1, \mathbf{x}_2, \cdots, \mathbf{x}_k$ are **linearly dependent** on $a \leq t \leq b$ if there exist constants $c_1, c_2, \cdots, c_k$, not all zero, such that

$$c_1\mathbf{x}_1 + c_2\mathbf{x}_2 + \cdots + c_k\mathbf{x}_k = \mathbf{0}$$

for all t in the interval (a, b). We say that the k functions $\mathbf{x}_1, \mathbf{x}_2, \cdots, \mathbf{x}_k$ are **linearly independent** on $a \leq t \leq b$ if they are not linearly dependent there. That is, $\mathbf{x}_1, \mathbf{x}_2, \cdots, \mathbf{x}_k$ are linearly independent on $a \le t \le b$ if

$$c_1 \mathbf{x}_1 + c_2 \mathbf{x}_2 + \cdots + c_k \mathbf{x}_k = \mathbf{0}$$

for all t in (a, b) implies $c_1 = c_2 = \ldots = c_n = 0$.

Example 4 The vectors

$$\mathbf{x}_1 = \begin{bmatrix} e^t \\ 0 \\ e^t \end{bmatrix}, \quad \mathbf{x}_2 = \begin{bmatrix} e^{-2t} \\ e^{-2t} \\ 0 \end{bmatrix}, \quad \mathbf{x}_3 = \begin{bmatrix} 3e^t - 2e^{-2t} \\ -2e^{-2t} \\ 3e^t \end{bmatrix}$$

are linearly dependent because

$$3\mathbf{x}_1 - 2\mathbf{x}_2 - \mathbf{x}_3 = 0.$$

Example 5 The vectors

$$\mathbf{x}_1 = \begin{bmatrix} 0 \\ 0 \\ 1 \end{bmatrix} e^t, \quad \mathbf{x}_2 = \begin{bmatrix} 1 \\ 0 \\ 1 \end{bmatrix} e^{-t}, \quad \text{and} \quad \mathbf{x}_3 = \begin{bmatrix} 1 \\ 2 \\ 3 \end{bmatrix} e^{3t}$$

are linearly independent because if

$$c_1\mathbf{x}_1 + c_2\mathbf{x}_2 + c_3\mathbf{x} = \begin{bmatrix} 0 & e^{-t} & e^{3t} \\ 0 & 0 & 2e^{3t} \\ e^t & e^{-t} & 3e^{3t} \end{bmatrix} \begin{bmatrix} c_1 \\ c_2 \\ c_3 \end{bmatrix} = \begin{bmatrix} 0 \\ 0 \\ 0 \end{bmatrix}$$

for all x, then we can solve this equation for the c_i since our matrix is invertible to obtain $c_1 = c_2 = c_3 = 0$. (We could have solved this by Cramer's rule (determinant is not zero), Gaussian elimination, or any other method to obtain this conclusion, too.)

With this concept of linear independence in place, we can state an extremely useful theorem.

THEOREM 5.1.3
The homogeneous system (5.4) always possesses n solution vectors that are linearly independent. Further, if $\mathbf{x}_1, \mathbf{x}_2, \ldots, \mathbf{x}_n$ are n linearly independent solutions of (5.4), then every solution $\mathbf{x}$ of (5.4) can be expressed

as a linear combination

$$c_1\mathbf{x}_1(t) + c_2\mathbf{x}_2(t) + \ldots + c_n\mathbf{x}_n(t) \tag{5.6}$$

of these n linearly independent solutions by proper choice of the constants $c_1, c_2, \ldots, c_n$. Expression (5.6) is called the **general solution** of (5.4) and is defined on (a, b), the interval on which solutions exist and are unique.

We thus have that the solutions $\mathbf{x}_1, \ldots, \mathbf{x}_n$ can be combined to give us any solution we desire. As before, the concept of a **fundamental set of solutions** gives us the set necessary to write the general solution.

Three Necessary and Sufficient Conditions for a Fundamental Set of Solutions of (5.4)

1. The number of vectors (elements) in this set must be the same as the number of first-order ODEs in system (5.4).
2. Each vector $\mathbf{x}_i$ in this set must be a solution to system (5.4).
3. The vectors must be linearly independent.

As before, we note that a fundamental set of solutions is *not* unique. Once we have a fundamental set of solutions, we can construct all possible solutions from it.

Testing for linear independence is usually the challenging part and we can use a familiar tool to help us.

Definition 5.1.3

Let

$$\mathbf{x}_1 = \begin{bmatrix} x_{11}(t) \\ x_{21}(t) \\ \vdots \\ x_{n1}(t) \end{bmatrix}, \quad \mathbf{x}_2 = \begin{bmatrix} x_{12}(t) \\ x_{22}(t) \\ \vdots \\ x_{n2}(t) \end{bmatrix}, \quad \mathbf{x}_n = \begin{bmatrix} x_{1n}(t) \\ x_{2n}(t) \\ \vdots \\ x_{nn}(t) \end{bmatrix}$$

be n real vector functions of t. The determinant

$$W(t) = W(\mathbf{x}_1, \mathbf{x}_2, \ldots, \mathbf{x}_n)(t) = \begin{vmatrix} x_{11}(t) & x_{12}(t) & \ldots & x_{1n}(t) \\ x_{21}(t) & x_{22}(t) & \ldots & x_{2n}(t) \\ \vdots & \vdots & \ddots & \vdots \\ x_{n1}(t) & x_{n2}(t) & \ldots & x_{nn}(t) \end{vmatrix}$$

is called the Wronskian.

THEOREM 5.1.4

Let $x_1, x_2, \ldots, x_n$ be defined as in Definition 5.1.3.
1. If $W(t_0) \neq 0$ for some $t_0 \in (a, b)$, then it follows that $x_1, x_2, \ldots, x_n$ are linearly independent on (a, b).
2. If $x_1, x_2, \ldots, x_n$ are linearly dependent on (a, b), then $W(t) = 0$, for all $t \in (a, b)$.

THEOREM 5.1.5

Let $x_1, x_2, \ldots, x_n$ be defined as in Definition 5.1.3. Suppose that the x_i are each a solution of (5.4). Then exactly one of the following statements is true:
1. $W(t) \neq 0$, for all $t \in (a, b)$
2. $W(t) = 0$, for all $t \in (a, b)$
Moreover, $W(t) \neq 0$ for all $t \in (a, b)$ if and only if the $\{x_i\}$ are linearly independent on (a, b). Similarly, $W(t) = 0$ for all $t \in (a, b)$ if and only if the $\{x_i\}$ are linearly dependent on (a, b).

Example 6 The vectors

$$x_1 = \begin{bmatrix} 0 \\ 0 \\ 1 \end{bmatrix} e^{-2t}, \quad x_2 = \begin{bmatrix} 1 \\ 0 \\ 1 \end{bmatrix} e^t, \quad \text{and} \quad x_3 = \begin{bmatrix} 5 \\ 10 \\ 3 \end{bmatrix} e^{3t}$$

are linearly independent because

$$\begin{vmatrix} 0 & e^t & 5e^{3t} \\ 0 & 0 & 10e^{3t} \\ e^{-2t} & e^t & 3e^{3t} \end{vmatrix} = 0 \cdot \begin{vmatrix} 0 & 10e^{3t} \\ e^t & 3e^{3t} \end{vmatrix} - e^t \cdot \begin{vmatrix} 0 & 10e^{3t} \\ e^{-2t} & 3e^{3t} \end{vmatrix} + 5e^{3t} \cdot \begin{vmatrix} 0 & 0 \\ e^{-2t} & e^t \end{vmatrix}$$

$$= -e^t(-10e^{3t}e^{-2t}) = 10e^{2t}, \tag{5.7}$$

which is never zero.

All of the theory so far has given us the necessary information to state a theorem for the nonhomogeneous linear system (5.2).

THEOREM 5.1.6

Let x_p be a solution of the nonhomogeneous system (5.2). Let

$$x_c = c_1 x_1 + x_2 x_2 + \ldots + c_n x_n$$

be the general solution of the corresponding homogeneous equation (5.4). Then every solution Φ of the nonhomogeneous system (5.2) can be expressed in the form

$$\Phi = x_c + x_p.$$

Just as with Theorem 4.1.2 for our nth-order equations, we have a theorem that says we only ever need to find a single particular solution because any other particular solution is a linear combination of the fundamental set of solutions for the homogeneous equation plus the particular solution.

THEOREM 5.1.7

Let x_p and X be two solutions of the nonhomogeneous equation $x' = Ax + F$ where A is $n \times n$. Then

$$X = x_p + c_1 x_1 + x_2 x_2 + \ldots + c_n x_n,$$

where $x_1, \ldots, x_n$ is a fundamental set of solutions for $x' = Ax$.

Problems

In Problems **1–8**, find the derivative and antiderivative of each matrix.

1. $\begin{bmatrix} \sin t & e^t \\ t^2 & 3t \end{bmatrix}$

2. $\begin{bmatrix} \cos^2 t & te^{-t} \\ 0 & 3t+1 \end{bmatrix}$

3. $\begin{bmatrix} \cos t & e^{3t} \\ t^2 & 0 \end{bmatrix}$

4. $\begin{bmatrix} e^{2t}\cos t & t^2 e^{-t} \\ 3 & \ln|t| \end{bmatrix}$

5. $\begin{bmatrix} \sin t & te^{-t} & e^{-t} \\ 0 & 0 & 2e^{3t} \\ e^t & t^2+e^{-t} & \cos t \end{bmatrix}$

6. $\begin{bmatrix} \cos t & e^{-3t} & e^t \\ t^2 & 0 & 2te^{-t} \\ e^{2t} & e^{-t} & 2-e^{3t} \end{bmatrix}$

7. $\begin{bmatrix} e^{-t}\sin t & -e^{3t} & e^{-3t} \\ \sqrt{t} & t^4 & \cos t \\ t^{4/3} & 0 & te^{3t} \end{bmatrix}$

8. $\begin{bmatrix} \tan t & 2+e^{-t} & e^{-t} \\ 0 & t+\sqrt{t} & 2e^{3t} \\ e^t & 3+e^t & \cos t \end{bmatrix}$

In Problems **9–15**, verify that given vectors are solutions to the equation $\frac{dx}{dt} = Ax$.

9. $x_1 = \begin{bmatrix} -1 \\ -2 \end{bmatrix} e^{4t}, \quad x_2 = \begin{bmatrix} 1 \\ 3 \end{bmatrix} e^t, \quad A = \begin{bmatrix} 10 & -3 \\ 18 & -5 \end{bmatrix}$

10. $x_1 = \begin{bmatrix} 1 \\ 1 \end{bmatrix} e^{-t}, \quad x_2 = \begin{bmatrix} 5 \\ 2 \end{bmatrix} e^{2t}, \quad A = \begin{bmatrix} 4 & -5 \\ 2 & -3 \end{bmatrix}$

11. $x_1 = \begin{bmatrix} 2 \\ 3 \end{bmatrix} e^{-2t}, \quad x_2 = \begin{bmatrix} 1 \\ 2 \end{bmatrix} e^{-3t}, \quad A = \begin{bmatrix} 1 & -2 \\ 6 & -6 \end{bmatrix}$

12. $x_1 = \begin{bmatrix} 1 \\ 1 \end{bmatrix} e^{3t}, \quad x_2 = \begin{bmatrix} -1 \\ 1 \end{bmatrix} e^t, \quad A = \begin{bmatrix} 2 & 1 \\ 1 & 2 \end{bmatrix}$

13. $x_1 = \begin{bmatrix} 1 \\ 0 \\ 0 \end{bmatrix} e^t, x_2 = \begin{bmatrix} 0 \\ 1 \\ 0 \end{bmatrix} e^{-t}, x_3 = \begin{bmatrix} -1 \\ 0 \\ 1 \end{bmatrix} e^{2t}, \quad A = \begin{bmatrix} 1 & 0 & -1 \\ 0 & -1 & 0 \\ 0 & 0 & 2 \end{bmatrix}$

14. $x_1 = \begin{bmatrix} 1 \\ 2 \\ 0 \end{bmatrix} e^{2t}, x_2 = \begin{bmatrix} 0 \\ 0 \\ 1 \end{bmatrix} e^t, x_3 = \begin{bmatrix} 1 \\ -1 \\ 1 \end{bmatrix} e^{-4t}, \quad A = \begin{bmatrix} -2 & 2 & 0 \\ 4 & 0 & 0 \\ \frac{-10}{3} & \frac{-5}{3} & 1 \end{bmatrix}$

15. $\mathbf{x}_1 = \begin{bmatrix} 1 \\ -3 \\ 1 \end{bmatrix} e^{-2t}, \mathbf{x}_2 = \begin{bmatrix} -1 \\ 2 \\ 0 \end{bmatrix} e^{-t}, \mathbf{x}_3 = \begin{bmatrix} 1 \\ 0 \\ -1 \end{bmatrix} e^t, \mathbf{A} = \begin{bmatrix} 1 & 1 & 0 \\ 6 & 2 & 6 \\ -6 & -3 & -5 \end{bmatrix}$

16. If $\mathbf{x}_1, \mathbf{x}_2$ are both solutions to $\mathbf{x}' = \mathbf{Ax}$, show that $c_1\mathbf{x}_1 + c_2\mathbf{x}_2$ is also a solution.

In Problems **17–24**, classify the following sets of vectors as linearly dependent or linearly independent.

17. $\left\{ \begin{bmatrix} 1 \\ 0 \end{bmatrix} e^t, \begin{bmatrix} 1 \\ 1 \end{bmatrix} e^t, \right\}$

18. $\left\{ \begin{bmatrix} -1 \\ 1 \end{bmatrix}, \begin{bmatrix} 3 \\ 2 \end{bmatrix} \right\}$

19. $\left\{ \begin{bmatrix} 3 \\ -1 \end{bmatrix}, \begin{bmatrix} -6 \\ 2 \end{bmatrix} \right\}$

20. $\left\{ \begin{bmatrix} 1 \\ -1 \end{bmatrix} e^{2t}, \begin{bmatrix} -2 \\ 2 \end{bmatrix} e^{-t}, \right\}$

21. $\left\{ \begin{bmatrix} -1 \\ 1 \end{bmatrix}, \begin{bmatrix} 2 \\ -1 \end{bmatrix} \right\}$

22. $\left\{ \begin{bmatrix} 1 \\ -1 \\ 0 \end{bmatrix}, \begin{bmatrix} 1 \\ 2 \\ 0 \end{bmatrix}, \begin{bmatrix} 0 \\ 0 \\ 3 \end{bmatrix} \right\}$

23. $\left\{ \begin{bmatrix} 3 \\ 1 \\ 0 \end{bmatrix} e^t, \begin{bmatrix} -3 \\ -1 \\ 0 \end{bmatrix} e^{2t}, \begin{bmatrix} 0 \\ 0 \\ 1 \end{bmatrix} e^{-t} \right\}$

24. $\left\{ \begin{bmatrix} 3 \\ -3 \\ 0 \end{bmatrix}, \begin{bmatrix} -1 \\ 1 \\ 1 \end{bmatrix}, \begin{bmatrix} 0 \\ 0 \\ 1 \end{bmatrix} \right\}$

In Problems **25–30**, determine if the following set forms a fundamental set of solutions for the given differential equation.

25. $\left\{ \begin{bmatrix} e^{3t}\sin(t) \\ e^{3t}\cos(t) \end{bmatrix}, \begin{bmatrix} e^{3t}\cos(t) \\ -e^{3t}\sin(t) \end{bmatrix}, \right\}, \quad \mathbf{x}' = \begin{bmatrix} 3 & 2 \\ -2 & 3 \end{bmatrix} \mathbf{x}$

26. $\left\{ \begin{bmatrix} e^{4t}\sin(2t) \\ e^{4t}\cos(2t) \end{bmatrix}, \begin{bmatrix} e^{4t}\cos(2t) \\ -e^{4t}\sin(2t) \end{bmatrix} \right\}, \quad \mathbf{x}' = \begin{bmatrix} 4 & 2 \\ -2 & 4 \end{bmatrix} \mathbf{x}$

27. $\left\{ \begin{bmatrix} -e^{5t} \\ e^{5t} \end{bmatrix}, \begin{bmatrix} e^t \\ e^t \end{bmatrix} \right\}, \quad \mathbf{x}' = \begin{bmatrix} 3 & -2 \\ -2 & 3 \end{bmatrix} \mathbf{x}$

28. $\left\{ \begin{bmatrix} e^{-t} \\ -e^{-t} \end{bmatrix}, \begin{bmatrix} 2e^{-t} \\ 0 \end{bmatrix} \right\}, \quad \mathbf{x}' = \begin{bmatrix} 1 & 2 \\ 0 & -1 \end{bmatrix} \mathbf{x}$

29. $\left\{ \begin{bmatrix} 1 \\ 0 \\ 0 \end{bmatrix} e^t, \begin{bmatrix} -1 \\ 1 \\ 0 \end{bmatrix} e^{-t}, \begin{bmatrix} 0 \\ 0 \\ 1 \end{bmatrix} e^{-2t} \right\}, \quad \mathbf{x}' = \begin{bmatrix} 1 & 2 & 0 \\ 0 & -1 & 0 \\ 0 & 0 & -2 \end{bmatrix} \mathbf{x}$

30. $\left\{ \begin{bmatrix} 0 \\ 0 \\ 1 \end{bmatrix} e^{-t}, \begin{bmatrix} 1 \\ 0 \\ \frac{1}{3} \end{bmatrix} e^t, \begin{bmatrix} -1 \\ 1 \\ -1 \end{bmatrix} e^{-t} \right\}, \quad \mathbf{x}' = \begin{bmatrix} 1 & 2 & 0 \\ 0 & -1 & 0 \\ 1 & 0 & -2 \end{bmatrix} \mathbf{x}$

5.2   Gaussian Elimination

We will give a little bit of "machinery" that will allow us to quickly solve systems of linear equations. Why? In Section 5.4, we will solve a system in the form

$$\mathbf{Ax} = \mathbf{0}$$

in order to help us solve the ODE system $\mathbf{x}' = \mathbf{Ax}$. For nonhomogeneous constant coefficient equations in which the forcing function is constant, we will need to solve a system in the form

$$\mathbf{Ax} = \mathbf{b}$$

in order to obtain the particular solution. Both of these equations can be solved via **Gaussian elimination** and we often just consider the second since the first is simply obtained when $\mathbf{b} = \mathbf{0}$. We assume the reader is familiar with Section B.1 of Appendix B in which some of the basics of matrices are covered. We finished that section with an equivalency theorem:

THEOREM 5.2.1

The following are equivalent characterizations of the $n \times n$ matrix $\mathbf{A}$:
(a) $\mathbf{A}$ is invertible.
(b) The system $\mathbf{Ax} = \mathbf{b}$ has a unique solution $\mathbf{x}$ for all $\mathbf{b}$ in $\mathbb{R}^n$.
(c) The system $\mathbf{Ax} = \mathbf{0}$ has $\mathbf{x} = \mathbf{0}$ as its unique solution.
(d) $\det(\mathbf{A}) \neq 0$.
(e) The n columns of $\mathbf{A}$ are linearly independent.

We will add to this list in this section.

In the simplest case, our matrix $\mathbf{A}$ will be **diagonal**:

$$\mathbf{A} = \begin{bmatrix} d_1 & 0 & \cdots & 0 \\ 0 & d_2 & \cdots & 0 \\ \vdots & 0 & \ddots & \vdots \\ 0 & 0 & \cdots & d_n \end{bmatrix}.$$

We could then solve $\mathbf{Ax} = \mathbf{b}$ by simply dividing the entry b_i of $\mathbf{b}$ by d_i for each i. However, we will usually not be so lucky. To solve a more general $\mathbf{Ax} = \mathbf{b}$ by hand, we begin with an example.

Example 1 Solve the given equations simultaneously by using multiples of each equation to reduce the system:

$$\begin{array}{rcll} -2x & +\, 4z = 2 & \quad (\text{eq1}) \\ 4x + 2y - 4z & = -2 & \quad (\text{eq2}) \\ 2x + 4y + 2z & = -2 & \quad (\text{eq3}) \end{array} \qquad (5.8)$$

to a solvable form.

Solution We want to systematically add and subtract equations from each other to achieve this. It is extremely important to remember that multiplying both sides of an equation by a (non-zero) number does not change the solution just as adding two equations together does not change the solution either. Thus performing the operations below gives us a simpler system with the *same solution set* as the original system. We will proceed by first making two entries in the same column equal and then subtracting the rows.

Let us fix the first equation and *subtract* multiples of it from the other two to try to eliminate x. If we multiply -2 times the first equation and subtract it from the second, we will eliminate the x variable in that equation. Likewise, if we multiply -1 times the first equation and subtract it from the third, we will eliminate the x from the third equation. Doing so gives

$$-2x + 4z = 2 \tag{5.9}$$
$$2y + 4z = 2 \quad (\text{eq2} - (-2\text{eq1}) \rightarrow \text{eq2a}) \tag{5.10}$$
$$4y + 6z = 0. \quad (\text{eq3} - (-\text{eq1}) \rightarrow \text{eq3a}) \tag{5.11}$$

If we now combine the new second and third equations in a similar way, we will obtain a form that we can easily solve. Thus we multiply the new second equation (eq2a) times 2 and subtract it from the new third equation (eq3a) and put it in the place of this last equation (now call it eq3b):

$$-2x + 4z = 2 \tag{5.12}$$
$$2y + 4z = 2 \tag{5.13}$$
$$-2z = -4. \quad (\text{eq3a} - 2\text{eq2a} \rightarrow \text{eq3b}) \tag{5.14}$$

We now use **back substitution** to solve the system, starting with the last row and working our way up. Solving the last equation gives us $z = 2$. We substitute this into the second equation and then see that $y = -3$. Substituting into the first equation gives $x = 3$.

We want to reexamine this last example, but now from the point of view of matrices and vectors. We consider the coefficient matrix with the right-hand side of (5.8) given as an additional column on the right. The resulting matrix is called the augmented matrix:

$$\begin{bmatrix} -2 & 0 & 4 & 2 \\ 4 & 2 & -4 & -2 \\ 2 & 4 & 2 & -2 \end{bmatrix}. \tag{5.15}$$

We then proceed to reduce the augmented matrix, just as we did with the system of equations. We let R_1, R_2, R_3 denote rows 1, 2, and 3, respectively, of our augmented matrix. We will use the notation $2R_1 - (-R_2) \rightarrow R_2$ to mean that we multiply row 2 by -1, subtract it from 2 times row 2, and then put the result in row 2 and this new row 2 will now be called R_2. Then beginning with (5.15)

$$R_2 - (-2R_1) \to R_2 \text{ gives} \quad \begin{bmatrix} -2 & 0 & 4 & 2 \\ 0 & 2 & 4 & 2 \\ 2 & 4 & 2 & -2 \end{bmatrix}, \quad (5.16)$$

$$R_3 - (-R_1) \to R_3 \text{ gives} \quad \begin{bmatrix} -2 & 0 & 4 & 2 \\ 0 & 2 & 4 & 2 \\ 0 & 4 & 6 & 0 \end{bmatrix}. \quad (5.17)$$

We note that the $(1,1)$ entry of the augmented matrix (-2 in this case) needed to be nonzero in order to eliminate the entries in the same column that lie below it. We call -2 a **pivot** and observe that we will always need to have a non-zero pivot in order to proceed. In the event that we obtain a 0 for a pivot element, we may interchange rows if it will help us proceed; if not, Gaussian elimination fails and the system has dependent rows. Continuing with our previous situation (the next pivot element is 2):

$$R_3 - 2R_2 \to R_3 \text{ gives} \quad \begin{bmatrix} -2 & 0 & 4 & 2 \\ 0 & 2 & 4 & 2 \\ 0 & 0 & -2 & -4 \end{bmatrix}. \quad (5.18)$$

With this last matrix, we can again use back substitution and calculate the solutions for x, y, z. We mentioned that the solution set is not changed if we (i) multiply a row by a constant or (ii) add a multiple of one row to another. The solution set is also not changed if we (iii) **interchange** two rows. These three operations are called **elementary row operations** and are used to reduce the original system to one that is easier to solve.

This process of zeroing out entries in our matrix, ultimately ending up with only zeros below the main diagonal of the matrix (i.e., an upper triangular matrix), is called **Gaussian elimination**. The element in the matrix that we use to "zero out" everything below it is called a **pivot**. It is sometimes convenient to make the pivots in (5.18) equal to one:

$$\begin{array}{l} \frac{-1}{2}R_1 \to R_1, \ \frac{1}{2}R_2 \to R_2, \\ \frac{-1}{2}R_3 \to R_3 \ \text{gives} \end{array} \quad \begin{bmatrix} 1 & 0 & -2 & -1 \\ 0 & 1 & 2 & 1 \\ 0 & 0 & 1 & 2 \end{bmatrix}. \quad (5.19)$$

Both (5.18) and (5.19) are upper triangular matrices, the only difference being that (5.19) has the number one as the first non-zero entry of each row. A matrix of this form is said to be in **row-echelon form**. In the event that some rows would end up with all zeros, row-echelon form requires that we move these rows to the end and keep the leading entry in this staircase layout. This should be believable because it is equivalent to switching the physical location of two equations and why should this change the solution? (It shouldn't.)

In Equation (5.18), we are able to see the pivots from this problem: $-2, 2, -2$. As it turns out, these give us a few additional facts:

> 1. $\det(\mathbf{A})$=product of the pivots (of $\mathbf{A}$)
> 2. $\mathbf{A}^{-1}$ exists only when all pivots are nonzero

Thus, $\det(\mathbf{A})$=8 and $\mathbf{A}$ is invertible. The latter of these can go into the equivalency statements of Theorem 5.2.1.

In the case when a unique solution exists, it is also possible to obtain the final answer by continuing the process from (5.19) instead of doing back substitution. In this situation, we reduce the row-echelon matrix to one that has only ones on the main diagonal, with zeros occurring in positions above and below the main diagonal in each column. The last row remains unchanged. The next step in this process, known as **Gauss-Jordan elimination**, is to work from left to right, eliminating all entries in each column that is not on the main diagonal. Normally, we would need to eliminate the entry in the first row and second column of the row-echelon matrix. But (5.19) already has this done and we thus proceed to the third column. We had

$$\begin{bmatrix} 1 & 0 & -2 & -1 \\ 0 & 1 & 2 & 1 \\ 0 & 0 & 1 & 2 \end{bmatrix}$$

so that

$$R_2 - 2R_3 \rightarrow R_2 \text{ gives } \begin{bmatrix} 1 & 0 & -2 & -1 \\ 0 & 1 & 0 & -3 \\ 0 & 0 & 1 & 2 \end{bmatrix} \quad (5.20)$$

$$R_1 - (-2R_3) \rightarrow R_1 \text{ gives } \begin{bmatrix} 1 & 0 & 0 & 3 \\ 0 & 1 & 0 & -3 \\ 0 & 0 & 1 & 2 \end{bmatrix}. \quad (5.21)$$

Now the final answer, $z = 2, y = -3, x = 3$, can be immediately seen because (5.21) is in **reduced-row echelon form**.

This example was one in which we have a **unique solution** but we will not always have this situation. In performing Gaussian elimination, there may be times when we get "stuck" at a certain point in the problem. Sometimes the problem can be overcome by interchanging rows while other times it cannot. If we really are stuck at a certain point then we will either have infinite solutions or no solution. Consider the following system written as augmented matrices and observe the two elementary row operations needed for the first two steps:

$$\begin{bmatrix} 1 & 2 & 1 & 4 \\ 2 & 4 & 1 & 2 \\ -1 & 1 & 2 & 5 \end{bmatrix} \begin{array}{l} R_2 - 2R_1 \rightarrow R_2 \\ R_3 - (-R_1) \rightarrow R_3, \end{array} \begin{bmatrix} 1 & 2 & 1 & 4 \\ 0 & 0 & -1 & -6 \\ 0 & 3 & 3 & 9 \end{bmatrix}. \quad (5.22)$$

We cannot eliminate the "3" in the last row as our next step because the pivot above it is zero. However, we can simply swap the second and third rows:

$$R_2 \leftrightarrow R_3 \quad \begin{bmatrix} 1 & 2 & 1 & 4 \\ 0 & 3 & 3 & 9 \\ 0 & 0 & -1 & -6 \end{bmatrix}. \tag{5.23}$$

This takes care of our problem and gives us an **upper triangular matrix** that can be solved by back substitution and gives $x = 4, y = -3, z = 6$, and our solution is unique.

Let us instead consider the following system and first two elementary row operations needed:

$$\begin{bmatrix} 1 & 2 & 1 & 4 \\ -1 & 1 & 2 & 5 \\ 4 & 2 & -2 & -2 \end{bmatrix} \begin{array}{c} R_2 - (-R_1) \rightarrow R_2 \\ R_3 - 4R_1 \rightarrow R_3, \end{array} \begin{bmatrix} 1 & 2 & 1 & 4 \\ 0 & 3 & 3 & 9 \\ 0 & -6 & -6 & -18 \end{bmatrix}. \tag{5.24}$$

The next step is to eliminate the "-6" in the last row and we do so with

$$R_3 - (-2R_2) \rightarrow R_3 \quad \begin{bmatrix} 1 & 2 & 1 & 4 \\ 0 & 3 & 3 & 9 \\ 0 & 0 & 0 & 0 \end{bmatrix}. \tag{5.25}$$

The last row drops out of the system and we simply need to solve the first two equations. This yields an infinite number of solutions: $x = -2 + z, y = 3 - z$ with z any real number.

Finally, we consider the following system and first two elementary row operations needed:

$$\begin{bmatrix} 1 & 2 & 1 & 4 \\ -1 & 1 & 2 & 5 \\ 2 & -2 & -4 & -2 \end{bmatrix} \begin{array}{c} R_2 - (-R_1) \rightarrow R_2 \\ R_3 - 2R_1 \rightarrow R_3, \end{array} \begin{bmatrix} 1 & 2 & 1 & 4 \\ 0 & 3 & 3 & 9 \\ 0 & -6 & -6 & -10 \end{bmatrix}. \tag{5.26}$$

The next step is to eliminate the "-6" in the last row and we do so with

$$R_3 - (-2R_2) \rightarrow R_3 \quad \begin{bmatrix} 1 & 2 & 1 & 4 \\ 0 & 3 & 3 & 9 \\ 0 & 0 & 0 & 8 \end{bmatrix}. \tag{5.27}$$

The last row doesn't make any sense and we thus say that the original system is not consistent and there is no solution.

Example 2: In Section B.1 of Appendix B, we consider the systems of equations

$$\begin{array}{ccc} x + 2y = 4 & 3x - 3y = -6 & -x + y = -1 \\ -x + y = 2, & -x + y = 2, & -x + y = 2. \\ \text{(a)} & \text{(b)} & \text{(c)} \end{array} \tag{5.28}$$

Write the reduced-row echelon forms of (5.28a)-(5.28c) and interpret the result.

For Equation (5.28a), we have

$$\begin{bmatrix} 1 & 2 & 4 \\ -1 & 1 & 2 \end{bmatrix} \longrightarrow \begin{bmatrix} 1 & 2 & 4 \\ 0 & 3 & 6 \end{bmatrix} \longrightarrow \begin{bmatrix} 1 & 0 & 0 \\ 0 & 1 & 2 \end{bmatrix}. \tag{5.29}$$

This gives us the solution $x = 0, y = 2$. For Equation (5.28b), we have

$$\begin{bmatrix} 3 & -3 & -6 \\ -1 & 1 & 2 \end{bmatrix} \longrightarrow \begin{bmatrix} 3 & -3 & -6 \\ 0 & 0 & 0 \end{bmatrix} \longrightarrow \begin{bmatrix} 1 & -1 & -2 \\ 0 & 0 & 0 \end{bmatrix}. \tag{5.30}$$

Our solution must satisfy $x - y = -2$. There are infinitely many solutions and we introduce a **parameter** t as an independent variable that will give us our solution:

$$x = -2 + t, \ y = t,$$

where t can be any real number. For Equation (5.28c), we have

$$\begin{bmatrix} -1 & 1 & -1 \\ -1 & 1 & 2 \end{bmatrix} \longrightarrow \begin{bmatrix} -1 & 1 & -1 \\ 0 & 0 & 4 \end{bmatrix}. \tag{5.31}$$

For this situation, we see that there is no way to satisfy both equations and we call this an inconsistent set of equations (no solution).

Example 3: Consider

$$x_1 + 2x_2 - x_3 + x_4 = 3$$
$$2x_1 + 4x_2 - x_3 + 3x_4 = 1.$$

Performing elementary row operations on the augmented matrix gives

$$\begin{bmatrix} 1 & 2 & -1 & 1 & 3 \\ 0 & 0 & 1 & 1 & -5 \end{bmatrix} \tag{5.32}$$

as the resulting echelon matrix. This system is consistent and with infinitely many solutions. We had previously defined the pivot elements as the first non-zero element in a given row for a matrix in echelon form. We further say that the columns with pivot elements are called the **basic variables** (or **leading variables**) while all the others are the **free variables**. In this example, x_1 and x_3 are the basic variables whereas x_2 and x_4 are free variables. The free variables are set to different (arbitrary) parameters, say $x_2 = s$ and $x_4 = t$. Our solution can then be written as

$$x_1 = -2 - 2s - 2t, \ x_3 = -5 - t$$

and any choice of s and t will give a solution.

Consider $\mathbf{A}\mathbf{x} = \mathbf{b}$ as

$$\begin{bmatrix} 1 & 2 \\ 4 & 3 \\ 3 & 1 \end{bmatrix} \begin{bmatrix} x_1 \\ x_2 \end{bmatrix} = \begin{bmatrix} b_1 \\ b_2 \\ b_3 \end{bmatrix}. \tag{5.33}$$

For $\mathbf{b} = [0 \ -5 \ -5]^T$, find the reduced-row echelon of the system and the resulting solution, if one exists. Then find the solution for (5.33) for the general $\mathbf{b}$ as written. What conditions are necessary for a solution to exist?

Performing row reduction gives

$$\begin{bmatrix} 1 & 2 & 0 \\ 4 & 3 & -5 \\ 3 & 1 & -5 \end{bmatrix} \longrightarrow \begin{bmatrix} 1 & 0 & -2 \\ 0 & 1 & 1 \\ 0 & 0 & 0 \end{bmatrix}. \tag{5.34}$$

Thus, $x_1 = -2, x_2 = 1$ is a solution. The first two columns of the reduced echelon form show that the columns are linearly independent and because the last row contains all zeros, we conclude that the vector $\mathbf{b}$ can be written as a linear combination of the two column vectors of $\mathbf{A}$.

When we consider the general $\mathbf{b}$, row reductions give

$$\begin{bmatrix} 1 & 2 & b_1 \\ 4 & 3 & b_2 \\ 3 & 1 & b_3 \end{bmatrix} \longrightarrow \begin{bmatrix} 1 & 0 & \frac{1}{5}(-3b_1 + 2b_2) \\ 0 & 1 & \frac{1}{5}(4b_1 - b_2) \\ 0 & 0 & b_1 - b_2 + b_3 \end{bmatrix}. \tag{5.35}$$

In order for a solution to exist, we thus need $b_1 - b_2 + b_3 = 0$. Thus any solution to (5.33) must lie in the plane $x - y + z = 0$, that is, the plane with normal vector $[-1 \ 1 \ -1]^T$.

In Problems **1–6**, write the systems of equations as an augmented matrix.

1. $\begin{cases} x_1 + 4x_2 = 3 \\ 2x_1 - 5x_2 = -1 \end{cases}$

2. $\begin{cases} 7x_1 - 3x_2 = 1 \\ 2x_2 = -5 \end{cases}$

3. $\begin{cases} 5x_1 + 8x_2 - 2x_3 = -1 \\ x_1 - 4x_2 + \frac{1}{2}x_3 = 0 \end{cases}$

4. $\begin{cases} 2x_1 - 3x_2 + x_3 = -7 \\ -x_1 + 27x_3 = 1 \end{cases}$

5. $\begin{cases} x_1 + 2x_2 - 7x_3 + 5x_4 = 1 \\ x_1 - x_3 + 3x_4 = -2 \\ 2x_1 + 3x_2 - 9x_3 + 3x_4 = 8 \end{cases}$

6. $\begin{cases} -x_1 + 2x_2 + 5x_3 + 2x_4 = 3 \\ x_1 - x_4 = 4 \\ 2x_1 + 3x_2 - x_3 + x_4 = -2 \end{cases}$

In Problems **7–12**, perform back substitution to solve the system.

7. $\begin{cases} x_1 + 4x_2 = 3 \\ \qquad 2x_2 = -4 \end{cases}$
   8. $\begin{cases} x_1 - 3x_2 = 5 \\ \qquad 2x_2 = 6 \end{cases}$

9. $\begin{bmatrix} 2 & 6 & -2 & -1 \\ 0 & -3 & 1 & 0 \\ 0 & 0 & 2 & 2 \end{bmatrix}$
   10. $\begin{bmatrix} 1 & 0 & -2 & -1 \\ 0 & -2 & 1 & 0 \\ 0 & 0 & 2 & 2 \end{bmatrix}$

11. $\begin{bmatrix} 1 & 3 & -6 & 0 \\ 0 & -2 & 1 & 1 \end{bmatrix}$
   12. $\begin{bmatrix} 2 & 1 & -2 & -3 \\ 0 & 1 & -2 & 1 \end{bmatrix}$

In Problems **13–20**, (i) use Gaussian elimination to reduce the matrix to echelon form and then back substitution to solve the system; (ii) use Gauss-Jordan elimination to solve the system.

13. $\begin{cases} 2x_1 + 3x_2 = 1 \\ 5x_1 + 7x_2 = 3 \end{cases}$
   14. $\begin{cases} x_1 + 2x_2 = 7 \\ 2x_1 + x_2 = 8 \end{cases}$

15. $\begin{cases} 2x_1 + 4x_2 = 6 \\ 3x_1 + 6x_2 = 9 \end{cases}$
   16. $\begin{cases} -2x_1 + 4x_2 = 3 \\ 4x_1 - 8x_2 = 1 \end{cases}$

17. $\begin{cases} x_1 - 3x_3 = -2 \\ 3x_1 + x_2 - 2x_3 = 5 \\ 2x_1 + 2x_2 + x_3 = 4 \end{cases}$
   18. $\begin{cases} x_1 + x_2 + 2x_3 = -1 \\ x_1 - 2x_2 + x_3 = -5 \\ 3x_1 + x_2 + x_3 = 3 \end{cases}$

19. $\begin{cases} x_1 + 2x_2 + 3x_3 = 3 \\ 2x_1 + 3x_2 + 8x_3 = 4 \\ 3x_1 + 2x_2 + 17x_3 = 1 \end{cases}$
   20. $\begin{cases} x_1 - 2x_2 + 2x_3 + x_4 = 2 \\ x_1 - 2x_2 + x_3 + x_4 = 4 \\ x_1 + x_2 + x_3 + x_4 = 1 \end{cases}$

5.3 Vector Spaces and Subspaces

We call a set V a vector space if its elements satisfy the same properties of addition and multiplication by scalars that geometric vectors do. More precisely, we have the following definition.

> **Definition 5.3.1**
> Let V be a set of elements (called **vectors**); this set, together with vector addition $+$ and scalar multiplication, is termed a vector space if for c_1, c_2 scalars and $\mathbf{u}, \mathbf{v}, \mathbf{w}$ in V the following axioms hold:
> (i) $\mathbf{u} + \mathbf{v} \in V$
> (ii) $c_1 \mathbf{v} \in V$
> (iii) $\mathbf{u} + \mathbf{v} = \mathbf{v} + \mathbf{u}$
> (iv) $\mathbf{u} + (\mathbf{v} + \mathbf{w}) = (\mathbf{u} + \mathbf{v}) + \mathbf{w}$
> (v) There is a zero vector, $\mathbf{0}$, such that $\mathbf{0} + \mathbf{u} = \mathbf{u}$
> (vi) Every vector has an additive inverse $-\mathbf{u}$ such that $\mathbf{u} + (-\mathbf{u}) = \mathbf{0}$
> (vii) $c_1(\mathbf{u} + \mathbf{v}) = c_1 \mathbf{u} + c_1 \mathbf{v}$
> (viii) $(c_1 + c_2)\mathbf{u} = c_1 \mathbf{u} + c_2 \mathbf{u}$
> (ix) $c_1(c_2 \mathbf{u}) = (c_1 c_2)\mathbf{u}$
> (x) $1\mathbf{u} = \mathbf{u}$

Example 1: Consider all the solutions of a given nth-order linear homogeneous differential equation. Together they form a vector space, specifically called the **solution space** of the given differential equation. We can easily check this by letting a, b be any real numbers and y_1, y_2, y_3 be solutions. Then Theorem 3.2.1, which says that linear combinations of solutions are still solutions for *any* scalars, shows that each of the above axioms is true. We leave it to the reader to show this. Thus the space of all solutions is a vector space.

For differential equations, we have seen that elements of the solution space (a vector space) of a given differential equation are solutions to this differential equation. The concept of vector space is one that is taken for granted in many courses that incorporate math and we'll illustrate the importance of knowing the vector space to which you are referring. In physics and engineering, we often consider a "vector" to be something with magnitude and direction. An example is

$$\mathbf{v}_1 = 2\hat{\imath} - 3\hat{\jmath}.$$

If you are told to sketch this vector, where would you draw it? Many of you would probably answer "in the x-y plane" and would go 2 units to the right and 3 units down from the origin to draw it. Now sketch the vector

$$\mathbf{v}_2 = 2\hat{\imath} - 3\hat{\jmath} + 0\hat{k}.$$

Is there any difference? Yes! The second vector, $\mathbf{v}_2$, really lives in the x-y-z space even though its z-component is 0. Mathematically, this is a significant distinction. In multivariable calculus, vectors are often written using the notation $\hat{\imath}, \hat{\jmath}, \hat{k}$. We will instead write vectors in a different form:

$$\mathbf{v}_1 = 2\hat{\imath} - 3\hat{\jmath} = \begin{bmatrix} 2 \\ -3 \end{bmatrix} \quad \text{and} \quad \mathbf{v}_2 = 2\hat{\imath} - 3\hat{\jmath} + 0\hat{k} = \begin{bmatrix} 2 \\ -3 \\ 0 \end{bmatrix}.$$

Vector $\mathbf{v}_1$ lives in $\mathbb{R}^2$ and elements in this space must have exactly two components both of which are real numbers (as opposed to complex numbers). On the other hand, vector $\mathbf{v}_2$ lives in $\mathbb{R}^3$ and elements in this space must have exactly three components, all of which are real numbers. In order to answer questions about these kinds of vectors, we need to specify the vector space to which we are referring.

Example 2: Consider all vectors of the form $\begin{bmatrix} x \\ y \end{bmatrix}$. With the normal componentwise addition and subtraction of vectors, we can easily check that all the axioms of a vector space are satisfied. The vector form represents the general form of a vector that lives in $\mathbb{R}^2$, which is simply the familiar x-y plane.

There are many examples of vector spaces but we will focus our efforts on understanding the specific example of a solution space for a given differential equation. We stated that all possible solutions of a linear homogeneous differential equation form the solution space for that differential equation. We have also described a fundamental set of solutions for a given nth-order linear homogeneous differential equation as being the elements (solutions) necessary to generate all possible solutions. This brings us to two more important topics, each of which is satisfied by this fundamental set of solutions.

Definition 5.3.2

A set of elements of a vector space is said to **span** the vector space if any vector in the space can be written as a linear combination of the elements in the spanning set.

It should be believable that if we don't have a set that spans the vector space, then we can possibly add more vectors so that it does span. In terms of differential equations, the following example illustrates this idea.

Example 3: Consider the set of vectors

$$\left\{ \begin{bmatrix} 1 \\ 0 \\ 0 \end{bmatrix}, \begin{bmatrix} 0 \\ 1 \\ 0 \end{bmatrix}, \begin{bmatrix} 2 \\ 3 \\ 0 \end{bmatrix} \right\}$$

in $\mathbb{R}^3$. This set of vectors does not span $\mathbb{R}^3$ because we can never obtain a non-zero entry in the 3rd component. Thus, the set does not span $\mathbb{R}^3$. If we include the vector $[0 \ 0 \ 1]^T$ in the set, then the resulting set would span $\mathbb{R}^3$.

Example 4: Consider the differential equation $y''(x) + y(x) = 0$. We can easily check that

$$\{\sin x, \sqrt{3}\sin x\}$$

are solutions but that this set *cannot* generate all possible solutions. For example, we could never write $2\cos x$ as a linear combination of these two elements (solutions). If we also include $2\cos x$, the resulting set

$$\{\sin x, \sqrt{3}\sin x, 2\cos x\}$$

does span the space. That is, all solutions of $y''(x) + y(x) = 0$ can be found by taking linear combinations of the solutions in $\{\sin x, \sqrt{3}\sin x, 2\cos x\}$.

At this point, we hope you are wondering why we don't just take the fundamental set of solutions as our spanning set in the case of a solution of a differential equation. The short answer is *we could* and *we will!* Regardless of the vector space, our ultimate goal is to find a set of vectors that spans the given space and is also **linearly independent**. Having a set with these two characteristics gives us the smallest possible set that can generate every element in the space.

Definition 5.3.3

A **basis** of a vector space is any collection of elements that spans the vector space and is also linearly independent.

Finding a spanning set sometimes requires adding more vectors to the given set, thus increasing the number of elements. Finding a linearly independent set sometimes requires discarding vectors that can be written as linear combinations of vectors in the given set, thus decreasing the number of elements. Finding a basis for a vector space is finding the balance between a set that has enough elements to span but not so many that there is any overlap or redundancy. We can state a nice theorem that relates these two concepts.

THEOREM 5.3.1

Let V be a vector space with a basis consisting of n vectors. Any set that spans V is either a basis of V (and thus has n vectors) or can be reduced to a basis of V by removing vectors. Any set that is linearly independent in V is either a basis of V (and thus has n vectors) or can be extended to a basis of V by inserting vectors.

We note that for this theorem, we do *not* say that we can add or remove any vectors we wish. In the case of a spanning set, it is necessary to only remove the ones that don't contribute anything new, while in the case of the linearly independent set we can only add vectors that are not already included.

The number of elements in the basis also tells us the size of the vector space. This may seem trivial in the case of a solution space—we know that for an nth-order linear homogeneous differential equation, any fundamental set of solutions, i.e., any basis, must have n elements. The size of a general vector space is often a very useful piece of information.

Definition 5.3.4

Let $\{\mathbf{u}_1, \mathbf{u}_2, \cdots, \mathbf{u}_n\}$ be a basis for a given vector space. The **dimension** of this vector space is equal to the number of elements in the basis, i.e., n in this case.

Example Consider the Cartesian plane, $\mathbb{R}^2$. The set

$$\left\{ v_1 = \begin{bmatrix} 1 \\ 0 \end{bmatrix}, \quad v_2 = \begin{bmatrix} 0 \\ 1 \end{bmatrix} \right\}$$

is a basis for the space. Why? If we consider any vector $\begin{bmatrix} a \\ b \end{bmatrix}$, we can write this as a linear combination of the two basis vectors, $av_1 + bv_2$. Thus the set spans the space. It should also be clear that $c_1v_1 + c_2v_2 = 0$ only when $c_1 = c_2 = 0$ and thus the two vectors are linearly independent.[1] This particular basis is known as the **standard basis** and its vectors are often denoted e_1, e_2.

We finish this section with one additional concept, that of a **subspace**. Removing elements from a basis yet still having a basis (but of a *different* space) gives rise to this notion of a subspace.

Definition 5.3.5

Let W be a subset of a vector space V. We call W a **subspace** if the following three properties hold:
(i) W contains the zero vector.
(ii) W is closed under addition; that is, if u, v are in W then so is $u + v$.
(iii) W is closed under scalar multiplication; that is, if c is a scalar and u a vector W then so is cu.

Example 6: Consider the two-dimensional Cartesian plane $\mathbb{R}^2$ and its standard basis. The y-axis is a subspace of $\mathbb{R}^2$ because any linear combination of vectors on this axis necessarily remains on this axis. In fact, any line through the origin is a subspace of $\mathbb{R}^2$.

Example 7: Consider the equation $y''' - y'' + y' - y = 0$. It is straightforward to check that a fundamental set of solutions is given by $\{\cos x, \sin x, e^x\}$. A subspace of this solution space can be created by taking $\{\cos x, \sin x\}$ as the basis for it. We know this particular subspace is also the solution space for the equation $y'' + y = 0$. Any solution in the subspace remains in the subspace. In order to leave the subspace, we must make a linear combination with an element *not* already in the subspace, in this case with e^x.

5.3.1   The Nullspace and Column Space

We conclude with some ideas that tie together the discussion of this section.[2] We began our discussion by considering $Ax=0$. We thus want to know what

---

[1] As mentioned earlier, a first step is to check that the vectors under consideration live in the proper space. Checking that you have a basis for a vector space is analogous to checking the conditions for a fundamental set of solutions. You need the following: (1) the same number of vectors as the size of the space, (2) the vectors need to be in the proper space, and (3) the vectors must be linearly independent.

[2] See any of the numerous linear algebra books, for example, Strang [47], for a more in-depth discussion.

vectors, when left-multiplied by $\mathbf{A}$, will give us the vector $\mathbf{0}$. While $\mathbf{x} = \mathbf{0}$ will always work, we are usually more interested in the non-zero vectors that will work.

Consider the following matrix equations:

$$\begin{bmatrix} 2 & -2 & 2 \\ 2 & -2 & 2 \\ 1 & -1 & 1 \end{bmatrix} \begin{bmatrix} 1 \\ 1 \\ 0 \end{bmatrix} = \begin{bmatrix} 0 \\ 0 \\ 0 \end{bmatrix} \text{ and } \begin{bmatrix} 2 & -2 & 2 \\ 2 & -2 & 2 \\ 1 & -1 & 1 \end{bmatrix} \begin{bmatrix} -1 \\ 0 \\ 1 \end{bmatrix} = \begin{bmatrix} 0 \\ 0 \\ 0 \end{bmatrix}. \quad (5.36)$$

Thus, the matrix $\mathbf{A}$ takes both $[1\ 1\ 0]^T$ and $[-1\ 0\ 1]^T$ to the zero vector $[0\ 0\ 0]^T$. Is this a property of the vectors or the matrix? Actually, it's both. The structure of the given matrix is such that any vector that is a linear combination of the above two vectors is also sent to the zero vector. And it's not hard to construct similar examples for an $m \times n$ matrix as well. For example, we can also see that

$$\begin{bmatrix} 2 & -2 & 2 \\ -3 & 3 & -3 \end{bmatrix} \begin{bmatrix} 0 \\ 1 \\ 1 \end{bmatrix} = \begin{bmatrix} 0 \\ 0 \end{bmatrix} \text{ and } \begin{bmatrix} 2 & -2 & 2 \\ -3 & 3 & -3 \end{bmatrix} \begin{bmatrix} 1 \\ 2 \\ 1 \end{bmatrix} = \begin{bmatrix} 0 \\ 0 \end{bmatrix}. \quad (5.37)$$

The following definition will clarify this situation:

Definition 5.3.6

The **nullspace** of an $m \times n$ matrix $\mathbf{A}$ is defined as

$$\text{null}(\mathbf{A}) = \{\mathbf{x} \in \mathbb{R}^n | \mathbf{Ax} = \mathbf{0}\}.$$

We note that $\text{null}(\mathbf{A})$ is a subspace of $\mathbb{R}^n$. This is easily seen because it satisfies the three conditions of a subspace:

(i) We know the zero vector is in $\text{null}(\mathbf{A})$ because $\mathbf{A0} = \mathbf{0}$.

(ii) If $\mathbf{x}, \mathbf{y}$ are two vectors in $\text{null}(\mathbf{A})$, then $\mathbf{A}(\mathbf{x} + \mathbf{y}) = \mathbf{Ax} + \mathbf{Ay}$ and thus $\mathbf{x} + \mathbf{y}$ is in $\text{null}(\mathbf{A})$.

(iii) For any scalar c and any vector $\mathbf{x}$ in $\text{null}(\mathbf{A})$, we have $\mathbf{A}(c\mathbf{x}) = c\mathbf{Ax} = c\mathbf{0} = \mathbf{0}$ and thus $c\mathbf{x}$ is in $\text{null}(\mathbf{A})$.

Because the three conditions are satisfied, we conclude that $\text{null}(\mathbf{A})$ is indeed a subspace of $\mathbb{R}^n$.

Sometimes the nullspace of $\mathbf{A}$ is referred to as the **kernel** of $\mathbf{A}$. We also observe that not all vectors are taken to zero. In the examples immediately preceding Definition 5.3.6, we can see that matrix A takes $[1\ 0\ 2]^T$ and $[1\ 0\ 1]^T$ to $[6\ -9]^T$ and $[4\ -6]^T$, respectively. What about these vectors that are *not* taken to zero upon left multiplication by a matrix? We have another definition.

Definition 5.3.7

The **column space** or **range** of an $m \times n$ matrix $\mathbf{A}$ is defined as

$$R(\mathbf{A}) = \{\mathbf{y} \in \mathbb{R}^m | \mathbf{A}\mathbf{x} = \mathbf{y}\}.$$

We note that $R(\mathbf{A})$ is a subspace of $\mathbb{R}^m$. It is also called the column space because it is the subspace of $\mathbb{R}^m$ that is spanned by the columns of $\mathbf{A}$. We also show this:

(i) We know the zero vector is in $R(\mathbf{A})$ because $\mathbf{A}\mathbf{0} = \mathbf{0}$.

(ii) If $\mathbf{y}_1, \mathbf{y}_2$ are two vectors in $R(\mathbf{A})$, then we know $\mathbf{A}\mathbf{x}_1 = \mathbf{y}_1$ and $\mathbf{A}\mathbf{x}_2 = \mathbf{y}_2$ for some vectors $\mathbf{x}_1, \mathbf{x}_2$ in $\mathbb{R}^n$. We then have $\mathbf{A}(\mathbf{x}_1+\mathbf{x}_2) = \mathbf{A}\mathbf{x}_1+\mathbf{A}\mathbf{x}_2 = \mathbf{y}_1+\mathbf{y}_2$ and thus $\mathbf{y}_1 + \mathbf{y}_2$ is in $R(\mathbf{A})$.

(iii) For any scalar c and any vector $\mathbf{y}$ in $R(\mathbf{A})$, we have $\mathbf{A}(\mathbf{x}) = \mathbf{y}$ for some $\mathbf{x}$. Then $\mathbf{A}(c\mathbf{x}) = c\mathbf{A}\mathbf{x} = c\mathbf{y}$ and thus $c\mathbf{y}$ is in $R(\mathbf{A})$.

Because the three conditions are satisfied, we conclude that $R(\mathbf{A})$ is indeed a subspace of $\mathbb{R}^m$.

At this point, we pause to define another important concept of linear algebra.

Definition 5.3.8

The **rank** of a matrix $\mathbf{A}$ is the number of linearly independent columns of $\mathbf{A}$.

This definition gives us the following useful result, where $\text{rank}(\mathbf{A}) = r$:

$$\dim(R(\mathbf{A})) = r. \tag{5.38}$$

Let's consider the matrix

$$\mathbf{A} = \begin{bmatrix} 2 & -2 & 2 & 2 \\ 2 & -2 & 2 & 1 \\ 1 & -1 & 1 & 0 \end{bmatrix}. \tag{5.39}$$

Bases from the nullspace and column space can be found easily using MAT-LAB, Maple, or Mathematica. We will proceed by hand. Since we want to find vectors in the nullspace, we don't need to consider the augmented matrix explicitly since the additional column would be all zeros. Performing elementary row operations on $\mathbf{A}$ gives

$$\begin{bmatrix} 2 & -2 & 2 & 2 \\ 0 & 0 & 0 & -1 \\ 0 & 0 & 0 & 0 \end{bmatrix}. \tag{5.40}$$

If we think of each of our columns as corresponding to a variable x_i, we observe that x_2 and x_3 are the free variables and our nullspace will thus have two vectors in its basis. Since any vector in the nullspace must right multiply this and give zero, we know that x_4 must be zero (because it's the lone entry in the 2nd row). The remaining equation that must be satisfied for vectors in the nullspace is thus $2x_1 - 2x_2 + 2x_3 = 0$. We set one of the free variables to one and the other to zero: $x_2 = 1$, $x_3 = 0$ and solve the first row for x_1. We then switch the values of our free variables: $x_2 = 0$, $x_3 = 1$ and solve again for x_1:

$$\begin{matrix} x_2 = 1 \\ x_3 = 0 \end{matrix} \Bigg\} \Rightarrow \begin{bmatrix} 1 \\ 1 \\ 0 \\ 0 \end{bmatrix} \quad \text{and} \quad \begin{matrix} x_2 = 0 \\ x_3 = 1 \end{matrix} \Bigg\} \Rightarrow \begin{bmatrix} -1 \\ 0 \\ 1 \\ 0 \end{bmatrix}. \tag{5.41}$$

If more free variables had been present, each of them takes a turn being set to one while the others are zero. We thus obtain a basis for our nullspace as

$$\text{null}(\mathbf{A}) = \left\{ \begin{bmatrix} 1 \\ 1 \\ 0 \\ 0 \end{bmatrix}, \begin{bmatrix} -1 \\ 0 \\ 1 \\ 0 \end{bmatrix} \right\}. \tag{5.42}$$

For the column space, the presence of two basic variables tells us that it will have two vectors in its basis. The pivot columns are the 1st and 4th and so a basis for the column space will be the 1st and 4th columns of the *original* matrix:

$$\text{R}(\mathbf{A}) = \left\{ \begin{bmatrix} 2 \\ 2 \\ 1 \end{bmatrix}, \begin{bmatrix} 2 \\ 1 \\ 0 \end{bmatrix} \right\}. \tag{5.43}$$

For our matrix $\mathbf{A}$, we have $m = 3$, $n = 4$. We should observe that we do have $\text{null}(\mathbf{A}) \subset \mathbb{R}^4$, $\text{R}(\mathbf{A}) \subset \mathbb{R}^3$, as we showed earlier. There is another worthwhile observation for this example:

$$\dim(\text{R}(\mathbf{A})) + \dim(\text{null}(\mathbf{A})) = 4.$$

The following is true in general and is a very important result in linear algebra:

> **THEOREM 5.3.2**
>
> For an $m \times n$ matrix $\mathbf{A}$, we have
>
> $$\dim(R(\mathbf{A})) + \dim(\text{null}(\mathbf{A})) = n. \qquad (5.44)$$

There is an interesting analogy between solving nonhomogeneous linear ODEs and solving the system $\mathbf{A}\mathbf{x} = \mathbf{b}$. Sections 1.3 and 5.1 describe how the linear ODE $\frac{dy}{dx} + P(x)y = Q(x)$ has the general solution of the form $y = y_h + y_p$ (and similarly for the vector version). For a linear system $\mathbf{A}\mathbf{x} = \mathbf{b}$, we can also think of the solution as being $\mathbf{x} = \mathbf{x}_h + \mathbf{x}_p$. If this is the case, then we need

$$\mathbf{A}(\mathbf{x}_h + \mathbf{x}_p) = (\mathbf{0} + \mathbf{b}).$$

Thus, we see that to get $\mathbf{x}_h$ we take the right-hand side equal to $\mathbf{0}$. In other words, it is the dependent columns of $\mathbf{A}$ that give rise to the homogeneous part or, equivalently, $\mathbf{x}_h$ is in the nullspace of $\mathbf{A}$.

Example 8: Write the solution to

$$\begin{bmatrix} 1 & 4 \\ 3 & 12 \\ -2 & -8 \end{bmatrix} \begin{bmatrix} x_1 \\ x_2 \end{bmatrix} = \begin{bmatrix} -1 \\ -3 \\ 2 \end{bmatrix}$$

in the form $\mathbf{x} = \mathbf{x}_h + \mathbf{x}_p$.

Solution

We know that $\mathbf{x}_h$ is obtained from the homogeneous equation

$$\begin{bmatrix} 1 & 4 \\ 3 & 12 \\ -2 & -8 \end{bmatrix} \mathbf{x}_h = \begin{bmatrix} 0 \\ 0 \\ 0 \end{bmatrix}.$$

Since the second column is a multiple of the first, we obtain

$$\mathbf{x}_h = C \cdot \begin{bmatrix} -4 \\ 1 \end{bmatrix}$$

for any constant C. A particular solution, by inspection (of, e.g., the first row in $\mathbf{A}\mathbf{x} = \mathbf{b}$), gives

$$\mathbf{x}_p = \begin{bmatrix} -1 \\ 0 \end{bmatrix}.$$

Thus, we can write the solution to the original system as

$$\mathbf{x} = \mathbf{x}_h + \mathbf{x}_p = C \cdot \begin{bmatrix} -4 \\ 1 \end{bmatrix} + \begin{bmatrix} -1 \\ 0 \end{bmatrix}.$$

Problems

In Problems **1–8**, show that the given set of vectors forms a basis for the specified vector space.

1. $\left\{ \begin{bmatrix} 2 \\ -1 \end{bmatrix}, \begin{bmatrix} -1 \\ 1 \end{bmatrix} \right\}$, $\mathbb{R}^2$

2. $\left\{ \begin{bmatrix} -1 \\ 1 \end{bmatrix}, \begin{bmatrix} 2 \\ 4 \end{bmatrix} \right\}$, $\mathbb{R}^2$

3. $\left\{ \begin{bmatrix} 10 \\ 1 \end{bmatrix}, \begin{bmatrix} 3 \\ 2 \end{bmatrix} \right\}$, $\mathbb{R}^2$

4. $\left\{ \begin{bmatrix} -2 \\ 0 \end{bmatrix}, \begin{bmatrix} 2 \\ 3 \end{bmatrix} \right\}$, $\mathbb{R}^2$

5. $\left\{ \begin{bmatrix} 1 \\ 1 \\ 1 \end{bmatrix}, \begin{bmatrix} 1 \\ 1 \\ 0 \end{bmatrix}, \begin{bmatrix} 1 \\ 0 \\ 0 \end{bmatrix} \right\}$, $\mathbb{R}^3$

6. $\left\{ \begin{bmatrix} 2 \\ 1 \\ -2 \end{bmatrix}, \begin{bmatrix} 1 \\ -1 \\ 1 \end{bmatrix}, \begin{bmatrix} 0 \\ 3 \\ 0 \end{bmatrix} \right\}$, $\mathbb{R}^3$

7. $\left\{ \begin{bmatrix} 1 \\ 0 \\ 0 \\ 2 \end{bmatrix}, \begin{bmatrix} 0 \\ 1 \\ 0 \\ -1 \end{bmatrix}, \begin{bmatrix} 3 \\ 2 \\ 0 \\ 1 \end{bmatrix}, \begin{bmatrix} 1 \\ 0 \\ 2 \\ 0 \end{bmatrix} \right\}$, $\mathbb{R}^4$

8. $\left\{ \begin{bmatrix} -1 \\ 1 \\ 1 \\ 1 \end{bmatrix}, \begin{bmatrix} 1 \\ -2 \\ 1 \\ 0 \end{bmatrix}, \begin{bmatrix} 0 \\ 3 \\ 1 \\ -1 \end{bmatrix}, \begin{bmatrix} 2 \\ 0 \\ -1 \\ 0 \end{bmatrix} \right\}$, $\mathbb{R}^4$

In Problems **9–21**, determine whether the given set of vectors forms a basis for the specified vector space.

9. $\left\{ \begin{bmatrix} 1 \\ -1 \end{bmatrix}, \begin{bmatrix} 1 \\ 2 \end{bmatrix} \right\}$, $\mathbb{R}^2$

10. $\left\{ \begin{bmatrix} 3 \\ -3 \end{bmatrix}, \begin{bmatrix} -2 \\ 2 \end{bmatrix} \right\}$, $\mathbb{R}^2$

11. $\left\{ \begin{bmatrix} 4 \\ 5 \end{bmatrix}, \begin{bmatrix} 7 \\ 8 \end{bmatrix} \right\}$, $\mathbb{R}^2$

12. $\left\{ \begin{bmatrix} 3 \\ 1 \end{bmatrix}, \begin{bmatrix} -1 \\ 1 \end{bmatrix} \right\}$, $\mathbb{R}^2$

13. $\left\{ \begin{bmatrix} 3 \\ -1 \end{bmatrix}, \begin{bmatrix} -1 \\ 2 \end{bmatrix}, \begin{bmatrix} -1 \\ 1 \end{bmatrix} \right\}$, $\mathbb{R}^3$

14. $\left\{ \begin{bmatrix} 1 \\ 0 \end{bmatrix}, \begin{bmatrix} 0 \\ -1 \end{bmatrix}, \begin{bmatrix} 3 \\ 5 \end{bmatrix} \right\}$, $\mathbb{R}^2$

15. $\left\{ \begin{bmatrix} 3 \\ 0 \\ 0 \end{bmatrix}, \begin{bmatrix} -1 \\ 0 \\ 1 \end{bmatrix}, \begin{bmatrix} 0 \\ 1 \\ 1 \end{bmatrix} \right\}$, $\mathbb{R}^3$

16. $\left\{ \begin{bmatrix} -1 \\ 1 \\ 0 \end{bmatrix}, \begin{bmatrix} 2 \\ -1 \\ 1 \end{bmatrix}, \begin{bmatrix} 0 \\ 0 \\ -1 \end{bmatrix} \right\}$, $\mathbb{R}^3$

17. $\left\{ \begin{bmatrix} 2 \\ 1 \\ 3 \end{bmatrix}, \begin{bmatrix} -1 \\ 0 \\ 0 \end{bmatrix}, \begin{bmatrix} 1 \\ 1 \\ -1 \end{bmatrix} \right\}$, $\mathbb{R}^3$

18. $\left\{ \begin{bmatrix} 2 \\ -1 \\ 1 \end{bmatrix}, \begin{bmatrix} 1 \\ 0 \\ 1 \end{bmatrix}, \begin{bmatrix} 3 \\ 1 \\ 1 \end{bmatrix} \right\}$, $\mathbb{R}^4$

19. $\left\{ \begin{bmatrix} 2 \\ -3 \\ 0 \\ 2 \end{bmatrix}, \begin{bmatrix} -1 \\ 1 \\ 0 \\ 5 \end{bmatrix}, \begin{bmatrix} 0 \\ 0 \\ 1 \\ 1 \end{bmatrix} \right\}$, $\mathbb{R}^4$

20. $\left\{ \begin{bmatrix} -1 \\ 1 \\ 1 \\ 1 \end{bmatrix}, \begin{bmatrix} 1 \\ -2 \\ 1 \\ 0 \end{bmatrix}, \begin{bmatrix} 0 \\ 3 \\ 1 \\ 0 \end{bmatrix}, \begin{bmatrix} 2 \\ 0 \\ -1 \\ 0 \end{bmatrix} \right\}$, $\mathbb{R}^4$

21. $\left\{ \begin{bmatrix} 1 \\ 0 \\ -1 \\ 1 \end{bmatrix}, \begin{bmatrix} 0 \\ 0 \\ -1 \\ 1 \end{bmatrix}, \begin{bmatrix} 2 \\ 2 \\ 1 \\ 2 \end{bmatrix}, \begin{bmatrix} 1 \\ 0 \\ 0 \\ 0 \end{bmatrix} \right\}, \mathbb{R}^4$

22. Show that $\mathbb{R}^2$ is a vector space.

23. Show that $\mathbb{R}^3$ is a vector space.

24. Show that the set of all 2×2 matrices with real entries, $\mathbb{R}^{2\times2}$, is a vector space.

25. Show that $\left\{ \begin{bmatrix} 1 & 0 \\ 0 & 0 \end{bmatrix}, \begin{bmatrix} 0 & 1 \\ 0 & 0 \end{bmatrix}, \begin{bmatrix} 0 & 0 \\ 1 & 0 \end{bmatrix}, \begin{bmatrix} 0 & 0 \\ 0 & 1 \end{bmatrix} \right\}$ is a basis for $\mathbb{R}^{2\times2}$ (see previous question).

26. Show that the plane $z = 0$ (i.e., the x-y plane) is a subspace of $\mathbb{R}^3$.

27. Show that the plane $x = 0$ (i.e., the y-z plane) is a subspace of $\mathbb{R}^3$.

28. Show that any plane passing through the origin in $\mathbb{R}^3$ is a subspace of $\mathbb{R}^3$.

29. Show that the set of $m \times n$ matrices forms a vector space with the previously defined rules for addition and scalar multiplication.

30. Show that the dimension of the vector space of 2×3 matrices is six.

Find bases for the column space and nullspace of the following matrices given in Problems **31–38**.

31. $\begin{bmatrix} 1 & 3 \\ 2 & 1 \end{bmatrix}$

32. $\begin{bmatrix} 1 & 1 \\ 3 & -1 \end{bmatrix}$

33. $\begin{bmatrix} 1 & 3 & 1 \\ 2 & 1 & -1 \end{bmatrix}$

34. $\begin{bmatrix} 1 & -3 & 1 \\ 2 & 0 & 3 \end{bmatrix}$

35. $\begin{bmatrix} 3 & 1 & 1 & 1 \\ 1 & 3 & -1 & -3 \end{bmatrix}$

36. $\begin{bmatrix} 2 & 1 & -1 & 0 \\ 2 & -1 & -1 & -2 \end{bmatrix}$

37. $\begin{bmatrix} -1 & 3 \\ 2 & -6 \\ -1 & 3 \end{bmatrix}$

38. $\begin{bmatrix} 4 & -6 \\ 8 & -12 \\ -2 & 3 \end{bmatrix}$

In Problems **39–43**, construct examples of the following matrices.

39. A 2×2 matrix that has a nullspace consisting only of the zero vector.

40. A 2×2 matrix that has a nullspace with a basis consisting of one non-zero vector.

41. A 3×3 matrix that has a nullspace consisting only of the zero vector.

42. A 3×3 matrix that has a nullspace with a basis consisting of one non-zero vector.

43. A 3×3 matrix that has a nullspace with a basis consisting of two non-zero vectors.

In Problems **44–46**, find bases of the nullspace and column space. Then repeat this with MATLAB, Maple, or Mathematica and comment on any differences.

44. $\begin{bmatrix} 2 & -1 & 1 \\ 2 & -2 & 4 \\ 6 & -4 & 6 \end{bmatrix}$

45. $\begin{bmatrix} 2 & -2 & 2 \\ 2 & -2 & 2 \\ 1 & -1 & 1 \end{bmatrix}$

46. $\begin{bmatrix} 2 & -1 & 3 & 1 \\ 2 & -2 & 2 & 1 \\ 0 & 1 & 1 & 0 \end{bmatrix}$

Another subspace can be obtained by considering the rows of $\mathbf{A}$, which are the same as the columns of $\mathbf{A}^T$. We thus define the **row space** of an $m \times n$ matrix $\mathbf{A}$ as

$$R(\mathbf{A}^T) = \{\mathbf{x} \in \mathbb{R}^n | \mathbf{A}^T \mathbf{y} = \mathbf{x}\}.$$

In a similar manner, we define the **left nullspace** of an $m \times n$ matrix $\mathbf{A}$ as

$$\text{null}(\mathbf{A}^T) = \{\mathbf{y} \in \mathbb{R}^m | \mathbf{y}^T \mathbf{A} = \mathbf{0}\}.$$

We note that the row space is a subspace of $\mathbb{R}^n$ and the left nullspace is a subspace of $\mathbb{R}^m$. Two useful results that follow for an $m \times n$ matrix $\mathbf{A}$ are

- $\dim(R(\mathbf{A}^T)) + \dim(\text{null}(\mathbf{A}^T)) = m.$

- dimension of row space = dimension of column space = rank.

In Problems **47–50**, find bases of the nullspace and column space. Then repeat this with MATLAB, Maple, or Mathematica and comment on any differences.

47. $\begin{bmatrix} 2 & -2 & 2 \\ 2 & -2 & 2 \\ 1 & -1 & 1 \end{bmatrix}$
48. $\begin{bmatrix} 2 & -1 & 1 \\ 2 & -2 & 4 \\ 6 & -4 & 6 \end{bmatrix}$
49. $\mathbf{A} = \begin{bmatrix} 2 & -2 & 2 & 2 \\ 2 & -2 & 2 & 2 \\ 1 & -1 & 1 & 0 \end{bmatrix}$

50. $\mathbf{A} = \begin{bmatrix} 2 & -1 & 3 & 1 \\ 2 & -2 & 2 & 1 \\ 0 & 1 & 1 & 0 \end{bmatrix}$

5.4 Eigenvalues and Eigenvectors

We have seen that we can write the linear system of equations

$$\frac{dx}{dt} = ax + by$$
$$\frac{dy}{dt} = cx + dy$$

in matrix-vector notation $\mathbf{x}' = \mathbf{A}\mathbf{x}$. If $x, A \in \mathbb{R}$, then this equation is easily solved by separation of variables, with solution $x = c_1 e^{At}$. When we considered nth order linear homogeneous equations with constant coefficients, we similarly guessed a solution of the form $y = e^{rx}$, substituted the result into the equation, and obtained the characteristic (or auxiliary) equation; solving for the roots gave r-values that made $y = e^{rx}$ a solution. Thus even though $\mathbf{x} \in \mathbb{R}^n, \mathbf{A} \in \mathbb{R}^{n \times n}$, we might still hope that we can find an exponential solution. Analogously to what we've done before, we assume a solution of the form

$$\mathbf{x} = \mathbf{v}e^{\lambda t},$$

where $\mathbf{v}$ is a vector (same size as $\mathbf{x}$) and λ is a scalar. (Either of these may be complex.) If $\mathbf{v}e^{\lambda t}$ is a solution, it must satisfy the original differential equation $\mathbf{x}' = \mathbf{A}\mathbf{x}$. Substitution gives

$$\lambda e^{\lambda t} \mathbf{v} = \mathbf{A} e^{\lambda t} \mathbf{v}. \tag{5.45}$$

Because $e^{\lambda t} \neq 0$, we can divide by it to obtain the equation

$$\mathbf{A}\mathbf{v} = \lambda \mathbf{v}. \tag{5.46}$$

Thus,

$\mathbf{v}e^{\lambda t}$ *is a solution to* $\mathbf{x}' = \mathbf{A}\mathbf{x}$ *if we can find* $\mathbf{v}, \lambda$ *such that* $\mathbf{A}\mathbf{v} = \lambda \mathbf{v}$.

In this section, we will consider the consequence of a square matrix acting on a vector $\mathbf{v}$ and yielding a constant multiple of the same vector $\mathbf{v}$. In symbols, we have

$$\mathbf{A}\mathbf{v} = \lambda \mathbf{v}, \tag{5.47}$$

where λ is a constant, called a **scale factor**. The scale factor λ modifies the length of the vector $\mathbf{v}$.

Example 1: Consider the matrix

$$A = \begin{bmatrix} -5 & 8 \\ -4 & 7 \end{bmatrix}$$

with

$$\lambda_1 = -1, \mathbf{v}_1 = \begin{bmatrix} 2 \\ 1 \end{bmatrix}, \quad \text{and} \quad \lambda_2 = 3, \mathbf{v}_2 = \begin{bmatrix} 1 \\ 1 \end{bmatrix}.$$

Verify that $\mathbf{A}\mathbf{v}_i = \lambda \mathbf{v}_i$ for $i = 1, 2$.

Solution

We multiply to obtain

$$\underbrace{\begin{bmatrix} -5 & 8 \\ -4 & 7 \end{bmatrix}}_{A} \underbrace{\begin{bmatrix} 2 \\ 1 \end{bmatrix}}_{\mathbf{v}_1} = \begin{bmatrix} -2 \\ -1 \end{bmatrix} = \underbrace{-1}_{\lambda_1} \underbrace{\begin{bmatrix} 2 \\ 1 \end{bmatrix}}_{\mathbf{v}_1} \tag{5.48}$$

and

$$\underbrace{\begin{bmatrix} -5 & 8 \\ -4 & 7 \end{bmatrix}}_{A} \underbrace{\begin{bmatrix} 1 \\ 1 \end{bmatrix}}_{\mathbf{v}_2} = \begin{bmatrix} 3 \\ 3 \end{bmatrix} = \underbrace{3}_{\lambda_2} \underbrace{\begin{bmatrix} 1 \\ 1 \end{bmatrix}}_{\mathbf{v}_2}. \tag{5.49}$$

The previous example showed two situations where the *same* matrix multiplies two different vectors and only changes them by scaling or stretching them. Note that multiplication of *any* vector by this matrix does *not* necessarily simply stretch it. For example,

$$\begin{bmatrix} -5 & 8 \\ -4 & 7 \end{bmatrix} \begin{bmatrix} 1 \\ -2 \end{bmatrix} = \begin{bmatrix} -21 \\ -18 \end{bmatrix},$$

which cannot be written as a product of a scalar and $(1 \quad -2)^T$.

A vector $\mathbf{v}$ that satisfies $\mathbf{Av} = \lambda \mathbf{v}$ for a given matrix $\mathbf{A}$ is called an **eigenvector** of $\mathbf{A}$ and the factor by which it is multiplied, λ, is called the **eigenvalue** of the matrix corresponding to the particular eigenvector. The eigenvalues and eigenvectors of a given matrix give us tremendous insight into the behavior of the matrix in situations as seemingly different as raising a matrix to a power to solving a system of differential equations!

We will now consider how to find the eigenvalues and eigenvectors associated with a matrix $\mathbf{A}$. Let us consider this same matrix

$$\mathbf{A} = \begin{bmatrix} -5 & 8 \\ -4 & 7 \end{bmatrix}.$$

We want to find λ and $\mathbf{v}$ so that

$$\mathbf{A}\mathbf{v} = \lambda \mathbf{v}.$$

Equivalently,

$$(\mathbf{A} - \lambda\mathbf{I})\mathbf{v} = \mathbf{0}.$$

Non-trivial solutions occur only when the matrix $(\mathbf{A} - \lambda\mathbf{I})$ is singular. From Theorem B.1.2, this happens when

$$\det(\mathbf{A} - \lambda\mathbf{I}) = 0.$$

This determinant gives us a polynomial in λ, called the **characteristic polynomial for A**, and when we set it equal to zero, we have the **characteristic equation** for $\mathbf{A}$. In our specific example, calculating the characteristic equation gives

$$\lambda^2 - 2\lambda - 3 = 0.$$

This characteristic equation is solved to obtain the characteristic roots, or eigenvalues, keeping in mind that we allow for complex roots as well as real roots. Because we allow for complex roots, the Fundamental Theorem of Algebra guarantees that there are exactly n roots when the characteristic polynomial is of degree n. As you may be aware, there is no general formula that exists for finding the roots of a polynomial that is degree 5 or higher. Having the help of MATLAB, Maple, or Mathematica to find the eigenvalues or the approximations of them will be very useful. Again referring to our specific example, the characteristic equation can be rewritten as

$$(\lambda - 3)(\lambda + 1) = 0,$$

which gives roots of $\lambda_1 = -1$, $\lambda_2 = 3$.

Once we have the eigenvalues, we take each one in turn and plug it into the equation $(\mathbf{A} - \lambda\mathbf{I})\mathbf{v} = \mathbf{0}$ and solve for $\mathbf{v}$. For this example,

$$(\mathbf{A} - \lambda_1\mathbf{I})\mathbf{v}_1 = \mathbf{0} \Longrightarrow \begin{bmatrix} -5 - (-1) & 8 \\ -4 & 7 - (-1) \end{bmatrix} \begin{bmatrix} v_{11} \\ v_{21} \end{bmatrix} = \begin{bmatrix} 0 \\ 0 \end{bmatrix}$$

$$\Longrightarrow -4v_{11} + 8v_{21} = 0.$$

You might think of any number of possible combinations that will make this last equality true. In fact, any v_{11}, v_{21} that satisfy $v_{11}/v_{21} = 2$ will work. Thus the eigenvectors are not unique in magnitude. Another way to think of this is that *any non-zero multiple of an eigenvector is still an eigenvector.* For example, if $\mathbf{v}$ is an eigenvector, so are $-\mathbf{v}$ and $2\mathbf{v}$. The simplest choice is perhaps $v_{11} = 2$, $v_{21} = 1$ which then gives

$$\mathbf{v}_1 = \begin{bmatrix} 2 \\ 1 \end{bmatrix}$$

as the eigenvector corresponding to the eigenvalue $\lambda_1 = -1$. We could do a similar calculation to obtain

$$\mathbf{v}_2 = \begin{bmatrix} 1 \\ 1 \end{bmatrix}$$

as the eigenvector corresponding to the eigenvalue $\lambda_2 = 3$.

For those who have read Section 5.3, we define

$$E_\lambda = \{v | (\mathbf{A} - \lambda\mathbf{I})\mathbf{v} = \mathbf{0}\} \tag{5.50}$$

and say that E_λ is the eigenspace of λ and is indeed a subspace. We also note that $\mathbf{v}$ is in the nullspace of $(\mathbf{A} - \lambda\mathbf{I})\mathbf{v}$.

We note that the above method worked in a straightforward fashion because our eigenvalues were real and distinct. If they were complex, we can still find eigenvectors but it is a bit more work as we will see later in this section. If we have eigenvalues that are repeated, we can *try* to plug them in as we did above, and this will give us at least one eigenvector but we may or may not be able to find more. Sometimes we will need to obtain a **generalized eigenvector**; this will be discussed in Section 5.6.

Example 2: A matrix $\mathbf{A}$ is said to be **symmetric** if $\mathbf{A}^T = \mathbf{A}$. Find the eigenvalues and eigenvectors of the symmetric matrix

$$\mathbf{A} = \begin{bmatrix} 2 & 3 \\ 3 & 2 \end{bmatrix}.$$

Solution

Calculating the characteristic equation, we find that $(2 - \lambda)^2 - 9 = 0$, so that $\lambda = 5, -1$. We take each eigenvalue in turn to find its corresponding eigenvector:

$$\lambda_1 = 5: \qquad (\mathbf{A} - \lambda_1 \mathbf{I})\mathbf{v}_1 = \mathbf{0} \tag{5.51}$$

$$\Rightarrow \begin{bmatrix} 2-5 & 3 \\ 3 & 2-5 \end{bmatrix} \begin{bmatrix} v_{11} \\ v_{21} \end{bmatrix} = \begin{bmatrix} 0 \\ 0 \end{bmatrix} \tag{5.52}$$

$$\Rightarrow -3v_{11} + 3v_{21} = 0. \tag{5.53}$$

Choosing $v_{11} = v_{21} = 1$ is a good choice. Repeating with the other eigenvalue:

$$\lambda_2 = -1: \qquad (\mathbf{A} - \lambda_2 \mathbf{I})\mathbf{v}_2 = \mathbf{0} \tag{5.54}$$

$$\Rightarrow \begin{bmatrix} 2-(-1) & 3 \\ 3 & 2-(-1) \end{bmatrix} \begin{bmatrix} v_{12} \\ v_{22} \end{bmatrix} = \begin{bmatrix} 0 \\ 0 \end{bmatrix} \tag{5.55}$$

$$\Rightarrow 3v_{12} + 3v_{22} = 0. \tag{5.56}$$

We can choose $v_{12} = 1, v_{22} = -1$ as one possible pair. Thus we have

$$\left\{ 5, \begin{bmatrix} 1 \\ 1 \end{bmatrix} \right\} \quad \text{and} \quad \left\{ -1, \begin{bmatrix} -1 \\ 1 \end{bmatrix} \right\}.$$

A special property of any symmetric matrix is that its eigenvalues and eigenvectors are *always* real-valued.

We revisit the equivalence theorem of the last section and add one more result. We will conclude our brief introduction to eigenvectors and eigenvalues with two important theorems. We will have occasion in later sections to make use of both of these results. The first theorem is an important and useful characterization of an invertible $n \times n$ matrix $\mathbf{A}$.

THEOREM 5.4.1

The following are equivalent characterizations of the $n \times n$ matrix $\mathbf{A}$:
(a) $\mathbf{A}$ is invertible.
(b) The system $\mathbf{A}\mathbf{x} = \mathbf{b}$ has a unique solution $\mathbf{x}$ for each $\mathbf{b}$ in $\mathbb{R}^n$.
(c) The system $\mathbf{A}\mathbf{x} = \mathbf{0}$ has $\mathbf{x} = \mathbf{0}$ as its unique solution.
(d) $\det(\mathbf{A}) \neq 0$.
(e) The n columns of $\mathbf{A}$ form a basis for $\mathbb{R}^n$.
(f) 0 is not an eigenvalue of $\mathbf{A}$.

Regarding (e) in this theorem, we note that because the columns of $\mathbf{A}$ form a basis of $\mathbb{R}^n$, they automatically span $\mathbb{R}^n$ *and* are linearly independent. We also remind the reader that there is never a unique basis for a given vector space. In fact, it will often be useful to convert from one basis to another and we will do so when we analyze the stability of solutions in future sections.

We refer the reader to Appendix C for a discussion of change of bases and coordinates.

The next theorem connects the trace and determinant of an $n \times n$ matrix $\mathbf{A}$ with the eigenvalues of $\mathbf{A}$. This theorem will be used in Chapter 6 when we wish to analyze the qualitative behavior of a linear system.

THEOREM 5.4.2

Let λ_i denote the eigenvalues of an $n \times n$ matrix $\mathbf{A}$. Then

$$\mathrm{Tr}(\mathbf{A}) = \sum_i \lambda_i, \qquad \det(\mathbf{A}) = \prod_i \lambda_i.$$

We note that Theorem 5.4.2 does not say that the eigenvalues lie on the diagonal. Sometimes this will be the case (for example, with triangular or diagonal matrices) but usually the eigenvalues do not appear in the matrix. Nevertheless, we now have a way to calculate the trace and determinant in terms of eigenvalues.

Eigencoordinates

Although we were able to obtain the solution to the system in terms of the variables x and y, it is sometimes easier to solve the system in the coordinate system of the eigenvectors (called the **eigencoordinates**). We refer the reader to Section B.4.1 in Appendix B for a more general discussion of coordinate transformation. We again consider the system of Example 1:

$$\frac{dx}{dt} = -5x + 8y$$

$$\frac{dy}{dt} = -4x + 7y. \tag{5.57}$$

Rather than convert the system to a second-order equation or examine it in the phase plane, we choose to find a **linear change of coordinates** that will make the system **uncoupled**, i.e., where each equation will only have functions and derivatives of one variable. If we create a matrix $\mathbf{V}$ with the columns being the eigenvectors $\mathbf{v}_1 = [2 \ \ 1]^T$, $\mathbf{v}_2 = [1 \ \ 1]^T$, then we can write our original variables x, y in terms of the eigencoordinates u_1, u_2:

$$\begin{bmatrix} x \\ y \end{bmatrix} = \begin{bmatrix} 2 & 1 \\ 1 & 1 \end{bmatrix} \begin{bmatrix} u_1 \\ u_2 \end{bmatrix}, \ \text{i.e., } \mathbf{x} = \mathbf{V}\mathbf{u}. \tag{5.58}$$

However, it is the opposite relation that will allow us to write our equation in eigencoordinates. The matrix $\mathbf{V}$ is invertible since it has eigenvectors as its columns. In order to solve for the eigencoordinates, we will need the inverse of a 2×2 matrix, given in Appendix B as (B.27) and here as (5.59):

$$\begin{bmatrix} a & b \\ c & d \end{bmatrix}^{-1} = \frac{1}{ad - bc} \begin{bmatrix} d & -b \\ -c & a \end{bmatrix}. \tag{5.59}$$

We then left-multiply by $\mathbf{V}^{-1}$ to obtain

$$\begin{bmatrix} u_1 \\ u_2 \end{bmatrix} = \begin{bmatrix} 1 & -1 \\ -1 & 2 \end{bmatrix} \begin{bmatrix} x \\ y \end{bmatrix}. \tag{5.60}$$

We can also write the eigenvalue-eigenvector equations, $\mathbf{A}\mathbf{v}_i = \lambda_i \mathbf{v}_i$ for $i = 1, 2$, as a matrix-vector system:

$$\mathbf{AV} = \mathbf{V\Lambda} \tag{5.61}$$

where we note that the order on the right is indeed $\mathbf{V\Lambda}$ for the matrix $\mathbf{\Lambda}$ that has the eigenvalues λ_1, λ_2 on the diagonal. Just as we did above we can left-multiply by $\mathbf{V}^{-1}$, this time to obtain

$$\mathbf{V}^{-1}\mathbf{AV} = \mathbf{\Lambda}. \tag{5.62}$$

Combining the above discussion together: we had our original system as $\mathbf{x}' = \mathbf{Ax}$ which in eigencoordinates is $\mathbf{Vu}' = \mathbf{AVu}$, which we left-multiply by $\mathbf{V}^{-1}$ to obtain

$$\mathbf{V}^{-1}\mathbf{Vu}' = \mathbf{V}^{-1}\mathbf{AVu}. \tag{5.63}$$

We then substitute $\mathbf{V}^{-1}\mathbf{AV} = \mathbf{\Lambda}$ to obtain

$$\mathbf{u}' = \mathbf{\Lambda u}. \tag{5.64}$$

Thus, we have written our system of differential equations $\mathbf{x}' = \mathbf{Ax}$ as an equivalent system in its eigencoordinates: $\mathbf{u}' = \mathbf{\Lambda u}$. We then have

$$\mathbf{u}' = \mathbf{V}^{-1}\mathbf{AVu} = \mathbf{\Lambda u} \tag{5.65}$$

$$= \begin{bmatrix} 1 & -1 \\ -1 & 2 \end{bmatrix} \begin{bmatrix} -5 & 8 \\ -4 & 7 \end{bmatrix} \begin{bmatrix} 2 & 1 \\ 1 & 1 \end{bmatrix} \mathbf{u} = \begin{bmatrix} -1 & 0 \\ 0 & 3 \end{bmatrix} \begin{bmatrix} u_1 \\ u_2 \end{bmatrix}. \tag{5.66}$$

Although this change of variables was a lot of work, (5.66) can be easily solved since each is a very simple separable equation. The solution is

$$\begin{bmatrix} u_1 \\ u_2 \end{bmatrix} = \begin{bmatrix} c_1 e^{-t} \\ c_2 e^{3t} \end{bmatrix}. \tag{5.67}$$

If our initial condition in the original system is $x(0) = x_0, y(0) = y_0$, then (5.60) together with (5.67) gives

$$\begin{bmatrix} c_1 \\ c_2 \end{bmatrix} = \begin{bmatrix} x_0 - y_0 \\ -x_0 + 2y_0 \end{bmatrix}. \tag{5.68}$$

Thus

$$\begin{bmatrix} u_1 \\ u_2 \end{bmatrix} = \begin{bmatrix} (x_0 - y_0)\, e^{-t} \\ (-x_0 + 2y_0)\, e^{3t} \end{bmatrix}. \tag{5.69}$$

In terms of our original variables, recalling that (5.58) states $\mathbf{x} = \mathbf{V}\mathbf{u}$, we have

$$\begin{bmatrix} x \\ y \end{bmatrix} = \mathbf{V}\mathbf{u} = \begin{bmatrix} 2 & 1 \\ 1 & 1 \end{bmatrix} \begin{bmatrix} (x_0 - y_0)\, e^{-t} \\ (-x_0 + 2y_0)\, e^{3t} \end{bmatrix}$$

$$= \begin{bmatrix} 2\,(x_0 - y_0)\, e^{-t} + (-x_0 + 2y_0)\, e^{3t} \\ (x_0 - y_0)\, e^{-t} + (-x_0 + 2y_0)\, e^{3t} \end{bmatrix}. \tag{5.70}$$

In studying differential equations from this approach, we observe that a proper choice of coordinates will often greatly simplify the system.

In Example 2, we considered a symmetric matrix. Another useful property of a symmetric matrix can be seen when we try to diagonalize it via eigenco-ordinates.

Example 3: Consider the matrix of the previous example: $\mathbf{A} = \begin{bmatrix} 2 & 3 \\ 3 & 2 \end{bmatrix}$.

Normalize each eigenvector so that it has length one. Then find matrices $\mathbf{V}$ and Λ such that $\mathbf{V}^{-1}\mathbf{A}\mathbf{V} = \Lambda$. Can you see another way to calculate $\mathbf{V}^{-1}$?

Solution

To normalize each eigenvector, we divide by its length and we put a ^ above the vector:

$$\hat{\mathbf{v}}_1 = \frac{\mathbf{v}_1}{\|\mathbf{v}_1\|} = \frac{[1 \ \ 1]^T}{\sqrt{2}} \tag{5.71}$$

$$\hat{\mathbf{v}}_2 = \frac{\mathbf{v}_2}{\|\mathbf{v}_2\|} = \frac{[-1 \ \ 1]^T}{\sqrt{2}}. \tag{5.72}$$

The matrix $\mathbf{V}$ we seek is the matrix with these normalized eigenvectors as columns:

$$\mathbf{V} = \begin{bmatrix} \frac{1}{\sqrt{2}} & -\frac{1}{\sqrt{2}} \\ \frac{1}{\sqrt{2}} & \frac{1}{\sqrt{2}} \end{bmatrix} = \frac{1}{\sqrt{2}} \begin{bmatrix} 1 & -1 \\ 1 & 1 \end{bmatrix},$$

where we factored out the $\frac{1}{\sqrt{2}}$ that was common in each term. The inverse is given by (5.59):

$$\mathbf{V}^{-1} = \frac{1}{\frac{1}{2} + \frac{1}{2}} \begin{bmatrix} \frac{1}{\sqrt{2}} & \frac{1}{\sqrt{2}} \\ -\frac{1}{\sqrt{2}} & \frac{1}{\sqrt{2}} \end{bmatrix} = \frac{1}{\sqrt{2}} \begin{bmatrix} 1 & 1 \\ -1 & 1 \end{bmatrix}$$

with $\Lambda = \begin{bmatrix} 5 & 0 \\ 0 & -1 \end{bmatrix}$. We see that $\mathbf{V}^{-1} = \mathbf{V}^T$ for our example.

For a general $n \times n$ matrix $\mathbf{Q}$ in which $\mathbf{Q}^{-1} = \mathbf{Q}^T$, we call this matrix **orthogonal**. The columns of an orthogonal matrix satisfy two key properties: (1) $\mathbf{q}_i \perp \mathbf{q}_j$ for $i \neq j$, (2) $\|\mathbf{q}_i\| = 1$. The first step in our previous example was to normalize the eigenvectors so that (2) would be satisfied. Regarding (1), it turns out that the eigenvectors of every symmetric matrix are **orthogonal** (i.e., **perpendicular**) to each other. Thus for symmetric matrices, we can always diagonalize with an orthogonal matrix

$$\mathbf{A} = \mathbf{Q}\mathbf{\Lambda}\mathbf{Q}^T.$$

Because symmetric matrices arise in many real-world examples, these are useful facts to remember.

Complex Eigenvalues

Our approach when we have complex eigenvalues is no different—except that our algebra now requires the use of complex numbers. See Section 3.5 for a brief review of complex numbers, if necessary.

Example Consider the matrix

$$\begin{bmatrix} 2 & -3 \\ 3 & 2 \end{bmatrix}.$$

From our above discussion, we know that the eigenvalues are $2 \pm 3i$. But if we didn't know/see this, we would proceed just as before. That is, we would calculate

$$\det(\mathbf{A} - \lambda\mathbf{I}) = 0 \implies (2 - \lambda)^2 + 9 = 0 \implies \lambda^2 - 4\lambda + 13 = 0.$$

This equation is quadratic and so we obtain

$$\lambda = \frac{4 \pm \sqrt{4^2 - 4(1)(13)}}{2(1)} = \frac{4 \pm \sqrt{-36}}{2} = \frac{4 \pm 6i}{2} = 2 \pm 3i.$$

To calculate the eigenvectors, we again use the equation $(\mathbf{A} - \lambda\mathbf{I})\mathbf{v} = \mathbf{0}$ to find the eigenvector by solving for $\mathbf{v}$. Unlike the real case, however, we only need to find *one* of the eigenvectors—it turns out that the second eigenvector is the complex conjugate of the first! For this example, let's use $\lambda_1 = 2 + 3i$ to find $\mathbf{v}_1$:

$$(\mathbf{A} - \lambda_1\mathbf{I})\mathbf{v}_1 = \mathbf{0} \implies \begin{bmatrix} 2 - (2 + 3i) & -3 \\ 3 & 2 - (2 + 3i) \end{bmatrix} \begin{bmatrix} v_{11} \\ v_{21} \end{bmatrix} = \begin{bmatrix} 0 \\ 0 \end{bmatrix}$$

$$\begin{bmatrix} -3i & -3 \\ 3 & -3i \end{bmatrix} \begin{bmatrix} v_{11} \\ v_{21} \end{bmatrix} = \begin{bmatrix} 0 \\ 0 \end{bmatrix}. \tag{5.73}$$

As before, the two rows of $(\mathbf{A} - \lambda_1\mathbf{I})$ are constant multiples of each other—the second row is just the first row multiplied by i. Thus we have

$$-3iv_{11} - 3v_{21} = 0 \implies -iv_{11} = v_{21}.$$

We are free to choose any values for v_{11}, v_{21} that will make this equation true as long as both are not zero. It is probably easiest to choose $v_{11} = 1$, which immediately gives $v_{21} = -i$. Thus

$$\lambda_1 = 2 + 3i \text{ has eigenvector } \mathbf{v}_1 = \begin{bmatrix} 1 \\ -i \end{bmatrix}.$$

The second eigenvalue-eigenvector pair is just the complex conjugate of the first:

$$\lambda_2 = 2 - 3i \text{ has eigenvector } \mathbf{v}_2 = \begin{bmatrix} 1 \\ i \end{bmatrix}.$$

As you might hope/expect, it is *not* a coincidence that the eigenvectors are complex conjugates of each other. That is, for complex eigenvalues, we will always have

$$\mathbf{v}_2 = \overline{\mathbf{v}_1}.$$

Example 5: Consider

$$\mathbf{A} = \begin{bmatrix} -1 & -2 \\ 2 & -1 \end{bmatrix}.$$

Find the eigenvalues and eigenvectors of $\mathbf{A}$.

Solution
We can easily calculate the eigenvalues as $\lambda_1 = -1 + 2i, \lambda_2 = -1 - 2i$. The first eigenvalue thus gives

$$\begin{aligned} \mathbf{0} &= (\mathbf{A} - \lambda_1 \mathbf{I})\mathbf{v}_1 \\ &= \begin{bmatrix} -1 - (-1 + 2i) & -2 \\ 2 & -1 - (-1 + 2i) \end{bmatrix} \begin{bmatrix} v_{11} \\ v_{21} \end{bmatrix} \\ &= \begin{bmatrix} -2i & -2 \\ 2 & -2i \end{bmatrix} \begin{bmatrix} v_{11} \\ v_{21} \end{bmatrix}. \end{aligned}$$

At this point, in the case of real distinct eigenvalues, we always had the situation where both rows gave us the same information. That is also the case now and we can see this by multiplying the second row by $-i$, for example. Thus it doesn't matter which row we choose, so let's consider the first row:

$$-2iv_{11} - 2v_{21} = 0.$$

Our only restriction is that we can't have $v_{11} = v_{21} = 0$. One possibility is thus $v_{11} = 1, v_{21} = -i$ and this gives us the eigenvector ($\mathbf{v}_1$) corresponding to $\lambda_1 = -1 + 2i$. If we repeat the above steps for $\lambda_2 = -1 - 2i$, we could obtain $v_{12} = 1, v_{22} = i$. Thus we have

$$\mathbf{v}_1 = \begin{bmatrix} 1 \\ -i \end{bmatrix}, \quad \mathbf{v}_2 = \begin{bmatrix} 1 \\ i \end{bmatrix}.$$

We close by pointing out that for triangular and diagonal matrices, the eigenvalues were exactly the diagonal entries. For a real-valued matrix of the form,

$$\begin{bmatrix} a & -b \\ b & a \end{bmatrix}, \quad a,b \in \mathbb{R},$$

the eigenvalues are $a \pm ib$. We will revisit matrices of this form in Section B.3.

* * * * * * * * * * *

In Problems **1–18**, calculate the characteristic equation of the given matrix by hand. Then find the eigenvalues and corresponding eigenvectors. The eigenvalues are real.

1. $\begin{bmatrix} 4 & -5 \\ 2 & -3 \end{bmatrix}$ 2. $\begin{bmatrix} 4 & -2 \\ 1 & 1 \end{bmatrix}$ 3. $\begin{bmatrix} 1 & -2 \\ 4 & -5 \end{bmatrix}$ 4. $\begin{bmatrix} 4 & -3 \\ 2 & -1 \end{bmatrix}$

5. $\begin{bmatrix} -4 & 7 \\ 0 & -1 \end{bmatrix}$ 6. $\begin{bmatrix} 6 & -4 \\ 3 & -1 \end{bmatrix}$ 7. $\begin{bmatrix} 2 & 1 \\ 1 & 2 \end{bmatrix}$ 8. $\begin{bmatrix} 7 & -6 \\ 12 & -10 \end{bmatrix}$

9. $\begin{bmatrix} 1 & 3 \\ 3 & 1 \end{bmatrix}$ 10. $\begin{bmatrix} 9 & -10 \\ 2 & 0 \end{bmatrix}$ 11. $\begin{bmatrix} 1 & 2 \\ -1 & 4 \end{bmatrix}$ 12. $\begin{bmatrix} -2 & 1 \\ 2 & -1 \end{bmatrix}$

13. $\begin{bmatrix} 4 & 1 \\ 3 & 2 \end{bmatrix}$ 14. $\begin{bmatrix} 0 & 1 \\ 3 & 2 \end{bmatrix}$ 15. $\begin{bmatrix} 2 & 2 \\ 2 & 2 \end{bmatrix}$ 16. $\begin{bmatrix} 2 & 1 \\ 3 & 0 \end{bmatrix}$

17. $\begin{bmatrix} 7 & 6 \\ 2 & -4 \end{bmatrix}$ 18. $\begin{bmatrix} -2 & 4 \\ 2 & 5 \end{bmatrix}$

In Problems **19–28**, calculate the characteristic equation of the given matrix by hand. Then find the eigenvalues and corresponding eigenvectors. The eigenvalues are complex (nonreal).

19. $\begin{bmatrix} 0 & -1 \\ 1 & 0 \end{bmatrix}$ 20. $\begin{bmatrix} -2 & -3 \\ 3 & -2 \end{bmatrix}$ 21. $\begin{bmatrix} 3 & -2 \\ 5 & -3 \end{bmatrix}$ 22. $\begin{bmatrix} 1 & -2 \\ 5 & -1 \end{bmatrix}$

23. $\begin{bmatrix} -1 & -5 \\ 1 & -3 \end{bmatrix}$ 24. $\begin{bmatrix} -1 & -5 \\ 2 & 5 \end{bmatrix}$ 25. $\begin{bmatrix} -3 & -5 \\ 1 & -7 \end{bmatrix}$ 26. $\begin{bmatrix} -1 & -2 \\ 5 & -3 \end{bmatrix}$

27. $\begin{bmatrix} 1 & -2 \\ 3 & -1 \end{bmatrix}$ 28. $\begin{bmatrix} 1 & -8 \\ 1 & 3 \end{bmatrix}$

In Problems **29–34**, use MATLAB, Maple, or Mathematica to find the eigenvalues and eigenvectors of the given matrix.

29. $\begin{bmatrix} -2 & 2 & 0 \\ 4 & 0 & 0 \\ -10/3 & 5/3 & 1 \end{bmatrix}$ 30. $\begin{bmatrix} 3 & 1 & 0 \\ 2 & 4 & 0 \\ 2/3 & 4/3 & 1 \end{bmatrix}$ 31. $\begin{bmatrix} 1 & -5 & -1 \\ 1 & 3 & 1 \\ 0 & 0 & 1 \end{bmatrix}$

32. $\begin{bmatrix} -1 & -1 & 1 \\ 1 & -1 & 0 \\ 0 & 0 & 1 \end{bmatrix}$ 33. $\begin{bmatrix} \frac{-2}{3} & \frac{-2}{3} & 0 \\ \frac{-4}{3} & \frac{-4}{3} & 0 \\ \frac{2}{3} & \frac{-1}{3} & -1 \end{bmatrix}$ 34. $\begin{bmatrix} 1 & 0 & 0 \\ 1 & 0 & -1 \\ 0 & 2 & 0 \end{bmatrix}$

In Problems **35–39**, use the given substitutions to rewrite the equations as an uncoupled system. Then write the solution in terms of u_1, u_2 and use this to find the solution in terms of x, y.

35. $u_1 = x - y$, $u_2 = -x + 2y$ for $\begin{cases} x' = x + 2y \\ y' = -x + 4y \end{cases}$

36. $u_1 = 2x - y$, $u_2 = -x + y$ for $\begin{cases} x' = x - 2y \\ y' = 4x - 5y \end{cases}$

37. $u_1 = \frac{1}{4}x$, $u_2 = -\frac{1}{4}x + y$ for $\begin{cases} x' = 5x \\ y' = 2x - 3y \end{cases}$

38. $u_1 = -\frac{1}{2}x + \frac{3}{2}y$, $u_2 = \frac{1}{2}x - \frac{1}{2}y$ for $\begin{cases} x' = 4x - 6y \\ y' = 2x - 4y \end{cases}$

39. $u_1 = \frac{3}{2}x - \frac{1}{2}y$, $u_2 = \frac{-1}{2}x + \frac{1}{2}y$ for $\begin{cases} x' = -2x + y \\ y' = -3x + 2y \end{cases}$

In Problems **40–42**, use $\mathbf{A} = \mathbf{V}\mathbf{\Lambda}\mathbf{V}^{-1}$ to efficiently calculate powers of a matrix: $\mathbf{A}^n = \mathbf{V}\mathbf{\Lambda}^n\mathbf{V}^{-1}$.

40. $\begin{bmatrix} 1 & 2 \\ -1 & 4 \end{bmatrix}^{10}$ **41.** $\begin{bmatrix} 2 & 1 \\ 1 & 2 \end{bmatrix}^{30}$ **42.** $\begin{bmatrix} 3 & 0 & 0 \\ 0 & 3 & 0 \\ 8 & -4 & -1 \end{bmatrix}^{20}$

A **skew-symmetric** matrix $\mathbf{A}$ is defined as one that satisfies $\mathbf{A}^T = -\mathbf{A}$. For a matrix with real-valued entries, a skew-symmetric matrix always has complex eigenvalues. In Problems **43–45**, find the eigenvalues of the given skew-symmetric matrices.

43. $\mathbf{A} = \begin{bmatrix} 0 & -1 \\ 1 & 0 \end{bmatrix}$ **44.** $\mathbf{A} = \begin{bmatrix} 0 & 3 \\ -3 & 0 \end{bmatrix}$ **45.** $\mathbf{A} = \begin{bmatrix} 0 & 1 & -1 \\ -1 & 0 & -1 \\ 1 & 1 & 0 \end{bmatrix}$

5.5 A General Method, Part I: Solving Systems with Real and Distinct or Complex Eigenvalues

In this section, we look at a general method for solving systems of first-order homogeneous linear differential equations with constant coefficients. It also provides a theoretical framework and makes one aware of what types of solutions are expected from such a linear system. Practically speaking, however, solving systems of three or more equations involves finding roots of polynomial equations of degree 3 or higher. Except in special cases, we need to use some technological tool to find these solutions. If we are using a computer algebra system, then we might as well just use it to solve the system and skip the method of this section. This is especially true if the graph of the solution is the only item of interest. On the other hand, if one is interested in the functional form of the solution, frequently the form of the solution given by a computer algebra system is large, cumbersome, and hard to study. Even simplification routines may not help much. In this case, the method of this section is valuable, even if we are using a machine to perform the individual steps. From a philosophical point of view, it is important that we know how to solve a problem, even though we may let a machine do the work.

Our discussion of eigenvalues and eigenvectors was motivated as a way to solve the constant coefficient, homogeneous version of our general system of linear equations (5.1), written in matrix vector form as: $\mathbf{x}' = \mathbf{A}\mathbf{x}$. As a brief recap, we guessed solutions of the form $\mathbf{x} = c\mathbf{v}e^{\lambda t}$ for some constant $c \neq 0$. Then the solution must satisfy the differential equation so that

$$c\lambda\mathbf{v}e^{\lambda t} = c\mathbf{A}\mathbf{v}e^{\lambda t}.$$

We can divide by $ce^{\lambda t}$ (because it is never zero) and then rewrite the equation as

$$\mathbf{A}\mathbf{v} = \lambda\mathbf{v}. \tag{5.74}$$

Now we are in the realm of an eigenvalue problem. In particular, the only way to have solutions other than $\mathbf{v} = \mathbf{0}$ is to have the matrix $A - \lambda I$ be singular, which means that $\det(A - \lambda I) = 0$. Evaluating this determinant yields an nth degree polynomial in λ which has n, possibly repeating and possibly complex (non-real) eigenvalues $\lambda_1, \ldots, \lambda_n$. Each of these eigenvalues is substituted for λ in Equation (5.74), which in turn is solved for its corresponding eigenvector $\mathbf{v}_i$. Based on our discussion in Section 5.4, we know that this last equation says that $\mathbf{v}e^{\lambda t}$ will be a solution of $\mathbf{x}' = \mathbf{A}\mathbf{x}$ when λ is an eigenvalue of $\mathbf{A}$ and $\mathbf{v}$ is the corresponding eigenvector. Thus one solution is

$$\mathbf{x}(t) = \mathbf{v}_i e^{\lambda_i t} c_i$$

where c_i is an arbitrary constant. In the case when all the eigenvalues are real and distinct, we can state a general theorem.

THEOREM 5.5.1

Consider the system $\mathbf{x}' = \mathbf{A}\mathbf{x}$, where the coefficient matrix $\mathbf{A}$ has n distinct real eigenvalues. Let $\lambda_1, \lambda_2, \cdots, \lambda_n$ be these eigenvalues with $\mathbf{v}_1, \mathbf{v}_2, \cdots, \mathbf{v}_n$ as the corresponding eigenvectors. Then the $c_i\mathbf{v}_i e^{\lambda_i t}$, $i = 1, 2, \cdots, n$ are linearly independent and the general solution is given by

$$\mathbf{x} = c_1\mathbf{v}_1 e^{\lambda_1 t} + c_2\mathbf{v}_2 e^{\lambda_2 t} + \cdots + c_n\mathbf{v}_n e^{\lambda_n t}$$

and is defined for $t \in (-\infty, \infty)$.

Example 1 Consider the system

$$x' = 5x - y$$
$$y' = 3y.$$

Find the general solution.

In matrix form this equation is

$$\mathbf{x}' = \begin{bmatrix} 5 & -1 \\ 0 & 3 \end{bmatrix} \mathbf{x}$$

so that the eigenvalues of

$$A = \begin{bmatrix} 5 & -1 \\ 0 & 3 \end{bmatrix}$$

are found from

$$\det(A - \lambda I) = \begin{vmatrix} 5 - \lambda & -1 \\ 0 & 3 - \lambda \end{vmatrix} = (5 - \lambda)(3 - \lambda) = 0.$$

The eigenvalues are

$$\lambda_1 = 3 \text{ and } \lambda_2 = 5.$$

Now an eigenvector $\mathbf{v}_1$ corresponding to $\lambda_1 = 3$ is found from

$$\begin{bmatrix} 2 & -1 \\ 0 & 0 \end{bmatrix} \begin{bmatrix} v_{11} \\ v_{21} \end{bmatrix} = \begin{bmatrix} 0 \\ 0 \end{bmatrix}$$

which gives $2v_{11} = v_{21}$. Letting $v_{11} = 1$, we obtain the eigenvector

$$\mathbf{v}_1 = \begin{bmatrix} 1 \\ 2 \end{bmatrix}.$$

Similarly, an eigenvector $\mathbf{v}_2$ corresponding to $\lambda = 5$ satisfies $(\mathbf{A} - \lambda_2 \mathbf{I})\mathbf{v}_2 = \mathbf{0}$:

$$\begin{bmatrix} 0 & -1 \\ 0 & -2 \end{bmatrix} \begin{bmatrix} v_{12} \\ v_{22} \end{bmatrix} = \begin{bmatrix} 0 \\ 0 \end{bmatrix}$$

so that $v_{22} = 0$. We are free to choose $\mathbf{v}_{12}$ to be anything except 0. So that if we let $v_{12} = 1$, then

$$\mathbf{v}_2 = \begin{bmatrix} 1 \\ 0 \end{bmatrix}.$$

So the general solution is

$$x(t) = c_1 \begin{bmatrix} 1 \\ 2 \end{bmatrix} e^{3t} + c_2 \begin{bmatrix} 1 \\ 0 \end{bmatrix} e^{5t}.$$

Example 2: Consider the following system:

$$\frac{dx}{dt} = -2x - y - 2z$$

$$\frac{dy}{dt} = -4x - 5y + 2z$$

$$\frac{dz}{dt} = -5x - y + z.$$

Calculate the eigenvalues and eigenvectors. Then use Theorem 5.5.1 to write the general solution.

We can calculate the eigenvalues either by hand or with the computer. Doing so gives

$$\lambda_1 = 3, \ \lambda_2 = -6, \ \lambda_3 = -3.$$

We can similarly find the eigenvectors either by hand or with the computer and we obtain

$$\mathbf{v}_1 = \begin{bmatrix} -1 \\ 1 \\ 2 \end{bmatrix}, \quad \mathbf{v}_2 = \begin{bmatrix} 1 \\ 2 \\ 1 \end{bmatrix}, \quad \mathbf{v}_3 = \begin{bmatrix} 1 \\ -1 \\ 1 \end{bmatrix}$$

as the respective eigenvectors. We can then write the general solution as

$$\mathbf{x} = c_1 e^{3t} \begin{bmatrix} -1 \\ 1 \\ 2 \end{bmatrix} + c_2 e^{-6t} \begin{bmatrix} 1 \\ 2 \\ 1 \end{bmatrix} + c_3 e^{-3t} \begin{bmatrix} 1 \\ -1 \\ 1 \end{bmatrix}$$

or, alternatively, we could write the general solution as

$$\begin{aligned} x(t) &= -c_1 e^{3t} + c_2 e^{6t} + c_3 e^{-3t} \\ y(t) &= c_1 e^{3t} + 2c_2 e^{6t} - c_3 e^{-3t} \\ z(t) &= 2c_1 e^{3t} + c_2 e^{6t} + c_3 e^{-3t}. \end{aligned}$$

In continuing our discussion of the solution of $\mathbf{x}' = \mathbf{A}\mathbf{x}$, we now look at the case when the eigenvalues of $\mathbf{A}$ are complex. In Section 5.4, we learned that complex eigenvectors always result when eigenvalues are complex. Recall that we obtained the eigenvalue-eigenvector equation because we assumed solutions of the form $\mathbf{v}e^{\lambda t}$. Thus, if we started with $\mathbf{x}' = \mathbf{A}\mathbf{x}$ and found complex eigenvalues $\lambda_1 = \alpha + i\beta, \lambda_2 = \alpha - i\beta$ (where α, β are both real) with corresponding (complex) eigenvectors $\mathbf{v}_1, \mathbf{v}_2$, then $e^{\lambda_1 t}\mathbf{v}_1$ and $e^{\lambda_2 t}\mathbf{v}_2$ are both solutions to $\mathbf{x}' = \mathbf{A}\mathbf{x}$.

Although this is true, the non-appealing aspect is that both eigenvectors are complex-valued even though our original system had only real entries. We had a similar situation occurring in Section 3.6 when we obtained complex-valued solutions to a second-order equation. In that section, we used Euler's formula to write a real-valued solution in terms of sines and cosines and that is exactly what we will do now:

$$\begin{aligned} e^{\lambda_1 t}\mathbf{v}_1 &= e^{(\alpha+i\beta)t}\mathbf{v}_1 = e^{\alpha t}(\cos \beta t + i \sin \beta t)\mathbf{v}_1 \\ e^{\lambda_2 t}\mathbf{v}_2 &= e^{(\alpha-i\beta)t}\mathbf{v}_2 = e^{\alpha t}(\cos \beta t - i \sin \beta t)\mathbf{v}_2. \end{aligned}$$

Theorem 5.1.2 tells us that *any* linear combination of solutions to $\mathbf{x}' = \mathbf{A}(t)\mathbf{x}$ is still a solution. We also introduce real-valued vectors $\mathbf{a}, \mathbf{b}$ that represent the real and complex parts of the eigenvectors, i.e., $\mathbf{v}_1 = \mathbf{a} + i\mathbf{b}$ (and thus $\mathbf{v}_2 = \mathbf{a} - i\mathbf{b}$). Using these facts we can rewrite a complex-valued solution:

$$k_1 e^{(\alpha+i\beta)t}\mathbf{v}_1 + k_2 e^{(\alpha-i\beta)t}\mathbf{v}_2$$

$$= k_1 e^{\alpha t}(\cos \beta t + i \sin \beta t)\mathbf{v}_1 + k_2 e^{\alpha t}(\cos \beta t - i \sin \beta t)\mathbf{v}_2$$

$$= k_1 e^{\alpha t}(\cos \beta t + i \sin \beta t)(\mathbf{a} + i\mathbf{b}) + k_2 e^{\alpha t}(\cos \beta t - i \sin \beta t)(\mathbf{a} - i\mathbf{b})$$

$$= k_1 e^{\alpha t}\left[\mathbf{a}(\cos \beta t) + i\mathbf{a}(\sin \beta t) + i\mathbf{b}(\cos \beta t) - \mathbf{b}(\sin \beta t)\right]$$
$$+ k_2 e^{\alpha t}\left[\mathbf{a}(\cos \beta t) - i\mathbf{a}(\sin \beta t) - i\mathbf{b}(\cos \beta t) - \mathbf{b}(\sin \beta t)\right]$$

$$= e^{\alpha t}\left\{\left[k_1\mathbf{a}(\cos \beta t) - k_1\mathbf{b}(\sin \beta t) + k_2\mathbf{a}(\cos \beta t) - k_2\mathbf{b}(\sin \beta t)\right]\right.$$
$$\left. + i\left[k_1\mathbf{a}(\sin \beta t) + k_1\mathbf{b}(\cos \beta t) - k_2\mathbf{a}(\sin \beta t) - k_2\mathbf{b}(\cos \beta t)\right]\right\}$$

$$= e^{\alpha t}\left\{\left[(k_1 + k_2)\mathbf{a}(\cos \beta t) + (-k_2 - k_1)\mathbf{b}(\sin \beta t)\right]\right.$$
$$\left. + i\left[(k_1 - k_2)\mathbf{a}(\sin \beta t) + (k_1 - k_2)\mathbf{b}(\cos \beta t)\right]\right\}$$

$$= e^{\alpha t}\left\{\left[c_1\mathbf{a}(\cos \beta t) - c_1\mathbf{b}(\sin \beta t)\right] + \left[c_2\mathbf{a}(\sin \beta t) + c_2\mathbf{b}(\cos \beta t)\right]\right\},$$

where the constants c_1, c_2 are related to the k_i by $c_1 = k_1 + k_2$, $c_2 = i(k_1 - k_2)$. The constants are arbitrary and we thus have a real-valued formulation of the original solution.

We state a theorem for the 2×2 case summarizing these results.

THEOREM 5.5.2

Consider the 2×2 system $\mathbf{x}' = \mathbf{A}\mathbf{x}$ whose coefficient matrix has eigenvalues $\lambda_1 = \alpha + i\beta$, $\lambda_2 = \alpha - i\beta$, with α, β real numbers. Choose one of the eigenvectors and write it as $\mathbf{v}_j = \mathbf{a} + i\mathbf{b}$, where $\mathbf{a}, \mathbf{b}$ are real-valued vectors. Then

$$\mathbf{x}_1 = e^{\alpha t}\left[\mathbf{a}(\cos \beta t) - \mathbf{b}(\sin \beta t)\right]$$
$$\mathbf{x}_2 = e^{\alpha t}\left[\mathbf{a}(\sin \beta t) + \mathbf{b}(\cos \beta t)\right] \tag{5.75}$$

are two linearly independent solutions, defined for $-\infty < t < \infty$. The general solution is a linear combination of these two:

$$\mathbf{x} = e^{\alpha t}\left\{c_1\left[\mathbf{a}(\cos \beta t) - \mathbf{b}(\sin \beta t)\right] + c_2\left[\mathbf{a}(\sin \beta t) + \mathbf{b}(\cos \beta t)\right]\right\} \tag{5.76}$$

Example 3: Write the real-valued general solution to

$$\mathbf{x}' = \begin{bmatrix} -1 & -2 \\ 2 & -1 \end{bmatrix}\mathbf{x}.$$

This example was considered at the end of Section 5.4. We found that $\lambda_1 = -1 + 2i$, $\lambda_2 = -1 - 2i$, with corresponding eigenvectors

$$\mathbf{v}_1 = \begin{bmatrix} 1 \\ -i \end{bmatrix}, \quad \mathbf{v}_2 = \begin{bmatrix} 1 \\ i \end{bmatrix}.$$

We rewrite one of the eigenvectors in terms of the real vectors $\mathbf{a}, \mathbf{b}$:

$$\mathbf{v}_1 = \mathbf{a} + i\mathbf{b} \Longrightarrow \mathbf{a} = \begin{bmatrix} 1 \\ 0 \end{bmatrix}, \quad \mathbf{b} = \begin{bmatrix} 0 \\ -1 \end{bmatrix}.$$

With $\alpha = -1, \beta = 2$, Theorem 5.5.2 states that the general solution is

$$\mathbf{x} = e^{-t} \left\{ c_1 \left[\mathbf{a}(\cos 2t) - \mathbf{b}(\sin 2t) \right] + c_2 \left[\mathbf{a}(\sin 2t) + \mathbf{b}(\cos 2t) \right] \right\}$$

$$= e^{-t} \left\{ c_1 \left[\begin{bmatrix} 1 \\ 0 \end{bmatrix} (\cos 2t) - \begin{bmatrix} 0 \\ -1 \end{bmatrix} (\sin 2t) \right] \right.$$

$$\left. + c_2 \left[\begin{bmatrix} 1 \\ 0 \end{bmatrix} (\sin 2t) + \begin{bmatrix} 0 \\ -1 \end{bmatrix} (\cos 2t) \right] \right\}.$$

We finish this section by tying together the techniques we have employed.

THEOREM 5.5.3

Suppose that $\mathbf{x}_1, \mathbf{x}_2, \cdots, \mathbf{x}_n$ form a fundamental set of solutions for the linear homogeneous system $\mathbf{x}' = \mathbf{A}\mathbf{x}$. The general solution to this system is a linear combination of n linearly independent solutions

$$\mathbf{x} = c_1 \mathbf{x}_1 + c_2 \mathbf{x}_2 + \cdots + c_n \mathbf{x}_n.$$

We saw an analogous statement in Theorem 5.1.3 but at that time we did not know how to compute any solutions. The previous theorems show us how to obtain a fundamental set of solutions and then we apply Theorem 5.5.3. As we have seen, obtaining a fundamental set of solutions may be very easy or quite complicated.

In **Problems 1–17**, find the eigenvalues and eigenvectors (they are all real) of the coefficient matrix and write the general solution.

1. $\begin{cases} x' = -x - 2y \\ y' = -x \end{cases}$
2. $\begin{cases} x' = x + 2y \\ y' = -3y \end{cases}$
3. $\begin{cases} x' = -4x - y \\ y' = 6x + y \end{cases}$

4. $\begin{cases} x' = 3x + y \\ y' = x + 3y \end{cases}$
5. $\begin{cases} x' = 5x - y \\ y' = 3x + y \end{cases}$
6. $\begin{cases} x' = 3x + 9y \\ y' = x + 3y \end{cases}$

7. $\begin{cases} x' = -7x + 5y \\ y' = -10x + 8y \end{cases}$
8. $\begin{cases} x' = 4x + 3y \\ y' = x + 2y \end{cases}$
9. $\begin{cases} x' = 4x + y \\ y' = 3x + 2y \end{cases}$

10. $\begin{cases} x' = 3x + 4y \\ y' = x + 3y \end{cases}$
11. $\begin{cases} x' = 10x - 6y \\ y' = 12x - 8y \end{cases}$
12. $\begin{cases} x' = x + 4y \\ y' = x - 2y \end{cases}$

13. $\begin{cases} x' = -5x + 6y \\ y' = -3x + 4y \end{cases}$
14. $\begin{cases} x' = 3x + 4y \\ y' = 4x + 3y \end{cases}$
15. $\begin{cases} x' = -17x + 36y \\ y' = -6x + 13y \end{cases}$

16. $\begin{cases} x' = -2x + 2y \\ y' = 3x - 2y \end{cases}$
17. $\begin{cases} x' = 2x + 3y \\ y' = x + 2y \end{cases}$

For Problems **18–26**, find the eigenvalues and eigenvectors of the coefficient matrix. Then find the general solution (the eigenvalues are all complex [non-real]).

18. $\begin{cases} x' = -3y \\ y' = 3x \end{cases}$
19. $\begin{cases} x' = -2y \\ y' = 2x \end{cases}$
20. $\begin{cases} x' = 3x + 4y \\ y' = -4x + 3y \end{cases}$

21. $\begin{cases} x' = -4x + 2y \\ y' = -10x + 4y \end{cases}$
22. $\begin{cases} x' = 3x - 5y \\ y' = 2x - 3y \end{cases}$
23. $\begin{cases} x' = -7x - 4y \\ y' = 10x + 5y \end{cases}$

24. $\begin{cases} x' = x - y \\ y' = 5x - 3y \end{cases}$
25. $\begin{cases} x' = -3x - 4y \\ y' = x - 3y \end{cases}$
26. $\begin{cases} x' = x - 5y \\ y' = x - y \end{cases}$

In Problems **27–32**, use MATLAB, Maple, or Mathematica to find the eigenvalues and eigenvectors of the coefficient matrix. If the eigenvalues are real, write the general solution; if they are complex, use the computer to find the general solution.

27. $\begin{cases} x' = 3x + y \\ y' = 2x + 4y \\ z' = 3x - y + z \end{cases}$
28. $\begin{cases} x' = -2x + 2y \\ y' = 4x \\ z' = 3y + z \end{cases}$
29. $\begin{cases} x' = -x - 2y + z \\ y' = 2x - y + z \\ z' = z \end{cases}$

30. $\begin{cases} x' = -x + y - z \\ y' = -y \\ z' = x - z \end{cases}$
31. $\begin{cases} x' = 2x - 2y \\ y' = 3x - y \\ z' = z \end{cases}$
32. $\begin{cases} x' = -x + y - z \\ y' = -y \\ z' = x - 2z \end{cases}$

In Problems **33–34**, consider the equation $a_2 x''(t) + a_1 x'(t) + a_0 x(t) = 0$. Solve it using the methods of Section 3.6. Then use the methods of Section 3.4 to convert it to a system of two first-order equations in the variables $x(t)$ and $y(t) = x'(t)$. Solve the resulting system using the methods of this current section. Compare your two answers.

33. $a_2 = 1, a_1 = 4, a_0 = 3$ 34. $a_2 = 1, a_1 = 2, a_0 = -8$

35. *Social Mobility* [37] Consider a town of 30,000 families (economic units) who have been grouped into three economic brackets: lower, middle, and upper. Each year 9% of the lower move into the middle, and 12% of the middle move back to the lower; 8% of the middle move to the upper, and 10% of the upper move down to the middle; finally 3% of the lower move directly to the upper, and 4% of the upper move down to the lower. These transitions occur continuously throughout the year as people are hired, laid off, fired, promoted, retire, and change careers.

We assume that initially there are 10,000 lower, 12,000 middle, and 8,000 upper income families and consider the compartmental diagram for this situation is shown in Figure 5.1.

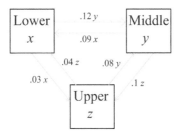

FIGURE 5.1: Compartmental diagram for income class model.
The system of equations is

$$\frac{dx}{dt} = -0.12x + 0.12y + 0.04z$$

$$\frac{dy}{dt} = 0.09x - 0.2y + 0.1z$$

$$\frac{dz}{dt} = 0.03x + 0.08y - 0.14z.$$

Find the general solution for this system and graph the solution.

36. Consider the previous application but with the coefficients changed slightly so the system is now governed by the equations

$$\frac{dx}{dt} = -0.14x + 0.01y + 0.04z$$

$$\frac{dy}{dt} = 0.11x - 0.09y + 0.1z$$

$$\frac{dz}{dt} = 0.03x + 0.08y - 0.14z.$$

Interpret the changes in the system compared to the previous one. Then use MATLAB, Maple, or Mathematica to find the general solution for this system and graph the solution.

5.6 A General Method, Part II: Solving Systems with Repeated Real Eigenvalues

We now continue our discussion of $\mathbf{x}' = \mathbf{A}\mathbf{x}$, begun in Section 5.5. We had assumed a solution of the form $c\mathbf{v}e^{\lambda t}$ and had concluded that $\lambda, \mathbf{v}$ must satisfy the eigenvalue equation

$$\mathbf{A}\mathbf{v} = \lambda\mathbf{v}.$$

We discussed the situation where we have distinct real eigenvalues. What if the eigenvalues were instead real and repeated? As we will see, sometimes we will be able to use the methods of the previous section while other times we will not. To address the latter case, we begin to develop the theory of the Jordan Form as it applies to our discussion of solving systems of differential equations and refer the reader to other texts, such as [52] and [23], for a more thorough treatment of the topic. Although not required, it is best if the reader is familiar with some of the terminology in Section 5.3.

We first consider the case of repeated real eigenvalues and use two specific matrices to show different scenarios.

Example 1: Find the eigenvalues and eigenvectors of

$$\mathbf{A} = \begin{bmatrix} 3 & 0 \\ 0 & 3 \end{bmatrix}, \quad \mathbf{B} = \begin{bmatrix} 3 & 2 \\ 0 & 3 \end{bmatrix}.$$

Solution

For the matrix $\mathbf{A}$, we can easily calculate the eigenvalues as $\lambda_{1,2} = 3, 3$. Substituting into $(\mathbf{A} - \lambda \mathbf{I})\mathbf{v} = \mathbf{0}$ gives

$$(\mathbf{A} - \lambda \mathbf{I})\mathbf{v} = \begin{bmatrix} 3 - 3 & 0 \\ 0 & 3 - 3 \end{bmatrix} \begin{bmatrix} v_{11} \\ v_{21} \end{bmatrix} = \mathbf{0}.$$

Because this is true for any choice of v_{11}, v_{21}, we can choose, for example, $v_{11} = 1, v_{21} = 0$ to give one eigenvector and then $v_{12} = 0, v_{22} = 1$ to give the other. Thus

$$\mathbf{v}_1 = \begin{bmatrix} 1 \\ 0 \end{bmatrix}, \quad \mathbf{v}_2 = \begin{bmatrix} 0 \\ 1 \end{bmatrix} \quad \text{for} \quad \mathbf{A} = \begin{bmatrix} 3 & 0 \\ 0 & 3 \end{bmatrix}.$$

For our matrix $\mathbf{B}$, we again have $\lambda_{1,2} = 3, 3$ and substitution into $(\mathbf{B} - \lambda \mathbf{I})\mathbf{v} = \mathbf{0}$ gives

$$\begin{bmatrix} 3 - 3 & 2 \\ 0 & 3 - 3 \end{bmatrix} \begin{bmatrix} v_{11} \\ v_{21} \end{bmatrix} = \begin{bmatrix} 0 & 2 \\ 0 & 0 \end{bmatrix} \begin{bmatrix} v_{11} \\ v_{21} \end{bmatrix} = \mathbf{0}.$$

The second row of the system gives an equation that is always true, whereas the first row gives us $0 \cdot v_{11} + 2 \cdot v_{21} = 0$. This can only happen when $v_{21} = 0$. We are free to choose v_{11} to be anything (non-zero) and so we let $v_{11} = 1$. Thus we have *one* eigenvector

$$\mathbf{v}_1 = \begin{bmatrix} 1 \\ 0 \end{bmatrix} \quad \text{for} \quad \mathbf{B} = \begin{bmatrix} 3 & 2 \\ 0 & 3 \end{bmatrix}.$$

Can we find another eigenvector? The short answer is "no." The remainder of this section will be dedicated to finding another vector that will help us.

In this situation, we see that $\mathbf{A}$ had *two* eigenvectors even though the eigenvalue was repeated. In general, if we can find n eigenvectors corresponding to n real eigenvalues (either distinct or repeated) for the coefficient matrix of a system of differential equations, we get lucky because we can write the general solution to $\mathbf{x}' = \mathbf{Ax}$ with the help of the following theorem.

THEOREM 5.6.1

Consider the system $\mathbf{x}' = \mathbf{Ax}$, where the coefficient matrix $\mathbf{A}$ has real eigenvalue λ repeated n times. Suppose we can find n eigenvectors $\mathbf{v}_1, \mathbf{v}_2, \cdots, \mathbf{v}_n$ corresponding to this eigenvalue. Then the $c_i \mathbf{v}_i e^{\lambda t}$, $i = 1, 2, \cdots, n$ are linearly independent and the general solution is given by

$$\mathbf{x} = c_1 \mathbf{v}_1 e^{\lambda t} + c_2 \mathbf{v}_2 e^{\lambda t} + \cdots + c_n \mathbf{v}_n e^{\lambda t}$$

and is defined for $t \in (-\infty, \infty)$.

Combined with Theorem 5.5.1, we see that as long as we can find n eigenvectors for a set of n real eigenvalues (regardless of whether they are distinct or repeated), we can obtain n linearly independent solutions with $c_i \mathbf{v}_i e^{\lambda t}$, $i = 1, 2, \cdots, n$. If we *cannot* find the same number of eigenvectors as the multiplicity of a repeated eigenvalue, such as the case with $\mathbf{B}$, we will need another tool.

If we put the above two matrices in the context of differential equations, Theorem 5.6.1 allows us to write the general solution of $\mathbf{x}' = \mathbf{Ax}$ as $\mathbf{x} = c_1 e^{3t} \mathbf{v}_1 + c_2 e^{3t} \mathbf{v}_2$ because we have two linearly independent eigenvectors $\mathbf{v}_1, \mathbf{v}_2$ but we cannot do the same for $\mathbf{B}$. The reader should recall that when we had a repeated root of the characteristic equation in solving an nth-order linear homogeneous constant-coefficient equation, we found a second solution by multiplying by t. Thus we had e^{rt}, te^{rt} as linearly independent solutions. Is it still the analogous situation here? To see if this is the case, let's assume that $\mathbf{x} = c_2 t e^{3t} \mathbf{v}$ is a solution to $\mathbf{x}' = \mathbf{Bx}$, then substitution gives

$$c_2(e^{3t} + 3te^{3t})\mathbf{v} = \mathbf{B}(c_2 t e^{3t} \mathbf{v})$$

$$\implies \frac{1}{te^{3t}}(e^{3t} + 3te^{3t})\mathbf{v} = \mathbf{Bv} \implies \mathbf{Bv} = \left[\frac{1}{t} + 3\right]\mathbf{v}.$$

We know that $\mathbf{v}$ is an eigenvector of $\mathbf{B}$ and this last equation is not true because the corresponding eigenvalue is 3. Thus $c_2 t e^{3t} \mathbf{v}$ is *not* a solution to $\mathbf{x}' = \mathbf{Bx}$. It turns out, however, that $te^{3t} \mathbf{v}$ is part of an additional solution but it is only a part of it.

If we consider the general situation where λ is a repeated eigenvalue of $\mathbf{x}' = \mathbf{Bx}$ and assume a solution of the form $\mathbf{x} = te^{\lambda t} \mathbf{v} + g(t)\mathbf{u}$, where $g(t), \mathbf{u}$ are to be determined, we would need it to satisfy the differential equation.

Substitution gives

$$(e^{\lambda t} + \lambda t e^{\lambda t})\mathbf{v} + g'(t)\mathbf{u} = \mathbf{B}(t e^{\lambda t}\mathbf{v} + g(t)\mathbf{u}).$$

Rearranging this gives

$$(e^{\lambda t})\mathbf{v} + g'(t)\mathbf{u} + \lambda t e^{\lambda t}\mathbf{v} = \mathbf{B}(t e^{\lambda t}\mathbf{v}) + \mathbf{B}(g(t)\mathbf{u})$$
$$\Longrightarrow (e^{\lambda t})\mathbf{v} + g'(t)\mathbf{u} - \mathbf{B}(g(t)\mathbf{u}) = \mathbf{B}(t e^{\lambda t}\mathbf{v}) - \lambda t e^{\lambda t}\mathbf{v}$$
$$\Longrightarrow (e^{\lambda t})\mathbf{v} - (\mathbf{B}g(t) - g'(t)\mathbf{I})\mathbf{u} = \underbrace{t e^{\lambda t}(\mathbf{B} - \lambda \mathbf{I})\mathbf{v}}_{= \, 0}$$

$$\Longrightarrow (\mathbf{B}g(t) - g'(t)\mathbf{I})\mathbf{u} = (e^{\lambda t})\mathbf{v}.$$

If we set $g(t) = e^{\lambda t}$ then this equation becomes

$$(\mathbf{B} - \lambda \mathbf{I})\mathbf{u} = \mathbf{v}, \tag{5.77}$$

and finding a vector $\mathbf{u}$ that satisfies this last equation would then give us an additional solution, which we could verify is linearly independent.

Definition 5.6.1

We say that $\mathbf{u} \neq \mathbf{0}$ is a **generalized eigenvector** of $\mathbf{A}$ associated to the eigenvalue λ if

$$(\mathbf{A} - \lambda \mathbf{I})^k \mathbf{u} = \mathbf{0} \tag{5.78}$$

for some integer $k > 0$. The index of the generalized eigenvector is the smallest k satisfying (5.78).

Our previous use of the word eigenvector, which we have always denoted $\mathbf{v}$, corresponds to the case $k = 1$ in (5.78):

$$(\mathbf{A} - \lambda \mathbf{I})^1 \mathbf{v} = \mathbf{0}.$$

The case of the matrix $\mathbf{B}$ from Example 1 corresponds to $k = 2$ in (5.78), which can be seen by expanding (5.77):

$$(\mathbf{B} - \lambda \mathbf{I})^2 \mathbf{u} = (\mathbf{B} - \lambda \mathbf{I})\underbrace{(\mathbf{B} - \lambda \mathbf{I})\mathbf{u}}_{} = (\mathbf{B} - \lambda \mathbf{I})\mathbf{v} = \mathbf{0}.$$

If we have more than one eigenvalue that repeats each with only one eigenvector (e.g., $\lambda_1 = 3$ occurs twice with eigenvector $\mathbf{v}_1$ and $\lambda_2 = -1$ occurs three times with eigenvector $\mathbf{v}_2$), we write a subscript on the $\mathbf{u}$ and use the same superscript as the eigenvalue in order to ensure that we associate the original eigenvector with its corresponding generalized eigenvector:

$$\{\mathbf{v}_i, \mathbf{u}_1^{(i)}, \mathbf{u}_2^{(i)}, \cdots, \mathbf{u}_{k-1}^{(i)}\}.$$

If there is no confusion, we will drop the superscript on the $\mathbf{u}$ and the subscript on the $\mathbf{v}$ for convenience. For $k > 1$, we see that the original eigenvector gives rise to a set of generalized eigenvectors:

$$(\mathbf{A} - \lambda\mathbf{I})\mathbf{u}_{k-1} = \mathbf{u}_{k-2}, \quad (\mathbf{A} - \lambda\mathbf{I})\mathbf{u}_{k-2} = \mathbf{u}_{k-3}, \quad \cdots,$$
$$(\mathbf{A} - \lambda\mathbf{I})\mathbf{u}_3 = \mathbf{u}_2, (\mathbf{A} - \lambda\mathbf{I})\mathbf{u}_2 = \mathbf{u}_1, (\mathbf{A} - \lambda\mathbf{I})\mathbf{u}_1 = \mathbf{v}. \qquad (5.79)$$

The set $\{\mathbf{v}, \mathbf{u}_1, \mathbf{u}_2, \cdots, \mathbf{u}_{k-1}\}$ satisfying (5.79) is called a **chain** of generalized eigenvectors. The chain is determined entirely by the choice of $\mathbf{v}$, which is referred to as the **bottom of the chain**. For those that have read Section 5.3, we define

$$\tilde{E}_\lambda = \{v|(\mathbf{A} - \lambda\mathbf{I})^k\mathbf{v} = \mathbf{0} \text{ for some } k\} \qquad (5.80)$$

as the generalized eigenspace of λ. From (5.50), we note that $E_\lambda \subseteq \tilde{E}_\lambda$. If λ is an eigenvalue of $\mathbf{A}$ of multiplicity m, then $\tilde{E}_\lambda$ is a subspace of dimension m.

Example 2 Consider the matrix

$$\mathbf{A} = \begin{bmatrix} 2 & 3 & 0 \\ 0 & 2 & -1 \\ 0 & 0 & 2 \end{bmatrix},$$

which has $\lambda = 2$ as an eigenvalue of multiplicity 3. Find the one eigenvector and the two generalized eigenvectors from (5.79).

Solution

To find the eigenvector, we solve $(\mathbf{A} - 2\mathbf{I})\mathbf{v} = \mathbf{0}$:

$$\begin{bmatrix} 0 & 3 & 0 \\ 0 & 0 & -1 \\ 0 & 0 & 0 \end{bmatrix} \begin{bmatrix} v_{11} \\ v_{21} \\ v_{31} \end{bmatrix} = \begin{bmatrix} 0 \\ 0 \\ 0 \end{bmatrix} \implies v_{21}, v_{31} = 0, v_{11} \neq 0,$$

and we can arbitrarily choose $v_{11} = 1$. There is no other linearly independent eigenvector—any other must be a multiple of $\mathbf{v}$. To find the two generalized eigenvectors that result from this we solve (5.79):

$$(\mathbf{A} - 2\mathbf{I})\mathbf{u}_1 = \mathbf{v} \implies u_{1,2} = \frac{1}{3}, \ u_{1,3} = 0, \ u_{1,1} \text{ arbitrary,}$$

and we choose $u_{1,1} = 0$ because this is simplest and will still satisfy $\mathbf{u}_1 \neq 0$. Similarly,

$$(\mathbf{A} - 2\mathbf{I})\mathbf{u}_2 = \mathbf{u}_1 \implies u_{2,2} = 0, \ u_{2,3} = \frac{-1}{3}, \ u_{2,1} \text{ arbitrary,}$$

and we choose $u_{2,1} = 0$ because this is simplest and will still satisfy $\mathbf{u}_2 \neq 0$. Thus the chain of generalized eigenvectors for this problem is

$$\left\{ \begin{bmatrix} 1 \\ 0 \\ 0 \end{bmatrix}, \begin{bmatrix} 0 \\ 1/3 \\ 0 \end{bmatrix}, \begin{bmatrix} 0 \\ 0 \\ -1/3 \end{bmatrix} \right\}.$$

We know that if a matrix has a full set of eigenvectors, then those eigenvectors are linearly independent. This was crucial for us to write a fundamental set of solutions of $x' = Ax$. We are fortunate that a similar thing holds for a set of chain vectors that arise from the eigenvector v.

THEOREM 5.6.2

Let v be an eigenvector of A corresponding to the repeated eigenvalue λ and let $\{v, u_1, u_2, \cdots, u_{k-1}\}$ be the resulting chain of generalized eigenvectors. Then $\{v, u_1, u_2, \cdots, u_{k-1}\}$ is linearly independent.

We now state the theorem for solving $x' = Ax$ when a chain of generalized eigenvectors is present.

THEOREM 5.6.3

Consider the system $x' = Ax$ for the $k \times k$ matrix A in which (i) λ is an eigenvalue of multiplicity k with a single eigenvector v and (ii) $\{v, u_1, \cdots, u_{k-1}\}$ is the corresponding chain of generalized eigenvectors. Set

$$x_1 = te^{\lambda t}v + e^{\lambda t}u_1$$

$$x_2 = \frac{t^2}{2!}e^{\lambda t}v + te^{\lambda t}u_1 + e^{\lambda t}u_2$$

$$\vdots$$

$$x_{k-1} = \frac{t^{k-1}}{(k-1)!}e^{\lambda t}v + \frac{t^{k-2}}{(k-2)!}e^{\lambda t}u_1 \cdots + te^{\lambda t}u_{k-2} + e^{\lambda t}u_{k-1}^{(k)}.$$

Then $e^{\lambda t}v, x_1, \cdots, x_{k-1}$ are linearly independent and the general solution to $x' = Ax$ can be written as

$$x = c_1 e^{\lambda t}v + c_2 x_1 + c_3 x_2 + \cdots + c_k x_{k-1}.$$

Remark 1: If we had multiple eigenvectors that gave rise to the full set of generalized eigenvectors, then the above theorem would apply to each set of eigenvectors and corresponding generalized eigenvectors.

Remark 2: If we also have distinct real eigenvalues, Theorem 5.5.1 applies to those eigenvalues and corresponding eigenvectors.

Example 3: Find the general solution of

$$x' = \begin{bmatrix} -1 & 2 & -4 \\ 0 & -1 & 0 \\ 0 & 0 & -1 \end{bmatrix} x.$$

Because the matrix is **upper triangular**, we can read the eigenvalues immediately from the diagonal and see that $\lambda = -1$ occurs three times. To find eigenvectors, we again need to solve $(\mathbf{A} - \lambda\mathbf{I})\mathbf{v} = \mathbf{0}$:

$$
\begin{bmatrix} -1-(-1) & 2 & -4 \\ 0 & -1-(-1) & 0 \\ 0 & 0 & -1-(-1) \end{bmatrix} \begin{bmatrix} v_{11} \\ v_{21} \\ v_{31} \end{bmatrix}
$$

$$
= \begin{bmatrix} 0 & 2 & -4 \\ 0 & 0 & 0 \\ 0 & 0 & 0 \end{bmatrix} \begin{bmatrix} v_{11} \\ v_{21} \\ v_{31} \end{bmatrix} = \begin{bmatrix} 0 \\ 0 \\ 0 \end{bmatrix}. \qquad (5.81)
$$

The last two rows do not restrict our choices of the entries of $\mathbf{v}_1$. The first row gives

$$
2v_{21} - 4v_{31} = 0 \Longrightarrow v_{21}/v_{31} = 2.
$$

We can, for example, choose $v_{21} = 2, v_{31} = 1$. We are free to do as we wish with the first entry v_{11}, say, $v_{11} = 0$. We could have also chosen $v_{21} = v_{31} = 0$ and then $v_{11} = 1$ (or any other non-zero number). Thus, we have two eigenvectors

$$
\mathbf{v}_1 = \begin{bmatrix} 0 \\ 2 \\ 1 \end{bmatrix}, \quad \mathbf{v}_2 = \begin{bmatrix} 1 \\ 0 \\ 0 \end{bmatrix}.
$$

We obtain the generalized eigenvector (or **chain vector**) by solving $(\mathbf{A} - (-1)\mathbf{I})\mathbf{u}_1 = \mathbf{v}_2$ for $\mathbf{u}_1$. (Note that we cannot solve this system if we put $\mathbf{v}_1$ on the right-hand side.) This gives us

$$
2u_{21} - 4u_{31} = 1 \quad \text{where} \quad \Longrightarrow u_{21} = 1/2, u_{31} = 0
$$

is one possibility. We are again free to choose u_{11} as anything, so we let $u_{11} = 0$. Thus

$$
\mathbf{u}_1 = \begin{bmatrix} 0 \\ 1/2 \\ 0 \end{bmatrix}.
$$

We can thus write the general solution as

$$
\mathbf{x} = c_1 e^{-t}\mathbf{v}_1 + c_2 e^{-t}\mathbf{v}_2 + c_3(te^{-t}\mathbf{v}_2 + e^{-t}\mathbf{u}_1)
$$

$$
= c_1 e^{-t} \begin{bmatrix} 0 \\ 2 \\ 1 \end{bmatrix} + c_2 e^{-t} \begin{bmatrix} 1 \\ 0 \\ 0 \end{bmatrix} + c_3 \left[te^{-t} \begin{bmatrix} 1 \\ 0 \\ 0 \end{bmatrix} + e^{-t} \begin{bmatrix} 0 \\ 1/2 \\ 0 \end{bmatrix} \right]. \qquad (5.82)
$$

We saw in Section 5.4, that being able to write $\mathbf{A}$ in its eigencoordinates greatly simplified our solution to the differential equation $\mathbf{x}' = \mathbf{A}\mathbf{x}$. In doing so, we used u_i to represent the eigencoordinates, solved the system $\mathbf{u}' = \mathbf{\Lambda}\mathbf{u}$ for $\mathbf{u}$, and converted back to our original variable via $\mathbf{x} = \mathbf{V}\mathbf{u}$. We will be able to state a similar result for matrices that are written in Jordan Canonical Form (sometimes called Jordan Normal Form).

We begin with the observation that if $\mathbf{A}$ is a $k \times k$ matrix, then the chain of generalized eigenvectors is a basis of $\mathbb{R}^k$. The Jordan Canonical Form of a matrix that we seek is the closest to diagonal that we will be able to find. For each eigenvector $\mathbf{v}_i$ that is the bottom of a chain of length k, we have an associated **Jordan block** of the form

$$\mathbf{J}_{\mathbf{v}_i} = \begin{bmatrix} \lambda_i & 1 & 0 & \cdots & 0 & 0 \\ 0 & \lambda_i & 1 & \cdots & 0 & \vdots \\ \vdots & & \ddots & \ddots & \vdots & 0 \\ 0 & 0 & \cdots & \lambda_i & 1 & 0 \\ 0 & 0 & \cdots & 0 & \lambda_i & 1 \\ 0 & 0 & \cdots & 0 & 0 & \lambda_i \end{bmatrix} \tag{5.83}$$

These Jordan blocks combine together to form a matrix in Jordan Canonical Form in which the only potential non-zero entries are on the diagonal and the **superdiagonal** lying just above this:

$$\mathbf{J} = \begin{bmatrix} J_{\mathbf{v}_1} & 0 & \cdots & 0 \\ 0 & J_{\mathbf{v}_2} & \cdots 0 & \vdots \\ \vdots & \ddots & \ddots & 0 \\ 0 & 0 & \cdots & J_{\mathbf{v}_j} \end{bmatrix} \tag{5.84}$$

If a given eigenvalue of multiplicity m has a full set of m eigenvectors, then the corresponding block will simply be diagonal (with 0 on the superdiagonal entries). In Section 5.4, we saw that we could diagonalize $\mathbf{A}$ with the matrix $\mathbf{V}$ in which the columns were the eigenvectors via $\mathbf{A} = \mathbf{V}\mathbf{\Lambda}\mathbf{V}^{-1}$. The result for generalized eigenvectors is analogous. We state the result for a single chain of generalized eigenvectors and will simply state that for the general Jordan Canonical Form, we will need to combine the various blocks together as in (5.84).

THEOREM 5.6.4

Suppose $\mathbf{A}$ is a $k \times k$ matrix with a single eigenvector $\mathbf{v}$ and a single eigenvalue λ of multiplicity k with $\{\mathbf{v}, \mathbf{u}_1, \mathbf{u}_2, \mathbf{u}_{k-1}\}$ being the resulting chain of generalized eigenvectors. Set

$$
P = \left[\begin{array}{c|c|c|c} v & u_1 & \cdots & u_{k-1} \end{array}\right] \text{ and } J = \begin{bmatrix} \lambda & 1 & 0 & \cdots & 0 \\ 0 & \lambda & 1 & \cdots & 0 \\ \vdots & & \ddots & \ddots & \vdots \\ 0 & 0 & \cdots & \lambda & 1 \\ 0 & 0 & \cdots & 0 & \lambda \end{bmatrix}, \quad (5.85)
$$

where P is a $k \times k$ matrix with the generalized eigenvectors as columns and J is a $k \times k$ Jordan block. Then

$$
A = PJP^{-1}. \tag{5.86}
$$

We consider an example to illustrate the theorem.

Example 4 Consider the matrix and chain of generalized eigenvectors from Example 2:

$$
A = \begin{bmatrix} 2 & 3 & 0 \\ 0 & 2 & -1 \\ 0 & 0 & 2 \end{bmatrix}, \left\{ \begin{bmatrix} 1 \\ 0 \\ 0 \end{bmatrix}, \begin{bmatrix} 0 \\ 1/3 \\ 0 \end{bmatrix}, \begin{bmatrix} 0 \\ 0 \\ -1/3 \end{bmatrix} \right\}.
$$

Verify that $A = PJP^{-1}$ where P, J are defined as in Theorem 5.6.4.

Solution

We write P and then calculate P^{-1}:

$$
P = \begin{bmatrix} 1 & 0 & 0 \\ 0 & \frac{1}{3} & 0 \\ 0 & 0 & \frac{-1}{3} \end{bmatrix} \Longrightarrow P^{-1} = \begin{bmatrix} 1 & 0 & 0 \\ 0 & 3 & 0 \\ 0 & 0 & -3 \end{bmatrix}.
$$

We then have

$$
\begin{bmatrix} 1 & 0 & 0 \\ 0 & \frac{1}{3} & 0 \\ 0 & 0 & \frac{-1}{3} \end{bmatrix} \begin{bmatrix} 2 & 1 & 0 \\ 0 & 2 & 1 \\ 0 & 0 & 2 \end{bmatrix} \begin{bmatrix} 1 & 0 & 0 \\ 0 & 3 & 0 \\ 0 & 0 & -3 \end{bmatrix} = \begin{bmatrix} 2 & 3 & 0 \\ 0 & 2 & -1 \\ 0 & 0 & 2 \end{bmatrix}.
$$

Now that we have a relationship between A and J, we can consider how to solve $w' = Jw$ (where we use w_i to represent the coordinates when we consider a system in its Jordan Canonical Form). We consider the 2×2 case as motivation:

$$
w' = Jw = \begin{bmatrix} \lambda & 1 \\ 0 & \lambda \end{bmatrix} x.
$$

We see that λ is the eigenvalue of J and there is only one corresponding eigenvector, $v = [1 \ 0]^T$. We can calculate the additional vector in the chain as $\mathbf{u}_1 = [0 \; 1]^T$. We now consider the individual differential equations,

$$w_1' = \lambda w_1 + w_2 \tag{5.87}$$
$$w_2' = \qquad \lambda w_2. \tag{5.88}$$

The last equation (5.88) can be solved easily as $w_2 = c_2 e^{\lambda t}$. We can then substitute this into (5.87) to obtain

$$w_1' = \lambda w_1 + c_2 e^{\lambda t}.$$

This equation is linear and can be solved with the methods of Section 1.3 to obtain $w_1 = c_1 e^{\lambda t} + c_2 t e^{\lambda t}$. Putting the solution in vector form gives

$$\mathbf{w} = \begin{bmatrix} w_1 \\ w_2 \end{bmatrix} = \begin{bmatrix} c_1 e^{\lambda t} + c_2 t e^{\lambda t} \\ c_2 e^{\lambda t} \end{bmatrix} = c_1 e^{\lambda t} \begin{bmatrix} 1 \\ 0 \end{bmatrix} + c_2 e^{\lambda t} \begin{bmatrix} t \\ 1 \end{bmatrix} \tag{5.89}$$

$$= c_1 e^{\lambda t} \begin{bmatrix} 1 \\ 0 \end{bmatrix} + c_2 e^{\lambda t} \begin{bmatrix} 0 \\ 1 \end{bmatrix} + c_2 t e^{\lambda t} \begin{bmatrix} 1 \\ 0 \end{bmatrix} \tag{5.90}$$

$$= c_1 e^{\lambda t} \mathbf{v} + c_2 (e^{\lambda t} \mathbf{u} + t e^{\lambda t} \mathbf{v}). \tag{5.91}$$

The method we used to solve the 2×2 case extends to any $k \times k$ Jordan block since the structure remains the same. This method of solution is the motivation for Theorem 5.6.3 (applying to a general $\mathbf{A}$) and its Corollary (applying to Jordan matrix $\mathbf{J}$).

THEOREM 5.6.5

Consider the system $\mathbf{w}' = \mathbf{J}\mathbf{w}$ for the $k \times k$ Jordan block $\mathbf{J}$ in which λ is an eigenvalue of multiplicity k with a single eigenvector $\mathbf{v}$. Then

$$\mathbf{v} = \begin{bmatrix} 1 \\ 0 \\ \vdots \\ 0 \end{bmatrix}, \mathbf{u}_1 = \begin{bmatrix} 0 \\ 1 \\ \vdots \\ 0 \end{bmatrix}, \cdots, \mathbf{u}_{k-1} = \begin{bmatrix} 0 \\ 0 \\ \vdots \\ 1 \end{bmatrix}$$

is a corresponding chain of generalized eigenvectors. The general solution can be written as

$$\begin{bmatrix} w_1 \\ w_2 \\ \vdots \\ w_k \end{bmatrix} = e^{\lambda t} \begin{bmatrix} 1 & t & \cdots & \frac{t^{k-2}}{(k-2)!} & \frac{t^{k-1}}{(k-1)!} \\ 0 & 1 & t & \cdots & \vdots \\ 0 & 0 & \ddots & t & \frac{t^2}{2} \\ \vdots & & & 1 & t \\ 0 & 0 & \cdots & 0 & 1 \end{bmatrix} \begin{bmatrix} c_1 \\ c_2 \\ \vdots \\ c_k \end{bmatrix}. \tag{5.92}$$

Note that the full set of generalized eigenvectors in Theorem 5.6.5 is just the **standard basis**, often denoted $\mathbf{e}_1, \mathbf{e}_2, \ldots, \mathbf{e}_n$. Theorems 5.6.5 and 5.6.6 give a convenient form for solving $\mathbf{x}' = \mathbf{A}\mathbf{x}$.

THEOREM 5.6.6

Consider the system $\mathbf{x}' = \mathbf{Ax}$ for the $k \times k$ matrix $\mathbf{A}$ in which (i) λ is an eigenvalue of multiplicity k with a single eigenvector $\mathbf{v}$ and (ii) $\{\mathbf{v}, \mathbf{u}_1, \cdots, \mathbf{u}_{k-1}\}$ is the corresponding chain of generalized eigenvectors. Set $\mathbf{P}$ to be the matrix containing the chain of generalized eigenvectors as its columns (as defined in Theorem 5.6.4). Then the general solution can be written

$$
\begin{bmatrix} x_1 \\ x_2 \\ \vdots \\ x_k \end{bmatrix} = \mathbf{P}e^{\lambda t}
\begin{bmatrix}
1 & t & \cdots & \frac{t^{k-2}}{(k-2)!} & \frac{t^{k-1}}{(k-1)!} \\
0 & 1 & t & \cdots & \vdots \\
0 & 0 & \ddots & t & \frac{t^2}{2} \\
\vdots & & & 1 & t \\
0 & 0 & \cdots & 0 & 1
\end{bmatrix}
\begin{bmatrix} c_1 \\ c_2 \\ \vdots \\ c_k \end{bmatrix}.
\tag{5.93}
$$

Example 5 Consider the matrices $\mathbf{A}$ and $\mathbf{P}$ from Example 4:

$$
\mathbf{A} = \begin{bmatrix} 2 & 3 & 0 \\ 0 & 2 & -1 \\ 0 & 0 & 2 \end{bmatrix}, \quad
\mathbf{P} = \begin{bmatrix} 1 & 0 & 0 \\ 0 & \frac{1}{3} & 0 \\ 0 & 0 & \frac{-1}{3} \end{bmatrix}.
$$

Use Theorem 5.6.6 to solve $\mathbf{x}' = \mathbf{Ax}$.

Solution

This is a direct application of Theorem 5.6.6:

$$
\begin{bmatrix} x_1 \\ x_2 \\ x_3 \end{bmatrix} = \mathbf{P}e^{\lambda t}
\begin{bmatrix} 1 & t & \frac{t^2}{2} \\ 0 & 1 & t \\ 0 & 0 & 1 \end{bmatrix}
\begin{bmatrix} c_1 \\ c_2 \\ c_3 \end{bmatrix}
\tag{5.94}
$$

$$
= e^{\lambda t}
\begin{bmatrix} 1 & 0 & 0 \\ 0 & \frac{1}{3} & 0 \\ 0 & 0 & \frac{-1}{3} \end{bmatrix}
\begin{bmatrix} 1 & t & \frac{t^2}{2} \\ 0 & 1 & t \\ 0 & 0 & 1 \end{bmatrix}
\begin{bmatrix} c_1 \\ c_2 \\ c_3 \end{bmatrix}
\tag{5.95}
$$

$$
= \begin{bmatrix} 1 & t & \frac{t^2}{2} \\ 0 & \frac{1}{3} & \frac{1}{3}t \\ 0 & 0 & \frac{-1}{3} \end{bmatrix}
\begin{bmatrix} c_1 \\ c_2 \\ c_3 \end{bmatrix}
= \begin{bmatrix} c_1 + c_2 t + c_3 \frac{t^2}{2} \\ c_2 \frac{1}{3} + c_3 \frac{1}{3} t \\ c_3 \frac{-1}{3} \end{bmatrix}.
\tag{5.96}
$$

We stated in (5.84) that we combine Jordan blocks together to give us the complete Jordan Canonical Form of a matrix. The size of each Jordan block $\mathbf{J}_{\mathbf{v}_i}$ in (5.84) is the dimension of each respective set of chain vectors and the total number of blocks in $\mathbf{J}$ is the total number of these sets. If we allow for complex eigenvalues, then every matrix can be decomposed into Jordan Canonical Form. However, for our study of differential equations, we will keep our focus on Jordan blocks that are of size 3×3 or lower, arising from real repeated eigenvalues, and will refer the interested reader to an advanced linear algebra text such as [23],[52] for further study. As such, we finish with a classification system that will allow us to apply Theorem 5.6.6.

THEOREM 5.6.7   Jordan Canonical Form

Let $\mathbf{A}$ be a matrix with real eigenvalue λ of multiplicity 1, 2, or 3.

1. If λ is an eigenvalue of $\mathbf{A}$ of multiplicity 1, then there is one corresponding 1×1 Jordan block.
2. If λ is an eigenvalue of $\mathbf{A}$ of multiplicity 2, then one of two possibilities occur:
   (a) there are two associated eigenvectors and thus two 1×1 Jordan blocks.
   (b) there is only one associated eigenvector with corresponding chain vector set $\{\mathbf{v}, \mathbf{u}\}$; there is one 2×2 Jordan block.
3. If λ is an eigenvalue of $\mathbf{A}$ of multiplicity 3, then one of three possibilities occur:
   (a) there are three associated eigenvectors and thus three 1×1 Jordan blocks.
   (b) there are two associated eigenvectors, one of which gives rise to a generalized eigenvector; denote these as $\mathbf{v}_1$ and $\{\mathbf{v}_2, \mathbf{u}_1^{(2)}\}$. Then $\mathbf{v}_1$ gives rise to one 1×1 Jordan block and $\{\mathbf{v}_2, \mathbf{u}_1^{(2)}\}$ gives rise to one 2×2 Jordan block.
   (c) there is only one associated eigenvector with corresponding chain vector set $\{\mathbf{v}, \mathbf{u}_1, \mathbf{u}_2\}$. This gives rise to one 3×3 Jordan block.

Example 6: In Example 3, we considered

$$\mathbf{x}' = \begin{bmatrix} -1 & 2 & -4 \\ 0 & -1 & 0 \\ 0 & 0 & -1 \end{bmatrix} \mathbf{x}$$

and found that it had two eigenvectors: $\mathbf{v}_1 = \begin{bmatrix} 0 \\ 2 \\ 1 \end{bmatrix}$, $\mathbf{v}_2 = \begin{bmatrix} 1 \\ 0 \\ 0 \end{bmatrix}$. We also found that $\mathbf{v}_2$ is the eigenvector that gives rise to the generalized eigenvector

$$\mathbf{u}_1 = \begin{bmatrix} 0 \\ \frac{1}{2} \\ 0 \end{bmatrix}.$$

Write the Jordan Canonical Form of this system and use Theorem 5.6.7 resolve the problem.

Solution
Because $\lambda = -1$ is of multiplicity 3 and we have two eigenvectors, we are

in 3(b) of Theorem 5.6.7. The Jordan block from $\mathbf{v}_1$ is simply the 1×1 block $\mathbf{J}_{\mathbf{v}_1} = -1$. The Jordan block from the chain of generalized eigenvectors $\{\mathbf{v}_2, \mathbf{u}_1^{(2)}\}$ is the 2×2 block

$$\mathbf{J}_{\mathbf{v}_2} = \begin{bmatrix} -1 & 1 \\ 0 & -1 \end{bmatrix}.$$

Combining these together gives

$$\mathbf{J} = \begin{bmatrix} \mathbf{J}_{\mathbf{v}_1} & 0 \\ 0 & \mathbf{J}_{\mathbf{v}_2} \end{bmatrix} = \left[\begin{array}{c|cc} -1 & 0 & 0 \\ \hline 0 & -1 & 1 \\ 0 & 0 & -1 \end{array} \right]$$

$$= \begin{bmatrix} -1 & 0 & 0 \\ 0 & -1 & 1 \\ 0 & 0 & -1 \end{bmatrix}, \tag{5.97}$$

where we initially put the horizontal and vertical lines inside the matrix to highlight its block structure. In order to apply Theorem 5.6.6, we put the generalized eigenvectors in the same order as the corresponding block in $\mathbf{J}$. Since we put the 1×1 block first, the first column of $\mathbf{P}$ will be $\mathbf{v}_1$ with $\mathbf{v}_2, \mathbf{u}_1^{(2)}$ being the second and third columns:

$$\mathbf{P} = \begin{bmatrix} 0 & 1 & 0 \\ 2 & 0 & \frac{1}{2} \\ 1 & 0 & 0 \end{bmatrix}.$$

We apply Theorem 5.6.6 to the two Jordan blocks to obtain

$$\begin{bmatrix} x_1 \\ x_2 \\ x_3 \end{bmatrix} = \mathbf{P} \left[\begin{array}{c|cc} e^{-t} & 0 & 0 \\ \hline 0 & e^{-t} & te^{-t} \\ 0 & 0 & e^{-t} \end{array} \right] \begin{bmatrix} c_1 \\ c_2 \\ c_3 \end{bmatrix} \tag{5.98}$$

$$= \begin{bmatrix} 0 & 1 & 0 \\ 2 & 0 & \frac{1}{2} \\ 1 & 0 & 0 \end{bmatrix} \begin{bmatrix} e^{-t} & 0 & 0 \\ 0 & e^{-t} & te^{-t} \\ 0 & 0 & e^{-t} \end{bmatrix} \begin{bmatrix} c_1 \\ c_2 \\ c_3 \end{bmatrix} \tag{5.99}$$

$$= \begin{bmatrix} 0 & e^{-t} & te^{-t} \\ 2e^{-t} & 0 & \frac{1}{2}e^{-t} \\ e^{-t} & 0 & 0 \end{bmatrix} \begin{bmatrix} c_1 \\ c_2 \\ c_3 \end{bmatrix} \tag{5.100}$$

$$= c_1 e^{-t} \begin{bmatrix} 0 \\ 2 \\ 1 \end{bmatrix} + c_2 e^{-t} \begin{bmatrix} 1 \\ 0 \\ 0 \end{bmatrix} + c_3 e^{-t} \begin{bmatrix} t \\ \frac{1}{2} \\ 0 \end{bmatrix}. \tag{5.101}$$

This answer is the same as the one we previously obtained:

$$\mathbf{x} = c_1 e^{-t} \mathbf{v}_1 + c_2 e^{-t} \mathbf{v}_2 + c_3 (te^{-t} \mathbf{v}_2 + e^{-t} \mathbf{u}_1).$$

Problems

In Problems **1–12**, find the eigenvalues and eigenvector of the coefficient matrix by hand (the eigenvalues are all repeated with only one eigenvector). Use the methods of this section to obtain a generalized eigenvector. Then use Theorem 5.6.3 and Theorem 5.5.3 to write the general solution.

1. $\begin{cases} x' = x + 2y \\ y' = y \end{cases}$

2. $\begin{cases} x' = -3x + y \\ y' = -3y \end{cases}$

3. $\begin{cases} x' = 7x - 2y \\ y' = 8x - y \end{cases}$

4. $\begin{cases} x' = -3x + y \\ y' = -4x + y \end{cases}$

5. $\begin{cases} x' = -2x + y \\ y' = -x \end{cases}$

6. $\begin{cases} x' = y \\ y' = -x + 2y \end{cases}$

7. $\begin{cases} x' = -x + 4y \\ y' = -y \end{cases}$

8. $\begin{cases} x' = -2x - y \\ y' = x - 4y \end{cases}$

9. $\begin{cases} x' = x - 2y \\ y' = 2x - 3y \end{cases}$

10. $\begin{cases} x' = -3x - y \\ y' = x - 5y \end{cases}$

11. $\begin{cases} x' = -x - 2y \\ y' = 2x + 3y \end{cases}$

12. $\begin{cases} x' = 7x + y \\ y' = -x + 5y \end{cases}$

In Problems **13–21**, use MATLAB, Maple, or Mathematica (or by hand, if instructed) to find the eigenvalues and eigenvector of the coefficient matrix (there is a repeated eigenvalue with only one corresponding eigenvector). Use the methods of this section to obtain a generalized eigenvector. Then use Theorem 5.6.3 and Theorem 5.5.3 to write the general solution.

13. $\begin{cases} x' = x - y - z \\ y' = x + 3y + z \\ z' = z \end{cases}$

14. $\begin{cases} x' = 2x + y \\ y' = y \\ z' = x + 3y + 2z \end{cases}$

15. $\begin{cases} x' = -x + y + z \\ y' = 3y + z \\ z' = -4y - z \end{cases}$

16. $\begin{cases} x' = 5x - 4y \\ y' = x + 2z \\ z' = 2y + 5z \end{cases}$

17. $\begin{cases} x' = 2x \\ y' = 2y \\ z' = x + y + 2z \end{cases}$

18. $\begin{cases} x' = 2x + y + z \\ y' = y - z \\ z' = x + 2y + 3z \end{cases}$

19. $\begin{cases} x' = y \\ y' = -2y - z \\ z' = x + y - z \end{cases}$

20. $\begin{cases} x' = 4x - y \\ y' = 4z \\ z' = 2x - y + 2z \end{cases}$

21. $\begin{cases} x' = x + z \\ y' = 2x - y - 3z \\ z' = -x + y + 3z \end{cases}$

In Problems **22–23**, write the general solution for $\mathbf{x}' = \mathbf{Ax}$, where $\mathbf{A}$ is the given matrix.

22. $\mathbf{A} = \begin{bmatrix} 1 & 0 & 0 & 0 & 0 & 0 \\ 0 & 3 & 0 & 0 & 0 & 0 \\ 0 & 0 & 3 & 0 & 0 & 0 \\ 0 & 0 & 0 & 3 & 1 & 0 \\ 0 & 0 & 0 & 0 & 3 & 1 \\ 0 & 0 & 0 & 0 & 0 & 3 \end{bmatrix}$

23. $\mathbf{A} = \begin{bmatrix} 1 & 0 & 0 & 0 & 0 & 0 \\ 0 & 2 & 1 & 0 & 0 & 0 \\ 0 & 0 & 2 & 0 & 0 & 0 \\ 0 & 0 & 0 & 3 & 1 & 0 \\ 0 & 0 & 0 & 0 & 3 & 1 \\ 0 & 0 & 0 & 0 & 0 & 3 \end{bmatrix}$

24. Find the Jordan normal form of

$$
\mathbf{A} = \begin{bmatrix}
7 & -8 & 2 & 5 & 3 & -7 \\
8 & -11 & 3 & 7 & 5 & -10 \\
-9 & 15 & -1 & -8 & -6 & 12 \\
0 & 0 & 0 & 2 & 0 & 2 \\
5 & -8 & 2 & 4 & 5 & -7 \\
-6 & 10 & -2 & -5 & -4 & 10
\end{bmatrix}.
$$

You may use MATLAB, Maple, or Mathematica to find the eigenvectors and help you find the generalized eigenvectors. How many Jordan blocks do you have? What is the size of each one? Now experiment with the built-in commands in each of the three that will find the Jordan normal form for you. Is the computer answer different from yours? Explain.

5.7 Matrix Exponentials

As with most of the sections in this chapter, we have been considering the first-order linear system (5.1), which can be written in matrix notation as

$$\mathbf{x}' = \mathbf{A}\mathbf{x}. \tag{5.102}$$

In Section 5.4, we saw that $\mathbf{x} = e^{\lambda t}\mathbf{v}$ is a solution to (5.102) provided that λ is an eigenvalue of $\mathbf{A}$ with $\mathbf{v}$ as the corresponding eigenvector. We know that solutions are unique, but is this the only form in which we can write the solution? The answer is "no" for reasons we will see shortly. For the moment let us *suppose that it makes sense to write*

$$e^{\mathbf{A}t}, \tag{5.103}$$

where $\mathbf{A}$ is the matrix in (5.102). And further suppose that it makes sense to take the derivative of this function in the same way we take a derivative of the function e^{at} when a is a constant; that is, $(e^{at})' = ae^{at}$. We thus *suppose that it also makes sense to take the derivative as follows:*

$$\frac{d}{dt}\left[e^{\mathbf{A}t}\right] = \mathbf{A}e^{\mathbf{A}t}. \tag{5.104}$$

If we assume a solution of the form $\mathbf{x} = e^{\mathbf{A}t}$, then

$$\mathbf{x}' = \mathbf{A}e^{\mathbf{A}t}$$

and substitution into (5.102) shows that both sides of the equation are indeed the same for all t. By our previous definitions, this means that $e^{\mathbf{A}t}$ is a solution. We thus have another form of a solution provided it makes sense to exponentiate a matrix and then take its derivative.

Formal Definition and Properties

We make the following definition of a **matrix exponential** to help us.

Definition 5.7.1

For any square matrix **A**, define

$$e^{\mathbf{A}} = \mathbf{I} + \frac{\mathbf{A}}{1!} + \frac{\mathbf{A}^2}{2!} + \cdots = \sum_{k=0}^{\infty} \frac{\mathbf{A}^k}{k!}. \tag{5.105}$$

For our purposes, we often consider $e^{\mathbf{A}t}$. Because a scalar times a matrix is just another matrix, the above definition gives us

$$e^{\mathbf{A}t} = \mathbf{I} + \frac{\mathbf{A}t}{1!} + \frac{\mathbf{A}^2 t^2}{2!} + \cdots = \sum_{k=0}^{\infty} \frac{\mathbf{A}^k t^k}{k!}. \tag{5.106}$$

The reader should note that this definition of the matrix exponential is in the form of an infinite series. It turns out that this series converges for any matrix **A** and thus it makes sense to talk about $e^{\mathbf{A}}$ or $e^{\mathbf{A}t}$ for any square matrix **A**.

In the case of a diagonal matrix, our computation is particularly easy as we see in the next example.

Example 1: Compute $e^{\mathbf{A}t}$ for $\mathbf{A} = \begin{bmatrix} 2 & 0 \\ 0 & -1 \end{bmatrix}$.

Solution

In order to apply (5.106), we need to calculate the successive powers of **A**:

$$\mathbf{A} = \begin{bmatrix} 2 & 0 \\ 0 & -1 \end{bmatrix}, \quad \mathbf{A}^2 = \begin{bmatrix} 4 & 0 \\ 0 & 1 \end{bmatrix}, \quad \cdots, \mathbf{A}^k = \begin{bmatrix} 2^k & 0 \\ 0 & (-1)^k \end{bmatrix}.$$

Applying (5.106) gives

$$e^{\mathbf{A}t} = \sum_{k=0}^{\infty} \frac{t^k}{k!} \begin{bmatrix} 2^k & 0 \\ 0 & (-1)^k \end{bmatrix} = \begin{bmatrix} \sum_{k=0}^{\infty} \frac{(2t)^k}{k!} & 0 \\ 0 & \sum_{k=0}^{\infty} \frac{(-1t)^k}{k!} \end{bmatrix} = \begin{bmatrix} e^{2t} & 0 \\ 0 & e^{-t} \end{bmatrix}.$$

This example about a 2×2 matrix can be generalized for a diagonal matrix of any size:

$$\text{for } \mathbf{A} = \begin{bmatrix} d_1 & 0 & \cdots & 0 \\ 0 & d_2 & \cdots & 0 \\ \vdots & & \ddots & \\ 0 & 0 & \cdots & d_n \end{bmatrix}, \text{ we have } e^{\mathbf{A}t} = \begin{bmatrix} e^{d_1 t} & 0 & \cdots & 0 \\ 0 & e^{d_2 t} & \cdots & 0 \\ \vdots & & \ddots & \\ 0 & 0 & \cdots & e^{d_n t} \end{bmatrix}.$$
$$\tag{5.107}$$

We state a few useful results that will help give us insight into the solution of a differential equation.

Let $\mathbf{A}, \mathbf{B}, \mathbf{P}$ be $n \times n$ matrices with $\mathbf{P}$ invertible. Then
(a) $e^{-\mathbf{A}} = (e^{\mathbf{A}})^{-1}$.
(b) If $\mathbf{A} = \mathbf{PBP}^{-1}$, then $e^{\mathbf{A}} = \mathbf{P}e^{\mathbf{B}}\mathbf{P}^{-1}$.
(c) If $\mathbf{A}$ and $\mathbf{B}$ commute (that is, $\mathbf{AB} = \mathbf{BA}$), then $e^{\mathbf{A}+\mathbf{B}} = e^{\mathbf{A}}e^{\mathbf{B}}$.

Example 2 Show that $e^{\mathbf{A}(t-t_0)} = e^{\mathbf{A}t}e^{-\mathbf{A}t_0}$.

Solution
Part (c) of the above theorem applies if we can show that $\mathbf{A}t$ and $-\mathbf{A}t_0$ commute. Using some of the properties of matrices, we have

$$(\mathbf{A}t)(-\mathbf{A}t_0) = (\mathbf{A}\mathbf{A})(-tt_0) = (\mathbf{A}\mathbf{A})(-t_0t) = (-\mathbf{A}t_0)(\mathbf{A}t),$$

which shows that the matrices do commute. Part (c) of the above theorem then gives us our desired conclusion.

Although the theorem gives some useful properties, it does not tell us how to compute the series for the matrix exponential. In Section 5.4, we saw that if $\mathbf{A}$ has a full set of eigenvectors we can put the eigenvectors as the columns of a matrix $\mathbf{V}$ and the relation $\mathbf{A} = \mathbf{V}\mathbf{\Lambda}\mathbf{V}^{-1}$ holds. Thus, we could calculate $e^{\mathbf{A}}$ when $\mathbf{A}$ possesses a full set of eigenvectors through (b).

Example 3 Consider the matrix

$$\mathbf{A} = \begin{bmatrix} -2 & -1 & -2 \\ -4 & -5 & 2 \\ -5 & -1 & 1 \end{bmatrix}.$$

Find the eigenvalues and eigenvectors and then use Theorem 5.7.1(b) to find $e^{\mathbf{A}}$.

Solution
We can readily calculate the eigenvalue-eigenvector pairs as

$$\left\{ 3, \begin{bmatrix} -1 \\ 1 \\ 2 \end{bmatrix} \right\}, \left\{ -6, \begin{bmatrix} 1 \\ 2 \\ 1 \end{bmatrix} \right\}, \left\{ -3, \begin{bmatrix} 1 \\ -1 \\ 1 \end{bmatrix} \right\}.$$

Thus the matrix is diagonalizable with $\mathbf{V}$ containing the eigenvectors as columns. Then we have

$$e^{\mathbf{A}} = \mathbf{V}e^{\mathbf{\Lambda}}\mathbf{V}^{-1} \tag{5.108}$$

$$= \begin{bmatrix} -1 & 1 & 1 \\ 1 & 2 & -1 \\ 2 & 1 & 1 \end{bmatrix} \begin{bmatrix} e^3 & 0 & 0 \\ 0 & e^{-6} & 0 \\ 0 & 0 & e^{-3} \end{bmatrix} \begin{bmatrix} -1 & 1 & 1 \\ 1 & 2 & -1 \\ 2 & 1 & 1 \end{bmatrix}^{-1} \tag{5.109}$$

$$= \begin{bmatrix} \frac{1}{3}e^3 + \frac{1}{3}e^{-6} + \frac{1}{3}e^{-3} & \frac{1}{3}e^{-6} - \frac{1}{3}e^3 & -\frac{1}{3}e^{-3} + \frac{1}{3}e^{-3} \\ -\frac{1}{3}e^3 + \frac{2}{3}e^{-6} - \frac{1}{3}e^{-3} & \frac{2}{3}e^{-6} + \frac{1}{3}e^{-3} & \frac{1}{3}e^3 - \frac{1}{3}e^{-3} \\ -\frac{2}{3}e^3 + \frac{1}{3}e^{-6} + \frac{1}{3}e^{-3} & \frac{1}{3}e^{-6} - \frac{1}{3}e^{-3} & \frac{2}{3}e^3 + \frac{1}{3}e^{-3} \end{bmatrix}. \tag{5.110}$$

The step of calculating $\mathbf{V}^{-1}$ can be done by hand using Appendix B.2 or one of the computer programs.

If we are able to write the matrix in Jordan Canonical Form via $\mathbf{A} = \mathbf{PJP}^{-1}$, then we can similarly apply Theorem 5.7.1(b) to find $e^{\mathbf{A}}$; however, $e^{\mathbf{J}}$ is not nearly as easy to calculate as $e^{\mathbf{A}}$. We refer the interested reader to the exercises for practice with these types of matrices. For a general matrix, the following theorem is useful for calculating the exponential of a matrix.

THEOREM 5.7.2

Let $\mathbf{A}$ be an $n \times n$ matrix.

a) There exist functions $\alpha_1(t), \alpha_2(t), \ldots, \alpha_n(t)$, such that

$$e^{\mathbf{A}t} = \alpha_1(t)\mathbf{A}^{n-1}t^{n-1} + \alpha_2(t)\mathbf{A}^{n-2}t^{n-2} + \cdots + \alpha_{n-1}(t)\mathbf{A}t + \alpha_n(t)\mathbf{I}. \tag{5.111}$$

b) For the polynomial (in r)

$$p(r) = \alpha_1(t)r^{n-1} + \alpha_2(t)r^{n-2} + \cdots + \alpha_{n-1}(t)r + \alpha_n(t), \tag{5.112}$$

if λ is an eigenvalue of $\mathbf{A}$, then

$$e^{\lambda} = p(\lambda),$$

so that $e^{\lambda t} = p(\lambda t)$.

c) If λ is an eigenvalue of multiplicity k, then

$$e^{\lambda} = \frac{dp(r)}{dr}\bigg|_{r=\lambda}, \quad e^{\lambda} = \frac{d^2 p(r)}{dr^2}\bigg|_{r=\lambda}, \quad \cdots, \quad e^{\lambda} = \frac{d^{k-1}p(r)}{dr^{k-1}}\bigg|_{r=\lambda}. \tag{5.113}$$

We calculate $e^{\mathbf{A}t}$ by first applying parts (b) and (c) of the above theorem to generate a set of linear equations in α_i. We then solve for these α_i and substitute into the formula in (a).

COROLLARY 5.7.1

For a 2×2 matrix $\mathbf{A} = \begin{bmatrix} a & -b \\ b & a \end{bmatrix}$, we have

$$e^{\mathbf{A}} = e^a \begin{bmatrix} \cos b & -\sin b \\ \sin b & \cos b \end{bmatrix} \quad \text{and thus} \quad e^{\mathbf{A}t} = e^{at} \begin{bmatrix} \cos bt & -\sin bt \\ \sin bt & \cos bt \end{bmatrix}.$$

Example 5.4

$$\text{Find } e^{\mathbf{A}t} \text{ if } \mathbf{A} = \begin{bmatrix} -1 & -2 & 3 \\ 0 & 2 & -1 \\ 0 & 0 & 2 \end{bmatrix}.$$

Solution

This is the case of $n = 3$ in part (a) of Theorem 5.7.2:

$$e^{\mathbf{A}t} = \alpha_1(t)\mathbf{A}^2 t^2 + \alpha_2(t)\mathbf{A}t + \alpha_3(t)\mathbf{I}$$

$$= \alpha_1 t^2 \begin{bmatrix} -1 & -2 & 3 \\ 0 & 2 & -1 \\ 0 & 0 & 2 \end{bmatrix}^2 + \alpha_2 t \begin{bmatrix} -1 & -2 & 3 \\ 0 & 2 & -1 \\ 0 & 0 & 2 \end{bmatrix} + \alpha_3 \begin{bmatrix} 1 & 0 & 0 \\ 0 & 1 & 0 \\ 0 & 0 & 1 \end{bmatrix}$$

$$= \begin{bmatrix} \alpha_1 t^2 - \alpha_2 t + \alpha_3 & -2\alpha_1 t^2 - 2\alpha_2 t & 5\alpha_1 t^2 + 3\alpha_2 t \\ 0 & 4\alpha_1 t^2 + 2\alpha_2 t + \alpha_3 & -4\alpha_1 t^2 - \alpha_2 t \\ 0 & 0 & 4\alpha_1 t^2 + 2\alpha_2 t + \alpha_3 \end{bmatrix}, \quad (5.114)$$

where we have stopped writing the dependence of α_i on t for notational convenience. The eigenvalues of $\mathbf{A}$ are easily seen to be $\lambda_1 = -1, \lambda_2 = 2, \lambda_3 = 2$. We thus set up a system of three equations to calculate $\alpha_1, \alpha_2, \alpha_3$: part (b) of Theorem 5.7.2 for λ_1 and part (c) for the repeated eigenvalues λ_2, λ_3. For the latter situation, we have

$$p(r) = \alpha_1 r^2 + \alpha_2 r + \alpha_3, \quad p'(r) = 2\alpha_1 r + \alpha_2.$$

This gives us

$$p(2t): \quad e^{2t} = \alpha_1(2t)^2 + \alpha_2(2t) + \alpha_3$$
$$p'(2t): \quad e^{2t} = 2\alpha_1(2t) + \alpha_2.$$

For the non-repeated eigenvalue, we have

$$p(-t): \quad e^{-t} = \alpha_1(-t)^2 + \alpha_2(-t) + \alpha_3.$$

We solve this system of equations, for example using MATLAB, Maple, or

Mathematica, to obtain

$$\alpha_1 = \frac{3te^{2t} - e^{2t} + e^{-t}}{9t^2}$$

$$\alpha_2 = -\frac{3te^{2t} - 4e^{2t} + 4e^{-t}}{9t}$$

$$\alpha_3 = \frac{-6te^{2t} + 5e^{2t} + 4e^{-t}}{9}.$$

Substituting these values into (5.114), we see that

$$e^{\mathbf{A}t} = \frac{1}{3}\begin{bmatrix} 3e^{-t} & 2e^{-t} - 2e^{2t} & +2te^{2t} + \frac{7}{3}e^{2t} - \frac{7}{3}e^{-t} \\ 0 & 3e^{2t} & -3te^{2t} \\ 0 & 0 & 3e^{2t} \end{bmatrix}. \tag{5.115}$$

We will consider one more example before showing how to use the matrix exponential to solve systems of differential equations.

Example 5: Find $e^{\mathbf{A}t}$ if $\mathbf{A} = \begin{bmatrix} -1 & -2 \\ 2 & -1 \end{bmatrix}$.

Solution
The above corollary gives us

$$e^{\mathbf{A}t} = e^{-t}\begin{bmatrix} \cos 2t & -\sin 2t \\ \sin 2t & \cos 2t \end{bmatrix}.$$

In determining the exponential of a matrix in the above discussion, we can use our computer programs to solve the system of equations that resulted by application of Theorem 5.7.2. For pedagogical reasons, it is useful and important to understand how the exponential of a matrix can actually be computed and the algebraic manipulations can be done with the programs. However, if our ultimate goal is to solve and/or gain some insight into the solution of a given differential equation, the matrix exponential is simply a tool that is at our disposal and we can simply use the built-in computer commands to calculate the exponential of a given matrix.

The Derivative of the Matrix Exponential

We motivated the matrix exponential by saying that if we could define $e^{\mathbf{A}t}$ and the derivative of it (in the "normal" manner), we would have found another expression of the solution of $\mathbf{x}' = \mathbf{A}\mathbf{x}$. We formalized $e^{\mathbf{A}t}$ and we now show that we can take its derivative in the "normal" manner.

$$\frac{d}{dt}\left[e^{\mathbf{A}t}\right] = \mathbf{A}e^{\mathbf{A}t} = e^{\mathbf{A}t}\mathbf{A}.$$

THEOREM 5.7.3

Remark: Note that this theorem shows that $\mathbf{A}$ commutes with $e^{\mathbf{A}t}$.

Proof: The proof of this is straightforward and uses the definition of limits and series formulation of the matrix exponential discussed in the previous section.

$$\begin{aligned}
\frac{d}{dt}\left[e^{\mathbf{A}t}\right] &= \lim_{h\to 0}\left[\frac{e^{\mathbf{A}(t+h)} - e^{\mathbf{A}t}}{h}\right] \\
&= \lim_{h\to 0}\left[\frac{e^{\mathbf{A}t}e^{\mathbf{A}h} - e^{\mathbf{A}t}}{h}\right] \quad \text{by Theorem 5.7.1} \\
&= e^{\mathbf{A}t}\lim_{h\to 0}\left[\frac{e^{\mathbf{A}h} - \mathbf{I}}{h}\right] \\
&= e^{\mathbf{A}t}\lim_{h\to 0}\left[\frac{\left[\mathbf{I} + \dfrac{\mathbf{A}h}{1!} + \dfrac{\mathbf{A}^2h^2}{2!} + \dfrac{\mathbf{A}^3h^3}{3!} + \cdots\right] - \mathbf{I}}{h}\right] \\
&= e^{\mathbf{A}t}\lim_{h\to 0}\left[\frac{\mathbf{A}}{1!} + \frac{\mathbf{A}^2h}{2!} + \frac{\mathbf{A}^3h^2}{3!} + \cdots\right] \\
&= e^{\mathbf{A}t}\mathbf{A}. \tag{5.116}
\end{aligned}$$

This proves the first equality. For the second, we only need to observe that $\mathbf{A}$ commutes with each term of the series of $e^{\mathbf{A}t}$. That is,

$$\begin{aligned}
\mathbf{A}e^{\mathbf{A}t} &= \mathbf{A}\left[\mathbf{I} + \frac{\mathbf{A}h}{1!} + \frac{\mathbf{A}^2h^2}{2!} + \frac{\mathbf{A}^3h^3}{3!} + \cdots\right] \\
&= \left[\mathbf{A}\mathbf{I} + \frac{\mathbf{A}^2h}{1!} + \frac{\mathbf{A}^3h^2}{2!} + \frac{\mathbf{A}^4h^3}{3!} + \cdots\right] \\
&= \left[\mathbf{I}\mathbf{A} + \frac{\mathbf{A}h\mathbf{A}}{1!} + \frac{\mathbf{A}^2h^2\mathbf{A}}{2!} + \frac{\mathbf{A}^3h^3\mathbf{A}}{3!} + \cdots\right] \\
&= \left[\mathbf{I} + \frac{\mathbf{A}h}{1!} + \frac{\mathbf{A}^2h^2}{2!} + \frac{\mathbf{A}^3h^3}{3!} + \cdots\right]\mathbf{A} \\
&= e^{\mathbf{A}t}\mathbf{A}. \tag{5.117}
\end{aligned}$$

The theorem is thus proved. ∎

Example 6 Show that $e^{\mathbf{A}t}$ is a solution of $\mathbf{x}' = \mathbf{A}\mathbf{x}$ for $\mathbf{A} = \begin{bmatrix} -1 & -2 \\ 2 & -1 \end{bmatrix}$.

Solution
This matrix was considered in Example 5 of this section and we found that

$$e^{\mathbf{A}t} = e^{-t} \begin{bmatrix} \cos 2t & -\sin 2t \\ \sin 2t & \cos 2t \end{bmatrix}.$$

For the derivative, we have

$$\left[e^{\mathbf{A}t} \right]' = -e^{-t} \begin{bmatrix} \cos 2t & -\sin 2t \\ \sin 2t & \cos 2t \end{bmatrix} + e^{-t} \begin{bmatrix} -2\sin 2t & -2\cos 2t \\ 2\cos 2t & -2\sin 2t \end{bmatrix}$$

$$= e^{-t} \begin{bmatrix} -\cos 2t - 2\sin 2t & \sin 2t - 2\cos 2t \\ -\sin 2t + 2\cos 2t & -\cos 2t - 2\sin 2t \end{bmatrix}. \tag{5.118}$$

Calculating $\mathbf{A}e^{\mathbf{A}t}$, we have

$$\mathbf{A}e^{\mathbf{A}t} = \begin{bmatrix} -1 & -2 \\ 2 & -1 \end{bmatrix} e^{-t} \begin{bmatrix} \cos 2t & -\sin 2t \\ \sin 2t & \cos 2t \end{bmatrix}$$

$$= e^{-t} \begin{bmatrix} -\cos 2t - 2\sin 2t & \sin 2t - 2\cos 2t \\ -\sin 2t + 2\cos 2t & -\cos 2t - 2\sin 2t \end{bmatrix}, \tag{5.119}$$

which shows that $e^{\mathbf{A}t}$ is a solution.

Thus for a homogeneous constant coefficient system of equations $\mathbf{x}' = \mathbf{A}\mathbf{x}$, we have another method of finding a solution. In the next section we will see how this relates to our previously obtained solutions (in Sections 5.5 and 5.6), and then will extend this knowledge to handle situations where the coefficient matrix is not constant or is nonhomogeneous.

Fundamental Matrices
We consider the homogeneous system in the form

$$\mathbf{x}' = \mathbf{A}\mathbf{x}.$$

We learned in Theorem 5.5.3 that if the vectors $\mathbf{x}_1(t), \mathbf{x}_2(t), \cdots, \mathbf{x}_n(t)$ form a fundamental set of solutions on (a, b), we can write the general solution as a linear combination of these vectors. Equivalently, the general solution is sometimes written as

$$\mathbf{x}(t) = \Phi(t)\mathbf{c} \tag{5.120}$$

where

$$\Phi(t) = \begin{bmatrix} \mathbf{x}_1(t) & \mathbf{x}_2(t) & \cdots & \mathbf{x}_n(t) \end{bmatrix} \quad \text{and} \quad \mathbf{c} = \begin{bmatrix} c_1 \\ c_2 \\ \vdots \\ c_n \end{bmatrix}. \tag{5.121}$$

The matrix $\Phi(t)$ is called a **fundamental matrix** and the vertical lines separate the columns. If we are given an initial condition for the system $\mathbf{x}(t_0) = \mathbf{x}_0$, where $a < t_0 < b$, we must have that

$$\mathbf{x}_0 = \Phi(t_0)\mathbf{c}.$$

Because the matrix has linearly independent columns regardless of t (because the columns are the vectors in the fundamental set of solutions), we know $\Phi(t)$ is always invertible (nonsingular). Thus

$$\mathbf{c} = \Phi^{-1}(t_0)\mathbf{x}_0.$$

Substituting into (5.120) gives

$$\mathbf{x} = \Phi(t)\Phi^{-1}(t_0)\mathbf{x}_0. \tag{5.122}$$

What is the significance of this? In solving systems of equations using the matrix exponential, we had not yet considered a problem with an initial condition. Noting that $\mathbf{c}e^{\mathbf{A}t}$ is also a solution to $\mathbf{x}' = \mathbf{A}\mathbf{x}$ for any constant vector $\mathbf{c}$ and also noting that $e^{\mathbf{A}0} = \mathbf{I}$, we can see that

$$\mathbf{x}' = \mathbf{A}\mathbf{x}, \ \mathbf{x}(t_0) = \mathbf{x}_0 \text{ has solution } e^{\mathbf{A}t}\mathbf{x}_0.$$

Because solutions must be unique for the constant coefficient system, we have

$$e^{\mathbf{A}t} = \Phi(t)\Phi^{-1}(t_0). \tag{5.123}$$

This is another way to calculate the matrix exponential, even though we would want to avoid using this method in practice due to the computation involved in calculating a matrix inverse. More importantly, this equality shows us how the vectors in a fundamental set of solutions relate to the matrix exponential.

Example Use the fundamental matrix to find the exponential of:

$$\mathbf{A} = \begin{bmatrix} 3 & 1 \\ 2 & 4 \end{bmatrix}.$$

Solution We find the eigenvalue-eigenvector pairs of $\mathbf{A}$ are

$$\left\{ 2, \begin{bmatrix} -1 \\ 1 \end{bmatrix} \right\}, \left\{ 5, \begin{bmatrix} 1 \\ 2 \end{bmatrix} \right\}.$$

Then we can calculate Φ as

$$\Phi(t) = \begin{bmatrix} -e^{2t} & e^{5t} \\ e^{2t} & 2e^{5t} \end{bmatrix}.$$

To compute e^{At}, we need to find $\Phi^{-1}(t_0)$ and we take $t_0 = 0$ for convenience (any t_0 will work since the columns are linearly independent for all t):

$$\Phi(0) = \begin{bmatrix} -1 & 1 \\ 1 & 2 \end{bmatrix} \Longrightarrow \Phi^{-1}(0) = \frac{1}{-3}\begin{bmatrix} 2 & -1 \\ -1 & -1 \end{bmatrix}.$$

Finally, we have

$$e^{At} = \Phi\Phi^{-1}(0) = \frac{1}{3}\begin{bmatrix} 2e^{2t} + e^{5t} & -e^{2t} + e^{5t} \\ -2e^{2t} + 2e^{5t} & e^{2t} + 2e^{5t} \end{bmatrix}. \tag{5.124}$$

Problems

In Problems **1–12**, let **A** be the given matrix and then calculate e^{At} without using the built-in computer commands.

1. $\begin{bmatrix} 3 & 0 \\ 0 & -2 \end{bmatrix}$ 2. $\begin{bmatrix} -1 & 0 \\ 0 & 2 \end{bmatrix}$ 3. $\begin{bmatrix} 3 & -2 \\ 0 & -6 \end{bmatrix}$ 4. $\begin{bmatrix} 2 & 1 \\ 0 & 1 \end{bmatrix}$

5. $\begin{bmatrix} 1 & -2 \\ 4 & -5 \end{bmatrix}$ 6. $\begin{bmatrix} 2 & 1 \\ 1 & 2 \end{bmatrix}$ 7. $\begin{bmatrix} 0 & 1 \\ 1 & 0 \end{bmatrix}$ 8. $\begin{bmatrix} 1 & -1 \\ 1 & 1 \end{bmatrix}$

9. $\begin{bmatrix} 3 & 2 \\ -2 & 3 \end{bmatrix}$ 10. $\begin{bmatrix} 1 & -2 \\ 2 & 1 \end{bmatrix}$ 11. $\begin{bmatrix} -1 & -3 \\ 3 & -1 \end{bmatrix}$ 12. $\begin{bmatrix} -2 & 3 \\ -3 & -2 \end{bmatrix}$

In Problems **13–15**, let **A** be the given matrix and then find e^{At} using the built-in computer commands.

13. $\begin{bmatrix} -1 & 1 & 2 \\ 0 & 1 & 0 \\ 1 & 0 & 0 \end{bmatrix}$ 14. $\begin{bmatrix} 1 & 0 & 0 \\ 4 & 3 & -3 \\ 2 & -1 & 1 \end{bmatrix}$ 15. $\begin{bmatrix} 2 & 3 & 2 & -3 \\ -1 & 2 & 1 & 2 \\ 0 & 0 & -3 & 0 \\ -1 & 3 & 0 & 1 \end{bmatrix}$

In Problems **16–18**, let **J** be the given matrix and then calculate $\mathbf{J}^2$, $\mathbf{J}^3$, $\mathbf{J}^4$.

16. $\begin{bmatrix} 2 & 1 \\ 0 & 2 \end{bmatrix}$ 17. $\begin{bmatrix} -1 & 1 \\ 0 & -1 \end{bmatrix}$ 18. $\begin{bmatrix} a & 1 \\ 0 & a \end{bmatrix}$

19. Find $\mathbf{J}^n$ for $\mathbf{J} = \begin{bmatrix} a & 1 \\ 0 & a \end{bmatrix}$.

In Problems **20–27**, let A be the given matrix and then find e^{At}. Then verify that this answer is a solution to the equation $\mathbf{x}' = \mathbf{Ax}$.

20. $\begin{bmatrix} 3 & -2 \\ 0 & -6 \end{bmatrix}$ 21. $\begin{bmatrix} 1 & -2 \\ 4 & -5 \end{bmatrix}$ 22. $\begin{bmatrix} 2 & 1 \\ 1 & 2 \end{bmatrix}$ 23. $\begin{bmatrix} 0 & 1 \\ 1 & 0 \end{bmatrix}$

24. $\begin{bmatrix} 1 & -1 \\ 1 & 1 \end{bmatrix}$ 25. $\begin{bmatrix} 3 & 2 \\ -2 & 3 \end{bmatrix}$ 26. $\begin{bmatrix} 1 & -2 \\ 1 & 3 \end{bmatrix}$ 27. $\begin{bmatrix} 5 & 0 \\ 2 & -3 \end{bmatrix}$

The **Cayley-Hamilton theorem** states that any square matrix satisfies its own characteristic equation. That is, if

$$\det(\mathbf{A} - \lambda\mathbf{I}) = a_0\lambda^n + a_1\lambda^{n-1} + \cdots + a_{n-2}\lambda^2 + a_{n-1}\lambda + a_n \tag{5.125}$$

is the characteristic polynomial of $\mathbf{A}$, then

$$a_0\mathbf{A}^n + a_1\mathbf{A}^{n-1} + \cdots + a_{n-2}\mathbf{A}^2 + a_{n-1}\mathbf{A} + a_n\mathbf{I} = \mathbf{0}. \tag{5.126}$$

(Theorem 5.7.2 was obtained from this theorem.) Show the Cayley-Hamilton theorem holds for the matrices of Problems **28–39** (cf. Problems **1–12** of this section).

28. $\begin{bmatrix} 3 & 0 \\ 0 & -2 \end{bmatrix}$ **29.** $\begin{bmatrix} -1 & 0 \\ 0 & 2 \end{bmatrix}$ **30.** $\begin{bmatrix} 3 & -2 \\ 0 & -6 \end{bmatrix}$ **31.** $\begin{bmatrix} 2 & 1 \\ 0 & 1 \end{bmatrix}$

32. $\begin{bmatrix} 1 & -2 \\ 4 & -5 \end{bmatrix}$ **33.** $\begin{bmatrix} 2 & 1 \\ 1 & 2 \end{bmatrix}$ **34.** $\begin{bmatrix} 0 & 1 \\ 1 & 0 \end{bmatrix}$ **35.** $\begin{bmatrix} 1 & -1 \\ 1 & 1 \end{bmatrix}$

36. $\begin{bmatrix} 3 & 2 \\ -2 & 3 \end{bmatrix}$ **37.** $\begin{bmatrix} 1 & -2 \\ 2 & 1 \end{bmatrix}$ **38.** $\begin{bmatrix} -1 & -3 \\ 3 & -1 \end{bmatrix}$ **39.** $\begin{bmatrix} -2 & 3 \\ -3 & -2 \end{bmatrix}$

5.8 Solving Linear Nonhomogeneous Systems of Equations

The reader should recall that Section 3.4 showed that *any* nth-order equation can be converted to a system of n first-order equations. Thus the methods that we consider in this section apply to any linear differential equation, regardless of whether it has constant or variable coefficients or whether it is homogeneous or nonhomogeneous. In this section, we consider first-order systems of the form

$$\mathbf{x}' = \mathbf{A}(t)\mathbf{x} + \mathbf{F}(t), \tag{5.127}$$

where $\mathbf{F}$ is a vector with continuous functions as its entries.

Constant Coefficient, Constant Forcing

We begin by considering (5.127) when $\mathbf{A}$ has constant entries and the forcing function $\mathbf{F}$ is also constant. The key for this problem is to remember that we need to obtain any particular solution and then add it to our homogeneous solution. One particular solution will occur when $\mathbf{x}' = \mathbf{0}$:

$$\mathbf{0} = \mathbf{A}\mathbf{x} + \mathbf{F}, \Longrightarrow \mathbf{A}\mathbf{x} = -\mathbf{F}. \tag{5.128}$$

This is simply $\mathbf{A}\mathbf{x} = \mathbf{b}$ when $\mathbf{b} = -\mathbf{F}$ and can be solved with Gaussian Elimination. Once we have obtained a particular solution, we can find the homogeneous solution with the methods of Sections 5.5, 5.6, or 5.7.

Example 1 Solve

$$\mathbf{x}' = \begin{bmatrix} -2 & 0 & 4 \\ 4 & 2 & -4 \\ 2 & 4 & 2 \end{bmatrix} \mathbf{x} + \begin{bmatrix} -2 \\ 2 \\ 4 \end{bmatrix}.$$

Solution

We first look for a particular solution by setting $\mathbf{x}' = \mathbf{0}$ and solve:

$$\mathbf{Ax} = -\mathbf{F} \implies \begin{bmatrix} -2 & 0 & 4 \\ 4 & 2 & -4 \\ 2 & 4 & 2 \end{bmatrix} \mathbf{x} = \begin{bmatrix} 2 \\ -2 \\ -4 \end{bmatrix}. \tag{5.129}$$

We can solve with Gaussian elimination to obtain the solution $\mathbf{x}_p = \begin{bmatrix} 5 \\ -5 \\ 3 \end{bmatrix}$.

To obtain the homogeneous solution, we consider $\mathbf{x}' = \mathbf{Ax}$ and use Section 5.5 or 5.7 to write the solution as

$$\mathbf{x}_h = C_1 e^{2t} \begin{bmatrix} 1 \\ \frac{-1}{2} \\ 1 \end{bmatrix} + C_2 \begin{bmatrix} \sin(2t) \\ -\sin(2t) \\ \frac{1}{2}\cos(2t) + \frac{1}{2}\sin(2t) \end{bmatrix}$$

$$+ C_3 \begin{bmatrix} \cos(2t) \\ -\cos(2t) \\ \frac{-1}{2}\sin(2t) + \frac{1}{2}\cos(2t) \end{bmatrix}, \tag{5.130}$$

with the general solution being $\mathbf{x} = \mathbf{x}_h + \mathbf{x}_p$.

Constant Coefficient and Diagonalizable Matrix A

At the end of Section 5.4, we discussed how switching to eigencoordinates allows us to solve a diagonal system in an easier form. (Appendix B.4.1 also discusses constant matrices that can be *diagonalized* in more detail.) In the situations where this can be done, we write the eigenvectors as columns of a matrix $\mathbf{V}$ and observe that $\mathbf{\Lambda} = \mathbf{V}^{-1}\mathbf{AV}$ is a diagonal matrix. If we define a new variable $\mathbf{y}$ by

$$\mathbf{x} = \mathbf{Vy},$$

then the constant coefficient nonhomogeneous system (5.127) can be written as

$$\mathbf{Vy}' = \mathbf{AVy} + \mathbf{F}(t) \implies \mathbf{y}' = \mathbf{V}^{-1}\mathbf{AVy} + \mathbf{V}^{-1}\mathbf{F}(t) = \mathbf{\Lambda y} + \mathbf{V}^{-1}\mathbf{F}(t).$$

This gives a system of *uncoupled equations*, each of which is linear and first-order and thus can be solved by methods of Section 1.3.

Example 2: Solve

$$\mathbf{x}' = \begin{bmatrix} 1 & -4 \\ -2 & -1 \end{bmatrix} \mathbf{x} + \begin{bmatrix} -\sin t \\ e^t \end{bmatrix}.$$

We can use our computer programs to find the eigenvalues and eigenvectors as

$$\lambda_1 = -3, \quad \lambda_2 = 3, \quad \mathbf{v}_1 = \begin{bmatrix} 1 \\ 1 \end{bmatrix}, \quad \mathbf{v}_2 = \begin{bmatrix} -2 \\ 1 \end{bmatrix}.$$

Creating our matrix whose columns are these eigenvectors and calculating its inverse gives

$$\mathbf{V} = \begin{bmatrix} 1 & -2 \\ 1 & 1 \end{bmatrix} \implies \mathbf{V}^{-1} = \frac{1}{3} \begin{bmatrix} 1 & 2 \\ -1 & 1 \end{bmatrix}.$$

By defining a new variable $\mathbf{y} = \mathbf{V}^{-1}\mathbf{x}$, our system becomes

$$\begin{aligned}
\mathbf{y}' &= \mathbf{V}^{-1}\mathbf{A}\mathbf{V}\mathbf{y} + \mathbf{V}^{-1}\mathbf{F}(t) \\
&= \begin{bmatrix} -3 & 0 \\ 0 & 3 \end{bmatrix} \mathbf{y} + \frac{1}{3} \begin{bmatrix} 1 & 2 \\ -1 & 1 \end{bmatrix} \begin{bmatrix} -\sin t \\ e^t \end{bmatrix} \\
&= \begin{bmatrix} -3 & 0 \\ 0 & 3 \end{bmatrix} \mathbf{y} + \frac{1}{3} \begin{bmatrix} -\sin t + 2e^t \\ \sin t + e^t \end{bmatrix}.
\end{aligned} \qquad (5.131)$$

This gives the uncoupled system of linear first-order equations

$$y_1' + 3y_1 = \frac{1}{3}(-\sin t + 2e^t)$$

$$y_2' - 3y_1 = \frac{1}{3}(\sin t + e^t).$$

This gives

$$y_1 = \frac{1}{30}(\cos t - 3\sin t) + \frac{1}{6}e^t + c_1 e^{-3t}$$

$$y_2 = \frac{1}{30}(-\cos t - 3\sin t) - \frac{1}{6}e^t + c_2 e^{3t}.$$

We still need to write this answer in terms of the original variables. Doing so gives us

$$\begin{aligned}
\mathbf{x} = \mathbf{V}\mathbf{y} &= \begin{bmatrix} 1 & -2 \\ 1 & 1 \end{bmatrix} \begin{bmatrix} y_1 \\ y_2 \end{bmatrix} = \begin{bmatrix} y_1 - 2y_2 \\ y_1 + y_2 \end{bmatrix} \\
&= \begin{bmatrix} \frac{1}{10}(\cos t + \sin t) + \frac{1}{2}e^t + e^{-3t}c_1 - 2e^{3t}c_2 \\ \frac{-1}{5}\sin t + e^{-3t}c_1 + e^{3t}c_2 \end{bmatrix} \\
&= c_1 e^{-3t} \begin{bmatrix} 1 \\ 1 \end{bmatrix} + c_2 e^{3t} \begin{bmatrix} -2 \\ 1 \end{bmatrix} \\
&\quad + e^t \begin{bmatrix} 1/2 \\ 0 \end{bmatrix} + \cos t \begin{bmatrix} 1/10 \\ 0 \end{bmatrix} + \sin t \begin{bmatrix} 1/10 \\ -1/5 \end{bmatrix}. \qquad (5.132)
\end{aligned}$$

The form of this solution should look familiar in that it is composed of the complementary solution (i.e., the solution to the system when $\mathbf{F}(t) = 0$) and a particular solution that is determined by the function $\mathbf{F}(t)$.

Matrix Exponential

In Section 1.3 we saw that a linear system of the form $x' + ax = f(t)$ has the solution

$$ce^{at} + e^{at} \int e^{-at} f(t)dt.$$

It can be shown that an analogous formulation of this solution exists in the case of (5.127). Thus $\mathbf{x}' = \mathbf{A}\mathbf{x} + \mathbf{F}(t)$ has solution

$$\mathbf{x} = e^{\mathbf{A}t}\mathbf{c} + e^{\mathbf{A}t} \int e^{-\mathbf{A}t}\mathbf{F}(t)dt. \tag{5.133}$$

(It is easiest to understand the derivation of this solution in terms of a general fundamental matrix and the method of variation of parameters, which we examine next.) If we also have an initial condition, the solution can be written to take this into account. Thus

$$\mathbf{x}' = \mathbf{A}\mathbf{x} + \mathbf{F}(t), \quad \mathbf{x}(t_0) = \mathbf{x}_0 \tag{5.134}$$

has solution

$$\mathbf{x} = e^{\mathbf{A}(t-t_0)}\mathbf{x}_0 + e^{\mathbf{A}t} \int_{t_0}^{t} e^{-\mathbf{A}s}\mathbf{F}(s)ds. \tag{5.135}$$

We again note that the form of this solution is composed of a complementary solution and a particular solution.

Example 3: Use the matrix exponential to solve

$$\mathbf{x}' = \begin{bmatrix} 1 & -4 \\ -2 & -1 \end{bmatrix} \mathbf{x} + \begin{bmatrix} -\sin t \\ e^t \end{bmatrix}.$$

Solution
We can calculate the matrix exponential by the methods of the previous section or by using the computer. If we choose the latter, we see that

$$e^{\mathbf{A}t} = \frac{1}{3} \begin{bmatrix} e^{-3t} + 2e^{3t} & 2e^{-3t} - 2e^{3t} \\ e^{-3t} - e^{3t} & 2e^{-3t} + e^{3t} \end{bmatrix}.$$

We also need to calculate $\int e^{-\mathbf{A}t}\mathbf{F}(t)dt$. We note that $e^{-\mathbf{A}t}$ is easily calculated from $e^{\mathbf{A}t}$ by simply replacing t with $-t$. Then we have

$$\int e^{-\mathbf{A}t} \mathbf{F}(t) dt$$

$$= \int \frac{1}{3} \begin{bmatrix} e^{3t} + 2e^{-3t} & 2e^{3t} - 2e^{-3t} \\ e^{3t} - e^{-3t} & 2e^{3t} + e^{-3t} \end{bmatrix} \begin{bmatrix} -\sin t \\ e^t \end{bmatrix}$$

$$= \int \frac{1}{3} \begin{bmatrix} -2e^{-3t}\sin t - e^{3t}\sin t + 2e^{4t} - 2e^{-2t} \\ -e^{3t}\sin t + e^{-3t}\sin t + e^{-2t} + 2e^{4t} \end{bmatrix}$$

$$= \frac{1}{30} \begin{bmatrix} 2e^{-3t}\cos t + 6e^{-3t}\sin t + e^{3t}\cos t - 3e^{3t}\sin t + 5e^{4t} + 10e^{-2t} \\ e^{3t}\cos t - 3e^{3t}\sin t - e^{-3t}\cos t - 3e^{-3t}\sin t - 5e^{-2t} + 5e^{4t} \end{bmatrix}.$$

We then left multiply by $e^{\mathbf{A}t}$ and obtain, after much simplification,

$$\frac{1}{10} \begin{bmatrix} \cos t + \sin t + 5e^t \\ -2\sin t \end{bmatrix}.$$

We add this last vector to the product of $e^{\mathbf{A}t}$ with an arbitrary vector. In the formula, we gave the arbitrary constant vector as $\mathbf{c}$ but for comparison purposes (with the previous example), we let $\mathbf{k}$ be our arbitrary constant vector.

$$\mathbf{x} = \frac{1}{3} \begin{bmatrix} e^{-3t} + 2e^{3t} & 2e^{-3t} - 2e^{3t} \\ e^{-3t} - e^{3t} & 2e^{-3t} + e^{3t} \end{bmatrix} \begin{bmatrix} k_1 \\ k_2 \end{bmatrix} + \frac{1}{10} \begin{bmatrix} \cos t + \sin t + 5e^t \\ -2\sin t \end{bmatrix}$$

$$= \frac{1}{3} \begin{bmatrix} (k_1 + 2k_2)e^{-3t} + (2k_1 - 2k_2)e^{3t} \\ (k_1 + 2k_2)e^{-3t} + (-k_1 + k_2)e^{3t} \end{bmatrix} + \frac{1}{10} \begin{bmatrix} \cos t + \sin t + 5e^t \\ -2\sin t \end{bmatrix}.$$

This solution is completely correct. However, so as to have the solution in the form given in Example 1, we let $c_1 = (k_1 + 2k_2)/3$ and $c_2 = (-k_1 + k_2)/3$. Then our answer can be written as

$$\mathbf{x} = c_1 e^{-3t} \begin{bmatrix} 1 \\ 1 \end{bmatrix} + c_2 e^{3t} \begin{bmatrix} -2 \\ 1 \end{bmatrix}$$

$$+ e^t \begin{bmatrix} 1/2 \\ 0 \end{bmatrix} + \cos t \begin{bmatrix} 1/10 \\ 0 \end{bmatrix} + \sin t \begin{bmatrix} 1/10 \\ -1/5 \end{bmatrix}. \tag{5.136}$$

Non-Constant Coefficient Systems

When we considered an nth-order equation with non-constant coefficients, we stated that there was no general formula for calculating a fundamental set of solutions. However, *if we somehow obtained* a fundamental set of solutions, we could solve the nonhomogeneous equation by using variation of parameters. The *method of variation of parameters* extends to first-order systems of equations. Unfortunately, so does the fact that there is no general formula for obtaining a fundamental set of solutions. Thus we will be able to solve a nonhomogeneous problem only if we somehow can obtain a fundamental set of solutions. In the case when the coefficients of the matrix are periodic, there is a technique called *Floquet theory* that can help us in this quest for a fundamental set. But in general, we will not be so lucky. Let us begin with a system of the form

$$\mathbf{x}' = \mathbf{A}(t)\mathbf{x} + \mathbf{F}(t). \tag{5.137}$$

We will assume that we somehow are able to obtain a fundamental matrix $\Phi(t)$ for the homogeneous system (i.e., when $\mathbf{F}(t) = \mathbf{0}$). Analogous to Section 4.5, we use variation of parameters to try to obtain a particular solution. That is, we try to find a vector $\mathbf{u}(t)$ such that

$$\mathbf{x}_p = \Phi(t)\mathbf{u}(t)$$

is a (particular) solution. We note that the dimensions of Φ and $\mathbf{u}$ require that the order of the multiplication is as given here. We will need the derivative of this to substitute into (5.137). Calculating this gives

$$\mathbf{x}_p' = \Phi'(t)\mathbf{u}(t) + \Phi(t)\mathbf{u}'(t),$$

and we again note that the order of the multiplication matters. Substitution into (5.137) gives us

$$\Phi'(t)\mathbf{u}(t) + \Phi(t)\mathbf{u}'(t) = \mathbf{A}(t)\Phi(t)\mathbf{u}(t) + \mathbf{F}(t). \tag{5.138}$$

We can arrange this as

$$[\Phi'(t) - \mathbf{A}(t)\Phi(t)]\,\mathbf{u}(t) + \Phi(t)\mathbf{u}'(t) = \mathbf{F}(t).$$

Because $\Phi(t)$ is a solution, this simplifies to

$$\Phi(t)\mathbf{u}'(t) = \mathbf{F}(t), \tag{5.139}$$

which we can solve for $\mathbf{u}'(t)$ since $\Phi(t)$ is a fundamental matrix (and thus is invertible):

$$\mathbf{u}'(t) = \Phi^{-1}(t)\mathbf{F}(t). \tag{5.140}$$

We can, in theory, integrate both sides of this last equation and thus obtain $\mathbf{u}(t)$. In practice, this integral may be difficult to evaluate but we will still be able to write our solution with an integral in it. Integrating gives

$$\mathbf{u}(t) = \int \Phi^{-1}(t)\mathbf{F}(t)\ dt, \tag{5.141}$$

and thus the particular solution is of the form

$$\mathbf{x}_p(t) = \Phi(t) \int \Phi^{-1}(t)\mathbf{F}(t)\ dt. \tag{5.142}$$

The general solution to (5.137) can be written as

$$\mathbf{x}(t) = \Phi(t)\mathbf{c} + \Phi(t) \int \Phi^{-1}(t)\mathbf{F}(t)\ dt. \tag{5.143}$$

We now use variation of parameters to solve the problems of Examples 1 and 2.

Example 3 Use variation of parameters to solve

$$\mathbf{x}' = \begin{bmatrix} 1 & -4 \\ -2 & -1 \end{bmatrix} \mathbf{x} + \begin{bmatrix} -\sin t \\ e^t \end{bmatrix}.$$

Solution

In Example 1, we found the eigenvalues and eigenvectors to be

$$\lambda_1 = -3, \quad \lambda_2 = 3, \quad \mathbf{v}_1 = \begin{bmatrix} 1 \\ 1 \end{bmatrix}, \quad \mathbf{v}_2 = \begin{bmatrix} -2 \\ 1 \end{bmatrix}.$$

We can use these to create a fundamental matrix

$$\Phi = \begin{bmatrix} e^{-3t} & -2e^{3t} \\ e^{-3t} & e^{3t} \end{bmatrix}.$$

The inverse is easily calculated because we are in the 2×2 case:

$$\Phi^{-1}(t) = \frac{1}{3} \begin{bmatrix} e^{3t} & 2e^{3t} \\ -e^{-3t} & e^{-3t} \end{bmatrix}.$$

Applying the formula for the solution and getting some help from our computer programs, we have

$$\Phi^{-1}(t)\mathbf{F}(t) = \frac{1}{3} \begin{bmatrix} -e^{3t}\sin t + 2e^{4t} \\ e^{-3t}\sin t + e^{-2t} \end{bmatrix}$$

$$\implies \int \Phi^{-1}(t)\mathbf{F}(t)\, dt = \frac{1}{30} \begin{bmatrix} e^{3t}\cos t - 3e^{3t}\sin t + 5e^{4t} \\ -e^{-3t}\cos t - 3e^{-3t}\sin t - 5e^{-2t} \end{bmatrix}.$$

Left multiplication of this result by $\Phi(t)$ and lots of simplification give us

$$\Phi(t) \int \Phi^{-1}(t)\mathbf{F}(t)\, dt = \frac{1}{10} \begin{bmatrix} \cos t + \sin t + 5e^t \\ -2\sin t \end{bmatrix}.$$

Thus, from (5.143), we can write our general solution as

$$\mathbf{x} = \begin{bmatrix} e^{-3t} & -2e^{3t} \\ e^{-3t} & e^{3t} \end{bmatrix} \begin{bmatrix} c_1 \\ c_2 \end{bmatrix} + \frac{1}{10} \begin{bmatrix} \cos t + \sin t + 5e^t \\ -2\sin t \end{bmatrix}$$

$$= c_1 e^{-3t} \begin{bmatrix} 1 \\ 1 \end{bmatrix} + c_2 e^{3t} \begin{bmatrix} -2 \\ 1 \end{bmatrix}$$

$$+ \cos t \begin{bmatrix} 1/10 \\ 0 \end{bmatrix} + \sin t \begin{bmatrix} 1/10 \\ -1/5 \end{bmatrix} + e^t \begin{bmatrix} 1/2 \\ 0 \end{bmatrix}. \tag{5.144}$$

As expected, this is the same answer as we obtained in Examples 1 and 2.

In Problems **1–6**, solve the following homogeneous systems with one (or more, if instructed) of the methods of this section.

1. $\mathbf{x}' = \begin{bmatrix} -3 & -2 \\ 3 & 4 \end{bmatrix} \mathbf{x}$ **2.** $\mathbf{x}' = \begin{bmatrix} -5 & -2 \\ 1 & -2 \end{bmatrix} \mathbf{x}$ **3.** $\mathbf{x}' = \begin{bmatrix} 4 & 0 \\ -1 & 2 \end{bmatrix} \mathbf{x}$

4. $\mathbf{x}' = \begin{bmatrix} 3 & 2 \\ -1 & 0 \end{bmatrix} \mathbf{x}$ **5.** $\mathbf{x}' = \begin{bmatrix} 1 & 0 & -2 \\ 1 & 2 & 5 \\ 0 & 0 & -1 \end{bmatrix} \mathbf{x}$ **6.** $\mathbf{x}' = \begin{bmatrix} -1 & 2 & 0 \\ 0 & 1 & 0 \\ 4 & 0 & 3 \end{bmatrix} \mathbf{x}$

In Problems **7–14**, solve the following nonhomogeneous systems with one (or more, if instructed) of the methods of this section.

7. $\mathbf{x}' = \begin{bmatrix} -3 & -2 \\ 3 & 4 \end{bmatrix} \mathbf{x} + \begin{bmatrix} 2 \\ 1 \end{bmatrix}$ **8.** $\mathbf{x}' = \begin{bmatrix} -5 & -2 \\ 1 & -2 \end{bmatrix} \mathbf{x} + \begin{bmatrix} 1 \\ 3 \end{bmatrix}$

9. $\mathbf{x}' = \begin{bmatrix} 4 & 0 \\ -1 & 2 \end{bmatrix} \mathbf{x} + \begin{bmatrix} \sin t \\ 0 \end{bmatrix}$ **10.** $\mathbf{x}' = \begin{bmatrix} 3 & 2 \\ -1 & 0 \end{bmatrix} \mathbf{x} + \begin{bmatrix} 0 \\ \cos t \end{bmatrix}$

11. $\mathbf{x}' = \begin{bmatrix} 1 & 0 & -2 \\ 1 & 2 & 5 \\ 0 & 0 & -1 \end{bmatrix} \mathbf{x} + \begin{bmatrix} e^t \\ \sin t \\ 0 \end{bmatrix}$ **12.** $\mathbf{x}' = \begin{bmatrix} -1 & 2 & 0 \\ 0 & 1 & 0 \\ 4 & 0 & 3 \end{bmatrix} \mathbf{x} + \begin{bmatrix} 0 \\ e^t \\ \cos t \end{bmatrix}$

13. $\mathbf{x}' = \begin{bmatrix} -1 & 0 & 3 \\ 0 & 1 & 0 \\ 1 & 0 & 1 \end{bmatrix} \mathbf{x} + \begin{bmatrix} \sin t \\ 0 \\ e^t \end{bmatrix}$ **14.** $\mathbf{x}' = \begin{bmatrix} 0 & 2 & 1/2 \\ 1 & 1 & 1/2 \\ 2 & 0 & 2 \end{bmatrix} \mathbf{x} + \begin{bmatrix} 0 \\ e^t \\ \cos t \end{bmatrix}$

Chapter 5 Review

In Problems **1–8**, determine whether the statement is true or false. If it is true, give reasons for your answer. If it is false, give a counterexample or other explanation of why it is false.

1. All linear systems of differential equations can be solved by first finding the eigenvalues and eigenvectors of the coefficient matrix and then taking a linear combination of $e^{\lambda_i t} \mathbf{v}_i$.

2. If the vectors $\mathbf{x}_1, \mathbf{x}_2, \ldots, \mathbf{x}_n$ are constant vectors (i.e., with no dependence on t), then the Wronskian equal to zero implies that the set of vectors is linearly dependent.

3. Every 2×3 matrix has the zero vector as the only element in its column space.

4. Every 3×2 matrix with linearly independent columns has only the zero vector in its nullspace.

5. A linear system of differential equations with repeated complex eigenvalues cannot be solved.

6. Consider an nth-order linear differential equation (studied in Chapter 4) that is written as a system of first-order equations (via the methods of Section 3.4). The solution obtained to the nth-order equation by the methods of Chapter 4 is mathematically equivalent to the solution obtained by solving the system via methods of Chapter 5.

7. One calculates the derivative of a matrix by taking the derivative of each element of the matrix.

8. A basis can be found so that every matrix can be diagonalized or written in Jordan normal form with respect to this basis.

In Problems **9–16**, determine whether the given set forms a fundamental set of solutions for the given differential equation.

9. $\left\{ \begin{bmatrix} e^t \\ 2e^t \end{bmatrix}, \begin{bmatrix} e^{-t} \\ 3e^{-t} \end{bmatrix} \right\}$, $\mathbf{x}' = \begin{bmatrix} 5 & -2 \\ 12 & -5 \end{bmatrix} \mathbf{x}$

10. $\left\{ \begin{bmatrix} 3e^{2t} \\ 2e^{2t} \end{bmatrix}, \begin{bmatrix} e^{-3t} \\ e^{-3t} \end{bmatrix} \right\}$, $\mathbf{x}' = \begin{bmatrix} 1 & -2 \\ 2 & -3 \end{bmatrix} \mathbf{x}$

11. $\left\{ \begin{bmatrix} e^{3t} \\ 2e^{3t} \end{bmatrix}, \begin{bmatrix} -e^{6t} \\ e^{6t} \end{bmatrix} \right\}$, $\mathbf{x}' = \begin{bmatrix} 5 & -1 \\ -2 & 4 \end{bmatrix} \mathbf{x}$

12. $\left\{ \begin{bmatrix} 1 \\ 1 \end{bmatrix} e^{-4t}, \begin{bmatrix} -1 \\ 1 \end{bmatrix} e^{2t} \right\}$, $\mathbf{x}' = \begin{bmatrix} -1 & -3 \\ -3 & -1 \end{bmatrix} \mathbf{x}$

13. $\left\{ \begin{bmatrix} 2 \\ 1 \end{bmatrix} e^{t}, \begin{bmatrix} 1 \\ 3 \end{bmatrix} e^{-2t} \right\}$, $\mathbf{x}' = \begin{bmatrix} 3 & 1 \\ 2 & -1 \end{bmatrix} \mathbf{x}$

14. $\left\{ \begin{bmatrix} 1 \\ 2 \end{bmatrix} e^{-4t}, \begin{bmatrix} 1 \\ -2 \end{bmatrix} e^{-8t} \right\}$, $\mathbf{x}' = \begin{bmatrix} -6 & 1 \\ 4 & -6 \end{bmatrix} \mathbf{x}$

15. $\left\{ \begin{bmatrix} 1 \\ -1 \\ 1 \end{bmatrix} e^{5t}, \begin{bmatrix} 1 \\ 1 \\ 1 \end{bmatrix} e^{t}, \begin{bmatrix} 1 \\ 1 \\ -1 \end{bmatrix} e^{-t} \right\}$, $\mathbf{x}' = \begin{bmatrix} 2 & -2 & 1 \\ -3 & 3 & 1 \\ 3 & -2 & 0 \end{bmatrix} \mathbf{x}$

16. $\left\{ \begin{bmatrix} 1 \\ -1 \\ 1 \end{bmatrix} e^{5t}, \begin{bmatrix} 1 \\ 1 \\ 1 \end{bmatrix} e^{t}, \begin{bmatrix} -1 \\ 1 \\ 1 \end{bmatrix} e^{2t} \right\}$, $\mathbf{x}' = \begin{bmatrix} 4 & -2 & -1 \\ -1 & 3 & -1 \\ 1 & -2 & 2 \end{bmatrix} \mathbf{x}$

In Problems **17–25**, determine whether the given set of vectors forms a basis for the specified vector space.

17. $\left\{ \begin{bmatrix} 2 \\ 1 \end{bmatrix}, \begin{bmatrix} 4 \\ -1 \end{bmatrix} \right\}, \mathbb{R}^2$

18. $\left\{ \begin{bmatrix} 1 \\ -1 \end{bmatrix}, \begin{bmatrix} 3 \\ 4 \end{bmatrix} \right\}, \mathbb{R}^2$

19. $\left\{ \begin{bmatrix} 5 \\ -2 \end{bmatrix}, \begin{bmatrix} -2 \\ -3 \end{bmatrix} \right\}, \mathbb{R}^2$

20. $\left\{ \begin{bmatrix} 2 \\ 4 \end{bmatrix}, \begin{bmatrix} -4 \\ -8 \end{bmatrix} \right\}, \mathbb{R}^2$

21. $\left\{ \begin{bmatrix} 1 \\ 1 \\ 0 \end{bmatrix}, \begin{bmatrix} 2 \\ 0 \\ -1 \end{bmatrix}, \begin{bmatrix} 4 \\ 2 \\ -1 \end{bmatrix} \right\}, \mathbb{R}^3$

22. $\left\{ \begin{bmatrix} 1 \\ 1 \\ 0 \end{bmatrix}, \begin{bmatrix} 2 \\ 0 \\ -1 \end{bmatrix}, \begin{bmatrix} 3 \\ 2 \\ 2 \end{bmatrix} \right\}, \mathbb{R}^3$

23. $\left\{ \begin{bmatrix} 1 \\ 1 \\ 0 \end{bmatrix}, \begin{bmatrix} 3 \\ 0 \\ -1 \end{bmatrix}, \begin{bmatrix} 1 \\ -2 \\ -1 \end{bmatrix}, \begin{bmatrix} 1 \\ 3 \\ -1 \end{bmatrix} \right\}, \mathbb{R}^3$

24. $\left\{ \begin{bmatrix} 1 \\ 1 \\ -1 \\ 1 \end{bmatrix}, \begin{bmatrix} 2 \\ 0 \\ -1 \\ -3 \end{bmatrix}, \begin{bmatrix} 2 \\ 2 \\ 1 \\ 2 \end{bmatrix}, \begin{bmatrix} 1 \\ 0 \\ 2 \\ 0 \end{bmatrix} \right\}, \mathbb{R}^4$

25. $\left\{ \begin{bmatrix} 1 \\ 0 \\ -1 \\ 2 \end{bmatrix}, \begin{bmatrix} 0 \\ 0 \\ -1 \\ 1 \end{bmatrix}, \begin{bmatrix} 2 \\ 2 \\ 1 \\ 2 \end{bmatrix}, \begin{bmatrix} 1 \\ 0 \\ 0 \\ 0 \end{bmatrix} \right\}, \mathbb{R}^4$

In Problems **26–31**, find bases for the nullspace and column space of the given matrix. Ask your instructor if this should be done by hand or with MATLAB, Maple, or Mathematica.

26. $\begin{bmatrix} -1 & 3 \\ 2 & -6 \\ -1 & 1 \end{bmatrix}$

27. $\begin{bmatrix} -2 & 4 \\ 2 & -4 \\ -1 & 2 \end{bmatrix}$

28. $\begin{bmatrix} -2 & 0 & 2 \\ -4 & -2 & 2 \\ 1 & 4 & 3 \end{bmatrix}$

29. $\begin{bmatrix} -1 & 2 & 2 \\ -1 & 1 & 2 \end{bmatrix}$

30. $\begin{bmatrix} 5 & -2 & 3 \\ -2 & 0 & 2 \end{bmatrix}$

31. $\begin{bmatrix} -1 & 2 & 2 \\ -1 & 7 & 8 \\ 2 & 1 & 2 \end{bmatrix}$

In Problems **32–47**, (i) find the eigenvalues and eigenvector of the coefficient matrix by hand; (ii) find two linearly independent solutions; (iii) then use Theorem 5.5.3 to find the general solution. Problems **32–39** have real, distinct roots and Problems **40–47** have complex (non-real) roots.

32. $\begin{cases} x' = 2x \\ y' = x + y \end{cases}$

33. $\begin{cases} x' = x + y \\ y' = 3x - y \end{cases}$

34. $\begin{cases} x' = 6x - y \\ y' = 5x \end{cases}$

35. $\begin{cases} x' = x + 2y \\ y' = -6x - 6y \end{cases}$

36. $\begin{cases} x' = -11x + 6y \\ y' = -18x + 10y \end{cases}$

37. $\begin{cases} x' = y \\ y' = -2x - 3y \end{cases}$

38. $\begin{cases} x' = -10x - y \\ y' = -4x - 10y \end{cases}$

39. $\begin{cases} x' = -6x + y \\ y' = -5x \end{cases}$

40. $\begin{cases} x' = 8y \\ y' = -2x \end{cases}$

41. $\begin{cases} x' = x + 3y \\ y' = -3x + y \end{cases}$

42. $\begin{cases} x' = 7x - 4y \\ y' = 4x + 7y \end{cases}$

43. $\begin{cases} x' = -y \\ y' = 16x \end{cases}$

44. $\begin{cases} x' = -2x - y \\ y' = 13x + 4y \end{cases}$

45. $\begin{cases} x' = 5x + 10y \\ y' = -x - y \end{cases}$

46. $\begin{cases} x' = -x + 4y \\ y' = -2x + 3y \end{cases}$

47. $\begin{cases} x' = 3x + 2y \\ y' = -2x + 5y \end{cases}$

In Problems **48–50**, use the given substitutions to rewrite the equations as an uncoupled system. Then write the solution in terms of u_1, u_2 and use this to find the solution in terms of x, y.

48. $u_1 = -5x + 2y, u_2 = 3x - y$ for $\begin{cases} x' = -7x + 2y \\ y' = -15x + 4y \end{cases}$

49. $u_1 = x - y, u_2 = -x + 2y$ for $\begin{cases} x' = -6x + 8y \\ y' = -4x + 6y \end{cases}$

50. $u_1 = 2x - y, u_2 = -2x + 2y$ for $\begin{cases} x' = 4x - y \\ y' = 2x + y \end{cases}$

In Problems **51–56**, find the eigenvalues and eigenvectors of the following matrices. Get clarification from your instructor as to whether you should find these by hand or by using MATLAB, Maple, or Mathematica.

51. $\begin{bmatrix} 4 & -8 & -10 \\ -1 & 6 & 5 \\ 1 & -8 & -7 \end{bmatrix}$ **52.** $\begin{bmatrix} 0 & -2 & 0 \\ 1 & 3 & 0 \\ -1 & -2 & 1 \end{bmatrix}$ **53.** $\begin{bmatrix} 2 & -2 & 1 \\ -3 & 3 & 1 \\ 3 & -2 & 0 \end{bmatrix}$

54. $\begin{bmatrix} 1 & -2 & 2 \\ -4 & 3 & 2 \\ 4 & -2 & -1 \end{bmatrix}$ **55.** $\begin{bmatrix} 1 & -2 & -1 \\ -1 & 0 & -1 \\ 1 & 1 & 2 \end{bmatrix}$ **56.** $\begin{bmatrix} 4 & -2 & -1 \\ -1 & 3 & -1 \\ 1 & -2 & 2 \end{bmatrix}$

For Problems **57–62**, (i) find the eigenvalues and eigenvector of the coefficient matrix by hand (the eigenvalues are all repeated with only one eigenvector); (ii) obtain a generalized eigenvector; (iii) then use Theorem 5.6.3 and Theorem 5.5.3 to write the general solution.

57. $\begin{cases} x' = -x + y \\ y' = -y \end{cases}$ **58.** $\begin{cases} x' = x + y \\ y' = -9x - 5y \end{cases}$ **59.** $\begin{cases} x' = 7x - 4y \\ y' = 9x - 5y \end{cases}$

60. $\begin{cases} x' = -4x + 12y \\ y' = -3x + 8y \end{cases}$ **61.** $\begin{cases} x' = 3x + 2y \\ y' = 3y \end{cases}$ **62.** $\begin{cases} x' = x + 4y \\ y' = -x + 5y \end{cases}$

In Problems **63–68**, use MATLAB, Maple, or Mathematica to find the eigenvalues and eigenvectors of the coefficient matrix (there is a complex pair). Then use the same program to find the general solution.

63. $\begin{cases} x' = -x - 2y + z \\ y' = 2x - y + z \\ z' = z \end{cases}$ **64.** $\begin{cases} x' = 2x - y - 2z \\ y' = x + y + z \\ z' = z \end{cases}$ **65.** $\begin{cases} x' = -x + y - z \\ y' = -y \\ z' = x - z \end{cases}$

66. $\begin{cases} x' = -x + z \\ y' = -y + z \\ z' = -y - z \end{cases}$ **67.** $\begin{cases} x' = -3x + y + 3z \\ y' = -5x - y + 3z \\ z' = -4y + 2z \end{cases}$ **68.** $\begin{cases} x' = -x + y - z \\ y' = -y \\ z' = x - 2z \end{cases}$

For Problems **69–80**, find the eigenvalues and eigenvector (all are real-valued, some repeated roots) of the coefficient matrix either by hand or using MATLAB, Maple, or Mathematica. Then use the methods of Sections 5.5 or 5.6 to obtain three linearly independent solutions and write the general solution.

69. $\begin{cases} x' = 2x - 2y + z \\ y' = -3x + 3y + z \\ z' = 3x - 2y \end{cases}$ **70.** $\begin{cases} x' = 4x - 8y - 10z \\ y' = -x + 6y + 5z \\ z' = x - 8y - 7z \end{cases}$

71. $\begin{cases} x' = x - 2y + 2z \\ y' = -4x + 3y + 2z \\ z' = 4x - 2y - z \end{cases}$ **72.** $\begin{cases} x' = -11x + 4y \\ y' = 8x - 25y \\ z' = -6x + 12y - 9z \end{cases}$

73. $\begin{cases} x' = -2y \\ y' = x + 3y \\ z' = -x - 2y + z \end{cases}$

74. $\begin{cases} x' = 2x - 16y + 6z \\ y' = -3x + 4y - 3z \\ z' = -4x + 19y - 8z \end{cases}$

75. $\begin{cases} x' = -x + z \\ y' = -3y + z \\ z' = -y - z \end{cases}$

76. $\begin{cases} x' = -8x - 8y - 2z \\ y' = -13x + 14y - z \\ z' = 12x - 15y - 6z \end{cases}$

77. $\begin{cases} x' = -5x + 7y + 14z \\ y' = 4x - 14y - 4z \\ z' = 5x + 8y + 4z \end{cases}$

78. $\begin{cases} x' = -2x + y + 2z \\ y' = 3x - 6y - 6z \\ z' = -2x + y - z \end{cases}$

79. $\begin{cases} x' = 7x + y + 2z \\ y' = 3x + 3y - 6z \\ z' = -2x + y + 8z \end{cases}$

80. $\begin{cases} x' = -5x + 2y + z \\ y' = -x + y + 2z \\ z' = -2x - 4y - 5z \end{cases}$

In Problems **81–88**, let $\mathbf{A}$ be the given matrix and obtain $e^{\mathbf{A}t}$ (either by hand or with the built-in computer commands). Then verify that this answer is a solution to the equation $\mathbf{x}' = \mathbf{A}\mathbf{x}$.

81. $\begin{bmatrix} 2 & 0 \\ 0 & 1 \end{bmatrix}$ **82.** $\begin{bmatrix} -3 & 0 \\ 0 & 4 \end{bmatrix}$ **83.** $\begin{bmatrix} 3 & 2 \\ 0 & 1 \end{bmatrix}$ **84.** $\begin{bmatrix} 1 & 0 \\ 1 & -1 \end{bmatrix}$

85. $\begin{bmatrix} 3 & 1 \\ 2 & 2 \end{bmatrix}$ **86.** $\begin{bmatrix} 3 & -1 \\ 2 & 0 \end{bmatrix}$ **87.** $\begin{bmatrix} 3 & -1 \\ 1 & 3 \end{bmatrix}$ **88.** $\begin{bmatrix} -2 & -3 \\ 3 & -2 \end{bmatrix}$

In Problems **89–93**, solve the following systems of equations.

89. $\begin{cases} x' = x + y + 3 \\ y' = 3x - y + 1 \end{cases}$ **90.** $\begin{cases} x' = 2x + e^t \\ y' = x + y \end{cases}$ **91.** $\begin{cases} x' = 6x - y + 1 \\ y' = 5x + \sin t \end{cases}$

92. $\begin{cases} x' = 2x - 2y + z \\ y' = -3x + 3y + z \\ z' = 3x - 2y \end{cases}$ **93.** $\begin{cases} x' = 4x - 8y - 10z + e^t \\ y' = -x + 6y + 5z + e^t \\ z' = x - 8y - 7z + e^t \end{cases}$

Chapter 5 Computer Labs

Chapter 5 Computer Lab: MATLAB

Code on solving the linear system $\mathbf{A}\mathbf{x} = \mathbf{b}$ with the built-in functions and with elementary row operations can be found at the end of Appendix B.

MATLAB Example 1: Enter the following code that calculates the eigenvalues and eigenvectors of the matrix $\begin{bmatrix} 3 & 0 & -2 \\ 0 & 1 & 2 \\ 0 & -2 & 1 \end{bmatrix}$.

```
>> clear all
>> A=[3, 0, -2; 0, 1, 2; 0, -2, 1]
>> lambda=eig(A) %calculates eigenvalues of A
>> [v,d]=eig(A) %calculates eigenvalues AND eigenvectors of A
>> %eigenvectors given as columns of v
>> %corresponding eigenvalues are on diagonal of d
>> lambda=eig(sym(A)) %uses sym to find and display eigenvalues
>> [v,d]=eig(sym(A))
>> d(1,1) %first eigenvalue of A
>> v(:,1) %corresponding first eigenvector of A
>> A*v(:,1)
>> d(1,1)*v(:,1) %verifies that A*v=lambda*v for first
>> % eigenvalue-eigenvector pair; the last two
>> %answers should be identical
>> A*v(:,3)
>> d(3,3)*v(:,3) %verifies that A*v=lambda*v for third
```

MATLAB Example 2: Enter the following code that solves the ODE system

$$\begin{bmatrix} x' \\ y' \\ z' \end{bmatrix} = \begin{bmatrix} -2 & -1 & -2 \\ -4 & -5 & 2 \\ -5 & -1 & 1 \end{bmatrix} \begin{bmatrix} x \\ y \\ z \end{bmatrix}$$

subject to the IC $x(0) = 2$, $y(0) = -1$, $z(0) = -3$ and then plots the solutions vs. t. The eigenvalues of the system are real.

```
>> clear all
>> A=[-2,-1,-2; -4,-5,2; -5,-1,1]
>> [v,d]=eig(sym(A))
>> syms t c1 c2 c3
>> %Type the line below on a single line, not with a <return>.
>> soln=[exp(d(1,1)*t)*v(:,1),exp(d(2,2)*t)*v(:,2),
   exp(d(3,3)*t)*v(:,3)]
>> %general soln is soln*[c1; c2; c3]
>> soln0=subs(soln,t,0)
>> x0=[2; -1; -3]
>> cvals=soln0\x0
>> v1n=v(:,1)*cvals(1)
>> v2n=v(:,2)*cvals(2)
>> v3n=v(:,3)*cvals(3)
>> Vmat=[v1n,v2n,v3n]
>> xvec=[exp(d(1,1)*t); exp(d(2,2)*t); exp(d(3,3)*t)]
>> soln=Vmat*xvec %the solution to the IVP
```

```
>> h1=ezplot(soln(1),[0,2]);
>> set(h1,'Color','m','LineStyle',':','LineWidth',2)
>> hold on
>> h2=ezplot(soln(2),[0,2]);
>> set(h2,'Color','b','LineStyle','-','LineWidth',2)
>> h3=ezplot(soln(3),[0,2]);
>> set(h3,'Color','k','LineWidth',2)
>> axis([0 2 -400 400])
>> xlabel('t')
>> ylabel('Solution values')
>> legend('x(t)','y(t)','z(t)')
>> hold off
```

MATLAB Example 3: Enter the following code that solves the ODE system

$$\begin{bmatrix} x' \\ y' \end{bmatrix} = \begin{bmatrix} -1 & -2 \\ 2 & -1 \end{bmatrix} \begin{bmatrix} x \\ y \end{bmatrix}$$

in which the eigenvalues of the system are complex.

```
>> clear all
>> syms t c1 c2
>> A=[-1,-2; 2,-1]
>> [v,d]=eig(sym(A))
>> real(v(:,1))
>> a_vec=real(v(:,1))
>> b_vec=imag(v(:,1))
>> alpha=real(d(1,1))%real part of 1st eigenvalue
>> beta=imag(d(1,1))%imaginary part of 1st eigenvalue
>> solnx1=exp(alpha*t)*(a_vec*cos(beta*t)-b_vec*sin(beta*t))
>> solnx2=exp(alpha*t)*(a_vec*sin(beta*t)+b_vec*cos(beta*t))
>> %solnx1 and solnx2 are from Eq.(5.75)
>> soln=c1*solnx1+c2*solnx2
```

MATLAB Example 4: Enter the following code that solves the ODE system

$$\begin{bmatrix} x' \\ y' \\ z' \end{bmatrix} = \begin{bmatrix} -1 & 2 & -4 \\ 0 & -1 & 0 \\ 0 & 0 & -1 \end{bmatrix} \begin{bmatrix} x \\ y \\ z \end{bmatrix}$$

in which the eigenvalues of the system are real and generalized eigenvectors are needed.

```
>> clear all
>> syms c1 c2 c3 t
```

```
>>  A=sym([-1,2,-4; 0,-1,0; 0,0,-1])
>>  [V,E]=eig(A)
    %output for MATLAB 7.0.1 was V(:,1)=[1,0; 0,2; 0,1]
>>  eqA1=A-E(1,1)*eye(3,3)
>>  equ_v1=eqA1\V(:,1) %soln to eqA1*equ_v1=V(:,1)
>>  equ_v2=eqA1\V(:,2) %soln to eqA1*equ_v2=V(:,2)
    %only V(:,1) yields an answer and so we work with equ_v1
>>  soln=c1*exp(E(1,1)*t)*V(:,2)+c2*exp(E(1,1)*t)*V(:,1)
    +c3*(t*exp(E(1,1)*t)*V(:,1)+exp(E(1,1)*t)*(equ_v1))
```

MATLAB Example 5: Enter the following code that uses the built-in commands to find (i) the nullspace and column space of the matrix **A**, (ii) the exponential of the matrix **B**t, and (iii) the reduced-row echelon form of **A**, **B**, and **C**. See Appendix B for code that reduces a matrix to reduced-row echelon form using elementary row operations.

$$\mathbf{A} = \begin{bmatrix} 2 & -2 & 2 & 2 \\ 2 & -2 & 2 & 2 \\ 1 & -1 & 1 & 0 \end{bmatrix}, \quad \mathbf{B} = \begin{bmatrix} -1 & -2 & 3 \\ 0 & 2 & -1 \\ 0 & 0 & 2 \end{bmatrix}, \quad \mathbf{C} = \begin{bmatrix} 2 & 6 & 0 & 4 \\ -1 & 1 & 4 & 2 \\ 1 & -1 & -4 & -2 \end{bmatrix}.$$

```
>>  A=[2, -2, 2, 2; 2, -2, 2, 2; 1, -1, 1, 0]
>>  null(A) %orthonormal basis for nullspace, which is
    %useful for numerical computation
>>  null(A,'r') %this is also a basis but is easier to
    %work with by hand
>>  colspace(sym(A)) %finds a basis for the column space
>>  B=[-1,-2,3;0,2,-1;0,0,2]    %defines the matrix B
>>  syms t;  %defines t as a symbolic variable
>>  expm(B*t)  %calculates the matrix exponential
>>  C=[2,6,0,4; -1,1,4,2; 1,-1,-4,-2]   %defines the matrix C
>>  rref(A)
>>  rref(B)
>>  rref(C)
```

MATLAB Exercises

1. Enter the commands given in MATLAB Example 1 and submit both your input and output.

2. Enter the commands given in MATLAB Example 2 and submit both your input and output.

3. Enter the commands given in MATLAB Example 3 and submit both your input and output.

4. Enter the commands given in MATLAB Example 4 and submit both your input and output.

5. Enter the commands given in MATLAB Example 5 and submit both your input and output.

6. Find the eigenvalues and eigenvectors of $\begin{bmatrix} 3 & 1 & 0 \\ 2 & 4 & 0 \\ 3 & -1 & 1 \end{bmatrix}$.

7. Find the eigenvalues and eigenvectors of the following matrices:

$$A = \begin{bmatrix} 2 & 1 \\ 1 & 2 \end{bmatrix}, \qquad B = \begin{bmatrix} 1 & 3 \\ 3 & 1 \end{bmatrix}$$

8. Find the eigenvalues and eigenvectors of the following matrices:

$$A = \begin{bmatrix} 3 & -2 \\ 5 & -3 \end{bmatrix}, \qquad B = \begin{bmatrix} -1 & -5 \\ 1 & -3 \end{bmatrix}$$

9. Find the eigenvalues and eigenvectors of the following matrices:

$$A = \begin{bmatrix} -2 & 2 & 0 \\ 4 & 0 & 0 \\ -10/3 & 5/3 & 1 \end{bmatrix}, \qquad B = \begin{bmatrix} 1 & -5 & -1 \\ 1 & 3 & 1 \\ 0 & 0 & 1 \end{bmatrix}$$

10. Solve the ODE system in which the eigenvalues are real with a full set of eigenvectors: $\begin{cases} x' = -4x - y \\ y' = 6x + y \end{cases}$

11. Solve the ODE system in which the eigenvalues are real with a full set of eigenvectors: $\begin{cases} x' = 5x - y \\ y' = 3x + y \end{cases}$

12. Solve the ODE system in which the eigenvalues are real with a full set of eigenvectors: $\begin{cases} x' = 3x + y \\ y' = 2x + 4y \\ z' = 3x - y + z \end{cases}$

13. Solve the ODE system in which the eigenvalues are real with a full set of eigenvectors: $\begin{cases} x' = -x - 2y + z \\ y' = 2x - y + z \\ z' = z \end{cases}$

14. Solve the ODE system, $\begin{bmatrix} x' \\ y' \end{bmatrix} = \begin{bmatrix} 5 & -1 \\ 0 & 3 \end{bmatrix} \begin{bmatrix} x \\ y \end{bmatrix}$, in which the eigenvalues are real with a full set of eigenvectors, subject to the IC $x(0) = 0$, $y(0) = 2$ and then plot the solutions vs. t.

15. Solve the ODE system in which the eigenvalues are complex: $\begin{cases} x' = -4x + 2y \\ y' = -10x + 4y \end{cases}$

16. Solve the ODE system in which the eigenvalues are complex: $\begin{cases} x' = -7x - 4y \\ y' = 10x + 5y \end{cases}$

17. Solve the ODE system in which you will need to first find the generalized eigenvectors: $\begin{cases} x' = 7x - 2y \\ y' = 8x - y \end{cases}$

18. Solve the ODE system in which you will need to first find the generalized eigenvectors: $\begin{cases} x' = -2x + y \\ y' = -x \end{cases}$

19. Solve the ODE system in which you will need to first find the generalized eigenvectors: $\begin{cases} x' = x - y - z \\ y' = x + 3y + z \\ z' = z \end{cases}$

20. Solve the ODE system in which you will need to first find the generalized eigenvectors: $\begin{cases} x' = -x + y + z \\ y' = 3y + z \\ z' = -4y - z \end{cases}$

21. Find the (i) nullspace, (ii) column space, and (iii) reduced-row echelon form of $\begin{bmatrix} 2 & 6 & -2 & -1 \\ 0 & -3 & 1 & 0 \\ 0 & 0 & 2 & 2 \end{bmatrix}$

22. Find the (i) nullspace, (ii) column space, and (iii) reduced-row echelon form of $\begin{bmatrix} 1 & 3 & -6 & 0 \\ 0 & -2 & 1 & 1 \end{bmatrix}$

23. Find the (i) nullspace, (ii) column space, and (iii) reduced-row echelon form of $\begin{bmatrix} 2 & 3 & -3 & 0 \\ 0 & 1 & -1 & 2 \\ 3 & 1 & 0 & -2 \end{bmatrix}$

24. Find the (i) nullspace, (ii) column space, and (iii) reduced-row echelon form of $\begin{bmatrix} -2 & 4 & -3 & 0 \\ 1 & 1 & -1 & 5 \\ 5 & 1 & 0 & -5 \end{bmatrix}$

25. Find the (i) nullspace, (ii) column space, and (iii) reduced-row echelon form of $\begin{bmatrix} -2 & 4 & 0 & 2 \\ 1 & 1 & 3 & 2 \\ 1 & -1 & 1 & 0 \end{bmatrix}$

26. Find the (i) nullspace, (ii) column space, and (iii) reduced-row echelon form of $\begin{bmatrix} 2 & 6 & 0 & 4 \\ -1 & 1 & 4 & 2 \\ 1 & -1 & -4 & -2 \end{bmatrix}$

Chapter 5 Computer Lab: Maple

Maple Example 1: Enter the following code that calculates the eigenvalues and eigenvectors of the matrix $\begin{bmatrix} 3 & 0 & -2 \\ 0 & 1 & 2 \\ 0 & -2 & 1 \end{bmatrix}$.

restart
with(LinearAlgebra) :
A := Matrix([[3, 0, −2], [0, 1, 2], [0, −2, 1]])
#We first calculate the eigenvalues by finding the roots of the characteristic
*   polynomial. Then we show the quicker way to do this using the built-in*
*   Maple functions.*
eq1 := CharacteristicPolynomial(A, λ)
solve(eq1, λ)
Eigenvalues(A)   #calculates eigenvalues more efficiently
Eigenvectors(A)   #calculates eigenvalues AND eigenvectors with the
*   eigenvectors being the columns of the matrix*
eq2 := Eigenvectors(A, output = 'list')   #Calculates eigenvalues and
*   eigenvectors; answer is a set with elements (i) the eigenvalue,*
*   (ii) its multiplicity, (iii) the corresponding set of eigenvector(s)*
lambda1 := eq2[1][1]   #This is the first eigenvalue
eq2[1][3]   #This is the SET that contains the corresponding eigenvector
v1 := eq2[1][3][1]   #This is the actual eigenvector
Multiply(A, v1)
lambda1 · v1   #These last two answers should be identical
lambda3 := eq2[3][1]   #This is the third eigenvalue
v3 := eq2[3][3][1]   #This is the corresponding eigenvector
Multiply(A, v3)
lambda3 · v3   #These last two answers should be identical

Maple Example 2: Enter the following code that solves the ODE system

$$\begin{bmatrix} x' \\ y' \\ z' \end{bmatrix} = \begin{bmatrix} -2 & -1 & -2 \\ -4 & -5 & 2 \\ -5 & -1 & 1 \end{bmatrix} \begin{bmatrix} x \\ y \\ z \end{bmatrix}$$

subject to the IC $x(0) = 2$, $y(0) = -1$, $z(0) = -3$ and then plots the solutions vs. t.

restart
with(LinearAlgebra) :
A := Matrix([[−2, −1, −2], [−4, −5, 2], [−5, −1, 1]])
eqEV := Eigenvectors(A)
soln1a := e^{t·eqEV[1][1]} · Column(eqEV[2], 1)   #e is from palette!
soln1b := e^{t·eqEV[1][2]} · Column(eqEV[2], 2)

$soln1c := e^{t \cdot eqEV[1][3]} \cdot Column(eqEV[2], 3)$
$gensoln := c_1 \cdot soln1a + c_2 \cdot soln1b + c_3 \cdot soln1c$
$ODE0 := subs(t = 0, \{gensoln[1] = 2, gensoln[2] = -1, gensoln[3] = -3\})$
$cvals := solve(ODE0, \{c_1, c_2, c_3\})$
$xsoln := subs(cvals, gensoln[1])$
$ysoln := subs(cvals, gensoln[2])$
$zsoln := subs(cvals, gensoln[3])$
$plot\,([xsoln, ysoln, zsoln], t = 0..2, -400..400, linestyle = [1, 2, 3],$
$\quad legend = ["x(t)", "y(t)", "z(t)"])$

Maple Example 3: Enter the following code that solves the ODE system

$$\begin{bmatrix} x' \\ y' \end{bmatrix} = \begin{bmatrix} -1 & -2 \\ 2 & -1 \end{bmatrix} \begin{bmatrix} x \\ y \end{bmatrix}$$

in which the eigenvalues of the system are complex.

$restart$
$with(LinearAlgebra):$
$A := Matrix([[-1, -2], [2, -1]])$
$eq1 := Eigenvectors(A)$
$eq1[1][1]$
$eq1[1][2]$
$avec := Re(Column(eq1[2], 1))$
$bvec := Im(Column(eq1[2], 1))$
$\alpha := Re(eq1[1][1])$
$\beta := Im(eq1[1][1])$
$solnx1 := e^{\alpha \cdot t} \cdot (avec \cdot \cos(\beta \cdot t) - bvec \cdot \sin(\beta \cdot t))$
$solnx2 := e^{\alpha \cdot t} \cdot (avec \cdot \sin(\beta \cdot t) + bvec \cdot \cos(\beta \cdot t))$
$soln := c_1 \cdot solnx1 + c_2 \cdot solnx2$

Maple Example 4: Enter the following code that solves the ODE system

$$\begin{bmatrix} x' \\ y' \\ z' \end{bmatrix} = \begin{bmatrix} -1 & 2 & -4 \\ 0 & -1 & 0 \\ 0 & 0 & -1 \end{bmatrix} \begin{bmatrix} x \\ y \\ z \end{bmatrix}$$

in which the eigenvalues of the system are real and generalized eigenvectors are needed.

$restart$
$with(LinearAlgebra):$
$A := Matrix([[-1, 2, -4], [0, -1, 0], [0, 0, -1]])$
$\quad \#Maple\ 16\ gave\ [[0,1,0],[2,0,0],[1,0,0]]\ as\ matrix\ of\ eigenvectors$

$eq1 := Eigenvectors(A)$

$A1 := A - DiagonalMatrix(eq1[1])$

$v_1 := Column(eq1[2], 1)$

$u_{v1} := LinearSolve(A1, v_1)$

    #with the above order of eigenvectors, Maple 18 gave an error because the
    system is inconsistent and can't be solved; 2nd column works in next line

$v_2 := Column(eq1[2], 2)$

$u_{v2} := LinearSolve(A1, v_2)$

$u_1 := subs(\_t0_1 = 0, \_t0_3 = 0, u_{v2})$

$soln := c_1 \cdot e^{eq1[1][1]\cdot t} \cdot v_1 + c_2 \cdot e^{eq1[1][1]\cdot t} \cdot v_2 + c_3 \cdot (t \cdot e^{eq1[1][1]\cdot t} \cdot v_2 + e^{eq1[1][1]\cdot t} \cdot u_1)$

Maple Example 5: Enter the following code that uses the built-in commands to find (i) the nullspace and column space of the matrix **A**, (ii) the exponential of the matrix **B**t, and (iii) the reduced-row echelon form of **A**, **B**, and **C**. See Appendix B for code that reduces a matrix to reduced-row echelon form using elementary row operations.

$$\mathbf{A} = \begin{bmatrix} 2 & -2 & 2 & 2 \\ 2 & -2 & 2 & 2 \\ 1 & -1 & 1 & 0 \end{bmatrix}, \quad \mathbf{B} = \begin{bmatrix} -1 & -2 & 3 \\ 0 & 2 & -1 \\ 0 & 0 & 2 \end{bmatrix}, \quad \mathbf{C} = \begin{bmatrix} 2 & 6 & 0 & 4 \\ -1 & 1 & 4 & 2 \\ 1 & -1 & -4 & -2 \end{bmatrix}.$$

$restart$

$with(LinearAlgebra) :$

$A := Matrix([[2, -2, 2, 2], [2, -2, 2, 2], [1, -1, 1, 0]])$

$NullSpace(A)$

$ColumnSpace(A)$

$B := Matrix([[-1, -2, 3], [0, 2, -1], [0, 0, 2]])$

$MatrixExponential(B \cdot t)$

$C := Matrix([[2, 6, 0, 4], [-1, 1, 4, 2], [1, -1, -4, -2]])$

$ReducedRowEchelonForm(A)$

$ReducedRowEchelonForm(B)$

$ReducedRowEchelonForm(C)$

Maple Exercises

1. Enter the commands given in Maple Example 1 and submit both your input and output.

2. Enter the commands given in Maple Example 2 and submit both your input and output.

3. Enter the commands given in Maple Example 3 and submit both your input and output.

4. Enter the commands given in Maple Example 4 and submit both your input and output.

5. Enter the commands given in Maple Example 5 and submit both your input and output.

6. Find the eigenvalues and eigenvectors of $\begin{bmatrix} 3 & 1 & 0 \\ 2 & 4 & 0 \\ 3 & -1 & 1 \end{bmatrix}$.

7. Find the eigenvalues and eigenvectors of the following matrices:
$$\mathbf{A} = \begin{bmatrix} 2 & 1 \\ 1 & 2 \end{bmatrix}, \qquad \mathbf{B} = \begin{bmatrix} 1 & 3 \\ 3 & 1 \end{bmatrix}$$

8. Find the eigenvalues and eigenvectors of the following matrices:
$$\mathbf{A} = \begin{bmatrix} 3 & -2 \\ 5 & -3 \end{bmatrix}, \qquad \mathbf{B} = \begin{bmatrix} -1 & -5 \\ 1 & -3 \end{bmatrix}$$

9. Find the eigenvalues and eigenvectors of the following matrices:
$$\mathbf{A} = \begin{bmatrix} -2 & 2 & 0 \\ 4 & 0 & 0 \\ -10/3 & 5/3 & 1 \end{bmatrix}, \qquad \mathbf{B} = \begin{bmatrix} 1 & -5 & -1 \\ 1 & 3 & 1 \\ 0 & 0 & 1 \end{bmatrix}$$

10. Solve the ODE system in which the eigenvalues are real with a full set of eigenvectors: $\begin{cases} x' = -4x - y \\ y' = 6x + y \end{cases}$

11. Solve the ODE system in which the eigenvalues are real with a full set of eigenvectors: $\begin{cases} x' = 5x - y \\ y' = 3x + y \end{cases}$

12. Solve the ODE system in which the eigenvalues are real with a full set of eigenvectors: $\begin{cases} x' = 3x + y \\ y' = 2x + 4y \\ z' = 3x - y + z \end{cases}$

13. Solve the ODE system in which the eigenvalues are real with a full set of eigenvectors: $\begin{cases} x' = -x - 2y + z \\ y' = 2x - y + z \\ z' = z \end{cases}$

14. Solve the ODE system, $\begin{bmatrix} x' \\ y' \end{bmatrix} = \begin{bmatrix} 5 & -1 \\ 0 & 3 \end{bmatrix} \begin{bmatrix} x \\ y \end{bmatrix}$, in which the eigenvalues are real with a full set of eigenvectors, subject to the IC $x(0) = 0$, $y(0) = 2$ and then plot the solutions vs. t.

15. Solve the ODE system in which the eigenvalues are complex: $\begin{cases} x' = -4x + 2y \\ y' = -10x + 4y \end{cases}$

16. Solve the ODE system in which the eigenvalues are complex: $\begin{cases} x' = -7x - 4y \\ y' = 10x + 5y \end{cases}$

17. Solve the ODE system in which you will need to first find the generalized eigenvectors: $\begin{cases} x' = 7x - 2y \\ y' = 8x - y \end{cases}$

18. Solve the ODE system in which you will need to first find the generalized eigenvectors: $\begin{cases} x' = -2x + y \\ y' = -x \end{cases}$

19. Solve the ODE system in which you will need to first find the generalized eigenvectors: $\begin{cases} x' = x - y - z \\ y' = x + 3y + z \\ z' = z \end{cases}$

20. Solve the ODE system in which you will need to first find the generalized eigenvectors: $\begin{cases} x' = -x + y + z \\ y' = 3y + z \\ z' = -4y - z \end{cases}$

21. Find the (i) nullspace, (ii) column space, and (iii) reduced-row echelon form of $\begin{bmatrix} 2 & 6 & -2 & -1 \\ 0 & -3 & 1 & 0 \\ 0 & 0 & 2 & 2 \end{bmatrix}$

22. Find the (i) nullspace, (ii) column space, and (iii) reduced-row echelon form of $\begin{bmatrix} 1 & 3 & -6 & 0 \\ 0 & -2 & 1 & 1 \end{bmatrix}$

23. Find the (i) nullspace, (ii) column space, and (iii) reduced-row echelon form of $\begin{bmatrix} 2 & 3 & -3 & 0 \\ 0 & 1 & -1 & 2 \\ 3 & 1 & 0 & -2 \end{bmatrix}$

24. Find the (i) nullspace, (ii) column space, and (iii) reduced-row echelon form of $\begin{bmatrix} -2 & 4 & -3 & 0 \\ 1 & 1 & -1 & 5 \\ 5 & 1 & 0 & -5 \end{bmatrix}$

25. Find the (i) nullspace, (ii) column space, and (iii) reduced-row echelon form of $\begin{bmatrix} -2 & 4 & 0 & 2 \\ 1 & 1 & 3 & 2 \\ 1 & -1 & 1 & 0 \end{bmatrix}$

26. Find the (i) nullspace, (ii) column space, and (iii) reduced-row echelon form of $\begin{bmatrix} 2 & 6 & 0 & 4 \\ -1 & 1 & 4 & 2 \\ 1 & -1 & -4 & -2 \end{bmatrix}$

Chapter 5 Computer Lab: Mathematica

Mathematica Example 1: Enter the following code that calculates the eigenvalues and eigenvectors of the matrix $\begin{bmatrix} 3 & 0 & -2 \\ 0 & 1 & 2 \\ 0 & -2 & 1 \end{bmatrix}$.

```
Quit[]
A = {{3, 0, -2}, {0, 1, 2}, {0, -2, 1}}
A//MatrixForm
MatrixForm[A]
eq1 = CharacteristicPolynomial[A, λ]
Solve[eq1==0, λ]
lam = Eigenvalues[A]
evs = Eigenvectors[A]
evs[[1]] (*first eigenvector*)
A.evs[[1]] (*matrix times eigenvector*)
lam[[1]] evs[[1]] (*should be same as above*)
A.evs[[3]]
lam[[3]] evs[[3]] (*should be same as above*)
```

Mathematica Example 2: Enter the following code that solves the ODE system

$$\begin{bmatrix} x' \\ y' \\ z' \end{bmatrix} = \begin{bmatrix} -2 & -1 & -2 \\ -4 & -5 & 2 \\ -5 & -1 & 1 \end{bmatrix} \begin{bmatrix} x \\ y \\ z \end{bmatrix}$$

subject to the IC $x(0) = 2$, $y(0) = -1$, $z(0) = -3$ and then plots the solutions vs. t.

```
Quit[]
(A = {{-2, -1, -2}, {-4, -5, 2}, {-5, -1, 1}})//MatrixForm
   (*Note the extra parenthesis before //MatrixForm *)
   (*This allows calculations to be done on A*)
eqEV = Eigensystem[A]
soln1a = e^(t eqEV[[1]][[1]]) eqEV[[2]][[1]]
soln1b = e^(t eqEV[[1]][[2]]) eqEV[[2]][[2]]
soln1c = e^(t eqEV[[1]][[3]]) eqEV[[2]][[3]]
gensoln = c1 soln1a + c2 soln1b + c3 soln1c
ODE0 = ReplaceAll[{gensoln[[1]]==2, gensoln[[2]]==-1,
   gensoln[[3]]==-3}, t→0]
cvals = Solve[ODE0, {c1, c2, c3}]
cvals[[1]]
xsoln = ReplaceAll[gensoln[[1]], cvals[[1]]]
ysoln = ReplaceAll[gensoln[[2]], cvals[[1]]]
zsoln = ReplaceAll[gensoln[[3]], cvals[[1]]]
```

```
Plot[{xsoln, ysoln, zsoln}, {t, 0, 2}, PlotRange→{-400, 400},
  PlotStyle→{Dashed, Dotted, DotDashed},
  PlotLegends→{"x(t)", "y(t)", "z(t)"}]
```

Mathematica Example 3: Enter the following code that solves the ODE system

$$\begin{bmatrix} x' \\ y' \end{bmatrix} = \begin{bmatrix} -1 & -2 \\ 2 & -1 \end{bmatrix} \begin{bmatrix} x \\ y \end{bmatrix}$$

in which the eigenvalues of the system are complex.

```
Quit[]
A = {{-1, -2}, {2, -1}}
MatrixForm[A]
eq1 = Eigensystem[A]
eq1[[1]][[1]]
eq1[[1]][[2]]
eq1[[2]][[1]]
avec = Re[eq1[[2]][[1]]]
bvec = Im[eq1[[2]][[1]]]
α = Re[eq1[[1]][[1]]]
β = Im[eq1[[1]][[1]]]
solnx1 = e^α t (avec Cos[β t] - bvec Sin[β t])
solnx2 = e^α t (avec Sin[β t] + bvec Cos[β t])
soln = c1 solnx1 + c2 solnx2
Column[soln]
```

Mathematica Example 4: Enter the following code that solves the ODE system

$$\begin{bmatrix} x' \\ y' \\ z' \end{bmatrix} = \begin{bmatrix} -1 & 2 & -4 \\ 0 & -1 & 0 \\ 0 & 0 & -1 \end{bmatrix} \begin{bmatrix} x \\ y \\ z \end{bmatrix}$$

in which the eigenvalues of the system are real and generalized eigenvectors are needed.

```
Quit[]
A = {{-1, 2, -4}, {0, -1, 0}, {0, 0, -1}}
MatrixForm[A]
eq1 = Eigensystem[A]
  (*Mathematica 10 gave {{-1,-1,-1},{{0,2,1},{1,0,0},{0,0,0}}}*)
A1 = A - DiagonalMatrix[eq1[[1]]]
v1 = eq1[[2]][[1]]
uv1 = LinearSolve[A1, v1] (*no solution*)
```

```
v2 = eq1[[2]][[2]]
uv2 = LinearSolve[A1, v2]
soln = c1 e^{eq1[[1]] t} v1 + c2 e^{eq1[[1]] t} v2 + c3 (t e^{eq1[[1]] t} v2 + e^{eq1[[1]] t} uv2)
```

Mathematica Example 5: Enter the following code that uses the built-in commands to find (i) the nullspace and column space of the matrix **A**, (ii) the exponential of the matrix **B**t, and (iii) the reduced-row echelon form of **A**, **B**, and **C**. See Appendix B for code that reduces a matrix to reduced-row echelon form using elementary row operations.

$$\mathbf{A} = \begin{bmatrix} 2 & -2 & 2 & 2 \\ 2 & -2 & 2 & 2 \\ 1 & -1 & 1 & 0 \end{bmatrix}, \quad \mathbf{B} = \begin{bmatrix} -1 & -2 & 3 \\ 0 & 2 & -1 \\ 0 & 0 & 2 \end{bmatrix}, \quad \mathbf{C} = \begin{bmatrix} 2 & 6 & 0 & 4 \\ -1 & 1 & 4 & 2 \\ 1 & -1 & -4 & -2 \end{bmatrix}.$$

```
Quit[]
A = {{2, -2, 2, 2}, {2, -2, 2, 2}, {1, -1, 1, 0}}
NullSpace[A]
MatrixForm[Transpose[RowReduce[Transpose[A]]]]
 (*non-zero columns in the above result are basis vectors
   for the column space*)
B = {{-1, -2, 3}, {0, 2, -1}, {0, 0, 2}}
MatrixExp[B t]//MatrixForm
C1 = {{2, 6, 0, 4}, {-1, 1, 4, 2}, {1, -1, -4, -2}}
RowReduce[A]//MatrixForm
RowReduce[B]//MatrixForm
RowReduce[C1]//MatrixForm
```

Mathematica Exercises

1. Enter the commands given in Mathematica Example 1 and submit both your input and output.

2. Enter the commands given in Mathematica Example 2 and submit both your input and output.

3. Enter the commands given in Mathematica Example 3 and submit both your input and output.

4. Enter the commands given in Mathematica Example 4 and submit both your input and output.

5. Enter the commands given in Mathematica Example 5 and submit both your input and output.

6. Find the eigenvalues and eigenvectors of $\begin{bmatrix} 3 & 1 & 0 \\ 2 & 4 & 0 \\ 3 & -1 & 1 \end{bmatrix}$.

7. Find the eigenvalues and eigenvectors of the following matrices:
$$A = \begin{bmatrix} 2 & 1 \\ 1 & 2 \end{bmatrix}, \qquad B = \begin{bmatrix} 1 & 3 \\ 3 & 1 \end{bmatrix}$$

8. Find the eigenvalues and eigenvectors of the following matrices:
$$A = \begin{bmatrix} 3 & -2 \\ 5 & -3 \end{bmatrix}, \qquad B = \begin{bmatrix} -1 & -5 \\ 1 & -3 \end{bmatrix}$$

9. Find the eigenvalues and eigenvectors of the following matrices:
$$A = \begin{bmatrix} -2 & 2 & 0 \\ 4 & 0 & 0 \\ -10/3 & 5/3 & 1 \end{bmatrix}, \qquad B = \begin{bmatrix} 1 & -5 & -1 \\ 1 & 3 & 1 \\ 0 & 0 & 1 \end{bmatrix}$$

10. Solve the ODE system in which the eigenvalues are real with a full set of eigenvectors: $\begin{cases} x' = -4x - y \\ y' = 6x + y \end{cases}$

11. Solve the ODE system in which the eigenvalues are real with a full set of eigenvectors: $\begin{cases} x' = 5x - y \\ y' = 3x + y \end{cases}$

12. Solve the ODE system in which the eigenvalues are real with a full set of eigenvectors: $\begin{cases} x' = 3x + y \\ y' = 2x + 4y \\ z' = 3x - y + z \end{cases}$

13. Solve the ODE system in which the eigenvalues are real with a full set of eigenvectors: $\begin{cases} x' = -x - 2y + z \\ y' = 2x - y + z \\ z' = z \end{cases}$

14. Solve the ODE system, $\begin{bmatrix} x' \\ y' \end{bmatrix} = \begin{bmatrix} 5 & -1 \\ 0 & 3 \end{bmatrix} \begin{bmatrix} x \\ y \end{bmatrix}$, in which the eigenvalues are real with a full set of eigenvectors, subject to the IC $x(0) = 0$, $y(0) = 2$ and then plot the solutions vs. t.

15. Solve the ODE system in which the eigenvalues are complex: $\begin{cases} x' = -4x + 2y \\ y' = -10x + 4y \end{cases}$

16. Solve the ODE system in which the eigenvalues are complex: $\begin{cases} x' = -7x - 4y \\ y' = 10x + 5y \end{cases}$

17. Solve the ODE system in which you will need to first find the generalized eigenvectors: $\begin{cases} x' = 7x - 2y \\ y' = 8x - y \end{cases}$

18. Solve the ODE system in which you will need to first find the generalized eigenvectors: $\begin{cases} x' = -2x + y \\ y' = -x \end{cases}$

19. Solve the ODE system in which you will need to first find the generalized
    eigenvectors: $\begin{cases} x' = x - y - z \\ y' = x + 3y + z \\ z' = z \end{cases}$

20. Solve the ODE system in which you will need to first find the generalized
    eigenvectors: $\begin{cases} x' = -x + y + z \\ y' = 3y + z \\ z' = -4y - z \end{cases}$

21. Find the (i) nullspace, (ii) column space, and (iii) reduced-row echelon
    form of $\begin{bmatrix} 2 & 6 & -2 & -1 \\ 0 & -3 & 1 & 0 \\ 0 & 0 & 2 & 2 \end{bmatrix}$

22. Find the (i) nullspace, (ii) column space, and (iii) reduced-row echelon
    form of $\begin{bmatrix} 1 & 3 & -6 & 0 \\ 0 & -2 & 1 & 1 \end{bmatrix}$

23. Find the (i) nullspace, (ii) column space, and (iii) reduced-row echelon
    form of $\begin{bmatrix} 2 & 3 & -3 & 0 \\ 0 & 1 & -1 & 2 \\ 3 & 1 & 0 & -2 \end{bmatrix}$

24. Find the (i) nullspace, (ii) column space, and (iii) reduced-row echelon
    form of $\begin{bmatrix} -2 & 4 & -3 & 0 \\ 1 & 1 & -1 & 5 \\ 5 & 1 & 0 & -5 \end{bmatrix}$

25. Find the (i) nullspace, (ii) column space, and (iii) reduced-row echelon
    form of $\begin{bmatrix} -2 & 4 & 0 & 2 \\ 1 & 1 & 3 & 2 \\ 1 & -1 & 1 & 0 \end{bmatrix}$

26. Find the (i) nullspace, (ii) column space, and (iii) reduced-row echelon
    form of $\begin{bmatrix} 2 & 6 & 0 & 4 \\ -1 & 1 & 4 & 2 \\ 1 & -1 & -4 & -2 \end{bmatrix}$

Chapter 5 Projects

Project 5A: Transition Matrix and Stochastic Processes

We now consider a **transition matrix** in stochastic processes, which describes the (transition) probability of one state going to the next. We make the assumption that the system going from one state to the next only depends on the current state of the system and not on the myriad of possibilities for how one could have arrived at the given configuration. This is a reasonable assumption and is a key characteristic of *Markov chains*, studied in the theory of probability. If we consider a system with 3 states, we write

$$\mathbf{T} = \begin{bmatrix} p_{11} & p_{21} & p_{31} \\ p_{12} & p_{22} & p_{32} \\ p_{13} & p_{23} & p_{33} \end{bmatrix},$$

where

p_{ij} is the probability of moving from state i to state j.

For example, p_{23} is the probability of the system moving from state 2 to state 3. Note that the probability p_{ij} is located in the $(\mathbf{T})_{ji}$ position of the matrix. If we begin with an initial state vector $\mathbf{x}_0$, we can calculate the next state as $\mathbf{x}_1 = \mathbf{T}\mathbf{x}_0$. Similarly, the next state of the system is $\mathbf{x}_2 = \mathbf{T}\mathbf{x}_1 = \mathbf{T}^2\mathbf{x}_0$, and so on. The long-term behavior of the system can be described by examining $\mathbf{x}_n = \mathbf{T}^n\mathbf{x}_0$ for large n.

Application:
Consider a town of 30,000 families (economic units) who have been grouped into three economic brackets: lower, middle, and upper. Each year there is a 9% chance that the lower move into the middle, and 12% chance that the middle move back to the lower; there is an 8% chance that the middle move to the upper, and 10% chance that the upper move down to the middle; finally there is a 3% chance that the lower move directly to the upper, and 4% chance that the upper move down to the lower. These transitions occur continuously throughout the year as people are hired, laid off, fired, promoted, retire, and change careers.

Assume that initially there are 10,000 lower, 12,000 middle, and 8,000 upper income families and consider the state diagram for this situation shown in Figure 5.2.

Write the transition matrix of this system. Assuming that n is measured in years, find the number of families in each state after 10 years; after 20 years; after 50 years. (Remember to use $\mathbf{A} = \mathbf{V}\mathbf{\Lambda}\mathbf{V}^{-1}$ for efficiently raising a matrix to a high power.)

Now find the eigenvalues and eigenvectors of the transition matrix. *Calculate* $\mathbf{T}^\infty\mathbf{x}_0$. Give a mathematical explanation for what is happening.

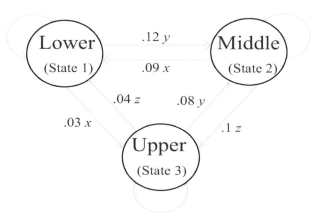

FIGURE 5.2: General state diagram for income class model of Project 5A.

Project 5B: Signal Processing

In Signal Processing, one is interested in observing and measuring a system with intent of applying a control that will make the system go to a known state. If $\mathbf{x}$ is the true state of the system and $\mathbf{y}$ the observed state, then the basic linear equations that one tries to analyze can be written

$$\frac{d\mathbf{x}}{dt} = \mathbf{A}\mathbf{x} + \mathbf{F}\mathbf{u}$$
$$\mathbf{y} = \mathbf{B}\mathbf{x} + \mathbf{C}\mathbf{u}, \qquad (5.145)$$

where $\mathbf{A}, \mathbf{F}, \mathbf{B}, \mathbf{C}$ are $n \times n$ matrices and $\mathbf{u}, \mathbf{x}, \mathbf{y}$ are n-dimensional vectors. We are interested in seeing how this linear system responds to an exponential input. This project investigates solutions of (5.145).

(1) Suppose the input signal is of the form $\mathbf{u} = e^{\mathbf{S}t}$ for diagonal matrix $\mathbf{S}$ with $s_{jj} \neq \lambda(\mathbf{A})$ for any $j = 1, \ldots, n$, where $\lambda(\mathbf{A})$ are the eigenvalues of $\mathbf{A}$. Show that

$$\mathbf{x}(t) = e^{\mathbf{A}t}\mathbf{x}(0) + e^{\mathbf{A}t}\int_0^t e^{(\mathbf{S}-\mathbf{A})\tau}\mathbf{F}d\tau.$$

(2) Show that this can be simplified to give

$$\mathbf{x}(t) = e^{\mathbf{A}t}\left(\mathbf{x}(0) - (\mathbf{S} - \mathbf{A})^{-1}\mathbf{F}\right) + (\mathbf{S} - \mathbf{A})^{-1}\mathbf{F}e^{\mathbf{S}t}$$

and conclude that the observed state of (5.145) can be written

$$\mathbf{y}(t) = \mathbf{B}e^{\mathbf{A}t}\left(\mathbf{x}(0) - (\mathbf{S} - \mathbf{A})^{-1}\mathbf{F}\right) + \left(\mathbf{C} + \mathbf{B}(\mathbf{S} - \mathbf{A})^{-1}\mathbf{F}\right)e^{\mathbf{S}t}.$$

Now let's solve this system again but assuming a general input $\mathbf{u}$.

Chapter 6

Geometric Approaches and
Applications of Systems of
Differential Equations

In this chapter, we examine systems of the form

$$\frac{d\mathbf{x}}{dt} = \mathbf{f}(t, \mathbf{x}, \mathbf{p}), \qquad (6.1)$$

where $\mathbf{x} \in \mathbb{R}^n$, $\mathbf{f}$ is an n-dimensional function (possibly nonlinear), $\mathbf{p} \in \mathbb{R}^m$ is a vector of parameters, and t is the independent variable. We have seen a system like this before with $n = 1$; in Sections 2.3–2.4, we considered

$$x' = rx(a - x)(x - b),$$

which is in the form of (6.1) with $n = 1$, $\mathbf{p} = [r \ \ a \ \ b]^T$ so that $m = 3$.

6.1 An Introduction to the Phase Plane

This section considers an example where both analytical and graphical methods provide useful insight into the behavior of solutions. Some of the topics in linear algebra found in Sections 5.4 and Appendix B are particularly helpful.

To begin our study, we will consider systems of two linear homogeneous first-order differential equations. These systems are easily solved by hand with the methods of the previous chapter, and the behaviors of their solutions are readily categorized. However, they also lend themselves to a useful graphical interpretation. We begin by considering systems of the form

$$x' = ax + by$$
$$y' = cx + dy \qquad (6.2)$$

where a, b, c, and d are constants and $x = x(t)$, $y = y(t)$.

To motivate our study of such a system, we consider an undamped mass-spring system, also known as a **simple harmonic oscillator**, which we studied previously in Section 3.7:

$$mx'' + kx = 0.$$

Using the methods of Section 3.4, we solve for the highest-order derivative to obtain

$$x'' = \frac{-k}{m}x.$$

Making the substitution $y = x'$ so that $y' = x''$ allows us to write the system in terms of the variables x, y:

$$x' = y$$
$$y' = \frac{-k}{m}x. \tag{6.3}$$

This is an example of a second-order system (6.2) with $a = d = 0$, $b = 1$, $c = -k/m$. At this point, we will *not* use numerical methods to analyze the behavior of solutions. Indeed, we could easily have solved this system because it is homogeneous with constant coefficients. Instead, we will use a graphical approach, analogous to phase line analysis of first-order autonomous equations. We view solutions in the **phase plane**, that is, the x-y plane where x and y are the two variables in this system. The right-hand sides of (6.3), and more generally (6.2), tell us how the system is changing at the given point and the change is given by (x', y'). Thus, to each point (x, y) in the phase plane, we will associate a vector (x', y') that describes the change. The collection of these vectors makes up what we call the **vector field**. Once we specify an initial condition, our motion is determined for all t.

For the equation of the simple harmonic oscillator, we can easily calculate slope lines according to the quadrant, keeping in mind that m and k are always positive. We refer the reader to Figure 6.1 as we go from quadrants 4 down to quadrant 1:

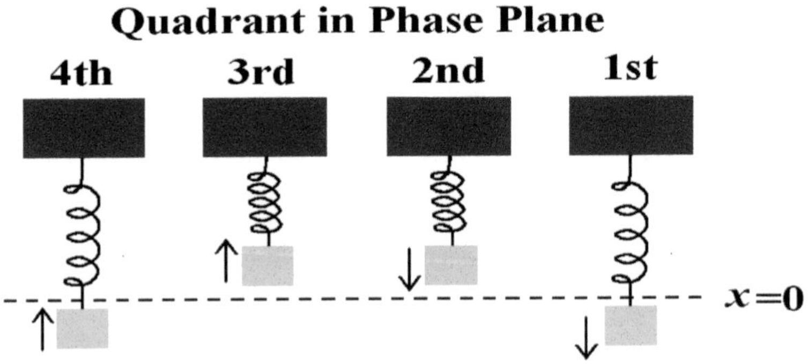

FIGURE 6.1:   Simple harmonic oscillator and the phase plane.

4th quadrant: $x > 0$, $y < 0$. Thus, (6.3) gives $x' < 0$ and $y' < 0$ and the solution in the phase plane is moving to the left (because $x' < 0$) and down (because $y' < 0$). In terms of the physical picture, the spring length is extended past its rest length but is beginning to compress as the block moves up.

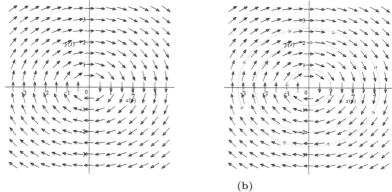

(a) **(b)**

FIGURE 6.2: Simple harmonic oscillator with $m = 2$, $k = 3$: **(a)** Vector field; **(b)** vector field with orbits corresponding to the initial conditions $(-1, 0)$, $(0, 2)$, and $(3, 1)$.

3rd quadrant: $x < 0$, $y < 0$. We have that $x' < 0$ and $y' > 0$ and the solution is moving to the left (because $x' < 0$) and up (because $y' > 0$). In terms of the physical picture, the spring length is shorter than its rest length and is continuing to compress as the block moves up.

2nd quadrant: $x < 0$, $y > 0$. We have that $x' > 0$ and $y' > 0$ and the solution is moving to the right (because $x' > 0$) and up (because $y' > 0$). In terms of the physical picture, the spring length is shorter than its rest length but is beginning to extend as the block moves down.

1st quadrant: $x > 0$, $y > 0$. We can substitute values to see that $x' > 0$ and $y' < 0$ and the solution is moving to the right (because $x' > 0$ and thus x is increasing) and down (because $y' < 0$). In terms of the physical picture, the spring length is extended past its rest length and is continuing to extend as the block moves down.

The vector field for the simple harmonic oscillator with $m = 2, k = 3$ is given in Figure 6.2(a). Equally useful is the inclusion of solutions in this picture. We will often refer to the solutions as **trajectories**. We should be careful here, though, because we are technically *not* looking at solutions since solutions live in the three-dimensional t-x-y space. What we are viewing are the **orbits** of this equation. As long as t does not explicitly appear in our equations, i.e., the system is autonomous (constant coefficient and homogeneous for linear equations), it is usually safe to think of the curves we draw as solutions. As we mentioned, given any initial condition, we can begin to trace out its path because we know that the vector field is tangent to the orbit at each point in the phase plane. Figure 6.2(b) shows two orbits of this system.

The phase plane with the orbits for various initial conditions drawn in is called the **phase portrait** of the system. Note that even if we do not superimpose the vector field, we still refer to it as the phase portrait. The reader should compare Figures 6.1 and 6.2 until convinced that they are describing the same situation. Remember that the horizontal axis represents the *position* and the vertical axis represents the *velocity* of the mass.

Every single point in the phase plane describes a part of a given orbit of our system. Although the origin may look boring (the zero vector!), we actually give it a special name: **equilibrium point**. This word has the same context as it did when we discussed autonomous first-order equations. It simply means a point where there is no change in the x or y value of the system, that is, $x'(t) = 0$ and $y'(t) = 0$. We note that for any constant coefficient, homogeneous, first-order system of the form given in Equation (5.1), the origin is *always* an equilibrium point.

Classification of Equilibrium Solutions

Eigenvalues and eigenvectors first arose in our discussion as the number λ and vector $\mathbf{v}$ that allowed $e^{\lambda t}\mathbf{v}$ to be a solution of the vector form of 6.2, $\mathbf{x}' = \mathbf{A}\mathbf{x}$ with $\mathbf{A} = \begin{bmatrix} a & b \\ c & d \end{bmatrix}$. We calculated the characteristic equation:

$$0 = \det(\mathbf{A} - \lambda\mathbf{I}) = \det\begin{bmatrix} a - \lambda & b \\ c & d - \lambda \end{bmatrix} = (a - \lambda)(d - \lambda) - bd$$

$$= \lambda^2 - (a + d)\lambda + (ad - bc). \tag{6.4}$$

We set $\beta = a + d$ and observe that this is the **trace** of the coefficient matrix and similarly for $\gamma = ad - bc$ as the **determinant**. The roots of the characteristic equation $\lambda^2 - \beta\lambda + \gamma$ are easily seen to be

$$\lambda_{1,2} = \frac{\beta \pm \sqrt{\beta^2 - 4\gamma}}{2}. \tag{6.5}$$

We can then write the general solution as

$$\mathbf{x} = c_1 e^{\lambda_1 t}\mathbf{v_1} + c_2 e^{\lambda_2 t}\mathbf{v_2}$$

as we did in Section 5.5. (Repeated eigenvalues that require Section 5.6 will be discussed shortly.) Our interest here is in understanding the qualitative behavior of the solution, especially as $t \to \infty$.

Example 1 Consider

$$x' = -5x + 8y$$
$$y' = -4x + 7y.$$

We can easily calculate the eigenvalue and eigenvector pairs as

$$\left\{ -1, \begin{bmatrix} 2 \\ 1 \end{bmatrix} \right\} \text{ and } \left\{ 3, \begin{bmatrix} 1 \\ 1 \end{bmatrix} \right\},$$

with general solution

$$\begin{bmatrix} x \\ y \end{bmatrix} = c_1 e^{-t}\begin{bmatrix} 2 \\ 1 \end{bmatrix} + c_2 e^{3t}\begin{bmatrix} 1 \\ 1 \end{bmatrix}.$$

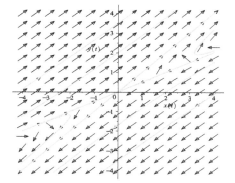

FIGURE 6.3: Phase portrait for Example 1. The origin is a saddle. The stable direction is along the $[2 \ 1]^T$ eigenvector and the unstable direction is along the $[1 \ 1]^T$ eigenvector.

The first pair says that along the vector $[2 \ 1]^T$, which lies along the line $y = x/2$, the solution is decreasing because $\lambda_1 = -1 < 0$. The second pair says that along the vector $[1 \ 1]^T$, which lies along the line $y = x$, the solution is increasing because $\lambda_2 = 3 > 0$. Our computer programs can generate the phase portrait, including the eigenvectors. Solutions are attracted to the origin along the first eigenvector (the stable direction) but are repelled from the origin along the second eigenvector (the unstable direction). Note that the Existence and Uniqueness theorem (Theorem 5.1.1) tells us that solutions do not cross each other. Because the system has constant coefficients, this means that trajectories cannot cross either. We can make an interesting mathematical observation: As $t \longrightarrow \infty$, all solutions approach the line $y = x$; as $t \longrightarrow -\infty$, all solutions approach the line $y = x/2$. This type of behavior is termed a **saddle point**; see Figure 6.3.

Example 2 Consider the system:

$$x' = -3x + 4y$$
$$y' = -4x - 3y.$$

We can calculate, either by hand or on the computer, that the eigenvalues are complex and are $\lambda_{1,2} = -3 \pm 4i$. We can write the real-valued solution as

$$\begin{bmatrix} x \\ y \end{bmatrix} = e^{-3t} \left\{ c_1 \begin{bmatrix} \sin 4t \\ \cos 4t \end{bmatrix} + c_2 \begin{bmatrix} \cos 4t \\ -\sin 4t \end{bmatrix} \right\}.$$

We know solutions rotate because of the complex eigenvalues and decay as $t \to \infty$ because of the e^{-3t}. If we consider $[1 \ 0]^T$, we can observe that it is rotated *clockwise* (via left multiplication by $\mathbf{A}$, since $[x' \ y']^T = [-3 \ -3]^T$).

Thus solutions spiral clockwise into the origin and we say the origin is a **stable spiral**; see Figure 6.4. See Appendix B.3 for an additional discussion and interpretation via **linear transformations**.

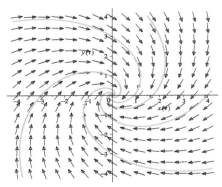

FIGURE 6.4:   Phase portrait for Example 2. The origin is a stable spiral. Note how each vector is rotated clockwise, as seen by left multiplication by **A**.

System (6.2) has equilibrium points whenever there is no change in either variable, i.e., whenever both derivatives are zero. As long as the system is non-degenerate, the only equilibrium is the origin $(0,0)$. If all the trajectories move toward an equilibrium point, it is said to be **stable** and is sometimes called a **sink**; if they all move away, it is said to be **unstable** and is sometimes called a **source**; if some trajectories move toward and others move away, it is called a **saddle**; and if all trajectories orbit around the equilibrium point, it is called a **center**. Further, the stable and unstable equilibria may exhibit spiraling behaviors in which case they are called **stable spirals** or **unstable spirals** (also called **spiral sinks** and **spiral sources**, respectively). Sometimes we will use the word **node** (e.g., **stable node/nodal sink** or **unstable node/nodal source**) to indicate that no spiraling is occurring. These results are summarized in the following theorem.

THEOREM 6.1.1

Consider (6.2). Let $\beta = a + d$ and $\gamma = ad - bc$; then
a. If $\gamma < 0$, then the origin is a saddle.
b. If $\gamma > 0$ and $\beta < 0$, then the origin is stable.
c. If $\gamma > 0$ and $\beta > 0$, then the origin is unstable.
d. If $\gamma > 0$ and $\beta = 0$, then the origin is a center.
e. If $\beta^2 - 4\gamma < 0$, then a sink or source is a stable spiral or an unstable spiral, respectively.

Figures 6.5 and 6.6 summarize these behaviors, based on the values of β and γ. We note that the straight lines in the case of the nodes and saddle are given by the **eigenvectors**. Also, the **borderline cases** can be considered as well:

(i) $\beta^2 - 4\gamma = 0$ corresponds to a repeated eigenvalue; if two eigenvectors exist, it is called a **star** whereas if only one eigenvector exists, then the general solution is given by the methods of Section 5.6 and the origin is called a **degenerate node**;

(ii) $\gamma = 0$ corresponds to at least one zero eigenvalue and the origin is either contained in a line of equilibria (one zero eigenvalue) or a plane of eigenvalues (if $\mathbf{A} = \mathbf{0}$); these are called **non-isolated equilibrium points**.

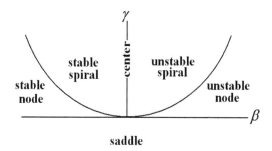

FIGURE 6.5: Classification of equilibria in the β-γ plane; see Theorem 6.1.1.

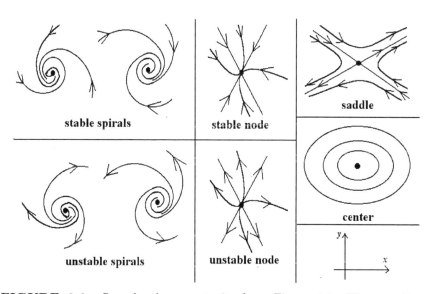

FIGURE 6.6: Sample phase portraits from Figure 6.5. The coordinate axes are given in the lower right corner of this figure.

Often our primary interest is in the behavior of solutions, rather than the solutions themselves. From the above discussion, notice that by looking at the system of equations and evaluating β and γ, we are able to predict the behavior of the system.

Example 3 Consider the system

$$x' = 2x + y$$
$$y' = -3x + 4y.$$

Here $\beta = 6$ and $\gamma = 8 + 3 = 11$, so $\beta^2 - 4\gamma = 36 - 44 = -8$. The origin is the only equilibrium point. From Figure 6.5 we note that since β and γ are both positive, we have either an unstable node or an unstable spiral. Since $\beta^2 - 4\gamma < 0$, the trajectories spiral away from the equilibrium and we see that we have an unstable spiral. If we mutliply $\mathbf{A}$ with $[1 \ \ 0]^T$, we see that the rotation is clockwise.

Example 4 Consider the simple harmonic oscillator from before:

$$x' = y$$
$$y' = \frac{-k}{m}x. \qquad (6.6)$$

Here $\beta = 0$ and $\gamma = k/m > 0$. The origin is the only equilibrium point and it's a center with trajectories encircling it in closed curves. The direction of rotation is again clockwise.

Three or More Equations

In the case of a 2×2 system, it is easy to use Theorem 6.1.1 to characterize the stability of the equilibria (the origin). For a general system of n first-order equations, we can again speak of equilibria of the system and we again have the types mentioned in Theorem 6.1.1. Except in very special cases, the origin will again be the only equilibrium solution that we need to consider. In the two-dimensional case, we talked about viewing the trajectories in the phase plane. For a system of n equations, the trajectories now live in n dimensional **phase space** and we can only see (at most) three dimensions at a time. We can, however, characterize the stability according to the eigenvalues of the matrix. It is important to note that the following theorem also holds for a 2×2 system and can easily be derived from Theorem 6.1.1 by writing the eigenvalues in terms of the trace and determinant.

THEOREM 6.1.2

Let $\{\lambda_1, \lambda_2, \cdots, \lambda_n\}$ be the n (real or complex [non-real], possibly re-peated) eigenvalues of the coefficient matrix $\mathbf{A}$ of a given linear constant-coefficient homogeneous system.

a. If the real part of the eigenvalue $\text{Re}(\lambda_i) < 0$ for all i, then the equilibrium point is stable.

b. If the real part of the eigenvalue $\text{Re}(\lambda_i) < 0$ for at least one i and the real part of the eigenvalue $\text{Re}(\lambda_j) > 0$ for at least one j, then the equilibrium point is a saddle.

c. If the real part of the eigenvalue $\text{Re}(\lambda_i) > 0$ for all i, then the equilibrium point is unstable.

d. If any of the eigenvalues are complex (non-real), then the stable or unstable equilibria is a spiral; if all of the eigenvalues are real, it is a node;

e. If a pair of complex conjugate eigenvalues λ_i, $\overline{\lambda_i}$ satisfy $\text{Re}(\lambda_i) = 0$, then the equilibrium is a center in the plane containing the corresponding eigenvectors.

As with two dimensions, we refer to equilibrium solutions as stable spirals (or spiral sinks), stable nodes (or nodal sinks), unstable spirals (or spiral sources), unstable nodes (or nodal sources), and saddles.

This theorem also applies to a system of two equations and it was straight-forward to view the phase portraits in the phase plane. In higher dimensions, we can have combinations of the different behaviors. For example, we may have an unstable spiral in one plane even though solutions may be attracted to this plane! Sketching phase portraits becomes much more difficult although the pictures from Figure 6.6 are still the generic ones that may result.

Example 6. Consider the following system:

$$x' = -2x - y - 2z$$
$$y' = -4x - 5y + 2z$$
$$z' = -5x - y + z.$$

Calculate the eigenvalues and use Theorem 6.1.2 to determine the stability of the equilibrium point (the origin).

Solution

We can calculate the eigenvalues either by hand or with the computer. Doing so gives

$$\lambda_1 = 3, \ \lambda_2 = -6, \ \lambda_3 = -3.$$

Part (b) of Theorem 6.1.2 shows that the origin is a saddle.

• • • • • • • • • • • •

Problems

In Problems **1–15**, (i) use Theorem 6.1.1 to determine the stability of the origin (it's the only equilibrium solution) for the given systems of equations; (ii) use a computer algebra system to draw the vector field; (iii) either by hand or with the computer, sketch some orbits to verify your conclusion in (i).

1. $\begin{cases} x' = -3x + 6y \\ y' = -2x + 5y \end{cases}$
2. $\begin{cases} x' = x - y \\ y' = 3x - 4y \end{cases}$
3. $\begin{cases} x' = x + 2y \\ y' = -x + 4y \end{cases}$

4. $\begin{cases} x' = 2x - 2y \\ y' = 5x - 2y \end{cases}$
5. $\begin{cases} x' = -3x + y \\ y' = 2x + y \end{cases}$
6. $\begin{cases} x' = y \\ y' = 3x + 2y \end{cases}$

7. $\begin{cases} x' = 4x + y \\ y' = 3x + 2y \end{cases}$
8. $\begin{cases} x' = -2y \\ y' = 4x + 5y \end{cases}$
9. $\begin{cases} x' = x - 8y \\ y' = x + 3y \end{cases}$

10. $\begin{cases} x' = 3x - 5y \\ y' = 5x + 3y \end{cases}$
11. $\begin{cases} x' = -x - 5y \\ y' = x - 3y \end{cases}$
12. $\begin{cases} x' = -3x - 5y \\ y' = x - 6y \end{cases}$

13. $\begin{cases} x' = 5x - 6y \\ y' = 2x - y \end{cases}$
14. $\begin{cases} x' = 5x \\ y' = -2x + 3y \end{cases}$
15. $\begin{cases} x' = 7x + 6y \\ y' = 2x - 4y \end{cases}$

16. Consider the simple harmonic oscillator with damping: $x'' + bx' + kx = 0$, where $b, k > 0$. Convert this to a system of 2 first-order equations and then answer questions (i)–(iii) above.

17. Use Theorem 6.1.1 to find values for b and d so that the equilibrium solution (the origin) of $\begin{cases} x' = x + by \\ y' = x + dy \end{cases}$ is an

(a) unstable node,   (b) stable node,   (c) unstable spiral,
(d) stable spiral,   (e) saddle point,   (f) center.

18. We previously showed that we can reduce a second-order equation to a system of 2 first-order equations and thus analyze with methods of this section. Consider (6.2) and follow the steps here to show how Section 3.6 can also give us Theorem 6.1.1.
(i) Differentiate $x' = ax + by$ to obtain x'' and substitute $y' = cx + dy$ into this equation.
(ii) Then substitute $by = x' - ax$ into this, collect like terms, and rearrange to obtain $x'' - (a + d)x' + (ad - bc)x = 0$.
(iii) Observe that $\beta = a + d$ is the *trace* of the coefficient matrix, and $\gamma = ad - bc$ is the determinant.
(iv) Show the characteristic equation of the second-order differential equation is $\lambda^2 - \beta\lambda + \gamma = 0$ and use the quadratic formula to verify the conclusions of Theorem 6.1.1 in terms of eigenvalues. (Note: we used r_1, r_2 in Chapter 4.)

19. Use Problem **18** to rewrite $\begin{cases} x' = -5x + 8y \\ y' = -4x + 7y. \end{cases}$ as a second-order equation.

Then use the Section 3.6 to find the general solution.

For Problems **20–27**, use Theorem 6.1.2 to classify the stability of the origin (the only equilibrium solution).

20. $\begin{cases} x' = -8x - 8y - 2z \\ y' = -13x + 14y - z \\ z' = 12x - 15y - 6z \end{cases}$

21. $\begin{cases} x' = -13x + 4y - 8z \\ y' = 8x + y + 4z \\ z' = 42x - 6y + 21z \end{cases}$

22. $\begin{cases} x' = 11x - 6y + 10z \\ y' = -5x + 4y - 5z \\ z' = -13x + 9y - 12z \end{cases}$

23. $\begin{cases} x' = 2x - 2y + z \\ y' = -3x + 3y + z \\ z' = 3x - 2y \end{cases}$

24. $\begin{cases} x' = -x + z \\ y' = -3y + z \\ z' = -y - z \end{cases}$

25. $\begin{cases} x' = x - 2y + 2z \\ y' = -4x + 3y + 2z \\ z' = 4x - 2y - z \end{cases}$

26. $\mathbf{x}' = \begin{bmatrix} -1 & -1 & 1 \\ 1 & -1 & 0 \\ 0 & 0 & 1 \end{bmatrix} \mathbf{x}$

27. $\mathbf{x}' = \begin{bmatrix} 1 & -5 & -1 \\ 1 & 3 & 1 \\ 0 & 0 & 1 \end{bmatrix} \mathbf{x}$

6.2 Nonlinear Equations and Phase Plane Analysis

Up to this point, we have studied a system of first-order linear differential equations with constant coefficients. We will now consider the more general situation where the equations are first-order but are *nonlinear*. There are numerous books devoted to the study of such equations and we are merely "scratching the surface" of this topic. For an extremely accessible yet thorough excursion into *nonlinear dynamics*, the interested reader should examine the book by Strogatz given in the references [48].

Let's consider the following two differential equations:

$$x' = f(t, x, y)$$
$$y' = g(t, x, y). \tag{6.7}$$

When f and g are nonlinear functions, it will be rare when we can actually find an exact solution to these equations. In such cases we must resort to graphical or numerical analysis and interpretation of the behavior of the solutions. To better understand this new approach, we will consider an *autonomous* nonlinear system of the form

$$x' = f(x, y)$$
$$y' = g(x, y), \tag{6.8}$$

where time, t, is not explicit. We begin with an existence and uniqueness theorem for a general system.

THEOREM 6.2.1

Consider the system given in (6.7) with initial condition $x(t_0) = x_0, y(t_0) = y_0$. If $\partial f/\partial x$, $\partial f/\partial y$, $\partial g/\partial x$, $\partial g/\partial y$ are all continuous on some rectangular region $R = \{(x,y) | a < x < b,\ c < y < d\}$ containing the point (t_0, x_0, y_0), then there exists a unique solution to (6.7) defined on $(t_0 - \tau, t_0 + \tau)$ for some $\tau > 0$.

We shall assume that all partial derivatives are continuous for the remainder of this section.

Equilibria

Just as with linear systems in Section 6.1, we are interested in finding the equilibria[1] of the system (6.8) of first-order differential equations. We know that the solution (x, y) is at equilibrium when it is not changing, i.e., when $x' = 0$ and $y' = 0$. In order to find equilibria, we need to consider the curves $f(x, y) = 0$ and $g(x, y) = 0$ in the phase plane. Any curve of the form $h(x, y) = k$, where k is a constant, is called an **isocline** or **level curve** of the function h. In other words, a curve is an isocline or level curve of a function if the function takes the same value at every point on the curve. In the special case when $k = 0$, the curve $h(x, y) = k = 0$ is called a **nullcline**. In the system (6.8), $f(x, y) = 0$ and $g(x, y) = 0$ are the nullclines since they take the value zero. From the discussion above we see that the intersection point of the two nullclines is an equilibrium. More formally (x^*, y^*) is an equilibrium of (6.8) if $f(x^*, y^*) = 0$ and $g(x^*, y^*) = 0$.

Example 1: Consider

$$x' = y - x^2$$
$$y' = y - x. \tag{6.9}$$

Here, $f(x, y) = y - x^2$ and $g(x, y) = y - x$. Therefore the nullclines are $y = x^2$ and $y = x$. These curves intersect at two points $(0, 0)$ and $(1, 1)$. From the definition we can conclude that these are the equilibrium solutions (x^*, y^*).

Directions of flow

We also interpreted the solutions and trajectories geometrically in the phase plane for our linear system. When the equations are nonlinear, we still consider the phase plane but now allow for the possibility of *multiple* equilibria. The vector field in the phase plane helps us to picture the behavior of the

---

[1] Recall that equilibria are also referred to as constant solutions, critical points, fixed points, and steady-state solutions.

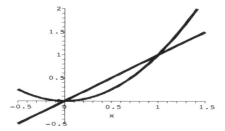

FIGURE 6.7: Nullclines of system (6.9).

solutions. One may think of the vector field as a flowing body of water. If we drop a stick in the water, it moves in a path determined by the flow vectors. A computer-generated vector field gives a more accurate sense of the flow of trajectories; it's easier to observe, e.g., that as a trajectory crosses a nullcline, the direction it moves changes; see Figure 6.8(a). Figure 6.8(b) shows the nullclines and vector field superimposed to illustrate horizontal and vertical flow along each nullcline. In practice, we will not superimpose nullclines and vector fields because only trajectories are plotted on vector fields and having nullclines over these would be very confusing.

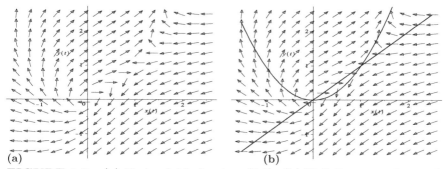

FIGURE 6.8: (a) Vector field of system (6.9). (b) Nullclines superimposed on vector field to show the horizontal or vertical flow along each nullcline. In practice we will not superimpose the nullclines because we will only plot trajectories (solutions) on the vector field; see Figure 6.9.

From a phase plane drawing, saddle points are evident; see Figure 6.9. However, it is difficult to tell if the equilbrium is a spiral sink, a center, or a spiral source. We discuss two methods used to classify our equilibrium points. The first is to use a differential-equations solver to find either exact or approximate solutions for initial points in various locations (i.e., for different initial conditions) and observe the behavior these solutions exhibit. For a review of the behavior near equilibrium points, see Figure 6.6. The second method is to do a local analysis of the solutions around the equilibrium point to classify the stability of the equilibrium. Such an approach requires us to linearize our equation or system and consider the behavior of the linearized system. The topology of the linearized system will be the same as the original system as long as the eigenvalues of the linearized system do not have zero real part.

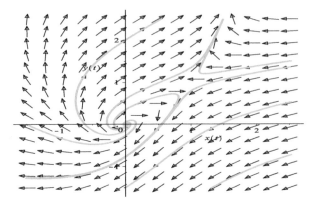

FIGURE 6.9: Phase portrait: trajectories superimposed onto vector field of system (6.9).

Linearization

Linearization of a differential equation is the replacement of the equation of a curve or surface by the appropriate tangent line or tangent plane centered at a point. From calculus, we know that a differentiable function $F(x)$ can be approximated near a point by its tangent line, $F(x) \approx F(a) + F'(a)(x-a)$. In calculus of several variables, this is extended to a plane tangent to a surface at a point (x, y). The surface $h(x, y)$ is approximated by the linear Taylor expansion:

$$h(x, y) \approx h(x^*, y^*) + \frac{\partial h}{\partial x}(x^*, y^*)(x - x^*) + \frac{\partial h}{\partial y}(x^*, y^*)(y - y^*).$$

This is usually done to study the behavior of a system near that point, in our case the equilibrium point. Linearization of a set of equations yields a system of linear differential equations. Therefore, we use the methods learned in Section 6.1 to analyze and determine the behavior of solutions close to the point of interest.

Even though linearization can be done at any point, we linearize only around equilibria. Let (x^*, y^*) be equilibria of the system (6.8),

$$\frac{dx}{dt} = f(x, y)$$

$$\frac{dy}{dt} = g(x, y),$$

and let $(x^*, y^*) = 0$ be an equilibrium, i.e., $f(x^*, y^*) = 0$ and $g(x^*, y^*) = 0$.

We use the notation $f_x(a, b)$ to denote the partial derivative of f with respect to x evaluated at (a, b), and $f_y(a, b)$ is the partial of f with respect to y. Similar notation is used for the function g.

The functions f and g at (x^*, y^*) are approximated by Taylor expanding $f(x, y)$ about (x^*, y^*) and omitting any higher-order terms:

$$\frac{dx}{dt} = f(x, y)$$
$$\approx f(x^*, y^*) + f_x(x^*, y^*)(x - x^*) + f_y(x^*, y^*)(y - y^*)$$
$$\frac{dy}{dt} = g(x, y)$$
$$\approx g(x^*, y^*) + g_x(x^*, y^*)(x - x^*) + g_y(x^*, y^*)(y - y^*).$$

Notice $f(x^*, y^*) = 0$ and $g(x^*, y^*) = 0$. Also

$$\frac{d(x - x^*)}{dt} = \frac{dx}{dt} \quad \text{and} \quad \frac{d(y - y^*)}{dt} = \frac{dy}{dt}$$

so we have the following pair of equations:

$$\frac{d(x - x^*)}{dt} \approx f_x(x^*, y^*)(x - x^*) + f_y(x^*, y^*)(y - y^*)$$
$$\frac{d(y - y^*)}{dt} \approx g_x(x^*, y^*)(x - x^*) + g_y(x^*, y^*)(y - y^*).$$

It is simpler to change coordinate systems and translate the equilibrium to the origin. Let $u = x - x^*$ and $v = y - y^*$. Then we have

$$\frac{du}{dt} \approx f_x(x^*, y^*)u + f_y(x^*, y^*)v$$
$$\frac{dv}{dt} \approx g_x(x^*, y^*)u + g_y(x^*, y^*)v. \tag{6.10}$$

This is a linear system which we rewrite as

$$\begin{bmatrix} \frac{du}{dt} \\ \frac{dv}{dt} \end{bmatrix} = \begin{bmatrix} f_x(x^*, y^*) & f_y(x^*, y^*) \\ g_x(x^*, y^*) & g_y(x^*, y^*) \end{bmatrix} \begin{bmatrix} u \\ v \end{bmatrix}.$$

The matrix of partial derivatives is sometimes called the **Jacobian matrix**, and we use the notation

$$J(x, y) = \begin{bmatrix} f_x(x, y) & f_y(x, y) \\ g_x(x, y) & g_y(x, y) \end{bmatrix}. \tag{6.11}$$

This matrix evaluated at the equilibirum, $J(x^*, y^*)$, is then used with Theorem 6.1.1 to determine the nature of the solution.

This linearization process may seem involved, but much of the complication is in the development of the equations. In practice it is only necessary to compute the four partial derivatives at the equilibrium and then use Theorem 6.1.1 (or equivalently, Theorem 6.1.2) to determine the nature of the solution.

Example 2: Consider the system (6.9) from Example 1:

$$\frac{dx}{dt} = y - x^2$$

$$\frac{dy}{dt} = y - x.$$

Find the equilibria and determine their stability via linearization.

Solution

We need to solve the right-hand sides of the equations simultaneously. Solving the second equation gives $y = x$. We then substitute into the the first equation to get

$$x - x^2 = 0 \Longrightarrow x = 0, 1.$$

Thus we have two equilibria, $(0,0)$ and $(1,1)$. Calculating their stability via linearization requires use of the Jacobian. We can easily calculate it as

$$\mathbf{J} = \begin{bmatrix} -2x & 1 \\ -1 & 1 \end{bmatrix}.$$

Evaluating the Jacobian at the respective equilibria gives

$$J(0,0) = \begin{bmatrix} -2x & 1 \\ -1 & 1 \end{bmatrix}_{(0,0)} = \begin{bmatrix} 0 & 1 \\ -1 & 1 \end{bmatrix},$$

so $\beta = \text{Tr}(J(0,0)) = 1$ and $\gamma = \det(J(0,0)) = 1$; thus $(0,0)$ is a spiral source. We also have

$$J(1,1) = \begin{bmatrix} -2 & 1 \\ -1 & 1 \end{bmatrix},$$

so $\beta = \text{Tr}(J(1,1)) = -1$ and $\gamma = \det(J(1,1)) = -1$; thus $(1,1)$ is a saddle point. See Figure 6.9.

6.2.1 Systems of More Than Two Equations

Nonlinear systems of equations are an active area of research as these equations occur in many areas of the sciences.

$$x_1' = f_1(t, x_1, x_2, \ldots, x_n)$$
$$x_2' = f_2(t, x_{1,2}, \ldots, x_n)$$
$$\vdots$$
$$x_n' = f_n(t, x_{1,2}, \ldots, x_n). \tag{6.12}$$

It will be extremely rare when we can actually find a closed form solution to these equations. As with two equations, we must then rely heavily on graphical, approximation, or numerical techniques. The existence and uniqueness theorem for this general system is analogous to the one for two equations—we need to check that each of the functions f_i and partial derivatives $\partial f_i/\partial x_j$ are continuous in order to guarantee existence and uniqueness. We will often be concerned with these equations when they are **autonomous**, that is, there is not explicit t in the problem. The following theorem gives the **linear stability** in a neighborhood of an equilibrium solution. If a system has multiple equilibrium solutions, the theorem may be applied to each equilibrium solution. We often write (6.12) in its vector notation for convenience:

$$\mathbf{x}' = \mathbf{f}(\mathbf{x}).$$

THEOREM 6.2.2

Let $\mathbf{x}' = \mathbf{f}(\mathbf{x})$ be a nonlinear system of n first-order equations with $\mathbf{x}^*$ as an equilibrium solution and $\mathbf{f}$ a sufficiently smooth vector function. Let $\mathbf{J}$ be the Jacobian (the matrix of partial derivatives) evaluated at this equilibrium solution:

$$\mathbf{J}(\mathbf{x}^*) = \begin{bmatrix} \partial f_1/\partial x_1 & \partial f_1/\partial x_2 & \cdots & \partial f_1/\partial x_n \\ \partial f_2/\partial x_1 & \partial f_2/\partial x_2 & \cdots & \partial f_2/\partial x_n \\ \vdots & & \ddots & \\ \partial f_n/\partial x_1 & \partial f_n/\partial x_2 & \cdots & \partial f_n/\partial x_n \end{bmatrix}_{\mathbf{x}=\mathbf{x}^*}. \tag{6.13}$$

Let $\{\lambda_1, \lambda_2, \cdots, \lambda_n\}$ be the n (real or complex, possibly repeated) eigenvalues of the Jacobian matrix.
a. If the real part of the eigenvalue $\mathrm{Re}(\lambda_i) < 0$ for all i, then the equilibrium is stable.
b. If the real part of the eigenvalue $\mathrm{Re}(\lambda_i) < 0$ for at least one i and $\mathrm{Re}(\lambda_j) > 0$ for at least one j, then the equilibrium is a saddle.
c. If the real part of the eigenvalue $\mathrm{Re}(\lambda_i) > 0$ for all i, then the equilibrium is unstable.
d. If any of the eigenvalues are complex, then the stable or unstable equilibria is a spiral; if all of the eigenvalues are real, it is a node.
e. If a pair of complex conjugate eigenvalues λ_i, $\overline{\lambda_i}$ satisfy $\mathrm{Re}(\lambda_i) = 0$, then the equilibrium is a linear center in the plane containing the corresponding eigenvectors.

THEOREM 6.2.3

Let $\mathbf{x}' = \mathbf{f}(\mathbf{x})$ be a nonlinear system of n first-order equations with $\mathbf{x}^*$ as an equilibrium solution and $\mathbf{f}$ a sufficiently smooth vector function. If $\mathrm{Re}(\lambda_i) \neq 0$ for all i, then the predictions given by the linear stability results of Theorem 6.2.2 hold for the equilibrium solution in the nonlinear system.

The significance of Theorem 6.2.3 cannot be understated. We found an equilibrium solution $\mathbf{x}^*$ and linearized about it. That is, we considered only the linear terms near this equilibrium solution. The results of the theorem allow us to conclude that only looking at linear terms near the equilibrium solution is sufficient to give us accurate stability predictions, as long as the real part of all eigenvalues is nonzero. This should be believable because adding nonlinear terms could possibly change the stability in these borderline cases. Alternative techniques beyond the scope of this book are needed to address these situations.

Thus the techniques used for autonomous systems are very familiar: finding equilibrium solutions, linearizing the system about the equilibria, determining the linear stability of the equilibria, and constructing phase portraits with the help of a computer program. Indeed, our plan of attack in order to understand the behavior of the solutions was identical for a system of two equations. The main difference here has to do with the structure of the space in which trajectories live. Things were very nice in two dimensions in that we could characterize many things about equilibria. Once we introduce a third (or more) dimension(s), very strange things can happen. The mathematical subject of **chaos** arose because of the kind of this strange behavior that can occur. We refer the interested reader to other books for an introduction.

Example 3: Find equilibrium solutions for the system

$$x' = -6x + 6y$$
$$y' = 36x - y - xz$$
$$z' = -3z + xy. \tag{6.14}$$

Then use Theorems 6.2.2 and 6.2.3 to classify the stability of the equilibrium solutions.

Solution
We use our three computer software packages to help us with these and give the code at the end of the example. We find three equilibria in the system:

$$(x^*, y^*, z^*) = (0, 0, 0), \left(\sqrt{105}, \sqrt{105}, 35\right), \left(-\sqrt{105}, -\sqrt{105}, 35\right).$$

In order to determine the stability of the equilibria, we first need to calculate the Jacobian matrix of the system:

$$\mathbf{J} = \begin{bmatrix} -6 & 6 & 0 \\ 36 - z & -1 & -x \\ y & x & -3 \end{bmatrix}, \tag{6.15}$$

and then substitute in the respective equilibrium points. We have

$$(0, 0, 0): \quad \lambda_1 = -3, \ \lambda_{2,3} = \frac{-7 \pm \sqrt{889}}{2},$$

which shows that the origin is a saddle point. According to Theorem 6.2.3, we can conclude that the origin is also a saddle in the original system (since $\text{Re}(\lambda_i) \neq 0$). For the second equilibrium, we have

$$(\sqrt{105}, \sqrt{105}, 35): \quad \lambda_1 = -10, \ \lambda_{2,3} = \pm i3\sqrt{14}.$$

According to Theorem 6.2.2, this second equilibrium solution is a linear center in two directions and stable in the third direction.[2] According to Theorem 6.2.3, we can only conclude that we have a linear center—it is possible that we have a nonlinear center in the full nonlinear system but it is also possible that the inclusion of the nonlinear terms makes this equilibrium solution either a stable spiral or an unstable spiral. For the third equilibrium, we have

$$(-\sqrt{105}, -\sqrt{105}, 35): \quad \lambda_1 = -10, \ \lambda_{2,3} = \pm i3\sqrt{14}.$$

This again gives the prediction of a linear center in two directions and stable in the third direction and doesn't allow us to conclude anything about the full system.

The system of the previous example is actually a very well-known system that has been studied extensively due to the seemingly unpredictable behavior of solutions with close initial conditions. Depending on the coefficients of the original equations, we can have between one and three equilibria and we have the possibility of trajectories wandering endlessly without approaching an equilibrium solution. See Problem **15** for another look at this system.

Problems

For Problems **1–9**, (i) find the equilibria of the given system; (ii) use linearization and Theorem 6.2.2 to classify the stability of the equilibria; (iii) use MATLAB, Maple, or Mathematica to draw the vector field of the system; (iv) sketch trajectories on the vector field for various initial conditions (either by hand or with the computer). You should verify that your answers from parts (iii) and (iv) agree with your predictions in parts (i) and (ii).

1. $\begin{cases} x' = y \\ y' = 4 - x^2 \end{cases}$
2. $\begin{cases} x' = y - 1 \\ y' = x^2 - y \end{cases}$
3. $\begin{cases} x' = y - x \\ y' = x^2 + 2y \end{cases}$

4. $\begin{cases} x' = y^2 - x \\ y' = x - 3y \end{cases}$
5. $\begin{cases} x' = y \\ y' = x^3 - x \end{cases}$
6. $\begin{cases} x' = \sin(x) - y \\ y' = y^2 - \frac{1}{4} \end{cases}$

7. $\begin{cases} x' = y^2 - 1 \\ y' = x^2 - y \end{cases}$
8. $\begin{cases} x' = y + x \\ y' = x^3 - 8y \end{cases}$
9. $\begin{cases} x' = x(3 - x) - 2xy \\ y' = y(2 - y) - xy \end{cases}$

---

[2]Locally, the plane is defined by the two eigenvectors of the linear center and the third direction is determined by the additional eigenvectors. More generally, we can define a **stable subspace**, **unstable subspace**, and **center subspace** that divide the space.

10. A well-known equation is the van der Pol oscillator, which models a triode valve where the resistance depended on the applied current [16]:

$$x'' + \epsilon x'(x^2 - 1) + x = 0, \tag{6.16}$$

where $x = x(t)$ and $\epsilon > 0$ is a constant. Using the methods of Section 3.4, convert this equation to a system of two first-order equations. Do part (i) and (ii) above for a general ϵ. Then do parts (iii) and (iv) with $\epsilon = 0.1$. Repeat steps (iii) and (iv) for $\epsilon = 10$, compare your phase portraits, and comment on any differences you see.

11. *We now reconsider Problem 2 from Section 2.6.* In a first physics course, students derive the equation of motion for a frictionless simple pendulum as

$$m\theta'' + g \sin \theta = 0, \tag{6.17}$$

where θ is the angle that the pendulum makes with the vertical. We will *not* assume that θ is small. For convenience, set $m = g$.
(a) Convert (6.17) to a system of two first-order equations.
(b) Find the equilibria for $\theta \in [-2\pi, 2\pi]$ and classify stability.
(c) Graph the nullclines on the phase plane.
(d) Use MATLAB, Maple, or Mathematica to sketch the vector field for the system of two first-order equations.
(e) Plot trajectories for various initial conditions and obtain a phase portrait similar to Figure 2.27 in Section 2.6. Again interpret the qualitatively different motions of the pendulum, keeping in mind that the pendulum is allowed to whirl over the top.

12. Now consider the simple pendulum with damping:

$$\theta'' + 0.3\theta' + \theta = 0.$$

Repeat parts (a)–(d) in the previous problem. Then plot trajectories for various initial conditions and compare with the phase portrait of the undamped motion. Interpret your picture and the differences between the two phase portraits.

13. *We now reconsider Project 4A of Chapter 4.* One formulation of the forced Duffing equation is

$$x'' + bx' + kx + \delta x^3 = F_0 \sin(\omega t), \tag{6.18}$$

where $x = x(t)$. When $\delta = 0$, the equation reduces to that of the forced mass on a spring of Section 4.7. Repeat steps 2-5 of this project but now also plot the trajectories in the phase plane. Because this system is nonautonomous, you will not be able to use `pplane`. Besides your explanations, be sure to address why the apparent crossing of solutions in the phase plane is not a violation of the Existence and Uniqueness theorem.

14. By Taylor expanding about the equilibrium point (x^*, y^*, z^*) and keeping only linear terms show that the three-dimensional system

$$
\begin{aligned}
x' &= f(x, y, z) \\
y' &= g(x, y, z) \quad \text{has Jacobian} \quad
\mathbf{J} =
\begin{bmatrix}
\frac{\partial f}{\partial x} & \frac{\partial f}{\partial y} & \frac{\partial f}{\partial z} \\
\frac{\partial g}{\partial x} & \frac{\partial g}{\partial y} & \frac{\partial g}{\partial z} \\
\frac{\partial h}{\partial x} & \frac{\partial h}{\partial y} & \frac{\partial h}{\partial z}
\end{bmatrix}_{(x^*, y^*, z^*)}. \\
z' &= h(x, y, z)
\end{aligned}
$$

15. The Lorenz system can be written in the form

$$
\begin{aligned}
x' &= -\sigma x + \sigma y \\
y' &= rx - y - xz \\
z' &= -\phi z + xy,
\end{aligned}
\tag{6.19}
$$

where σ, r, ϕ are positive parameters. The system arose as a model of convective rolls in the atmosphere. Lorenz studied the parameter values $\sigma = 10, \phi = 8/3$ and examined how the behavior of solutions changed as r increased.

(a) Determine the equilibria and their stability for $r = 0.5$.
(b) Determine the equilibria and their stability for $r = 2$.
(c) Determine the equilibria and their stability for $r = 25$.
(d) Plot trajectories in x-y-z **phase space** for parts (a), (b), (c). Go from $t = 0$ to $t = 100$ for three different initial conditions.

6.3   Bifurcations

The system of equations (6.8) with an explicit parameter vector $\mathbf{p}$,

$$
\begin{aligned}
x' &= f(x, y, \mathbf{p}) \\
y' &= g(x, y, \mathbf{p}),
\end{aligned}
$$

can have bifurcations as the parameter $\mathbf{p}$ varies, just as we did with first-order autonomous equations in Section 2.3. Recall that in 1-D, we said that a **bifurcation** occurs when we have a qualitative change in the number or stability of equilibrium solutions. This is true in higher dimensions but we may also have the appearance (or disappearance) of periodic solutions as well.

Recall that Theorem 6.2.2 characterized the stability of an equilibrium solution according to its eigenvalue. Part (a) specifically required $\operatorname{Re}(\lambda_i) < 0$ for all i in order to characterize the solution as stable. If one of λ_i has its real part change from negative to positive, the solution will lose its stability.

Consider only the eigenvalue(s), λ, that is involved in the bifurcation. In order for the $\operatorname{Re}(\lambda)$ to change sign, we must either have

(a) $\lambda = 0$ bifurcations can be either **saddlenode**, **transcritical**, or **pitchfork** bifurcations

(b) $\lambda = \pm ib$ bifurcations, where $b \in \mathbb{R}$, are known as **Hopf** bifurcations

These bifurcations are known as **local bifurcations** as they involve a change in the behavior of the system near an equilibrium point and only near this point. We have seen the names of the $\lambda = 0$ in 1-D but the $\lambda = \pm ib$ (pure imaginary) name, **Hopf bifurcation**, is new to us. Besides having a pure imaginary eigenvalue, we will also have the appearance or disappearance of an isolated periodic trajectory called a **limit cycle**. (By isolated, we simply mean that we can find an $\epsilon > 0$ such that if we begin within ϵ of the limit cycle, we will not encounter another closed orbit.)

$\lambda = 0$ bifurcations

We can have saddlenode bifurcations in which a saddle point and a node (either stable or unstable) coalesce and disappear as a parameter is varied; we can have transcritical bifurcations in which a saddle and a node exchange stability as one passes through the other as a parameter is varied; we can have subcritical pitchfork bifurcations (involving the birth/death of two unstable equilibria) and supercritical pitchfork bifurcations (involving the birth/death of two stable equilibria) in which two equilibria are born out of one but with different stability than the one from which they were born. These types of bifurcations occur when one eigenvalue of the point (i.e., equilibrium solution) undergoing the bifurcation is zero.

We consider the basic form for a supercritical pitchfork bifurcation:

$$
\begin{aligned}
x' &= rx - x^3 \\
y' &= -y
\end{aligned}
\tag{6.20}
$$

where $r \in \mathbb{R}$. The equilibria are $(x^*, y^*) = (0,0), (\sqrt{r}, -0), (\sqrt{r}, 0)$, where the last two equilibria only exist when $r > 0$. As in the first-order equation, we have three cases to consider: $r < 0, r = 0$, and $r > 0$. We could check the stability analytically using the linearization previously discussed. The Jacobian is easily calculated as

$$
\mathbf{J} = \begin{bmatrix} r - 3x^2 & 0 \\ 0 & -1 \end{bmatrix}.
$$

Evaluating at the equilibrium points gives for $r < 0$

$$
J(0,0) = \begin{bmatrix} r & 0 \\ 0 & -1 \end{bmatrix}.
$$

Thus $\beta = \text{Tr}(J(0,0)) = r - 1$ and $\gamma = \det(J(0,0)) = -r$ so $(0,0)$ is a spiral source. Because we are in the case when $r < 0$, we see that $\beta < 0$ and $\gamma > 0$ and thus the origin is stable.

For $r = 0$, we have

$$J(0,0) = \begin{bmatrix} 0 & 0 \\ 0 & -1 \end{bmatrix}.$$

Thus $\beta = \mathrm{Tr}(J(0,0)) = -1$ and $\gamma = \det(J(0,0)) = 0$ and the prediction is for a linear center. This is a borderline case and thus the conclusion may be affected by the nonlinear terms that were ignored.

For $r > 0$, we have three equilibria:

$$J(0,0) = \begin{bmatrix} r & 0 \\ 0 & -1 \end{bmatrix} \quad \text{and} \quad J(\pm\sqrt{r},0) = \begin{bmatrix} r - 3r & 0 \\ 0 & -1 \end{bmatrix}.$$

Thus $\beta = \mathrm{Tr}(J(0,0)) = r - 1$ and $\gamma = \det(J(0,0)) = -r$ and because $r > 0$, we know this is a saddle point. The Jacobian is the same for the other two equilibria and we have that $\beta = \mathrm{Tr}(J(\pm\sqrt{r},0)) = -2r - 1$ and $\gamma = \det(J(0,0)) = 2r$, which shows that both points are stable equilibrium points.

Thus, for $r < 0$ the origin is the only equilibrium solution and it is stable. For $r > 0$, two additional equilibria exist that were born out of the origin (when $r = 0$, they are located at $(0,0)$) and both are stable, while the origin has now changed its stability; see Figure 6.10.

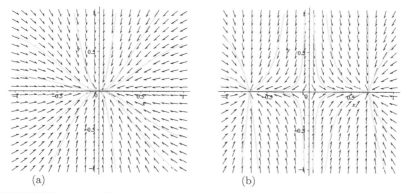

(a)   (b)

FIGURE 6.10: Equations 6.20. (a) Phase plane plot of trajectories for $r = -0.5$; there is only one equilibrium, $(0,0)$, and it is stable. (b) Phase plane plot of trajectories for $r = 0.5$; there are three equilibria, $(-\sqrt{5},0)$, $(0,0)$, $(\sqrt{5},0)$. The equilibrium $(0,0)$ is a saddle and $(\pm\sqrt{5},0)$ are stable nodes.

To see this, we calculate the characteristic equation of the Jacobian evaluated at one of the equilibria (in this example, all three give the same answer):

$$\det(\mathbf{J}(0,0) - \lambda\mathbf{I}) = (r - \lambda)(-1 - \lambda) = \lambda^2 + \lambda(1 - r) - r = 0.$$

We substitute $\lambda = 0$ and see that $r = 0$ is the parameter value for one of the above four bifurcations. We could then use linearization near the equilibria before and after the bifurcation value to see if there were a qualitative change in the number or stability of the equilibria.

Example 1: Consider the system

$$x' = rx - x^2$$
$$y' = -y, \tag{6.21}$$

for $r \in \mathbb{R}$. Determine the type of $\lambda = 0$ bifurcation that occurs and classify it according to the change in the number or stability of the equilibria. Use the computer to corroborate your conclusions.

Solution
We calculate the equilibria as $(0,0)$ and $(r,0)$. It seems that there is always two equilibria except when $r = 0$ in which case we will only have $(0,0)$. The Jacobian is seen to be

$$J(x,y) = \begin{bmatrix} r - 2x & 0 \\ 0 & -1 \end{bmatrix},$$

with characteristic polynomial $\lambda^2 + (1 - r + 2x)\lambda - r + 2x = 0$. To find the $\lambda = 0$ bifurcations, we set $\lambda = 0$ in this equation and then plug in each of the equilibria

Equilibrium $(0,0)$:	$-r = 0$	(6.22)
Equilibrium $(r,0)$:	$-r + 2r = 0.$	(6.23)

Both equations are the same: $r = 0$. This should not be a surprise given that the equilibrium points are the same when $r = 0$. Thus we expect a $\lambda = 0$ bifurcation to occur when $r = 0$. To determine the type of bifurcation, we examine what happens for r-values before and after this value, i.e., for $r < 0$ and $r > 0$. Then

$$J(0,0) = \begin{bmatrix} r & 0 \\ 0 & -1 \end{bmatrix}, \tag{6.24}$$

and thus $(0,0)$ is stable for $r < 0$ and a saddle for $r > 0$ whereas for

$$J(r,0) = \begin{bmatrix} -r & 0 \\ 0 & -1 \end{bmatrix}, \tag{6.25}$$

which shows that $(r,0)$ is a saddle for $r < 0$ and stable for $r > 0$. Similar to 1-D, we call this a transcritical bifurcation; see Figure 6.11.

The other two types of 1-D bifurcations seen in Section 2.3, saddlenode bifurcations and subcritical pitchfork bifurcations, are again seen in two or more dimensions and occur when $\lambda = 0$ in the characteristic equation. It is often useful to track the location of the equilibrium as a function of one of

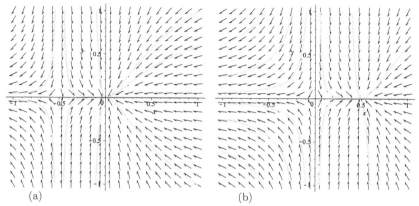

(a)   (b)

FIGURE 6.11: (a) Phase plane plot of trajectories for $r = -0.5$; $(0,0)$ is stable and $(r,0)$ is a saddle. (b) Phase plane plot of trajectories for $r = 0.5$; $(0,0)$ is now a saddle and $(r,0)$ is stable. The system thus underwent a transcritical bifurcation at $r = 0$.

the parameters. For example, if we consider the Equations (6.20) that gave us a supercritical pitchfork bifurcation:

$$x' = rx - x^3$$
$$y' = -y,$$

we found the equilibria to be $(0,0)$ and $(\pm\sqrt{r},0)$. The $(0,0)$ equilibrium is stable when $r < 0$ and is unstable (a saddle) if $r > 0$. The remaining two equilibria do not exist for $r < 0$ and are stable for $r > 0$. Because $y^* = 0$ for all equilibria, we will plot r vs. x^*. This type of picture is called a **bifurcation diagram** and the typical convention is to denote stability with a stable curve and instability with a dashed curve as shown in Figure 6.12(a). If we consider the Equations (6.21) that gave us a transcritical bifurcation:

$$x' = rx - x^2$$
$$y' = -y,$$

we found the equilibria to be $(0,0)$ and $(r,0)$. The $(0,0)$ equilibrium is stable when $r < 0$ and is unstable (a saddle) if $r > 0$. In contrast, the $(r,0)$ equilibrium is unstable (a saddle) when $r < 0$ and is stable when $r > 0$. We again plot r vs. x^* and observe the bifurcation diagram in Figure 6.12(b).

Hopf bifurcation

The other type of local bifurcation that can now occur to change the stability of an equilibrium point is a **Hopf bifurcation**. These also come in supercritical and subcritical flavors. This bifurcation occurs when the real part of a complex pair of eigenvalues becomes zero ($\lambda = \pm ib$, with $b \neq 0$ and $b \in \mathbb{R}$)

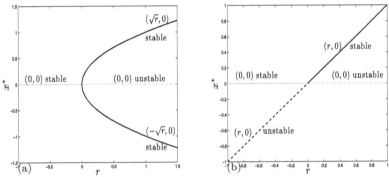

FIGURE 6.12: Bifurcation diagrams: the x-location and stability of the equilibria as a function of the parameter r. (a) The supercritical pitchfork bifurcation of Equations (6.20); (b) The transcritical bifurcation of Equations (6.21).

as a parameter is varied, and either before or after the bifurcation value we have the presence of a limit cycle.

Example 2 Consider the system

$$\dot{x} = y + rx \tag{6.26}$$
$$\dot{y} = -x + ry - x^2 y \tag{6.27}$$

where $r \in \mathbb{R}$. Show that the origin undergoes a **supercritical Hopf bifurcation** when $r = 0$.

Solution
The origin is clearly an equilibrium point and we determine its stability via the Jacobian:

$$J(0,0) = \begin{bmatrix} r & 1 \\ -1 & r \end{bmatrix}. \tag{6.28}$$

The characteristic equation is then $\lambda^2 - 2r\lambda + 1 + r^2 = 0$. In order to have a Hopf bifurcation, we need $\lambda = \pm ib$ with $b \in \mathbb{R}$ and $b \neq 0$. Substituting this condition gives

$$-b^2 + 1 + r^2 - 2irb = 0.$$

In order for this to hold, we equate real and imaginary parts on both sides of the equation:

$$-b^2 + 1 + r^2 = 0 \qquad \text{real part} \tag{6.29}$$
$$-2irb = 0 \qquad \text{imaginary part.} \tag{6.30}$$

The second equation gives $r = 0$ as one condition (since $b \neq 0$). Substitution of this into the first gives $b^2 = 1$ and since this satisfies both $b \in \mathbb{R}$ and $b \neq 0$, we conclude that we *may* have a Hopf bifurcation when $r = 0$. The additional requirement we need is the appearance/disappearance of a limit cycle as r passes through 0. We enter this system into one of our computer systems and choose a variety of initial conditions for a pair of parameter values close to the bifurcation value, say $r = \pm 0.1$. In Figure 6.13, we observe that the origin is a stable spiral for $r = -0.1$. In contrast, we see that the origin is an unstable spiral for $r = 0.1$ *and* there is a **limit cycle** surrounding the origin. Thus our system underwent a Hopf bifurcation when $r = 0$. The Hopf bifurcation is called *supercritical* because the limit cycle involved is stable; that is, trajectories starting close to it (whether inside or outside) eventually approach it.

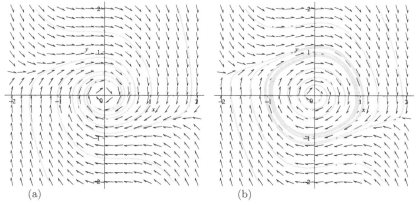

(a)   (b)

FIGURE 6.13:   (a) Phase plane plot of trajectories for $r = -0.1$; $(0, 0)$ is a stable spiral. (b) Phase plane plot of trajectories for $r = 0.1$; $(0, 0)$ is now an unstable spiral and a stable limit cycle is present.

The **limit cycle** is an isolated closed trajectory (*not* a center) and is a periodic solution. While we must have a switch in stability of the equilibrium point, it is the presence of the limit cycle on only one side of the parameter bifurcation value that makes this a Hopf bifurcation and the stability of the limit cycle that determines the type of Hopf bifurcation. We can also have an **unstable limit cycle** in which case an initial condition starting close to it (whether inside or outside) was repelled by it. If an unstable limit cycle is present in a Hopf bifurcation, we say that it is **subcritical**. In either case, the limit cycle increases in amplitude as we move further from the bifurcation of the parameter value and will shrink to a radius of zero as we approach the bifurcation value. If we do not have a limit cycle that appears on either side, we call the situation a **degenerate Hopf bifurcation**. There are other types of *global* bifurcations that can also create (or destroy) limit cycles but that do not exclusively occur in a neighborhood of an equilibrium point.

Example 3 Consider the system

$$\dot{x} = py \tag{6.31}$$
$$\dot{y} = -x + ry + x^2 + xy + y^2 \tag{6.32}$$

where $p, r \in \mathbb{R}$. Show that the origin undergoes a **subcritical Hopf bifurcation** when $r = 0, p > 0$.

Solution

The origin is clearly an equilibrium point. We note that $(1,0)$ is also an equilibrium point but it is not involved in the Hopf bifurcation so we ignore it for now. We determine the stability of $(0,0)$ via the Jacobian:

$$J(0,0) = \begin{bmatrix} 0 & p \\ -1 & r \end{bmatrix}. \tag{6.33}$$

The characteristic equation is then $\lambda^2 - r\lambda + p = 0$. In order to have a Hopf bifurcation, we need $\lambda = \pm ib$ with $b \in \mathbb{R}$ and $b \neq 0$. Substituting this condition gives

$$-b^2 + p - irb = 0.$$

In order for this to hold, we equate real and imaginary parts on both sides of the equation:

$$-b^2 + p = 0 \qquad \text{real part} \tag{6.34}$$
$$-irb = 0 \qquad \text{imaginary part.} \tag{6.35}$$

The second equation gives $r = 0$ as one condition (since $b \neq 0$). Substitution of this into the first gives $b^2 = p$, which will only be satisfied when $p > 0$ since we know that $b \in \mathbb{R}$ and $b \neq 0$. Thus we conclude that we *may* have a Hopf bifurcation when $r = 0$. The additional requirement we need is the appearance/disappearance of a limit cycle as r passes through 0 for $p > 0$. Up to this point, our bifurcations have only involved one parameter. Because our conditions now involve two parameters, r and p, we say that we have a **two parameter family** of differential equations. While we can have subcritical Hopf bifurcations that only involve one parameter, other bifurcations (that we have not yet encountered) may need 2 or more parameters in order to occur.

We enter this system into one of our computer systems and choose a variety of initial conditions for a pair of parameter values close to the bifurcation value, say $r = \pm 0.04$ and $p = .2$. (There are numerous other choices that would work.) In Figure 6.13, we observe that the origin is a stable spiral for $r = -0.04$ and is an unstable spiral for $r = 0.04$. However, any trajectory with an initial condition starting just to the left of $(1,0)$ in either picture becomes unbounded. While this is an acceptable conclusion when the origin is unstable, it does not make sense when the origin is stable. To see this, consider the initial condition $(.6, 0)$. In both pictures, this trajectory becomes unbounded. However, if we consider the initial condition $(.1, 0)$, we observe that it become unbounded for $r = 0.04$ but clearly approaches the origin when $r = -0.04$. If we trace the trajectory *backward* in time for $r = -0.04$, we see that it came from an **unstable limit cycle**. Thus we conclude that our system has undergone a subcritical Hopf bifurcation.

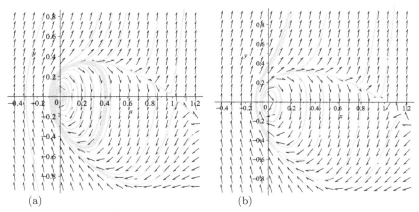

(a)   (b)

FIGURE 6.14: (a) Phase plane plot of trajectories for $r = -0.04, p = 0.2$; $(0, 0)$ is a stable spiral and an unstable limit cycle is present. (b) Phase plane plot of trajectories for $r = 0.04, p = 0.2$; $(0, 0)$ is now an unstable spiral. Although the trajectories in both (a) and (b) spiral in the same way, the stability of the origin helps us "observe" and infer the unstable limit cycle in (a), passing through about $(.4, 0)$.

From Example 3, we can see that it often takes trial and error in finding numerical values that will confirm our analytical conclusions. Observing limit cycles can be a balance between parameter values that are too close to the bifurcation values to observe this phenomenon and parameter values that are too far away and other things have already changed. It turns out that we still have an unstable limit cycle for $r = 0.05, p = 0.2$ but it has already disappeared when $r = 0.06$ through an entirely different bifurcation—a **homoclinic bifurcation** in which the limit cycle grew in amplitude but was destroyed when it touched the other equilibrium point $(1, 0)$.

We briefly present an alternative approach to determining the presence of a Hopf bifurcation in a 2-dimensional system. Recall Theorem 6.1.1 and Figure 6.5 that determined conditions on a linear system. We know that a supercritical or subcritical Hopf bifurcation occurs when the spiral equilibrium point changes stability. Thus, we must be crossing the $\beta = 0$ line for $\gamma > 0$. Revisiting Example 2 of this section, we had

$$J(0,0) = \begin{bmatrix} r & 1 \\ -1 & r \end{bmatrix},$$

so that our alternative approach requires

$$\beta = 2r = 0, \qquad \gamma = r^2 + 1 > 0.$$

This gives the same condition $r = 0$ as previously found (the equation on γ is always satisfied). Revisiting Example 3 of this section, we had

$$J(0,0) = \begin{bmatrix} 0 & p \\ -1 & r \end{bmatrix},$$

so that our alternative approach requires

$$\beta = r = 0, \qquad \gamma = p > 0.$$

This gives the same conditions $r = 0$ and $p > 0$ as previously found. In both cases, we still need to numerically find the limit cycle on one side of the bifurcation value but this approach is sometimes easier.

We can also have $\lambda = 0$ and Hopf bifurcations occurring in the same problem and we explore this more in the problems.

•   •   •   •   •   •   •   •   •   •   •

Problems

In Problems **1–8**, determine the type of bifurcation that occurs. Confirm your analytical results with qualitatively different phase portraits before and after the bifurcation. If instructed, draw the bifurcation diagram as well.

1. $\begin{cases} x' = r - x^2 \\ y' = -y \end{cases}$
        2. $\begin{cases} x' = y - x^2 \\ y' = rx - y \end{cases}$

3. $\begin{cases} x' = rx - x^2 \\ y' = x - y \end{cases}$
        4. $\begin{cases} x' = ry - x^2 \\ y' = x - y \end{cases}$

5. $\begin{cases} x' = rx - x^3 \\ y' = -y \end{cases}$
        6. $\begin{cases} x' = rx + x^3 \\ y' = -y \end{cases}$

7. $\begin{cases} x' = rx + 2y \\ y' = -2x + ry + x^2 y \end{cases}$
        8. $\begin{cases} x' = rx + 2y \\ y' = -2x + ry - x^2 y \end{cases}$

9. Consider the oscillator described by

$$x'' + x'(x^2 - \epsilon) + x = 0, \tag{6.36}$$

where $x = x(t)$ and $\epsilon \in \mathbb{R}$. Show this system undergoes a Hopf bifurcation when $\epsilon = 0$. What type of Hopf bifurcation is it?

10. Another famous nonlinear differential equation is the double-well oscilla-tor

$$x'' + \epsilon x' - x + x^3 = 0.$$

Using the methods of Section 3.4, convert this equation to a system of two first-order equations. Find and classify any $\lambda = 0$ or Hopf bifurcations that occur. Draw phase portraits for $\epsilon = 0, 0.25, 1$ and describe any differences.

11. Consider the system given by $\begin{cases} x' = -by + f(x,y) \\ y' = bx + g(x,y) \end{cases}$ with $f(0) = g(0) = 0$
and $f_x(0) = f_y(0) = g_x(0) = g_y(0) = 0$ and the quantity

$$\mu = \frac{1}{16}[f_{xxx} + f_{xyy} + g_{xxy} + g_{yyy}] \qquad (6.37)$$

$$+ \frac{1}{16b}[f_{xy}(f_{xx} + f_{yy}) - g_{xy}(g_{xx} + g_{yy}) - f_{xx}g_{xx} + f_{yy}g_{yy}],$$

where the partial derivatives are all evaluated at $(0,0)$. It is shown in
[27] that the system will undergo a supercritical Hopf bifurcation if $\mu < 0$
and a subcritical Hopf bifurcation if $\mu > 0$. Use this to show that the
$x' = y + rx$, $y' = -x + ry - x^2y$ undergoes a supercritical Hopf bifurcation
when $r = 0$.

12. Consider the Lorenz system

$$x' = -\sigma x + \sigma y$$
$$y' = rx - y - xz$$
$$z' = -\phi z + xy,$$

where σ, r, ϕ are positive parameters.
(a) Show that the origin is always an equilibrium solution with eigenvalues

$$\lambda_1 = -\phi, \lambda_{2,3} = -\frac{(\sigma+1) \pm \sqrt{\sigma^2 - 2\sigma + 1 + 4r\sigma}}{2}.$$

(b) Show that the two additional equilibria are given by $(x^*, y^*, z^*) = (\sqrt{\phi(r-1)}, \sqrt{\phi(r-1)}, r-1), (-\sqrt{\phi(r-1)}, -\sqrt{\phi(r-1)}, r-1)$.
(c) Show that the origin undergoes a supercritical pitchfork bifurcation
at $r = 1$.
(d) Show that each of these additional equilibria undergoes a subcritical
Hopf bifurcation when $r_c = \frac{\sigma(\sigma + \phi + 3)}{\sigma - \phi - 1}$. With $\sigma = 10, \phi = 8/3$, look
for the unstable limit cycle in this three-dimensional system by starting
close to the equilibrium solution for r slightly smaller than r_c. You will
see that initial conditions close enough to the point will approach it while
those just slightly bigger will go away from it.

6.4 Epidemiological Models

This section deals with the interaction of groups of people in an effort
to better understand the spread of a certain disease. There are a myriad of

mathematical models that can be used to describe the spread and transmission of a wide range of diseases. Most of the framework in epidemiology consists of dividing the population under consideration into various classes. Each class defines the state of the individual in reference to the disease being modeled. These classes typically consist of three groups: those that are **susceptible** to the disease, those that are **infected** with the disease, and those that are **recovered** from the disease. A model with only these three classes is called a Susceptible-Infected-Recovered (SIR) model. The type of mathematics that is used or implemented in the model is determined by the rules governing or describing the movement of individuals from one class to another. In the SIR model, individuals move from the susceptible class to the infected class to the recovered class depending on their interactions with infected individuals and on their bodies' ability to fight the disease. There are many variations of this model. For example, the recovered individuals may be permanently immune from the disease (e.g., measles) or they may have temporary immunity or no immunity and be susceptible again to the disease (e.g., syphilis). In the latter case, there are no recovered individuals as everyone who has been cured becomes immediately susceptible. However, in the case of temporary immunity the recovered individuals eventually become susceptible again.

In this section, we will formulate the well-known SIR model and some variations, discuss the rates of transmission and recovery, and mathematically analyze the resulting model. We will find the equilibria of the system, determine their stability, and examine whether any bifurcations are possible. The topics from Section 6.2 will help us understand the effect of a disease on the population and give us insight into the key parameters driving the spread of this disease.

A Model without Vital Dynamics

We consider a simple model in which individuals do not enter or leave the system through natural birth or death (or immigration or emigration). A disease that spreads rapidly through a population is a good candidate for this framework. Pioneering work in epidemiology was done in 1927 by Kermack and McKendrick [25] in their study of the Bombay plague of 1906. They divided the population into $S(t)$ = number of susceptible individuals, $I(t)$ = number of recovered individuals, and $R(t)$ = number of removed (or deceased) individuals. They made three basic assumptions about the disease:

1. It traveled quickly through the population and thus no people were able to leave or enter the system. There are no births, deaths, and no immigration or emigration. In epidemiological terms, this is a model *without vital dynamics*.

2. When an infected individual encounters a susceptible individual, there is a probability, β, that the susceptible individual will get the disease. This occurs in proportion to the numbers of individuals in the infected and

susceptible classes. This is referred to as **standard incidence**.[3]

3. Infected individuals recover at a constant rate, α.[4]

With these assumptions, we can write the following set of equations:

$$\frac{dS}{dt} = -\beta \frac{I}{N} S$$

$$\frac{dI}{dt} = \beta \frac{I}{N} S - \alpha I$$

$$\frac{dR}{dt} = \alpha I, \tag{6.38}$$

where $N = S + I + R$ is the total population and $S(t), I(t), R(t) \geq 0$; see Figure 6.15.

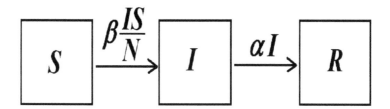

FIGURE 6.15: Flowchart for SIR model without vital dynamics. This diagram illustrates the movement of individuals from one class to another.

We will typically refer to β as the transmission rate because it is the average number of contacts per unit time sufficient for transmission of the disease, that is, $\beta = \dfrac{\text{contacts}}{\text{unit time}} \times \dfrac{\text{probability of transmission}}{\text{contact}}$. Sometimes β is also referred to as the average number of **adequate contacts**. Thus the average number of contacts that a susceptible individual has with infectives per unit time is given by $\beta I/N$. The number of new infections per unit time in a susceptible community with S individuals is $\left[\frac{\beta I}{N}\right] S$. Since α is the recovery rate, the average length of time an individual will remain infected is given by $1/\alpha$. If we multiply the average number of adequate contacts β by the average infectious period $1/\alpha$, we get the "average number of adequate contacts of a typical infective during the infectious period" (from Hethcote [21]). It is

---

[3]For smaller populations it is sometimes better to consider a contact rate proportional to the total population and this is referred to as **mass action**—the interaction term in (6.38) would then be $\beta S I$.

[4]If we let $u(t)$ represent the proportion of infected individuals remaining at t time units, then we can write $u' = -\alpha u$. The solution is $u(t) = u(0)e^{-\alpha t}$ and thus the fraction that is still infective after t time units is $e^{-\alpha t}$. If we think of this in terms of probability, we see this is an exponential distribution and thus the average length of the infective period is $1/\alpha$.

sometimes useful to consider only fractions of a population and we can do so with this model by introducing new variables:[5] $s = S/N, i = I/N, r = R/N$. This is just a special case of non-dimensionalization from Section 2.4. In the current model (6.38), we can divide both sides of the three equations by the total population N and thus obtain

$$\frac{ds}{dt} = -\beta si$$

$$\frac{di}{dt} = \beta is - \alpha i$$

$$\frac{dr}{dt} = \alpha i. \tag{6.39}$$

The variables under consideration will satisfy $0 \le s, i, r \le 1$ and $s + i + r = 1$. Model (6.39) is often easier to deal with both in terms of interpreting the results and from a numerical point of view. It allows us to consider relative changes in the population size. Mathematically, the two formulations are equivalent. Homework Problem **6** requires the reader to go through this derivation.

Since the total population is constant we have

$$s' + i' + r' = 0,$$

and we can reduce the dimension of our system to two by substituting $r = 1 - s - i$. We thus arrive at the equivalent yet simpler formulation

$$\frac{ds}{dt} = -\beta si$$

$$\frac{di}{dt} = \beta is - \alpha i. \tag{6.40}$$

Our solution can now be examined in $\mathbb{R}^2$ rather than $\mathbb{R}^3$. We could solve these equations analytically by considering $di/ds = i'/s'$ but we will instead use the linearization methods covered in Section 6.2. We can observe that there is an entire line of equilibria with

$$i^* = 0, s^* = \text{arbitrary}.$$

This makes sense as whatever fraction of susceptible people that remain when the final infected person recovers will always remain susceptible. We are interested in understanding how the disease spreads across a completely susceptible population and thus $s^* = 1$ is the equilibrium with which we will be concerned. This equilibrium point in which there are no infected individuals is

---

[5]We use i as the fraction of infective individuals and caution the reader to *not* confuse it with the imaginary $i = \sqrt{-1}$, which will sometimes arise as an eigenvalue in these problems.

known as the **disease-free equilibrium** (*DFE*). We calculate the Jacobian matrix of (6.40) as

$$\begin{bmatrix} -\beta i & -\beta s \\ \beta i & \beta s - \alpha \end{bmatrix} \tag{6.41}$$

and evaluate it at the *DFE* to obtain

$$\begin{bmatrix} 0 & -\beta \\ 0 & \beta - \alpha \end{bmatrix}. \tag{6.42}$$

We can use the trace and determinant to determine its stability. Here we have

$$\text{Tr}(J(1,0)) = \beta - \alpha, \qquad \det(J(1,0)) = 0$$

for $(0,0)$. The latter condition shows we will have a borderline case. Even so, we will proceed in order to gain an understanding of how the solution behaves in the remaining eigendirection. If $\text{Tr} < 0$, then we are borderline between a saddle and stable equilibria, whereas if $\text{Tr} > 0$, then we are borderline between a saddle and an unstable equilibria. We summarize with

$$DFE \text{ is stable when } \beta < \alpha$$
$$DFE \text{ is unstable when } \beta > \alpha, \tag{6.43}$$

acknowledging that we still have a direction with no change (this will not happen when we introduce births and deaths). Recalling that β is average number of **adequate contacts** for disease transmission and α is the recovery rate of infected individuals, we see that this stability condition for the *DFE* is believable. *The disease will spread if the average number of adequate contacts β is larger than the recovery rate α.* Mathematical epidemiologists often consider the **basic reproductive number**, R_0, which gives the average number of infections caused by one infected individual over his/her period of infection as an equivalent method of determining if a disease will spread. Thus $R_0 < 1$ says that, on average, an infected individual infects less than 1 individual (for example, 10 infected individuals might only infect 7 others before recovering). Thus the disease will eventually die out. Similarly, $R_0 > 1$ says that, on average, an infected individual infects more than 1 individual (for example, 5 infected individuals might infect 8 others before recovering). In this latter case, the disease will spread through the population. If $R_0 > 1$, we say that there is an **epidemic**.

The mathematical epidemiologists examine the stability conditions for the *DFE* and determine which condition or conditions will first cause it to become unstable. When considering R_0, the stability conditions are rewritten in terms of R_0 such that

$$R_0 < 1 \iff DFE \text{ equilibrium is stable}$$
$$R_0 > 1 \iff DFE \text{ equilibrium is unstable.} \tag{6.44}$$

In our case, there is only one condition that will cause a change in stability. Our goal is to manipulate that expression so that $R_0 = 1$ corresponds to the switch in stability. In our current example, we can manipulate (6.43) to obtain

$$R_0 = \frac{\beta}{\alpha}. \tag{6.45}$$

Following the conditions in (6.44), the DFE is stable if $\frac{\beta}{\alpha} < 1$ and unstable if $\frac{\beta}{\alpha} > 1$, which is just what we previously found.[6]

In the exercises, you will show that when a small number of infective individuals are introduced into a susceptible population,

$$\frac{\beta}{\alpha} \approx \frac{\ln\left(\frac{s(0)}{s^*}\right)}{1 - s^*}. \tag{6.46}$$

While it is usually not hard to determine the average length of the infection for a given disease, without this approximation it can often be difficult to determine the number of adequate contacts.

Example 1:   Consider a geographically isolated college campus and suppose 95% of the students are susceptible to the influenza virus at the beginning of the school year [9]. By the end of the year, many had become sick with the flu and only 42% were still susceptible after the flu had run its course on campus. Estimate the basic reproductive number for this flu and determine if there was an epidemic. How might this epidemic have been prevented?

Solution

Since 42% of the individuals were still susceptible after the flu was gone, we see that $s^* = .42$. Similarly, we see that $s(0) = .95$. Thus we see that

$$R_0 = \frac{\beta}{\alpha} \approx \frac{\ln\frac{s(0)}{s^*}}{1 - s^*} = \frac{\ln\frac{.95}{.42}}{1 - .42} = 1.41,$$

and there was indeed an epidemic on the campus. Preventing the epidemic would require us to alter some of the values in the formula for R_0. For example, if we were to restrict the interactions of the students (which may or may not be practical), we could change the contact rate, thereby lowering R_0. We also could have **vaccinated** individuals at the start of the year. This would lower the initial number of susceptible individuals and would also reduce the number of students that had the flu. In this latter situation, we should note that we can't selectively vaccinate those students who will end up getting sick because we don't know in advance who will get sick.

---

[6]Problem 9 examines an equivalent formulation of (6.44)-(6.45) when we don't normalize the equations: DFE stable $\iff \frac{S(0)}{N} < \frac{\beta}{\alpha}$.

A Model with Vital Dynamics

We now consider a somewhat more realistic model by allowing natural births and deaths in the population to occur. We again let S represent the number of susceptible individuals, I represent the number of infected individuals, and R represent the number of recovered individuals. We assume that the death rate and birth rate are the same so that there is no change in the overall population. Denote this rate as $\mu > 0$ and let $N = S + I + R$ again represent the total population. We assume that all individuals are born susceptible to the disease even when they are born from an infected individual. The number of births from the susceptible class, the infected class, and the recovered class are defined as μS, μI, and μR, respectively. Thus μN is the total number of births or newcomers into the susceptible class. Similarly the number of deaths in each class per unit time are given by μS, μI, and μR, so that the population level remains constant. If we again make assumptions 2 and 3 from the previous model, our diagram is shown in Figure 6.16.

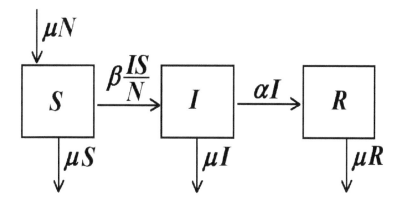

FIGURE 6.16: Flowchart for SIR model with vital dynamics.

The corresponding equations are

$$\frac{dS}{dt} = \mu N - \mu S - \beta \frac{I}{N} S$$
$$\frac{dI}{dt} = \beta \frac{I}{N} S - \alpha I - \mu I$$
$$\frac{dR}{dt} = \alpha I - \mu R. \tag{6.47}$$

This is called a **SIR model with vital dynamics**. Because N is constant, we can reduce the system to only two equations. If we consider the normalized populations of $s = S/N, i = I/N, r = R/N$, we can rewrite the governing equations as

$$\frac{ds}{dt} = \mu - \mu s - \beta i s$$

$$\frac{di}{dt} = \beta i s - (\alpha + \mu) i. \tag{6.48}$$

(The reader should convince him/herself that this can be done; see homework Problem **7**.) We have two equilibria (s^*, i^*) for this system,

$$(1, 0) \text{ and } \left(\frac{\alpha + \mu}{\beta}, \frac{\mu}{\alpha + \mu} - \frac{\mu}{\beta} \right). \tag{6.49}$$

We have seen the first equilibrium point before—it's the DFE in which there is no disease and everyone is susceptible. The latter is new to us—it is called an **endemic equilibria (EE)** and represents the population levels if the disease persists.

It is very important to observe that the EE only makes sense biologically if $i^* > 0$, i.e., if $\dfrac{\mu}{\alpha + \mu} > \dfrac{\mu}{\beta}$. The Jacobian of the system can be calculated as

$$J(s^*, i^*) = \begin{bmatrix} -\beta i^* - \mu & -\beta s^* \\ \beta i^* & \beta s^* - (\alpha + \mu) \end{bmatrix}. \tag{6.50}$$

Evaluating the Jacobian at the DFE gives

$$J_{DFE} \begin{bmatrix} -\mu & -\beta \\ 0 & \beta - (\alpha + \mu) \end{bmatrix}. \tag{6.51}$$

We calculate the eigenvalues here instead of using the trace and determinant because the former is easier in this case since we have a triangular matrix. The eigenvalues for the DFE are

$$\lambda_1 = -\mu, \quad \lambda_2 = \beta - (\alpha + \mu).$$

Since $\mu > 0$ the DFE is a stable node if $\beta < (\alpha + \mu)$ and is a saddle point if $\beta > (\alpha + \mu)$. Based on these stability conditions the DFE switches stability when $\beta = \alpha + \mu$. From this we can conclude that the basic reproductive number is

$$R_0 = \frac{\beta}{\alpha + \mu}. \tag{6.52}$$

This says that the disease will die out as long as the contact rate is less than the recovery rate plus the death rate (i.e., $R_0 < 1$). This should be believable because in order for the disease to persist, there should be at least as many people coming into the infective class as there are leaving it (due to death or recovery).

We noted above that the EE will only exist if $i^* > 0$, i.e., if $\frac{\mu}{\alpha + \mu} > \frac{\mu}{\beta}$. This condition can be rearranged to say that

$$i^* > 0 \iff \beta > \alpha + \mu.$$

This is exactly the condition for the instability of the *DFE*. It is no coincidence that the *EE* becomes biologically relevant at the instant when the *DFE* becomes unstable! We leave it as an exercise for the reader to show that

EE is a saddle point (and not biologically relevant) when $R_0 < 1$, and

EE is stable (and biologically relevant) when $R_0 > 1$.   (6.53)

We can take sample parameter values to show some of the plots. In Figure 6.17, the parameter values are chosen as $\beta = .3, \alpha = .4, \mu = .05$. This would correspond to an average infectious period of $\frac{1}{\alpha} = 2.5$ days and an average lifetime of $\frac{1}{\mu} = 20$ days.[7] These values give a basic reproductive number that is less than 1 and thus the *EE* is not biologically relevant. We note in Figure 6.17 that while mathematically we can choose any values for s and i, their biological meaning requires us to only choose values where $s + i \le 1$.

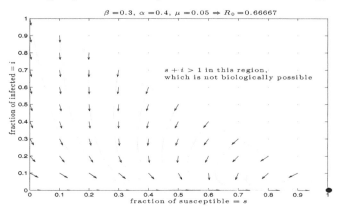

FIGURE 6.17: Only the *DFE* is biologically relevant. $R_0 < 1$ and the *DFE* is stable.

If we increase the contact rate to $\beta = .6$, we obtain an *EE* that is biologically relevant and stable. The *DFE* is a saddle for these parameter values and $R_0 > 1$. The previously used value of $\mu = .02$ is necessary in order to graphically see the equilibria above the horizontal axis. (The reader should refer to Equation (6.49) to calculate the location of the *EE* for these parameter values. However, even for realistic values of μ, the *EE* will exist and be stable whenever $R_0 > 1$. See Figure 6.18, which contains the solutions in the phase plane and the corresponding time series plots for one of the solution curves.

We must again impose the mathematical restriction that $s + i \le 1$ since our total normalized population cannot exceed 1 (and we know that $r = 1 - s - i$). In Figure 6.18(a), we see that solutions are spiraling into the *EE*. Mathematically, we can show that the *EE* is a spiral sink. Biologically, the damped

---

[7]The value $\mu = .05$ is an unrealistically low value that is used only to be consistent with Figure 6.18. The reason will be justified then.

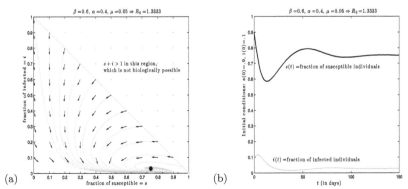

FIGURE 6.18: (a) Phase plane plot of numerous solutions; the *DFE* is located at $(s^*, i^*) = (1, 0)$; the *EE* is at $(s^*, i^*) = (.75, .027778)$. (b) Time series plot for the initial condition $s(0) = .9, i(0) = .1$.

oscillations we see in Figure 6.18(b) can be interpreted as the effect of the mass action law together with the vital dynamics: the more susceptibles there are, the greater the likelihood of adequate contacts thus resulting in more infections. This results in a decrease of susceptibles in the next time period. The reduction in the fraction of susceptibles that results from more infections makes it less likely for an infective individual to have an adequate contact with a susceptible and infect. This will then allow for more susceptibles in the next time period and so on until a steady state is reached.

We go back again to the basic reproductive number to give one additional interpretation of it. Let's rewrite this as

$$R_0 = \beta \cdot \left(\frac{1}{\alpha + \mu} \right).$$

We said that β can be interpreted as the average number of susceptibles infected by infectious individuals per unit time. By a similar argument given in the footnote in the derivation of Equation (6.38), we can interpret

$$\frac{1}{\alpha + \mu}$$

as the average length of the infectious period. Thus R_0 is the number of infections caused by an infected individual during her/his period of infectiousness. The interested reader is again encouraged to examine the texts mentioned at the beginning of this section for a more in-depth look at mathematical epidemiology.

Bifurcations in Epidemiological Systems

Recall that a bifurcation is a qualitative change in the system, often due to a change in the number or the stability of equilibria. Using the information in Sections 2.3.2 and 6.3, we see that the *DFE* and *EE* underwent a transcritical bifurcation where the *EE* passed through the DFE as it became biologically relevant and the two switched stability. This is typically what happens in an epidemiological system. For larger epidemiological systems involving more classes, the *EE* is often very difficult to obtain as a closed-form expression. In such cases, the existence of an *EE* is often deduced from the switch in stability of the *DFE* through a transcritical bifurcation! The analytical results can then be confirmed numerically.

In plotting the transcritical bifurcation curves, we plot each one of the equilibrium points' coordinates, s^* or i^*, as a function of one specific parameter. When it is feasible, we consider R_0 as a function of the parameter we are varying and plot R_0 as the horizontal component (instead of a specific parameter of the model). This allows us to see that the transcritical bifurcation occurs when R_0 goes through 1, with the *EE* becoming biologically relevant and stable. For example, consider the coordinates of the *EE* in the model given in Equation (6.48),

$$(s^*, i^*) = \left(\frac{\alpha + \mu}{\beta}, \frac{\mu}{\alpha + \mu} - \frac{\mu}{\beta} \right).$$

We can rearrange terms and manipulate the expressions to make them explicit functions of R_0:

$$(s^*, i^*) = \left(\frac{1}{R_0}, \frac{\mu}{\beta} (R_0 - 1) \right). \tag{6.54}$$

This gives a line in the R_0-i plane. If we also plot the curve produced by the i^* coordinate of the *DFE*, $i^* = 0$ (a horizontal line in the R_0-i plane), we obtain the plot in Figure 6.19.

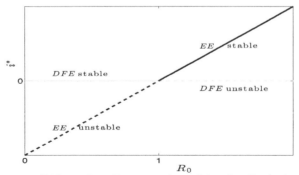

FIGURE 6.19: Bifurcation diagram plotted in the R_0-i plane. A transcritical bifurcation occurs because the *DFE* and *EE* switch stability as the *EE* becomes biologically relevant.

While the transcritical bifurcation is the typical way for the *EE* to gain stability, it is sometimes part of another bifurcation curve (a saddlenode bi-furcation curve). In these situations, we can have a **backward bifurcation** and the *EE* can exist even when $R_0 < 1$. This means that continuously vary-ing the parameters back to their preepidemic values may not be enough to get rid of the epidemic. This phenomenon is known as **hysteresis** and is also observed in many physical systems [48].

Mathematical Modeling

We take a moment to consider various branches of *mathematics* that can be used to study the spread of epidemics in a population. As with all mathemat-ical models, there are limitations and not all details can be incorporated into the model. We need to make many assumptions in order to create a man-ageable mathematical model that captures the key components of the disease under consideration and allows for new insight of the disease.

- *Ordinary differential equations* assume a continuous population variable and thus work well when there is a large population of people. Much information can be obtained from these models but answers do *not* have to be whole numbers. This is the specific branch that we studied in this section.

- *Partial differential equations* are commonly used for large populations when we want to keep track of multiple characteristics of a person that may change. For example, we may consider a SIR model with *age structure* in which we also keep track of the age of the individuals within each class.

- *Difference equation* models are often used to describe epidemics in which either the population or time is considered discrete (e.g., may be sea-sonal). For example, one may look at generations of mosquitoes that carry a certain disease and examine how the disease passes through the population over many generations. These models can often be used if we do not want to consider a continuous population variable but want whole numbers instead. These models can also arise as a *discretization* of one of the continuous models above. For example, if we wrote down the formulas for solving the system using the fourth-order Runge-Kutta method for a SIR model, we would have a difference equation model.

- *Stochastic processes* may be used for models in which we want to con-sider a discrete population and time. A key difference between this type of mathematical framework and the previous is the assumption of *ran-domness* in one or more of the parameters. That is, at each step there is a probability that something will happen but it is not guaranteed. In this type of model, beginning with the same initial state of the system may yield different results every time. This is in direct contrast to the previous three *deterministic* approaches in which a given initial state will always result in the same outcome.

- models use graph theory to understand how individuals are connected with each other and thereby gain a better understanding of how the disease spreads through a given community. By focusing on key individuals that are well connected to others in the population, the spread of the disease can be controlled.

- models are becoming increasingly common because of the large-scale computer simulations that arise from them and their ability to differentiate between individuals in the same class. In this type of model, a set of characteristics is used to describe each individual and thus each movement and interaction can be tracked. Based on these descriptions, a large-scale computer simulation can be done by considering a city of 500,000 people and watching how the interactions of the people can affect the spread of disease. These simulations have become popular with the threatened use of biological weapons in heavily populated areas.

Each type of modeling has its place and one needs to carefully choose the appropriate mathematical framework from which to model and analyze a given epidemic. The interested reader should examine Hethcote [21], Daley and Gani [15], and and others for additional details of the subject of epidemiology. In this book, we consider only models of ordinary differential equations. These are perhaps the simplest to understand and the ones in which we can carry the mathematics the furthest. Even though a number of assumptions will be required to formulate these models, the results can often shed tremendous insight on the behavior of individuals, their interactions with each other, and the overall effect of the disease on the population. These models can also give researchers insight on how to control the spread of a disease; mathematical models can sometimes indicate whether it is possible to eradicate a disease from the population (e.g., as happened with smallpox) or whether there is little or no chance of doing so. Variations of the basic SIR model are explored in the Problems, such as in Chapter Review Problem **34** that explores the *SIS* model (where recovered individuals are again susceptible) and Problem **13** in this section that explores the *SEIR* model (where an additional class of individuals exist that have contracted the disease but cannot yet infect others, that is, they are not yet infectious).

Problems

1. Consider Example 1. Assuming that the infective period of the flu is about 3.5 days, determine the adequate number of contacts required for the disease to spread.

2. Again consider Example 1 and suppose that a vaccination strategy is implemented at the beginning of the school year. Determine what per-

centage of students must receive the vaccination in order to make $R_0 < 1$. You may assume that if 10% of the students are vaccinated, then only $.95 \times .9$ were initially susceptible and $.42 \times .1$ would not have developed the flu even though they received the vaccination. Is this assumption reasonable? Explain.

3. In this exercise, we use (6.40) to obtain (6.46).
   (a) Find an implicit solution to (6.40) in which time is not explicit by solving $di/ds = i'/s'$. Write this solution in the form $f(s, i) = c$.
   (b) Observe that your implicit solution $f(s, i) = c$ describes the evolution of the disease for a given initial condition. Show that $f(s(0), i(0)) = f(s^*, i^*)$.
   (c) Use the initial approximations $s(0) \approx 1$ and $i(0) \approx 0$, and the limiting values (after the disease has passed) $i^* = 0$, $s^* = 0$, along with your answers in parts (a) and (b) to obtain

$$\beta/\alpha \approx \frac{\ln\left(\frac{s(0)}{s^*}\right)}{1 - s^*}.$$

It is important to check that a given epidemiological model is a **well-posed** one. That is, there should exist a **forward invariant set** (a region of phase space) in which a solution in the region cannot leave it for any $t > t_0$. In the case of a fixed population, we need to have the total number of individuals remain constant and none of the variables should ever become negative-valued. Use this information to answer the following two questions.

4. (a) By substituting $s = 0$ into (6.40), show that no individuals will be able to become infected.
   (b) By substituting $i = 0$ into (6.40), show that no individuals will be able to become infected without infected individuals.
   (c) Use $s + i \leq 1$ and your results from (a) and (b) to conclude that

$$\{(s, i)|0 \leq s, 0 \leq i, s + i \leq 1\}$$

is a forward invariant set and the model is thus well posed in the first quadrant.

5. (a) By substituting $s = 0$ into (6.48), show that the susceptible population cannot become negative.
   (b) By substituting $i = 0$ into (6.48), show that no individuals will be able to become infected without infected individuals.
   (c) Use $s + i \leq 1$ and your results from (a) and (b) to conclude that

$$\{(s, i)|0 \leq s, 0 \leq i, s + i \leq 1\}$$

is a forward invariant set and the model is thus well posed in the first quadrant.

6. Derive the normalized model without vital dynamics, (6.39).

7. Derive the normalized (reduced) model with vital dynamics, (6.48).

8. Consider an illness that is passing through an isolated college campus that is on the quarter system. Assume that students that get the illness can function okay for a while but then often have to drop out for the remainder of the quarter because they are falling too behind in their work. Thus we have susceptible individuals, infected individuals, and drop-outs (that are removed from the population and are assumed to no longer have contact with any enrolled student). Our model can be written as the SIR model without vital dynamics

$$\frac{dS}{dt} = -\beta \frac{I}{N} S$$
$$\frac{dI}{dt} = \beta \frac{I}{N} S - \alpha I$$
$$\frac{dR}{dt} = \alpha I, \tag{6.55}$$

where S and I have their normal meaning and R stands for those students that are removed from the population, i.e., that have dropped out. Our goal is to normalize (6.55).

(a) Define $N = S + I$ as the currently enrolled population and $\tilde{N} = S + I + R$ as the total population. Decide if the variables N and $\tilde{N}$ are constant or dependent on time. Explain why $s = S/N$, $i = I/N$, and $r = R/\tilde{N}$ are the appropriate choices for the new variables. (Note that the denominator in the r-variable is different.)

(b) Explain why $N/\tilde{N} = (1 - r)$.

(c) Use the definitions from part (a) to rewrite model (6.55) as

$$\frac{ds}{dt} = -\beta si + \alpha is$$
$$\frac{di}{dt} = \beta is - \alpha i + \alpha i^2$$
$$\frac{dr}{dt} = \alpha i(1 - r). \tag{6.56}$$

Note that this is *not* the same as (6.39).

(d) Analyze this new system, (6.56).

This normalization technique has also been used in a model of college drinking by Almada et al. [2].

9. Consider the basic SIR model without vital dynamics (6.38). This problem derives an alternate form of the basic reproductive number in the situation where we don't normalize the variables.

(a) Reduce the system to two equations (in I and R).

(b) Find the one equilibrium solution and classify its stability.

(c) Use (b) to derive an equivalent formulation for stability of the DFE:

$$\frac{S(0)}{N} < \frac{\alpha}{\beta}.$$

10. Consider the model given by (6.47) with a constant influx, Λ, of people into the population

$$\frac{dS}{dt} = \Lambda - \mu S - \beta \frac{I}{N} S$$

$$\frac{dI}{dt} = \beta \frac{I}{N} S - \alpha I - \mu I$$

$$\frac{dR}{dt} = \alpha I - \mu R. \tag{6.57}$$

By adding the equations together, obtain an expression for $N'(t)$, where $N = S + I + R$. Solve this equation and describe the population level as $t \to \infty$.

11. Consider the model given by (6.47) in which susceptible individuals can be vaccinated at a rate ν

$$\frac{dS}{dt} = \mu N - (\mu + \nu)S - \beta \frac{I}{N} S$$

$$\frac{dI}{dt} = \beta \frac{I}{N} S - \alpha I - \mu I$$

$$\frac{dR}{dt} = \alpha I - \mu R + \nu S. \tag{6.58}$$

(a) Show that the population level remains constant.
(b) By considering only the first two equations (justified by your result in part (a), compute the DFE.
(c) Determine the stability of the DFE.
(d) From your results of part (c), find an expression for R_0 and give an interpretation of it.

12. An article by researchers at Princeton in early 2014 suggested that Facebook will lose as many as 80% of its users by 2017 [14]. They modified the basic SIR model without vital dynamics to include what they called **infectious recovery dynamics**:

$$\frac{dS}{dt} = -\beta S \frac{I}{N}$$

$$\frac{dI}{dt} = \beta S \frac{I}{N} - \nu I \frac{R}{N}$$

$$\frac{dR}{dt} = \nu I \frac{R}{N}, \tag{6.59}$$

where S represents individuals who have never used Facebook, I are those who currently use Facebook, and R are those who previously used Facebook but have now quit.

(a) Show that overall population in (6.59) remains constant.

(b) Write an equivalent system of two equations that does not contain S.

(c) Show that $(I^*, R^*) = (0, R(0)), (N, 0)$ are the two equilibria and classify their stability.

(d) Use (c) to show that the stability condition for $(0, R(0))$ is given by

$$\frac{S(0)}{R(0)} < \frac{\alpha}{\beta}$$

and compare your result to that of Problem 9.

(e) Assuming that the stability condition you obtained in (d) holds, determine what is predicted to happen to the Facebook population as $t \to \infty$.

(f) The researchers obtained the estimate of "80% by 2017" by fitting the model to the data in order to estimate the parameters β and ν. If instructed, find the article and comment on the shortcomings of the model, including from where the infectious and recovery rates were estimated as well as whether a model without vital dynamics is the best choice of models.

13. Many diseases, such as tuberculosis, HIV, and Ebola, have an **exposed period** (or **latent period**) in which the susceptibles have contracted the disease but have not yet developed symptoms and cannot transmit the disease. After a period of time, they become infectious and are then able to transmit the diseases to susceptible people.

$$\frac{dS}{dt} = -\beta S \frac{I}{N} + \mu N - \mu S$$

$$\frac{dE}{dt} = \beta S \frac{I}{N} - \mu E - \delta E$$

$$\frac{dI}{dt} = \delta E - \mu I - \gamma I$$

$$\frac{dR}{dt} = \gamma I - \mu R. \tag{6.60}$$

(a) Show that the total population $N = S + E + I + R$ remains constant.

(b) Calculate a simpler system involving only three equations.

(c) Find R_0 for this simpler system and give an interpretation of it.

14. An HIV model by Perelson [41] proposed the following model for the dynamics of blood (lymphocyte) concentrations:

$$\frac{dT}{dt} = s + pT \left(1 - \frac{T}{T_{max}}\right) - kVT - \alpha T$$

$$\frac{dI}{dt} = kVT - \delta I$$

$$\frac{V}{dt} = N\delta I - \gamma V,$$

where T is the uninfected T-cell population, I is the infected T-cell population, and V is the virus concentration. For the parameters, $\alpha > 0$ is the rate of death per uninfected cells, δ is the rate of death per infected cells, γ is the rate of death of the virus, k is the infection rate constant, s represents the external production of T cells, p is the growth rate (for the assumed logistic growth), and N is the number of virions produced by each infected cell during its lifetime. With no virus, the T-cell equation has a stable steady state $\overline{T} = \dfrac{T_{max}}{2p} \left[p - \alpha + \sqrt{(p-\alpha)^2 + \dfrac{4sp}{T_{max}}} \right]$. The basic reproductive number is known to be

$$R_0 = \frac{k}{\gamma}\overline{T}N.$$

(a) Derive this quantity and give a biological interpretation of it.
(b) Numerically investigate the number and stability of the equilibria for $R_0 < 1$ and $R_0 > 1$.

6.5   Models in Ecology

This section deals with additional applications of systems of nonlinear differential equations to populations of animals. As with epidemiological models, there are numerous mathematical approaches that may be used and we will consider only models with ordinary differential equations. We begin with the classical model describing the interaction between two species: a predator and its prey. This model was originally proposed by A. J. Lotka and V. Volterra in the 1920s and is now referred to the Lotka-Volterra model. Although this model is simplistic and it is easy to find weaknesses in it, it is still a valuable example to study. Many populations in the wild oscillate as do a number of other phenomena such as the auto industry (in fact there is a whole industrial sector called cyclicals). The predator-prey model is a theoretical model in the sense that while it may not predict correct numbers, it proves that conditions can be formulated which lead to stable oscillatory behaviors.

Lotka-Volterra Predator Prey Model

We begin with the model assumptions, simplifications, and notation:

1. There are two species interacting: a prey species x and a predator species y. For the purposes of this model no other species interact with these two.

2. In the absence of the predator, the prey exhibits pure exponential growth. In particular $\frac{dx}{dt} = ax$ where $a > 0$. Implicit in this assumption is that there is sufficient food and space to allow the prey species to grow indefinitely.

3. In the absence of the prey, the predator species dies out exponentially. In particular, $\frac{dy}{dt} = -dy$ where $d > 0$. Thus although it is not explicitly mentioned, there is other food for the predators, but not enough to sustain the population. Thus it dies out over several years rather than over a month or two.

4. When the two species are in the presence of each other, the predators kill the prey in such a way that the predator population increases at a rate proportional to the product of the number of predators and the number of prey (i.e., xy). Similarly the prey population is decreased by an amount proportional to the product of the population sizes. This term captures the likelihood of an encounter.

This xy term perhaps deserves some further discussion. Naively it makes sense that if either the number of predators or the number of prey increase, the number of interactions and hence deaths increase also. But why xy and not, say, $x^2\sqrt{y}$, or some other such function with the same properties? Originally this term was borrowed from chemistry models of rates of reactions where molecules in solution interacted by randomly bumping into one another [16]. The predator-prey assumptions yield the following system of equations:

$$\frac{dx}{dt} = ax - bxy = x(a - by)$$

$$\frac{dy}{dt} = cxy - dy = y(cx - d). \tag{6.61}$$

If $b = c$, this would mean that every eaten x corresponds to a new y. Thus, we should have $b > c > 0$ unless x and y have been rescaled so that this restriction is not necessary. We are interested in the long-term behavior of the system and calculate the equilibria of the system:

$$(0,0), \qquad \text{and} \qquad \left(\frac{d}{c}, \frac{a}{b}\right).$$

In order to determine the stability of these solutions, we calculate the Jacobian:

$$J(x,y) = \begin{bmatrix} a - by & -bx \\ cy & cx - d \end{bmatrix}$$

and evaluate it at the respective equilibrium points. For the origin, we have

$$J(0,0) = \begin{bmatrix} a & 0 \\ 0 & -d \end{bmatrix}.$$

Here $\text{Tr}(J(0,0)) = a - d$ and $\det(J(0,0)) = -ad$. Since a and b are both positive, $\det(J(0,0)) < 0$, and hence $(0,0)$ is a saddle point. At $\left(\frac{d}{c}, \frac{a}{b}\right)$ we have

$$J\left(\frac{d}{c}, \frac{a}{b}\right) = \begin{bmatrix} 0 & \frac{-bd}{c} \\ \frac{ca}{b} & 0 \end{bmatrix}.$$

Thus $\text{Tr}(J\left(\frac{d}{c}, \frac{a}{b}\right)) = 0$ and $\det(J\left(\frac{d}{c}, \frac{a}{b}\right)) = ad > 0$. Referring to Theorem 6.1.1, we have $\beta = 0$ and $\gamma > 0$ which indicates that this equilibrium is a linear center. The $\left(\frac{d}{c}, \frac{a}{b}\right)$ solution has trajectories circling about it in a counterclockwise direction; see Figure 6.20(a),(b). As this is a theoretical model, these are more or less arbitrarily chosen parameters. Figure 6.20(a) shows the predator and prey solution curves plotted as a function of time. Figure 6.20(b) plots these solutions on an xy phase plane.

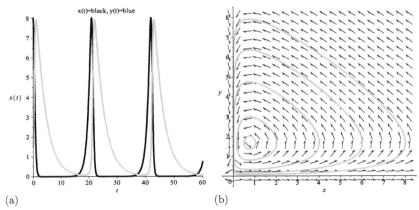

(a)   (b)

FIGURE 6.20: Numerical solutions of system (6.61) with $a = 1$, $b = 0.6$, $c = 0.4$, and $d = 0.3$. (a) Predator and prey populations as a function of time. (b) Predator-prey trajectories plotted in the phase plane.

The phase plane diagram verifies that the equilibrium is a linear center and is in fact a **nonlinear center**.[8] Observe the following behavior on both graphs. When the predator population is low, the prey population starts to rise. Soon the predator population rises to the point that the prey population starts to drop. Even though the prey population is dropping, for a time the prey population is still high enough that the predator population continues to rise. At a point, the prey population drops so low that the predator population can no longer be maintained and it starts to drop. Both populations drop until the predator population is low enough that the prey population begins to grow again and the cycle repeats. From the graphs, one can even determine the predicted period of the cycle.

The work up to now has been a general technique which applies to any system of two differential equations. The analysis from this point on depends on the particular form of these equations. In particular, the linearized equations are combined to form a single separable differential equation. This is used to

---

[8]A linear center refers to one predicted by the Jacobian but where the inclusion of linear terms may change this result. A nonlinear center refers to a center that remains that way when the full system with nonlinear terms is considered.

analytically determine the shape of the trajectories near a center for several important cases.

Going back to the information on linearization we see that if we translate the coordinate system, so that an equilibrium (x^*, y^*) is moved to the origin using the equations $u = x - x^*$ and $v = y - y^*$, the system behaves near the origin according to the linearized system of equations. In particular, we had

$$\frac{du}{dt} = f_x(x^*, y^*)u + f_y(x^*, y^*)v$$

$$\frac{dv}{dt} = g_x(x^*, y^*)u + g_y(x^*, y^*)v$$

in Equation (6.10). In the case of the predator-prey model, $u = x - \frac{d}{c}$, $v = y - \frac{a}{b}$, and

$$\frac{du}{dt} = -\left(\frac{bd}{c}\right)v,$$

$$\frac{dv}{dt} = \left(\frac{ac}{b}\right)u.$$

The chain rule from calculus gives

$$\frac{du}{dv} = \frac{du}{dt}\frac{dt}{dv} \implies \frac{du}{dv} = -\frac{b^2 d}{ac^2}\frac{v}{u}. \tag{6.62}$$

The key to this analysis is that (6.62) is a separable differentiable equation. Thus

$$ac^2 u\, du = -db^2 v\, dv.$$

Integrating both sides yields

$$\frac{ac^2 u^2}{2} = -\frac{b^2 dv^2}{2} + C_1.$$

Rearranging gives

$$\frac{ac^2 u^2}{2} + \frac{b^2 dv^2}{2} = C_1 \quad \text{or} \quad \frac{u^2}{b^2 d} + \frac{v^2}{ac^2} = C.$$

Substituting to get back the original variables yields

$$\frac{\left(x - \frac{d}{c}\right)^2}{b^2 d} + \frac{\left(y - \frac{a}{b}\right)^2}{ac^2} = C.$$

This is, of course, the equation for an ellipse centered at the equilibrium with axes parallel to the coordinate axes. The implication is that close to the equilibrium, the trajectories look like ellipses, which confirms our prior determination that the fixed point was a center. Indeed, close to the equilibrium the trajectories are ellipses, further away they still oscillate about the equilibrium, but with a more complicated shape. Interestingly the original nonlinear predator-prey equations can be solved in closed form to give the implicit solution $(y^a\, e^{-by})(x^c\, e^{-dx}) = K$, and the reader is referred to Olinick's book [40] for a derivation. Olinick's book was written in 1978, before computers became commonplace, and there are some interesting discussions about how to graph solutions such as $(y^a\, e^{-by})(x^c\, e^{-dx}) = K$ using drafting techniques.

Many modeling books report a famed data set of Canadian lynxes and hares that appear to support this model. However, there has been much controversy over this data set. The data represent the trapping for fur, and there is a question about how well trapping data represent the true census (trends in fur prices might play a role in the number of animals trapped). An observation that at one point the rise and fall in the lynx population preceded the rise and fall in the hare population resulted in a paper by Gilpin entitled *Do Hares Eat Lynx?* [17]. Hall points out, in a paper [19] that should be read by all aspiring modelers, that the data set represents hares from eastern Canada and lynx from western Canada (lynx furs were worth shipping but hare furs were not). In fact these populations were not interacting at all. Further, hare populations were seen to oscillate on Anticosti Island in a manner similar to hares on the mainland, but no lynx lived on the island. Great care must be taken when modeling if you are using data you have not collected yourself.

In natural settings, it is very hard to observe a true predator-prey relationship due to other factors that affect the populations. There are, however, examples of the predator-prey phenomena observed in experimental settings. One performed in 1957 by Huffaker studied two species of mite: one fed on oranges, the other fed on the other species of mite. The results agree favorably with the predictions of the predator-prey model.

We now consider a modification of this model that uses a modified logistic function to model the prey. Recall the logistic function from Section 2.4:

$$\frac{dN}{dt} = aN\left(1 - \frac{N}{K}\right),$$

where N is the population, a is the intrinsic growth rate, and K is the carrying capacity. This models the population as growing exponentially for small populations and then approaching its carrying capacity. However, if we consider that small populations might have more difficulty in finding a mate, then a better model would be

$$\frac{dN}{dt} = aN^2\left(1 - \frac{N}{K}\right).$$

This next example incorporates this idea into the classic Lotka-Volterra model [39], [48].

Example 1 Consider a modified version of the classic Lotka-Volterra equation:

$$\frac{dN}{dt} = aN^2 \left(1 - \frac{N}{K}\right) - bPN$$

$$\frac{dP}{dt} = P(cN - d), \tag{6.63}$$

where N is the prey population, P is the predator population, a is the intrinsic growth rate and K the carrying capacity of the prey, and b, c, d have the same meaning as discussed earlier. With five parameters in the model, it can become unwieldy but the technique of non-dimensionalization from Section 2.4 allows us to rewrite the system as

$$\frac{dx}{d\tau} = x^2 (1 - x) - xy$$

$$\frac{dy}{d\tau} = \alpha y(x - \delta), \tag{6.64}$$

as you will show in the Problem **7** with $x = \frac{N}{K}$, $y = \frac{bP}{aK}$, $\tau = aKt$, $\alpha = \frac{c}{a}$, and $\delta = \frac{d}{cK}$.

We find equilibria of the system (6.64) to be $(0,0)$, $(1,0)$, and $(\delta, \delta(1 - \delta))$. Calculating the eigenvalues of $J(0,0)$ gives $\lambda_{1,2} = 0, -\alpha\delta$. An eigenvalue of 0 does not allow us to conclude anything about the stability of the origin but we can overcome this by considering the cases $x = 0$ and then $y = 0$. Taking the former shows that the predator population will die off in the absence of prey (eigenvalue $-\alpha\delta$). If we take $y = 0$ our model reduces to $x' = x^2(1 - x)$, which we can analyze on the phase line to conclude that for small initial population, it will increase to 1. Thus, $(0,0)$ is a saddle.

For the equilibrium $(1,0)$, the eigenvalues of $J(1,0)$ are $\lambda_{1,2} = -1, \alpha(1 - \delta)$. We know that $\delta > 0$ so this point will be a saddle when $\delta < 1$ and stable when $\delta > 1$.

For the final equilibrium point, it will exist biologically only when $\delta < 1$ since $\alpha > 0$. When it does exist biologically, we can see whether a Hopf bifurcation will be present. We use the alternative approach using the trace and determinant of the Jacobian shown at the end of Section 6.3, which requires $\beta = 0$ and $\gamma > 0$:

$$\beta = \delta(1 - 2\delta) = 0, \qquad \gamma = \delta^2 \alpha(1 - \delta) > 0.$$

Thus, we may have a Hopf bifurcation when $\delta = \frac{1}{2}$ (with the other condition $\delta < 1$ not adding anything). If we look at phase portraits, we see that we have a stable limit cycle when we are just below $\delta = \frac{1}{2}$; see Figure 6.21. Thus we have a supercritical Hopf bifurcation. Interpreting the plots of Figure 6.21(b), we see that the population will reach a stable solution for any choice of $x, y > 0$ and will approach this equilibrium in an oscillatory fashion. For certain parameters, Figure 6.21(a) shows that our populations may continue to oscillate (stable limit cycle). This is different from the basic Lotka-Volterra model in which we had an infinite number of nonlinear centers surrounding the equilibrium point. Although not shown here, our analytical work also allows us to conclude that if $\delta = \frac{d}{cK} > 1$, that is, if the death rate of the prey is too fast (d), or if the conversion rate of prey to predators is too slow (c), or if the carrying capacity of the prey is too small, then the predator population will no longer exist and will die off.

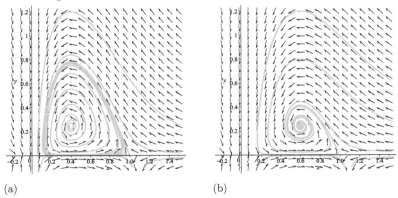

(a)   (b)

FIGURE 6.21: Supercritical Hopf bifurcation in system (6.63). Negative trajectories are shown for mathematical completeness even though they do not make biological sense. (a) A stable limit cycle exists for $\alpha = 2$, $\delta = 0.4$. (b) The stable limit cycle has disappeared for $\alpha = 2$, $\delta = 0.6$ and $(\delta, \delta(1 - \delta))$ has become unstable.

Lotka-Volterra Competition Model

Besides modeling predator-prey systems, we can use our approach to model **competitive systems** as well. We revisit our previous assumptions and modify them as follows:

1. There are two species interacting for the same food source: species x and species y. For the purposes of this model no other species interact with these two.

2. In the absence of the other, each species will exhibit *logistic* growth.

3. When the two species are in the presence of each other, only one will eat with the other seeking food elsewhere.

Our modified Lotka-Volterra competition model can be written as

$$\frac{dx}{dt} = a\,x \left(1 - \frac{x}{K_1} \right) - b\,xy$$

$$\frac{dy}{dt} = c\,y \left(1 - \frac{y}{K_2} \right) - d\,xy. \tag{6.65}$$

We highlight that interaction has a negative effect on both populations. We can find equilibria and stability in a straightforward manner and we summarize the trivial ones here (where either x or y or both are zero):

$$(0,0): \text{Eigenvalues } \lambda_{1,2} = a, c \implies \text{always unstable} \tag{6.66}$$

$$(0,K_2): \text{Eigenvalues } \lambda_{1,2} = -c, (a - bK_2) \tag{6.67}$$
$$\implies \text{stable for } a < bK_2, \text{ saddle for } a > bK_2$$

$$(K_1,0): \text{Eigenvalues } \lambda_{1,2} = -a, (c - dK_1) \tag{6.68}$$
$$\implies \text{stable for } c < dK_1, \text{ saddle for } c > dK_1.$$

For the non-trivial equilibrium

$$\left(\frac{K_1 c(a - bK_2)}{ac - bdK_1K_2}, \frac{K_2 a(c - dK_1)}{ac - bdK_1K_2} \right), \tag{6.69}$$

we calculate the trace and determinant instead:

$$\beta = -\frac{ac(a - bK_2 + c - dK_1)}{ac - bdK_1K_2}, \quad \gamma = \frac{ac(a - bK_2)(c - dK_1)}{ac - bdK_1K_2}. \tag{6.70}$$

In order for it to be biologically meaningful, we require

$$a - bK_2, \quad c - dK_1, \text{ and } \quad ac - bdK_1K_2 \tag{6.71}$$

to all have the same sign, either positive or negative. In order for this equilibrium to be stable, we need $\beta < 0$ and $\gamma > 0$. We observe that the numerator of β contains the sum of the first two quantities in (6.71) while the numerator contains the third. In order to have $\beta < 0$ we thus need all the quantities in (6.71) to be positive. All positive quantities in (6.71) will also make $\gamma > 0$. Conversely, all negative quantities in (6.71) will make both $\beta > 0$ and $\gamma < 0$ and thus the non-trivial equilibrium point will either be a saddle or stable when it is biologically relevant. We can also observe that it will become biologically relevant in a transcritical bifurcation with either $(K_1, 0)$ or $(0, K_2)$ and may enter as either a stable node or a saddle depending on the sign of the terms in (6.71). Note that we will always have at least one of $(K_1, 0)$, $(0, K_2)$, and $\left(\frac{K_1 c(a - bK_2)}{ac - bdK_1K_2}, \frac{K_2 a(c - dK_1)}{ac - bdK_1K_2} \right)$ stable. See Figure 6.22 for the four cases that we have. Thus, we may have stable co-existence in certain situations or we may only have one species that survives. Stable co-existence requires both $a > bK_2$ and $c > dK_1$. In words, we would need to have a larger intrinsic growth rate of both species, or interaction terms with a smaller (in magnitude) coefficient, or a smaller carrying capacity of both species.

Many variations and combinations of the predator-prey and competition models can be done and some of these extensions and related problems are explored in the Problems.

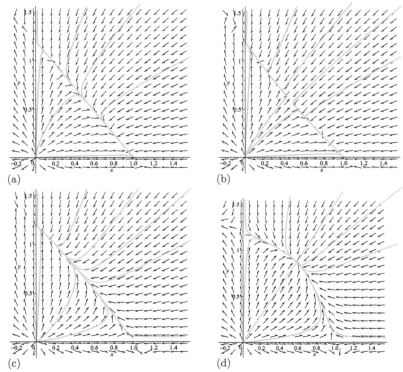

(a)   (b)

(c)   (d)

FIGURE 6.22: Possible phase portraits for logistic competition model (6.65). We set $b = d = K_1 = 1$, $K_2 = 1.2$ for all the plots. (a) With $a = 1.3$, $c = .9$, only $(K_1, 0)$ is stable and the non-trivial equilibria is not biologically relevant. (b) With $a = 1.1$, $c = .9$, both $(K_1, 0)$ and $(0, K_2)$ are stable and the non-trivial point is biologically relevant but is a saddle. (c) With $a = 1.1$, $c = 1.3$, only $(0, K_2)$ is stable and the non-trivial point is again not biologically relevant. (d) With $a = 2.1$, $c = 2.1$, both $(K_1, 0)$ and $(0, K_2)$ are saddles and the non-trivial point is biologically relevant and stable.

●   ●   ●   ●   ●   ●   ●   ●   ●   ●   ●   ●

Problems

In Problems **1–6**, consider the given Lotka-Volterra models, with $x, y \geq 0$, and do the following:

(a) Give a possible interpretation of the equations based on the information in this section.

(b) Find the equilibria of this system.

(c) Use the Jacobian analysis to determine the stability of the equilibria.

(d) Find all bifurcations that occur in this system, if applicable.

(e) Use a computer program to generate a phase portrait for various initial conditions.

(f) Interpret the results in terms of the two variables and any parameters.

1. $\begin{cases} x' = x(3-x) - 2xy \\ y' = y(2-y) - xy \end{cases}$ 2. $\begin{cases} x' = x(3-x) - xy \\ y' = y(2-y) - xy \end{cases}$

3. $\begin{cases} x' = x(3-x) - xy \\ y' = -2y + xy \end{cases}$ 4. $\begin{cases} x' = 3x - bxy, \quad b > 0 \\ y' = 2y - xy \end{cases}$

5. $\begin{cases} x' = x(1-x) - xy \\ y' = y(x-d), \quad d > 0 \end{cases}$ 6. $\begin{cases} x' = x(b-x) - 2xy \\ y' = y(1-y) + xy, \quad b > 0 \end{cases}$

7. In Example 1, we claimed that (6.63) could be rescaled to (6.64) with the substitutions $x = \dfrac{N}{K}$, $y = \dfrac{bP}{aK}$, $\tau = aKt$, $\alpha = \dfrac{c}{a}$, and $\delta = \dfrac{d}{cK}$. Show this.

In Problems **8–9**, do (a)–(f) above for these two models of an arms race. Consider two economically competing nations which we call Purple and Green. Both nations desire peace and hope to avoid war, but they are not pacifistic. They will not go out of their way to launch aggression, but they will not sit idly by if their country is attacked. They believe in self-defense and will fight to protect their nation and their way of life. Both nations feel that the maintenance of a large army and the stockpiling of weapons are purely "defensive" gestures when they do it, but at least somewhat "offensive" when the other side does it. Since the two nations are in competition, there is an underlying sense of "mutual fear." The more one nation arms, the more the other nation is spurred to arm. These models are known as the *Richardson's Arms Race model* in honor of Lewis F. Richardson [42], who considered this model in 1939 for the combatants of World War I.

8. Let $x(t)$ and $y(t)$ represent the yearly rates of armament expenditures of the two nations in some standardized monetary unit. We assume that each country adjusts the rate of increase or decrease of its armaments in response to the level of the other's. The simplest assumption is that each nation's rate is directly proportional to the expenditure of the other nation. We further assume that excessive armament expenditures present a drag on the nation's economy so that the actual level of expenditure reduces the rate of change of the expenditure. The simplest way to model this is to assume that the rate of change for a nation is directly and negatively proportional to its own expenditure. We thus arrive at Richardson's Model:

$$\frac{dx}{dt} = ay - mx \tag{6.72}$$

$$\frac{dy}{dt} = bx - ny \tag{6.73}$$

where a, b, m, and n are positive constants.

9. Consider (6.72)-(6.73) with the additional consideration of underlying grievances of each country toward the other. To model this, we introduce two additional constant terms, r and s, to the equations (6.72) and (6.73)

and obtain

$$\frac{dx}{dt} = ay - mx + r \tag{6.74}$$

$$\frac{dy}{dt} = bx - ny + s. \tag{6.75}$$

A positive value of r or s indicates that there is a grievance of one country toward the other which causes an increase in the rate of arms expenditures. If r or s is negative, then there is an underlying feeling of good will, so there is a decrease in the rate of arms expenditures. For simplicity, assume that r, s have the same sign.

Chapter 6 Review

In Problems **1–6**, determine whether the statement is true or false. If it is true, give reasons for your answer. If it is false, give a counterexample or other explanation of why it is false.

1. Linearization of a nonlinear system about an equilibrium solution reduces it to a linear system in the neighborhood of this equilibrium solution.

2. The phase plane can be drawn for any 2-dimensional autonomous system that can be written in the form $\mathbf{x}' = \mathbf{A}\mathbf{x}$.

3. For the 2-dimensional system $\mathbf{x}' = \mathbf{f}(t, \mathbf{x})$, we may have trajectories that cross in the x-y plane even if the Existence and Uniqueness Theorem is satisfied.

4. The stability of an equilibrium point in 3+ dimensions can be determined by the signs of the trace and determinant of the Jacobian matrix.

5. For a Hopf bifurcation to occu, all we need to observe is the real part of a pair of eigenvalues switching sign.

6. SIR models do not allow for births and deaths due to natural causes.

For each of Problems **7–20**, do the following parts: (i) consider the given matrix as the coefficient matrix $\mathbf{A}$ of the system $\mathbf{x}' = \mathbf{A}\mathbf{x}$ and use Theorem 6.1.1 to determine the classification of the origin; (ii) use one of the computer programs to draw the vector field for each system $\mathbf{x}' = \mathbf{A}\mathbf{x}$ and some trajectories for various initial conditions.

7. $\begin{bmatrix} 3 & 1 \\ 1 & 3 \end{bmatrix}$
8. $\begin{bmatrix} 1 & 0 \\ 4 & 3 \end{bmatrix}$
9. $\begin{bmatrix} 5 & 1 \\ -1 & 5 \end{bmatrix}$
10. $\begin{bmatrix} 2 & 2 \\ -1 & 0 \end{bmatrix}$

11. $\begin{bmatrix} 4 & 1 \\ -10 & -2 \end{bmatrix}$
12. $\begin{bmatrix} -1 & 2 \\ 1 & 0 \end{bmatrix}$
13. $\begin{bmatrix} 5 & -6 \\ 3 & -4 \end{bmatrix}$
14. $\begin{bmatrix} -10 & 24 \\ -4 & 10 \end{bmatrix}$

15. $\begin{bmatrix} 5 & 1 \\ -17 & -3 \end{bmatrix}$
16. $\begin{bmatrix} 4 & 1 \\ -20 & -4 \end{bmatrix}$
17. $\begin{bmatrix} 2 & 3 \\ -1 & 2 \end{bmatrix}$
18. $\begin{bmatrix} 1 & -1 \\ 2 & 4 \end{bmatrix}$

19. $\begin{bmatrix} 5 & -6 \\ 2 & -1 \end{bmatrix}$ 20. $\begin{bmatrix} -5 & 2 \\ -2 & -3 \end{bmatrix}$

For Problems **21–26**, use Theorem 6.1.2 to classify the stability of the origin (the only equilibrium solution).

21. $\begin{cases} x' = -11x + 4y \\ y' = 8x - 25y \\ z' = -6x + 12y - 9z \end{cases}$
    22. $\begin{cases} x' = 2x - 16y + 6z \\ y' = -3x + 4y - 3z \\ z' = -4x + 19y - 8z \end{cases}$

23. $\begin{cases} x' = 4x - 8y - 10z \\ y' = -x + 6y + 5z \\ z' = x - 8y - 7z \end{cases}$
    24. $\begin{cases} x' = -2y \\ y' = x + 3y \\ z' = -x - 2y + z \end{cases}$

25. $\begin{cases} x' = -17x + 4y + 2z \\ y' = 5x - 25y + z \\ z' = 3x + 12y - 12z \end{cases}$
    26. $\begin{cases} x' = 6x - 16y + 10z \\ y' = -5x + 4y - 5z \\ z' = -8x + 19y - 12z \end{cases}$

For Problems **27–30**, (i) find the equilibria of the given system; (ii) use linearization and Theorem 6.2.2 to classify the stability of the equilibria; (iii) use MATLAB, Maple, or Mathematica to draw the vector field of the system; (iv) sketch trajectories on the vector field for various initial conditions (either by hand or with the computer). You should verify that your answers from parts (iii) and (iv) agree with your predictions in parts (i) and (ii).

27. $\begin{cases} x' = y^2 - 1 \\ y' = x \end{cases}$
    28. $\begin{cases} x' = x + 1 \\ y' = x + y^2 \end{cases}$

29. $\begin{cases} x' = x(y - 1) \\ y' = x - y^2 + 2y \end{cases}$
    30. $\begin{cases} x' = y^3 - y \\ y' = x - y \end{cases}$

For Problems **31–33**, determine the type of bifurcation that occurs. Confirm your analytical results with qualitatively different phase portraits before and after the bifurcation. If instructed, draw the bifurcation diagram as well.

31. $\begin{cases} x' = x - r \\ y' = x + y^2 \end{cases}$
    32. $\begin{cases} x' = x(x - r) \\ y' = x^2 - y \end{cases}$
    33. $\begin{cases} x' = x(r + y^2) \\ y' = x - y \end{cases}$

34. Consider the *SIS* epidemic model with vital dynamics

$$S' = \mu N - \beta S \frac{I}{N} + \alpha I - \mu S$$

$$I' = \beta S \frac{I}{N} - \alpha I - \mu I, \tag{6.76}$$

where $S, I \geq 0$, $\beta, \alpha, \mu > 0$ are parameters, and $N = S + I$.
(a) Give a biological interpretation of the model and an example of diseases that it might reasonably model.
(b) Show that the population remains constant for all t.
(c) Find all equilibria and classify their stability.
(d) Find the basic reproductive number and interpret it.
(e) Eliminate the S equation (because of the constant population) and solve I on the phase line. Do the results differ from the above analysis?

35. Consider the *SIS* epidemic model with vital dynamics and constant influx of population

$$S' = \Lambda - \beta S \frac{I}{N} + \alpha I - \mu S$$

$$I' = \beta S \frac{I}{N} - \alpha I - \mu I, \tag{6.77}$$

where $S, I \geq 0$, $\beta, \alpha, \mu > 0$ are parameters, $\Lambda > 0$, and $N = S + I$.

(a) By adding the equations together, describe the population level N^* as $t \to \infty$.

(b) Consider the **limiting system** obtained by substituting N^* from (a) into the system of equations and consider only the I equation. Why is this allowed?

(c) Determine the equilibria of the system and classify their stability.

(d) Find the basic reproductive number and interpret it.

36. Consider the predator-prey model with an Allee effect in the prey (x) population:

$$x' = x \left(x - \frac{1}{4} \right) (1 - x) - xy$$

$$y' = y(x - r), \tag{6.78}$$

where $x, y \geq 0$ and $r > 0$ is a parameter.

(a) Give a biological interpretation of the parameter r.

(b) Find and classify the stability of the equilibria.

(c) Give graphical evidence that a Hopf bifurcation occurs at $r = \frac{5}{8}$. Is it subcritical or supercritical?

37. Use MATLAB, Maple, or Mathematica to examine the Rössler system [48].

$$x' = -y - z$$
$$y' = x + ay$$
$$z' = b + z(x - c).$$

For $a = b = .2, c = 2.5$, determine the equilibria and their stability. Use the computer to draw the phase portrait. Describe what you see. Then repeat these steps for $a = b = .2, c = 3.5$. Again, describe what you see.

Chapter 6 Computer Labs

Chapter 6 Computer Lab: MATLAB

MATLAB Example 1: Enter the following code that demonstrates how to graph the vector field for the mass on a spring without friction using the values $m = 2$, $k = 3$. Thus we consider the second-order ODE $2x''(t) + 3x(t) = 0$, which we must rewrite as the first-order system

$$\frac{dx}{dt} = y$$

$$\frac{dy}{dt} = \frac{-3}{2}x \qquad (6.79)$$

```
>>  clear all
>>  [X,Y]=meshgrid(-2:.5:2,-2:.5:2);
>>  DX=Y;
>>  DY=(-3/2)*X;
>>  DW=sqrt(DX.^2+DY.^2); %used to normalize vector lengths
>>  quiver(X,Y,DX./DW,DY./DW,.5);
>>  xlabel('x'); ylabel('y');
>>  axis([-2 2 -2 2])
>>  title('vector field for mass-spring system')
```

MATLAB Example 2: Enter the following code that draws the phase portrait for the ODE system

$$x' = y - x^2$$
$$y' = y - x.$$

This example may be supplemented/replaced with **pplane**, given at the end of this MATLAB computer lab.

We first create the function **ExamplePP.m** which contains the equations:

```
function f=ExamplePP(xn,yn)
%
%The original system is
%x'(t)=y-x^2 and y'(t)=y-x
%We let yn(1)=x, yn(2)=y, xn=t
%
f= [yn(2)-yn(1).^2; yn(2)-yn(1)];
```

Then create a script named **Ch6MatlabEx2.m** with the following code that uses **ode45** or **RK4.m** to numerically solve the system for a given set of initial conditions.

```
%begin script
t0=0; tf=10;
```

```
IC1=[3,1];
[t,y]=RK4(@ExamplePP,[t0,tf],IC1,.05);
IC2=[-1,0];
[t2,y2]=RK4(@ExamplePP,[t0,tf],IC2,.05);
IC3=[-0.5,0.9];
[t3,y3]=RK4(@ExamplePP,[t0,tf],IC3,.05);
IC4=[3,-.5];
[t4,y4]=RK4(@ExamplePP,[t0,tf],IC4,.05);
IC5=[-0.15,0];
[t5,y5]=RK4(@ExamplePP,[t0,tf],IC5,.05);
IC6=[-1,2];
[t6,y6]=RK4(@ExamplePP,[t0,tf],IC6,.05);


% Now plot the numerical solutions we found above.
plot(y(:,1),y(:,2))
axis([-4 3 -4 6])
hold on
plot(y2(:,1),y2(:,2))
plot(y3(:,1),y3(:,2))
plot(y4(:,1),y4(:,2))
plot(y5(:,1),y5(:,2))
plot(y6(:,1),y6(:,2))
[X,Y]=meshgrid(-4:.5:3,-4:.5:6);
DX=Y-X.^2;
DY=Y-X;
DW=sqrt(DX.^2+DY.^2);
quiver(X,Y,DX./DW,DY./DW,.5);
xlabel('x'); ylabel('y');
title('phase plane example')
hold off
%end of script
```

Save the file Ch6MatlabEx2.m and click the *Run* icon in MATLAB editor
window.

MATLAB Example 3: Enter the following code that calculates the equilibria of

$$x' = -6x + 6y$$
$$y' = 36x - y - xz$$
$$z' = -3z + xy$$

and determines their stability.

```
>> clear all
>> syms x y z %defines the variables as symbolic
>> f=[-6*x+6*y,36*x-y-x*z,-3*z+x*y] %rhs of equations
>> [x1,y1,z1]=solve(f)
>> %Note that x1 has the 3 x-coordinates, y1 has
>> %the 3 y-coords, and z1 has the 3 z-coords
>> equil1=[x1(1,1), y1(1,1), z1(1,1)] %1st equil
>> equil2=[x1(2,1), y1(2,1), z1(2,1)] %2nd equil
>> equil3=[x1(3,1), y1(3,1), z1(3,1)] %3rd equil
>> jac1=jacobian(f,[x,y,z])
   %calculates jacobian
>> J1=subs(jac1,{x,y,z},equil1) %substitutes
   %1st equil into jacobian
>> J2=subs(jac1,{x,y,z},equil2) %substitutes
   %2nd equil into jacobian
>> J3=subs(jac1,{x,y,z},equil3) %substitutes
   %3rd equil into jacobian
>> eigenvals1=eig(J1) %computes eigenvalues for 1st equil
>> eigenvals2=eig(J2) %computes eigenvalues for 2nd equil
>> eigenvals3=eig(J3) %computes eigenvalues for 3rd equil
```

MATLAB Example 4: Enter the following code that finds the $\lambda = 0$ bifurcations of the SIR model with vital dynamics. To do so, consider the following two equations, where $x = \frac{S}{N}$, $y = \frac{I}{N}$, and recall that we only need to consider the first two equations since N is constant. We still have $\beta, \gamma, \mu > 0$.

$$x' = \mu - \beta xy - \mu x$$
$$y' = \beta xy - (\mu + \gamma)y. \tag{6.80}$$

```
>> clear all
>> syms mu beta alpha x y x1 y1 lambda
>> f=[mu-beta*x*y-mu*x; beta*x*y-(mu+alpha)*y]%rhs of equations
>> [x1,y1]=solve(f,x,y)
>> equil1=[x1(1,1),y1(1,1)] %DFE
>> equil2=[x1(2,1),y1(2,1)] %EE
>> J=jacobian(f,[x,y]) %calculates jacobian
>> J0=subs(J,[x,y],equil1)%jacobian at DFE
>> chpoly=charpoly(J0,lambda) %characteristic polynomial at DFE
>> chpoly0=subs(chpoly,lambda,0) %possible lambda=0 bifns
>> params=[beta,mu,alpha]
>> paramvals1=[.4,.25,.2]
>> eq3a=subs(chpoly0,params,paramvals1)
>> eig(subs(J0,params,paramvals1))
>> equil2a=subs(equil2,params,paramvals1)
```

```
>> eigenvals2a=eig(subs(J,[x,y,params],[equil2a,paramvals1]))
>> eval(eigenvals2a)
>> paramvals2=[.5,.25,.2]
>> eq4a=subs(chpoly0,params,paramvals2)
>> eig(subs(J0,params,paramvals2))
>> equil2b=subs(equil2,params,paramvals2)
>> eigenvals2b=eig(subs(J,[x,y,params],[equil2b,paramvals2]))
>> eval(eigenvals2b)
>> simplify(subs(equil2,beta,mu+alpha))
```

Exercises for MATLAB are given after the following discussion of `pplane`.

pplane in MATLAB

For a system of only two equations, a very user-friendly software supplement called **pplane** exists for MATLAB and it is freely available for educational use. See <http://math.rice.edu/~dfield/>. There are two to three programs that you will need to download (depending on your version of MATLAB) and install in your working directory. `pplane` is much easier to implement than either of the above methods. It can also numerically find equilibria, determine stability of them, plot nullclines, and many other things. The drawback is that it only works for a system of two autonomous equations, whereas the above methods work for any number of equations. We give a brief introduction here.

Once you have placed the relevant programs in your working directory, type `pplane8`. A new window should pop up; see Figure 6.23. (Note that this is for MATLAB R2013a. Download the appropriate files for other versions of MATLAB, e.g., `pplane7` is for MATLAB 7.)

We again consider the equations of the previous example and enter them in the `pplane` window as

$$x' = y-x^2$$
$$y' = y-x$$

In the display window, set minimum $x = -4$, maximum $x = 3$, minimum $y = -4$, and maximum $y = 6$.

In the direction field box, make sure *arrows* is marked. Click *proceed* and observe the direction field. Click once to trace a trajectory forward and backward in time. Repeat a few times.

Now go to *solutions* $\longrightarrow$ *find an equilibrium point*. Then click with the mouse where you expect to see an equilibrium. Observe the Jacobian is given as are the corresponding eigenvalues and eigenvectors. (You can click *display the linearization* to see what happens near the fixed point.)

In the display window, graph t vs. x and t vs. y by going to *graph* $\longrightarrow$ *both*. Click on a trajectory that you want to see plotted.

Now re-plot the picture but with *lines*, *nullclines*, and *none* checked (three separate plots). Experiment with *number of field points per row or column* to observe how this changes the display window.

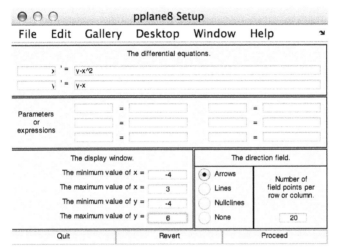

FIGURE 6.23: Pop-up window for `pplane` program used to obtain numerical solutions of two first-order equations. Note this was done with MATLAB R2013a and the corresponding program is `pplane8`.

MATLAB Exercises

Turn in both the commands that you enter for the exercises below as well as the output/figures. These should all be in one document. Please highlight or clearly indicate all requested answers. Some of the questions will require you to modify the above MATLAB code to answer them.

1. Enter the commands given in MATLAB Example 1 and submit both your input and output.

2. Enter the commands given in MATLAB Example 2 and submit both your input and output.

3. Enter the commands given in MATLAB Example 3 and submit both your input and output.

4. Enter the commands given in MATLAB Example 4 and submit both your input and output.

5. Plot the vector field for the first-order ODE system $\begin{cases} x' = -5x + 8y \\ y' = -4x + 7y. \end{cases}$

6. Use `pplane` or modify the code from Examples 1 and 2 to draw phase portrait for the ODE system $\begin{cases} x' = -4x - y \\ y' = 6x + y. \end{cases}$

7. Use `pplane` or modify the code from Examples 1 and 2 to draw phase portrait for the ODE system $\begin{cases} x' = 5x - y \\ y' = 3x + y. \end{cases}$

8. Use `pplane` or modify the code from Examples 1 and 2 to draw phase portrait for the ODE system $\begin{cases} x' = -4x + 2y \\ y' = -10x + 4y. \end{cases}$

9. Use `pplane` or modify the code from Examples 1 and 2 to draw phase portrait for the ODE system $\begin{cases} x' = -7x - 4y \\ y' = 10x + 5y. \end{cases}$

10. Determine the equilibria and their stability analytically for the ODE system $\begin{cases} x' = y, \\ y' = 4 - x^2. \end{cases}$ Then confirm your results with a phase portrait.

11. Determine the equilibria and their stability analytically for the ODE system $\begin{cases} x' = y, \\ y' = 4 - x^2. \end{cases}$ Then confirm your results with a phase portrait for the ODE system.

12. Determine the equilibria and their stability analytically for the ODE system $\begin{cases} x' = y - x, \\ y' = x^2 + 2y. \end{cases}$ Then confirm your results with a phase portrait for the ODE system.

13. Determine the equilibria and their stability analytically for the ODE system $\begin{cases} x' = y, \\ y' = x^3 - x. \end{cases}$ Then confirm your results with a phase portrait for the ODE system.

14. Determine the equilibria and their stability analytically for the ODE system $\begin{cases} x' = y^2 - 1, \\ y' = x^2 - y. \end{cases}$ Then confirm your results with a phase portrait for the ODE system.

15. Determine the $\lambda = 0$ bifurcation that occurs for the ODE system $\begin{cases} x' = r - x^2, \\ y' = -y, \end{cases}$ and the value of r at which it occurs.

16. Determine the $\lambda = 0$ bifurcation that occurs for the ODE system $\begin{cases} x' = y - 2x, \\ y' = r + x^2 - y, \end{cases}$ and the value of r at which it occurs.

17. Determine the $\lambda = 0$ bifurcation that occurs for the ODE system $\begin{cases} x' = rx - x^2, \\ y' = x - y, \end{cases}$ and the value of r at which it occurs.

18. Determine the $\lambda = 0$ bifurcation that occurs for the ODE system $\begin{cases} x' = rx - x^3, \\ y' = -y, \end{cases}$ and the value of r at which it occurs.

19. Determine the Hopf bifurcation that occurs for the ODE system $\begin{cases} x' = rx + 2y, \\ y' = -2x + ry + x^2y, \end{cases}$ and the value of r at which it occurs.

20. For the Lotka-Volterra model $\begin{cases} x' = x(3 - x - 2y), \\ y' = y(2 - x - y), \end{cases}$ find all the biologically relevant equilibria and determine their stability. Then draw the phase portrait.

21. Consider the Lotka-Volterra model $\begin{cases} x' = x(1 - x) - xy \\ y' = y(x - d), \quad d > 0 \end{cases}$ where x is the prey, y is the predator, and $d > 0$ is a parameter. Determine the $\lambda = 0$ bifurcation that occurs.

Chapter 6 Computer Lab: Maple

Maple Example 1: Enter the following code that demonstrates how to graph the vector field for the mass on a spring without friction using the values $m = 2$, $k = 3$. Thus we consider the second-order ODE $2x''(t) + 3x(t) = 0$, which we must rewrite as the first-order system

$$\frac{dx}{dt} = y$$
$$\frac{dy}{dt} = \frac{-3}{2}x \qquad (6.81)$$

restart
with(DEtools) :
eq1 := $x'(t) = y(t)$
eq2 := $y'(t) = \dfrac{-3}{2} \cdot x(t)$

DEplot ([*eq1, eq2*], [$x(t), y(t)$], $t = 0..3, x = -2..2, y = -2..2, linecolor = black,$
    dirgrid $= [12, 12]$, *title* $=$ "vector field for mass-spring system",
    arrows $=$ *medium*)

Maple Example 2: Enter the following code that draws the phase portrait for the ODE system

$$x' = y - x^2$$
$$y' = y - x.$$

restart
with(plots) : with(DEtools) :
eq1 := x'(t) = y(t) - x(t)²
eq2 := y'(t) = y(t) - x(t)
IC := [[x(0) = 3, y(0) = 1], [x(0) = -1, y(0) = 0], [x(0) = -.5, y(0) = .5],
  [x(0) = 3, y(0) = 2], [x(0) = -.15, y(0) = 0], [x(0) = -.15, y(0) = 2]]
DEplot([eq1, eq2], [x(t), y(t)], t = 0..10, x = -4..3, y = -4..6, IC, stepsize = .05,
  title = "phase plane example", linecolor = black, method = classical[rk4])
phaseportrait([eq1, eq2], [x(t), y(t)], t = 0..10, IC, stepsize = .05,
  scene = [x(t), y(t)], title = "phase plane example", linecolor = black,
  method = classical[rk4])

Maple Example 3: Enter the following code that calculates the equilibria of

$$x' = -6x + 6y$$
$$y' = 36x - y - xz$$
$$z' = -3z + xy$$

and determines their stability.

restart
with(LinearAlgebra) : with(VectorCalculus) :
eq1a := -6 · x + 6 · y
eq1b := 36 · x - y - x · z
eq1c := -3 · z + x · y
eq2 := solve({eq1a, eq1b, eq1c}, {x, y, z})
eq2a := allvalues(eq2[2])
J := Jacobian([eq1a, eq1b, eq1c], [x, y, z])
eq3a := subs(eq2[1], J)
eq3b := Eigenvalues(eq3a)
evalf(eq3b)
eq4a := subs(eq2a[1], J)
eq4b := Eigenvalues(eq4a)
eq5a := subs(eq2a[2], J)
eq5b := Eigenvalues(eq5a)

Maple Example 4: Enter the following code that finds the $\lambda = 0$ bifurcations of the SIR model with vital dynamics. To do so, consider the following two equations, where $x = \frac{S}{N}$, $y = \frac{I}{N}$, and recall that we only need to consider the first two equations since N is constant. We still have $\beta, \gamma, \mu > 0$.

$$x' = \mu - \beta xy - \mu x$$
$$y' = \beta xy - (\mu + \gamma)y. \tag{6.82}$$

restart
with(LinearAlgebra) : with(VectorCalculus) :
eq1a := $\mu - \beta \cdot x \cdot y - \mu \cdot x$
eq1b := $\beta \cdot x \cdot y - (\mu + \alpha) \cdot y$
eq2 := solve({eq1a, eq1b}, {x, y})
J := Jacobian([eq1a, eq1b], [x, y])
J0 := subs(eq2[1], J)
chpoly := CharacteristicPolynomial(J0, λ) = 0
chpoly0 := subs($\lambda = 0$, chpoly)
#The last equation is the condition when a "$\lambda = 0$" bifurcation may occur.
 This gives, as we know, the condition $\beta = \alpha + \mu$ (i.e., $R_0 = 1$). We now
 choose 2 sets of parameter values, for < 0 and > 0, and check stability.
params1 := {$\beta = .4, \mu = .25, \alpha = .2$}
eq3a := subs(params1, lhs(chpoly0))
equil1 := eq2[1]
Eigenvalues(subs(params1, J0))
equil2a := subs(params1, eq2[2])
Eigenvalues(subs(equil2a, params1, J))
params2 := {$\beta = .5, \mu = .25, \alpha = .2$}
eq4a := subs(params2, lhs(chpoly0))
Eigenvalues(subs(params2, J0))
equil2b := subs(params2, eq2[2])
Eigenvalues(subs(equil2b, params2, J))
subs($\beta = \mu + \alpha$, eq2[2]) # The EE is the same as the DFE
#end of code

The last line shows us that the two equilibria are the same when $\beta = \alpha + \mu$. The lines from *params1* until just before *params2* show that for $\beta < \mu + \alpha$, the DFE is stable whereas the EE is not biologically relevant and is unstable. The lines from *params2* until the second to last one show that for $\beta > \mu + \alpha$, the DFE is unstable whereas the EE is biologically relevant and is now stable. Taken together, we can conclude that a transcritical bifurcation occurred between the two equilibria.

Maple Exercises
Turn in both the commands that you enter for the exercises below as well as

the output/figures. These should all be in one document. Please highlight or clearly indicate all requested answers. Some of the questions will require you to modify the above Maple code to answer them.

1. Enter the commands given in Maple Example 1 and submit both your input and output.

2. Enter the commands given in Maple Example 2 and submit both your input and output.

3. Enter the commands given in Maple Example 3 and submit both your input and output.

4. Enter the commands given in Maple Example 4 and submit both your input and output.

5. Plot the vector field for the first-order ODE system $\begin{cases} x' = -5x + 8y \\ y' = -4x + 7y. \end{cases}$

6. Modify the code from Examples 1 and 2 to draw phase portrait for the ODE system $\begin{cases} x' = -4x - y \\ y' = 6x + y. \end{cases}$

7. Modify the code from Examples 1 and 2 to draw phase portrait for the ODE system $\begin{cases} x' = 5x - y \\ y' = 3x + y. \end{cases}$

8. Modify the code from Examples 1 and 2 to draw phase portrait for the ODE system $\begin{cases} x' = -4x + 2y \\ y' = -10x + 4y. \end{cases}$

9. Modify the code from Examples 1 and 2 to draw phase portrait for the ODE system $\begin{cases} x' = -7x - 4y \\ y' = 10x + 5y. \end{cases}$

10. Determine the equilibria and their stability analytically for the ODE system $\begin{cases} x' = y, \\ y' = 4 - x^2. \end{cases}$ Then confirm your results with a phase portrait.

11. Determine the equilibria and their stability analytically for the ODE system $\begin{cases} x' = y, \\ y' = 4 - x^2. \end{cases}$ Then confirm your results with a phase portrait for the ODE system.

12. Determine the equilibria and their stability analytically for the ODE system $\begin{cases} x' = y - x, \\ y' = x^2 + 2y. \end{cases}$ Then confirm your results with a phase portrait for the ODE system.

13. Determine the equilibria and their stability analytically for the ODE system $\begin{cases} x' = y, \\ y' = x^3 - x. \end{cases}$ Then confirm your results with a phase portrait for the ODE system.

14. Determine the equilibria and their stability analytically for the ODE system $\begin{cases} x' = y^2 - 1, \\ y' = x^2 - y. \end{cases}$ Then confirm your results with a phase portrait for the ODE system.

15. Determine the $\lambda = 0$ bifurcation that occurs for the ODE system $\begin{cases} x' = r - x^2, \\ y' = -y, \end{cases}$ and the value of r at which it occurs.

16. Determine the $\lambda = 0$ bifurcation that occurs for the ODE system $\begin{cases} x' = y - 2x, \\ y' = r + x^2 - y, \end{cases}$ and the value of r at which it occurs.

17. Determine the $\lambda = 0$ bifurcation that occurs for the ODE system $\begin{cases} x' = rx - x^2, \\ y' = x - y, \end{cases}$ and the value of r at which it occurs.

18. Determine the $\lambda = 0$ bifurcation that occurs for the ODE system $\begin{cases} x' = rx - x^3, \\ y' = -y, \end{cases}$ and the value of r at which it occurs.

19. Determine the Hopf bifurcation that occurs for the ODE system $\begin{cases} x' = rx + 2y, \\ y' = -2x + ry + x^2 y, \end{cases}$ and the value of r at which it occurs.

20. For the Lotka-Volterra model $\begin{cases} x' = x(3 - x - 2y), \\ y' = y(2 - x - y), \end{cases}$ find all the biologically relevant equilibria and determine their stability. Then draw the phase portrait.

21. Consider the Lotka-Volterra model $\begin{cases} x' = x(1 - x) - xy \\ y' = y(x - d), \quad d > 0 \end{cases}$ where x is the prey, y is the predator, and $d > 0$ is a parameter. Determine the $\lambda = 0$ bifurcation that occurs.

Chapter 6 Computer Lab: Mathematica

Mathematica Example 1: Enter the following code that demonstrates how to graph the vector field for the mass on a spring without friction using the values $m = 2$, $k = 3$. Thus we consider the second-order ODE $2x''(t) + 3x(t) = 0$, which we must rewrite as the first-order system

$$\frac{dx}{dt} = y$$

$$\frac{dy}{dt} = \frac{-3}{2}x \qquad (6.83)$$

```
Quit[]
VectorPlot[{y, - 3x/2}, {x, -2, 2}, {y, -2, 2},
  VectorScale→{Small, Small, None}, Axes→Automatic,
  AxesLabel→{"x", "y"}, VectorPoints→10]
```

Mathematica Example 2: Enter the following code that draws the phase portrait for the ODE system

$$x' = y - x^2$$
$$y' = y - x.$$

```
Quit[]
dex[t_] = y[t] - x[t]^2
dey[t_] = y[t] - x[t]
ICx={x[0]==3,x[0]==-1,x[0]==-.5,x[0]==3,x[0]==-.15,x[0]==-.8}
ICy={y[0]==1,y[0]==0,y[0]==.5,y[0]==2,y[0]==0,y[0]==0}
soln1=NDSolve[{x'[t]==dex[t],ICx[[1]], y'[t]==dey[t],ICy[[1]]},
  {x, y}, {t, 0, 3}, StartingStepSize→.05,
  Method→{FixedStep, Method→ExplicitRungeKutta}]
soln2 = NDSolve[{x'[t]==dex[t], ICx[[2]], y'[t]==dey[t],
  ICy[[2]]}, {x, y}, {t, -1, 1}, StartingStepSize→.05,
  Method→{FixedStep, Method→ExplicitRungeKutta}]
soln3 = NDSolve[{x'[t]==dex[t], ICx[[3]], y'[t]==dey[t],
  ICy[[3]]}, {x, y}, {t, -1, 10}, StartingStepSize→.05,
  Method→{FixedStep, Method→ExplicitRungeKutta}]
soln4 = NDSolve[{x'[t]==dex[t], ICx[[4]], y'[t]==dey[t],
  ICy[[4]]}, {x, y}, {t, 0, 10}, StartingStepSize→.05,
  Method→{FixedStep, Method→ExplicitRungeKutta}]
soln5 = NDSolve[{x'[t]==dex[t], ICx[[5]], y'[t]==dey[t],
  ICy[[5]]}, {x, y}, {t, -1, 8}, StartingStepSize→.05,
  Method→{FixedStep, Method→ExplicitRungeKutta}]
p1 = ParametricPlot[{Evaluate[{x[t], y[t]}/.soln1],
  Evaluate[{x[t], y[t]}/.soln2], Evaluate[{x[t], y[t]}/.soln3],
  Evaluate[{x[t], y[t]}/.soln4], Evaluate[{x[t], y[t]}/.soln5]},
  {t, 0, 7.5},PlotStyle→{Thickness[0.015]},
  PlotRange→{{-3, 4}, {-3, 5}}];
p2 = VectorPlot[{y - x^2, y - x}, {x, -3, 3}, {y, -3, 5},
  VectorScale→{Small, Small, None}, Axes→Automatic,
  AxesLabel→{"x", "y"}, VectorPoints→10];
Show[p1, p2, PlotLabel→"Phase plane plot"]
fx = Function[{x, y}, y - x^2];
fy = Function[{x, y}, y - x];
```

```
StreamPlot[{fx[x, y], fy[x, y]}, {x, -3, 4}, {y, -3, 3}]
```

Mathematica Example 3: Enter the following code that calculates the equilibria of

$$x' = -6x + 6y$$
$$y' = 36x - y - xz$$
$$z' = -3z + xy$$

and determines their stability.

```
Quit[]
eq1x = -6 x + 6 y
eq1y = 36 x - y - x z
eq1z = -3 z + x y
eq2 = Solve[eq1x==0 && eq1y==0 && eq1z==0, {x, y, z}]
  (*sometimes FullSimplify helps at this step*)
(J = D[{eq1x,eq1y,eq1z}, {{x,y,z}}])//MatrixForm (*Jacobian*)
  (*Note the extra parenthesis before //MatrixForm *)
  (*This allows calculations to be done on J*)
(J1 = J/.eq2[[1]])//MatrixForm
eig1 = FullSimplify[Eigenvalues[J1]]
N[eig1]
(J2 = J/.eq2[[2]])//MatrixForm
eig2 = FullSimplify[Eigenvalues[J2]]
N[eig2]
(J3 = J/.eq2[[3]])//MatrixForm
eig3 = FullSimplify[Eigenvalues[J3]]
```

Mathematica Example 4: Enter the following code that finds the $\lambda = 0$ bifurcations of the SIR model with vital dynamics. To do so, consider the following two equations, where $x = \frac{S}{N}$, $y = \frac{I}{N}$, and recall that we only need to consider the first two equations since N is constant. We still have $\beta, \gamma, \mu > 0$.

$$x' = \mu - \beta xy - \mu x$$
$$y' = \beta xy - (\mu + \gamma)y. \tag{6.84}$$

```
Quit[]
eq1a = μ - β x y - μ x
eq1b = β x y - (μ + α) y
eq2 = Solve[{eq1a==0, eq1b==0}, {x, y}]
(J = D[{eq1a, eq1b}, {{x, y}}])//MatrixForm
(J0 = J/.eq2[[1]])//MatrixForm
chpoly = CharacteristicPolynomial[J0, λ]
```

```
chpoly0 = ReplaceAll[chpoly==0, λ→0]
  (*The last equation is the condition when a "λ=0"
    bifurcation may occur. This gives, as we know, the condition
    beta= alpha + mu (i.e., R0 = 1).  We now choose 2 sets of
    parameter values, for < 0 and > 0, and check stability.*)
params1 = {β→.4, μ→.25, α→.2}
First[chpoly0] (*left-hand-side*)
Last[chpoly0] (*right-hand-side*)
eq3a = ReplaceAll[First[chpoly0], params1]
equil1 = eq2[[1]]
ReplaceAll[Eigenvalues[J0], params1]
equil2a = ReplaceAll[eq2[[2]], params1]
Eigenvalues[ReplaceAll[ReplaceAll[J, params1], equil2a]]
params2 = {β→.5, μ→.25, α→.2}
eq4a = ReplaceAll[First[chpoly0], params2]
Eigenvalues[ReplaceAll[J0, params2]]
equil2b = ReplaceAll[eq2[[2]], params2]
Eigenvalues[ReplaceAll[ReplaceAll[J, params2], equil2b]]
Simplify[ReplaceAll[eq2[[2]],β→μ + α]]
(*The EE is the same as the DFE*)
(*end of the code*)
```

The last line shows us that the two equilibria are the same when $\beta = \alpha + \mu$. The lines from *params1* until just before *params2* show that for $\beta < \mu + \alpha$, the DFE is stable whereas the EE is not biologically relevant and is unstable. The lines from *params2* until the second to last one show that for $\beta > \mu + \alpha$, the DFE is unstable whereas the EE is biologically relevant and is now stable. Taken together, we can conclude that a transcritical bifurcation occurred between the two equilibria.

Mathematica Exercises

Turn in both the commands that you enter for the exercises below as well as the output/figures. These should all be in one document. Please highlight or clearly indicate all requested answers. Some of the questions will require you to modify the above Mathematica code to answer them.

1. Enter the commands given in Mathematica Example 1 and submit both your input and output.

2. Enter the commands given in Mathematica Example 2 and submit both your input and output.

3. Enter the commands given in Mathematica Example 3 and submit both your input and output.

4. Enter the commands given in Mathematica Example 4 and submit both your input and output.

5. Plot the vector field for the first-order ODE system $\begin{cases} x' = -5x + 8y \\ y' = -4x + 7y. \end{cases}$

6. Use `pplane` or modify the code from Examples 1 and 2 to draw phase portrait for the ODE system $\begin{cases} x' = -4x - y \\ y' = 6x + y. \end{cases}$

7. Use `pplane` or modify the code from Examples 1 and 2 to draw phase portrait for the ODE system $\begin{cases} x' = 5x - y \\ y' = 3x + y. \end{cases}$

8. Use `pplane` or modify the code from Examples 1 and 2 to draw phase portrait for the ODE system $\begin{cases} x' = -4x + 2y \\ y' = -10x + 4y. \end{cases}$

9. Use `pplane` or modify the code from Examples 1 and 2 to draw phase portrait for the ODE system $\begin{cases} x' = -7x - 4y \\ y' = 10x + 5y. \end{cases}$

10. Determine the equilibria and their stability analytically for the ODE system $\begin{cases} x' = y, \\ y' = 4 - x^2. \end{cases}$ Then confirm your results with a phase portrait.

11. Determine the equilibria and their stability analytically for the ODE system $\begin{cases} x' = y, \\ y' = 4 - x^2. \end{cases}$ Then confirm your results with a phase portrait for the ODE system.

12. Determine the equilibria and their stability analytically for the ODE system $\begin{cases} x' = y - x, \\ y' = x^2 + 2y. \end{cases}$ Then confirm your results with a phase portrait for the ODE system.

13. Determine the equilibria and their stability analytically for the ODE system $\begin{cases} x' = y, \\ y' = x^3 - x. \end{cases}$ Then confirm your results with a phase portrait for the ODE system.

14. Determine the equilibria and their stability analytically for the ODE system $\begin{cases} x' = y^2 - 1, \\ y' = x^2 - y. \end{cases}$ Then confirm your results with a phase portrait for the ODE system.

15. Determine the $\lambda = 0$ bifurcation that occurs for the ODE system $\begin{cases} x' = r - x^2, \\ y' = -y, \end{cases}$ and the value of r at which it occurs.

16. Determine the $\lambda = 0$ bifurcation that occurs for the ODE system $\begin{cases} x' = y - 2x, \\ y' = r + x^2 - y, \end{cases}$ and the value of r at which it occurs.

17. Determine the $\lambda = 0$ bifurcation that occurs for the ODE system $\begin{cases} x' = rx - x^2, \\ y' = x - y, \end{cases}$ and the value of r at which it occurs.

18. Determine the $\lambda = 0$ bifurcation that occurs for the ODE system $\begin{cases} x' = rx - x^3, \\ y' = -y, \end{cases}$ and the value of r at which it occurs.

19. Determine the Hopf bifurcation that occurs for the ODE system $\begin{cases} x' = rx + 2y, \\ y' = -2x + ry + x^2y, \end{cases}$ and the value of r at which it occurs.

20. For the Lotka-Volterra model $\begin{cases} x' = x(3 - x - 2y), \\ y' = y(2 - x - y), \end{cases}$ find all the biologically relevant equilibria and determine their stability. Then draw the phase portrait.

21. Consider the Lotka-Volterra model $\begin{cases} x' = x(1 - x) - xy \\ y' = y(x - d), \quad d > 0 \end{cases}$ where x is the prey, y is the predator, and $d > 0$ is a parameter. Determine the $\lambda = 0$ bifurcation that occurs.

Chapter 6 Projects

Project 6A: An MSEIR Model [21]

Consider a model for a disease for which infection confers permanent immunity (e.g., measles, rubella, mumps, and chicken pox) and for which a mother's immunity gives her newborn infant *passive* immunity from the disease; that is, some of the mother's antibodies will protect the newborn from the disease for, say, the first six months of her/his life. Verify that Figure 6.24 gives rise to the following system of differential equations:

$$\frac{dM}{dt} = b(N - S) - \epsilon M - \mu M$$

$$\frac{dS}{dt} = bS + \epsilon M - \beta S \frac{I}{N} - \mu S$$

$$\frac{dE}{dt} = \beta S \frac{I}{N} - \delta E - \mu E$$

$$\frac{dI}{dt} = \delta E - \gamma I - \mu I$$

$$\frac{dR}{dt} = \gamma I - \mu R, \tag{6.85}$$

where $N = M + S + E + I + R$ is the total population.
    Show that

$$\frac{dN}{dt} = (b - \mu)N,$$

where $N = M + S + E + I + R$ is the total population. Is the total population constant or changing? Rewrite the differential equation in its fractional form (with $m = \frac{M}{N}, s = \frac{S}{N}, e = \frac{E}{N}, i = \frac{I}{N}, r = \frac{R}{N}$) and eliminate s from the system to obtain

$$\frac{dm}{dt} = (\mu + q)(e + i + r) - \epsilon m$$

$$\frac{de}{dt} = \beta(1 - m - e - i - r)i - (\delta + \mu + q)e$$

$$\frac{di}{dt} = \delta e - (\gamma + \mu + q)i$$

$$\frac{dr}{dt} = \gamma i - (\mu + q)r, \tag{6.86}$$

where $q = b - \mu$ is the difference between the birth and death rates. Show that
$$\{(m, e, i, r) | 0 \leq m, 0 \leq e, 0 \leq i, 0 \leq r, m + e + i + r \leq 1\}$$

is a positively invariant domain. Determine conditions on the stability of the *DFE* and show that the basic reproductive number is given by

$$R_0 = \frac{\beta\delta}{(\gamma + \mu + q)(\delta + \mu + q)}.$$

Interpret the meaning of R_0 in terms of the parameters of the system. Try to find the unique endemic equilibrium of the system that lies in the invariant region. Determine its stability and when it is biologically relevant. For chicken pox, $1/\epsilon = 6$ months, $1/\delta = 14$ days, and $1/\gamma = 7$ days; choose reasonable values for the birth and death rates in (6.86). Choose two different β-values that will give qualitatively different behavior in the system (only the *DFE* and then another that gives both an *EE* and a *DFE*). Give plots for these two different situations similar to that of Figure 6.18.

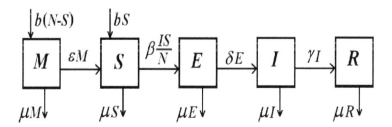

FIGURE 6.24: Flowchart for *MSEIR* model with exponentially changing size.

Project 6B: Routh-Hurwitz Criteria

In this project we explore an alternative way to examine whether $\text{Re}(\lambda) < 0$ for all λ_i. Although this project may be done by hand, it may help in easing some of the computation to have MATLAB, Maple, or Mathematica available. In the context of this chapter, this will allow us to determine the conditions for the local asymptotic stability of an equilibrium point via the Routh-Hurwitz critera. These criteria give conditions that guarantee the local asymptotic stability of the linearized system at zero. The conditions are on the coefficients of its characteristic polynomial. The Routh-Hurwitz criteria is the standard tool for cases where the eigenvalues are difficult to compute explicitly [16]. "Difficult to compute explicitly" usually means bigger than a 2×2 matrix. Especially for epidemiological models, this critera will help us determine stability of the point.

For the Routh-Hurwitz criteria, consider the nth-order system which has been linearized at an equilibrium $\mathbf{u}*$, which without loss of generality is assumed to be zero. Hence, the linearization is given by the system

$$\frac{d\mathbf{u}}{dt} = A\mathbf{u}$$

where $\mathbf{u} \in \mathbb{R}^n$ and $\mathbf{A}$ is an $n \times n$ matrix (Jacobian of $\mathbf{u}^* = 0$). The characteristic polynomial of this system can be written as

$$p(\lambda) = \lambda^n + a_1\lambda^{n-1} + a_2\lambda^{n-2} + ... + a_n,$$

where $a_i \in \mathbb{R}$ for $i = 1, ..., n$. Without loss of generality, we can assume that $a_n \neq 0$ otherwise $p(\lambda)$ has a zero root, i.e., $\lambda = 0$ is a solution, and the problem reduces to finding the non-zero roots of an $(n-1)$st degree polynomial with $a_{n-1} \neq 0$ and so on. In order for $\mathbf{u}^*$ to be stable we need $\text{Re}(\lambda_i) < 0$ for all $i = 1, ..., n$. These conditions are derived from the determinants of the principal minors, Δ_i, $i = 1, .., n$, of the Hurwitz matrix (see, e.g., [16]),

$$\mathbf{H}_n = \begin{bmatrix} a_1 & 1 & 0 & 0 & 0 & 0 & 0 & ... \\ a_3 & a_2 & a_1 & 1 & 0 & 0 & 0 & ... \\ a_5 & a_4 & a_3 & a_2 & a_1 & 1 & 0 & ... \\ a_7 & a_6 & a_5 & a_4 & a_4 & a_2 & 1 & ... \\ \vdots & & & & & \ddots & \end{bmatrix}$$

where the (i, j)th entry is

$$\begin{array}{lll} a_{2i-j} & \text{for} & 0 < 2i - j \leq n, \\ 1 & \text{for} & 2i = j, \\ 0 & \text{for} & 2i < j \ \text{ or } \ 2i > n + j. \end{array}$$

(Notice the main diagonal has the a_i and observe the contents of each row based on this.)

Routh-Hurwitz Criteria [30]:
For the polynomial $p(\lambda)$ to have only roots with negative real parts, it is necessary and sufficient that $\Delta_i > 0$, for $i = 1, \ldots, n$, where

$$\Delta_1 = a_1, \quad \Delta_2 = \begin{vmatrix} a_1 & 1 \\ a_3 & a_2 \end{vmatrix}, \quad \Delta_3 = \begin{vmatrix} a_1 & 1 & 0 \\ a_3 & a_2 & a_1 \\ a_5 & a_4 & a_3 \end{vmatrix}, \cdots, \Delta_n = a_n \Delta_{n-1}.$$

A Simple Example: Consider the characteristic polynomial $p(\lambda) = \lambda^3 + \lambda^2 + \lambda + 1$. Then

$$H_3 = \begin{bmatrix} a_1 & 1 & 0 \\ a_3 & a_2 & a_1 \\ 0 & 0 & a_3 \end{bmatrix} = \begin{bmatrix} 1 & 1 & 0 \\ 1 & 1 & 1 \\ 0 & 0 & 1 \end{bmatrix} \Rightarrow \Delta_1 = 1, \Delta_2 = 0, \Delta_3 = 0,$$

which shows that $p(\lambda)$ is unstable. We could have stopped at $\Delta_2 = 0$ because one violation of $\Delta_i > 0$ is enough.

More Realistic Example: van der Pol Oscillators Coupled via a Bath
Consider the system [12] given by

$$x''(t) - \epsilon(1 - x(t)^2)x'(t) + x(t) = \gamma(z(t) - x(t)),$$
$$y''(t) - \epsilon(1 - y(t)^2)y'(t) + y(t) = \gamma(z(t) - y(t)),$$
$$z'(t) = \gamma(x(t) - z(t)) + \gamma(y(t) - z(t)),$$

where γ and ϵ are nonnegative. The origin is an equilibrium point and we will determine its stability based on a relation of the parameters γ-ϵ.
(1) Calculate the Jacobain about the origin and show that the characteristic polynomial is

$$\begin{aligned} p(\lambda) = {} & \lambda^5 + (-2\epsilon + 2\gamma)\lambda^4 + (\epsilon^2 + 2 + 2\gamma - 4\epsilon\gamma)\lambda^3 \\ & + (2\epsilon^2\gamma + 2\gamma^2 - 2\epsilon - 2\epsilon\gamma + 4\gamma)\lambda^2 \\ & + (1 + \lambda^2 - 2\epsilon\gamma^2 + 2\gamma - 4\epsilon\gamma)\lambda + 2\gamma(1 + \gamma). \end{aligned} \tag{6.87}$$

(3) Verify the Hurwitz matrix for (6.87) is

$$\mathbf{H}_5 = \begin{bmatrix} a_1 & 1 & 0 & 0 & 0 \\ a_3 & a_2 & a_1 & 1 & 0 \\ a_5 & a_4 & a_3 & a_2 & a_1 \\ 0 & 0 & a_5 & a_4 & a_3 \\ 0 & 0 & 0 & 0 & a_5 \end{bmatrix}, \tag{6.88}$$

and show that the $\Delta_i, i = 1, 2, 3, 4, 5$ for our system are

$$\Delta_1 = 2(\gamma - \epsilon), \qquad \Delta_2 = (2 - 8\epsilon)\gamma^2 + (-2\epsilon + 8\epsilon^2)\gamma - 2\epsilon^3 - 2\epsilon,$$

$$\Delta_3 = -8\epsilon\gamma^4 + (4 - 16\epsilon + 20\epsilon^2 - 16\epsilon^3)\gamma^3$$

$$+ (-16\epsilon^3 - 4\epsilon + 16\epsilon^2 + 16\epsilon^4)\gamma^2 + (4\epsilon^4 - 4\epsilon - 4\epsilon^5 - 12\epsilon^3)\gamma + 4\epsilon^4,$$

$$\Delta_4 = 4\epsilon(2\epsilon^2\gamma - \epsilon - \epsilon\gamma - 4\epsilon\gamma^2 + 2\gamma^2)$$

$$(\epsilon^3\gamma^2 + 2\epsilon^3\gamma - 2\epsilon^2\gamma^3 - 5\epsilon^2\gamma^2 - 2\epsilon^2\gamma - \epsilon^2 + 2\epsilon\gamma^3 + 2\epsilon\gamma^2 - \gamma^4),$$

$$\Delta_5 = 8\gamma(1 + \gamma)\epsilon(2\epsilon^2\gamma - \epsilon - \epsilon\gamma - 4\epsilon\gamma^2 + 2\gamma^2)$$

$$(\epsilon^3\gamma^2 + 2\epsilon^3\gamma - 2\epsilon^2\gamma^3 - 5\epsilon^2\gamma^2 - 2\epsilon^2\gamma - \epsilon^2 + 2\epsilon\gamma^3 + 2\epsilon\gamma^2 - \gamma^4).$$

(5) By applying the Routh-Hurwitz criteria to expressions for Δ_2 and Δ_4, we now determine the stability of the origin in the entire first quadrant of the (γ, ϵ) plane.

(a) Consider the curve(s) that separate the $\Delta_i > 0$ region, for $i = 2, 4$. Obtain an analytic expression for these curves by considering the solution set of each expression. Solve $\Delta_2 = 0$ to obtain the curve C_1 given by

$$C_1: \quad \epsilon = -\frac{B}{6} + \frac{2(-4\gamma^2 + 3\gamma + 3)}{3B} + \frac{4\gamma}{3} \tag{6.89}$$

where

$$B = \left(64\gamma^3 + 36\gamma^2 + 144\gamma + 12\sqrt{96\gamma^5 + 153\gamma^4 - 12\gamma^3 + 132\gamma^2 + 36\gamma + 12}\right)^{\frac{1}{3}}.$$

(b) Solve $\Delta_4 = 0$ to obtain the curve C_2 given by

$$C_2: \quad \epsilon = \frac{4\gamma^2 + 1 - \sqrt{16\gamma^4 - 8\gamma^3 + 9\gamma^2 + 2\gamma + 1}}{4\gamma}. \tag{6.90}$$

(c) Conclude that above the curve C_1 in the first quadrant of the (γ, ϵ) plane, $\Delta_2 < 0$. The curve C_1 lies between the line $\epsilon = 0$ and the curve C_2. Below the C_2 curve, $\Delta_4 < 0$. Thus for $\epsilon > 0$ and $\gamma > 0$ there is always a negative Δ_i for some i and by the Routh-Hurwitz criteria the origin is always an unstable equilibrium point in the original system.

The interested reader should attempt to apply this criteria to some 3×3 (or larger!) systems that are known to have complicated eigenvalues for additional practice with this method.

Chapter 7

Laplace Transforms

7.1 Introduction

Consider the first-order electric circuit with an inductor and resistor only shown in Figure 7.1 with the switch in closed position (including V_0 in the circuit) and in open position (so that V_0 is not part of the circuit). Suppose that the switch begins in the closed position so that constant voltage V_0 is applied for $0 \le t < 1$, then a step down in voltage from V_0 to 0 occurs at $t = 0$ as the switch is opened, and thus a constant applied voltage 0 is applied for $t > 1$:

$$LI' + RI = \begin{cases} V_0, & 0 \le t < 1 \\ 0, & t > 1 \end{cases} \tag{7.1}$$

where I is the current in the circuit, L measures the inductance, and R measures the resistance, and V_0 is the constant applied voltage.

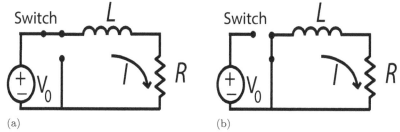

(a)   (b)

FIGURE 7.1: (a) *RL* circuit with switch is in the closed position so that V_0 is included in the circuit. (b) *RL* circuit with the switch is in the open position so that V_0 does not contribute.

We previously considered this equation for a continuous forcing function, but now our forcing function is a **piecewise continuous function**. We can separate this into two problems, each of whose solution we can easily find:

$$0 \le t < 1 : \quad LI' + RI = V_0 \Longrightarrow I = \tfrac{V_0}{R} + e^{\frac{-Rt}{L}}\left(I(0) - \tfrac{V_0}{R}\right)$$

$$t > 1 : \quad LI' + RI = 0 \Longrightarrow I = I(1^+)e^{\frac{-R}{L}(t-1)}.$$

The initial condition $I(1^+)$ in the $t > 1$ solution can be found examining $\lim_{t \to 1^-} I(t)$ in the first solution. Since that solution is continuous, we can plug

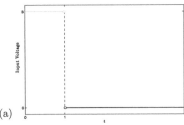
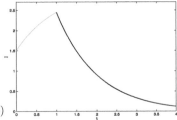

FIGURE 7.2: (a) Plot of input forcing function with $V_0 = 3$ in (7.1). (b) Solution to (7.1) with $V_0 = 3$, $R = 1$, $L = 1$, $I(0) = 1.5$. The blue solution is for $0 \leq t < 1$ while the black solution is for $t \geq 1$.

in $t = 1$ and rewrite the general solution as

$$
I(t) = \begin{cases} \frac{V_0}{R} + e^{\frac{-Rt}{L}}\left(I(0) - \frac{V_0}{R}\right), & 0 \leq t < 1 \\ \left[\frac{V_0}{R} + e^{\frac{-R}{L}}\left(I(0) - \frac{V_0}{R}\right)\right]e^{\frac{-R}{L}(t-1)}, & t > 1. \end{cases} \tag{7.2}
$$

Figure 7.2 shows the plot of the solution on $[0, 4]$.

This type of problem seemed simple enough in its set-up but the jump at $t = 1$ resulted in our solving two IVPs and then rewriting one of the initial conditions in terms of the original solution. As we will see in the ensuing sections, Laplace transforms will be a way to avoid solving this problem twice by first transforming the problem from the time domain to the *frequency domain* and solving a much simpler problem before transforming back to the time domain. While in the frequency domain we will additionally be able to observe the long-term behavior of the solutions with minimal effort. But first we will need to develop some of the mathematical machinery that will be used in our study.

Unit Step Function

The forcing function of our motivational equation (7.1) is a piecewise continuous function. It is a bit special, however, in that it is *constant* where it is continuous.

Definition 7.1.1

A function f is said to be piecewise continuous on a finite interval $a \leq t \leq b$ if this interval can be divided into a finite number of subintervals such that

1. f is continuous in the interior of each of these subintervals.

2. $f(t)$ approaches a limit (finite) as t approaches either endpoint of the subintervals from its interior.

More specifically, if f is piecewise continuous on $a \leq t \leq b$ and t_0 is an endpoint of one of the subintervals, then the right- and left-hand limit exists, that is,

$$f(t_0^-) = \lim_{t \to t_0^-} f(t) \quad \text{and} \quad f(t_0^+) = \lim_{t \to t_0^+} f(t) \text{ are finite and exist.}$$

Example 1 Consider the function f defined by

$$f(t) = \begin{bmatrix} -1, & 0 < t < 2 \\ 1, & t > 2. \end{bmatrix}$$

Solution

The function f is piecewise continuous, for at $t = 2$ we have

$$f(2^-) = \lim_{t \to 2^-} f(t) = -1$$
$$f(2^+) = \lim_{t \to 2^+} f(t) = 1.$$

We now consider a very special piecewise continuous function, called the **unit step function**

$$U_0(t) = \begin{cases} 0, & t < 0 \\ 1, & t > 0, \end{cases}$$

which is also referred to as a **Heaviside function.** We note that for some applications it is convenient to define $U_0(0) = \frac{1}{2}$, which is the average of $U_0(0^-)$ and $U_0(0^+)$. However, in calculating Laplace transforms, we will see that the value of the function at the point of discontinuity is not relevant but rather it is the values to the left and right of this discontinuity. *We thus don't include a value of the function at the discontinuity in order to stress this point and we will view the function as instantaneously changing its values at the point of discontinuity.* The subscript represents the location of the discontinuity and we can horizontally translate this curve to define the **shifted unit step function**. For each real number $a \geq 0$, we have

$$U_a(t) = \begin{cases} 0, & t < a \\ 1, & t > a. \end{cases}$$

We can think of this as a fundamental building block for piecewise continuous functions. We will often refer to both $U_0(t)$ and $U_a(t)$ as unit step functions. Using our previous notation, we can again talk about the limits to the left and right of $t = a$ for our unit step function and write

$$U_a(a^-) = \lim_{t \to a^-} = 0 \quad \text{and} \quad U_a(a^+) = \lim_{t \to a^+} = 1. \tag{7.3}$$

Example 2 Graph the unit step function $U_0(t)$ and

$$\begin{cases} 5, \ t < 2 \\ 0, \ t > 2. \end{cases} \tag{7.4}$$

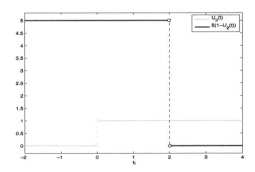

FIGURE 7.3: Graphs of $U_0(t)$ and (7.4).

Then rewrite (7.4) using $U_a(t)$.

Solution

We can graph both functions as shown in Figure 7.3.

We see two differences: there are each zero on different sides of the discontinuity and the non-zero portion is not one. For the first part, we observe that

$$1 - U_2(t) = \begin{cases} 1, & t < 2 \\ 0, & t > 2. \end{cases}$$

Multiplying by 5 will thus give us our desired result:

$$5(1 - U_2) = \begin{cases} 5, & t < 2 \\ 0, & t > 2. \end{cases}$$

Example 3: Rewrite the **box function**

$$f(t) = \begin{cases} 0, & 0 < t < 2 \\ 3, & 2 < t < 5 \\ 0, & t > 5 \end{cases}$$

using shifted unit step functions.

Solution

This is just the linear combination $3U_2(t) - 3U_5(t)$, that is,

$$3U_2(t) - 3U_5(t) = \begin{cases} 0, & t < 2 \\ 3, & t > 2 \end{cases} - \begin{cases} 0, & t < 5 \\ 3, & t > 5 \end{cases}$$

$$= \begin{cases} 0, & 0 < t < 2 \\ 3, & 2 < t < 5 \\ 0, & t > 5. \end{cases}$$

Unit Impulse Function

Considering what happens in electrical and mechanical systems when subjected to an impulse (a large but finite input over a very short time period) is often a problem of great interest. Suppose we consider the following problem.

Example 4: Consider the first-order electric circuit with capacitor and resistor. Figure 7.4(a) shows the switch in the open position (where V_s does not contribute) and Figure 7.4(b) shows the switch in the closed position (where V_s is now included in the circuit). At time $t = 0$, the switch is moved from the closed to the open position so that only R and C are in the circuit and there remains an initial charge $Q(0)$ in the circuit. However, suppose that a **voltage spike**, V_s, is applied and lasts from $t = 1$ to $t = 1.1$, represented by the quick closing and then re-opening of the switch. Find the voltage for any time $t > 0$.

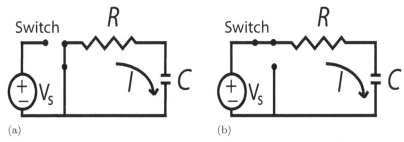

(a)　　　　　　　　　　(b)

FIGURE 7.4: RC circuit of Example 4; recall that $I = \frac{dQ}{dt}$. (a) The switch is in the open position so that V_s does not contribute. (b) The switch is in the closed position so that V_s is now included in the circuit.

Solution

The autonomous part of the differential equation is that of an RC circuit and the forcing is due to the voltage spike:

$$RQ' + \frac{Q}{C} = \begin{cases} 0, & 0 < t < 1 \\ V_s, & 1 < t < 1.1 \\ 0, & 1.1 < t \end{cases} \tag{7.5}$$

where R measures the resistance, C measures the capacitance, Q is the charge on the capacitor, and V_s is the voltage spike. This involves solving *three* differential equations and we could rewrite the forcing as a combination of step functions:

$$V_s(U_1 - U_{1.1}) = \begin{cases} 0, & 0 < t < 1 \\ V_s, & 1 < t < 1.1 \\ 0, & 1.1 < t \end{cases}$$

For comparison in a later section, the solution can be written as

$$\underline{0 < t < 1:} \quad RQ' + \tfrac{Q}{C} = 0 \Rightarrow Q = Q(0)e^{\frac{-t}{RC}}$$

$$\underline{1 < t < 1.1:} \quad RQ' + \tfrac{Q}{C} = V_s \Rightarrow Q = V_sC - e^{\frac{-t}{RC}}\left(V_sCe^{\frac{1}{RC}} - Q(0)\right) \quad (7.6)$$

$$\underline{t > 1.1:} \quad RQ' + \tfrac{Q}{C} = 0 \Rightarrow Q = e^{\frac{1.1-t}{RC}}\left(V_sC + Q(0)e^{\frac{-1.1}{RC}} - V_sCe^{\frac{-.1}{RC}}\right)$$

as you will show in Problem **12**, and where we again note that we will not define the function at the point of discontinuity. We can plot this solution for specific values as can be seen in Figure 7.5.

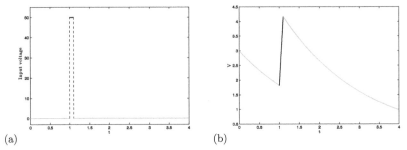

(a)   (b)

FIGURE 7.5: (a) The input voltage into the RC circuit: a voltage spike with $V_s = 50$ for t in $(1, 1.1)$ and 0 otherwise. (b) Solution to (7.5) with $Q(0) = 3$, $R = 2$, $C = 1$, $V_s = 50$.

In Example 4, the voltage spike was a tall and narrow box function. For the differential equation $P(D)x = f(t)$, we consider the **impulse** of the force $f(t)$ that is centered at t_0 as the area under the curve of the spike:

$$I(t) = \int_{-\infty}^{\infty} f(t)\, dt. \quad (7.7)$$

In our Example 4, this area was $50 \times 0.1 = 5$ (ignoring units). Mathematically, we might wonder how much difference our solution would be if the voltage spike lasted from $t = 1$ to 1.05 or even 1.01 instead. To do so, consider the specific box function:

$$f(t) = \frac{1}{2\alpha}\left(U_{t_0 - \alpha}(t) - U_{t_0 + \alpha}(t)\right)$$

$$= \begin{bmatrix} \dfrac{1}{2\alpha}, & \text{for } t_0 - \alpha < t < t_0 + \alpha \\ 0, & \text{otherwise} \end{bmatrix} \quad (7.8)$$

with α a constant. For the f given in (7.8), we can calculate this area as

$$
\begin{aligned}
I(t) &= \int_{t_0-\alpha}^{t_0+\alpha} f(t)\, dt \\
&= \int_{-\infty}^{\infty} f(t)\, dt \\
&= 1,
\end{aligned}
\tag{7.9}
$$

where we were able to extend the integral over all t-values because $f = 0$ outside of $[-\alpha, \alpha]$. We are most interested in what happens for small α and we thus define

$$
\delta(t - t_0) = \lim_{\alpha \to 0} \frac{1}{2\alpha} \left(U_{t_0-\alpha}(t) - U_{t_0+\alpha}(t) \right).
\tag{7.10}
$$

The function $\delta(t - t_0)$ is called an idealized **unit impulse function**. It is widely used in physics as well where it goes by another name: the **Dirac delta function**. We have thus constructed a rather interesting function, properly called a **generalized function**, that we now formalize with a definition.

Definition 7.1.2

The (idealized) unit impulse function δ is the function that satisfies

$$
\delta(t - t_0) = \begin{cases} 0, & \text{for } t \neq t_0 \\ \infty, & \text{for } t = t_0 \end{cases}
\tag{7.11}
$$

and

$$
\int_{-\infty}^{\infty} \delta(t - t_0)\, dt = 1.
\tag{7.12}
$$

Using some of our previous notation to denote what happens to the left and to the right of t_0, we can also write (7.11) as

$$
\delta(0^-) = 0, \quad \delta(0^+) = 0, \quad \delta(0) = \infty.
\tag{7.13}
$$

We motivated the unit impulse function in Example 4 as a brief (in time) force and found the explicit solution. We will see in the ensuing sections that Laplace transforms will allow us to solve differential equations "quickly" with step functions, box functions, or impulse functions and we will come back and compare the idealized impulse with the finite impulse of Example 3. Before moving on, however, we will state a few useful theorems involving the Dirac delta function.

THEOREM 7.1.1

Suppose that $g(t)$ is a bounded continuous function. Then

$$\int_{-\infty}^{\infty} \delta(t - t_0) g(t)\, dt = g(t_0).$$

One final result shows a relation between the unit step function (Heaviside function) and the idealized unit impulse function (Dirac delta function).

Consider the unit step function

$$U_0(t) = \begin{cases} 0 & t < 0 \\ 1 & t > 0 \end{cases}$$

and the idealized unit impulse function centered at the origin:

$$\delta(t) = \begin{cases} 0, & \text{for } t \neq 0 \\ \infty, & \text{for } t = 0 \end{cases}, \quad \text{with} \quad \int_{-\infty}^{\infty} \delta(t)\, dt = 1.$$

Then

$$\frac{d}{dt}[U_0(t)] = \delta(t). \tag{7.14}$$

Remark: We stated that we refer to $\delta(t)$ as a generalized function and thus we often refer to the above notation as a **generalized derivative**.

Recall that we have chosen $U_0(0)$ as undefined. In this case or if we had defined $U_0(0) = \frac{1}{2}$, we could use the typical definition of the derivative. However, we instead choose to introduce the equivalent **centered difference for the derivative** of $f(x)$:

$$f'(x) = \lim_{h \to 0} \frac{f(x + h) - f(x - h)}{2h}. \tag{7.15}$$

Although the relation between the Heaviside and Dirac delta function may seem a bit strange, we apply the definition of the derivative to $U_0(t)$ for $t \neq 0$:

$$\frac{d}{dt}[U_0(t)] = \lim_{h \to 0} \frac{U_0(t + h) - U_0(t - h)}{2h}$$

$$= \begin{cases} \lim_{h \to 0} \dfrac{0 - 0}{2h} = \lim_{h \to 0} \dfrac{0}{2h} = \lim_{h \to 0} 0 = 0, & t < 0 \\[3mm] \lim_{h \to 0} \dfrac{1 - 1}{2h} = \lim_{h \to 0} \dfrac{0}{2h} = \lim_{h \to 0} 0 = 0, & t > 0 \end{cases} \tag{7.16}$$

and observe that this matches exactly with $\delta(t) = 0$ for $t \neq 0$. For the derivative when $t = 0$, we have

$$\frac{d}{dt}[U_0(0)] = \lim_{h \to 0} \frac{U_0(h) - U_0(-h)}{h}$$

$$= \begin{cases} \lim_{h \to 0^-} \dfrac{0 - 1}{h} = \lim_{h \to 0^-} \dfrac{-1}{h} = \infty \\ \lim_{h \to 0^+} \dfrac{1 - 0}{h} = \lim_{h \to 0^+} \dfrac{1}{h} = \infty. \end{cases} \qquad (7.17)$$

This matches the definition for the Dirac delta function at 0: $\delta(0) = \infty$. This will come up when we are applying the Laplace transform to the unit step function.

Brief Overview

In the remainder of this chapter we will develop properties of the Laplace transform, which will be very useful in certain applications. While we considered first-order circuit equations in the examples of this section, we could also consider the equation

$$m\frac{d^2x}{dt^2} + b\frac{dx}{dt} + kx = F(t)$$

for the position $x(t)$ of a mass on a spring. We saw in Chapter 4 this mass-spring system and considered some of its applications. In all of the problems we considered previously, the forcing function $F(t)$ was continuous. In many practical applications, however, $F(t)$ may have discontinuities. For instance, the mass may be periodically struck, imparting a force $F(t)$ at time t. In this case, the methods we used in Chapter 4 are awkward and the solutions may be cumbersome. It is just these kinds of applications for which the Laplace transform is useful.

The Laplace transform will be formally defined in the next section, but to set the stage and motivate what is about to come, we will let $\mathcal{L}$ represent the Laplace transform. We will see shortly that the Laplace transform of a function $f(t)$ produces a function $F(s)$, that is,

$$F(s) = \mathcal{L}\{f(t)\}$$

where F is a function of a new independent variable s.

After learning some of the basics of computing Laplace transforms of some basic functions, we will see that the Laplace transform converts a *differential* equation $P(D)x = f(t)$ with its initial conditions into an *algebraic* equation in $Y(s) = \mathcal{L}\{x(t)\}$. It is often the case that such an algebraic equation is easier to solve than the differential equation. We will develop a method of converting from the solution of the algebraic equation to the solution of the differential equation; the original variable $x(t)$ is obtained from the inverse Laplace transform $x(t) = \mathcal{L}^{-1}\{Y(s)\}$. The ideas discussed here are schematically shown in the following diagram.

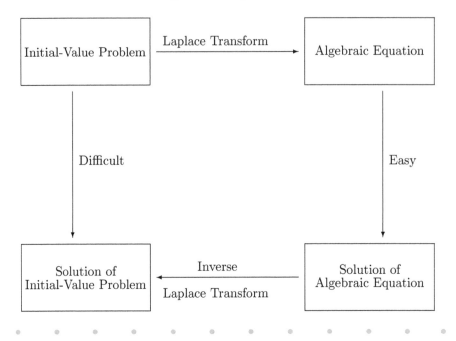

In Problem **1–11**, write the given piecewise continuous functions using shifted unit step functions.

1. $f(t) = \begin{cases} 2, & 0 < t < 4 \\ 0, & t > 4 \end{cases}$

2. $f(t) = \begin{cases} 0, & 0 < t < 4 \\ 3, & t > 4 \end{cases}$

3. $f(t) = \begin{cases} 0, & 0 < t < 6 \\ 5, & t > 6 \end{cases}$

4. $f(t) = \begin{cases} 4, & 0 < t < 10 \\ 0, & t > 10 \end{cases}$

5. $f(t) = \begin{cases} 0, & 0 < t < 5 \\ 2, & 5 < t < 7 \\ 0, & t > 7 \end{cases}$

6. $f(t) = \begin{cases} 0, & 0 < t < 3 \\ -6, & 3 < t < 9 \\ 0, & t > 9 \end{cases}$

7. $f(t) = \begin{cases} t, & 0 < t < 3 \\ 3, & t > 3 \end{cases}$

8. $f(t) = \begin{cases} 2t, & 0 < t < 5 \\ 10, & t > 5 \end{cases}$

9. $f(t) = \begin{cases} t, & 0 < t < 1 \\ 1, & 1 < t < 2 \\ 0, & t > 2 \end{cases}$

10. $f(t) = \begin{cases} t, & 0 < t < 1 \\ t^2, & 1 < t < 2 \\ 0, & t < 2 \end{cases}$

11. $f(t) = \begin{cases} 1, & 0 < t < 2 \\ 2, & 2 < t < 3 \\ 3, & 3 < t < 5 \\ 0, & t > 5 \end{cases}$

12. Derive the solutions of Equation (7.6).

In Problems **13-22**, use Theorem 7.1.1 to evaluate the given integral.

13. $\displaystyle\int_{-\infty}^{\infty} \delta(t)\cos t\, dt$

14. $\displaystyle\int_{-\infty}^{\infty} \delta\left(t - \frac{\pi}{2}\right)\sin t\, dt$

15. $\displaystyle\int_{-\infty}^{\infty} \delta(t-2)e^{-t}\, dt$

16. $\displaystyle\int_{-\infty}^{\infty} \delta(t-5)e^{-t}\sin t\, dt$

17. $\displaystyle\int_{-\infty}^{\infty} \delta(t-2)\frac{t^2}{1+t^2}\, dt$

18. $\displaystyle\int_{-\infty}^{\infty} \delta(t-1)\arctan t\, dt$

19. $\displaystyle\int_{0}^{3} (1+e^{-t})\delta(t-2)\, dt$

20. $\displaystyle\int_{-2}^{1} (1+e^{-t})\delta(t-2)\, dt$

21. $\displaystyle\int_{-1}^{2} \cos 2t\,\delta(t)\, dt$

22. $\displaystyle\int_{-3}^{2} (e^{2t}+t)\delta(t+2)\, dt$

7.2   Fundamentals of the Laplace Transform

We now formally introduce the Laplace transform of a function f.

Definition 7.2.1

Let f be a real-valued function of the real variable t, defined for $t > 0$. Consider the function F defined by

$$F(s) = \int_{0^-}^{\infty} e^{-st} f(t)\, dt \tag{7.18}$$

for all values of s for which this integral exists. The function F is called the Laplace transform of the function f. We denote the Laplace transform F of f by $\mathcal{L}\{f\}$ and denote $F(s)$ by $\mathcal{L}\{f(t)\}$.

Remark 1: The lower limit of integration is given as 0^- to account for special functions such as the idealized unit impulse, $\delta(t)$. If there is no problem with the limit at 0, we will just write 0 for the lower limit (instead of 0^-). While this distinction differs from many of the current popular introductory differential equations texts, we refer the reader to [36] for a more detailed discussion of why this distinction is crucial from both a mathematical and engineering perspective.

Remark 2: Since the integral in (7.18) is improper, it may not exist for every function f for which we might wish to compute the Laplace transform. Conditions will often have to be imposed upon f in order for this integral to exist.

Remark 3: In this definition, the integral is with respect to the variable t, with s a "dummy" variable for the integral. Thus, the Laplace transform of $f(t)$ will produce a function of s. We will allow s to be complex; however, many results we obtain will only depend on the real part of s. Thus, for convenience, we will often consider a real-valued s and will specifically note when we intend s to have a non-zero imaginary part.

Using the definition, it is not difficult to calculate the Laplace transform of familiar functions. We now present several examples to illustrate this concept.

Example 1: For the function $f(t) = 1$ for all $t > 0$, the Laplace transform is

$$\mathcal{L}\{1\} = \int_0^\infty e^{-st} \cdot 1\, dt = \int_0^\infty e^{-st}\, dt = \lim_{k \to \infty} \left.\frac{-1}{s} e^{-st}\right|_0^k$$

$$= \lim_{k \to \infty} -\frac{1}{s} e^{-sk} - \frac{-1}{s} = \frac{1}{s} \quad \text{for all } s > 0$$

so

$$\mathcal{L}\{1\} = \frac{1}{s}, \quad \text{for } s > 0,$$

which is a function of s as expected.

Example 2: Let $f(t) = t$, $t > 0$; then integrating by parts gives the Laplace transform as

$$\mathcal{L}\{t\} = \int_0^\infty e^{-st} \cdot t\, dt = \int_0^\infty t\, e^{-st}\, dt$$

$$= \lim_{k \to \infty} \left.-\frac{t}{s} e^{-st}\right|_0^k + \frac{1}{s} \int_0^\infty e^{-st}\, dt$$

$$= \lim_{k \to \infty} \left.-\frac{1}{s^2} e^{-st}\right|_0^k = \frac{1}{s^2}.$$

Hence

$$\mathcal{L}\{t\} = \frac{1}{s^2}, \quad \text{for } s > 0.$$

Example 3: For the function $f(t) = e^{at}$, for $t > 0$, the Laplace transform is found from

$$\mathcal{L}\{e^{at}\} = \int_0^\infty e^{-st} e^{at}\, dt = \int_0^\infty e^{(a-s)t}\, dt$$

$$= \lim_{k \to \infty} \left.\frac{1}{a-s} e^{(a-s)t}\right|_0^k = \lim_{k \to \infty} \frac{1}{a-s} e^{(a-s)k} - \frac{1}{a-s}$$

so that if $s > a$, then $e^{(a-s)k} \to 0$ as $k \to \infty$, thus

$$\mathcal{L}\{e^{at}\} = \frac{1}{s-a} \quad \text{if } s > a.$$

We mention at this point that the three examples so far have been with a real-valued s. In this next example, we revisit Example 1 for a complex-valued s.

Example 1 revisited. Consider again the function $f(t) = 1$ for all $t > 0$ and let $s = \alpha + i\beta$. The Laplace transform is

$$\mathcal{L}\{1\} = \int_0^\infty e^{-st} \cdot 1 \, dt$$

$$= \lim_{k \to \infty} \left. \frac{-1}{s} e^{-st} \right|_0^k$$

$$= \lim_{k \to \infty} \left. -\frac{e^{-\alpha t}(\cos(\beta t) + i \sin(\beta t))}{s} \right|_0^k .$$

Because cos and sin are bounded functions, we see that this limit will converge if $\alpha > 0$ and will diverge if $\alpha < 0$. Thus, for complex s, we have

$$\mathcal{L}\{1\} = \frac{1}{s}, \quad \text{for} \quad \text{Re}(s) > 0,$$

which agrees with our results in Example 1 when we only considered a real-valued s.

Example 4. For $f(t) = \sin bt, \ t > 0$, the Laplace transform is found from

$$\mathcal{L}\{\sin bt\} = \int_0^\infty e^{-st} \sin bt \, dt.$$

Integrating by parts twice,

$$\int e^{-st} \sin bt \, dt = \frac{-b \cos bt e^{-st}}{s^2 + b^2} - \frac{s \sin bt \, e^{-st}}{s^2 + b^2} \qquad \text{for } s > 0.$$

Thus

$$\int_0^\infty e^{-st} \sin bt \, dt = \lim_{k \to \infty} \int_0^k e^{-st} \sin bt \, dt$$

$$= \left[\lim_{k \to \infty} \frac{-b \cos bt \, e^{-st}}{s^2 + b^2} - \left. \frac{s \sin bt \, e^{-st}}{s^2 + b^2} \right] \right|_0^k$$

$$= \lim_{k \to \infty} \left[\frac{-b}{s^2 + b^2} (\cos bk \, e^{-sk}) - \frac{s}{s^2 + b^2} \sin bk \, e^{-sk} \right]$$

$$\quad - \left[\frac{-b}{s^2 + b^2} + \frac{0}{s^2 + b^2} \right]$$

$$= \frac{b}{s^2 + b^2},$$

so

$$\mathcal{L}\{\sin bt\} = \frac{b}{s^2 + b^2}, \quad s > 0.$$

A similar calculation shows that

$$\mathcal{L}\{\cos bt\} = \frac{s}{s^2 + b^2} \quad \text{for } s > 0.$$

Example 5: We previously considered the unit step function, defined as

$$U_a(t) = \begin{cases} 0 & t < a \\ 1 & t > a. \end{cases}$$

In our definition of the Laplace transform, we wrote the lower limit as 0^- in order to incorporate any discontinuities into the transform. The unit step function U_a, with $a = 0$ a possibility, satisfies the conditions of the existence theorem for Laplace transforms so that $\mathcal{L}\{U_a(t)\}$ exists. Thus

$$\mathcal{L}\{U_a(t)\} = \int_{0^-}^{\infty} e^{-st} U_a(t)dt = \int_{0^-}^{a^-} e^{-st} 0\, dt + \int_{a^-}^{\infty} e^{-st}\, dt$$

$$= \lim_{k \to \infty} \int_{a^-}^{k} e^{-st}\, dt = \lim_{k \to \infty} \left. -\frac{e^{-st}}{s} \right|_{a^-}^{k}$$

$$= \lim_{k \to \infty} \frac{-e^{-sk} + e^{-sa}}{s} = \frac{e^{-sa}}{s}.$$

Thus we have

$$\mathcal{L}\{U_a(t)\} = \frac{e^{-as}}{s}, \quad s > 0.$$

Note that for $a > 0$, we include the discontinuity with a^-. We can use unit step functions to express other step functions.

Using the familiar property of the integral that the integral of a sum is the sum of the integrals, we say that the Laplace transform is *linear*.

THEOREM 7.2.1   Linearity

Let f_1 and f_2 be functions whose Laplace transform exists and let C_1 and C_2 be constants. Then,

$$\mathcal{L}\{C_1 f_1(t) + C_2 f_2(t)\} = C_1 \mathcal{L}\{f_1(t)\} + C_2 \mathcal{L}\{f_2(t)\}.$$

We previously showed that

$$f(t) = \begin{cases} 0 & 0 < t < 2 \\ 3 & 2 < t < 5 \\ 0 & t > 5 \end{cases} = 3U_2(t) - 3U_5(t).$$

Then

$$\mathcal{L}\{f(t)\} = \mathcal{L}\{3U_2(t) - 3U_5(t)\} = 3\mathcal{L}\{U_2(t)\} - 3\mathcal{L}\{U_5(t)\}$$

$$= \frac{3}{s}e^{-2s} - \frac{3}{s}e^{-5s} = \frac{3}{s}(e^{-2s} - e^{-5s}).$$

Use the linearity property of the Laplace transform and Euler's formula to obtain $\mathcal{L}\{\sin bt\}$ and $\mathcal{L}\{\sin bt\}$.

From Example 3, we have that $\mathcal{L}\{e^{at}\} = \frac{1}{s-a}$ if $s > a$. Thus, we can write

$$\mathcal{L}\{\cos(bt) + i\sin(bt)\} = \mathcal{L}\{e^{ibt}\}$$

$$= \frac{1}{s - ib} = \frac{1}{s - ib} \cdot \frac{s + ib}{s + ib}$$

$$= \frac{s + ib}{s^2 + b^2}. \tag{7.19}$$

We obtain our desired transforms with

$$\mathcal{L}\{\cos(bt)\} = \operatorname{Re}(\mathcal{L}\{e^{ibt}\}) = \frac{s}{s^2 + b^2}$$

$$\mathcal{L}\{\sin(bt)\} = \operatorname{Im}(\mathcal{L}\{e^{ibt}\}) = \frac{b}{s^2 + b^2}.$$

Another useful property concerning the nth derivative of the Laplace transform of a function $f(t)$ is given in the next theorem.

THEOREM 7.2.2

Suppose the real function f is integrable on every finite closed subinterval of $a \leq t < \infty$, with the Laplace transform F given by

$$F(s) = \int_{0-}^{\infty} e^{-st} f(t)\, dt.$$

Then
$$\mathcal{L}\{t^n f(t)\} = (-1)^n \frac{d^n}{ds^n}[F(s)].$$

Example 7: To find $\mathcal{L}\{t^2 \sin bt\}$, we have

$$\mathcal{L}\{t^2 \sin bt\} = (-1)^2 \frac{d^2}{ds^2} F(s)$$

where

$$F(s) = \mathcal{L}\{\sin bt\} = \frac{b}{s^2 + b^2}$$

and

$$\frac{dF}{ds} = \frac{-2bs}{(s^2 + b^2)^2}$$

$$\frac{d^2F}{ds^2} = \frac{6bs^2 - 2b^3}{(s^2 + b^2)^3},$$

so that

$$\mathcal{L}\{t^2 \sin bt\} = \frac{6bs^2 - sb^3}{(s^2 + b^2)^3}.$$

In each of these examples, we have seen that the integral (7.18) exists for some (possibly infinite) range of values of s. Does the Laplace transform always exist? The answer is "No," as we will shortly see. We will now investigate conditions for which the transform does exist.

Existence of the Laplace Transform

We will introduce a few ideas to help us determine when the Laplace transform exists. The next definition will allow us to determine a class of functions for which the transform always exists. The next definition will allow us to determine how fast a function grows.

Definition 7.2.2

A function f is said to be of exponential order if there exists a constant α and positive constants t_0 and M such that

$$|f(t)| < M e^{\alpha t}$$

for all $t > t_0$ at which $f(t)$ is defined. More explicitly, if f is of exponential order corresponding to some definite constant α, then we say that f is of exponential order $e^{\alpha t}$.

In other words, we say f is of exponential order if a constant α exists such that the product $e^{-\alpha t}|f(t)|$ is bounded for all sufficiently large values of t, so that the values of $|f(t)|$ cannot become infinite more rapidly than a multiple of some exponential function $e^{\alpha t}$. Note that if f is of exponential order $e^{\alpha t}$, then f is also of exponential order $e^{\beta t}$ for any $\beta > \alpha$. Now clearly every bounded function is of exponential order, with constant $\alpha = 0$. So, for instance, $\sin bt$ and $\cos bt$ are of exponential order. The next few examples consider some combinations of familiar functions.

For the function $f(t) = e^{at} \sin bt$, $f(t)$ is of exponential order with constant $\alpha = a$, since

$$e^{-\alpha t}|f(t)| = e^{-at}e^{at}|\sin bt| = |\sin bt|$$

which is bounded for all t.

Consider the function $f(t) = t^n$, where $n > 0$. Then

$$e^{-\alpha t}|f(t)| = e^{-\alpha t}t^n.$$

Now for any $\alpha > 0$,

$$\lim_{t \to \infty} e^{-\alpha t}t^n = 0.$$

This means there exists $M > 0$ and $t_0 > 0$ so that

$$e^{-\alpha t}|f(t)| = e^{-\alpha t}t^n < M \quad \text{for } t > t_0.$$

Thus $f(t) = t^n$ is of exponential order, with the constant α equal to any positive number.

The function $f(t) = e^{t^2}$ is *not* of exponential order, as

$$e^{-\alpha t}|f(t)| = e^{-\alpha t + t^2}$$

which is unbounded as $t \to \infty$ no matter the value of α.

To apply these ideas, we need a familiar result from calculus. This result gives a condition for the existence of an improper integral.

THEOREM 7.2.3   Comparison Test for Improper Integrals

Let g and G be real functions such that

$$0 \le g(t) \le G(t) \quad \text{on } a \le t < \infty.$$

Suppose

$$\int_a^\infty G(t)\,dt \quad \text{exists,}$$

and g is integrable on every finite closed subinterval of $a \le t < \infty$. Then

$$\int_a^\infty g(t)\,dt \quad \text{exists.}$$

The function G is sometimes called a **dominating function** or **majorizing function**. The comparison test gives the next result since $g(t) \le |g(t)|$ for any function $g(t)$.

THEOREM 7.2.4

Suppose the real function g is integrable on every finite closed subinterval of $a \le t < \infty$ and suppose

$$\int_a^\infty |g(t)|\,dt \quad \text{exists.}$$

Then

$$\int_a^\infty g(t)\,dt \quad \text{exists.}$$

With these two results in place, we can now state and prove an existence theorem for Laplace transforms.

THEOREM 7.2.5

Let f be a real function that has the following properties:
(1) f is piecewise continuous in every finite closed interval $0 \le t \le b, b > 0$
(2) f is of exponential order, that is, there exists $\alpha, M > 0$ and $t_0 > 0$ so that

$$e^{-\alpha t}|f(t)| < M \quad \text{for } t > t_0.$$

Then the Laplace transform

$$\mathcal{L}\{f\} = \int_{0^-}^\infty e^{-st} f(t)\,dt$$

of f exists for $s > \alpha$.

Proof: We have

$$\int_{0^-}^{\infty} e^{-st} f(t)\, dt = \int_{0^-}^{t_0} e^{-st} f(t)\, dt + \int_{t_0}^{\infty} e^{-st} f(t)\, dt.$$

Now by hypothesis (1), the first integral on the right exists.
Considering the second integral, by hypothesis (2) we have

$$e^{-st}|f(t)| < e^{-st} M e^{-\alpha t} = M e^{-(s-\alpha)t} \text{ for } t > t_0.$$

Also,

$$\int_{t_0}^{\infty} M e^{-(s-\alpha)t}\, dt = \lim_{k \to \infty} \left. \frac{-M e^{-(s-\alpha)t}}{s-\alpha} \right|_{t_0}^{k}$$

$$= \lim_{k \to \infty} \left(\frac{M}{s-\alpha} \right) \left(e^{-(s-\alpha)t_0} - e^{-(s-\alpha)k} \right)$$

$$= \left[\frac{M}{s-\alpha} \right] e^{-(s-\alpha)t_0} \text{ if } s > \alpha.$$

Thus

$$\int_{t_0}^{\infty} M e^{-(s-\alpha)t}\, dt \quad \text{exists for } s > \alpha.$$

Now by hypothesis (1), $e^{-st}|f(t)|$ is integrable on every finite closed subinterval of $t_0 \le t < \infty$ so applying the comparison test with

$$g(t) = e^{-st}|f(t)|$$

and

$$G(t) = M e^{-(s-\alpha)t}$$

we see that

$$\int_{t_0}^{\infty} e^{-st}|f(t)|\, dt \quad \text{exists for } s > \alpha.$$

That is,

$$\int_{t_0}^{\infty} |e^{-st} f(t)|\, dt \quad \text{exists for } s > \alpha.$$

Thus,

$$\int_{t_0}^{\infty} e^{-st} f(t)\, dt \quad \text{also exists if } s > \alpha.$$

So the Laplace transform of f exists for $s > \alpha$. ∎

If we look at the proof, we have actually shown that if f is as stated, then

$$\int_{t_0}^{\infty} e^{-st}|f(t)|\, dt \quad \text{exists if } s > \alpha.$$

Further, hypothesis 1 shows that

$$\int_{0^-}^{t_0} e^{-st}|f(t)|\, dt \text{ exists.}$$

Thus

$$\int_{0^-}^{\infty} e^{-st}|f(t)|\, dt \quad \text{exists if } s > \alpha,$$

so not only does $\mathcal{L}\{f(t)\}$ exist, but so does $\mathcal{L}\{|f(t)|\}$ for $s > \alpha$. More precisely, we say that $\int_{0^-}^{\infty} e^{-st}f(t)\, dt$ is **absolutely convergent** for $s > \alpha$.

It should be pointed out that the conditions on f described in the hypothesis of the theorem are not necessary for the existence of $\mathcal{L}\{f\}$; however, there are functions f that *do not* satisfy the hypothesis of the theorem but for which $\mathcal{L}\{f\}$ exists. For instance, suppose we replace hypothesis 1 with the less restrictive condition that f is piecewise continuous in every finite closed interval $a \leq t \leq b$ where $a > 0$, and is such that $t^n f(t)$ remains bounded as $t \to 0^+$ for some $0 < n < 1$. Then, provided hypothesis 2 remains satisfied, it can be shown that $\mathcal{L}\{f\}$ still exists.

Example 11: Consider the function $f(t) = t^{-\frac{1}{3}}$, $t > 0$; then it can be shown that $\mathcal{L}\{f\}$ exists. Although f does not satisfy the hypothesis of the theorem $[f(t) \to \infty$ as $t \to 0^+]$ it *does* satisfy the less restrictive requirement stated above (with $n = \frac{2}{3}$) and is indeed of exponential order.

Laplace Transform of Derivatives

Since we seek to apply Laplace transforms to differential equations, we need to know how they affect the derivative of a function.

THEOREM 7.2.6   Differentiation

Let f be a real function that is continuous for $t \geq 0$ and of exponential order $e^{\alpha t}$ and suppose f' is piecewise continuous in every finite closed interval $0 \leq t \leq b$. Then

$$\mathcal{L}\{f'\} \text{ exists for } s > \alpha \text{ and}$$

$$\mathcal{L}\{f'(t)\} = s\mathcal{L}\{f(t)\} - f(0^-). \tag{7.20}$$

Recall that we use 0^- as the lower limit of integration and this arises in the above formula. We often refer to $f(0^-)$ as a **pre-initial condition**.

Proof: Since

$$\mathcal{L}\{f'(t)\} = \int_{0^-}^{\infty} e^{-st}f'(t)\, dt,$$

integrating by parts gives

$$\mathcal{L}\{f'(t)\} = \lim_{k\to\infty} e^{-st} f(t)\Big|_{0^-}^{k} + s\int_{0^-}^{\infty} e^{-st} f(t)\, dt$$

$$= \lim_{k\to\infty} e^{-sk} f(k) - f(0^-) + s\mathcal{L}\{f(t)\}$$

$$= s\mathcal{L}\{f(t)\} - f(0^-) \qquad \text{(since } f \text{ is of exponential order 1).} \blacksquare$$

The relation between the pre-initial condition given in (7.20) and the **post-initial-value** $f(0^+)$ is given by

$$\lim_{s\to\infty} sF(s) = f(0^+), \tag{7.21}$$

where we specify that the limit is taken along the positive real axis since s may be complex. As mentioned earlier, we will often write 0 instead of 0^- if there is not a problem at 0.

Example If $f(t) = \sin^2 at$, then $f'(t) = 2a\sin at \cos at$ so that

$$\mathcal{L}\{2a\sin at \cos at\} = s\mathcal{L}\{\sin^2 at\}$$

$$= s\left(\frac{2a^2}{s(s^2 + 4a^2)}\right) = \frac{2a^2}{s^2 + 4a^2}.$$

In a very similar manner, it is also possible to consider higher-order derivatives.

THEOREM 7.2.7 **Higher Order Derivatives**

Let f be a real function having a continuous $(n-1)^{\text{st}}$ derivative $f^{(n-1)}$ for $t \geq 0$ and suppose $f, f', \ldots, f^{(n-1)}$ are all of exponential order e^{at}. Further suppose $f^{(n)}$ is piecewise continuous. Then $\mathcal{L}\{f^{(n)}\}$ exists for $s > \alpha$ and

$$\mathcal{L}\{f^{(n)}(t)\} = s^n \mathcal{L}\{f(t)\} - s^{(n-1)} f(0^-)$$
$$- s^{(n-2)} f'(0^-) - s^{(n-3)} f''(0^-) - \cdots - f^{(n-1)}(0^-).$$

The proof of Theorem 7.2.7 follows from induction on the previous theorem.

Example To find $\mathcal{L}\{\sin bt\}$, we proceed as follows:

$$\mathcal{L}\{f''(t)\} = s^2 \mathcal{L}\{f(t)\} - sf(0) - f'(0), \tag{7.22}$$

so that we have

$$f(t) = \sin bt \text{ so } f(0) = 0.$$

Further,

$$f'(t) = b\cos bt \text{ gives } f'(0) = b \text{ and } f''(t) = -b^2 \sin bt.$$

This yields

$$\mathcal{L}\{-b^2 \sin bt\} = s^2 \mathcal{L}\{\sin bt\} - b, \quad \text{applying (7.22).}$$

Simplifying

$$(s^2 + b^2)\mathcal{L}\{\sin bt\} = b$$

and solving for $\mathcal{L}\{\sin bt\}$ we have

$$\mathcal{L}\{\sin bt\} = \frac{b}{s^2 + b^2}$$

which agrees with Example 4 of this section.

Problems

In Problems **1–11**, use the definition of the Laplace transform to find $\mathcal{L}\{f(t)\}$ for each of the given functions $f(t)$. If $\mathcal{L}\{f(t)\}$ exists, give the domain for $F(s) = \mathcal{L}\{f(t)\}$.

1. $f(t) = t^2$

2. $f(t) = 3e^{-2t}$

3. $f(t) = t - 3$

4. $f(t) = te^{-t}$

5. $f(t) = |t - 2|$

6. $f(t) = te^{t\sqrt{t}}$

7. $f(t) = \sinh t$

8. $f(t) = 28e^{13t}$

9. $f(t) = \begin{cases} 2, & 0 < t < 4 \\ 0, & t > 4 \end{cases}$

10. $f(t) = \begin{cases} t, & 0 < t < 1 \\ 1, & 1 < t < 2 \\ 0, & t > 2 \end{cases}$

11. $f(t) = \begin{cases} t, & 0 < t < 1 \\ t^2, & 1 < t < 2 \\ 0, & t > 2 \end{cases}$

In Problems **12–32**, using the properties of the Laplace transform developed in this section, find the Laplace transform $\mathcal{L}\{f(t)\}$ for each of the given functions $f(t)$.

12. $f(t) = t + \sin t$

13. $f(t) = t - 5$

14. $f(t) = |t - 5|$

15. $f(t) = \cos^2 t$

16. $f(t) = \cos^2 at$; a, a constant

17. $f(t) = \sin t \sin 2t$

18. $f(t) = \sin at \sin bt$; a, b constants

19. $f(t) = \sin^3 t$

20. $f(t) = \sin^2 t \cos t$

21. $f(t) = \begin{cases} 0, & 0 < t < 4 \\ 3, & t > 4 \end{cases}$

22. $f(t) = \begin{cases} 0, & 0 < t < 6 \\ 5, & t > 6 \end{cases}$

23. $f(t) = \begin{cases} 4, & 0 < t < 10 \\ 0, & t > 10 \end{cases}$

24. $f(t) = \begin{cases} 0, & 0 < t < 5 \\ 2, & 5 < t < 7 \\ 0, & t > 7 \end{cases}$

25. $f(t) = \begin{cases} 0, & 0 < t < 3 \\ -6, & 3 < t < 9 \\ 0, & t > 9 \end{cases}$

26. $f(t) = \begin{cases} 1, & 0 < t < 2 \\ 2, & 2 < t < 3 \\ 3, & 3 < t < 5 \\ 0, & t > 5 \end{cases}$

27. $f(t) = \begin{cases} 1, & 0 < t < 2 \\ 0, & 2 < t < 6 \\ 1, & t > 6 \end{cases}$

28. $f(t) = \begin{cases} t, & 0 < t < 3 \\ 3, & t > 3 \end{cases}$

29. $f(t) = \begin{cases} 2t, & 0 < t < 5 \\ 10, & t > 5 \end{cases}$

30. $f(t) = \begin{cases} 0, & 0 < t < 1 \\ e^{-t}, & t > 1 \end{cases}$

31. $f(t) = \begin{cases} 0, & 0 < t < 4 \\ t - 4, & 4 < t < 7 \\ 3, & t > 7 \end{cases}$

32. $f(t) = \begin{cases} 6, & 0 < t < 1 \\ 8 - 2t, & 1 < t < 3 \\ 2, & t > 3 \end{cases}$

From a table of integrals

$$\int e^{au} \sin bu \, du = e^{au} \frac{a \sin bu - b \cos bu}{a^2 + b^2}$$

$$\int e^{au} \cos bu \, du = e^{au} \frac{a \cos bu + b \sin bu}{a^2 + b^2}.$$

Use the integrals to find the Laplace transform $\mathcal{L}\{f(t)\}$ of the following functions in Problems **33**–**34**.

33. $f(t) = \cos \alpha t$

34. $f(t) = \sin \alpha t$

35. Another useful property of Laplace transforms can be given as follows: if $\mathcal{L}\{f(t)\} = F(s)$ exists for some $s > a$ for a function $f(t)$ in which $f(t)/t$ is bounded for $t > 0$, then

$$\mathcal{L}\left\{\frac{1}{t}f(t)\right\} = \int_s^\infty F(y)\, dy. \tag{7.23}$$

(a) Verify (7.23) by writing $F(y) = \int_0^\infty e^{-yt} f(t)\, dt$ and integrating both sides from $y = s$ to $y = \infty$.

(b) Using the definition of the Laplace transform, observe the troubles in evaluating $\mathcal{L}\left\{\frac{\sin t}{t}\right\}$.

(c) Using (7.23) evaluate $\mathcal{L}\left\{\frac{\sin t}{t}\right\}$.

36. The gamma function $\Gamma(\alpha)$ is defined as

$$\Gamma(\alpha) = \int_0^\infty x^{\alpha-1} e^{-x}\, dx \quad \text{for } \alpha > 0,$$

which simplifies to $\Gamma(n) = (n-1)!$ for natural numbers. We will explore further properties of the gamma function in the next chapter. In this problem, using the definition, show that

$$\mathcal{L}\{t^\alpha\} = \frac{\Gamma(\alpha+1)}{s^{\alpha+1}} \quad \text{for } \alpha > -1.$$

37. Using the results of Problem **36**, find the following Laplace transforms:

(a) $\mathcal{L}\left\{\frac{1}{\sqrt{t}}\right\}$ 　　　(b) $\mathcal{L}\{\sqrt{t}\}$ 　　　(c) $\mathcal{L}\{t^{3/2}\}$.

38. Let $f(t)$ and $g(t)$ be of exponential order. Show that their sum $f(t)+g(t)$ is also of exponential order.

39. Let $f(t)$ and $g(t)$ be piecewise continuous on $a \leq t \leq b$. Show that their sum $f(t) + g(t)$ is also piecewise continuous on $a \leq t \leq b$.

40. Using the results of Problems **38** and **39** show

$$\mathcal{L}\{f(t) + g(t)\} = \mathcal{L}\{f(t)\} + \mathcal{L}\{g(t)\}.$$

41. Show that the function $f(t) = 3\sin(e^{t^2})$ is of exponential order but that its derivative is not.

7.3　The Inverse Laplace Transform

Thus far our work has been centered upon the problem: Given a function f, find its Laplace transform $\mathcal{L}\{f(t)\}$. The next question for us to pursue is given a function F, find a function f whose Laplace transform is the given F. We introduce the notation $\mathcal{L}^{-1}\{F\}$ to denote such a function f, i.e.,

$$f(t) = \mathcal{L}^{-1}\{F(s)\},$$

which means $\mathcal{L}\{f(t)\} = F(s)$. $\mathcal{L}^{-1}$ is called the **inverse Laplace transform.** There are now three questions that we need to consider:

1. Given a function F, does an inverse transform exist?
2. If it exists, is it unique?
3. How do you find it?

These questions have different levels of import in our considerations here, from very theoretical mathematics to very practical. We will first answer question 2.

THEOREM 7.3.1　　　(Uniqueness)

Let f and g be two functions that are continuous for $t \geq 0$ and that have the same Laplace transform, that is,

$$\mathcal{L}\{f(t)\} = \mathcal{L}\{g(t)\},$$

then 　　　　　　　　　　$f(t) = g(t)$ 　for all $t \geq 0$.

Remark: In probability theory the uniqueness theorem plays an essential role in the theory of moment generating functions (m.g.f.) where the uniqueness is essential for determining the distribution of a random variable.

We must emphasize that in the uniqueness theorem the function is continuous, as the next example clarifies.

Example Consider

$$g(t) = \begin{cases} 1 & 0 < t < 3 \\ 2 & t = 3 \\ 1 & t > 3. \end{cases}$$

Then

$$\mathcal{L}\{g(t)\} = \int_0^\infty e^{-st} g(t)\, dt = \int_0^3 e^{-st}\, dt + \int_3^\infty e^{-st}\, dt$$

$$= \frac{1}{s} \quad \text{if } s > 0.$$

This, however, is also the Laplace transform of 1, i.e.,

$$\mathcal{L}\{1\} = \frac{1}{s} \quad \text{now clearly } g(t) \neq 1 \quad \text{for all } t.$$

Thus, this discontinuous function g is also an inverse transform of F defined by

$$F(s) = \frac{1}{s}.$$

However, the only continuous inverse transform of $F(s) = \frac{1}{s}$ is $f(t) = 1$ for all t so that the *unique* continuous inverse transform is

$$\mathcal{L}^{-1}\left\{\frac{1}{s}\right\} = 1.$$

This is how we address uniqueness. *We require continuity of the inverse transform.*

Question: How do we compute inverse Laplace transforms?

This is an important question for us. However, we will not directly compute inverse transforms in this text. We will refer to Table 7.1 to give us some often-used transforms. We note, however, that the uniqueness property allows us to conclude that *the inverse Laplace transform is also a linear operator.* That is,

$$\mathcal{L}^{-1}\{C_1 F_1(s) + C_2 F_2(s)\} = \mathcal{L}^{-1}\{C_1 F_1(s)\} + \mathcal{L}^{-1}\{C_2 F_2(s)\}. \qquad (7.24)$$

As with all tables, certain preliminary manipulations might be required to put a given $F(s)$ into a form present in the table. Often partial fraction decomposition is required. Of course, with MATLAB, Maple, or Mathematica, our job is much easier as will be seen at the end of the chapter.

Table 7.1: Table of Laplace Transforms

$f(t) = \mathcal{L}^{-1}\{F(s)\}$	$F(s) = \mathcal{L}\{f(t)\}$	Covered in text
1. 1	$\dfrac{1}{s}$	Sec. 7.2, Example 1
2. e^{at}	$\dfrac{1}{s-a}$	Sec. 7.2, Example 3
3. $\sin bt$	$\dfrac{b}{s^2+b^2}$	Sec. 7.2, Example 4
4. $\cos bt$	$\dfrac{s}{s^2+b^2}$	Sec. 7.2, Example 4
5. $\sinh bt$	$\dfrac{b}{s^2-b^2}$	Sec. 7.2, Prob. 7
6. $\cosh bt$	$\dfrac{s}{s^2-b^2}$	Chap. Rev. Prob. 13
7. $t^n (n=1,2,\ldots)$	$\dfrac{n!}{s^{n+1}}$	Chap. Rev. Prob. 19
8. $t^\alpha (\alpha > -1)$	$\dfrac{\Gamma(\alpha+1)}{s^{\alpha+1}}$	Sec. 7.2, Prob. 36
9. $t^n e^{at} (n=1,2,\ldots)$	$\dfrac{n!}{(s-a)^{n+1}}$	Chap. Rev. Prob. 20
10. $t\sin bt$	$\dfrac{2bs}{(s^2+b^2)^2}$	Chap. Rev. Prob. 15
11. $t\cos bt$	$\dfrac{s^2-b^2}{(s^2+b^2)^2}$	Chap. Rev. Prob. 16
12. $e^{-at}\sin bt$	$\dfrac{b}{(s+a)^2+b^2}$	Sec. 7.4, Example 2
13. $e^{-at}\cos bt$	$\dfrac{s+a}{(s+a)^2+b^2}$	Chap. Rev. Prob. 14
14. $\dfrac{\sin bt - bt\cos bt}{2b^3}$	$\dfrac{1}{(s^2+b^2)^2}$	Chap. Rev. Prob. 18
15. $\dfrac{t\sin bt}{2b}$	$\dfrac{s}{(s^2+b^2)^2}$	Chap. Rev. Prob. 17
16. $U_a(t)$	$\dfrac{e^{-as}}{s}$	Sec. 7.2, Example 5
17. $U_a(t)f(t-a)$	$e^{-as}F(s)$	Sec. 7.4, Thrm. 7.4.2
18. $e^{at}f(t)$	$F(s-a)$	Sec. 7.4, Thrm. 7.4.1
19. $\delta(t-t_0)$	e^{-st_0}	Sec. 7.4, Thrm. 7.4.3
20. $\ln t$	$-\dfrac{\gamma + \ln s}{s}$, $\gamma \approx 0.5772$ is Euler's constant;	See Sec. 8.5

Example 2 Find $\mathcal{L}^{-1}\left\{\dfrac{12}{s^4} + \dfrac{1}{s-5}\right\}$.

Solution

Because of the linearity of the inverse transform, we can consider each of the terms in the sum separately. In addition, the s^4 in the denominator of the first term should make us think of entry 7 in the table and the second term should make us think of entry 2 in the table. Thus we write

$$\mathcal{L}^{-1}\left\{\frac{12}{s^4} + \frac{1}{s-5}\right\} = 2\mathcal{L}^{-1}\left\{\frac{3!}{s^4}\right\} + \mathcal{L}^{-1}\left\{\frac{1}{s-5}\right\}$$

$$= 2t^3 + e^{5t}. \tag{7.25}$$

Example 3 Find $\mathcal{L}^{-1}\left\{\dfrac{1}{s^2 + 6s + 13}\right\}$.

Solution

We look for

$$F(s) = \frac{1}{as^2 + bs + c}$$

in the table but there is no such critter; we do, however, find

$$\frac{b}{(s+a)^2 + b^2}.$$

Thus we rewrite

$$s^2 + 6s + 13 = s^2 + 6s + 9 + 4 = (s+3)^2 + 4.$$

Hence

$$\frac{1}{s^2 + 6s + 13} = \frac{1}{(s+3)^2 + 4} = \frac{1}{2}\frac{2}{(s+3)^2 + 2^2}.$$

So

$$\mathcal{L}^{-1}\left\{\frac{1}{s^2 + 6s + 13}\right\} = \frac{1}{2}\mathcal{L}^{-1}\left\{\frac{2}{(s+3)^2 + 2^2}\right\}$$

$$= \frac{1}{2}e^{-3t}\sin 2t.$$

Example 4 Compute $\mathcal{L}^{-1}\left\{\dfrac{1}{s(s^2+1)}\right\}$.

Solution

By partial fractions,

$$\frac{1}{s(s^2+1)} = \frac{A}{s} + \frac{Bs+c}{s^2+1}$$

so that
$$A = 1, \quad B = -1, \quad \text{and } C = 0.$$

Thus
$$\frac{1}{s(s^2 + 1)} = \frac{1}{s} - \frac{s}{s^2 + 1},$$

which means
$$\mathcal{L}^{-1}\left\{\frac{1}{s(s^2 + 1)}\right\} = \mathcal{L}^{-1}\left\{\frac{1}{s}\right\} - \mathcal{L}^{-1}\left\{\frac{s}{s^2 + 1}\right\}$$
$$= 1 - \cos t.$$

7.3.1   Laplace Transform Solution of Linear Differential Equations

Armed with the ability to calculate some of the basic inverse Laplace Transforms, we now look at a few examples in which we apply them to solve initial-value problems consisting of an nth order linear differential equation with a specified initial condition.

The Method:
Consider the nth-order linear differential equation with constant coefficients:
$$a_n \frac{d^n y}{dt^n} + a_{n-1}\frac{d^{n-1}y}{dt^{n-1}} + \cdots + a_1 \frac{dy}{dt} + a_0 y = b(t)$$

which satisfies the initial conditions
$$y(0) = c_0, y'(0) = c_1, \ldots, y^{(n-1)}(0) = c_{n-1}.$$

The existence theorem in Chapter 4 ensures that this initial-value problem has a unique solution. Thus if we now take the Laplace transform of the differential equation,
$$a_n \mathcal{L}\left\{\frac{d^n y}{dt^n}\right\} + a_{n-1}\mathcal{L}\left\{\frac{d^{n-1}y}{dt^{n-1}}\right\} + \cdots + a_1\mathcal{L}\left\{\frac{dy}{dt}\right\} + a_0 \mathcal{L}\{y(t)\} = \mathcal{L}\{b(t)\}.$$

Each Laplace transform is given as
$$\mathcal{L}\left\{\frac{d^n y}{dt^n}\right\} = s^n \mathcal{L}\{y(t)\} - s^{n-1}y(0) - s^{n-2}y^1(0) - \cdots - y^{(n-1)}(0)$$
$$= s^n \mathcal{L}\{y(t)\} - c_0 s^{n-1} - c_1 s^{n-2} - \cdots - c_{n-1}$$

$$\mathcal{L}\left\{\frac{d^{n-1}y}{dt^{n-1}}\right\} = s^{n-1}\mathcal{L}\{y(t)\} - s^{n-2}y(0) - s^{n-3}y^1(0) - \cdots - y^{(n-2)}(0)$$

$$= s^{n-1}\mathcal{L}\{y(t)\} - c_0 s^{n-2} - c_1 s^{n-3} - \cdots - c_{n-2}$$

$$\vdots$$

$$\mathcal{L}\left\{\frac{dy}{dt}\right\} = s\mathcal{L}\{y(t)\} - y(0) = s\mathcal{L}\{y(t)\} - c_0.$$

Now letting $Y(s) = \mathcal{L}\{y(t)\}$ and $B(s) = \mathcal{L}\{b(t)\}$ we obtain

$$[a_n s^n + a_{n-1}s^{n-1} + \cdots + a_1 s + a_0]Y(s) - c_0[a_n s^{n-1} + a_{n-1}s^{n-2} + \cdots + a_1]$$
$$- c_1[a_n s^{n-2} + a_{n-1}s^{n-3} + \cdots + a_2]$$

$$-$$

$$\vdots$$

$$- c_{n-2}[a_n s + a_{n-1}] - c_{n-1}a_n = B(s).$$

This is an algebraic equation in the "unknown" $Y(s)$. We solve this equation for $Y(s)$ so that

$$y(t) = \mathcal{L}^{-1}\{Y(s)\}.$$

The idea presented here is one of the main features of the Laplace transform. It allows a differential equation to be expressed as an algebraic equation that can be solved easily. Once the algebraic equation is solved, the inverse Laplace transform is applied to recover the solution of the differential equation.

Example 7.14   Solve the initial-value problem

$$\frac{dy}{dt} - 2y = e^{5t}, \quad y(0) = 3.$$

Solution:

Taking Laplace transforms of both sides of the differential equation gives

$$\mathcal{L}\left\{\frac{dy}{dt}\right\} - 2\mathcal{L}\{y(t)\} = \mathcal{L}\{e^{5t}\}.$$

Now

$$\mathcal{L}\left\{\frac{dy}{dt}\right\} = s\mathcal{L}\{y(t)\} - y(0),$$

letting

$$Y(s) = \mathcal{L}\{y(t)\}$$

so

$$\mathcal{L}\left\{\frac{dy}{dt}\right\} = sY(s) - 3.$$

Thus

$$sY(s) - 3 - 2Y(s) = \mathcal{L}\{e^{5t}\} = \frac{1}{(s-5)},$$

so

$$(s-2)Y(s) - 3 = \frac{1}{s-5}.$$

Solving for $Y(s)$ gives

$$(s-2)Y(s) = \frac{1}{s-5} + 3$$

as

$$Y(s) = \frac{3s - 14}{(s-5)(s-2)}.$$

Now we find $y(t)$,

$$y(t) = \mathcal{L}^{-1}\left\{\frac{3s - 14}{(s-5)(s-2)}\right\}.$$

Applying partial fractions

$$\frac{3s - 14}{(s-5)(s-2)} = \frac{A}{s-5} + \frac{B}{s-2}$$

gives

$$3s - 14 = A(s-2) + B(s-5)$$

so that

$$A = \frac{1}{3}, \quad B = \frac{8}{3}.$$

This implies

$$\mathcal{L}^{-1}\left\{\frac{3s - 14}{(s-2)(s-5)}\right\} = \frac{1}{3}\mathcal{L}^{-1}\left\{\frac{1}{s-5}\right\} + \frac{8}{3}\mathcal{L}^{-1}\left\{\frac{1}{s-2}\right\}.$$

Now

$$\mathcal{L}^{-1}\left\{\frac{1}{s-2}\right\} = e^{2t} \quad \text{and} \quad \mathcal{L}^{-1}\left\{\frac{1}{s-5}\right\} = e^{5t}.$$

Thus

$$y(t) = \frac{8}{3}e^{2t} + \frac{1}{3}e^{5t}$$

is the solution of the initial-value problem.

Example 6: Solve the initial-value problem.

$$\frac{d^2y}{dt^2} - 2\frac{dy}{dt} - 8y = 0$$

$$y(0) = 3, \quad y'(0) = 6.$$

Taking the Laplace transform of both sides of the differential equation gives

$$\mathcal{L}\left\{\frac{d^2y}{dt^2}\right\} - 2\mathcal{L}\left\{\frac{dy}{dt}\right\} - 8\mathcal{L}\{y\} = \mathcal{L}\{0\}.$$

Now

$$\mathcal{L}\left\{\frac{d^2Y}{dt^2}\right\} = s^2Y(s) - sy(0) - y'(0)$$

where

$$Y(s) = \mathcal{L}\{y(t)\};$$

also

$$\mathcal{L}\left\{\frac{dy}{dt}\right\} = sY(s) - y(0).$$

Thus

$$s^2Y(s) - 3s - 6 - 2(sY(s) - 3) - 8Y(s) = 0,$$

which is

$$[s^2 - 2s - 8]Y(s) - 3s = 0.$$

Solving for $Y(s)$ gives

$$Y(s) = \frac{3s}{s^2 - 2s - 8},$$

which is

$$Y(s) = \frac{3s}{(s - 4)(s + 2)}.$$

We find $Y(t)$ by applying the inverse transform

$$Y(t) = \mathcal{L}^{-1}\left\{\frac{3s}{(s - 4)(s + 2)}\right\}.$$

Using partial fractions

$$\frac{3s}{(s - 4)(s + 2)} = \frac{A}{s - 4} + \frac{B}{s + 2}$$

gives the solution

$$A = 2, \quad B = 1.$$

Thus

$$\mathcal{L}^{-1}\left\{\frac{3s}{(s - 4)(s + 2)}\right\} = 2\mathcal{L}^{-1}\left\{\frac{1}{s - 4}\right\} + \mathcal{L}^{-1}\left\{\frac{1}{s + 2}\right\}$$

$$= 2e^{4t} + e^{-2t}.$$

So the solution of the initial problem is

$$Y(t) = 2e^{4t} + e^{-2t}.$$

Laplace transforms can sometimes be used to solve initial-value problems involving linear differential equations with nonconstant coefficients. However, Laplace transforms do not provide a general method for solving equations with nonconstant coefficients.

Example 7: Solve $y'' + 4ty' - 8y = 4$ with the initial conditions $y(0) = 0, y'(0) = 0$.

Solution

Applying the Laplace transform to both sides of the equation gives

$$\mathcal{L}\{y''\} + 4\mathcal{L}\{ty'\} - 8\mathcal{L}\{y\} = \mathcal{L}\{4\}. \tag{7.26}$$

Now recall

$$\mathcal{L}\{t^n f(t)\} = (-1)^n \frac{d^n}{ds^n}\mathcal{L}\{f(t)\}$$

so that

$$\mathcal{L}\{ty'\} = -\frac{d}{ds}\mathcal{L}\{y'\} = -\frac{d}{ds}[sY(s) - y(0)]$$
$$= -sY'(s) - Y(s).$$

Applying this to Equation (7.26), we have

$$s^2 Y(s) - sy(0) - y'(0) + 4[-sY'(s) - Y(s)] - 8Y(s) = \frac{4}{s}.$$

Simplifying, we obtain

$$(s^2 - 12)Y(s) - 4sY'(s) = \frac{4}{s},$$

which is a first-order differential that can be solved for $Y(s)$. Rewriting this equation as

$$Y'(s) + \left(\frac{3}{s} - \frac{s}{4}\right)Y(s) = \frac{-1}{s^2}$$

we see that this is a first-order linear differential equation. An integrating factor for this equation is

$$u(s) = s^3 e^{-s^2/8},$$

and upon applying this integrating factor from Section 1.6 (or other method), we obtain the solution

$$Y(s) = \frac{4}{s^3} + \frac{ce^{s^2/8}}{s^3}.$$

Now if $y(t) = \mathcal{L}^{-1}\{Y(s)\}$ exists, then

$$\lim_{s \to \infty} Y(s) = 0.$$

Thus, here $c = 0$, so

$$Y(s) = \frac{4}{s^3}.$$

Using this

$$y(t) = \mathcal{L}^{-1}\{Y(s)\} = 2\mathcal{L}^{-1}\left\{\tfrac{2}{s}\right\}$$
$$= 2t^2.$$

The next example gives the general solution for a mass-spring system in terms of the Laplace transform. The latter part of the example considers a discontinuous forcing function.

Example   Mass on a Spring (revisited)
As we saw in Chapter 4, the motion of a mass suspended on a spring satisfies the differential equation

$$m\frac{d^2x}{dt^2} + b\frac{dx}{dt} + kx = f(t)$$

where m is the mass, b is a damping coefficient, k the spring constant, and $f(t)$ is a forcing function. The displacement of the mass also satisfies the initial conditions $x(0) = x_0$ and $x'(0) = v_0$. So applying the Laplace transform to the differential equation gives

$$\mathcal{L}\left\{m\frac{d^2x}{dt^2} + b\frac{dx}{dt} + kx\right\} = \mathcal{L}\{f(t)\}.$$

So

$$m\mathcal{L}\left\{\frac{d^2x}{dt^2}\right\} + b\mathcal{L}\left\{\frac{dx}{dt}\right\} + k\mathcal{L}\{x\} = \mathcal{L}\{f(t)\}$$

which is

$$m[s^2\mathcal{L}\{x\} - x_0 s - v_0] + b[s\mathcal{L}\{x\} - x_0] + k\mathcal{L}\{x\} = \mathcal{L}\{f(t)\}.$$

Letting $X = \mathcal{L}\{x\}$ and rewriting this expression gives

$$(ms^2 + bs + k)X - mx_0 s - mv_0 - bx_0 = \mathcal{L}\{f(t)\}.$$

Thus

$$X = \frac{\mathcal{L}\{f(t)\} + m(x_0 s + v_0) + bx_0}{ms^2 + bs + k}.$$

Applying the inverse transform gives

$$x(t) = \mathcal{L}^{-1}\left\{\frac{\mathcal{L}\{f(t)\} + m(x_0 s + v_0) + bx_0}{ms^2 + bs + k}\right\}.$$

Of course, this is a very general expression for the displacement function $x(t)$ of a mass on a spring. A specific example is considered next.

Example 9: Consider the result of the Example 8 for the case of forced undamped motion described by

$$x'' + 4x = \sin 3t$$

with $x(0) = x'(0) = 0$. Then since $\mathcal{L}\{f(t)\} = \mathcal{L}\{\sin 3t\} = \frac{3}{s^2+9}$ and $m = 1, b = 0, k = 4, x_0 = 0, v_0 = 0$ we have

$$x(t) = \mathcal{L}^{-1}\left\{\frac{\frac{3}{s^2+9}}{s^2+4}\right\} = \mathcal{L}^{-1}\left\{\frac{3}{(s^2+9)(s^2+4)}\right\}.$$

Applying partial fractions gives

$$\frac{3}{(s^2+9)(s^2+4)} = \frac{3}{5}\left(\frac{1}{s^2+4}\right) - \frac{3}{5}\left(\frac{1}{s^2+9}\right)$$

$$= \frac{3}{10}\left(\frac{2}{s^2+4}\right) - \frac{1}{5}\left(\frac{3}{s^2+9}\right).$$

Now

$$\mathcal{L}^{-1}\left\{\frac{2}{s^2+4}\right\} = \sin 2t \quad \text{and} \quad \mathcal{L}^{-1}\left\{\frac{3}{s^2+9}\right\} = \sin 3t.$$

Thus

$$x(t) = \frac{3}{10}\sin 2t - \frac{1}{5}\sin 3t$$

is the displacement.

● ● ● ● ● ● ● ● ● ● ● ●

Problems

Find $\mathcal{L}^{-1}\{F(s)\}$ for each of the functions $F(s)$ in Problems **1–16**.

1. $F(s) = \dfrac{3}{s^2+4}$

2. $F(s) = \dfrac{5s}{s^2-9}$

3. $F(s) = \dfrac{5}{(s-3)^2}$

4. $F(s) = \dfrac{5s}{s^2+4s+4}$

5. $F(s) = \dfrac{s+2}{s^2+4s+7}$

6. $F(s) = \dfrac{s+6}{s^2+8s+20}$

7. $F(s) = \dfrac{s+1}{s^3+2s}$

8. $F(s) = \dfrac{s+1}{(s^2+9)^2}$

9. $F(s) = \dfrac{2s+7}{(s+3)^4}$

10. $F(s) = \dfrac{8(s+1)}{(2s-1)^3}$

11. $F(s) = \dfrac{s-2}{s^2+5s+6}$

12. $F(s) = \dfrac{5s+8}{s^2+3s-10}$

13. $F(s) = \dfrac{5s+6}{s^2+9}e^{-\pi s}$

14. $F(s) = \dfrac{2s+9}{s^2+4s+13}e^{-3s}$

15. $F(s) = \dfrac{e^{-4s}-e^{-7s}}{s^2}$

16. $F(s) = \dfrac{2-e^{-3s}}{s^2+9}$

For each of the Problems **17–31**, use Laplace transforms to solve each of the initial-value problems.

17. $y' - y = e^{3t}$, $y(0) = 2$ **18.** $y' + y = 3e^{2t}$, $y(0) = -1$

19. $y'' + 4y = 0$, $y(0) = 2, y'(0) = 3$ **20.** $y'' + 16y = 0$, $y(0) = 7, y'(0) = 0$

21. $y'' + 9y = 4$, $y(0) = 0, y'(0) = 6$ **22.** $y'' - y' - 6y = 0$, $y(0) = 2, y'(0) = 6$

23. $y'' - 5y' + 6y = 0$, $y(0) = 1, y'(0) = 3$

24. $y'' + y' - 12y = 0$, $y(0) = 4, y'(0) = -1$

25. $y'' + 2y' + 5y = 0$, $y(0) = 2, y'(0) = 4$

26. $y'' + 2y' + y = te^{-3t}$, $y(0) = 2$, $y'(0) = 0$

27. $y'' - 8y' + 15y = 6te^{4t}$, $y(0) = 5$, $y'(0) = 4$

28. $y'' - y' - 2y = 13e^{-t} \sin 3t$, $y(0) = 0$, $y'(0) = 3$

29. $y'' + 3y' - 10y = 50e^{2t} \cos t$, $y(0) = -1$, $y'(0) = 1$

30. $y''' - 5y'' + 7y' - 3y = 20 \sin t$, $y(0) = 0$, $y'(0) = 0$, $y''(0) = -2$

31. $y''' - 6y'' + 11y' - 6y = 19te^{4t}$, $y(0) = -1$, $y'(0) = 0$, $y''(0) = -6$

7.4   Translated Functions, Delta Function, and Periodic Functions

Now that we have established the existence of the Laplace transform we will examine some additional properties and functions that will be very useful in certain applications.

Our mass on a spring was sometimes forced with functions such as $e^{at} \sin(bt)$. We thus consider a useful property of the Laplace transform that will allow us to deal with this.

THEOREM 7.4.1   Translation Property

Suppose f is such that $\mathcal{L}\{f\}$ exists for $s > \alpha$. Then for any constant a,

$$\mathcal{L}\{e^{at} f(t)\} = F(s - a)$$

for any $s > \alpha + a$, where $F(s) = \mathcal{L}\{f(t)\}$.

Proof: Let

$$F(s) = \mathcal{L}\{f(t)\} = \int_{0^-}^{\infty} e^{-st} f(t)\, dt,$$

so

$$F(s - a) = \int_{0^-}^{\infty} e^{-(s-a)t} f(t)\, dt = \int_{0^-}^{\infty} e^{-st} e^{at} f(t)\, dt = \mathcal{L}\{e^{at} f(t)\}. \blacksquare$$

The translation property aids in many calculations of $\mathcal{L}\{f(t)\}$ as can be seen in the following examples.

Example 1. Find $\mathcal{L}\{e^{at} t\}$.

Solution

Now

$$\mathcal{L}\{t\} = \frac{1}{s^2} \quad \text{for } s > 0$$

so

$$\mathcal{L}\{e^{at}t\} = \frac{1}{(s-a)^2} \quad \text{for } s > a.$$

Example 2: Find $\mathcal{L}\{e^{at}\sin bt\}$.

Solution

We have

$$\mathcal{L}\{\sin bt\} = \frac{b}{s^2 + b^2}$$

so

$$\mathcal{L}\{e^{at}\sin bt\} = \frac{b}{(s-a)^2 + b^2}.$$

Another useful property of the unit step function in connection with the Laplace transform is the translation of a function $f(t)$ in a positive direction. That is, consider a new function defined by

$$\begin{cases} 0, & 0 < t < a \\ f(t-a), & t > a. \end{cases}$$

Then since

$$U_a(t) = \begin{cases} 0, & 0 < t < a \\ 1, & t > a \end{cases}$$

we have

$$U_a(t)f(t-a) = \begin{cases} 0, & 0 < t < a \\ f(t-a), & t > a. \end{cases}$$

Thus for Laplace transforms of this function we have the following result.

THEOREM 7.4.2

Suppose f is a function which has a Laplace transform F, i.e.,

$$F(s) = \int_{0-}^{\infty} e^{-st}f(t)\,dt.$$

Then the translated function

$$U_a(t)f(t-a) = \begin{cases} 0, & 0 < t < a \\ f(t-a), & t > a \end{cases}$$

has Laplace transform

$$\mathcal{L}\{U_a(t)f(t-a)\} = e^{-as}\mathcal{L}\{f(t)\}.$$

Proof: We have

$$\mathcal{L}\{U_a(t)f(t-a)\} = \int_{0^-}^{\infty} e^{-st}U_a(t)f(t-a)\,dt$$

$$= \int_{0^-}^{a} e^{-st}\cdot 0\,dt + \int_{a}^{\infty} e^{-st}f(t-a)\,dt$$

$$= \int_{a}^{\infty} e^{-st}f(t-a)\,dt.$$

We now change variables, letting $\tau = t - a$. Then

$$\int_{a}^{\infty} e^{-st}f(t-a)\,dt = \int_{0^-}^{\infty} e^{-s(\tau+a)}f(\tau)\,d\tau = e^{-sa}\int_{0^-}^{\infty} e^{-s\tau}f(\tau)\,d\tau$$

$$= e^{-sa}\mathcal{L}\{f(t)\}. \blacksquare$$

Example Find the Laplace transform of

$$g(t) = \begin{cases} 0, & 0 < t < 5 \\ t - 3, & t > 5. \end{cases}$$

Solution

We must express $t - 3$ for $t > 5$ in terms of $t - 5$. Thus $t - 3 = t - 5 + 2$ so

$$g(t) = \begin{cases} 0, & 0 < t < 5 \\ t - 5 + 2, & t > 5 \end{cases}$$

which is

$$U_5(t)f(t-5) = \begin{cases} 0, & 0 < t < 5 \\ (t-5)+2, & t > 5 \end{cases}$$

where $f(t) = t + 2$. Thus we need

$$\mathcal{L}\{f(t)\} = \mathcal{L}\{t + 2\} = \mathcal{L}\{t\} + \mathcal{L}\{2\}$$

$$= \frac{1}{s^2} + \frac{2}{s}.$$

This gives

$$\mathcal{L}\{g(t)\} = \mathcal{L}\{U_s(t)f(t-5)\} = e^{-5s}\mathcal{L}\{f(t)\}$$

$$= e^{-5s}\left(\frac{1}{s^2} + \frac{2}{s}\right).$$

Example Find the Laplace transform of

$$g(t) = \begin{cases} 0, & 0 < t < \frac{\pi}{2} \\ \sin t, & t > \frac{\pi}{2}. \end{cases}$$

Solution

We must express $\sin t$ in terms of $t - \frac{\pi}{2}$. Noting that $\sin t = \cos(t - \frac{\pi}{2})$, we have

$$g(t) = \begin{cases} 0, & 0 < t < \frac{\pi}{2} \\ \cos(t - \frac{\pi}{2}), & t > \frac{\pi}{2}. \end{cases}$$

Hence

$$U_{\frac{\pi}{2}} f\left(t - \frac{\pi}{2}\right) = \begin{cases} 0, & 0 < t < \frac{\pi}{2} \\ \cos(t - \frac{\pi}{2}), & t > \frac{\pi}{2} \end{cases}$$

where $f(t) = \cos t$. Thus, since

$$\mathcal{L}\{\cos t\} = \frac{s}{s^2 + 1}$$

we have

$$\mathcal{L}\{g(t)\} = \mathcal{L}\left\{U_{\frac{\pi}{2}}(t)f\left(t - \frac{\pi}{2}\right)\right\} = e^{-\frac{\pi}{2}s}\left(\frac{s}{s^2 + 1}\right).$$

Example 5: Find $\mathcal{L}^{-1}\left\{e^{-4s}\left(\frac{2}{s^2} + \frac{5}{s}\right)\right\}$.

Solution

This is of the form $\mathcal{L}^{-1}\{e^{-as}F(s)\}$ where $a = 4$ with $F(s) = \frac{2}{s^2} + \frac{5}{s}$. Thus

$$\mathcal{L}^{-1}\{e^{-as}F(s)\} = U_a(t)f(t - a)$$

where

$$f(t) = \mathcal{L}^{-1}\{F(s)\} = \mathcal{L}^{-1}\left\{\frac{2}{s^2} + \frac{5}{s}\right\}$$

$$= 2\mathcal{L}^{-1}\left\{\frac{1}{s^2}\right\} + 5\mathcal{L}^{-1}\left\{\frac{1}{s}\right\}$$

$$= 2t + 5 \cdot 1 = 2t + 5.$$

So $f(t - 4) = 2(t - 4) + 5 = 2t - 3$. Hence

$$\mathcal{L}^{-1}\{e^{-4s}F(s)\} = U_4(t)f(t - 4) = U_4(t)(2t - 3)$$

$$= \begin{cases} 0, & 0 < t < 4 \\ 2t - 4, & t > 4. \end{cases}$$

Example 6: We now reconsider our motivational example given by Equation (7.1) from Section 7.1 in which the first-order electric circuit with an inductor and resistor has applied step voltage:

$$LI' + RI = \begin{cases} V_0, & 0 \le t < 1 \\ 0, & t > 1. \end{cases} \tag{7.27}$$

We rewrite the forcing function as a step function and then take the Laplace transform of both sides, $\mathcal{L}\{LI' + RI\} = \mathcal{L}\{V_0(1 - U_1(t))\}$, to obtain

$$L[sY(s) - Y(0)] + RY(s) = V_0 \frac{1 - e^{-s}}{s}.$$

Solving for $Y(s)$ gives

$$Y(s) = \frac{V_0(1 - e^{-s})}{s(sL + R)} + \frac{LY(0)}{sL + R}.$$

We then need to find the inverse transform to obtain the solution in the original variable. We could either plug into one of the computer algebra systems or re-write slightly and then apply the rules that we have learned thus far:

$$I(t) = \mathcal{L}^{-1}\{Y(s)\} = \frac{V_0}{L} \mathcal{L}^{-1}\left\{\frac{(1 - e^{-s})}{s(s + \frac{R}{L})}\right\} + \mathcal{L}^{-1}\left\{\frac{Y(0)}{s + \frac{R}{L}}\right\}$$

$$= \frac{V_0}{R}\left(1 - e^{\frac{-Rt}{L}} - U_1(t)\left(1 - e^{\frac{-R(t-1)}{L}}\right)\right) + I(0)e^{\frac{-Rt}{L}}. \qquad (7.28)$$

Rearranging shows that this is exactly the solution we obtained in (7.2).

In Section 7.1, we motivated the study of the Laplace transform with a step forcing function that we just showed in Example 4 was easier to solve using the Laplace transform. In order to re-consider Example 4 of Section 7.1 with a Dirac delta forcing function, we first need to know how to take its Laplace transform.

The Laplace transform of $\delta(t - t_0)$ can easily be found:

$$\mathcal{L}\{\delta(t - t_0)\} = \int_{0^-}^{\infty} \delta(t - t_0)e^{-st}\, dt$$

$$= \int_{-\infty}^{\infty} \delta(t - t_0)e^{-st}e^{-st}\, dt$$

$$= e^{-st_0}. \qquad \text{(by Theorem 7.1.1)} \qquad (7.29)$$

We note that our definition of the Laplace transform with 0^- as the lower limit was crucial in obtaining this result. An alternative derivation from the definition $\delta_a(t - t_0)$ is pursued in the Problems. We state this result here for reference.

THEOREM 7.4.3

For $t_0 \geq 0$,

$$\mathcal{L}\{\delta(t - t_0)\} = e^{-st_0}.$$

Example 7: We again consider the first-order electric circuit with capacitor and resistor from Example 4 in Section 7.1, $RQ' + \frac{Q}{C} = f(t)$, in which no voltage is applied from the initial $t = 0$ except for a voltage spike, V_s, that is applied from $t = 1$ to $t = 1.1$. Use the Dirac delta function to find the voltage for any time $t > 0$.

Solution

As we derived in Section 7.1, the area under the Dirac delta function is 1 even though it has zero width. The voltage spike in our example is of height V_s and width 0.1, which gives an area of $0.1V_s$. If we center our delta function in the middle of our interval, our voltage spike takes the form

$$f(t) = 0.1V_s\delta(t - 1.05). \tag{7.30}$$

Taking the Laplace transform of $RQ' + \frac{Q}{C} = 0.1V_s\delta(t - 1.05)$ gives

$$R[sY(s) - Y(0)] + \frac{Y(s)}{C} = .1V_se^{-s\cdot1.05}.$$

Solving for $Y(s)$ gives

$$Y(s) = \frac{\frac{.1V_s}{R}e^{-s\cdot1.05} + Y(0)}{s + \frac{1}{RC}}.$$

We then need to find the inverse transform to obtain the solution in the original variable. Doing so gives

$$Q(t) = \mathcal{L}^{-1}\{Y(s)\} = \frac{0.1V_s}{R}\mathcal{L}^{-1}\left\{\frac{e^{-s\cdot1.05}}{s + \frac{1}{RC}}\right\} + \mathcal{L}^{-1}\left\{\frac{Y(0)}{s + \frac{1}{RC}}\right\}$$

$$= \frac{0.1V_s}{R}U_{1.05}(t)e^{\frac{-(t-1.05)}{RC}} + Q(0)e^{\frac{-t}{RC}}. \tag{7.31}$$

Rearranging shows that this is almost the same solution we obtained in (7.2). The main difference is the presence of a continuous rise in charge in (7.2) whereas this is represented by a step function in (7.31). Figure 7.6 shows that the solution using the idealized impulse gives an excellent approximation to the solution using a voltage spike of finite width. Of course, the solution via the Laplace transform was much quicker. The other item of note is the presence of V_sC in (7.2) whereas we have $\frac{V_s}{R}$ in (7.31). This may seem inconsistent until we observe that the product RC is a constant for the circuit and is called the **time constant**.[1] That is, $RC = \text{const} \Rightarrow V_s C = V_s \cdot \frac{\text{const}}{R} = \text{const} \cdot \frac{V_s}{R}$. With this knowledge, the two formulas are equivalent.

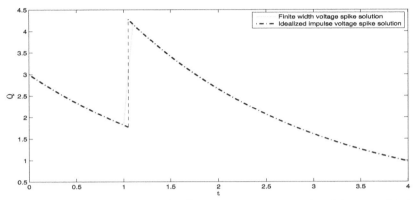

FIGURE 7.6: The solution (7.2) with voltage spike of width 0.1 and height V_s as given in Section 7.1 is in blue. The solution (7.31) with idealized impulse function of the same area, $0.1V_s\delta(t-1.05)$. In both graphs, $Q(0) = 3$, $R = 2$, $C = 1$, $V_s = 50$.

Example 3 Consider the forced undamped motion described by

$$x'' + 9x = 3\delta(t - \pi)$$

with $x(0) = 1, x'(0) = 0$. Then since $\mathcal{L}\{F(t)\} = 3\mathcal{L}\{\delta(t-\pi)\} = 3e^{-\pi s}$ and $b = 0, m = 9, x_0 = 1, v_0 = 0$ we have

$$x(t) = \mathcal{L}^{-1}\left\{\frac{3e^{-\pi s} + 9s}{9s^2 + 1}\right\}.$$

Now using the translation property, we have

$$x(t) = \cos 3t + \sin 3(t - \pi)u(t - \pi)$$

which can be written as

$$x(t) = \begin{cases} \cos 3t, & t < \pi \\ \cos 3t - \sin 3t, & \pi < t. \end{cases}$$

---

[1]Although it didn't come up yet, we note that the time constant in an RL circuit is $\frac{R}{L}$.

We have now considered a step function and delta function as well as periodic function such as $\sin bt$ and $\cos bt$. However, forcing function don't need to be smooth periodic functions, and we could imagine a periodic step function or a periodic impulse. We thus need to address how to deal with periodic functions. Recall a function $f(x)$ is **periodic** with period p if

$$f(x + p) = f(x) \quad \text{for all} \quad x.$$

THEOREM 7.4.4

Suppose f is a periodic function of period P for which a Laplace transform $\mathcal{L}\{f(t)\}$ exists. Then

$$\mathcal{L}\{f(t)\} = \frac{\int_0^P e^{-st} f(t) \, dt}{1 - e^{-Ps}}.$$

Proof: Since

$$\mathcal{L}\{f(t)\} = \int_0^\infty e^{-st} f(t) \, dt = \int_0^P e^{-st} f(t) \, dt + \int_P^\infty e^{-st} f(t) \, dt$$

it follows that we compute

$$\int_P^\infty e^{-st} f(t) \, dt = \int_0^\infty e^{-s(\tau + P)} f(\tau + P) \, d\tau = e^{-sP} \int_0^\infty e^{-st} f(t) \, dt$$

$$= e^{-sP} \mathcal{L}\{f(t)\}.$$

Thus

$$\mathcal{L}\{f(t)\} = \int_0^P e^{-st} f(t) \, dt + e^{-sP} \mathcal{L}\{f(t)\}$$

so that

$$\mathcal{L}\{f(t)\} = \frac{1}{1 - e^{-sP}} \int_0^P e^{-st} f(t) \, dt. \ \blacksquare$$

Example 9: Find the Laplace transform of

$$f(t) = \begin{cases} 1, & 0 < t < 2 \\ -1, & 2 < t < 4 \end{cases}$$

and $f(t + 4) = f(t)$ for $t \geq 4$.

Solution

This is a periodic function with period $P = 4$ as seen in the graph shown in

Figure 7.7. So

$$\mathcal{L}\{f(t)\} = \frac{\int_0^4 e^{-st} f(t)\, dt}{1 - e^{-4s}}$$

$$= \frac{1}{1 - e^{-4s}} \left[\int_0^2 e^{-st}\, dt + \int_2^4 e^{-st}(-1)\, dt \right]$$

$$= \frac{1}{1 - e^{-4s}} \left[\left.\frac{-1}{s} e^{-st}\right|_0^2 + \left.\frac{e^{-st}}{s}\right|_2^4 \right]$$

$$= \frac{1}{1 - e^{-4s}} \left(\frac{1}{s}\right) [-e^{-2s} + 1 + e^{-4s} - e^{-2s}]$$

$$= \frac{1 - e^{-2s}}{s(1 + e^{-2s})}.$$

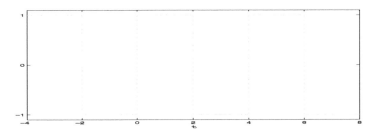

FIGURE 7.7:   The piecewise defined periodic function of Example 9.

Problems

Find $\mathcal{L}\{f(t)\}$ for each of the functions $f(t)$ in Problems **1–12**.
1. $f(t) = te^{2t}$
2. $f(t) = te^{-10t}$
3. $f(t) = t^3 e^{-2t}$
4. $f(t) = e^{2t} \sin 3t$
5. $f(t) = e^{4t} \cos 2t$
6. $f(t) = tU_2(t)$
7. $f(t) = (2t + 1)U_1(t)$
8. $f(t) = \delta(t - 1)$
9. $f(t) = \delta(t - \pi)$
10. $f(t) = \delta(t - \frac{3\pi}{2})$
11. $f(t) = \delta(t - 4 + \pi)$
12. $f(t) = \delta(t - 10)$

In Problems **13–22**, find $\mathcal{L}^{-1}\{Y(s)\}$ for each of the functions $Y(s)$.

13. $Y(s) = \dfrac{1}{s^2 - 6s + 10}$

14. $Y(s) = \dfrac{3}{s^2 + 4s + 13}$

15. $Y(s) = \dfrac{2s + 3}{s^2 + 2s + 5}$

16. $Y(s) = \dfrac{2s + 1}{4s^2 + 4s + 5}$

17. $Y(s) = \dfrac{s}{(s + 2)^2}$

18. $Y(s) = \dfrac{s}{(s - 2)^2}$

19. $Y(s) = \dfrac{e^{-3s}}{s^2}$

20. $Y(s) = \dfrac{e^{-s}}{s + 3}$

21. $Y(s) = \dfrac{e^{-2s}}{s(s+3)}$ **22.** $Y(s) = e^{-3s}\left(\dfrac{2}{s+3} - \dfrac{4}{s}\right)$

23. Determine a value of t_0 so that $\displaystyle\int_0^1 \sin^2[\pi(t - t_0)]\delta\left(t - \dfrac{1}{2}\right) dt = \dfrac{3}{4}$.

24. If $\displaystyle\int_1^5 t^n \delta(t - 2)\, dt = 8$ what is the value of the exponent n?

25. Sketch the graph of the function $f(t)$ defined by $f(t) = \displaystyle\int_{0^-}^t \delta(v)\, dv$, $0 < t < \infty$. Can the graph be obtained from a unit step function?

In Problems **26–31**, solve the given initial-value problems using Laplace Transforms.

26. $x' + x = \begin{cases} 1, & 0 \le t < 2 \\ -2, & t > 2, \end{cases}$ with $x(0) = 0$

27. $x'' - 5x' + 6x = U_1(t),\quad x(0) = 0, x'(0) = 1$

28. $x'' - x = -20\delta(t - 3),\quad x(0) = 1, x'(0) = 0$

29. $x'' + 4x = \delta(t - 4\pi),\quad x(0) = 1, x'(0) = 0$

30. $x'' + x = \delta(t - 2\pi)\cos t,\quad x(0) = 0, x'(0) = 1$

31. $x'' + 2x' + 2x = \delta(t - \frac{\pi}{2}) + \cos t,\quad x(0) = 0, x'(0) = 1$

32. The differential equation

$$v\frac{dy}{dt} = r_i f(t) - r_o y$$

is used to describe the absorption of a drug by a body organ of volume v. The function $y(t)$ is the concentration of the drug in the organ's fluid at time t, and r_i and r_o are the rates of fluid flow into and out of the organ, respectively. The function $f(t)$ is the concentration of the drug entering the organ.

(a) If $y(0) = 0$ and $r_i = r_o$ find $y(t)$ if $f(t) = t[1 - U_a(t)]$.

(b) If $y(0) = 1$ and $r_i \ne r_o$ find $y(t)$ if $f(t) = t[1 - U_a(t)]$.

33. A particular forced vibration of a mass m at the end of a vertical spring with spring constant k is described by the differential equation

$$m\frac{d^2x}{dt^2} + kx(t) = 1 + U_1(t) - 2U_2(t).$$

(a) Describe what $f(t)$ represents concerning the motion of the top of the spring.

(b) Solve the differential equation for $m = 1$ and $k = 4$ with the initial conditions $x(0) = x'(0) = 0$.

(c) Plot the solution on $0 \le t \le 10$. Considering the driving function $1 + U_1(t) - 2U_2(t)$, what do you observe?

34. *Alternative derivation of Theorem 7.4.3.* Define

$$\delta_\alpha(t - t_0) = \frac{1}{2\alpha}(U_{t_0-\alpha}(t) - U_{t_0+\alpha}(t))$$

so that $\lim_{\alpha \to 0} \delta_\alpha(t-t_0) = \delta(t-t_0)$ as in Equation (7.10). As can be justified by the bounded convergence theorem, we have

$$\mathcal{L}\{\delta(t - t_0)\} = \mathcal{L}\{\lim_{\alpha \to 0} \delta_\alpha(t - t_0)\} = \lim_{\alpha \to 0} \mathcal{L}\{\delta_\alpha(t - t_0)\}. \tag{7.32}$$

(a) Show that $\mathcal{L}\{\delta_\alpha(t - t_0)\} = e^{-st_0}\left(\dfrac{e^{\alpha s} - e^{-\alpha s}}{2\alpha s}\right)$.

(b) Use (7.32) and your results from (a) to conclude Theorem 7.4.3.

7.5 The s-Domain and Poles

In signal processing and control systems, one often measures the impulse response and frequency response in order to gain information about the system that gave rise to the observed signal. This ties in with the Laplace transform, which we have seen transforms functions from the time domain (with variable t) to the frequency domain (with variable s). Using inverse Laplace transforms, we saw how we can obtain solutions in our original variables. However, we can actually determine very useful information about the behavior of the solution in the time-domain without transforming back to it. To see this, consider Example 8 of Section 7.3 in which we considered the general equation for a forced mass on a spring:

$$m\frac{d^2x}{dt^2} + b\frac{dx}{dt} + kx = f(t)$$

where m is the mass, b is a damping coefficient, k the spring constant, and $f(t)$ is a forcing function. We let the initial conditions be written as $x(0) = x_0$ and $x'(0) = v_0$ so that the solution, via the Laplace transform, is

$$x(t) = \mathcal{L}^{-1}\{X(s)\} = \mathcal{L}^{-1}\left\{\frac{\mathcal{L}\{f(t)\} + m(x_0s + v_0) + bx_0}{ms^2 + bs + k}\right\}. \tag{7.33}$$

A quick comparison with (7.33) shows that the characteristic equation, written with variable s instead of r, is in the denominator. Thus, the roots of the characteristic polynomial are the zeros of the denominator of $X(s)$. We must keep in mind, though, that sometimes our roots were complex. Thus, we need to think of $X(s)$ as a complex function. When we first defined the Laplace transform, we stated that s should be considered complex even if many of the results could be stated for a real-valued s.

We also note that the initial conditions will not affect the long-term behavior of the system nor will they determine whether the system has oscillations or not. However, in order to understand how an input $E(t)$ affects a given system, we suppose that the initial conditions correspond to the homogeneous system *starting from rest under this input at some definite time* (for example, $t = 0$). Engineers often consider the following form of a forced linear time-invariant ODE:

$$\underbrace{a_n \frac{d^n x}{dt^n} + a_{n-1} \frac{d^{n-1} x}{dt^{n-1}} + \ldots + a_1 \frac{dx}{dt} + a_0 x}_{P(D)x} = f(t) \qquad (7.34)$$

for $a_n, a_{n-1}, \ldots, a_0 \in \mathbb{R}$ and $P(D)$ the operator notation with the specific forcing function

$$f(t) = \underbrace{b_m \frac{d^m E}{dt^m} + b_{m-1} \frac{d^{m-1} E}{dt^{m-1}} + \ldots + b_1 \frac{dE}{dt} + b_0 E}_{N(D)E} \qquad (7.35)$$

where $E(t)$ is the input to the system, $b_m, b_{m-1}, \ldots, b_0 \in \mathbb{R}$, and $N(D)$ is the operator notation. We will also assume $n \geq m$, since this will occur in most physical examples [45]. Then we define the **transfer function** of the linear time-invariant ODE (7.34)-(7.35) as

$$H(s) = \frac{\mathcal{L}\{x(t)\}}{\mathcal{L}\{E(t)\}}. \qquad (7.36)$$

If we think of the solution of the homogeneous equation as giving the normal response of the system (e.g., the mass-spring system), then the transfer function is the ratio of the Laplace transforms of any normal response to the input that produces it [45]. We can simplify (7.36) by observing that the Laplace transform of (7.34)-(7.35) with "from rest" initial conditions gives

$$\mathcal{L}\left\{ \underbrace{a_n \frac{d^n x}{dt^n} + \ldots + a_1 \frac{dx}{dt} + a_0 x}_{(a_n s^n + \cdots + a_1 s + a_0)\mathcal{L}\{x(t)\}} \right\} = \mathcal{L}\left\{ \underbrace{b_m \frac{d^m E}{dt^m} + \ldots + b_1 \frac{dE}{dt} + b_0 E}_{(b_m s^m + \cdots + b_1 s + b_0)\mathcal{L}\{E(t)\}} \right\},$$

that is, $P(s)\mathcal{L}\{x(t)\} = N(s)\mathcal{L}\{E(t)\}$, so that (7.36) becomes

$$H(s) = \frac{N(s)}{P(s)} = \frac{b_m s^m + b_{m-1} s^{m-1} + \cdots + b_1 s + b_0}{a_n s^n + a_{n-1} s^{n-1} + \cdots + a_1 s + a_0}. \qquad (7.37)$$

The transfer function can also give us the **frequency-response function** given (compare with Equation (4.74)) if we replace s with $i\omega$. Engineers often refer to the ratio $\frac{b_m}{a_n}$ as the **gain factor**.

Example 1: Consider the mass-spring systems with damping b from the dashpot:

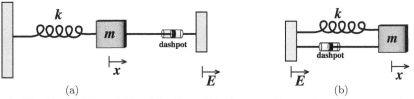

(a)   (b)

In Problems 11 and 12 of Section 4.7, the equations of motion were shown
to be (a) $mx'' + bx' + kx = bE'$ and (b) $mx'' + bx' + kx = bE' + kE$. Find
the transfer functions for each of these.

Solution

(a) Comparing the forcing terms to (7.35), we find that $b_0 = 0$, $b_1 = b$. Thus,
(7.37) gives us the transfer function

$$H(s) = \frac{bs}{ms^2 + bs + k}.$$

(b) Comparing with (7.35) we see that $b_0 = k$ and $b_1 = b$. Thus, (7.37) gives
us the transfer function

$$H(s) = \frac{bs + k}{ms^2 + bs + k}.$$

Engineers will often utilize the transfer function to gain information about
the solution of the system rather than using the inverse Laplace transform to
convert the function from the s-domain back to the t-domain. For example, in
the mass-spring system it will often be preferable to gain information about
the original system from the transfer function we obtained in Example 1 rather
than through the inverse Laplace transform (7.33). *This is the approach we
use in this section.*

We have seen the **singularities** (i.e., the zeros in the denominators) of real-
valued rational functions in calculus (or before) and we examined whether we
had a vertical asymptote at this value or if a factor in the numerator actually
cancelled with it so that we just had a hole in the graph. Since we will assume
a complex-valued rational function, we briefly state some useful terminology.

As we did in Section 3.5, we typically let w and z be the variables represent-
ing complex numbers and $N(z)$ and $N(w)$ will be complex-valued functions.
In this section, s will also be a complex variable. If N and P are two complex-
valued polynomials, then $\frac{N}{P}$ is a complex-valued rational function. We say
that $\frac{N}{P}$ is in **reduced form** if there are no factors in common.

Example 2 Consider

$$\frac{2}{s-3}, \qquad \frac{s+2}{s^2+4s+3}, \qquad \frac{s+1}{s^2+4s+3}.$$

We note that $s^2+4s+3 = (s+3)(s+1)$. Then $\frac{2}{s-3}$ and $\frac{s+2}{s^2+4s+3}$ are in reduced
form. The third rational function is not in reduced form but canceling the

common factor allows us to re-write it in reduced form $\frac{1}{s+3}$.

Definition 7.5.1

Let $\frac{N(s)}{P(s)}$ be a rational complex-valued function in reduced form. A **pole** of $\frac{N}{P}$ is defined as any value of s such that $P(s) = 0$.

Besides its singularities, we will also examine the **zeros** of the function, i.e., the roots of the numerators. These will be important in obtaining information about the role a given forcing function has in the long-term behavior of the system.

Example 3: Consider the rational functions

$$(a)\frac{2}{s-3}, \qquad (b)\frac{s+2}{s^2+9}, \qquad (c)\frac{s+1}{s^2+4s+3}, \qquad (d)\frac{s(s+1)}{s^2+10s+25}.$$

Then (a) has a single pole at s=3 and no zeros; (b) has two poles—one at $s = -3i$ and one at $s = 3i$ and one zero at $s = -2$. We saw that (c) reduces so there is only one pole at $s = -3$ and no zeros. Finally, (d) has two poles at $s = -5$ and zeros at $s = 0, -1$.

Example 4: Consider the mass-spring systems considered in Example 1. Find the poles and zeros of the transfer function assuming $x(0) = x'(0) = 0$.

Solution

In that example, we obtained the transfer functions

$$(a)\ \ H(s) = \frac{bs}{ms^2+bs+k}, \qquad (b)\ \ H(s) = \frac{bs+k}{ms^2+bs+k}.$$

Both have poles at $s = \frac{-b}{2m} \pm \frac{1}{2m}\sqrt{b^2 - 4mk}$. For (a), we have a zero at $s = 0$ while for (b) we have a zero at $s = -\frac{k}{b}$.

The zeros, s_z, tell us that forcing functions of the form e^{ts_z} will have *no effect* on the system. In (a), this means forcing the system with $E =$ constant will have no effect on the long-term solution. In (b), forcing the system with $E = e^{-kt/b}$ will not have any effect on the system.

For the poles we observe that they are exactly the roots of the characteristic equation and thus will determine those solutions that will grow versus decay and will even determine which decay fastest. For the physical mass-spring system it was always the case that the real part of the roots were negative and thus solutions always approached zero as $t \to \infty$. Although for a general transfer function $H(s)$ it is not the case that the real part of a pole must be negative, it will always be the case that the long-term behavior of the homogeneous solution can be obtained from the poles of the transfer function.

Example 6 Consider

$$(a)\ e^{-2t}, \qquad (b)\ e^{-4t} + e^{-t}, \qquad (c)\ e^{t/5}, \qquad (d)\ e^{-3t}\sin(2t).$$

We know that (a), (b), and (d) are solutions that decay to zero whereas (c) grows without bound. Moreover, we can observe that (d) converges to zero fastest, followed by (a) with (b) decaying slowest. Notice that it is the right-most root that determines the speed of the decay.

We can view an analogous example in the s-domain.

Example 6 Consider the transfer functions

$$(a)\ \frac{1}{s+2}, \qquad (b)\ \frac{1}{(s+4)(s+1)}, \qquad (c)\ \frac{1}{5s-1}, \qquad (d)\ \frac{1}{s^2+6s+13}$$

from four different forced ODEs with rest initial conditions. Determine the long-term behavior of the solutions in the t-domain. For those that will give solutions that decay, state which will decay the fastest.

Solution

The poles of these functions are the same roots of the characteristic equation that gave rise to the functions in Example 5. Thus, (a) has a pole at $s = -2$, (b) at $s = -1, -4$, (c) at $s = \frac{1}{5}$, and (d) at $s = -3 \pm 2i$. Our conclusions are the same as in Example 5.

Notice that it is only the real part of the pole that will affect the behavior of the solution of the homogeneous system. We can state this succinctly as follows.

THEOREM 7.5.1

Consider the reduced form of the complex-valued transfer function $H(s)$ defined in (7.37). If $\text{Re}(s_i) < 0$ for every pole s_i, then the solution $x(t)$ for the homogeneous equation (i.e., $f(t) = 0$ in (7.34)) in the t-domain decays exponentially to zero as $t \to \infty$. If $\text{Re}(s_i) > 0$ for at least one pole s_i, then the solution $x(t)$ for the homogeneous equation in the t-domain grows exponentially as $t \to \infty$. If any poles of $H(s)$ satisfy $\text{Im}(s_i) \neq 0$, then the solution $x(t)$ will have oscillatory motion.

Although we began with the mass-spring system as our motivational example, the LRC circuit is of a similar form

$$L\frac{d^2Q}{dt^2} + R\frac{dQ}{dt} + \frac{1}{C}Q = \mathcal{E} \quad \text{or equivalently} \quad L\frac{d^2I}{dt^2} + R\frac{dI}{dt} + \frac{1}{C}I = \mathcal{E}',$$

where L is the inductance, R is the resistance, C is the capacitance, Q is the charge, $I = \frac{dQ}{dt}$ is the current, and $\mathcal{E}$ is the input voltage. Engineers use things such as oscilloscopes to measure the amplitude and frequency of a signal and then use this information to help determine the characteristics of the electric circuit. It is also common in physical applications to create a **Bode plot** which is simply a graph of the magnitude of the gain plotted as a function of the frequency (magnitude Bode plot) or the phase of the gain plotted as a function of the frequency (phase Bode plot).

We finish with an example of the pole-zero plot of a transfer function. This is a convenient way to graphically see the zeros and poles of a given differential equation.

Example 7: For the ODE $(D^2 + 3D + 2)(D^2 + 4D + 5)(x) = E'' + 4E$, find (a) the transfer function and (b) the forcing function that will not have any effect on the system. Then (c) draw the location of these poles in the complex plane and use this to determine the long-term behavior in the t-domain.

Solution

(a) The transfer function is $H(s) = \dfrac{s^2 + 4}{(s^2 + 3s + 2)(s^2 + 6s + 10)}$ with zeros at $s = \pm 2i$ and poles at $s = -1, -2, -3 \pm i$. (b) From the zeros of $H(s)$, the forcing function e^{2ti} will not have an effect on the system. In other words, any forcing of the form $A \sin 2t + B \cos 2t$ for constants A, B will not affect the sytem. (c) The pole-zero plot is given below; since the poles all satisfy $\text{Re}(s) < 0$, the solutions in the t-domain will decay to zero.

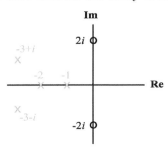

FIGURE 7.8: The locations of the zeros and poles in the complex plane for the transfer function in Example 7. The poles, marked as a blue ×, correspond to a solution that will decay. The black o are the zeros and give the roots of forcing functions that will not affect the system.

Problems

In Problems 1–8, determine the poles and zeros of the rational complex-valued functions.

1. $\dfrac{3}{s-4}$ 2. $\dfrac{\pi}{s-\pi}$ 3. $\dfrac{s+2}{s^2+4}$ 4. $\dfrac{s+2}{s^2-4}$

5. $\dfrac{s+2}{s^2+3s+2}$ 6. $\dfrac{s}{s^2+4s+4}$ 7. $\dfrac{s+3}{s^2+2s+2}$ 8. $\dfrac{s-1}{s^2+4s+5}$

In Problems **9–14**, find the transfer function for the following forced differential equations where the initial state of each is from rest.

9. $x'' + 3x' + 4x = 2E'' + E' + E$ 10. $x'' + 4x + 5 = E' + 32E$

11. $I'' + 4I' + 13I = E'' + E$ 12. $I'' + 2I' + 17I = E''$

13. $RCQ' + Q = E$ 14. $LI' + RI = E$

In Problems **15–24**, compute the transfer function of the given ODE, where $E = E(t)$. Use the poles in the complex plane to determine the long-term behavior of the solution of the original ODE in the t-domain. Assume "from rest" initial conditions.

15. $x'' + 9x = E$ 16. $x'' + 10x' + 25x = 3E$

17. $x'' + 3x' + 2x = 2E' + E$ 18. $x'' + 5x' + 6x = E'' + E$

19. $x'' + 4x = E'' - 4E$ 20. $x'' + x = E'' + 2E' + E$

21. $x'' + 2x' + 2x = E'' - E$ 22. $x'' + 4x' + 5x = E'' + 4E' + 4E$

23. $(D^2 + 2D + 2)(D^2 - 1)(x) = E''' - 2E''$

24. $(D^2 + 4D + 5)(D^2 - 2D + 3)(x) = E''' - E'$

In Problems **25–32**, suppose the transfer function of an ODE results in the given function. (a) Draw the pole-zero plots in the complex plane. Determine (b) the long-term behavior of the solution in the t-domain and (c) the forcing functions that will have no effect on the system.

25. $\dfrac{1}{s(s^2+4)}$ 26. $\dfrac{3s-4}{(s-3)(s-1)}$ 27. $\dfrac{s+3}{s^2+4s+13}$ 28. $\dfrac{s+2}{s^2+2s+10}$

29. $\dfrac{s^2-1}{(s^2+1)^2}$ 30. $\dfrac{6s}{(s^2+9)^2}$ 31. $\dfrac{s+1}{(s+1)^2+4}$ 32. $\dfrac{s}{(s+2)^2+3}$

In Problems **33–36**, obtain the frequency-response function in the following manner: (a) find the transfer function for the forced differential equations as given where the initial state of each is from rest; (b) then substitute $s = i\omega$. Compare with Equation (4.74) and Problems **13** and **14** in Section 4.7.

33. $x'' + 2\zeta\omega_n x' + \omega_n^2 x = \frac{F_0}{m}E$ 34. $I'' + \frac{R}{L}I' + \frac{1}{LC}I = \frac{F_0}{L}E$

35. $mx'' + bx' + kx = bE'$ 36. $mx'' + bx' + kx = bE' + kE$

7.6   Solving Linear Systems Using Laplace Transforms

We can apply the Laplace transform method to find the solution of a first-order system

$$a_1 \frac{dx}{dt} + a_2 \frac{dy}{dt} + a_3 x + a_4 y = \beta_1(t)$$

$$b_1 \frac{dx}{dt} + b_2 \frac{dy}{dt} + b_3 x + b_4 y = \beta_2(t) \qquad (7.38)$$

where $a_1, a_2, a_3, a_4, b_1, b_2, b_3$, and b_4 are constants and $\beta_1(t), \beta_2(t)$ are known functions. In addition, x and y satisfy the initial conditions

$$x(0) = c_1 \quad \text{and} \quad y(0) = c_2$$

where c_1 and c_2 are constants.

The procedure we employ is a straightforward extension of the method we used to solve an ordinary differential equation. Letting

$$X(s) = \mathcal{L}\{x(t)\} \quad \text{and} \quad Y(s) = \mathcal{L}\{y(t)\}$$

we apply the Laplace transform to Equation (7.38) to obtain a system of algebraic equations in $X(s)$ and $Y(s)$. This system of algebraic equations is then solved for $X(s), Y(s)$. Once these are obtained, we find (if possible)

$$x(t) = \mathcal{L}^{-1}\{X(s)\} \quad \text{and} \quad y(t) = \mathcal{L}^{-1}\{Y(s)\}.$$

Example 1: Use Laplace transforms to solve the system

$$\frac{dx}{dt} - 6x + 3y = 8e^t$$

$$\frac{dy}{dt} - 2x - y = 4e^t$$

that satisfies the initial conditions $x(0) = -1, y(0) = 0$.

Solution

Applying the Laplace transform to both sides of the system gives

$$\mathcal{L}\left\{\frac{dx}{dt}\right\} - 6\mathcal{L}\{x(t)\} + 3\mathcal{L}\{y(t)\} = \mathcal{L}\{8e^t\}$$

$$\mathcal{L}\left\{\frac{dy}{dt}\right\} - 2\mathcal{L}\{x(t)\} - \mathcal{L}\{y(t)\} = \mathcal{L}\{4e^t\}$$

and letting $X(s) = \mathcal{L}\{x(t)\}$ and $Y(s) = \mathcal{L}\{y(t)\}$ gives

$$\mathcal{L}\left\{\frac{dx}{dt}\right\} = sX(s) - x(0) = sX(s) + 1$$

$$\mathcal{L}\left\{\frac{dy}{dt}\right\} = sY(s) - y(0) = sY(s).$$

So, using the Laplace transform table, we have

$$sX(s) + 1 - 6X(s) + Y(s) = \frac{8}{s-1}$$

$$sY(s) - 2X(s) - Y(s) = \frac{4}{s-1}$$

which simplifies to

$$(s-6)X(s) + 3Y(s) = \frac{-s+9}{s-1}$$

$$-2X(s) + (s-1)Y(s) = \frac{4}{s-1}.$$

Now solving this system of algebraic equations for $X(s)$ and $Y(s)$ gives

$$X(s) = \frac{-s+7}{(s-1)(s-4)} \quad \text{and} \quad Y(s) = \frac{2}{(s-1)(s-4)}.$$

Thus

$$x(t) = \mathcal{L}^{-1}\{X(s)\} = \mathcal{L}^{-1}\left\{\frac{-s+7}{(s-1)(s-4)}\right\}$$

and

$$y(t) = \mathcal{L}^{-1}\{Y(s)\} = \mathcal{L}^{-1}\left\{\frac{2}{(s-1)(s-4)}\right\}.$$

Applying partial fractions we have

$$\frac{-s+7}{(s-1)(s-4)} = \frac{A}{s-1} + \frac{B}{s-4}$$

which gives $A = -2$ and $B = 1$, so

$$x(t) = -2\mathcal{L}^{-1}\left\{\frac{1}{s-1}\right\} + \mathcal{L}^{-1}\left\{\frac{1}{s-4}\right\}.$$

Using the Laplace transform table, we find $x(t)$ as

$$x(t) = -2e^t + e^{4t}$$

and in a very similar manner, we find $y(t)$ as

$$y(t) = \frac{-2}{3}e^t + \frac{2}{3}e^{4t}.$$

Example 2: Solve the system

$$x' = -x + y + 25 \sin t$$
$$y' = -x - 3y$$

with the initial conditions $x(0) = -1, y(0) = -1$.

Solution

If we take the Laplace transform of both sides of this system we obtain

$$\mathcal{L}\{x'\} = -\mathcal{L}\{x\} + \mathcal{L}\{y\} + 25\mathcal{L}\{\sin t\}$$
$$\mathcal{L}\{y'\} = -\mathcal{L}\{x\} - 3\mathcal{L}\{y\}.$$

Letting $X(s) = \mathcal{L}\{x(t)\}$ and $Y(s) = \mathcal{L}\{y(t)\}$, then this becomes

$$sX(s) + 1 = -X(s) + Y(s) + \frac{25}{s^2 + 1}$$
$$sY(s) + 1 = -X(s) - 3Y(s).$$

Solving these linear equations for $X(s)$ and $Y(s)$ gives

$$X(s) = -(3 + s)Y(s) - 1$$

and

$$Y(s) = -\frac{s}{(s+2)^2} - \frac{25}{(s+2)^2(s^2+1)}.$$

Now applying partial fractions to the second term here,

$$\frac{25}{(s+2)^2(s^2+1)} = \frac{A}{s+2} + \frac{B}{(s+2)^2} + \frac{Cs+D}{(s^2+1)},$$

gives

$$A + C = 0$$
$$2A + B + 4C + D = 0$$
$$A + 4C + 4D = 0$$
$$2A + B + 4D = 25.$$

Solving this system gives $A = 4, B = 5, C = -4,$ and $D = 3$.
Thus

$$\mathcal{L}^{-1}\left\{\frac{25}{(s+2)^2(s^2+1)}\right\} = 4e^{-2t} + 5te^{-2t} - 4\cos t + 3\sin t.$$

Combining the above yields the explicit form of the solution as

$$y(t) = -e^{-2t}(3t + 5) + 4\cos t - 3\sin t.$$

We find $x(t)$ by solving $y' = -x - 3y$ for x as

$$x(t) = -3y - y'$$

so that

$$x(t) = e^{-2t}(3t + 8) - 9\cos t + 13\sin t.$$

Problems

In each of the following Problems **1–12**, use the Laplace transform to find the solution of the linear system that satisfies the given initial conditions. Assume $x = x(t)$, $y = y(t)$.

1. $\begin{cases} x' - 2y = 0 \\ y' + x - 3y = 2 \\ x(0) = 3, \ y(0) = 0 \end{cases}$

2. $\begin{cases} x' + y = 3e^{2t} \\ y' + x = 0 \\ x(0) = 2, \ y(0) = 0 \end{cases}$

3. $\begin{cases} x' - 2x - 3y = 0 \\ y' + x + 2y = t \\ x(0) = -1, \ y(0) = 0 \end{cases}$

4. $\begin{cases} x' - 4x + 2y = 2t \\ y' - 8x + 4y = 1 \\ x(0) = 3, \ y(0) = 5 \end{cases}$

5. $\begin{cases} x' - 5x + 2y = 3e^{4t} \\ y' - 4x + y = 0 \\ x(0) = 3, \ y(0) = 0 \end{cases}$

6. $\begin{cases} x' + x + y = 5e^{2t} \\ y' - 5x - y = -3e^{2t} \\ x(0) = 3, \ y(0) = 2 \end{cases}$

7. $\begin{cases} 2x' + y' - x - y = e^{-t} \\ x' + y' + 2x + y = e^{t} \\ x(0) = 2, \ y(0) = 1 \end{cases}$

8. $\begin{cases} 2x' + y' + x + 5y = 4t \\ x' + y' + 2x + 2y = 2 \\ x(0) = 3, \ y(0) = -4 \end{cases}$

9. $\begin{cases} 2x' - 2x + y' = 1 \\ x' - 3x + y' - 3y' = 2 \\ x(0) = 0, \ y(0) = 0 \end{cases}$

10. $\begin{cases} x' + 2x + y' = 16e^{-2t} \\ 2x' + 3y' + 5y = 15 \\ x(0) = 4, \ y(0) = -3 \end{cases}$

11. $\begin{cases} x' - y' = -e^{t} \\ 2x' - 2y' - y = 8 \\ x(0) = -1, \ y(0) = -10 \end{cases}$

12. $\begin{cases} x'' + 2x - 4y' = 0 \\ x' + y'' - 4y = 0 \\ x(0) = -4, \ y(0) = 1 \\ x'(0) = 8, \ y'(0) = 2 \end{cases}$

13. **(a)** Letting, as usual, $X(s) = \mathcal{L}\{x(t)\}$ and $Y(s) = \mathcal{L}\{y(t)\}$, the Laplace transform converts the initial-value problem $\begin{cases} 2x' + y' - y = t \\ x' + y' = t^2 \end{cases}$ with $x(0) = 1$ and $y(0) = 0$ into the system $\begin{cases} sX + (s-1)Y = 2 + \frac{1}{s^2} \\ sX + sY = 1 + \frac{2}{s^3}. \end{cases}$

(b) Solve the system in part (a) for $Y(s)$ as $Y(s) = \dfrac{4 - s}{s^3(s+1)}$. Then use partial fractions and the inverse Laplace transform to show that $y(t) = 5 - 5t + 2t^2 - 5e^{-t}$.

(c) Show that $X(s) + Y(s) = \frac{1}{s} + \frac{2}{s^4}$ and solve for $x(t)$.

7.7   The Convolution

Sometimes inverse Laplace transforms and the use of the table of transforms can be simplified by using the convolution theorem that we give below. We first define the convolution of two functions f and g.

Definition 7.7.1

Let f and g be two functions that are piecewise continuous on every finite closed interval $0 \le t \le b$ and of exponential order. The function denoted by $f * g$ and defined by

$$f(t) * g(t) = \int_{0^-}^{t} f(\tau)g(t - \tau)\, d\tau \qquad (7.39)$$

is called the convolution of the functions f and g.

The convolution operation $*$ is commutative, that is, $f * g = g * f$. To see this, we need only change variables in (7.39). Letting $u = t - \tau$, we have

$$f(t) * g(t) = \int_{0^-}^{t} f(\tau)g(t - \tau)\, d\tau$$

$$= -\int_{t}^{0^-} f(t - \mu)g(u)\, du$$

$$= \int_{0^-}^{t} g(u)f(t - \mu)\, du$$

$$= g(t) * f(t).$$

Example 1: Let $f(t) = e^{-at}$ and $g(t) = \sin bt$. Then we have

$$f(t) * g(t) = \int_{0}^{t} e^{-a\tau} \sin b(t - \tau)\, d\tau$$

$$= \int_{0^-}^{t} e^{-a(t-\tau)} \sin b\tau\, d\tau \qquad (\text{using } f * g = g * f)$$

$$= e^{-at} \int_{0^-}^{t} e^{at} \sin b\tau\, d\tau$$

$$= \frac{1}{a^2 + b^2}[a \sin bt - b \cos bt + be^{-at}].$$

We now state a useful fact about the convolution of two functions f and g.

PROPOSITION 7.7.1

If the functions f and g are piecewise continuous on a closed interval $0 \le t \le b$ and of exponential order e^{at}, then the convolution $f * g$ is also piecewise continuous on every finite closed interval $0 \le t \le b$ and of exponential order $e^{(a+\varepsilon)t}$ where $\varepsilon > 0$ is a constant.

We will not prove this proposition, but we note here that it guarantees the existence of $\mathcal{L}\{f * g\}$ when $\mathcal{L}\{f\}$ and $\mathcal{L}\{g\}$ exist.

THEOREM 7.7.1   Convolution

If f and g are piecewise continuous functions on every closed interval $0 \le t \le b$ and are of exponential order e^{at} then
$$\mathcal{L}\{f * g\} = \mathcal{L}\{f\}\mathcal{L}\{g\}$$
for $s > a$.

Proof: Applying the definition of the Laplace transform, $\mathcal{L}\{f * g\}$ is the function defined by
$$\int_{0-}^{\infty} e^{-st} \left[\int_{0-}^{t} f(\tau)g(t - \tau)\, d\tau \right] dt$$

which can be written as
$$\int_{0-}^{\infty} \int_{0}^{t} e^{-st} f(\tau)g(t - \tau)\, d\tau\, dt.$$

Now consider the region R_1 shown in Figure 7.9.

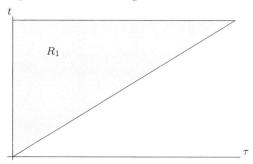

FIGURE 7.9: Region of integration R_1.

Thus
$$\int_{0-}^{\infty} \int_{0-}^{t} e^{-st} f(\tau)g(t - \tau)\, d\tau\, dt = \int \int_{R_1} e^{-st} f(\tau)g(t - \tau)\, d\tau\, dt.$$

So if we consider the change of variables $u = t - \tau, v = \tau$, which has Jacobian 1, the double integral over R_1 becomes the double integral
$$\int \int_{R_2} e^{-s(u+v)} f(v)g(u)\, du\, dv$$

where R_2 is the quarter plane defined by $u > 0, v > 0$ shown in Figure 7.10.

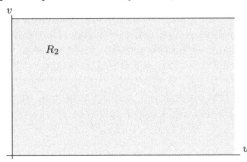

FIGURE 7.10: Region of integration R_2.

We then have

$$\iint_{R_2} e^{-s(u+v)} f(u)g(u)\,du\,dv = \int_{0-}^{\infty} \int_{0}^{\infty} e^{-s(u+v)} f(v)g(u)\,du\,dv$$

$$= \int_{0-}^{\infty} e^{-sv} f(v)\,dv \int_{0}^{\infty} e^{-su} g(u)\,du.$$

Indeed, since the integrals involved in this calculation are absolutely convergent for $s > a$, the operations we performed above are valid so we have shown

$$\mathcal{L}\{f * g\} = \mathcal{L}\{f\}\mathcal{L}\{g\} \quad \text{for } s > a. \quad \blacksquare$$

If we denote $\mathcal{L}\{f\}$ by F and $\mathcal{L}\{g\}$ by G then the conclusion of the last theorem can be written as

$$\mathcal{L}\{f(t) * g(t)\} = F(s)G(s).$$

Thus,

$$\mathcal{L}^{-1}\{F(s)G(s)\} = f(t) * g(t) = \int_{0-}^{t} f(\tau)g(t - \tau)\,d\tau \qquad (7.40)$$

and since $*$ is a commutative operation, we have

$$\mathcal{L}^{-1}\{F(s)G(s)\} = g(t) * f(t) = \int_{0-}^{t} g(\tau)f(t - \tau)\,d\tau. \qquad (7.41)$$

These remarks allow another method for computing inverse Laplace transforms. Suppose we are given a function H and need to find $\mathcal{L}^{-1}\{H(s)\}$. If we can express $H(s)$ as a product $F(s)G(s)$ where $\mathcal{L}^{-1}\{F(s)\} = f(t)$ and $\mathcal{L}^{-1}\{G(s)\} = g(t)$ are known, then we can apply (7.40) or (7.41) to determine $\mathcal{L}^{-1}\{H(s)\}$.

Example 2: Find $\mathcal{L}^{-1}\left\{\frac{1}{s(s^2+1)}\right\}$ using the convolution and the transform table.

Writing $\frac{1}{s(s^2+1)}$ as the product $F(s)G(s)$ where

$$F(s) = \frac{1}{s} \quad \text{and} \quad G(s) = \frac{1}{s^2+1}$$

we have by Table 7.1

$$f(t) = \mathcal{L}^{-1}\left\{\frac{1}{s}\right\} = 1 \text{ and } g(t) = \mathcal{L}^{-1}\left\{\frac{1}{s^2+1}\right\} = \sin t$$

so

$$\mathcal{L}^{-1}\left\{\frac{1}{s(s^2+1)}\right\} = f(t) * g(t) = \int_0^t 1 \cdot \sin(t-\tau)\, d\tau$$

or

$$\mathcal{L}^{-1}\left\{\frac{1}{s(s^2+1)}\right\} = g(t) * f(t) = \int_0^t \sin\tau \cdot 1\, d\tau.$$

Here the second integral is (slightly) more simple to evaluate, so we have

$$\mathcal{L}^{-1}\left\{\frac{1}{s(s^2+1)}\right\} = 1 - \cos t.$$

Note that we obtained this result earlier by means of a partial fractions decomposition.

Example 38 Newton's Law of Cooling (revisited)

Recall from Chapter 1 we considered Newton's Law of Cooling:

$$\frac{dT}{dt} = k[T(t) - T_s(t)],$$

which says that the rate of change of the temperature $T(t)$ of a body is proportional to the difference between the temperature of the body and its surroundings $T_s(t)$. Unlike our work in Chapter 1, here we assume T_s is not necessarily a constant. So, applying the Laplace transform to both sides of this differential equation gives

$$\mathcal{L}\left\{\frac{dT}{dt}\right\} = \mathcal{L}\{k[T(t) - T_s(t)]\}$$

which is

$$s\mathcal{L}\{T\} - T(0^-) = k\mathcal{L}\{T\} - k\mathcal{L}\{T_s\}.$$

Solving this equation for $\mathcal{L}\{T\}$ gives

$$\mathcal{L}\{T\} = \frac{T(0^-)}{s-k} - \frac{k}{s-k}\mathcal{L}\{T_s\}.$$

Using the convolution theorem we find the inverse Laplace transform as

$$T(t) = T(0^-)e^{kt} - \int_{0^-}^{t} kT_s(z)e^{k(t-z)}\,dz,$$

that is,

$$T(t) = T(0^-)e^{kt} - ke^{kt}\int_{0^-}^{t} T_s(z)e^{-kz}\,dz. \tag{7.42}$$

Equation (7.42) is the solution of Newton's Law of Cooling with a nonconstant surrounding temperature. If we take $T_s(t) = T_s$ a constant, as we did in Chapter 1, then Equation (7.42) becomes

$$\begin{aligned}
T(t) &= T(0^-)e^{kt} - ke^{kt}\int_{0}^{t} T_s e^{-kz}\,dz \\
&= T(0^-)e^{kt} - T_s e^{kt} + T_s \\
&= (T(0^-) - T_s)e^{kt} + T_s
\end{aligned}$$

which is the solution we obtained in Chapter 1; see Equation (1.4).

Although the idealized unit impulse is not a piecewise continuous function and thus doesn't satisfy the criteria of Theorem 7.7.1, there is a comparable Theorem addressing the convolution of generalized functions. Based on this, we state two convolution results here that may be of interest.

Consider the idealized unit impulse function given in Definition 7.1.2. Then the following convolution results hold:

$$\delta(t) * g(t) = \begin{cases} g(t), & t > 0 \\ 0, & t < 0 \end{cases} \tag{7.43}$$

and

$$\delta(t-a) * g(t) = \begin{cases} g(t-a), & t > a \\ 0, & t < a. \end{cases} \tag{7.44}$$

If we apply the formula from Theorem 7.7.1 or Definition 7.7.1, this should be believable and we refer the interested reader to [36] for further discussion.

Problems

In each of the following Problems **1–6**, compute the convolution $(f * g)(t)$ for the given pair of functions.

1. $f(t) = 1$, $g(t) = t^2$
2. $f(t) = \sin 2t$, $g(t) = e^{-t}$
3. $f(t) = t^3$, $g(t) = \sin 4t$
4. $f(t) = \cos 2t$, $g(t) = e^{-st}$
5. $f(t) = t$, $g(t) = \delta(t)$
6. $f(t) = \sin t$, $g(t) = \delta\left(t - \frac{\pi}{2}\right)$

Find $f(t) = \mathcal{L}^{-1}\{F(s)\}$ using the convolution theorem for each of the functions $F(s)$ in Problems **7–12**.

7. $F(s) = \dfrac{1}{s^2 + 5s + 6}$

8. $F(s) = \dfrac{1}{s^2 + 3s - 4}$

9. $F(s) = \dfrac{1}{s(s^2 + 9)}$

10. $F(s) = \dfrac{1}{s(s^2 + 4s + 13)}$

11. $F(s) = \dfrac{1}{s^3(s + 3)}$

12. $F(s) = \dfrac{1}{(s^2 + 4)(s + 7)}$

13. Show that the convolution operation $*$ is **associative**; that is

$$(f * (g * h))(t) = ((f * g) * h)(t)$$

for functions f, g, and h.

14. Show that the convolution operation $*$ satisfies the **distributive property over addition**; that is,

$$(f * (g + h))(t) = (f * g)(t) + (f * h)(t)$$

for functions f, g, and h.

15. Show that for any constant k, $((kf) * g)(t) = k(f * g)(t)$.

16. Show that $t^{-1/2} * t^{-1/2} = \pi$. Hint: Find $\mathcal{L}\{t^{-1/2}\}$.

17. Compute for a constant k

  (a) $\sin kt * \cos kt$ (b) $\sin kt * \sin kt$ (c) $\cos kt * \cos kt$

  (d) Using the results, compute $\mathcal{L}^{-1}\left\{\dfrac{ks}{(s^2 + k^2)^2}\right\}$.

Chapter 7 Review

In Problems **1–6**, determine whether the statement is true or false. If it is true, give reasons for your answer. If it is false, give a counterexample or other explanation of why it is false.

1. For any function $f(t)$, the Laplace transform $\mathcal{L}\{f(t)\}$ exists.

2. $\mathcal{L}\{C_1 f_1(t) + C_2 f_2(t)\} = C_1 \mathcal{L}\{f_1(t)\} C_2 \mathcal{L}\{f_2(t)\}$.

3. The translated function $[U_a(t)f(t - a) = \begin{cases} 0 & 0 < t < a \\ f(t - a) & t > a \end{cases}$ has Laplace transform $\mathcal{L}\{U_a(t)f(t - a)\} = e^{-as}\mathcal{L}\{f(t)\}$.

4. If $\mathcal{L}\{f(t)\} = F(s)$ exists for some $s > a$ for a function $f(t)$ in which $f(t)/t$ is bounded for $t > 0$, then $\mathcal{L}\left\{\dfrac{1}{t}f(t)\right\} = \displaystyle\int_s^\infty F(z)\,dz$.

5. If $\mathcal{L}\{f(t)\} = \mathcal{L}\{g(t)\}$, then $f(t) = g(t)$ for all $t \geq 0$.

6. The poles in the s-domain of the Laplace transformed ODE determine the long-term behavior of the system in the t-domain.

In Problems **7–12**, use the definition of the Laplace transform to find $\mathcal{L}\{f(t)\}$ for each of the given functions $f(t)$.

7. $f(t) = 3$ **8.** $f(t) = 5U_2(t)$ **9.** $f(t) = 2t^4$

10. $f(t) = |t - 2|$ **11.** $f(t) = t^2 e^{-t}$ **12.** $f(t) = e^t \cos t$

In Problems **13–18**, use the definition of the Laplace transform, Theorems in the chapter, or other techniques (but not Table 7.1) to find $\mathcal{L}\{f(t)\}$ for each of the given functions $f(t)$.

13. $f(t) = \cosh bt$ **14.** $f(t) = e^{-at} \cos bt$

15. $f(t) = t \sin bt$ **16.** $f(t) = t \cos bt$

17. $f(t) = \dfrac{t \sin bt}{2b}$ **18.** $f(t) = \dfrac{\sin bt - bt \cos bt}{2b^3}$

19. $f(t) = t^n, n = 1, 2, \ldots$ **20.** $f(t) = t^n e^{at}, n = 1, 2, \ldots$

In Problems **21–24**, evaluate the following integrals involving the Dirac delta function.

21. $\displaystyle\int_0^4 (1 + e^{-t})\delta(t - 3)\, dt$ **22.** $\displaystyle\int_{-5}^1 (1 + e^{-t})\delta(t - 2)\, dt$

23. $\displaystyle\int_0^{\frac{\pi}{3}} \tan t\delta\!\left(t - \frac{\pi}{4}\right) dt$ **24.** $\displaystyle\int_{-1}^2 \sin 2t\delta(t)\, dt$

In Problems **25–30**, find the poles of the given complex-valued rational function and use it to determine the long-term behavior of the solution in the t-domain. Then calculate $\mathcal{L}^{-1}\{F(s)\}$.

25. $F(s) = \dfrac{7s}{s^2 - 4}$ **26.** $F(s) = \dfrac{5}{s^2 + 9}$ **27.** $F(s) = \dfrac{5s}{s^2 + 6s + 9}$

28. $F(s) = \dfrac{6}{(s - 2)^2}$ **29.** $F(s) = \dfrac{3}{s^2 + 8s + 17}$ **30.** $F(s) = \dfrac{s - 3}{s^2 + 2s + 25}$

In Problems **31–34**, (a) compute the transfer function of the given ODE, where $E = E(t)$, and then (b) graph the pole-zero plot. (c) State the long-term behavior of the ODE in the t-domain and state the forcing functions that will not have any effect on the system. Assume "from rest" initial conditions.

31. $x'' + 5x' + 6x = E$ **32.** $x'' + 3x' + 2x = E'$

33. $(D^2 + 4D + 8)(D^2 + 1)(x) = 4E' + 8E$

34. $(D^2 + 6D + 13)(D + 1)(x) = E'' + 4E$

In Problems **35–44**, use Laplace transforms to solve the initial-value problem.

35. $x' - x = e^{4t}$, $x(0) = 6$

36. $x' + 9x = 0$, $x(0) = 2$

37. $x'' + 81x = 0$, $x(0) = 1$, $x'(0) = 2$

38. $x'' - x' - 12x = 0$, $x(0) = 1$, $x'(0) = 0$

39. $x'' + 2x' + 5y = 0$, $x(0) = 6$, $x'(0) = 0$

40. $x'' + 2x' + x = te^{-2t}$, $x(0) = 1$, $x'(0) = 0$

41. $x'' - 8x' + 15x = 5te^{2t}$, $x(0) = 5$, $x'(0) = 4$

42. $x'' - x' - 2x = 13e^{-t}\cos 4t$, $x(0) = 0$, $x'(0) = 3$

43. $x''' - 7x' - 6x = 13t$, $x(0) = 1$, $x'(0) = 2$, $x''(0) = -2$

44. $x'' + 3x' - 10x = e^t$, $x(0) = -1$, $x'(0) = 1$

45. Suppose an insect pest population grows at a rate proportional to its size with rate $r = 1/10$ (per day). In an attempt to eliminate the pest, a farmer applies a pesticide. The pesticide kills the insect at a rate that, in one day, decreases linearly with time from the maximum rate $P > 0$ to 0 (in units of 10,000 per day). As this is a rather potent pesticide, regulations allow the farmer to apply it only once. If the initial pest population is approximately 200,000 and the farmer applies a dosage $P = 30$, will the pest population be eliminated?

46. A forced vibration of a mass m at the end of a vertical spring, with spring constant k and no damping, is described by $mx''(t) + kx(t) = f(t)$, where

$$f(t) = 1 + U_1(t) - 2U_2(t).$$

Solve this differential equation for $m = 1, k = 8, x(0) = x'(0) = 0$. Plot the solution on the interval $0 \le t \le 3$ and discuss what you observe.

47. Using the definition of the Laplace transform, show that

$$\mathcal{L}\{e^{(a+bi)t}\} = \frac{1}{s - (a + bi)} \quad \text{for } s > a.$$

48. Prove that if f and g are piecewise continuous on $[a, b]$, then the product fg is piecewise continuous on $[a, b]$.

49. The differential equation for the current $I(t)$ in a series circuit with inductance L and resistance R is

$$LI' + RI = E(t)$$

where $E(t)$ is the applied voltage. If $I(0) = 0$ and E_0 is a constant, solve this equation if

(a) $E(t)$ is the square pulse $E(t) = E_0[U_1(t) - U_2(t)]$

(b) $E(t)$ is a single pulse of a sine wave $E(t) = (\sin t)[U_0(t) - U_\pi(t)]$

50. The **floor** or **unit staircase** function is defined as

$$f(t) = n \text{ if } n - 1 \le t < n,$$

for n a positive integer. Show that $f(t) = \sum\limits_{k=0}^{\infty} U(t-k)$. Using this result, and assuming that the Laplace transform exists for this infinite series, show using a geometric series that $\mathcal{L}\{f(t)\} = \dfrac{1}{s(1-e^{-s})}$.

51. Using the method in Problem **50**, show that the function

$$f(t) = \sum_{k=0}^{\infty} (-1)^k U(t-k)$$

has Laplace transform $\mathcal{L}\{f(t)\} = \dfrac{1}{s(1+e^{-s})}$.

52. Using Problem **51**, show that the **square wave** function

$$g(t) = \begin{cases} \alpha & n-1 \le t < n, n \text{ an odd integer} \\ \beta & n-1 \le t < n, n \text{ an even integer} \end{cases}$$

has Laplace transform $\mathcal{L}\{f(t)\} = \dfrac{\alpha + \beta e^{-s}}{s(1+e^{-s})}$.

In Problems **53–56**, use Laplace transforms to solve the following initial-value problems.

53. $\begin{cases} x' + 2x + y' = 8e^{-3t} \\ 2x' + 3y' + 5y = 12 \end{cases}$ with $x(0) = 2, y(0) = -3$

54. $\begin{cases} 2x' - 2x + y' = 12\sin t \\ x' - 3x + y' - 3y = 2 \end{cases}$ with $x(0) = 1, y(0) = 3$

55. $\begin{cases} x' - y' = -e^{2t} \\ 2x' - 2y' - y = 8 \end{cases}$ with $x(0) = -1, y(0) = -10$

56. $\begin{cases} x' - 2y' = -5e^{3t} \\ x' - 2y' - y = 5 \end{cases}$ with $x(0) = -10, y(0) = -1$

In Problems **57–60**, compute the convolution $(f * g)(t)$ for the given pair of functions.

57. $f(t) = 3, g(t) = t^3$

58. $f(t) = \sin 3t, g(t) = e^{-2t}$

59. $f(t) = 2t^4, g(t) = \cos 4t$

60. $f(t) = \cos t, g(t) = \delta\left(t - \frac{\pi}{2}\right)$

In Problems **61–65**, use the convolution to find $\mathcal{L}^{-1}\{F(t)\}$.

61. $F(s) = \dfrac{1}{s^2 + 7s + 12}$

62. $F(s) = \dfrac{1}{s^2 + 3s - 40}$

63. $F(s) = \dfrac{1}{s(s^2 + 4)}$

64. $F(s) = \dfrac{6}{s^2(s^2 + 36)}$

65. $F(s) = \dfrac{1}{(s^2 + 64)(s + 8)}$

Chapter 7 Computer Labs

Chapter 7 Computer Lab: MATLAB

MATLAB Example 1: Enter the code below to find, using *Symbolic Math Toolbox*, some basic Laplace transforms and inverse Laplace transforms.

```
>> clear all
>> syms s t a b
>> eq1=laplace(1,t,s) %Laplace transform of 1
>> eq2=laplace(sin(b*t),t,s)
>> f1=t*cos(b*t)
>> eq3=laplace(f1,t,s)
>> f2=exp(-a*t)*sin(b*t)
>> eq4=laplace(f2,t,s)
>> g1=1/(s^2+6*s+13)
>> eq5=ilaplace(g1,s,t) %inverse Laplace transform
>> g2=exp(-4*s)*(2/s^2+5/s)
>> eq6=ilaplace(g2,s,t)
```

MATLAB Example 2: Enter the code below to find, using *Symbolic Math Toolbox*, the Laplace transforms and inverse Laplace transforms or the unit step (Heaviside) function and idealized impulse (delta) function.

```
>> clear all
>> syms s t a b
>> f1=dirac(t-3)
>> eq0=int(f1,t,-inf,inf) %must be 1 by definition
>> eq1=laplace(f1,t,s)
>> eq2=ilaplace(eq1,s,t)
>> f2=heaviside(t)
>> eq3=diff(f2,t)
>> eq4=int(f2,t)
>> eq5=laplace(f2,t,s)
>> eq6=ilaplace(eq5,s,t)
>> f3=(2*t-3)*heaviside(t-4) %Heaviside function
>> eq7=laplace(f3,t,s)
>> eq8=ilaplace(eq7,s,t)
```

MATLAB Example 3: Enter the code below, using *Symbolic Math Toolbox*, to calculate the frequency response function of the given differential equation via its transfer function given in (7.36)-(7.37).

$$\frac{d^2x}{dt^2} + 2\zeta\omega_n \frac{dx}{dt} + \omega_n^2 x = \frac{F_0}{m} E'.$$

```
>> clear all
>> syms zeta wn F0 m s omega
>> zeta=sym('zeta','real')
>> zeta=sym('zeta','positive')
>> omega=sym('omega','real')
>> omega=sym('omega','positive')
>> wn=sym(wn,'real')
>> wn=sym(wn,'positive')
>> F0=sym(F0,'real')
>> F0=sym(F0,'positive')
>> m=sym(m,'real')
>> m=sym(m,'positive')
>> %observe b1=F0/m
>> H=(F0*s/m)/(s^2+2*zeta*wn*s+wn^2)
>> g=subs(H,s,i*omega)
>> sqrt(simplify(g*conj(g)))
```

MATLAB Example 4: Enter the code below, using *Symbolic Math Toolbox*, to determine the long-term behavior of the solution of $(D^2 + 3D + 4)(D^2 + 7D + 2)(x) = E' + E$ using the location of the poles of its transfer function.

```
>> clear all
>> syms s
>> P=(s^2+3*s+4)*(s^2+7*s+2)
>> H=(s+1)/P
>> sol=solve(P,s)
>> eval(sol)
```

MATLAB Example 5: Enter the code below, using *Symbolic Math Toolbox*, to calculate the convolution and see the relation with Laplace transforms.

```
>> clear all
>> syms s t a b tau
>> g1=1/(s^2*(s^2+1))
>> eq1=ilaplace(g1,s,t)
>> eq2=int(tau*sin(t-tau),tau,0,t) %using convolution
>> f1=exp(-a*tau)*sin(b*(t-tau))
>> eq3=int(f1,tau,0,t)
>> simplify(laplace(eq3,t,s))
```

MATLAB Exercises

Turn in both the commands that you enter for the exercises below as well as the output/figures. These should all be in one document. Please highlight or clearly indicate all requested answers. Some of the questions will require you to modify the above MATLAB code to answer them.

1. Enter the commands given in MATLAB Example 1 and submit both your input and output.

2. Enter the commands given in MATLAB Example 2 and submit both your input and output.

3. Enter the commands given in MATLAB Example 3 and submit both your input and output.

4. Enter the commands given in MATLAB Example 4 and submit both your input and output.

5. Enter the commands given in MATLAB Example 5 and submit both your input and output.

6. Find the Laplace transform of $\cos t$.

7. Find the Laplace transform of $\sin 5t$.

8. Find the Laplace transform of $t \sin 5t$.

9. Find the Laplace transform of $\frac{\sin bt - bt \cos bt}{2b^3}$.

10. Find the Laplace transform of $e^{3t} \cos 4t$.

11. Find the Laplace transform of $e^{-2t} \sin 3t$.

12. Find the Laplace transform of $\ln t$.

13. Find the Laplace transform of $\delta(t - 1)$.

14. Find the Laplace transform of $t\delta(t - 2)$.

15. Find the Laplace transform of $U_0(t)$.

16. Find the Laplace transform of $tU_1(t)$.

17. Find the inverse Laplace transform of s^2.

18. Find the inverse Laplace transform of $s^3 - \frac{3}{s^2}$.

19. Find the inverse Laplace transform of $\frac{s^2 - b^2}{(s^2 + b^2)^2}$.

20. Find the inverse Laplace transform of $\frac{s}{(s^2 + b^2)^2}$.

21. Find the inverse Laplace transform of $\frac{2}{s+1} - \frac{3}{s^2}$.

22. Find the inverse Laplace transform of $e^{-s} - \frac{3}{s^3}$.

23. Find the inverse Laplace transform of $e^{-s}\left(s - \frac{3}{s^2+1}\right)$.

24. Find the inverse Laplace transform of $e^{-2s}\left(\frac{1}{s(s^2+4)} - \frac{3}{s^2}\right)$.

25. Find the inverse Laplace transform of $-\frac{\gamma + \ln s}{s}$.

26. Use the transfer function to calculate the frequency response function of $x'' + 2x' + x = E' + E$.

27. Use the transfer function to calculate the frequency response function of $x'' + 5x' + 10x = 2E'' + E$.

28. Use the transfer function to calculate the frequency response function of $4x'' + 4x' + x = E'$.

29. Use the transfer function to calculate the frequency response function of $x'' + 2x' + 2x = E' + 3E$.

30. Use the transfer function to calculate the frequency response function of $(D^2 + 2D + 1)(D^2 + 6D + 10)(x) = E''' + 2E$.

31. Use the transfer function to calculate the frequency response function of $(D^2 + 2D + 2)(D^2 + 10D + 29)(x) = E'''' + E''$.

32. Use the poles of the transfer function to determine the long-term behavior of solutions of $x'' + 2x' + x = E' + E$.

33. Use the poles of the transfer function to determine the long-term behavior of solutions of $x'' + 5x' + 10x = 2E'' + E$.

34. Use the poles of the transfer function to determine the long-term behavior of solutions of $4x'' + 4x' + x = E'$.

35. Use the poles of the transfer function to determine the long-term behavior of solutions of $x'' + 2x' + 2x = E' + 3E$.

36. Use the poles of the transfer function to determine the long-term behavior of solutions of $(D^2 + 2D + 1)(D^2 + 6D + 10)(x) = E''' + 2E$.

37. Use the poles of the transfer function to determine the long-term behavior of solutions of $(D^2 + 2D + 2)(D^2 + 10D + 29)(x) = E'''' + E''$.

38. Use the definition of the convolution, (7.39), to find $t * \sin(t)$.

39. Use the definition of the convolution, (7.39), to find $\cos t * \sin(t)$.

40. Use the definition of the convolution, (7.39), to find $\delta(t) * \cos(t)$.

41. Use the definition of the convolution, (7.39), to find $U_0(t) * e^t$.

Chapter 7 Computer Lab: Maple

Maple Example 1: Enter the code below to find some basic Laplace transforms and inverse Laplace transforms.

restart

with(*inttrans*)

eq1 := *laplace*(1, *t*, *s*)

eq2 := *laplace*(sin(*b* · *t*), *t*, *s*)

f1 := *t* · cos(*b* · *t*)

eq3 := *laplace*(*f1*, *t*, *s*)

f2 := e$^{-a \cdot t}$ · sin(*b* · *t*)

eq4 := *laplace*(*f2*, *t*, *s*)

$g1 := \dfrac{1}{s^2 + 6 \cdot s + 13}$

eq5 := *invlaplace*(*g1*, *s*, *t*)

$g2 := e^{-4 \cdot s} \cdot \left(\dfrac{2}{s^2} + \dfrac{5}{s} \right)$

eq6 := *invlaplace*(*g2*, *s*, *t*)

convert(*eq6*, '*piecewise*')

Maple Example 2: Enter the code below to find the Laplace transforms and inverse Laplace transforms or the unit step (Heaviside) function and idealized impulse (delta) function.

restart

with(*inttrans*)

f1 := Dirac(*t* − 3)

$eq0 := \displaystyle\int_{-\infty}^{\infty} f1 \, dt$

eq1 := *laplace*(*f1*, *t*, *s*)

eq2 := *invlaplace*(*eq1*, *s*, *t*)

f2 := Heaviside(*t*)

$eq3 := \dfrac{d}{dt} f2$

$eq4 := \displaystyle\int f2 \, dt$

eq5 := *laplace*(*f2*, *t*, *s*)

eq6 := *invlaplace*(*eq5*, *s*, *t*)

f3 := (2 · *t* − 3) · Heaviside(*t* − 4)

eq7 := *laplace*(*f3*, *t*, *s*)

eq8 := *invlaplace*(*eq7*, *s*, *t*)

Maple Example 3: Enter the code below to calculate the frequency response function of the given differential equation via its transfer function given in (7.36)-(7.37).

$$\frac{d^2 x}{dt^2} + 2\zeta\omega_n \frac{dx}{dt} + \omega_n^2 x = \frac{F_0}{m} E'.$$

restart

$$H := \frac{\dfrac{F_0}{m} \cdot s}{(s^2 + 2 \cdot zeta \cdot wn \cdot s + wn^2)}$$

$g := subs(s = I \cdot \omega, H)$

$gabs := simplify(evalc(abs(g)))$

Maple Example 4: Enter the code below to determine the long-term behavior of the solution of $(D^2 + 3D + 4)(D^2 + 7D + 2)(x) = E' + E$ using the location of the poles of its transfer function.

restart

$$H := \frac{s+1}{(s^2 + 3 \cdot s + 4) \cdot (s^2 + 7 \cdot s + 2)}$$

$solve(numer(H), s)$

$sol := solve(denom(H), s)$

$evalf(sol)$

Maple Example 5: Enter the code below to calculate the convolution and see the relation with Laplace transforms.

restart

$with(inttrans)$

$$g1 := \frac{1}{s^2 \cdot (s^2 + 1)}$$

$eq1 := invlaplace(g1, s, t)$

$$eq2 := \int_0^t \tau \cdot \sin(t - \tau) \, d\tau$$

$f1 := e - a \cdot \tau \cdot \sin(b \cdot (t - \tau))$

$$eq3 := \int_0^t f1 \, d\tau$$

$laplace(eq3, t, s)$

Maple Exercises

Turn in both the commands that you enter for the exercises below as well as the output/figures. These should all be in one document. Please highlight or clearly indicate all requested answers. Some of the questions will require you to modify the above Maple code to answer them.

1. Enter the commands given in Maple Example 1 and submit both your input and output.

2. Enter the commands given in Maple Example 2 and submit both your input and output.

3. Enter the commands given in Maple Example 3 and submit both your input and output.

4. Enter the commands given in Maple Example 4 and submit both your input and output.

5. Enter the commands given in Maple Example 5 and submit both your input and output.

6. Find the Laplace transform of $\cos t$.

7. Find the Laplace transform of $\sin 5t$.

8. Find the Laplace transform of $t \sin 5t$.

9. Find the Laplace transform of $\frac{\sin bt - bt \cos bt}{2b^3}$.

10. Find the Laplace transform of $e^{3t} \cos 4t$.

11. Find the Laplace transform of $e^{-2t} \sin 3t$.

12. Find the Laplace transform of $\ln t$.

13. Find the Laplace transform of $\delta(t - 1)$.

14. Find the Laplace transform of $t\delta(t - 2)$.

15. Find the Laplace transform of $U_0(t)$.

16. Find the Laplace transform of $tU_1(t)$.

17. Find the inverse Laplace transform of s^2.

18. Find the inverse Laplace transform of $s^3 - \frac{3}{s^2}$.

19. Find the inverse Laplace transform of $\frac{s^2 - b^2}{(s^2 + b^2)^2}$.

20. Find the inverse Laplace transform of $\frac{s}{(s^2 + b^2)^2}$.

21. Find the inverse Laplace transform of $\frac{2}{s+1} - \frac{3}{s^2}$.

22. Find the inverse Laplace transform of $e^{-s} - \frac{3}{s^3}$.

23. Find the inverse Laplace transform of $e^{-s}(s - \frac{3}{s^2+1})$.

24. Find the inverse Laplace transform of $e^{-2s}\left(\frac{1}{s(s^2+4)} - \frac{3}{s^2}\right)$.

25. Find the inverse Laplace transform of $-\frac{\gamma + \ln s}{s}$.

26. Use the transfer function to calculate the frequency response function of $x'' + 2x' + x = E' + E$.

27. Use the transfer function to calculate the frequency response function of $x'' + 5x' + 10x = 2E'' + E$.

28. Use the transfer function to calculate the frequency response function of $4x'' + 4x' + x = E'$.

29. Use the transfer function to calculate the frequency response function of $x'' + 2x' + 2x = E' + 3E$.

30. Use the transfer function to calculate the frequency response function of $(D^2 + 2D + 1)(D^2 + 6D + 10)(x) = E''' + 2E$.

31. Use the transfer function to calculate the frequency response function of $(D^2 + 2D + 2)(D^2 + 10D + 29)(x) = E''' + E''$.

32. Use the poles of the transfer function to determine the long-term behavior of solutions of $x'' + 2x' + x = E' + E$.

33. Use the poles of the transfer function to determine the long-term behavior of solutions of $x'' + 5x' + 10x = 2E'' + E$.

34. Use the poles of the transfer function to determine the long-term behavior of solutions of $4x'' + 4x' + x = E'$.

35. Use the poles of the transfer function to determine the long-term behavior of solutions of $x'' + 2x' + 2x = E' + 3E$.

36. Use the poles of the transfer function to determine the long-term behavior of solutions of $(D^2 + 2D + 1)(D^2 + 6D + 10)(x) = E''' + 2E$.

37. Use the poles of the transfer function to determine the long-term behavior of solutions of $(D^2 + 2D + 2)(D^2 + 10D + 29)(x) = E''' + E''$.

38. Use the definition of the convolution, (7.39), to find $t * \sin(t)$.

39. Use the definition of the convolution, (7.39), to find $\cos t * \sin(t)$.

40. Use the definition of the convolution, (7.39), to find $\delta(t) * \cos(t)$.

41. Use the definition of the convolution, (7.39), to find $U_0(t) * e^t$.

Chapter 7 Computer Lab: Mathematica

Mathematica Example 1: Enter the code below to find some basic Laplace transforms and inverse Laplace transforms.

```
Quit[]
eq1 = LaplaceTransform[1, t, s]
eq2 = LaplaceTransform[Sin[b t], t, s]
f1[t_] = t Cos[b t]
eq3 = LaplaceTransform[f1[t], t, s]
f2[t_] = e^{-a t} Sin[b t]
eq4 = LaplaceTransform[f2[t], t, s]
g1[s_] = 1 / (s^2 + 6 s + 13)
eq5 = InverseLaplaceTransform[g1[s], s, t]
FullSimplify[eq5]
g2[s_] = e^{-4 s} (2/s^2 + 5/s)
```

```
eq6 = InverseLaplaceTransform[g2[s], s, t]
```

Mathematica Example 2: Enter the code below to find the Laplace transforms and inverse Laplace transforms or the unit step (Heaviside) function and idealized impulse (delta) function.

```
Quit[]
f1[t_] = DiracDelta[t - 3]
eq0 = ∫_{-∞}^{∞} f1[t]ⅆt (*entered from the palette*)
eq1 = LaplaceTransform[f1[t], t, s]
eq2 = InverseLaplaceTransform[eq1, s, t]
f2[t_] = UnitStep[t]
eq3 = f2'[t]
eq4 = ∫ f2[t]ⅆt
eq5 = LaplaceTransform[f2[t], t, s]
eq6 = InverseLaplaceTransform[eq5, s, t]
f3[t_] = (2 t - 3) UnitStep[t - 4]
eq7 = LaplaceTransform[f3[t], t, s]
eq8 = InverseLaplaceTransform[eq7, s, t]
```

Mathematica Example 3: Enter the code below to calculate the frequency response function of the given differential equation via its transfer function given in (7.36)-(7.37).

$$\frac{d^2x}{dt^2} + 2\zeta\omega_n\frac{dx}{dt} + \omega_n^2 x = \frac{F_0}{m}E'.$$

```
Quit[]
H[s_] = (FO/m s)/(s^2 + 2 zeta wn s + wn^2)
g = ReplaceAll[H[s], s→i ω] (*i entered from palette*)
gabs = ComplexExpand[Abs[g]]
```

Mathematica Example 4: Enter the code below to determine the long-term behavior of the solution of $(D^2 + 3D + 4)(D^2 + 7D + 2)(x) = E' + E$ using the location of the poles of its transfer function.

```
Quit[]
H[s_] = (s + 1)/((s^2 + 3 s + 4) (s^2 + 7 s + 2))
Solve[Numerator[H[s]]==0, s]
sol = Solve[Denominator[H[s]]==0, s]
N[sol]
```

Mathematica Example 5: Enter the code below to calculate the convolution and see the relation with Laplace transforms.

```
Quit[]
```
$$g1[s\_] = \frac{1}{s^2(s^2+1)}$$
```
eq1 = InverseLaplaceTransform[g1[s], s, t]
```
$$eq2 = \int_0^t \tau \operatorname{Sin}[t - \tau] \, d\tau$$
$$f1[\tau\_] = e^{-a\,\tau} \operatorname{Sin}[b\,(t-\tau)]$$
$$eq3[t\_] = \int_0^t f1[\tau] \, d\tau$$
```
FullSimplify[LaplaceTransform[eq3[t], t, s]]
```

Mathematica Exercises

Turn in both the commands that you enter for the exercises below as well as the output/figures. These should all be in one document. Please highlight or clearly indicate all requested answers. Some of the questions will require you to modify the above Mathematica code to answer them.

1. Enter the commands given in Mathematica Example 1 and submit both your input and output.

2. Enter the commands given in Mathematica Example 2 and submit both your input and output.

3. Enter the commands given in Mathematica Example 3 and submit both your input and output.

4. Enter the commands given in Mathematica Example 4 and submit both your input and output.

5. Enter the commands given in Mathematica Example 5 and submit both your input and output.

6. Find the Laplace transform of $\cos t$.

7. Find the Laplace transform of $\sin 5t$.

8. Find the Laplace transform of $t \sin 5t$.

9. Find the Laplace transform of $\frac{\sin bt - bt \cos bt}{2b^3}$.

10. Find the Laplace transform of $e^{3t} \cos 4t$.

11. Find the Laplace transform of $e^{-2t} \sin 3t$.

12. Find the Laplace transform of $\ln t$.

13. Find the Laplace transform of $\delta(t - 1)$.

14. Find the Laplace transform of $t\delta(t - 2)$.

15. Find the Laplace transform of $U_0(t)$.

16. Find the Laplace transform of $tU_1(t)$.

17. Find the inverse Laplace transform of s^2.

18. Find the inverse Laplace transform of $s^3 - \frac{3}{s^2}$.

19. Find the inverse Laplace transform of $\frac{s^2 - b^2}{(s^2 + b^2)^2}$.

20. Find the inverse Laplace transform of $\frac{s}{(s^2 + b^2)^2}$.

21. Find the inverse Laplace transform of $\frac{2}{s+1} - \frac{3}{s^2}$.

22. Find the inverse Laplace transform of $e^{-s} - \frac{3}{s^3}$.

23. Find the inverse Laplace transform of $e^{-s}\left(s - \frac{3}{s^2+1}\right)$.

24. Find the inverse Laplace transform of $e^{-2s}\left(\frac{1}{s(s^2+4)} - \frac{3}{s^2}\right)$.

25. Find the inverse Laplace transform of $-\frac{\gamma + \ln s}{s}$.

26. Use the transfer function to calculate the frequency response function of $x'' + 2x' + x = E' + E$.

27. Use the transfer function to calculate the frequency response function of $x'' + 5x' + 10x = 2E'' + E$.

28. Use the transfer function to calculate the frequency response function of $4x'' + 4x' + x = E'$.

29. Use the transfer function to calculate the frequency response function of $x'' + 2x' + 2x = E' + 3E$.

30. Use the transfer function to calculate the frequency response function of $(D^2 + 2D + 1)(D^2 + 6D + 10)(x) = E''' + 2E$.

31. Use the transfer function to calculate the frequency response function of $(D^2 + 2D + 2)(D^2 + 10D + 29)(x) = E''' + E''$.

32. Use the poles of the transfer function to determine the long-term behavior of solutions of $x'' + 2x' + x = E' + E$.

33. Use the poles of the transfer function to determine the long-term behavior of solutions of $x'' + 5x' + 10x = 2E'' + E$.

34. Use the poles of the transfer function to determine the long-term behavior of solutions of $4x'' + 4x' + x = E'$.

35. Use the poles of the transfer function to determine the long-term behavior of solutions of $x'' + 2x' + 2x = E' + 3E$.

36. Use the poles of the transfer function to determine the long-term behavior of solutions of $(D^2 + 2D + 1)(D^2 + 6D + 10)(x) = E''' + 2E$.

37. Use the poles of the transfer function to determine the long-term behavior of solutions of $(D^2 + 2D + 2)(D^2 + 10D + 29)(x) = E''' + E''$.

38. Use the definition of the convolution, (7.39), to find $t * \sin(t)$.

39. Use the definition of the convolution, (7.39), to find $\cos t * \sin(t)$.

40. Use the definition of the convolution, (7.39), to find $\delta(t) * \cos(t)$.

41. Use the definition of the convolution, (7.39), to find $U_0(t) * e^t$.

Chapter 7 Projects

This project is a follow-up on some of the ideas presented in Section 6.4 and follows the work in Daley and Gani [15].

There are certain diseases such as typhoid and tuberculosis in which the infectives who are the source of the infection in a population appear healthy. Since the infectives appear healthy, they often go undetected in the population until the disease spread warrants a mass screening program. We call such infectives *carriers* to distinguish them from susceptibles, who, on being infected, may be quickly recognized by their symptoms and removed from the population. Once carriers are detected, they are removed from contact with the susceptibles.

The situation described here requires a different type of model than presented in Section 6.4. The infection of a susceptible through contact with a carrier results in the removal of the infected susceptible, while the number of carriers remains unaltered. The carrier population declines through a process that is independent of the susceptibles.

If we let $C(t)$ represent the size of the carrier population at time t and $S(t)$ be the size of the susceptible population at time t, then one simple model for this carrier-borne infection is

$$\frac{dS}{dt} = -\beta S(t)C(t)$$

$$\frac{dC}{dt} = -\gamma C(t) \tag{7.45}$$

with initial conditions $S(0) = s_0$ and $C(0) = c_0$, where β and γ are positive constants.

Solve this system of differential equations to find $S(t)$ and $C(t)$. For different values of β and γ and initial conditions s_0, c_0, plot your solutions and discuss the behavior of the susceptible and carrier populations.

In practice, when a carrier-borne disease is recognized in a population where it is not normally present, measures are taken to locate the source(s) of the infection. Suppose then that there are c_0 carriers initially in the population, and at some time later, say at t_0, the disease is recognized in the population. At this time, efforts are increased to remove the carriers from the population so that the carrier removal rate γ is increased to γ'. This modification changes the system (7.45) to

$$\frac{dS}{dt} = -\beta S(t)C(t)$$

$$\frac{dC}{dt} = \begin{cases} -\gamma C(t) & 0 < t \leq t_0 \\ -\gamma' C(t) & t > t_0 \end{cases} \tag{7.46}$$

with the same initial conditions $S(0) = s_0$ and $C(0) = c_0$ as before.

Using Laplace transforms, solve this system of differential equations to find $S(t)$ and $C(t)$. For the values of β, γ and initial conditions s_0, c_0 you used for the system (7.45), plot your solutions for different values of γ'. Discuss the behavior of the susceptible and carrier populations comparing your results to the results obtained for the system (7.45).

One possible modification of the system (7.46) that can be studied using Laplace transforms arises by allowing there to be a constant influx of new carriers to the carrier population. If we let α be the rate at which new carriers arrive in the population (α can be thought of as an immigration rate), and assume that this rate is independent of the populations of carriers and susceptibles, then we have the modified system

$$\frac{dS}{dt} = -\beta S(t)C(t)$$

$$\frac{dC}{dt} = \begin{cases} -\gamma C(t) + \alpha & 0 < t \le t_0 \\ -\gamma' C(t) + \alpha & t > t_0 \end{cases} \qquad (7.47)$$

with the same initial conditions $S(0) = s_0$ and $C(0) = c_0$ as before.

Solve this system using Laplace transforms and perform the same kind of analysis you did on the other systems with this system. What is the effect of the immigration rate α?

As a final extension and challenge, we can extend the carrier-borne model described by system (7.46) in a different manner. Consider the effect of allowing a small proportion k of the susceptibles infected to become carriers. This modification changes the differential equation for the carriers and gives the equation

$$\frac{dC}{dt} = -\gamma C(t) + k\beta S(t)C(t),$$

since the proportion of new carriers introduced to the carrier population arises from the infected susceptible population. Using this, the system (7.46) with the increased detection rate γ' becomes

$$\frac{dS}{dt} = -\beta S(t)C(t)$$

$$\frac{dC}{dt} = -\gamma C(t) + k\beta S(t)C(t). \qquad (7.48)$$

Observe that this system is similar to the SI model, Equation (6.40) considered in Section 6.4. Determine conditions on the stability of the *DFE* and find the basic reproductive number R_0. Interpret the meaning of R_0 in terms of the parameters of the system. Try to find the unique endemic equilibrium of the system that lies in the invariant region. Determine its stability and when it is biologically relevant.

If the system has initial conditions $S(0) = s_0$ and $C(0) = c_0$, show that the number of susceptibles $s_\infty = \lim_{t\to\infty} S(t)$ surviving the epidemic is the unique solution of the equation

$$k(s_0 - s_\infty) + c_0 = \frac{\gamma}{\beta} \ln\left(\frac{s_0}{s_\infty}\right).$$

Project 7B: Integral Equations

The convolution theorem has many useful theoretical applications in other branches of mathematics. We will now briefly consider an application closely related to differential equations.

An equation in which an unknown function occurs in a definite integral is called an *integral equation*. In some applications, the unknown function occurs not only in the integral, but in differentiated form in other terms of the same equation. In such cases, the equation is called an *integrodifferential equation*. In either case, if the definite integral is of the convolution type, then the convolution theorem may enable us to transform such an equation into an algebraic equation.

To illustrate this discussion, we will consider an integral that is of the convolution type. Consider an equation of the form

$$x(t) = f(t) + \int_0^t g(t - \tau)x(\tau)\, d\tau \tag{7.49}$$

in which f and g are *known* functions. Our goal is to find a function x for which (7.49) is true.

If we assume that the functions in (7.49) have transforms, we have

$$X(s) = F(s) + G(s)X(s),$$

which we can algebraically solve for $X(s)$ as

$$X(s) = \frac{F(s)}{1 - G(s)}. \tag{7.50}$$

Equation (7.50) is the transform of the solution function $x(t)$ of (7.49). To find $x(t)$ explicitly, we calculate $\mathcal{L}^{-1}\{X(s)\}$.

For example, we can find $x(t)$ so that

$$x(t) = t^3 + \int_0^t x(\tau)\sin(t - \tau)\, d\tau.$$

Here, in this equation, $f(t) = t^3$ and $g(t) = \sin t$ so that applying Laplace transforms yields the equation

$$X(s) = \frac{6}{s^4} + \frac{X(s)}{s^2 + 1}.$$

Solving this equation for $X(s)$ gives

$$X(s) = \frac{6}{s^4} + \frac{6}{s^6},$$

so that

$$x(t) = \mathcal{L}^{-1}\{X(s)\} = \mathcal{L}^{-1}\left\{\frac{6}{s^4} + \frac{6}{s^6}\right\}$$

$$= t^3 + \frac{t^5}{20}.$$

Solve the given integral equation using Laplace transforms.

1. $x(t) = t - \displaystyle\int_0^t (t - u)x(u)\,du$

2. $x(t) = 2 + \displaystyle\int_0^t x(u)\,du$

3. $x(t) = 5 - \displaystyle\int_0^t x(t - u)u\,du$

As a simple application, consider the current $i(t)$ in a single loop RLC circuit which satisfies the equation

$$L\frac{di(t)}{dt} + Ri(t) + \frac{1}{C}\int_0^t i(\tau)\,d\tau = e(t)$$

where R, L, and C are constants and $e(t)$ is a known driving function.

Convert this equation into an algebraic equation in the transform $I(s) = \mathcal{L}\{i(t)\}$ by taking the Laplace transform of both sides. If the constants R, L, and C are given and the function $e(t)$ has a transform, solve for $I(s)$ and find the current $i(t)$ from $\mathcal{L}^{-1}\{I(s)\}$.

For instance, in a single loop R-L-C circuit, we have

$$L = .01, \qquad R = 10, \qquad C = \frac{1}{2500}$$

and a constant driving function $e(t) = 30$. Using these values, find the current $i(t)$ if we assume $i(0) = 0$.

Chapter 8

Series Methods

In this chapter we will explore a different approach to solving an ordinary differential equation. We have developed many methods in the previous chapters for obtaining the solution to an ordinary differential equation; in this chapter we will present a method based upon the power series representation of a function.

In previous chapters, we have not discussed (in general) a method of solution for an ordinary differential equation with variable coefficients; series methods provide one possible method of solution.

8.1 Power Series Representations of Functions

We begin with a discussion of the basic properties of a power series.

A **power series** in $(x - a)$ is an infinite series of the form

$$\sum_{n=0}^{\infty} c_n(x - a)^n = c_0 + c_1(x - a) + c_2(x - a)^2 + \dots, \tag{8.1}$$

where $c_0, c_1, \dots$, are constants, called the **coefficients** (what else would they be called?) of the series; a is a constant, called the **center** of the series; and x is an independent variable. In particular, the power series centered at zero $(a = 0)$ has the form

$$\sum_{n=0}^{\infty} c_n x^n = c_0 + c_1 x + c_2 x^2 + \dots. \tag{8.2}$$

It is a *very* important first observation to note that all polynomials are (finite termed) power series.

A series of the form (8.1) can always be reduced to the form (8.2) by the substitution $X = x - a$. This substitution is merely a translation of the coordinate system. It is easy to see that the behavior of (8.2) near zero is exactly the same as the behavior of (8.1) near a. For this reason we need only study the properties of series of the form (8.2).

We say that (8.2) *converges* at a point $x = x_0$ if the infinite series (of real numbers) $\sum_{n=0}^{\infty} c_n x_0^n$ converges. More specifically, this means that the

$$\lim_{N \to \infty} \sum_{n=0}^{N} c_n x_0^n < \infty.$$

If this limit does not exist as a finite (real) number, then the power series *diverges* at $x = x_0$. It should be clear that the series (8.2) converges at $x = 0$, but are there other values for which the series converges?

Every power series (8.2) has an **interval of convergence** consisting of all values x for which the series converges. The interval of convergence is given in the form $|x| < R$, where R is called the **radius of convergence** of the power series (8.2). On its interval of convergence, a power series (8.2) *converges absolutely*, that is, the series

$$\sum_{n=0}^{\infty} |c_n| \, |x^n|$$

converges; outside the interval of convergence, $|x| > R$, the series (8.2) diverges. When $R = 0$, the interval of convergence consists only of $x = 0$; when $R = \infty$, the power series (8.2) converges absolutely for all x. We can often find the radius of convergence from the familiar ratio test we studied in calculus. We state it here for reference.

THEOREM 8.1.1   Ratio Test

Suppose for a numerical series

$$\sum_{n=1}^{\infty} c_n \tag{8.3}$$

that the limit of the ratio of successive terms is

$$\lim_{n \to \infty} \left| \frac{c_{n+1}}{c_n} \right| = L.$$

i) If $L < 1$ then the series (8.3) converges.
ii) If $L > 1$ then the series (8.3) diverges.
iii) If $L = 1$, either convergence or divergence is possible. The test is inconclusive.

Example 1: Determine the radius and interval of convergence for the series

$$\sum_{n=0}^{\infty} \frac{x^n}{(n+1)2^n}.$$

Substituting for each $x = x_0$ in this power series gives a numerical series with coefficients

$$c_n = \frac{x_0^n}{(n+1)2^n}$$

so that

$$L = \lim_{n \to \infty} \left| \frac{c_{n+1}}{c_n} \right|$$

$$= \lim_{n \to \infty} \left| \frac{\frac{x_0^{n+1}}{(n+2)2^{n+1}}}{\frac{x_0^n}{(n+1)2^n}} \right|$$

$$= \frac{|x_0|}{2} \lim_{n \to \infty} \frac{n+1}{n+2}$$

$$= \frac{|x_0|}{2}.$$

Now the ratio test requires $L < 1$ so

$$\frac{|x_0|}{2} < 1,$$

which is $-2 < x_0 < 2$, so the radius of convergence is 2 and the series converges (absolutely) in $(-2, 2)$.

Convergence of the power series is also possible at the endpoints of this interval. If $x = 2$, the series becomes

$$\sum_{n=0}^{\infty} \frac{1}{(n+1)}$$

which is the harmonic series and hence diverges. If $x = -2$, the series becomes

$$\sum_{n=0}^{\infty} \frac{(-1)^n}{(n+1)}$$

which converges by the alternating series test. Thus, the series converges absolutely for $-2 < x < 2$ and converges on $[-2, 2)$.

As we just saw in this last example, a power series may converge or diverge at the endpoints of its interval of convergence. In general, extra care must be taken when determining whether a power series with interval of convergence $|x| < R$ converges at one or both endpoints $x = \pm R$.

Example 2: Determine the radius of convergence of the power series

$$\sum_{n=0}^{\infty} n! x^n.$$

Solution
The radius of convergence is found by substituting $x = x_0$ into this power series and calculating the limit of successive terms of the corresponding numerical series. Specifically,

$$L = \lim_{n \to \infty} \left| \frac{c_{n+1}}{c_n} \right| = \lim_{n \to \infty} \left| \frac{(n+1)! x_0^{n+1}}{n! x_0^n} \right|$$

$$= |x_0| \lim_{n \to \infty} (n+1)$$

$$= \infty \text{ for } x_0 \neq 0.$$

The only value that will make $L < 1$ is $x_0 = 0$. Thus the power series converges only for $x_0 = 0$.

Example 3: Determine the radius of convergence of the power series

$$\sum_{n=0}^{\infty} \frac{x^n}{n!}.$$

Solution
The radius of convergence is found by computing

$$L = \lim_{n \to \infty} \left| \frac{c_{n+1}}{c_n} \right| = \lim_{n \to \infty} \left| \frac{\frac{x_0^{n+1}}{(n+1)!}}{\frac{x_0^n}{n!}} \right|$$

$$= |x_0| \lim_{n \to \infty} \frac{n!}{(n+1)!}$$

$$= |x_0| \lim_{n \to \infty} \frac{1}{n+1}$$

$$= 0.$$

Thus, $L < 1$ for all x_0, hence the power series converges for all real x_0.

The interval of convergence for a power series is crucial. For each x in this interval, the power series

$$\sum_{n=0}^{\infty} c_n x^n = c_0 + c_1 x + c_2 x^2 + \dots$$

converges. Thus, the series defines, in a very natural way, a *function* of x. Thus, on the interval of convergence, we have

$$f(x) = \sum_{n=0}^{\infty} c_n x^n$$

where $f(x)$ is the limit of the series at the value x, that is,

$$f(x) = \lim_{N \to \infty} \sum_{n=0}^{N} c_n x^n.$$

Now what is truly great about power series is that on their interval of convergence they behave nicely. In fact, the function

$$f(x) = \sum_{n=0}^{\infty} c_n x^n \tag{8.4}$$

is differentiable on $|x| < R$. Moreover, the derivative is obtained by *term-by-term* differentiation:

$$f'(x) = \left(\sum_{n=0}^{\infty} c_n x^n \right)' = \sum_{n=1}^{\infty} n c_n x^{n-1}, \tag{8.5}$$

and the termwise derivative $f'(x)$ has the *same* radius of convergence as the original series (8.4). This wonderful fact is a consequence of the absolute convergence of the series (8.4) on its interval of convergence $|x| < R$ and forms the basis upon which a method of solution of differential equations will be developed.

The series in (8.5) may again be differentiated term by term to obtain the second derivative $f''(x)$. In fact, we can repeat this process indefinitely, so that the function $f(x)$ defined by (8.4) is **infinitely differentiable**.

Observe that

$$f(0) = c_0,$$
$$f'(0) = c_1,$$
$$f''(0) = 2! c_2,$$

and that in general

$$f^{(n)}(0) = n! c_n.$$

Thus,

$$c_n = \frac{f^{(n)}(0)}{n!},$$

so that (8.4) becomes the **Maclaurin** series of $f(x)$,

$$f(x) = \sum_{n=0}^{\infty} \frac{f^{(n)}(0)}{n!} x^n. \tag{8.6}$$

Any function that has a Maclaurin series (8.6) or, more generally, a **Taylor series**,

$$f(x) = \sum_{n=0}^{\infty} \frac{f^{(n)}(a)}{n!} (x-a)^n, \tag{8.7}$$

which converges to $f(x)$ in the interval $|x| < R$ for series (8.6) or the interval $|x - a| < R$ for the series (8.7), is said to be **analytic** in the interval $|x| < R$ or $|x - a| < R$, respectively.

Now, applying these ideas it is possible to obtain power series representations of familiar functions.

Example 4: Find the Maclaurin series for

$$f(x) = e^x.$$

Solution

Since

$$f^{(n)}(x) = e^x \text{ for all } n \geq 0,$$

then

$$f^{(n)}(0) = 1 \text{ for all } n \geq 0.$$

Thus,

$$e^x = \sum_{n=0}^{\infty} \frac{f^{(n)}(0)}{n!} x^n = \sum_{n=0}^{\infty} \frac{x^n}{n!},$$

which is the series we considered above. We showed that this series converges absolutely for $|x| < \infty$. Thus,

$$e^x = 1 + x + \frac{x^2}{2} + \frac{x^3}{3!} + \cdots . \tag{8.8}$$

The Maclaurin series (8.8) gives an infinite termed "polynomial-like" representation for e^x. Using this series, the function e^x can be approximated by the first few terms. For instance, using just the first two terms

$$e^x \approx 1 + x.$$

Graphing the functions $e^x 1 + x, 1 + x + x^2/2$ and $1 + x + \frac{x^2}{2} + \frac{x^3}{6}$, as seen in Figure 8.1, higher-order polynomials yield a better approximation.

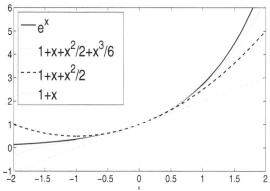

FIGURE 8.1:   Three Maclaurin polynomial approximations for e^x.

Now using the same procedure as in Example 4, we can give the following power series which are analytic for all x:

$$\sin x = \sum_{n=0}^{\infty} \frac{(-1)^n \, x^{2n+1}}{(2n+1)!} = x - \frac{x^3}{3!} + \frac{x^5}{5!} - \frac{x^7}{7!} + \cdots \qquad (8.9)$$

$$\cos x = \sum_{n=0}^{\infty} \frac{(-1)^n \, x^{2n}}{(2n)!} = 1 - \frac{x^2}{2!} + \frac{x^4}{4!} - \frac{x^6}{6!} + \cdots \qquad (8.10)$$

$$\sinh x = \sum_{n=0}^{\infty} \frac{x^{2n+1}}{(2n+1)!} = x + \frac{x^3}{3!} + \frac{x^5}{5!} + \frac{x^7}{7!} + \cdots \qquad (8.11)$$

$$\cosh x = \sum_{n=0}^{\infty} \frac{x^{2n}}{(2n)!} = 1 + \frac{x^2}{2!} + \frac{x^4}{4!} + \frac{x^6}{6!} + \cdots . \qquad (8.12)$$

Before we use these series to study differential equations, we consider a few useful properties.

THEOREM 8.1.2   Addition of power series

If the power series

$$c_0 + c_1 x + c_2 x^2 + \ldots$$

and

$$b_0 + b_1 x + b_2 x^2 + \ldots$$

converge for a given value of x, then for that same value of x we have

$$(b_0 + b_1 x + b_2 x^2 + \ldots) + (c_0 + c_1 x + c_2 x^2 + \ldots)$$
$$= (b_0 + c_0) + (b_1 + c_1)x + (b_2 + c_2)x^2 + \ldots .$$

Power series can be scaled by a constant multiple, as given next.

THEOREM 8.1.3   Multiplication by a constant

If the power series

$$c_0 + c_1 x + c_2 x^2 + \cdots$$

converges for a given value of x and a is a real number, then for that same value of x we have

$$a(c_0 + c_1 x + c_2 x^2 + \cdots) = ac_0 + ac_1 x + ac_2 x^2 + \cdots.$$

Using the distributive property, multiplication of power series is very familiar. It is just "infinite-termed" polynomial multiplication.

THEOREM 8.1.4   Multiplication of two power series

If the power series

$$c_0 + c_1 x + c_2 x^2 + \cdots$$

and

$$b_0 + b_1 x + b_2 x^2 + \cdots$$

converge for a given value of $|x| < R$, then for $|x| < R$ we have

$$(b_0 + b_1 x + b_2 x^2 + \ldots)(c_0 + c_1 x + c_2 x^2 + \cdots)$$
$$= b_0 c_0 + (b_0 c_1 + b_1 c_0)x + (b_0 c_2 + b_1 c_1 + b_2 c_0)x^2 + \cdots.$$

We have already noted how to differentiate a power series. Integration is similar.

THEOREM 8.1.5   Integration of power series

If the power series

$$c_0 + c_1 x + c_2 x^2 + \cdots$$

converges for a given value of $|x| < R$, then for $|x| < R$ we have

$$\int (c_0 + c_1 x + c_2 x^2 + \cdots)\, dx = c_0 x + \frac{c_1}{2}x^2 + \frac{c_2}{3}x^3 + \cdots.$$

Let's look at a few examples.

Example 5: In the exercises, you will show that the power series

$$\sum_{n=0}^{\infty} x^n = 1 + x + x^2 + x^3 + \cdots$$

converges to

$$\frac{1}{1-x}$$

for $|x| < 1$. This is the familiar geometric series. Now if we replace x by $-x$, we obtain the power series

$$1 - x + x^2 - x^3 + \cdots = \frac{1}{1+x}. \tag{8.13}$$

Replacing x by x^2 in this representation gives yet another power series

$$1 - x^2 + x^4 - x^6 + \cdots = \frac{1}{1+x^2}. \tag{8.14}$$

Integrating this power series term by term and noting that

$$\int \frac{1}{1+x^2}\, dx = \arctan x$$

gives the power series for the inverse tangent function as

$$\arctan x = x - \frac{x^3}{3} + \frac{x^5}{5} - \frac{x^7}{7} + \cdots. \tag{8.15}$$

Setting $x = 1$ in this power series representation gives a remarkable result

$$\frac{\pi}{4} = 1 - \frac{1}{3} + \frac{1}{5} - \frac{1}{7} + \cdots.$$

This result, known as **Gregory's series**, has been discovered (and rediscovered) by some of history's greatest mathematicians. We should note here that our argument is not yet completely valid. You should show that the series representation for $\arctan x$ converges for $x = 1$.

Example Show, using a power series approach, that

$$e^{x+y} = e^x e^y.$$

Solution The power series representation for e^x is given by (8.8), as

$$e^x = 1 + x + \frac{x^2}{2} + \frac{x^3}{3!} + \cdots$$

which converges for all real x. Thus, using the product of power series, we have

$$e^x e^y = \left(1 + x + \frac{x^2}{2} + \frac{x^3}{3!} + \cdots\right)\left(1 + y + \frac{y^2}{2} + \frac{y^3}{3!} + \cdots\right)$$

$$= \left(1 + y + \frac{y^2}{2} + \frac{y^3}{3!} + \cdots\right) + x\left(1 + y + \frac{y^2}{2} + \frac{y^3}{3!} + \cdots\right)$$

$$+ \frac{x^2}{2}\left(1 + y + \frac{y^2}{2} + \frac{y^3}{3!} + \cdots\right) + \frac{x^3}{3}\left(1 + y + \frac{y^2}{2} + \frac{y^3}{3!} + \cdots\right) + \cdots$$

$$= 1 + (x + y) + \frac{x^2}{2} + xy + \frac{y^2}{2} + \cdots$$

$$= 1 + (x + y) + \frac{(x + y)^2}{2} + \cdots$$

$$= e^{x+y}.$$

The last equality follows from recognizing the power series expansion for e^{x+y}.

Example 7: Show, using power series, that

$$\sin^2 x + \cos^2 x = 1.$$

Solution

Using the power series (8.9) for $\sin x$ we have

$$\sin^2 x = \left(x - \frac{x^3}{3!} + \frac{x^5}{5!} - \frac{x^7}{7!} + \cdots\right)\left(x - \frac{x^3}{3!} + \frac{x^5}{5!} - \frac{x^7}{7!} + \cdots\right)$$

$$= x^2 + \left(-\frac{1}{3!} - \frac{1}{3!}\right)x^4 + \cdots$$

$$= x^2 - \frac{x^4}{3} + \cdots .$$

Similarly,

$$\cos^2 x = \left(1 - \frac{x^2}{2!} + \frac{x^4}{4!} - \frac{x^6}{6!} + \cdots\right)\left(1 - \frac{x^2}{2!} + \frac{x^4}{4!} - \frac{x^6}{6!} + \cdots\right)$$

$$= 1 - x^2 + \frac{x^4}{3} + \cdots .$$

Thus

$$\sin^2 x + \cos^2 x = \left(x^2 - \frac{x^4}{3} + \cdots\right) + \left(1 - x^2 + \frac{x^4}{3} + \cdots\right)$$

$$= 1.$$

In addition to these power series manipulations, power series are also useful for computing limits.

Example 8: We know from calculus that

$$\lim_{x \to 0} \left(\frac{\sin x}{x} \right) = 1.$$

We can verify this result by using a power series. Since

$$\sin x = x - \frac{x^3}{3!} + \frac{x^5}{5!} - \frac{x^7}{7!} + \dots$$

it follows that

$$\frac{\sin x}{x} = 1 - \frac{x^2}{3!} + \frac{x^4}{5!} - \frac{x^6}{7!} + \dots \qquad \text{for } x \neq 0.$$

Thus, taking the limit gives

$$\lim_{x \to 0} \left(\frac{\sin x}{x} \right) = \lim_{x \to 0} \left(1 - \frac{x^2}{3!} + \frac{x^4}{5!} - \frac{x^6}{7!} + \dots \right)$$
$$= 1.$$

As we move through our work with power series, we see their nature as infinite, termed "polynomial-like" functions. Since polynomials are very friendly functions, power series must surely be friendly functions as well!

Before we consider power series methods of solving differential equations, we give one additional useful application.

Many integrals, even simple-looking ones, cannot be calculated in "closed form." Power series, being easy to integrate, offer a useful alternative approach.

Example 9: Approximate

$$\int_0^1 \sin x^2 \, dx.$$

Solution

Using the power series (8.9) for $\sin x$ with x^2 replacing x, we have

$$\sin x^2 = x^2 - \frac{x^6}{3!} + \frac{x^{10}}{5!} - \frac{x^{14}}{7!} + \dots$$

which converges for all x. Thus

$$\int_0^x \sin t^2 \, dt = \frac{x^3}{3} - \frac{x^7}{7 \times 3!} + \frac{x^{11}}{11 \times 5!} - \frac{x^{15}}{15 \times 7!} + \dots .$$

Now, this power series is not a representation for any elementary function. But, it is a perfectly good antiderivative of $\sin x^2$. Thus,

$$\int_0^1 \sin t^2 \, dt = \frac{x^3}{3} - \frac{x^7}{7 \times 3!} + \frac{x^{11}}{11 \times 5!} - \frac{x^{15}}{15 \times 7!} + \cdots \Big|_0^1$$

$$= \frac{1}{3} - \frac{1}{7 \times 3!} + \frac{1}{11 \times 5!} - \frac{1}{15 \times 7!} + \cdots$$

$$\approx \frac{1}{3} - \frac{1}{7 \times 3!} + \frac{1}{11 \times 5!} - \frac{1}{15 \times 7!}$$

$$\approx 0.3102681578.$$

Using a numerical integration routine, we have

$$\int_0^1 \sin t^2 \, dt \approx 0.31026830172$$

so that the series approximation is very good.

Problems

In Problems **1–4**, determine the interval and radius of convergence of the given series.

1. $\displaystyle\sum_{n=0}^{\infty} \frac{n^2 x^n}{10^n}$

2. $\displaystyle\sum_{n=0}^{\infty} \frac{3^n x^n}{(n+1)^2}$

3. $\displaystyle\sum_{n=1}^{\infty} \frac{(-1)^n x^{2n-1}}{(2n-1)!}$

4. $\displaystyle\sum_{n=1}^{\infty} \frac{2^n (x-3)^n}{n+3}$

Show, using a power series argument, the following identities in Problems **5–8**. In Problem **8**, assume that the power series representations for e^x, $\sin x$, and $\cos x$ are valid for complex numbers $z = x + iy$, where x and y are real and i is the imaginary number.

5. $\sin 2x = 2 \sin x \cos x$

6. $e^{-x} = \dfrac{1}{e^x}$ (Hint: Show $e^{-x} e^x = 1$.)

7. $\displaystyle\lim_{x \to 0} \frac{1 - \cos x}{x} = 0$

8. $e^{ix} = \cos x + i \sin x$ (Euler's formula)

9. Among the simplest and most useful power series, known as the geometric series, is

$$S(x) = 1 + x + x^2 + x^3 + \cdots = \sum_{n=0}^{\infty} x^n.$$

(a) For which real numbers x does $S(x)$ converge? Why?

(b) What function does the series converge to? Why?

10. Derive the power series representations (8.9)–(8.12). Show that each of these series is analytic for all x.

11. Consider the function $f(x) = \begin{cases} e^{-1/x^2}, & x \neq 0 \\ 0, & x = 0. \end{cases}$

    (a) Show that f has derivatives of all orders at $x = 0$ and that

    $$f'(0) = f''(0) = \ldots = 0.$$

    You might want to use the limit $f'(0) = \lim\limits_{x \to 0} \dfrac{f(x) - f(0)}{x}$.

    (b) Conclude that $f(x)$ does not have a Taylor series expansion at $x = 0$, even though it is infinitely differentiable there. Thus, explain why f is not analytic at $x = 0$.

12. Suppose that two functions $f(x)$ and $g(x)$ have Taylor series expansions

    $$f(x) = a_0 + a_1(x - c) + a_2(x - c)^2 + \ldots$$
    $$g(x) = b_0 + b_1(x - c) + b_2(x - c)^2 + \ldots$$

    for $|x - c| < r$. Show that
    (a) The Wronskian of f and g at c is $a_0 b_1 - a_1 b_0$.
    (b) f and g are linearly independent on $|x - c| < r$ if $W(f, g)(c) \neq 0$.

8.2 The Power Series Method

The fundamental assumption made in solving a linear differential equation of the form

$$f(x, y, y', y'', \ldots) = 0$$

by the power series method is that *the solution of the differential equation can be expressed in the form of a power series*, say

$$y = \sum_{n=0}^{\infty} c_n x^n = c_0 + c_1 x + c_2 x^2 + \cdots. \tag{8.16}$$

Since a power series can be differentiated, term by term, it follows that $y', y'', \ldots$ can be obtained as

$$y' = \sum_{n=1}^{\infty} n c_n x^{n-1} = c_1 + 2c_2 x + 3c_3 x^2 + \cdots \tag{8.17}$$

$$y'' = \sum_{n=2}^{\infty} n(n-1) c_n x^{n-2} = 2c_2 + 3 \cdot 2c_3 x + \cdots. \tag{8.18}$$

$$\vdots$$

These series can be substituted into the given differential equation and the terms involving like powers of x are then collected. Such an expression would be of the form

$$k_0 + k_1 x + k_2 x^2 + \ldots = \sum_{n=0}^{\infty} k_n x^n = 0, \tag{8.19}$$

where k_0, k_1, $k_2, \cdots$ are expressions involving the unknown coefficients c_0, c_1, $c_2, \cdots$. Now Equation (8.19) must hold for all values of x in the interval of convergence; so all the coefficients $k_0, k_1, k_2, \ldots$ must be zero, that is,

$$k_0 = 0, \quad k_1 = 0, \quad k_2 = 0, \quad \cdots .$$

Using these equations, we may be able to determine the coefficients $c_0, c_1, c_2, \ldots$ of the power series solution y. We illustrate this method in the next several examples.

Example 1: Solve the initial-value problem

$$y' = y + x^2$$

with $y(0) = 1$.

Solution
Using the Equations (8.16) and (8.17) in the differential equation $y' = y + x^2$ gives

$$c_1 + 2c_2 x + 3c_3 x^2 + \ldots = (c_0 + c_1 x + c_2 x^2 + \ldots) + x^2.$$

Collecting like powers of x we obtain

$$(c_1 - c_0) + (2c_2 - c_1)x + (3c_3 - c_2 - 1)x^2 + (4c_4 - c_3)x^3 + \ldots = 0.$$

Equating each of these coefficients to zero gives the equations

$$c_1 - c_0 = 0,$$

$$2c_2 - c_1 = 0,$$

$$3c_3 - c_2 - 1 = 0,$$

$$4c_4 - c_3 = 0,$$

$$\vdots$$

From these equations we have $c_0 = c_1$ and

$$c_2 = \frac{c_1}{2} = \frac{c_0}{2!},$$

$$c_3 = \frac{c_2 + 1}{3} = \frac{c_0 + 2}{3!},$$

$$c_4 = \frac{c_3}{4} = \frac{c_0 + 2}{4!}$$

$$\vdots$$

We now substitute these back into the power series solution (8.16) y to get

$$y = c_0 + c_0 x + \frac{c_0}{2!} x^2 + \frac{c_0 + 2}{3!} x^3 + \frac{c_0 + 2}{4!} x^4 + \cdots .$$

The initial condition $y(0) = 1$ gives

$$1 = c_0 + 0 + 0 + \ldots$$

so the series becomes

$$y = 1 + x + \frac{x^2}{2!} + \frac{3}{3!} x^3 + \frac{3}{4!} x^4 + \cdots .$$

This almost looks like an exponential power series. If you look at it hard enough, you will see (maybe) that adding and subtracting

$$2 \left(1 + x + \frac{x^2}{2!} \right)$$

gives

$$y = 3 \left(1 + x + \frac{x^2}{2!} + \frac{x^3}{3!} + \ldots \right) - 2 \left(1 + x + \frac{x^2}{2!} \right) .$$

Now we see that the series is the power series of e^x. Thus, the solution is

$$y = 3e^x - x^2 - 2x - 2.$$

This is our first example; we will not always be so lucky to recognize a power series expansion. Note that this equation is a first-order nonhomogeneous linear differential equation and we could have solved it as such.

Example 2: Solve the differential equation

$$y'' + y = 0.$$

Solution

Using the Equations (8.16) and (8.18), we have

$$(2c_2 + 3 \cdot 2c_3 x + \ldots) + (c_0 + c_1 x + c_2 x^2 + \ldots) = 0.$$

Collecting like powers of x gives

$$(2c_2 + c_0) + (3 \cdot 2c_3 + c_1)x + (4 \cdot 3c_4 + c_2)x^2 + \ldots = 0.$$

Each of these coefficients must be zero, so

$$2c_2 + c_0 = 0,$$

$$3 \cdot 2c_3 + c_1 = 0,$$

$$4 \cdot 3c_4 + c_2 = 0,$$

$$5 \cdot 4c_5 + c_3 = 0,$$

$$\vdots$$

Thus,

$$c_2 = -\frac{c_0}{2!}, \quad c_3 = -\frac{c_1}{3!}, \quad c_4 = -\frac{c_2}{4 \cdot 3} = \frac{c_0}{4!}, \quad c_5 = -\frac{c_3}{5 \cdot 4} = \frac{c_1}{5!}, \quad \ldots.$$

Using these coefficients in the power series (8.16) for y gives

$$y = c_0 + c_1 x - \frac{c_0}{2!}x^2 - \frac{c_1}{3!}x^3 + \frac{c_0}{4!}x^4 + \frac{c_1}{5!}x^5 + \cdots.$$

This can be written as the sum of two power series

$$y = c_0 \left(1 - \frac{x^2}{2!} + \frac{x^4}{4!} - \frac{x^6}{6!} + \cdots \right) + c_1 \left(x - \frac{x^3}{3!} + \frac{x^5}{5!} - \frac{x^7}{7!} + \cdots \right).$$

Here, again, we recognize each of the power series, the first series is that of $\cos x$ and the second is the one for $\sin x$. Hence, the solution is

$$y = c_0 \cos x + c_1 \sin x.$$

Note that this differential equation is a second-order homogeneous equation with constant coefficients.

You might now wonder why we did the power series approach on the last two examples, as each was solvable by our earlier methods. The next two examples will reveal one of the reasons we have introduced this method.

Example 3: Solve the differential equation

$$y'' + xy' + y = 0.$$

Solution
Using power series written in summation notation (for the sake of compactness) we have

$$\sum_{n=2}^{\infty} n(n-1)c_n x^{n-2} + x \sum_{n=1}^{\infty} n c_n x^{n-1} + \sum_{n=0}^{\infty} c_n x^n = 0. \qquad (8.20)$$

Now, to group coefficients of like powers of x, we need to collect the sums into one single sum. The procedure is akin to change of variables techniques in integration and is as follows. For the first sum

$$S_1 = \sum_{n=2}^{\infty} n(n-1)c_n x^{n-2},$$

we will obtain an exponent k by substituting $k = n - 2$. Thus, $n = k + 2$ and we see that $n = 2$ gives $k = 0$ while $n = \infty$ gives $k = \infty$. Thus we have

$$S_1 = \sum_{k=0}^{\infty} (k+2)(k+1)c_{k+2} x^k.$$

Next, we need to re-index the second series

$$S_2 = x \sum_{n=1}^{\infty} nc_n x^{n-1} = \sum_{n=1}^{\infty} nc_n x^n = \sum_{k=1}^{\infty} kc_k x^k = \sum_{k=0}^{\infty} kc_k x^k.$$

Note here that the last expression follows since $kc_k x^k = 0$ when $k = 0$. Finally, the third series becomes

$$S_3 = \sum_{k=0}^{\infty} c_k x^k.$$

So, the expression (8.20) becomes

$$\sum_{k=0}^{\infty} (k+2)(k+1)c_{k+2} x^k + \sum_{k=0}^{\infty} kc_k x^k + \sum_{k=0}^{\infty} c_k x^k = 0.$$

Gathering like powers of x gives the expression

$$\sum_{k=0}^{\infty} [(k+2)(k+1)c_{k+2} + (k+1)c_k] x^k = 0.$$

Setting the coefficients in the brackets to zero gives the general *recursion* formula

$$(k+2)(k+1)c_{k+2} + (k+1)c_k = 0.$$

This gives $c_{k+2} = -\dfrac{c_k}{k+2}$ so that

$$c_2 = -\frac{c_0}{2}, \quad c_3 = -\frac{c_1}{3}, \quad c_4 = -\frac{c_2}{4} = \frac{c_0}{2 \cdot 4}, \quad c_5 = -\frac{c_3}{5} = \frac{c_1}{3 \cdot 5}, \quad \cdots .$$

The power series for y can be written in the form

$$y = c_0 + c_1 x - \frac{c_0}{2} x^2 - \frac{c_1}{3} x^3 + \frac{c_0}{2 \cdot 4} x^4 + \frac{c_1}{3 \cdot 5} x^5 - \cdots .$$

Separating the terms that involve c_0 and c_1 gives

$$y = c_0 \left(1 - \frac{x^2}{2} + \frac{x^4}{2 \cdot 4} - \frac{x^6}{2 \cdot 4 \cdot 6} + \cdots \right)$$

$$+ c_1 \left(x - \frac{x^3}{3} + \frac{x^5}{3 \cdot 5} - \frac{x^7}{3 \cdot 5 \cdot 7} + \cdots \right).$$

There is no closed form for either of these series. This *does not* mean that this solution is not useful. Indeed, it is a perfectly good (and natural) solution. For instance, suppose that the differential equation satisfied the initial conditions $y(0) = 1$ and $y'(0) = 1$; then $c_0 = 1$ and $c_1 = 1$. The solution is

$$y = 1 + x - \frac{x^2}{2} - \frac{x^3}{3} + \frac{x^4}{8} - \frac{x^5}{15} + \cdots.$$

Graphing $1 + x, 1 + x - \frac{x^2}{2}, 1 + x - \frac{x^2}{2} - \frac{x^3}{3}$, and $1 + x - \frac{x^2}{2} - \frac{x^3}{3} + \frac{x^4}{8}$ as in Figure 8.2 gives an idea of the behavior of this series solution.

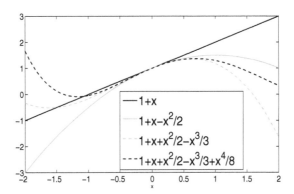

FIGURE 8.2: Four approximate solutions to the differential equation $y'' + xy' + y = 0$ with initial conditions $y(0) = 1$ and $y'(0) = 1$.

Example 4: Find a power series solution of the initial-value problem

$$(x^2 - 1)\frac{d^2y}{dx^2} + 3x\frac{dy}{dx} + xy = 0,$$

with $y(0) = 4$ and $y'(0) = 6$.

Solution

Since the initial conditions are prescribed at $x = 0$, we will seek a power series solution about this point. We assume a solution of the form

$$y = \sum_{n=0}^{\infty} c_n x^n.$$

This gives

$$\frac{dy}{dx} = \sum_{n=1}^{\infty} nc_n x^{n-1}$$

and

$$\frac{d^2y}{dx^2} = \sum_{n=2}^{\infty} n(n-1)c_n x^{n-2}$$

so that substituting these into the differential equation gives

$$(x^2 - 1)\sum_{n=2}^{\infty} n(n-1)c_n x^{n-2} + 3x\sum_{n=1}^{\infty} nc_n x^{n-1} + x\sum_{n=0}^{\infty} c_n x^n = 0.$$

Thus

$$\sum_{n=2}^{\infty} n(n-1)c_n x^n - \sum_{n=2}^{\infty} n(n-1)c_n x^{n-2} + 3\sum_{n=1}^{\infty} nc_n x^n + \sum_{n=0}^{\infty} c_n x^{n+1} = 0,$$

so that reindexing the second and fourth series gives

$$\sum_{n=2}^{\infty} n(n-1)c_n x^n - \sum_{n=0}^{\infty} (n+2)(n+1)c_{n+2} x^n + 3\sum_{n=1}^{\infty} nc_n x^n + \sum_{n=1}^{\infty} c_{n-1} x^n = 0.$$

Combining like powers of x and writing just a single series gives

$$-2c_2 + (c_0 + 3c_1 - 6c_3)x + \sum_{n=2}^{\infty}[-(n+2)(n+1)c_{n+2} + n(n+2)c_n + c_{n-1}]x^n = 0.$$

Comparing coefficients we have

$$-2c_2 = 0$$
$$c_0 + 3c_1 - 6c_3 = 0$$

and

$$-(n+2)(n+1)c_{n+2} + n(n+2)c_n + c_{n-1} = 0 \quad \text{for } n \geq 2.$$

This gives $c_2 = 0$ and $c_3 = \frac{1}{6}c_0 + \frac{1}{2}c_1$, and the recurrence relation

$$c_{n+2} = \frac{n(n+2)c_n + c_{n-1}}{(n+1)(n+2)} \quad \text{for } n \geq 2.$$

Applying this recurrence relation successively yields

$$c_4 = \frac{8c_2 + c_1}{12} = \frac{1}{12}c_1,$$
$$c_5 = \frac{15c_3 + c_2}{20} = \frac{1}{8}c_0 + \frac{3}{8}c_1$$

so that

$$y = c_0 + c_1 x + \left(\frac{c_0}{6} + \frac{c_1}{2}\right)x^3 + \frac{c_1}{12}x^4 + \left(\frac{c_0}{8} + \frac{3c_1}{8}\right)x^5 + \cdots .$$

This series can be written as

$$y = c_0\left(1 + \frac{1}{6}x^3 + \frac{1}{8}x^5 + \cdots\right) + c_1\left(x + \frac{1}{2}x^3 + \frac{1}{12}x^4 + \frac{3}{8}x^5 + \cdots\right).$$

Now applying the initial condition $y(0) = 4$ gives

$$c_0 = 4$$

and differentiating the series gives

$$y' = c_0\left(\frac{1}{2}x^2 + \frac{5}{8}x^4 + \cdots\right) + c_1\left(1 + \frac{3}{2}x^2 + \frac{1}{3}x^3 + \frac{15}{8}x^4 + \cdots\right)$$

so that $y'(0) = 6$ gives

$$c_1 = 6.$$

Thus, the power series solution that satisfies the initial-value problem is

$$y = 4\left(1 + \frac{1}{6}x^3 + \frac{1}{8}x^5 + \cdots\right) + 6\left(x + \frac{1}{2}x^3 + \frac{1}{12}x^4 + \frac{3}{8}x^5 + \cdots\right)$$

$$= 4 + 6x + \frac{11}{3}x^3 + \frac{1}{2}x^4 + \frac{11}{4}x^5 + \cdots .$$

Suppose now that the initial-value problem of Example 4 was replaced by the initial-value problem

$$(x^2 - 1)\frac{d^2y}{dx^2} + 3x\frac{dy}{dx} + xy = 0$$

with $y(5) = 4$ and $y'(5) = 6$.

This problem is virtually the same as Example 4; however, the initial-values in the problem are prescribed at $x = 5$, instead of $x = 0$. We seek a power series solution of the form

$$y = \sum_{n=0}^{\infty} c_n(x - 5)^n.$$

The simplest way to proceed here is to make the substitution $t = x - 5$. Here $x = t + 5$ and the initial-value problem becomes

$$((t+5)^2 - 1)\frac{d^2y}{dt} + 3(t+5)\frac{dy}{dt} + (t+5)y = 0$$

with $y(0) = 4$ and $y'(0) = 6$. We then seek a solution in powers of t of the form

$$y = \sum_{n=0}^{\infty} c_n t^n.$$

Thus, this problem is now very similar to Example 4 and it is suggested that the reader pursue the calculations to obtain the c_n's. Once the c_n's are found, the initial conditions $y(0) = 4$ and $y'(0) = 6$ can be applied; then replacing t by $x - 5$, the solution to the original initial-value problem is obtained.

Taylor and Maclaurin series can also be used to find power series solutions to differential equations. This is a slick method.

Example 6 Again, consider the differential equation

$$y' = y + x^2,$$

with the condition $y(0) = 1$.

Solution
Differentiating both sides of the differential equation repeatedly and successively evaluating each derivative at the initial-value of $x = 0$, we have

$$y'(0) = (y + x^2)|_{x=0} = y(0) = 1,$$
$$y''(0) = (y' + 2x)|_{x=0} = y'(0) = 1,$$
$$y'''(0) = (y'' + 2)|_{x=0} = y''(0) + 2 = 3,$$
$$y^{(4)}(0) = y'''|_{x=0} = y'''(0) = 3.$$
$$\vdots$$

Substituting these values into the Maclaurin series (8.2) gives

$$y(x) = \sum_{n=0}^{\infty} \frac{y^{(n)}(0)}{n!} x^n$$
$$= 1 + x + \frac{x^2}{2!} + \frac{3x^3}{3!} + \frac{3x^4}{4!} + \cdots,$$

which we obtained earlier.

Problems

In Problems **1–14**, solve the following differential equations using the power series method. Also solve the equations using one of the methods we have presented earlier, if possible.

1. $y' = y^2 - x$ with $y(0) = 1$.

2. $y' - 2y = x^2$ with $y(1) = 1$.

3. $y' = y + xe^y$ with $y(0) = 0$.

4. $y' = x + y$ with $y(0) = 1$.

5. $y' - x^2 = y^2$ with $y(0) = 1$.

6. $y' = x + \sin y$ with $y(0) = 0$.

7. $y' = e^x + \sin y$ with $y(0) = 0$.

8. $y' = x^2 + 2y^2$ with $y(0) = 1$.

9. $y' = x + \frac{1}{y}$ with $y(0) = 1$.

10. $y' = 1 + x\sin y$ with $y(0) = 0$.

11. $y' = x^2 + y^3$ with $y(1) = 1$.

12. $y' = x^3 + y^2$ with $y(1) = 1$.

13. $y'' + y = 1 + x + x^2$ with $y(0) = 1$ and $y'(0) = -1$.

14. $y'' - xy' + y^2 = 0$ with $y(0) = 1$ and $y'(0) = -1$.

In Problems **15–17**, find the general power series solution to the differential equation. Try to recognize the solution in terms of familiar functions. This may not be possible.

15. $(1 + x^2)y'' + xy' + xy = 0$.

16. $xy'' - x^2y' + (x^2 - 2)y = 0$, $y(0) = 0$ and $y'(0) = 1$.

17. $y'' - 2xy' - 2y = x$, $y(0) = 1$ and $y'(0) = -1/4$.

18. Sir George Biddell Airy (1801–1892) was Lucasian professor of mathematics, director of the observatory, and Plumian professor of astronomy at Cambridge University in England until 1835. He was then appointed director of the Greenwich Observatory (Astronomer Royal). He remained there until his retirement in 1881. He did much work in lunar and solar photography, planetary motion, optics, and other areas. The **Airy equation**

$$y'' - xy = 0$$

has applications in the theory of diffraction. Find the general solution of this equation.

19. The **Hermite equation** in honor of Charles Hermite (1822–1901) is

$$y'' - 2xy' + 2py = 0,$$

where p is a constant. This equation arises in quantum mechanics in connection with the **Schrödinger equation** for a harmonic oscillator. Show that if p is a positive integer, one of the two linearly independent solutions of the Hermite equation is a polynomial, called the **Hermite polynomial** $H_p(x)$.

8.3   Ordinary and Singular Points

The power series method we have considered in the previous section sometimes fails to yield a solution for one equation while working very well for an apparently similar equation.

Example 1: Consider the differential equation

$$x^2 y'' + axy' + by = 0.$$

Apply the power series method for each of the following cases:

Case 1: $a = -2$, $b = 2$.

In the differential equation, set

$$y = \sum_{n=0}^{\infty} c_n x^n, \quad y' = \sum_{n=1}^{\infty} n c_n x^{n-1}, \quad \text{and } y'' = \sum_{n=2}^{\infty} n(n-1) c_n x^{n-2}.$$

Then, since $n(n-1) = 0$ at $n = 0$ and 1,

$$
\begin{aligned}
x^2 y'' + axy' + by &= x^2 \sum_{n=0}^{\infty} n(n-1) c_n x^{n-2} + ax \sum_{n=0}^{\infty} n c_n x^{n-1} + b \sum_{n=0}^{\infty} c_n x^n \\
&= \sum_{n=0}^{\infty} n(n-1) c_n x^n + \sum_{n=0}^{\infty} a n c_n x^n + \sum_{n=0}^{\infty} b c_n x^n \\
&= \sum_{n=0}^{\infty} [n(n-1) + an + b] c_n x^n \\
&= 0.
\end{aligned}
\tag{8.21}
$$

Now substituting $a = -2$, $b = 2$ into (8.21) gives

$$\sum_{n=0}^{\infty} (n^2 - 3n + 2) c_n x^n = 0.$$

Equating the coefficients to zero gives

$$(n-2)(n-1) c_n = 0,$$

which means that $c_n = 0$ for all $n \neq 1$ or 2. Hence,

$$y = c_1 x + c_2 x^2$$

is the general solution when $a = -2$, $b = 2$.

Case 2: $a = -1$ and $b = 1$.

Substitution into (8.21) gives

$$\sum_{n=0}^{\infty} (n^2 - 2n + 1) c_n x^n = 0,$$

so
$$(n-1)^2 c_n = 0.$$

Thus, $c_n = 0$ for all $n \neq 1$, so that the solution is

$$y = c_1 x.$$

But here we note that the differential equation

$$x^2 y'' + axy' + by = 0$$

is second-order, so that its general solution involves *two* linearly independent solutions. The power series method has given us only one. We use the method of reduction of order to find the other solution.

Let $y = xv$ so that

$$x^2(xv)'' - x(xv)' + xv = 0,$$

and this equation becomes

$$x^3 v'' + x^2 v' = 0.$$

Now making the substitution $z = v'$, we obtain the first-order separable differential equation

$$x^3 \frac{dz}{dx} + x^2 z = 0,$$

so

$$z = \frac{c}{x}.$$

Thus,

$$v' = \frac{c}{x}, \text{ which gives } v = c \ln |x|.$$

Thus, the general solution is

$$y = Ax + Bx \ln |x|.$$

Note here that $\ln |x|$ is not defined at $x = 0$, so it is not a surprise that the power series method failed to obtain this term as part of the solution.

Case 3: $a = 1$ and $b = 1$.

Equation (8.21) becomes

$$\sum_{n=0}^{\infty} (n^2 + 1) c_n = 0.$$

Equating the coefficients to zero gives

$$(n^2 + 1) c_n = 0,$$

so

$$c_n = 0 \text{ for every } n, \text{ since } n^2 + 1 \neq 0.$$

Thus, the power series method fails completely in helping us find the general solution. The general solution is

$$y = A \cos \ln |x| + B \sin \ln |x|.$$

(Check it!)

We will now present conditions under which the power series method does work. One also notes here that the methods of Chapter 4, Section 3 apply.

THEOREM 8.3.1

There is a unique Maclaurin series $y(x)$ satisfying the initial-value problem

$$y'' + a(x)y' + b(x)y = 0,$$

with $y(0) = \alpha$ and $y'(0) = \beta$, provided $a(x)$ and $b(x)$ can each be represented by a Maclaurin series converging in an interval $|x| < R$. The power series $y(x)$ also converges in $|x| < R$.

This result guarantees that the power series method works whenever $a(x)$ and $b(x)$ are analytic (have convergent Maclaurin series) at $x = 0$. We need some terminology.

Definition 8.3.1

The point $x = x_0$ is called an ordinary point of the differential equation

$$y'' + a(x)y' + b(x)y = 0 \qquad (8.22)$$

when both $a(x)$ and $b(x)$ are analytic at $x = x_0$. If $x = x_0$ is not an ordinary point, it is called a singular point of the differential equation.

If we write each of the second-order homogeneous equations in the previous example in the form (8.22), we have

$$y'' + \frac{a}{x}y' + \frac{b}{x^2}y = 0.$$

We see here that neither of the terms

$$a(x) = \frac{a}{x} \text{ or } b(x) = \frac{b}{x^2}$$

are defined at $x = 0$, so that they fail to have a power series representation that converges in an open interval containing $x = 0$.

Consider the differential equation

$$(1 - x^2)y'' - xy' + 2y = 0.$$

Rewriting this equation gives

$$y'' - \frac{x}{1 - x^2}y' + \frac{2}{1 - x^2}y = 0,$$

for which we see that

$$a(x) = -\frac{x}{1 - x^2} \text{ and } b(x) = \frac{2}{1 - x^2}.$$

Each of these functions has a Maclaurin series

$$a(x) = -x(1 + x^2 + x^4 + \ldots)$$

and

$$b(x) = 2(1 + x^2 + x^4 + \ldots)$$

so that $x = 0$ is an ordinary point and the power series method will yield the general solution. Note here that $x = \pm 1$ are singular points of these functions, so that a Taylor series centered at $x = \pm 1$ may or may not exist.

This is a specific case of a special type of differential equation. The **Čebyšev differential equation** is given as

$$(1 - x^2)y'' - xy' + \mu^2 y = 0 \tag{8.23}$$

where μ is a constant.

Pafnuty Lvovich Čebyšev (1821–1894) (pronounced and often written as Chebyshev or Tchebyshev) was a mathematics professor at the University of St. Petersburg. He made numerous important contributions in the areas of probability theory, number theory, special functions, and mechanics. The Čebyšev differential equation (8.23) has some very interesting properties, some of which we now consider.

Example 2: Solve the differential equation

$$(1 - x^2)y'' - xy' + \mu^2 y = 0.$$

Solution
Rewriting the equation as

$$y'' - \frac{x}{1 - x^2}y' + \frac{\mu^2}{1 - x^2}y = 0$$

gives $x = \pm 1$ as singular points. Thus, on the interval $(-1, 1)$ letting

$$y(x) = \sum_{n=0}^{\infty} c_n x^n$$

and substituting into the differential equation gives

$$(1 - x^2) \sum_{n=2}^{\infty} c_n n(n-1)x^{n-2} - x \sum_{n=1}^{\infty} c_n n x^{n-1} + \mu^2 \sum_{n=0}^{\infty} c_n x^n = 0.$$

Simplifying this expression and rewriting gives

$$\sum_{n=0}^{\infty} [c_{n+2}(n+2)(n+1) - c_n(n^2 - \mu^2)]x^n = 0.$$

Comparing coefficients gives the recurrence relation

$$c_{n+2} = \frac{(n^2 - \mu^2)c_n}{(n+1)(n+2)} \quad \text{for } n = 0, 1, 2, \cdots . \tag{8.24}$$

This recurrence relation yields the even terms as

$$c_2 = \frac{-\mu^2 c_0}{2}, \quad c_4 = \frac{-(4-\mu)^2 \mu^2 c_0}{24}, \quad \cdots$$

and the odd terms

$$c_3 = \frac{(1 - \mu^2)c_1}{6}, \quad c_5 = \frac{(1 - \mu^2)(9 - \mu^2)c_1}{120}, \quad \cdots$$

so that the general solution is

$$y(x) = c_0 \left(1 + \frac{-\mu^2}{2}x^2 + \frac{-(4 - \mu^2)\mu^2}{24}x^4 + \cdots \right)$$
$$+ c_1 \left(x + \frac{1 - \mu^2}{6}x^3 + \frac{(1 - \mu^2)(9 - \mu^2)}{120}x^5 + \cdots \right).$$

If μ is an even integer, we see that all even coefficients having index greater than μ are zero. For instance, when $\mu = 4$ we have

$$c_6 = \frac{(4^2 - 4^2)c_4}{30} = 0$$

so that $c_8 = 0, c_{10} = 0 \cdots$. Thus when $\mu = 4$ the recurrence relation gives

$$c_2 = -8c_0, \quad c_4 = -c_2 = 8c_0, \quad c_6 = c_8 = c_{10} \ldots = 0.$$

Using these values, we have the fourth-degree polynomial

$$P(x) = c_0[1 - 8t^2 + 8t^4]$$

as a solution of Čebyšev's equation (8.23).

Similarly, if μ is an odd positive integer then we also obtain a polynomial of (8.23). These polynomial solutions of (8.23) for each nonnegative integer valued μ are known as *Čebyšev (Tchebyshev) polynomials of the first kind* and $T_n(x)$ is usually used to denote these nth degree Čebyšev polynomials. The first few Čebyšev polynomials are

$$T_0(x) = 1, \quad T_1(x) = x, \quad T_2(x) = 2x^2 - 1$$

$$T_3(x) = 4x^3 - 3x, \quad T_4(x) = 8x^4 - 8x^2 + 1,$$

and a graph of $T_2(x)$, $T_3(x)$, and $T_4(x)$ is shown in Figure 8.3. More properties and aspects of Čebyšev polynomials are considered in the Problems.

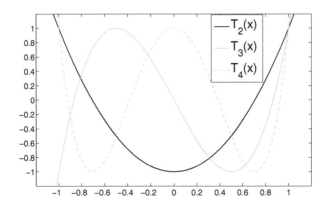

FIGURE 8.3: The Čebyšev polynomials $T_2(x)$, $T_3(x)$, and $T_4(x)$.

We next consider a method of solution about a singular point.

Solutions about Singular Points

Consider the second-order homogenous linear differential equation

$$f_0(x)\frac{d^2y}{dx^2} + f_1(x)\frac{dy}{dx} + f_2(x)y = 0 \tag{8.25}$$

and suppose that x_0 is a singular point. In this situation we are not assured that a power series solution

$$y = \sum_{n=0}^{\infty} c_n x^n$$

exists. We need to find a different type of solution, but how should we approach the problem?

To this end, we need to make a definition.

Definition 8.3.2

Consider the differential equation

$$f_0(x)\frac{d^2y}{dx^2} + f_1(x)\frac{dy}{dx} + f_2(x)y = 0$$

and assume that at least one of the functions

$$p(x) = \frac{f_1(x)}{f_0(x)}, \quad q(x) = \frac{f_2(x)}{f_0(x)}$$

is not analytic at x_0 so that x_0 is a singular point of the differential equation (8.25). If the functions defined by the products

$$(x - x_0)p(x) \quad \text{and} \quad (x - x_0)^2q(x) \tag{8.26}$$

are both analytic at x_0, then x_0 is called a *regular* singular point of the differential equation (8.25). If either or both of the functions defined by (8.26) are not analytic at x_0, then x_0 is called an *irregular* singular point of (8.25).

Example The differential equation

$$3x^2\frac{d^2y}{dx^2} - 4x\frac{dy}{dx} + (x + 2)y = 0 \tag{8.27}$$

can be written as

$$\frac{d^2y}{dx^2} - \frac{4}{3x}\frac{dy}{dx} + \frac{(x + 2)}{3x^2}y = 0,$$

so that

$$p(x) = \frac{-4}{3x} \quad \text{and} \quad q(x) = \frac{x + 2}{3x^2}.$$

Both fail to be analytic at $x = 0$. Thus, $x = 0$ is a singular point of the differential equation (8.27). Now the products are

$$xp(x) = \frac{-4}{3} \quad \text{and} \quad x^2q(x) = \frac{x + 2}{3},$$

both of which are analytic at $x = 0$. Thus $x = 0$ is a regular singular point of the differential equation (8.27).

Example The differential equation

$$x^2(x - 2)^2\frac{d^2y}{dx^2} + 2(x - 2)\frac{dy}{dx} + (x + 1)y = 0$$

can be written as

$$\frac{d^2y}{dx^2} + \frac{2}{x^2(x-2)}\frac{dy}{dx} + \frac{(x+1)}{x^2(x-2)^2}y = 0 \qquad (8.28)$$

so that

$$p(x) = \frac{2}{x^2(x-2)} \quad \text{and} \quad q(x) = \frac{x+1}{x^2(x-2)^2}$$

are not analytic at $x = 0$ and $x = 2$. Thus $x = 0$ and $x = 2$ are singular points of the differential equation (8.28).

Now at $x = 0$, we consider the products

$$xp(x) = \frac{2}{x(x-2)} \quad \text{and} \quad x^2q(x) = \frac{x+1}{(x-2)^2};$$

here $x^2q(x)$ is analytic at $x = 0$, while $xp(x)$ is *not* analytic at $x = 0$. Thus $x = 0$ is an irregular singular point of (8.28).

Next considering the point $x = 2$, the products

$$(x-2)p(x) = \frac{2}{x^2} \quad \text{and} \quad (x-2)^2q(x) = \frac{x+1}{x^2}$$

are *both* analytic at $x = 2$ so that $x = 2$ is a regular singular point of (8.28).

These ideas lead us to develop a method of obtaining a series solution when the differential equation has a regular singular point. We will develop this method in the next section.

● ● ● ● ● ● ● ● ● ● ● ●

Problems

For each of the following differential equations in Problems **1–6**, find all singular points. Determine if these points are regular or irregular.

1. $(x^2 - 3x)y'' + (x+2)y' + y = 0.$ **2.** $(x^3 + x^2)y'' + (x^2 - 2x)y' + 4y = 0.$
3. $x^2(x+2)y'' + xy' - (2x-1)y = 0$ **4.** $(x^5 + x^4 - 6x^3)y'' + x^2y' + (x-2)y = 0.$
5. $(2x+1)x^2y'' - (x+2)y' + 2e^x y = 0.$
6. $(x^4 - 2x^3 + x^2)y'' + 2(x-1)y' + x^2y = 0.$

7. The Čebyšev differential equation $(1-x^2)y'' - xy' + \mu^2 y = 0$ has the polynomial solution $T_n(x)$ when n is a nonnegative integer. Using the recurrence relation (8.24) find $T_5(x)$ and $T_6(x)$.

8. (a) Show, by substitution, that

$$y(x) = \cos(n \arccos(x))$$

is a solution of the Čebyšev differential equation

$$(1 - x^2)y'' - xy' + n^2y = 0$$

with n a nonnegative integer and $-1 < x < 1$.

(b) Show for $n = 0, 1, 2$ that the function

$$y(x) = \cos(n \arccos(x))$$

is a polynomial in x and that $T_n(x) = y(x)$. Show that

$$T_n(x) = \begin{cases} \cos(n \arccos(x)) & \text{for } |x| \leq 1 \\ \cosh(n \operatorname{arccos}(x)) & \text{for } |x| > 1. \end{cases}$$

8.4 The Method of Frobenius

We now proceed to give a result concerning solutions about regular singular points.

THEOREM 8.4.1

If x_0 is a regular singular point of the differential equation

$$f_0(x)\frac{d^2y}{dx^2} + f_1(x)\frac{dy}{dx} + f_2(x)y = 0,$$

then this differential equation has at least one nontrivial solution of the form

$$|x - x_0|^r \sum_{n=0}^{\infty} c_n(x - x_0)^n,$$

where r is a constant (that can be determined) and this solution is valid in some (deleted) interval $0 < |x - x_0| < R, R > 0$, about x_0.

Example 1: We have seen that $x = 0$ is a regular singular point of the differential equation

$$3x^2\frac{d^2y}{dx^2} - 4x\frac{dy}{dx} + (x + 2)y = 0;$$

the theorem assures us that there is (at least) one nontrivial solution of the form

$$|x|^r \sum_{n=0}^{\infty} c_n x^n$$

which is valid in the interval $0 < |x| < R$ about $x = 0$.

We know the differential equation

$$x^2(x-2)^2\frac{d^2y}{dx^2} + 2(x-2)\frac{dy}{dx} + (x+1)y = 0$$

has a regular singular point at $x = 2$ so that there is (at least) one solution of the form

$$|x-2|^r \sum_{n=0}^{\infty} c_n(x-2)^n,$$

which is defined for all values in $0 < |x-2| < R$.

Now, in the case $x = 0$, an irregular singular point of the differential equation, we cannot say that

$$|x|^r \sum_{n=0}^{\infty} c_n x^n$$

is a solution in any interval $0 < |x| < R$.

Question: How do we determine the coefficients c_n and the number r when x_0 is a regular singular point of the differential equation?

The method we consider is very similar to the preceding method of solution for ordinary points. The method we will now develop is often called the **method of Frobenius.**

Georg Frobenius was born in Charlottenburg, Germany in 1849. He received his doctorate at the University of Berlin (awarded with distinction) in 1870 under the supervision of Weierstrass. In 1874, after having taught at the secondary school level for several years, he was appointed to the University of Berlin as an extraordinary professor of mathematics. Frobenius made enormous contributions to group theory, particularly group representations, linear algebra, and analysis. In [20] Haubrich gives the following overview of Frobenius's work:

> The most striking aspect of his mathematical practice is his extraordinary skill at calculations. In fact, Frobenius tried to solve mathematical problems to a large extent by means of a calculative, algebraic approach. Even his analytical work was guided by algebraic and linear algebraic methods. For Frobenius, conceptual argumentation played a somewhat secondary role. Although he argued in a comparatively abstract setting, abstraction was not an end in itself. Its advantages to him seemed to lie primarily in the fact that it can lead to much greater clearness and precision.

Frobenius developed a method of determining the coefficients c_n and the number r when x_0 is a regular singular point. We will illustrate this method on the first differential equation we considered.

Example 8.8 Find a solution to the differential equation

$$3x^2 \frac{d^2y}{dx^2} - 4x\frac{dy}{dx} + (x+2)y = 0 \tag{8.29}$$

in some interval $0 < x < R$.

Solution Since $x = 0$ is a regular singular point, we know that

$$y = |x|^r \sum_{n=0}^{\infty} c_n x^n$$

is a solution on $0 < |x| < R$.

Thus for $0 < x < R$, we seek a solution of the form

$$y = x^r \sum_{n=0}^{\infty} c_n x^n$$

$$= \sum_{n=0}^{\infty} c_n x^{n+r}.$$

Now

$$\frac{dy}{dx} = \sum_{n=0}^{\infty} c_n (n+r) x^{n+r-1}$$

and

$$\frac{d^2y}{dx^2} = \sum_{n=0}^{\infty} c_n (n+r)(n+r-1) x^{n+r-2}$$

so that substitution into the differential equation (8.29) gives

$$3x^2 \frac{d^2y}{dx^2} - 4x\frac{dy}{dx} + (x+2)y$$

$$= 3x^2 \sum_{n=0}^{\infty} c_n (n+r)(n+r-1) x^{n+r-2}$$

$$-4x \sum_{n=0}^{\infty} c_n (n+r) x^{n+r-1} + (x+2) \sum_{n=0}^{\infty} c_n x^{n+r}$$

$$= 3 \sum_{n=0}^{\infty} c_n (n+r)(n+r-1) x^{n+r}$$

$$-4 \sum_{n=0}^{\infty} c_n (n+r) x^{n+r} + \sum_{n=0}^{\infty} c_n x^{n+r+1} + 2 \sum_{n=0}^{\infty} c_n x^{n+r}$$

$$= 0.$$

Simplification of this expression gives

$$\sum_{n-0}^{\infty}(3(n+r)(n+r-1)-4(n+r)+2)c_n x^{n+r} + \sum_{n=1}^{\infty} c_{n-1}x^{n+r} = 0$$

or equivalently

$$(3(r(r-1)-4r+2)c_0 x^r + \sum_{n=1}^{\infty}[(3(n+r)(n+r-1)-4(n+r)+2)c_n+c_{n-1}]x^{n+r} = 0.$$

Assuming $c_0 \neq 0$ and equating the coefficient of the lowest power of x to zero gives

$$(3r(r-1)-4r+2)c_0 = 0$$

or

$$3r(r-1)-4r+2 = 0.$$

This quadratic equation, known as the **indicial equation**, is

$$3r^2 - 7r + 2 = 0$$

and has roots

$$r_1 = \frac{1}{3} \quad \text{and} \quad r_2 = 2.$$

These roots are known as the **exponents** of the differential equation and are the only possible values for the constant r. Now equating the coefficients of higher powers of x to zero we obtain the recurrence relation

$$(3(n+r)(n+r-1)-4(n+r)+2)c_n + c_{n-1} = 0 \quad \text{for} \quad n \geq 1.$$

Now when $r_1 = \frac{1}{3}$ we obtain

$$\left(3\left(n+\frac{1}{3}\right)\left(n-\frac{2}{3}\right) - 4\left(n+\frac{1}{3}\right) + 2\right)c_n + c_{n-1} = 0 \quad \text{for} \quad n \geq 1$$

which simplifies to

$$n(3n-5)c_n + c_{n-1} = 0 \quad \text{for} \quad n \geq 1$$

or

$$c_n = \frac{-c_{n-1}}{n(3n-5)} \quad \text{for} \quad n \geq 1.$$

Thus

$$c_1 = \frac{c_0}{2},$$

$$c_2 = \frac{-c_1}{2} = \frac{-c_0}{4},$$

$$c_3 = \frac{-c_2}{12} = \frac{c_0}{48},$$

$$\vdots$$

so that

$$y_1 = c_0 x^{\frac{1}{3}} \left(1 + \frac{1}{2}x - \frac{1}{4}x^2 + \frac{1}{48}x^3 - \cdots \right)$$

is the solution corresponding to $r_1 = \frac{1}{3}$.

Similarly, $r_2 = 2$ gives the recurrence relation

$$(3(n+2)(n+1) - 4(n+2) + 2)c_n + c_{n-1} = 0 \quad \text{for} \quad n \geq 1$$

which simplifies as

$$n(3n+5)c_n + c_{n-1} = 0 \quad \text{for} \quad n \geq 1.$$

This gives

$$c_n = \frac{-c_{n-1}}{n(3n+5)} \quad \text{for} \quad n \geq 1$$

so that

$$c_1 = \frac{-c_0}{8}, \quad c_2 = \frac{-c_1}{22} = \frac{c_0}{176}, \quad c_3 = \frac{-c_2}{42} = \frac{-c_0}{7392}, \quad \cdots .$$

This gives

$$y_2 = c_0 x^2 \left(1 - \frac{1}{8}x + \frac{1}{176}x^2 - \frac{1}{7392}x^3 + \cdots \right)$$

as the solution corresponding to the root $r_2 = 2$.

These solutions, corresponding to $r_1 = \frac{1}{3}$ and $r_2 = 2$, respectively, are linearly independent, so the general solution of the differential equation (8.29) is

$$y = k_1 x^{\frac{1}{3}} \left(1 + \frac{1}{2}x - \frac{1}{4}x^2 + \frac{1}{48}x^3 - \cdots \right)$$

$$+ k_2 x^2 \left(1 - \frac{1}{8}x + \frac{1}{176}x^2 - \frac{1}{7392}x^3 + \cdots \right)$$

where k_1 and k_2 are arbitrary constants.

We now outline the method we presented in this example.

Outline of the Method of Frobenius

For x_0 a regular singular point of the differential equation

$$f_0(x)\frac{d^2y}{dx^2} + f_1(x)\frac{dy}{dx} + f_2(x)y = 0, \tag{8.30}$$

to find a solution valid in some interval $0 < |x - x_0| < R$,

1. Assume a solution of the form

$$y = (x - x_0)^r \sum_{n=0}^{\infty} c_n (x - x_0)^n$$

$$= \sum_{n=0}^{\infty} c_n (x - x_0)^{n+r}$$

where $c_0 \neq 0$.

2. Differentiate y term by term to obtain

$$\frac{dy}{dx} = \sum_{n=0}^{\infty} c_n (n + r)(x - x_0)^{n+r-1}$$

and

$$\frac{d^2 y}{dx^2} = \sum_{n=0}^{\infty} c_n (n + r)(n + r - 1)(x - x_0)^{n+r-2}$$

and substitute these expressions into the differential equation (8.30) for

$$y, \quad \frac{dy}{dx}, \quad \text{and} \quad \frac{d^2 y}{dx^2}.$$

3. Simplify the expression so that it is of the form

$$K_0(x - x_0)^{r+k} + K_1(x - x_0)^{r+k+1} + K_2(x - x_0)^{r+k+2} + \ldots = 0$$

where k is an integer and the coefficients $K_i, i = 0, 1, 2, \ldots$ are functions of r and some of the coefficients c_n.

4. To be valid for all x in $0 < |x - x_0| < R$,

$$K_0 = K_1 = K_2 = \ldots = 0.$$

5. Upon equating to zero the coefficient K_0 of the *lowest* power $r + k$ of $(x - x_0)$, we obtain a quadratic equation in r, called the **indicial equation** of the differential equation (8.30). The two roots are called the **exponents** of the differential equation, denoted by r_1 and r_2. Here r_1 and r_2 can be distinct real numbers, repeated real or complex. For convenience, r_1 is such that

$$\text{Re}(r_1) \geq \text{Re}(r_2).$$

6. Equate the remaining coefficients $K_1, K_2, \ldots$ to obtain a set of equations, involving r, which must be satisfied by the coefficients c_n in the series.

7. Substitute the root r_1 for r obtained in Step **6** and then determine the c_n which satisfy these conditions.

8. If $r_2 \neq r_1$, repeat Step **7** using r_2 instead of r_1.

9. The general solution is obtained as a linear combination of the solution(s) obtained in Steps **7** and **8**.

Example Use the method of Frobenius to find a solution of

$$xy'' + (1+x)y' - \frac{1}{16x}y = 0.$$

Solution

Rewrite this equation as

$$y'' + \frac{1+x}{x}y' - \frac{1}{16x^2}y = 0$$

so that

$$p(x) = \frac{1+x}{x} \quad \text{and} \quad q(x) = \frac{-1}{16x^2}.$$

Both fail to be analytic at $x = 0$, so $x = 0$ is a singular point. Now

$$xp(x) = 1 + x \quad \text{and} \quad x^2q(x) = \frac{-1}{16}$$

are both analytic at $x = 0$, thus $x = 0$ is a regular singular point. Thus, there is at least one solution of the form

$$y = \sum_{n=0}^{\infty} c_n(x-x_0)^{n+r} \quad \text{with} \quad c_0 \neq 0.$$

Differentiating gives

$$y' = \sum_{n=0}^{\infty} c_n(n+r)(x-x_0)^{n+r-1}$$

and

$$y'' = \sum_{n=0}^{\infty} c_n(n+r)(n+r-1)(x-x_0)^{n+r-2}$$

which yields upon substitution into the differential equation

$$x\sum_{n=0}^{\infty} c_n(n+r)(n+r-1)x^{n+r-2}$$

$$+ (1+x)\sum_{n=0}^{\infty} c_n(n+r)x^{n+r-1} - \frac{1}{16x}\sum_{n=0}^{\infty} c_n x^{n+r} = 0.$$

Rearrangement gives

$$\left[r(r-1) + r - \frac{1}{16}\right]c_0 x^{r-1} + \sum_{n=1}^{\infty} c_n(n+r)(n+r-1)x^{n+r-1}$$

$$+ \sum_{n=1}^{\infty} c_n(n+r)x^{n+r-1} + \sum_{n=0}^{\infty} c_n(n+r)x^{n+r}$$

$$- \sum_{n=1}^{\infty} \frac{1}{16}c_n x^{n+r-1} = 0,$$

and further simplification gives

$$\left[r(r-1)+r-\frac{1}{16}\right]c_0 x^{r-1}+\sum_{n=1}^{\infty}\left(\left[(n+r)(n+r-1)+(n+r)-\frac{1}{16}\right]c_n\right.$$

$$\left.+(n+r-1)c_{n-1}\right)x^{n+r-1}=0.$$

Equating the first coefficient to zero with $c_0 \neq 0$ gives the indicial equation

$$r(r-1)+r-\frac{1}{16}=0$$

which is

$$r^2-\frac{1}{16}=0.$$

This quadratic equation has roots $r_1 = \frac{1}{4}$ and $r_2 = -\frac{1}{4}$. Thus, the differential equation has two solutions

$$y_1 = \sum_{n=0}^{\infty} c_n x^{n+\frac{1}{4}} \quad \text{and} \quad y_2 = \sum_{n=0}^{\infty} c_n x^{n-\frac{1}{4}}.$$

Now, using these series, we can find the coefficients.

Beginning with the larger root $r_1 = \frac{1}{4}$ and equating the series coefficients to zero gives

$$\left[\left(n+\frac{1}{4}\right)\left(n-\frac{3}{4}\right)+\left(n+\frac{1}{4}\right)-\frac{1}{16}\right]c_n+\left(n-\frac{3}{4}\right)c_{n-1}=0$$

so that the coefficients c_n must satisfy the recurrence relation

$$c_n = \frac{(3-4n)c_{n-1}}{2(2n^2+n)} \quad \text{for} \quad n \geq 1.$$

Hence

$$c_1 = -\frac{1}{6}c_0, \; c_2 = -\frac{1}{4}c_1 = \frac{1}{24}c_0, \; c_3 = -\frac{1}{112}c_0, \; \cdots.$$

Using these gives

$$y_1 = c_0 x^{\frac{1}{4}}\left(1-\frac{1}{6}x+\frac{1}{24}x^2-\frac{1}{112}x^3+\dots\right).$$

Similarly, we use the root $r_2 = -\frac{1}{4}$ so that

$$\left[\left(n-\frac{1}{4}\right)\left(n-\frac{5}{4}\right)+\left(n-\frac{1}{4}\right)-\frac{1}{16}\right]c_n+\left(n-\frac{5}{4}\right)c_{n-1}=0.$$

This yields the recurrence relation

$$c_n = \frac{(5-4n)c_{n-1}}{2(2n^2-n)} \quad \text{for} \quad n \geq 1$$

that the coefficients c_n must satisfy. The values are

$$c_1 = \frac{1}{2}c_0, \quad c_2 = -\frac{1}{4}c_1 = -\frac{1}{8}c_0, \quad c_3 = \frac{7}{240}c_0, \cdots .$$

Using these coefficients gives the solution

$$y_2 = c_0 x^{-\frac{1}{4}} \left(1 + \frac{1}{2}x - \frac{1}{8}x^2 + \frac{7}{240}x^3 - \cdots \right).$$

Forming a linear combination of y_1 and y_2 gives a general solution of the differential equation as

$$y = C_1 y_1 + C_2 y_2$$
$$= C_1 x^{\frac{1}{4}} \left(1 - \frac{1}{6}x + \frac{1}{24}x^2 - \frac{1}{112}x^3 + \cdots \right)$$
$$+ C_2 x^{-\frac{1}{4}} \left(1 + \frac{1}{2}x - \frac{1}{8}x^2 + \frac{7}{240}x^3 - \cdots \right).$$

The previous calculations, while straightforward, can be quite tedious and there is a lot of room for algebraic error. You might imagine that it would be easier to use MATLAB, Maple, or Mathematica to find the coefficient of the assumed form of the series even with the extra complication of a regular singular point. It will be a bit more complicated as we will have to first deal with the indicial equation.

In the previous examples, we have found the indicial equation via direct substitution of the series solution into the differential equation. The method of Frobenius depends upon the roots of the indicial equation so it is of interest to obtain a general formula for this equation.

Proceeding here, we will assume for simplicity that $x = 0$ is a regular singular point of the differential equation

$$\frac{d^2 y}{dx^2} + p(x)\frac{dy}{dx} + q(x)y = 0. \tag{8.31}$$

Otherwise, if $x = x_0 \neq 0$ were a regular singular point, we could then write $X = x - x_0$. Then the functions

$$xp(x) \quad \text{and} \quad x^2 q(x)$$

are analytic at $x = 0$. Thus, both $xp(x)$ and $x^2 q(x)$ have convergent power series representations.

That is,

$$xp(x) = p_0 + p_1 x + p_2 x^2 + \cdots$$

and

$$x^2 q(x) = q_0 + q_1 x + q_2 x^2 + \cdots$$

or alternately

$$p(x) = \frac{p_0}{x} + p_1 + p_2 x + p_3 x^2 + \cdots$$

and

$$q(x) = \frac{q_0}{x^2} + \frac{q_1}{x} + q_2 + q_3 x + \cdots .$$

Now substitution of these series into the differential equation (8.31) gives

$$\left(\sum_{n=0}^{\infty} c_n (n+r)(n+r-1) x^{n+r-2} \right) + \left(\frac{p_0}{x} + p_1 + p_2 x + p_3 x^2 + \ldots \right)$$
$$\times \left(\sum_{n=0}^{\infty} c_n (n+r) x^{n+r-1} \right)$$
$$+ \left(\frac{q_0}{x^2} + \frac{q_1}{x} + q_2 + q_3 x + \ldots \right)$$
$$\times \left(\sum_{n=0}^{\infty} c_n x^{n+r} \right) = 0.$$

Expanding a little gives

$$\left(\sum_{n=0}^{\infty} c_n (n+r)(n+r-1) x^{n+r-2} \right)$$
$$+ \left(\sum_{n=0}^{\infty} c_n p_0 (n+r) x^{n+r-2} \right)$$
$$+ (p_1 + p_2 x + p_3 x^2 + \ldots) \left(\sum_{n=0}^{\infty} c_n (n+r) x^{n+r-1} \right)$$
$$+ \left(\sum_{n=0}^{\infty} c_n q_0 x^{n+r-2} \right)$$
$$+ \left(\frac{q_1}{x} + q_2 + q_3 x + \ldots \right) \left(\sum_{n=0}^{\infty} c_n x^{n+r} \right) = 0.$$

Thus, with $n = 0$, the coefficient of x^{r-2} is given as

$$-r c_0 + r^2 c_0 + r c_0 p_0 + c_0 q_0 = c_0 (r^2 + (p_0 - 1)r + q_0)$$
$$= c_0 (r(r-1) + p_0 r + q_0).$$

So for a differential equation of the form

$$\frac{d^2 y}{dx^2} + p(x) \frac{dy}{dx} + q(x) y = 0$$

with $x = 0$ a regular singular point, the indicial equation is given as

$$r(r-1) + p_0 r + q_0 = 0 \qquad (8.32)$$

or
$$r^2 + (p_0 - 1)r + q_0 = 0.$$

The values of r that satisfy this equation are called the *exponents of the differential equation* and are found from (8.32) using the quadratic formula. These exponents are

$$r_1 = \frac{1 - p_0 + \sqrt{1 - 2p_0 + p_0^2 - 4q_0}}{2} \quad \text{and} \quad r_2 = \frac{1 - p_0 - \sqrt{1 - 2p_0 + p_0^2 - 4q_0}}{2}$$

where
$$p_0 = \lim_{x \to 0} xp(x) \quad \text{and} \quad q_0 = \lim_{x \to 0} x^2 q(x).$$

Example Determine the indicial roots for the following differential equations.

(a) $y'' - \dfrac{1}{3x}y' + \dfrac{1}{3x^2}y = 0$ (b) $xy'' + y' - y = 0$.

Solution

(a) Here we note that $x = 0$ is a singular point of the differential equation which is regular since

$$xp(x) = -\frac{1}{3} \quad \text{and} \quad x^2 q(x) = \frac{1}{3}.$$

Now $p_0 = -\frac{1}{3}$ and $q_0 = \frac{1}{3}$ so that the indicial equation is

$$r(r - 1) + p_0 r + q_0 = r(r - 1) - \frac{1}{3}r + \frac{1}{3} = 0.$$

That is,
$$r^2 - \frac{4}{3}r + \frac{1}{3} = 0,$$

which has roots $r_1 = 1$ and $r_2 = \frac{1}{3}$.

(b) Rewriting this differential equation as

$$y'' + \frac{1}{x}y' - \frac{1}{x}y = 0$$

we see that $x = 0$ is a singular point since

$$p(x) = \frac{1}{x} \quad \text{and} \quad q(x) = \frac{-1}{x}$$

fail to be analytic at $x = 0$.

Now $xp(x) = 1$ and $x^2 q(x) = -x$ are both analytic at $x = 0$, so $x = 0$ is a regular singular point. Here

$$p_0 = \lim_{x \to 0} xp(x) = 1 \quad \text{and} \quad q_0 = \lim_{x \to 0} x^2 q(x) = 0$$

so the indicial equation is

$$r(r-1) + p_0 r + q_0 = r(r-1) + r = 0,$$

that is,

$$r^2 = 0.$$

So the roots here are $r_1 = r_2 = 0$.

Example 6: Determine the indicial roots for the following differential equations given in (a)–(b).

(a) $x^2 y'' + (\sin x)y' - (\cos x)y = 0$ (b) $x^2 y'' + \frac{1}{2}(x + \sin x)y' + y = 0$.

Solution

(a) Rewriting this differential equation as

$$y'' + \frac{\sin x}{x^2}y' - \frac{\cos x}{x^2}y = 0$$

gives that $x = 0$ is a singular point of the differential equation as

$$p(x) = \frac{\sin x}{x^2} \quad \text{and} \quad q(x) = \frac{-\cos x}{x^2}$$

fail to be analytic at $x = 0$.

Now

$$xp(x) = \frac{\sin x}{x} = \frac{1}{x}\left(x - \frac{x^3}{3!} + \frac{x^5}{5!} - \cdots\right)$$

$$= 1 - \frac{x^2}{3!} + \frac{x^4}{5!} - \cdots.$$

So $xp(x)$ is analytic at $x = 0$. Further $x^2 q(x) = -\cos x$ which is (clearly) also analytic at $x = 0$. Thus $x = 0$ is a regular singular point. Now

$$p_0 = \lim_{x \to 0} xp(x) = 1 \quad \text{and} \quad q_0 = \lim_{x \to 0} x^2 q(x) = -1,$$

which gives the indicial equation as

$$r(r-1) + p_0 r + q_0 = r(r-1) + r - 1 = 0$$

so

$$r^2 - 1 = 0.$$

The indicial roots for this differential equation are

$$r_1 = 1 \quad \text{and} \quad r_2 = -1.$$

(b) This differential equation can be written as

$$\frac{d^2y}{dx^2} + \frac{1}{2}\left(\frac{x + \sin x}{x^2}\right)\frac{dy}{dx} + \frac{1}{x^2}y = 0$$

so that $x = 0$ is a singular point as

$$p(x) = \frac{x + \sin x}{2x^2} \quad \text{and} \quad q(x) = \frac{1}{x^2}$$

are not analytic at $x = 0$.

To see that $x = 0$ is a regular singular point expand $xp(x)$ as

$$xp(x) = x\frac{(x + \sin x)}{2x^2} = \frac{x + \sin x}{2x} = \frac{1}{2} + \frac{\sin x}{2x}$$

$$= \frac{1}{2} + \frac{1}{2}\left(1 - \frac{x^2}{3!} + \frac{x^4}{5!} - \cdots\right)$$

so

$$p_0 = \lim_{x \to 0} xp(x) = 1.$$

Further, $x^2q(x) = 1$ which is analytic for all x. Now the indicial equation is

$$r(r-1) + p_0 r + q_0 = r(r-1) + r + 1 = 0$$

which is $r^2 + 1 = 0$ and has (complex!) roots

$$r_1 = i \quad \text{and} \quad r_2 = -i.$$

These examples give some of the possible situations that can arise when finding the indicial roots of the differential equation. The first example we considered has distinct real roots and applying the method of Frobenius is similar to the previous examples we have considered.

In the next example we consider a differential equation with distinct real roots that differ by an integer value. This particular case requires a bit of additional care.

Example 7: Find a general solution of

$$x^2\frac{d^2y}{dx^2} + (x^2 - 3x)\frac{dy}{dx} + 3y = 0.$$

Solution

In this differential equation

$$p(x) = \frac{x^2 - 3x}{x^2} \quad \text{and} \quad q(x) = \frac{3}{x^2}$$

which fail to be analytic at $x = 0$. However,

$$xp(x) = \frac{x^2 - 3x}{x} = x - 3$$

and

$$x^2 q(x) = 3,$$

which are both analytic at $x = 0$. Hence, $x = 0$ is a regular singular point.
Now

$$p_0 = \lim_{x \to 0} xp(x) = -3 \quad \text{and} \quad q_0 = \lim_{x \to 0} x^2 q(x) = 3$$

so the indicial equation is

$$r(r - 1) + p_0 r + q_0 = r(r - 1) - 3r + 3 = 0,$$

that is,

$$r^2 - 4r + 3 = 0.$$

The indicial roots are $r_1 = 3$ and $r_2 = 1$. These roots differ by a positive integer. We assume solutions of the form

$$y_1 = x^3 \sum_{n=0}^{\infty} c_n x^n \quad \text{and} \quad y_2 = x \sum_{n=0}^{\infty} c_n x^n.$$

Now

$$y_1' = \sum_{n=0}^{\infty} (n + 3) c_n x^{n+2}$$

and

$$y_1'' = \sum_{n=0}^{\infty} (n + 3)(n + 2) c_n x^{n+1}$$

so

$$x^2 \sum_{n=0}^{\infty} (n + 3)(n + 2) c_n x^{n+1} + (x^2 - 3x) \sum_{n=0}^{\infty} (n + 3) c_n x^{n+2}$$

$$+ 3 \sum_{n=0}^{\infty} c_n x^{n+3} = 0.$$

Simplifying and equating the higher powers of x to zero give

$$[(n + 3)(n + 2) - 3(n + 3) + 3] c_n + (n + 2) c_{n-1} = 0 \quad \text{for} \quad n \geq 1,$$

which is

$$n(n + 2) c_n + (n + 2) c_{n-1} = 0 \quad \text{for} \quad n \geq 1.$$

Thus, the recurrence relation that the coefficients c_n must satisfy is

$$c_n = -\frac{c_{n-1}}{n} \quad \text{for} \quad n \geq 1$$

so

$$c_1 = -c_0, \quad c_2 = -\frac{c_1}{2} = \frac{c_0}{2!}, \quad c_3 = -\frac{c_2}{3} = -\frac{c_0}{3!}.$$

Indeed, in general,

$$c_n = \frac{(-1)^n c_0}{n!}$$

so

$$y_1 = c_0 x^3 \left[1 - x + \frac{x^2}{2!} - \frac{x^3}{3!} + \ldots + \frac{(-1)^n x^n}{n!} + \ldots \right]$$

$$= c_0 x^3 e^{-x}.$$

For the root $r_2 = 1$, we obtain the recurrence formula in the same manner as above. Thus, for y_2 the coefficients must satisfy

$$n(n-2)c_n + nc_{n-1} = 0 \quad \text{for} \quad n \geq 1.$$

Now for $n \neq 2$, we have

$$c_n = -\frac{c_{n-1}}{n-2} \quad \text{for} \quad n \geq 1, n \neq 2.$$

So, for $n = 1$

$$c_1 = c_0$$

and for $n = 2$,

$$2 \cdot 0c_2 + 2c_1 = 0,$$

which is $c_1 = 0$. However, this implies $c_0 = 0$, a contradiction of our assumption. This contradiction implies that there is no solution y_2 of the form $y_2 = \sum_{n=0}^{\infty} c_n x^{n-1}$.

This can be emphasized further by considering the coefficients c_n for $n \geq 3$; from the condition $0 \cdot c_2 + 2c_1 = 0$, c_2 must be arbitrary, so for $n \geq 3$

$$c_3 = -c_2, \quad c_4 = \frac{-c_3}{2} = \frac{c_2}{2!}, \quad c_5 = \frac{-c_4}{3} = \frac{c_2}{3!} \ldots$$

so that using these values of the coefficients gives

$$y_2 = c_2 x \left[x^2 - x^3 + \frac{x^4}{2!} - \frac{x^5}{3!} + \ldots + \frac{(-1)^n x^{n+2}}{n!} + \ldots \right] \quad (8.33)$$

$$= c_2 x^3 \left[1 - x + \frac{x^2}{2!} - \frac{x^3}{3!} + \ldots + (-1)^n \frac{x^n}{n!} + \ldots \right]$$

$$= c_2 x^3 e^{-x}$$

which is the solution obtained for y_1.

This really should not be a surprise as the indicial roots r_1 and r_2 in this example differ by 2. [This value of 2 appears as the factor of x^2 in the expansion of y_2 in equation (8.33).]

So, in the case of the indicial roots differing by an integer, only one solution will arise from the method of Frobenius. To find another (nearly independent) solution of this differential equation, we employ the method of reduction.

Here, let $y = f(x)v$ where $f(x)$ is a known solution of the differential equation. In this case for y_1 we have, with $c_0 = 1$,

$$y = x^3 e^{-x} v.$$

From this, we obtain

$$\frac{dy}{dx} = x^3 e^{-x} \frac{dv}{dx} + (3x^2 e^{-x} - x^3 e^{-x})v$$

and

$$\frac{d^2 y}{dx^2} = x^3 e^{-x} \frac{d^2 v}{dx^2} + 2(3x^2 e^{-x} - x^3 e^{-x})\frac{dv}{dx}$$

$$+(x^3 e^{-x} - 6x^2 e^{-x} + 6xe^{-x})v.$$

Substituting into the differential equation and simplifying give

$$x\frac{d^2 v}{dx^2} + (3 - x)\frac{dv}{dx} = 0.$$

Letting

$$w = \frac{dv}{dx}$$

reduces the order and we obtain

$$x\frac{dw}{dx} + (3 - x)w = 0.$$

This first-order linear differential equation has solution (check it!)

$$w = x^{-3} e^x.$$

Thus

$$v = \int w\, dx = \int x^{-3} e^x\, dx.$$

Hence

$$y = y_2 = x^3 e^{-x} \int x^{-3} e^x\, dx,$$

which is a linearly independent solution from y_1.

Now it is interesting to note the series form of y_2; expanding e^x in a power series gives

$$y_2 = x^3 e^{-x} \int x^{-3} \left(1 + x + \frac{x^2}{2!} + \frac{x^3}{3!} + \frac{x^4}{4!} + \cdots \right) dx$$

$$= x^3 e^{-x} \int \left(x^{-3} + x^{-2} + \frac{1}{2}x^{-1} + \frac{1}{6} + \frac{1}{24}x + \cdots \right) dx$$

$$= x^3 e^{-x} \left[\frac{-1}{2x^2} - \frac{1}{x} + \frac{1}{2}\ln x + \frac{1}{6}x + \frac{1}{48}x^2 + \cdots \right]$$

where we integrated the series, term by term.

Now writing e^{-x} in series form gives

$$y_2 = \left(x^3 - x^4 + \frac{x^5}{2} - \frac{x^6}{6} + \cdots \right) \left(-\frac{1}{2x^2} - \frac{1}{x} + \frac{1}{6}x + \frac{1}{48}x^2 + \cdots \right)$$

$$+ \frac{1}{2}x^3 e^{-x} \ln x$$

$$= \left(-\frac{1}{2}x - \frac{1}{2}x^2 + \frac{3}{4}x^3 - \frac{1}{4}x^4 + \cdots \right) + \frac{1}{2}x^3 e^{-x} \ln x.$$

Note the logarithm here is of the form $cy_1 \ln x$.

The general solution of the differential equation is thus

$$y = C_1 y_1(x) + C_2 y_2(x)$$

where C_1 and C_2 are arbitrary constants.

In this last example, it was fortunate that we were able to express the solution y in closed form. This simplified the computations involved in finding the second solution y_2. The method of reduction still may be applied even if we cannot find a closed form for the first solution y_1. We would carry out the computations term by term in the series for y_1. These computations can be quite spectacular and are often very complicated.

We will now consider (briefly) the case of complex roots of the indicial equation.

Suppose that the indicial equation

$$r(r-1) + p_0 r + q_0 = 0$$

had complex roots r_1 and r_2. Here r_1 and r_2 occur as complex conjugates so that if $r_1 = a + bi$ then $r_2 = a - bi$ where a, b are real, $b \neq 0$. Now $r_1 - r_2 = 2bi$, so the difference of these roots is not an integer. Therefore, the solution will not contain a logarithmic term as we had previously. Thus, we assume a solution of the form

$$y_1 = x^{r_1} \sum_{n=0}^{\infty} c_n x^n.$$

But how do we treat a complex exponent $x^{r_1} = x^{a+bi}$?
   Recalling Euler's formula

$$e^{ix} = \cos x + i \sin x$$

gives

$$x^{a+bi} = x^a x^{bi} = x^a e^{(b \ln x)i}$$
$$= x^a [\cos(b \ln x) + i \sin(b \ln x)] \quad \text{for } x > 0.$$

Thus, the series for y_1 will have coefficients c_n that are complex. Writing

$$c_n = a_n + ib_n \quad \text{for } n \geq 0$$

formally gives the series

$$y_1(x) = x^{r_1} \sum_{n=0}^{\infty} (a_n + ib_n) x^n$$

$$= x^a [\cos(b \ln x) + i \sin(b \ln x)] \left[\sum_{n=0}^{\infty} a_n x^n + i \sum_{n=0}^{\infty} b_n x^n \right]$$

$$= x^a \left[\cos(b \ln x) \sum_{n=0}^{\infty} a_n x^n - \sin(b \ln x) \sum_{n=0}^{\infty} b_n x^n \right]$$

$$+ ix^a \left[\cos(b \ln x) \sum_{n=0}^{\infty} bn x^n + \sin(b \ln x) \sum_{n=0}^{\infty} a_n x^n \right].$$

We note here that both the real and imaginary parts of $y_1(x)$ are real functions. From this analysis, we see that differential equations with complex indicial roots have solutions that are complex valued. The solution $y_2(x)$ corresponding to the complex conjugate root r_2 is the conjugate of $y_1(x)$. It can be shown that $y_1(x)$ and $y_2(x)$ are linearly independent. We will close this discussion by considering an example of this case.

Example 8: Consider the differential equation

$$x^2 y'' + xy' + (1 - x)y = 0.$$

Rewriting this equation as

$$y'' + \frac{1}{x} y' + \frac{(1 - x)}{x^2} y = 0$$

gives that $x = 0$ is a singular point of the differential equation as

$$p(x) = \frac{1}{x} \quad \text{and} \quad q(x) = \frac{1-x}{x^2},$$

both which fail to be analytic at $x = 0$. However, $xp(x)$ and $xq(x)$ are analytic at $x = 0$, so $x = 0$ is a regular singular point.

Now

$$p_0 = \lim_{x \to 0} xp(x) = 1 \quad \text{and} \quad q_0 = \lim_{x^2 \to 0} xq(x) = 1,$$

so the indicial equation is

$$r(r-1) + p_0 r + q_0 = r(r-1) + r + 1$$
$$= r^2 + 1 = 0.$$

The indicial roots or exponents for this differential equation are

$$r_1 = i \quad \text{and} \quad r_2 = -i.$$

Thus, if we assume a solution of the form

$$y = x^r \sum_{n=0}^{\infty} c_n x^n$$

we have

$$y_1 = x^i \sum_{n=0}^{\infty} c_n x^n \quad \text{and} \quad y_2 = x^{-i} \sum_{n=0}^{\infty} c_n x^n.$$

Formally treating this imaginary exponent of x as one would treat a real exponent, we have upon differentiation of y_1 that

$$y_1' = \sum_{n=0}^{\infty} (n+i)c_n x^{n+i-1}$$

and

$$y_1'' = \sum_{n=0}^{\infty} (n+i)(n+i-1)c_n x^{n+i-2},$$

so substituting into the differential equation gives

$$x^2 \sum_{n=0}^{\infty} (n+i)(n+i-1)c_n x^{n+i-2} + x \sum_{n=0}^{\infty} (n+i)c_n x^{n+i-1} + (1-x)\sum_{n=0}^{\infty} c_n x^{n+i} = 0.$$

Equating coefficients to zero here gives the recurrence relation for the c_n's as

$$c_{n+1} = \frac{1}{(n+1+i)^2 + 1} c_n$$
$$= \frac{(n+1) - 2i}{(n+1)[(n+1)^2 + 4]} c_n \quad \text{for } n \geq 0.$$

Clearly, this recurrence relation will generate complex-valued coefficients in terms of c_0. To simplify our calculations we will take $c_0 = 1 + i$, solely as an illustration. Thus,

$$c_1 = \frac{1 - 2i}{5} c_0 = \left(\frac{1 - 2i}{5}\right)(1 + i) = \frac{3 - i}{2},$$

$$c_2 = \frac{1 - i}{8} c_1 = \left(\frac{1 - i}{8}\right)\left(\frac{3 - i}{2}\right) = \frac{1 - 2i}{20},$$

$$c_3 = \frac{3 - 2i}{39} c_2 = \left(\frac{3 - 2i}{39}\right)\left(\frac{1 - 2i}{20}\right) = \frac{-(1 + 8i)}{780},$$

$$\vdots$$

so that taking the real and imaginary parts we have

$$a_0 = 1, \ a_1 = \frac{3}{5}, \ a_2 = \frac{1}{20}, \ a_3 = \frac{-1}{780}, \ \cdots$$

and

$$b_0 = 1, \ b_1 = \frac{-1}{5}, \ b_2 = \frac{-1}{10}, \ b_3 = \frac{-2}{195}, \ \cdots .$$

Using these we obtain the solution

$$y_1(x) = \cos(\ln x)\left[1 + \frac{3}{5}x + \frac{1}{20}x^2 + \cdots\right] - \sin(\ln x)\left[1 - \frac{1}{5}x - \frac{1}{10}x^2 + \cdots\right]$$

$$+ \ i\cos(\ln x)\left[1 - \frac{1}{5}x - \frac{1}{10}x^2 + \cdots\right] + \sin(\ln x)\left[1 + \frac{3}{5}x + \frac{1}{20}x^2 + \cdots\right].$$

The solution $y_2(x)$ is found from the conjugate of $y_1(x)$.

Problems

For each of the following differential equations in Problems **1–12**, apply the method of Frobenius to obtain the solution.

1. $xy'' - y = 0$
2. $(x^2 + x)y'' - 2y' - 2y = 0$
3. $2x^2 y'' + 3xy' - (x + 1)y = 0$
4. $x^2 y'' - xy' + \left(2x^2 + \frac{5}{9}\right)y = 0$
5. $x^2 y'' + x(x + 1)y' - y = 0$
6. $(3x^2 + x^3)y'' - xy' + y = 0$
7. $2x^2 y'' + xy' + (x^2 - 1)y = 0$
8. $(x^2 + x^3)y'' + (x^2 - x)y' + y = 0$
9. $xy'' + y' + y = 0$
10. $xy'' - xy' - y = 0$
11. $(x^2 + x^3)y'' - (x^2 + x)y' + y = 0$
12. $x^2 y'' + (x^2 - x)y' + 2y = 0.$

8.5 Bessel Functions

We will now develop the theory of a type of function that occurs in connection with many problems in applied mathematics, physics, and engineering. Its diverse applications range from electric fields to heat conduction to application in abstract probability theory.

The differential equation

$$x^2 \frac{d^2 y}{dx^2} + x \frac{dy}{dx} + (x^2 - p^2)y = 0, \tag{8.34}$$

where p is a parameter, is called **Bessel's equation of order p.** Any solution of Bessel's equation of order p is called a **Bessel function of order p.** The differential equation (8.34) and its solutions are named in honor of Friedrich Wilhelm Bessel (1784–1846) who first studied some of their remarkable properties. [It should be noted here that the solutions of (8.34) are called Bessel functions, not the possessive Bessel's functions. This is common practice with Bessel functions but is not typical of mathematics in general. For example, Green's functions (possessive form) are solutions to another type of differential equation.]

Bessel Functions of Order Zero

If $p = 0$, Equation (8.34) becomes

$$x \frac{d^2 y}{dx^2} + \frac{dy}{dx} + xy = 0, \tag{8.35}$$

which is called **Bessel's equation of order zero.** We will first find solutions of this equation in an interval $0 < x < R$.

Noting that $x = 0$ is a regular singular point of (8.35), we assume a solution of the form

$$y = \sum_{n=0}^{\infty} c_n x^{n+r} \tag{8.36}$$

with $c_0 \neq 0$. Differentiating (8.36) twice and substituting into (8.35) gives

$$\sum_{n=0}^{\infty} (n+r)(n+r-1)c_n x^{n+r-1} + \sum_{n=0}^{\infty} (n+r)c_n x^{n+r-1} + \sum_{n=0}^{\infty} c_n x^{n+r+1} = 0.$$

Simplifying this expression, we obtain

$$\sum_{n=0}^{\infty} (n+r)^2 c_n x^{n+r-1} + \sum_{n=2}^{\infty} c_{n-2} x^{n+r-1} = 0,$$

which is

$$r^2 c_0 x^{r-1} + (1+r)^2 c_1 x^r + \sum_{n=2}^{\infty} [(n+r)^2 c_n + c_{n-2}] x^{n+r-1} = 0. \qquad (8.37)$$

Thus, upon equating to zero the coefficient of the lowest power of x in (8.37) gives the indicial equation

$$r^2 = 0,$$

which has (clearly) two equal roots $r_1 = r_2 = 0$. This gives, by equating to zero the coefficients of the higher powers of x in (8.37),

$$(1+r)^2 c_1 = 0, \qquad (8.38)$$

and

$$(n+r)^2 c_n + c_{n-2} = 0 \quad \text{for } n \geq 2. \qquad (8.39)$$

Taking $r = 0$ in (8.38) gives $c_1 = 0$ and, again letting $r = 0$ in (8.39), we have the recurrence

$$n^2 c_n + c_{n-2} = 0 \quad \text{for } n \geq 2,$$

or more usefully

$$c_n = -\frac{c_{n-2}}{n^2} \quad \text{for } n \geq 2.$$

Using this expression, we obtain successively

$$c_2 = \frac{-c_0}{2^2}, \quad c_3 = \frac{-c_1}{3^2} = 0, \quad c_4 = \frac{-c_2}{4^2} = \frac{c_0}{2^2 \cdot 4^2}, \quad c_5 = \frac{-c_3}{5^2} = 0, \cdots$$

so that we see all the odd-termed coefficients are zero and the general form for the even-termed coefficients is

$$c_{2n} = \frac{(-1)^n c_0}{2^2 4^2 6^2 \cdots (2n)^2} = \frac{(-1)^n c_0}{(n!)^2 2^{2n}}, \quad \text{for } n \geq 1.$$

Thus, when $r = 0$ in (8.36) the solution of Bessel's equation (8.34) is

$$y(x) = c_0 \sum_{n=0}^{\infty} \frac{(-1)^n}{(n!)^2} \left(\frac{x}{2}\right)^{2n}.$$

Setting the arbitrary constant $c_0 = 1$, we obtain the particular solution of Bessel's equation (8.34),

$$y(x) = \sum_{n=0}^{\infty} \frac{(-1)^n}{(n!)^2} \left(\frac{x}{2}\right)^{2n}.$$

This particular situation defines a function, denoted by J_0, and is called the **Bessel function of the first kind of order zero.** Thus,

$$J_0(x) = \sum_{n=0}^{\infty} \frac{(-1)^n}{(n!)^2} \left(\frac{x}{2}\right)^{2n} \qquad (8.40)$$

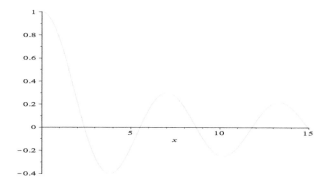

FIGURE 8.4: The Bessel function $J_0(x)$.

is a particular solution of (8.35). A graph of $J_0(x)$ is shown in Figure 8.4; note that $J_0(x)$ has a damped oscillatory behavior.

Since the indicial equation had two roots that were equal, we know that a solution of (8.35) that is linearly independent of J_0 is of the form

$$y = x \sum_{n=0}^{\infty} c_n^* x^n + J_0(x) \ln x$$

for $0 < x < R$ and constants c_n^*. Using the method of reduction of order, we see that this linearly independent solution y_2 is given by

$$y_2(x) = J_0(x) \int \frac{e^{-\int \frac{1}{x} dx}}{[J_0(x)]^2} dx = J_0(x) \int \frac{dx}{x[J_0(x)]^2}.$$

Now

$$
\begin{aligned}
[J_0(x)]^2 &= \left[1 - \frac{x^2}{4} + \frac{x^4}{64} - \frac{x^6}{2304} + \cdots\right]\left[1 - \frac{x^2}{4} + \frac{x^4}{64} - \frac{x^6}{2304} + \cdots\right] \\
&= 1 - \frac{x^2}{2} + \frac{3x^4}{32} - \frac{5x^6}{576} + \cdots
\end{aligned}
$$

so

$$\frac{1}{[J_0(x)]^2} = 1 + \frac{x^2}{2} + \frac{5x^4}{32} + \frac{23x^6}{576} + \cdots .$$

This gives

$$
\begin{aligned}
y_2(x) &= J_0(x) \int \frac{dx}{x[J_0(x)]^2} \\
&= J_0(x) \int \frac{1}{x}\left[1 + \frac{x^2}{2} + \frac{5x^4}{32} + \frac{23x^6}{576} + \cdots\right] dx \\
&= J_0(x) \int \left[\frac{1}{x} + \frac{x}{2} + \frac{5x^3}{32} + \frac{23x^5}{576} + \cdots\right] dx
\end{aligned}
$$

and simplifying further gives

$$y_2(x) = J_0(x)\left[\ln x + \frac{x^2}{4} + \frac{5x^4}{128} + \frac{23x^6}{3456} + \cdots\right]$$

$$= J_0(x)\ln x + J_0(x)\left[\frac{x^2}{4} + \frac{5x^4}{128} + \frac{23x^6}{3456} + \cdots\right]$$

$$= J_0(x)\ln x + \left[1 - \frac{x^2}{4} + \frac{x^4}{64} - \frac{x^6}{2304} + \cdots\right]\left[\frac{x^2}{4} + \frac{5x^4}{128} + \frac{23x^6}{3456} + \cdots\right]$$

$$= J_0(x)\ln x + \frac{x^2}{4} - \frac{3x^4}{128} + \frac{11x^6}{13824} + \cdots,$$

so we have found a form of the second solution, but in this form of the expansion we will have trouble finding the general coefficient c_{2n}^*.

Noting that

$$(-1)^2\frac{1}{2^2(1!)^2}(1) = \frac{1}{2^2} = \frac{1}{4},$$

$$(-1)^3\frac{1}{2^4(2!)^2}\left(1 + \frac{1}{2}\right) = -\frac{3}{2^42^2 \cdot 2} = -\frac{3}{128},$$

$$(-1)^4\frac{1}{2^6(3!)^2}\left(1 + \frac{1}{2} + \frac{1}{3}\right) = \frac{11}{2^62^2 \cdot 6} = \frac{11}{13824}.$$

It seems that, in general,

$$c_{2n}^* = \frac{(-1)^{n+1}}{2^{2n}(n!)^2}\sum_{k=1}^{n}\frac{1}{k} \quad \text{for } n \geq 1.$$

This is indeed the case, as can be shown (cf. Watson [51]). Using this fact, we have

$$y_2(x) = J_0(x)\ln x + \sum_{n=1}^{\infty}\frac{(-1)^{n+1}x^{2n}}{2^{2n}(n!)^2}\sum_{k=1}^{n}\frac{1}{k}. \tag{8.41}$$

Now $J_0(x)$ and $y_2(x)$ are linearly independent so that the general solution of (8.35) can be written as a linear combination of $J_0(x)$ and $y_2(x)$. However, it is more customary in the theory of Bessel functions to write a "special" linear combination of $J_0(x)$ and $y_2(x)$. This combination is defined as

$$\frac{2}{\pi}[y_2(x) - (\gamma - \ln 2)J_0(x)]$$

where γ is the number

$$\gamma = \lim_{n\to\infty}\left(1 + \frac{1}{2} + \frac{1}{3} + \cdots + \frac{1}{n} - \ln n\right)$$

$$\approx 0.5772,$$

which is known as **Euler's constant.**

This second solution of (8.35) is the function

$$Y_0(x) = \frac{2}{\pi}\left[J_0(x)\ln x + \sum_{n=1}^{\infty} \frac{(-1)^{n+1}x^{2n}}{2^{2n}(n!)^2}\sum_{k=1}^{n}\frac{1}{k} + (\gamma - \ln 2)J_0(x)\right]$$

or

$$Y_0(x) = \frac{2}{\pi}\left[\left(\ln\frac{x}{2} + \gamma\right)J_0(x) + \sum_{n=1}^{\infty}\frac{(-1)^{n+1}x^{2n}}{2^{2n}(n!)^2}\sum_{x=1}^{n}\frac{1}{k}\right].\tag{8.42}$$

The function $Y_0(x)$ is called the **Bessel function of the second kind of order zero.** The expression (8.42) is known as *Weber's form* of $Y_0(x)$. So, taking $Y_0(x)$ as the second solution of (8.35), the general solution of (8.35) is the linear combination

$$y = C_1 J_0(x) + C_2 Y_0(x) \quad \text{for } 0 < x < R,$$

where C_1 and C_2 are arbitrary constants, with $J_0(x)$ defined by (8.40) and $Y_0(x)$ defined by (8.42).

A graph of $Y_0(x)$ is shown in Figure 8.5, and both $J_0(x)$ and $Y_0(x)$ are shown on the same graph in Figure 8.6. Note that the zeros of $J_0(x)$ separate the zeros of $Y_0(x)$.

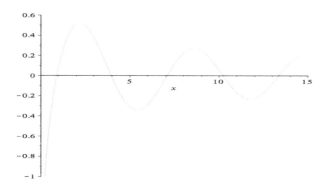

FIGURE 8.5: The Bessel function $Y_0(x)$.

Bessel Functions of Order p

We will now consider solutions of Bessel's equation

$$x^2\frac{d^2y}{dx^2} + x\frac{dy}{dx} + (x^2 - p^2)y = 0\tag{8.43}$$

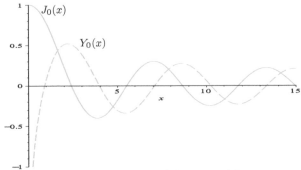

FIGURE 8.6: The Bessel functions $J_0(x)$ and $Y_0(x)$.

for real $p > 0$ and valid for $0 < x < R$. Studying Equation (8.43), we see that $x = 0$ is a regular singular point, so we can assume a solution

$$y = \sum_{n=0}^{\infty} c_n x^{n+r} \tag{8.44}$$

valid for $0 < x < R$ with $c_0 \neq 0$. Proceeding as we have done previously by differentiating (8.44) and substituting into (8.43) and simplifying, we obtain

$$(r^2 - p^2)c_0 x^r + [(r+1)^2 - p^2]c_1 x^{r+1} + \sum_{n=2}^{\infty}[[(n+r)^2 - p^2]c_n + c_{n-2}]x^{n+r} = 0. \tag{8.45}$$

Equating the coefficients of (8.45) to zero gives

$$r^2 - p^2 = 0, \quad \text{since } c_0 \neq 0 \tag{8.46}$$

$$[(r+1)^2 - p^2]c_1 = 0 \tag{8.47}$$

and

$$[(n+r)^2 - p^2]c_n + c_{n-2} = 0 \quad \text{for } n \geq 2. \tag{8.48}$$

Equation (8.46) is, of course, the indicial equation for the differential equation (8.43). The indicial equation has roots $r_1 = p > 0$ and $r_2 = -p$. Now if $r_1 - r_2 = 2p > 0$ is not a positive integer, then we know from the previous section that the differential equation (8.43) has two linearly independent solutions of the form (8.44). On the other hand, if $r_1 - r_2 = 2p$ is a positive integer, we are only certain of a solution in the form (8.44) to exist corresponding to the larger root $r_1 = p$. This is the solution we shall now proceed to obtain.

Thus, letting $r = r_1 = p$ in (8.47) gives

$$(2p+1)c_1 = 0,$$

but since $p > 0$, we have $c_1 = 0$. Letting $r = r_1 = p$ in (8.48) gives the recurrence relation

$$c_n = -\frac{c_{n-2}}{n(n+2p)}, \quad n \geq 2. \tag{8.49}$$

Now since $c_1 = 0$, (8.49) gives that all odd coefficients are zero; further we find that the even coefficients are given by

$$c_{2n} = \frac{(-1)^n c_0}{[2 \cdot 4 \dots (2n)][(2 + 2p)(4 + 2p) \dots (2n + 2p)]}$$
$$= \frac{(-1)^n c_0}{2^{2n} n! [(1 + p)(2 + p) \dots (n + p)]} \quad \text{for } n \geq 1.$$

So, the solution of the differential equation (8.43) corresponding to the larger root p is given by

$$y_1(x) = c_0 \sum_{n=0}^{\infty} \frac{(-1)^n x^{2n+p}}{2^{2n} n! [(1 + p)(2 + p) \dots (n + p)]}. \tag{8.50}$$

If p is a positive integer then (8.50) can be written as

$$y_1(x) = c_0 2^p p! \sum_{n=0}^{\infty} \frac{(-1)^n}{n!(n + p)!} \left(\frac{x}{2}\right)^{2n+p}. \tag{8.51}$$

What if p is not a positive integer? In this case, to express $y_1(x)$ in a form similar to (8.51) we need to introduce a function that generalizes the notion of the factorial.

For $\alpha > 0$ the **gamma function** is defined as

$$\Gamma(\alpha) = \int_0^{\infty} x^{\alpha-1} e^{-x} \, dx \tag{8.52}$$

which is a convergent improper integral for each value of $\alpha > 0$. Integrating (8.52) by parts one time gives the recurrence relation for $\alpha > 0$

$$\Gamma(\alpha) = (\alpha - 1)\Gamma(\alpha - 1), \tag{8.53}$$

so that if α were a positive integer, repeatedly applying (8.53) yields

$$\alpha! = \Gamma(\alpha + 1). \tag{8.54}$$

It is in this sense that if $\alpha > 0$ but not an integer, we use (8.54) to *define* $\alpha!$

Some of these properties may seem surprising. The gamma function, as do the Bessel functions, all belong to a broad class of functions that mathematicians, physicists, and engineers term **special functions.** Special functions are functions, usually defined in terms of a convergent power series or integral, that play a special role in the solution of some problems of practical importance. The gamma and Bessel functions we are studying in this section are perhaps new to you and their properties may seem surprising and unfamiliar. However, the class of special functions contains very familiar functions as well. One only needs to remember that the functions e^x, $\sin x$, and $\cos x$ are all *defined* as convergent power series [although this is *not* how you probably first learned of them], to realize that the "strangeness" of the properties of the gamma and Bessel functions is just an artifact of their newness to your collection of mathematical facts.

Special functions have been extensively studied. Indeed there are some excellent texts that develop the theory and relations of these and other special functions.

Returning now to our discussion of the gamma function, we see that so far, we have defined $\Gamma(\alpha)$ for $\alpha > 0$. It can be shown that

$$\Gamma\left(\frac{1}{2}\right) = \sqrt{\pi}$$

so that using (8.35) we can compute, for instance, $\Gamma(\frac{5}{2})$, that is,

$$\Gamma\left(\frac{5}{2}\right) = \frac{3}{2}\Gamma\left(\frac{3}{2}\right)$$

$$= \left(\frac{3}{2}\right)\left(\frac{1}{2}\right)\Gamma\left(\frac{1}{2}\right)$$

$$= \left(\frac{3}{4}\right)(\sqrt{\pi}) = \frac{3\sqrt{\pi}}{4} \approx 1.3293.$$

In this way, we could say $\left(\frac{3}{2}\right)! = \frac{3\sqrt{\pi}}{4} \approx 1.3293$. For other positive values of α, $\Gamma(\alpha)$ may need to be calculated numerically using (8.52) and a numerical integration routine.

For values of $\alpha < 0$, the integral (8.52) diverges so that $\Gamma(\alpha)$ is not defined for $\alpha < 0$. However, we can extend the definition of $\Gamma(\alpha)$ to $\alpha < 0$ by *demanding* that the recurrence relation (8.53) is valid for all values of α. In this way, $\Gamma(\alpha)$ becomes defined for every noninteger negative value of α; a graph of $\Gamma(x)$ is shown in Figure 8.7.

We digressed to the gamma function for a purpose. We are interested in the solution $y_1(x)$ of the differential equation (8.43) when p is not a positive integer. In this case, if we use the recurrence relation (8.53) repeatedly, we obtain

$$\Gamma(n + p + 1) = (n + p)(n + p - 1) \ldots (p + 1)\Gamma(p + 1).$$

So, for p not a positive integer, the solution given by (8.50) becomes

$$y_1(x) = c_0 \Gamma(p + 1) \sum_{n=0}^{\infty} \frac{(-1)^n x^{2n+p}}{2^{2n} n! \Gamma(n + p + 1)}$$

$$= c_0 2^p \Gamma(p + 1) \sum_{n=0}^{\infty} \frac{(-1)^n}{n! \Gamma(n + p + 1)} \left(\frac{x}{2}\right)^{2n+p}. \qquad (8.55)$$

Note that (8.55) reduces to (8.51) when $p = 0$.

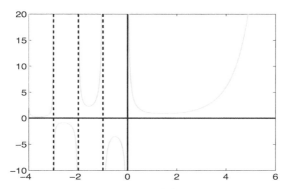

FIGURE 8.7: The gamma function $\Gamma(x)$.

If we let c_0 in (8.55) be given as

$$c_0 = \frac{1}{2^p \Gamma(p+1)}$$

we then obtain the particular solution of (8.34) known as the **Bessel function of the first kind of order** p. This function, often denoted J_p, is given as

$$J_p(x) = \sum_{n=0}^{\infty} \frac{(-1)^n}{n!\Gamma(n+p+1)} \left(\frac{x}{2}\right)^{2n+p}. \tag{8.56}$$

Taking $p = 0$ in (8.56), we see that (8.56) reduces to the Bessel function of the first kind of order zero given by (8.40). Now if $p = 1$, then Bessel's equation (8.34) becomes

$$x^2 \frac{d^2 y}{dx^2} + x \frac{dy}{dx} + (x^2 - 1) = 0, \tag{8.57}$$

which is Bessel's equation of order 1. Letting $p = 1$ in (8.56) gives the **Bessel function of the first kind of order one**, denoted J_1 and given as

$$J_1(x) = \sum_{n=0}^{\infty} \frac{(-1)^n}{n!(n+1)!} \left(\frac{x}{2}\right)^{2n+1}. \tag{8.58}$$

A graph of $J_1(x)$ is shown in Figure 8.8. Note the damped oscillatory behavior. Graphing both $J_0(x)$ and $J_1(x)$, as shown in Figure 8.9, shows that the positive roots of J_0 and J_1 separate each other.

This is in fact true for the function J_p, $p \geq 0$. It can be shown (c.f. Watson [51]) that J_p has a damped oscillatory behavior as $x \to \infty$ and that the positive roots of J_p and J_{p+1} separate each other.

We have obtained a solution to Bessel's equation (8.34) for every $p \geq 0$. We now find another linearly independent solution of (8.34). We have already given such a solution in the case $p = 0$. This is $Y_0(x)$ defined by (8.42). For

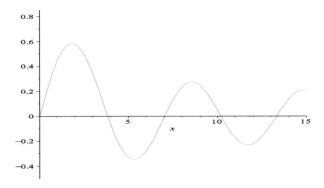

FIGURE 8.8: The Bessel function $J_1(x)$.

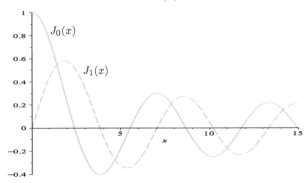

FIGURE 8.9: The Bessel functions $J_0(x)$ and $J_1(x)$.

$p > 0$, we have noted that if $2p$ is not a positive integer, then the differential equation (8.34) has a linearly independent solution of the form (8.44) corresponding to the smaller root $r_2 = -p$.

Taking $r = r_2 = -p$ in (8.47) gives

$$(-2p + 1)c_1 = 0 \tag{8.59}$$

and in (8.48) we have

$$c_n = -\frac{c_{n-2}}{n(n - 2p)}, \quad n \geq 2, \ n \neq 2p. \tag{8.60}$$

Studying the recurrence relation (8.60), we see that there are three distinct cases for the form of the solution $y = y_2(x)$. These cases are:

1. If $2p$ is not a positive integer then

$$y_2(x) = c_0 x^{-p}\left(1 + \sum_{n=1}^{\infty} \alpha_{2n} x^{2n}\right) \tag{8.61}$$

where c_0 is an arbitrary constant and the $\alpha_{2n}(n = 1, 2, \ldots)$ are (definite) constants.

2. If $2p$ is an odd positive integer then

$$y_2(x) = c_0(x)^{-p}\left(1 + \sum_{n=1}^{\infty} \beta_{2n} x^{2n}\right) + c_{2p} x^p \left(1 + \sum_{n=1}^{\infty} \gamma_{2n} x^{2n}\right) \qquad (8.62)$$

where c_0 and c_{2p} are arbitrary constants and $\beta_{2n}, \gamma_{2n} (n = 1, 2, \ldots)$ are (definite) constants.

3. If $2p$ is an even positive integer then

$$y_2(x) = c_{2p} x^p \left(1 + \sum_{n=1}^{\infty} \delta_{2n} x^{2n}\right) \qquad (8.63)$$

where c_{2p} is an arbitrary constant and the $\delta_{2n}(n = 1, 2, \ldots)$ are (definite) constants.

The solution defined in Case 1 is linearly independent of J_p. In Case 2 the solution defined is linearly independent of J_p if $c_{2p} = 0$. However, in Case 3, the solution is a constant multiple of J_p and hence is *not* linearly independent of J_p. So, if $2p$ is not a positive even integer, there exists a linearly independent solution of the form (8.44) corresponding to the smaller root $-p$. That is, if p is not a positive integer, Bessel's equation (8.34) has a solution of the form

$$y_2(x) = \sum_{n=0}^{\infty} c_{2n} x^{2n-p} \qquad (8.64)$$

which is linearly independent of J_p.

To determine the coefficients c_{2n} in (8.64), we note that (8.60) is obtained if p is replaced by $-p$ in the recurrence relation (8.49). This implies that a solution of the form (8.64) can be obtained from (8.56) by replacing p by $-p$. This gives the solution denoted by J_{-p} as

$$J_{-p}(x) = \sum_{n=0}^{\infty} \frac{(-1)^n}{n!\Gamma(n-p+1)} \left(\frac{x}{2}\right)^{2n-p}. \qquad (8.65)$$

Thus, if $p > 0$ is not an integer, the general solution of Bessel's equation of order p is given as the linear combination

$$y = c_1 J_p(x) + c_2 J_{-p}(x)$$

where c_1 and c_2 are arbitrary constants.

If p is a positive integer, the solution defined in Case 3 is not linearly independent of J_p as we have already pointed out. So, in this case a solution that is linearly independent of J_p is given by

$$Y_p(x) = x^{-p} \sum_{n=0}^{\infty} c_n^* x^n + C J_p(x) \ln x$$

where $C \neq 0$. This linearly independent solution can be found using the method of reduction of order, much the same way as we did previously for Y_0. In this manner, we obtain Y_p, so that just as in the case of Bessel's equation of order zero, it is customary to choose a special linear combination as the second solution of (8.34). This special combination, denoted Y_p, is defined as

$$Y_p(x) = \frac{2}{\pi} \left[\left(\ln \frac{x}{2} + \gamma \right) J_p(x) - \frac{1}{2} \sum_{n=0}^{p-1} \frac{(p-n-1)}{n!} \left(\frac{x}{2} \right)^{2n-p} \right.$$
$$\left. + \frac{1}{2} \sum_{n=0}^{\infty} (-1)^{n+1} \left(\sum_{k=1}^{n} \frac{1}{k} + \sum_{k=1}^{n+p} \frac{1}{k} \right) \left[\frac{1}{n!(n+p)!} \left(\frac{x}{2} \right)^{2n+p} \right] \right]$$

$$(8.66)$$

where γ is Euler's constant. The solution Y_p is called the **Bessel function of the second kind of order** p. This formulation of Y_p is sometimes called Weber's form of the Bessel function. A graph of $Y_1(x)$ is shown in Figure 8.10.

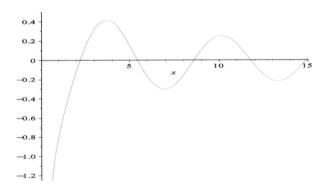

FIGURE 8.10:   The Bessel function $Y_1(x)$.

Problems

1. Show directly that the series for $J_0(x)$, given in (8.40), converges absolutely for all x.

2. Show directly that the series for $J_1(x)$,

$$J_1(x) = \sum_{n=0}^{\infty} \frac{(-1)^n}{n!(n+1)!} \left(\frac{x}{2}\right)^{2n+1},$$

converges absolutely for all x and that

$$J_0'(x) = -J_1(x).$$

3. Using the series for $J_p(x)$,

$$J_p(x) = \sum_{n=0}^{\infty} \frac{(-1)^n}{n!\Gamma(n+p+1)} \left(\frac{x}{2}\right)^{2n+p},$$

show directly that

$$\frac{d}{dx}[x^p J_p(kx)] = kx^p J_{p-1}(kx)$$

and

$$\frac{d}{dx}[x^{-p} J_p(kx)] = -kx^{-p} J_{p+1}(kx)$$

where k is a constant.

4. Show that $J_0(kx)$, where k is a constant, satisfies the differential equation

$$x\frac{d^2y}{dx^2} + \frac{dy}{dx} + k^2 xy = 0.$$

5. Show that the transformation

$$y = \frac{u(x)}{\sqrt{x}}$$

reduces the Bessel equation of order p

$$x^2\frac{d^2y}{dx^2} + x\frac{dy}{dx} + (x^2 - p^2)y = 0$$

to the form

$$\frac{d^2u}{dx^2} + \left[1 + \left(\frac{1}{4} - p^2\right)\frac{1}{x^2}\right]u = 0.$$

Using this, show that

$$y_1(x) = \frac{\cos x}{\sqrt{x}} \quad \text{and} \quad y_2(x) = \frac{\sin x}{\sqrt{x}}$$

are solutions of the Bessel equation of order one half.

6. By a suitable change of variables a differential equation sometimes can be transformed into a Bessel equation. Show that a solution of

$$x^2 y'' + (\alpha^2 \beta^2 x^{2\beta} + \frac{1}{4} - p^2 \beta^2)y = 0 \quad \text{for } x > 0$$

is given by $y = \sqrt{x}\, f(\alpha x^\beta)$ where $f(\cdot)$ is a solution of the Bessel equation of order p.

7. (a) Show that the differential equation

$$\frac{d^2 y}{dt^2} + e^{-2t} y = 0,$$

which models an aging spring, can be converted to an equation with solution $J_0(t)$ by using the change of variable $x = e^{-t}$.

(b) Predict the behavior of the solution by considering the limiting differential equation as $t \to \infty$.

Chapter 8 Review

In Problems **1–5**, determine whether the statement is true or false. If it is true, give reasons for your answer. If it is false, give a counterexample or other explanation of why it is false.

1. When $L = 1$, the ratio test gives useful information about the convergence of a series.

2. Power series "behave" a lot like polynomials.

3. Only closed-form solutions are useful as solutions of differential equations.

4. Every ordinary point is a singular point.

5. Every function has a convergent power series representation.

In Problems **6–8**, show, using a power series argument, the following identities.

6. $e^x e^{-x} = 1$ **7.** $\displaystyle\lim_{x \to 0} \frac{1 - \cos x}{x} = 0$ **8.** $1 - e^{-x} \approx x$ for small x

In Problems **9–18**, solve the following differential equations using the power series method.

9. $y' = 2y + 6, \ y(0) = 3$ **10.** $y' = x + 2y, \ y(0) = 3$

11. $y' = y + \cos x, \ y(0) = 1$ **12.** $y' = 3 + xy^2, \ y(0) = 5$

13. $y' = e^{2x} + x \sin y, \ y(0) = 0$ **14.** $y' = x^2 + 5y, y(0) = 6$

15. $y' = 2x^2 + y^4, \ y(0) = 2$ **16.** $y' = x^3 + y^2, \ y(0) = 4$

17. $y' - 2y = 2x^3, \ y(1) = 1$

18. $y'' + 2y = 4 + 2x + x^2, \ y(0) = 6, y'(0) = -2$

In Problems **19–21**, apply the method of Frobenius to obtain the solution to each of the following differential equations.

19. $x^2y'' - xy' + \left(x^2 + \frac{8}{9}\right)y = 0.$ **20.** $x^2y'' + xy' + \left(x^2 - \frac{1}{9}\right)y = 0.$
21. $3xy'' - (x-2)y' - 2y = 0.$

Chapter 8 Computer Labs

Chapter 8 Computer Lab: MATLAB

MATLAB Example 1: Enter the code below, using *Symbolic Math Toolbox*, to find the series expansion of e^x, $\sin x^2$, and the sum of $\frac{1}{k^2}, \frac{1}{x^k}$.

```
>> clear all
>> syms x k
>> f1=exp(x)
>> T=taylor(f1,'Order',8,'ExpansionPoint',0)
>> pretty(T)
>> f2=sin(x^2)
>> taylor(f2,'Order',8,'ExpansionPoint',0)
>> g1=1/k^2
>> symsum(g1,k,1,5) %partial sum only
>> symsum(g1,k,1,inf)
>> g2=1/x^k
>> symsum(g2,k,0,5)
>> symsum(g2,k,0,inf)
```

MATLAB Example 2: Enter the code below, using *Symbolic Math Toolbox*, to solve the ODE $y'' + xy' + y = 0$ by series expansion.

```
>> clear all
>> syms x c0 c1 c2 c3 c4 c5 c6
>> y=c0+c1*x+c2*x^2+c3*x^3+c4*x^4+c5*x^5
>> eqODE=diff(y,x,2)+x*diff(y,x)+y
>> eq1=collect(eqODE)
>> eq2=coeffs(eq1,x) %n must be large enough in y
>> [c2,c3,c4,c5]=solve(eq2(1),eq2(2),eq2(3),eq2(4),c2,c3,c4,c5)
>> ysoln=subs(y)
```

```
>> collect(collect(ysoln,c0),c1)
```

MATLAB Example 3: Enter the code below, using *Symbolic Math Toolbox,* to find a series solution for the ODE $xy'' + (x+1)y' - \dfrac{1}{16x}y = 0$, which has a regular singular point.

```
>> clear all
>> syms x c0 c1 c2 c3 c4 c5 r
>> y=c0*x^r+c1*x^(1+r)+c2*x^(2+r)+c3*x^(3+r)+c4*x^(4+r)
>> yn=c0*x^r+c1*x^(1+r)+c2*x^(2+r)+c3*x^(3+r)
>> eqODE=x*diff(y,x,2)+(x+1)*diff(y,x)-y/(16*x)
>> eq1=collect(eqODE)
   %It won't collect with x in denom or r in exponent.
   %We rewrite the expression to get coeffs we need
>> eq1a=collect(simplify(eq1),x) %lowest power of x is x^(1-r)
>> eq2=coeffs(eq1a,x) %first term is x^(1-r) power
>> xrcoeff=eq2(2) %coeff of x^r
>> xrp1coeff=eq2(3) %coeff of x^(r+1)
>> xrp2coeff=eq2(4) %coeff of x^(r+2)
>> eq3=solve(eq2(1),r)
>> r1=eq3(1)
>> r2=eq3(2)
>> cvals1=solve(xrcoeff,xrp1coeff,xrp2coeff,c1,c2,c3)
>> cvals=subs([cvals1.c1,cvals1.c2,cvals1.c3],r,r1)
>> y=subs(yn,[r,c1,c2,c3],[r1,cvals(1),cvals(2),cvals(3)])
>> cvals2=subs([cvals1.c1,cvals1.c2,cvals1.c3],r,r2)
>> y2=subs(yn,[r,c1,c2,c3],[r2,cvals2(1),cvals2(2),cvals2(3)])
```

MATLAB Exercises

Turn in both the commands that you enter for the exercises below as well as the output/figures. These should all be in one document. Please highlight or clearly indicate all requested answers. Some of the questions will require you to modify the above MATLAB code to answer them.

1. Enter the commands given in MATLAB Example 1 and submit both your input and output.

2. Enter the commands given in MATLAB Example 2 and submit both your input and output.

3. Enter the commands given in MATLAB Example 3 and submit both your input and output.

4. Find the Taylor series expansion of e^{2x} about $x = 0$ up to and including x^8 terms.

5. Find the Taylor series expansion of e^{2x} about $x = 1$ up to and including $(x-1)^8$ terms.

6. Find the Taylor series expansion of $\ln(x+1)$ about $x = 0$ up to and including x^7 terms.

7. Find the Taylor series expansion of $\ln x$ about $x = 1$ up to and including $(x-1)^7$ terms.

8. Find the Taylor series expansion of $\cos(x^2)$ about $x = 0$ up to and including x^8 terms.

9. Find the Taylor series expansion of $\cos(5x)$ about $x = 1$ up to and including $(x-1)^5$ terms.

10. Find the Taylor series expansion of $x^4 + 2x + 1$ about $x = 0$ up to and including x^4 terms.

11. Find the Taylor series expansion of $x^4 + 2x + 1$ about $x = 1$ up to and including $(x-1)^4$ terms.

12. Find the Taylor series expansion of $\dfrac{1}{x}$ about $x = -1$ up to and including $(x+1)^8$ terms.

13. Find the Taylor series expansion of $\dfrac{1}{x}$ about $x = 1$ up to and including $(x-1)^8$ terms.

14. Solve $y' - x^2 = y^2$ by series expansion.

15. Solve $y' - e^x + \sin y$ by series expansion.

16. Solve $y'' + xy' + y = 0$ by series expansion.

17. Solve $y' = x + \frac{1}{y}$, $y(0) = 1$ by series expansion.

18. Solve $y' = x^2 + y^3$, $y(1) = 1$ by series expansion.

19. Solve $y'' + y = 1 + x + x^2$, $y(0) = 1$, $y'(0) = -1$ by series expansion.

20. Find a series solution for the ODE $xy'' - y = 0$, which has a regular singular point.

21. Find a series solution for the ODE $2x^2 y'' + 3xy' - (x+1)y = 0$, which has a regular singular point.

22. Find a series solution for the ODE $x^2 y'' + x(x+1)y' - y = 0$, which has a regular singular point.

23. Find a series solution for the ODE $2x^2 y'' + xy' + (x^2 - 1)y = 0$, which has a regular singular point.

24. Find a series solution for the ODE $xy'' + y' + y = 0$, which has a regular singular point.

25. Find a series solution for the ODE $(x^2 + x^3)y'' - (x^2 + x)y' + y = 0$, which has a regular singular point.

Chapter 8 Computer Lab: Maple

Maple Example 1: Enter the code below to find the series expansion of e^x, $\sin x^2$, and the sum of $\dfrac{1}{k^2}, \dfrac{1}{x^k}$.

restart
$f1 := e^x$
$taylor(f1, x = 0, 8)$
$mtaylor(f1, x = 0, 8)$
$f2 := \sin(x^2)$
$mtaylor(f2, x = 0, 8)$
$g1 := \dfrac{1}{k^2}$
$sum(g1, k = 1..5)$
$$\sum_{k=1}^{5} g1$$
$$\sum_{k=1}^{\infty} g1$$
$g2 := \dfrac{1}{x^k}$
$$\sum_{k=0}^{5} g2$$
$$\sum_{k=0}^{\infty} g2$$

Maple Example 2: Enter the code below to solve the ODE $y'' + xy' + y = 0$ by series expansion.

restart
$eqODE := y''(x) + x \cdot y'(x) + y(x) = 0$
$$eq1 := y(x) = \sum_{k=0}^{n} c_k \cdot x^k$$
$eq1a := eval(subs(n = 5, eq1))$
$eq2 := eval(subs(eq1a, eqODE))$
$eq3 := collect(eq2, x)$
$eq3a := coeff(lhs(eq3), x, 0)$
$eq3b := coeff(lhs(eq3), x, 1)$
$eq3c := coeff(lhs(eq3), x, 2)$

$eq3d := coeff(lhs(eq3), x, 3)$

$eq4 := solve(\{eq3a, eq3b, eq3c, eq3d\}, \{c_2, c_3, c_4, c_5\})$

$eqsoln := subs(eq4, eval(subs(n = 5, eq1)))$

$eqsoln1 := collect(collect(eqsoln, c_0), c_1)$

Maple Example 3: Enter the code below to find a series solution for the ODE $xy'' + (x+1)y' - \dfrac{1}{16x}y = 0$, which has a regular singular point.

restart

$eqODE := x \cdot y''(x) + (1+x) \cdot y'(x) - \dfrac{y(x)}{16 \cdot x} = 0$

$eqy := y(x) = \displaystyle\sum_{k=0}^{n} c_k \cdot x^{k+r}$

$eq1 := eval(subs(n = 5, eqy))$

$eq2 := subs(eq1, eqODE)$

$eq3 := combine(expand(eq2), x)$

$eq4a := coeff(lhs(eq3), x^{r-1})$

$eq4b := solve(eq4a, r)$

$eq5a := coeff(lhs(eq3), x^r)$

$eq5b := coeff(lhs(eq3), x^{1+r})$

$eq5c := coeff(lhs(eq3), x^{2+r})$

$eqsoln1 := solve(subs(r = eq4b[1], \{eq5a, eq5b, eq5c\}), \{c_1, c_2, c_3\})$

$eqsoln2 := solve(subs(r = eq4b[2], \{eq5a, eq5b, eq5c\}), \{c_1, c_2, c_3\})$

$soln := eval(subs(n = 4, eqy))$

$soln1 := y_1(x) = subs(eqsoln1, r = eq4b[1], soln)$

$soln2 := y_2(x) = subs(eqsoln2, r = eq4b[2], soln)$

Maple Exercises

Turn in both the commands that you enter for the exercises below as well as the output/figures. These should all be in one document. Please highlight or clearly indicate all requested answers. Some of the questions will require you to modify the above Maple code to answer them.

1. Enter the commands given in Maple Example 1 and submit both your input and output.

2. Enter the commands given in Maple Example 2 and submit both your input and output.

3. Enter the commands given in Maple Example 3 and submit both your input and output.

4. Find the Taylor series expansion of e^{2x} about $x = 0$ up to and including x^8 terms.

5. Find the Taylor series expansion of e^{2x} about $x = 1$ up to and including $(x-1)^8$ terms.

6. Find the Taylor series expansion of $\ln(x+1)$ about $x = 0$ up to and including x^7 terms.

7. Find the Taylor series expansion of $\ln x$ about $x = 1$ up to and including $(x-1)^7$ terms.

8. Find the Taylor series expansion of $\cos(x^2)$ about $x = 0$ up to and including x^8 terms.

9. Find the Taylor series expansion of $\cos(5x)$ about $x = 1$ up to and including $(x-1)^5$ terms.

10. Find the Taylor series expansion of $x^4 + 2x + 1$ about $x = 0$ up to and including x^4 terms.

11. Find the Taylor series expansion of $x^4 + 2x + 1$ about $x = 1$ up to and including $(x-1)^4$ terms.

12. Find the Taylor series expansion of $\dfrac{1}{x}$ about $x = -1$ up to and including $(x+1)^8$ terms.

13. Find the Taylor series expansion of $\dfrac{1}{x}$ about $x = 1$ up to and including $(x-1)^8$ terms.

14. Solve $y' - x^2 = y^2$ by series expansion.

15. Solve $y' - e^x + \sin y$ by series expansion.

16. Solve $y'' + xy' + y = 0$ by series expansion.

17. Solve $y' = x + \frac{1}{y}$, $y(0) = 1$ by series expansion.

18. Solve $y' = x^2 + y^3$, $y(1) = 1$ by series expansion.

19. Solve $y'' + y = 1 + x + x^2$, $y(0) = 1$, $y'(0) = -1$ by series expansion.

20. Find a series solution for the ODE $xy'' - y = 0$, which has a regular singular point.

21. Find a series solution for the ODE $2x^2 y'' + 3xy' - (x+1)y = 0$, which has a regular singular point.

22. Find a series solution for the ODE $x^2 y'' + x(x+1)y' - y = 0$, which has a regular singular point.

23. Find a series solution for the ODE $2x^2 y'' + xy' + (x^2 - 1)y = 0$, which has a regular singular point.

24. Find a series solution for the ODE $xy'' + y' + y = 0$, which has a regular singular point.

25. Find a series solution for the ODE $(x^2 + x^3)y'' - (x^2 + x)y' + y = 0$, which has a regular singular point.

Chapter 8 Computer Lab: Mathematica

Mathematica Example 1: Enter the code below to find the series expansion of e^x, $\sin x^2$, and the sum of $\dfrac{1}{k^2}, \dfrac{1}{x^k}$.

```
Quit[]
f1[x_] = e^x
eq1 = Series[f1[x], {x, 0, 8}]
Normal[eq1]
f2[x_] = Sin[x^2]
Series[f2[x], {x, 0, 8}]//Normal
g1[k_] = 1/k^2
```
$$\sum_{k=1}^{5} g1[k]$$
$$\sum_{k=1}^{\infty} g1[k]$$
```
g2[k_] = 1/x^k
```
$$\sum_{k=0}^{5} g2[k]$$
$$\sum_{k=0}^{\infty} g2[k]$$

Mathematica Example 2: Enter the code below to solve the ODE $y'' + xy' + y = 0$ by series expansion.

```
Quit[]
y1[x_] = c0 + c1 x + c2 x^2 + c3 x^3 + c4 x^4 + c5 x^5
dey[x_] = y''[x] + x y'[x] + y[x]
eq2 = Simplify[ReplaceAll[dey[x], {y[x]→y1[x], y'[x]→y1'[x],
   y''[x]→y1''[x]}]]
eq3a = Coefficient[eq2, x, 0]
eq3b = Coefficient[eq2, x, 1]
eq3c = Coefficient[eq2, x, 2]
eq3d = Coefficient[eq2, x, 3]
eq4 = Solve[{eq3a==0,eq3b==0,eq3c==0,eq3d==0}, {c2, c3, c4,c5}]
eqsoln = ReplaceAll[y1[x], eq4]
eqsoln1 = Collect[Collect[eqsoln[[1]], c0], c1]
```

Mathematica Example 3: Enter the code below to find a series solution for the ODE $xy''+(x+1)y'-\dfrac{1}{16x}y = 0$, which has a regular singular point.

```
Quit[]
yr[x_] = c0 xʳ + c1 xʳ⁺¹ + c2 xʳ⁺² + c3 xʳ⁺³ + c4 xʳ⁺⁴
eqODE[x_] = x y''[x] + (1 + x) y'[x] - y[x]/(16 x)
eq1 = Expand[ReplaceAll[eqODE[x], {y[x]→yr[x], y'[x]→yr'[x],
  y''[x]→yr''[x]}]]
eq2a = Coefficient[eq1, xʳ⁻¹] (*compare to above output*)
eq2b = Solve[eq2a==0, r]
eq3a = Coefficient[Coefficient[eq1, xʳ], x, 0]
eq3b = Coefficient[Coefficient[eq1, xʳ⁺¹], x, 0]
eq3c = Coefficient[Coefficient[eq1, xʳ⁺²], x, 0]
eq3d = Coefficient[Coefficient[eq1, xʳ⁺³], x, 0]
eqsoln1 = Solve[ReplaceAll[{eq3a==0, eq3b==0, eq3c==0,
eq3d==0},eq2b[[1]]], {c1, c2, c3, c4}]
eqsoln2 = Solve[ReplaceAll[{eq3a==0, eq3b==0, eq3c==0,
eq3d==0},eq2b[[2]]], {c1, c2, c3, c4}]
y1[x_] = Collect[ReplaceAll[ReplaceAll[yr[x], eqsoln1],
eq2b[[1]]], c0]
y2[x_] = Collect[ReplaceAll[ReplaceAll[yr[x], eqsoln2],
eq2b[[2]]], c0]
```

Mathematica Exercises

Turn in both the commands that you enter for the exercises below as well as
the output/figures. These should all be in one document. Please highlight or
clearly indicate all requested answers. Some of the questions will require you
to modify the above Mathematica code to answer them.

1. Enter the commands given in Mathematica Example 1 and submit both
   your input and output.

2. Enter the commands given in Mathematica Example 2 and submit both
   your input and output.

3. Enter the commands given in Mathematica Example 3 and submit both
   your input and output.

4. Find the Taylor series expansion of e^{2x} about $x = 0$ up to and including
   x^8 terms.

5. Find the Taylor series expansion of e^{2x} about $x = 1$ up to and including
   $(x-1)^8$ terms.

6. Find the Taylor series expansion of $\ln(x+1)$ about $x = 0$ up to and
   including x^7 terms.

7. Find the Taylor series expansion of $\ln x$ about $x = 1$ up to and including $(x - 1)^7$ terms.

8. Find the Taylor series expansion of $\cos(x^2)$ about $x = 0$ up to and including x^8 terms.

9. Find the Taylor series expansion of $\cos(5x)$ about $x = 1$ up to and including $(x - 1)^5$ terms.

10. Find the Taylor series expansion of $x^4 + 2x + 1$ about $x = 0$ up to and including x^4 terms.

11. Find the Taylor series expansion of $x^4 + 2x + 1$ about $x = 1$ up to and including $(x - 1)^4$ terms.

12. Find the Taylor series expansion of $\dfrac{1}{x}$ about $x = -1$ up to and including $(x + 1)^8$ terms.

13. Find the Taylor series expansion of $\dfrac{1}{x}$ about $x = 1$ up to and including $(x - 1)^8$ terms.

14. Solve $y' - x^2 = y^2$ by series expansion.

15. Solve $y' - e^x + \sin y$ by series expansion.

16. Solve $y'' + xy' + y = 0$ by series expansion.

17. Solve $y' = x + \frac{1}{y}$, $y(0) = 1$ by series expansion.

18. Solve $y' = x^2 + y^3$, $y(1) = 1$ by series expansion.

19. Solve $y'' + y = 1 + x + x^2$, $y(0) = 1$, $y'(0) = -1$ by series expansion.

20. Find a series solution for the ODE $xy'' - y = 0$, which has a regular singular point.

21. Find a series solution for the ODE $2x^2y'' + 3xy' - (x + 1)y = 0$, which has a regular singular point.

22. Find a series solution for the ODE $x^2y'' + x(x + 1)y' - y = 0$, which has a regular singular point.

23. Find a series solution for the ODE $2x^2y'' + xy' + (x^2 - 1)y = 0$, which has a regular singular point.

24. Find a series solution for the ODE $xy'' + y' + y = 0$, which has a regular singular point.

25. Find a series solution for the ODE $(x^2 + x^3)y'' - (x^2 + x)y' + y = 0$, which has a regular singular point.

Chapter 8 Projects

There will be times when the method of Frobenius fails and we need to consider other approaches in our attempts to find series solutions to a given differential equation. Consider

$$x^2 y'' + (1 + 3x)y' + y = 0. \qquad (8.67)$$

Show that there is an irregular singular point at 0. Even though the method of Frobenius should only work at regular singular points, let's try it here. By assuming a solution of the form $y = \sum a_n x^{n+r}$, show that the indicial equation is $r = 0$. By substituting into (8.67), show that the coefficients satisfy $a_{n+1} = -(n+1)a_n$, $n = 0, 1, 2, \cdots$ and thus $a_n = (-1)^n n! a_0$. Conclude that one solution is thus

$$y_1 = a_0 \sum_{n=0}^{\infty} (-1)^n n! x^n. \qquad (8.68)$$

Now comes the strange part. Show that this series has a radius of convergence of zero and thus diverges for all $x \neq 0$. Since we assumed that a Frobenius series should have a non-zero radius of convergence, this means that (8.67) actually has no Frobenius series solution!

We often focus on series that converge; that is, a series that approaches the exact function as we add more and more terms. In the remainder of this project we will consider an *asymptotic series*, which has the property that it approaches the function at a specific point even though it is a divergent series. In summary, if $f_n(x)$ is the partial sum approximation to $f(x)$, then

$$\text{Convergent Series: } \lim_{n \to \infty} f_n(x) = f(x),$$

$$\text{Asymptotic Series: } \lim_{x \to x_0} f_n(x) = f(x_0).$$

For two functions $f(x)$ and $g(x)$, we use the notation

$$f(x) \sim g(x), \ x \to x_0$$

to mean that the relative error between f and g goes to zero as $x \to x_0$ and

$$f(x) \ll g(x), \ x \to x_0$$

to mean that $f(x)/g(x)$ goes to zero as $x \to x_0$. Our solution (8.68) is an asymptotic series solution to (8.67). We now outline how we can obtain a second solution. Substitute $y = e^S$ into (8.67) to obtain

$$x^2 S'' + x^2 (S')^2 + (1 + 3x)S' + 1 = 0. \tag{8.69}$$

Near an irregular singular point, it is usually the case that $x^2 S''$ can be neglected relative to $x^2 (S')^2$ and that $3xS'$ can be neglected relative to S'. We thus can write

$$x^2 (S')^2 + S' + 1 \sim 0, \quad x \to 0^+.$$

Solve this quadratic for S' and conclude that (because x is small)

$$S' \sim -1, \quad x \to 0^+,$$
$$S' \sim -x^{-2}, \quad x \to 0^+. \tag{8.70}$$

Integrate both of the above asymptotic relations. The first gives the solution already obtained. Use the result from the second integration to conclude that the *controlling factor in the leading behavior* is governed by $e^{1/x}$. To find the full leading behavior, substitute $S(x) = x^{-1} + C(x)$ (with $C(x) \ll x^{-1}$) into (8.69) to obtain

$$x^2 C'' + x^2 (C')^2 - (1 - 3x)C' - x^{-1} + 1 = 0.$$

We can again neglect certain terms as $x \to 0^+$ and obtain

$$c'(x) + x^{-1} \sim 0, \ x \to 0^+.$$

Solve this and conclude that the full leading behavior is given by

$$y_2 = c_2 x^{-1} e^{1/x}, \ x \to 0^+.$$

Choose an initial condition and verify, by numerically solving (8.67), that $y_1 + y_2$ agrees with this solution for small x.

Project 8B: Hypergeometric Functions

We will examine solutions of the differential equation

$$x(1 - x)y'' + [c - (a + b + 1)x]y' - aby = 0,$$

for $|x| < 1$ satisfying the initial condition, $y(0) = 1$.
To do so, we first state two helpful definitions. The first uses the gamma function, $\Gamma(\alpha)$, seen earlier in this chapter. We define

$$(\alpha)_n \overset{\text{def}}{=} \frac{\Gamma(\alpha + n)}{\Gamma(\alpha)} = \begin{cases} \alpha(\alpha + 1) \cdots (\alpha + n - 1), & n \geq 1 \\ 1, & n = 0. \end{cases}$$

With this definition of $(\alpha)_n$, we can define the following:

DEFINITION 8B

Let a, b, and c be real numbers, and suppose c is a nonnegative number.
Consider

$$F(a, b; c; x) \overset{\text{def}}{=} \sum_{j=0}^{\infty} \frac{(a)_j (b)_j}{n!(c)_j} x^n$$

on the open interval $(-1, 1)$. The series is a real analytic function called
a **hypergeometric function.**

THEOREM 8B

The hypergeometric function, $F(a, b; c; x)$, is the unique real analytic
solution to the hypergeometric differential equation

$$x(1 - x)\frac{\partial^2 y}{\partial x^2} + [c - (a + b + 1)x]\frac{\partial y}{\partial x} - aby = 0 \qquad (8.71)$$

for $|x| < 1$, satisfying the initial condition, $y(0) = 1$.

Prove this by letting $y(x) = \sum_{j=1}^{\infty} \alpha_j x^j$ with $\alpha_0 = 1$. It may help you
to make the observation

$$(n + 1)(x + n)\alpha_{n+1} = (a + n)(b + n)\alpha_n.$$

Show that the indicial roots of Equation (8.71) are $r_1 = 0$ and $r_2 = 1 - c$.
Using the Frobenius method with the root $r_1 = 0$, obtain a series
solution for (8.71).

Appendix A

An Introduction to MATLAB,
Maple, and Mathematica

This is meant as a crash course in MATLAB®, Maple, and Mathematica or brief refresher. All have fairly friendly "help" menus which the reader should consult when questions arise that are not answered here. All three packages also have *Student Versions* available for about the price of one hardback textbook. Typically, MATLAB is used for numerical problems, whereas Maple and Mathematica are used for symbolic manipulation. We will see throughout the text that this distinction is often blurred.

Computer laboratories in gray boxes (such as this) are given at the end of each chapter and provide code that will help the reader solve problems within the given chapter.

Initially, we want to plot functions of one variable. Sometimes this will be an explicit representation as in $y = f(x)$. But enough times the separation of variables process (or other method) leaves us with an implicit representation that can be written as $f(x, y) = c$. In order to plot solutions for either case we need to specify initial conditions. As we will see, MATLAB, Maple, and Mathematica have methods for handling this.

A.1 MATLAB

This code was written using MATLAB R2014a. Certain code, particularly in the *Symbolic Math Toolbox*, may not work in earlier versions.

In MATLAB, we execute the commands by either (i) typing them at the command line prompt ≫ and pressing <Enter> or (ii) typing the commands in a `script`, saving the `script`, and then executing the `script` by typing the name in the command window and pressing <Enter>. (Note that you DO NOT type in ≫ for any of these commands.) You cannot go back and correct a line that you typed. However, you can type in the correct one and re-execute it. For example,

```
>> x=0:1:5
>> y=x^2
```

gives the error

```
??? Error using  ^
Inputs must be a scalar and a square matrix.
To compute elementwise POWER, use POWER (.^) instead.
```

and you need to retype (or use the up arrow to recall the line) and then correct
it as

```
>> y=x.^2
```

and the last line is the one MATLAB will use until you redefine or clear the
variable. On occasion, it may confuse variables if you rename them multiple
times. The commands

```
>> clear y
```

and

```
>> clear all
```

will wipe the memory clean of the variable y and then of all variables.

As we saw above, we will proceed with the command window input. An
important thing to realize is that everything in MATLAB is a matrix. If
you have not yet encountered this word in your study of mathematics, just
think of a matrix as a structure with both rows and columns. A vector is
just considered a matrix of either one column or one row. And a scalar is
considered to be a matrix with exactly one row and one column.

Another main thing to remember about MATLAB is that you have the
choice of putting a semicolon ";" at the end of each line or of not putting
one. Putting the semicolon suppresses the output, whereas not putting the
semicolon shows you the results of the statement you just executed. It does
not matter whether you put a semicolon after a statement that plots or does
something to an already existing plot. The MATLAB commands given in
this text will be in **typewriter** font. Anything that follows a percentage sign
"%" is ignored by MATLAB. MATLAB is also case sensitive. That is, the
variables t and T are two different variables. Variables also continue to be
defined/hold a value until we overwrite them or clear them.

We again note at this point that if you type something incorrectly and get
an error message, you can use the up and down arrows on the keyboard to
scroll through previous commands that were entered. Sometimes an error
is because of a simple typo and using the up arrow to recall your previous
command and then correcting the mistake is often quite time saving. Also,
if you entered something wrong, you cannot cut the error from the command

window—you can only execute the correct command and remember that the most recent assignment of a variable is the one in MATLAB's memory.

If we want to plot the function $y = f(x)$ we need to know the x-values that we are going to plug into our function. Then we need to evaluate the function at these values. Sounds simple enough, right? Let's try to plot $f(x) = x^3 - x$ between the x-values of -2 and 2.

```
>> x=-2:.01:2;
>> y=x.^3 -x;
>> plot(x,y)
```

The first line can be read as "go from $x = -2$ to $x = 2$ in steps of .01." The step size of .01 was more or less arbitrary but will give us a smooth plot. The second line can be read as "take the entire vector x and cube *each* entry and then subtract the corresponding entry of x from this." Remember x is a vector (or a matrix) and it doesn't make mathematical sense to cube a vector—we really just want to cube each entry. The "." is used to tell MATLAB to do an operation *elementwise* and this applies to multiplication, division, and raising to a power. (Addition and subtraction are always done elementwise.) If we want to label the axes, title the graph, and change the x-y values that we see in the window to, say, $x \in [-\pi/2, \pi/2]$ and $y \in [-2, 2]$, we could do so with the commands

```
>> xlabel('x');
>> ylabel('y');
>> title('Plot of f(x)=x^3-x');
>> axis([-pi/2 pi/2 -2 2]);
```

We could also do things like plot the graph in a different color or superimpose another graph on it. Suppose we wanted to also graph the function $f(x) = x^2 - 1$ in the color green and on the same graph. We could use the commands

```
>> y1=x.^2 -1;
>> hold on
>> plot(x,y1,'g')
>> title('Plot of x^3-x and x^2 -1');
>> hold off
```

A.1.1   Some Helpful MATLAB Commands

MATLAB has a help section that can be accessed with the mouse or the keyboard. If you know the command you want to use but forget the proper syntax, the keyboard version is probably the way to go (e.g., if you want to know how to use the built-in MATLAB function ode45, you would type `help ode45` and information would appear in the command window). We summarize the commands in the list below.

`help <command>` - specify a command and the info about it will be given
`lookfor <topic>` - specify a topic and all commands that have related topics will be listed
`who` - lists current variables
`cd amssi` - changes the working directory to amssi (assuming it exists)
`pwd` - tells you which directory you are currently working in
`dir` - lists files in that directory
`clear all` - clears variables from memory
`format long` - lets you see more non-zero digits
`size(x)` - gives the size of the variable (matrix) x
`length(x)` - gives the length of the vector x
`zeros(n,m)` - an n x m matrix of zeros
`ones(n,m)` - an n x m matrix of ones
`eye(n,m)` - the n x m identity matrix
`rand(n,m)` - an n x m matrix whose entries are random numbers uniformly distributed in (0,1)
`randn(n,m)` - an n x m matrix whose entries are random numbers normally distributed with mean 0 and variance 1
`linspace(x1,x2,n)` - a vector with uniformly spaced points whose beginning value is x1 and the last value is x2 with a total of n points

A few other useful commands are the common function: `exp`, `log`, `sqrt`, `cos`, `sin`, `tan`, `atan`, and others. The `Help` menu in the top right of the window can help you find others.

Plotting in MATLAB

MATLAB gives numerous options for plotting. Some of the more often used ones are given here.

`figure` - creates a new graph window
`orient tall`, `orient portrait`, `orient landscape` - orients the picture in the desired manner for printing (portrait is the default)

Various line types, plot symbols, and colors can also be implemented in MATLAB with the command `plot(x,y,option)` where `option` is a character string made from one element from any or all the following three columns:

y	yellow	.	point	-	solid
m	magenta	o	circle	:	dotted
c	cyan	x	x-mark	-.	dashdot
r	red	+	plus	--	dashed
g	green	s	square	p	pentagram
w	white	d	diamond	*	star
k	black	v	triangle (down)	<	triangle (left)
b	blue	^	triangle (up)	>	triangle (right)

As an example, `plot(x,y,'m+:')` plots a magenta dotted line with a plus at each data point, whereas `plot(x,y,'bd')` plots a blue diamond at each data point but does not connect the points with a line.

The reader is encouraged to go through the following example in detail, making sure that the syntax of each line is understood.

Before getting into a detailed example, we pause to discuss MATLAB's `subplot` command. In the previous code, we graphed a function that used the entire figure window. There will be many times when this is an inefficient use of space and we would like to restrict ourselves to a smaller portion of the screen. For example, suppose we want to plot x and x^3 separately so that we can see how they appear separately and then plot them together. We can do this conveniently with `subplot`.

The command `subplot` simply divides the figure window into smaller sub-windows and then plots what follows in the given smaller window. The syntax for it is

$$\text{subplot}(\text{m},\text{n},\text{p})$$

where `m`, `n`, `p` are positive integers. If `m`, `n`, `p` are single digits, i.e., $0 <$`m`, `n`, `p`< 10, then we could also use the syntax `subplot(mnp)`. The figure is divided into `m` (horizontal) rows and `n` (vertical) columns. The `p` refers to the figure and the numbering starts with 1 as the top left picture. One of the beauties of `subplot` is that we can mix and match how we divide up the window as long as there is no contradiction. See Figure A.1 for two examples of this and the next example for an implementation of `subplot`.

If we wanted to superimpose figures, whether using `subplot` or not, we use the command `hold on` to tell MATLAB to superimpose everything that follows until we tell it to `hold off` for the given figure (or subfigure). Example 1 now illustrates some of these additional tools.

Example 1: Basic commands and built-in functions.
```
>> pi+exp(1)+log(4)
>> format long
>> pi+exp(3)+log(1/2)
>> x=-2:.1:pi/2;
>> eq1=x.^2-1+sin(x);
>> plot(x,eq1)
```

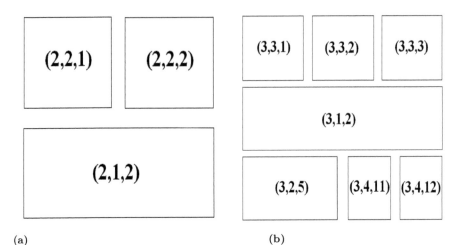

(a)   (b)

FIGURE A.1: Two examples of the subplot command. In (a), we could obtain two plots on top by dividing the figure into 2 rows and 2 columns and then obtain one plot on the bottom by dividing the figure into 2 rows and 1 column. In (b), we again divide up the picture so that it is consistent. E.g., we obtain the middle strip by dividing the picture into 3 rows and 1 column (with the middle strip being picture 2) and we could obtain the lower right one by dividing the picture into 3 rows and 4 columns (with the bottom right one being picture 12).

```
>> plot(x,eq1,'b--','LineWidth',3)
>> xlabel('x','Fontsize',12), ylabel('y','Fontsize',12)
>> legend('x^2-1+sin(x)')
>> title('A basic plot')
>> axis([-2 pi/2 -1.5 1.5])
>> f=sqrt(x+2)+log(x+2); %note that log(x) is ln(x) in MATLAB
>> plot(x,f)
>> plot(x,eq1,'b-',x,f,'m.')
>> xlabel('x','Fontsize',12), ylabel('y','Fontsize',12)
>> title('Superimposed plots')
>> legend('x^2-1+sin(x)','sqrt(x+2)+log(x+2)')
>> axis([-2 1 -1.5 2])
>> x=0:.1:3;
>> eq2=log(x);
>> eq3=exp(-x.^2);
>> plot(x,eq2,'b')
>> hold on
>> plot(x,eq3,'k--')
>> xlabel('x','Fontsize',12), ylabel('y','Fontsize',12)
>> axis([0 3 -2 2]), title('ln(x) vs exp(-x^2)')
>> hold off %anything that follows will not be superimposed
```

```
>> orient tall %makes picture take up full page when you print
```

Unlike in Maple and Mathematica, worksheets are not saved for future execution by default. To save your input in a file, try examining the commands `diary on` and `diary off` to see their usefulness. Alternatively, all the commands may be stored in an m-file to be executed.

There are many *toolboxes* that can accompany MATLAB. Although it is not necessary for the reader to have access to any of these in order to use many of the numerical capabilities mentioned here and throughout the text, the *Symbolic Math Toolbox* is often helpful as it will allow symbolic manipulation. For example, we will be able to solve for the roots of a polynomial or calculate a "nice" form of a basis of the nullspace of a matrix. The reader may or may not find it worthwhile to purchase this particular toolbox.

Example 2: Some commands that use MATLAB's *Symbolic Math Toolbox*

```
>> clear all
>> syms x
>> f(x)=x^4+2*x^3-8*x^2
>> f(3)
>> subs(f(x),x,3)
>> eq1=solve(f(x),x)
>> eq1(4)
>> factor(f(x))
>> g(x)=1-x^2
>> f(x)*g(x)
>> factor(f(x)*g(x))
>> f(x)/x
>> simplify(f(x)/x)
>> h(x)=x^2+2*x+4
>> solve(h(x),x)
>> eq2=solve(h(x),x)
>> eq2(1)
>> real(eq2(1))
>> imag(eq2(1))
>> 2*i+eq2(1)
>> eval(2*i+eq2(1))
>> eq1=x^4+3*x^2-7*x+6
>> solve(eq1)
>> %Next 2 lines find roots numerically w/o Symbolic Math Toolbox
>> p=[1 0 3 -7 6] % Coefficients of polynomial
>> roots(p)
```

A.1.2 Programming with a `script` and a `function` in MAT-LAB

The use of a `script` allows one to create functions or simply store a set of commands to be executed. One needs to be careful to save a `script` to the appropriate directory and/or be in the right directory in the command window. A `script` is simply a text file that MATLAB knows to execute because it has the extension 'm'. Clicking on *File→New→script* will pop up a new window in which you can type.

The easiest use of a `script` is to simply enter a list of commands that would normally be entered in the command window. E.g., we could create a `script` called `ExampleA1.m` and enter the following commands:

```
x=-2:.01:2;
y=x.^3 -x;
plot(x,y)
xlabel('x');
ylabel('y');
title('Plot of f(x)=x^3-x');
axis([-pi/2 pi/2 -2 2]);
y1=x.^2 -1;
hold on
plot(x,y1,'g')
```

In the command window, we would then type

```
>> ExampleA1;
```

and the result would be the same we obtained at the beginning of this appendix. That is, it would plot the function $x^3 - x$ and $x^2 - 1$ on the same graph and with the various colors, axes, etc. specified. Alternatively, we could execute the file by clicking on the green arrow in the top middle of the Toolbar in the `script` window.

A more important use of a `Script` is to create lines that have nested loops. Next we use a `Script` (or equivalently a `Function`) to plot the Taylor series expansion of $\sin(x)$ about $x = 0$ and compare with the actual function. More specifically, our function will take two numbers, N and $x1$, as inputs where $N + 1$ is the number of terms in the expansion and $x1$ is the interval over which we will compare the function with its series expansion. Let's call this file `SineTaylorSeries.m`. In a `Script` or `Function`, we would enter:

Example 3:

```
function f=SineTaylorSeries(N,x1)
%This function takes two inputs N and x1.  It compares the Taylor
%expansion of sine with (N+1) terms to the actual sine over
```

```
%the interval [-x1,x1].

x=-x1:0.1:x1;
SinePlot=zeros(size(x1));
for k=0:N
SinePlot = SinePlot + (-1)^k*x.^(2*k+1)/factorial(2*k+1);
end
plot(x,SinePlot,'b')
hold on
plot(x,sin(x),'r+')
hold off
end
```

and save the file as SineTaylorSeries.m. Then, in the command window we could enter

```
>> SineTaylorSeries(2,3)
>> SineTaylorSeries(3,6)
>> SineTaylorSeries(8,10)
```

We close this part by noting that it is not necessary that the variable use in the script agree with the variable use in the command window. The use of m-files will be very important in the implementation of the Runge-Kutta method given in the book.

Exercises
The remainder of this lab will make use of many MATLAB commands, including zoom, fzero, sqrt, subplot, axis, format long, legend, xlabel, ylabel, title. Turn in both the commands that you entered as well as the output/figures. Please highlight or clearly indicate all requested answers.

1. Enter the commands given in MATLAB Example 1 and submit both your input and output.

2. Enter the commands given in MATLAB Example 2 and submit both your input and output.

3. Enter the commands given in MATLAB Example 3 and submit both your input and output.

4. Determine which of e^{π} or $\pi^{e^{1}}$ is the larger value. Use format long to write your answers correct to 12 decimal places.

5. Determine which of $\sin(\ln(\sqrt{.7}))$ or $\ln(\sin(\sqrt{.7}))$ is the larger value. Use format long to write your answers correct to 12 decimal places.

6. Plot $y = \sin 2x$, $-3\pi \leq x \leq 3\pi$ dotted and red. Label the axes.

7. Plot $y = (1 - x)^2 - 2$, $-1 \leq x \leq 3$ dash-dot and cyan. Label the axes.

8. Plot $y = \tan x$, $-\frac{\pi}{2} < x < \frac{\pi}{2}$ with blue points. Label the axes.

9. Plot $y = e^{-x^2}$, $-5 \leq x \leq 5$ solid and magenta. Label the axes.

10. Plot $x = \ln t$, $.1 \leq t \leq 5$ and only show x-values from -3 to 2. Label the axes.

11. Plot $x = \sqrt{t}$, $0 \leq t \leq 4$ and use title and legend. Label the axes.

12. Use `subplot` to plot $y = \sin x, \sin 2x, \cos x, \cos 2x$, $-2\pi \leq x \leq 2\pi$ in four different subfigures (in the same window).

13. Superimpose the plots of $y = \tan^{-1} x, y = \sin\left(\frac{1}{x}\right)$ on $0.1 \leq x \leq 10$ in the same figure.

14. Superimpose the plots of $y = \sin x, y = \cos x$ on $0 \leq x \leq \pi$ in the same figure with a stepsize of .01. Use `zoom` to approximate the intersection point to 4 decimal places.

15. Use `fzero` to find the intersection point of $y = \sin x, y = x^2 - 1$ on $0 \leq x \leq \pi/2$ to 8 decimal places. (Hint: To find the intersection of $\sin x$ and $\cos x$ at $5\pi/4$ we could type `fzero(inline('sin(x)-cos(x)'),4)` (where 4 is an initial guess "close" to the intersection point) and the intersection point would be displayed numerically).

16. Use `fzero` to find the intersection point of $y = \tan x, y = x^2 + 1$ on $3\pi \leq x \leq 7\pi/2$ to 8 decimal places. (Use the hint above.)

17. We know from Calculus that the Taylor expansion of $\cos x$ about $x = 0$ is given by

$$\cos x = 1 - \frac{x^2}{2!} + \frac{x^4}{4!} - \cdots = \sum_{k=0}^{\infty} \frac{(-1)^k x^{2k}}{(2k)!}, \quad \text{for } |x| < \infty.$$

Write a `Function` or `Script` in MATLAB that will take N, $x1$, and $x2$ (respectively, the number of terms in this expansion and the interval of comparison $[x1, x2]$) as inputs and will compare the actual function value with its approximation. Use the interval $[0, 6\pi]$ and compare the actual function with its expansion of 6 terms, 15 terms, and 30 terms (3 separate plots). Limit your viewing window on the vertical axis to $[-1.5\ 1.5]$.

18. From Calculus, we know that the infinite series

$$\sum_{k=1}^{\infty} \frac{1}{k} = 1 + \frac{1}{2} + \frac{1}{3} + \frac{1}{4} + \frac{1}{5} + \frac{1}{6} + \cdots$$

diverges. Write a MATLAB `Function` or `Script` that will calculate the first N terms of this series. Run the code for $N = 100, 1000, 10,000$.

19. Find the roots of $x^2 + 3x + 2$.

20. Find the roots of $x^2 + 3x + 4$.

21. Find the roots of $2x^2 - 3x + 4$.

22. Find the roots of $x^2 + 4x + 4$.

23. Find the roots of $x^3 + 3x + 4$.

24. Find the roots of $x^4 + 2x^2 + 1$.

25. Find the roots of $x^4 + 2x^2 + 5$.

26. Find the roots of $x^4 + 2x^2 + x - 5$.

A.2 Maple

The code in this book was written using Maple 18 in Worksheet Mode. Certain code may not work in earlier versions.

In Maple, we execute commands by entering them in the command window and pressing <Enter>. Any errors that were made can be corrected on the same line that they were typed. For example, the input

$(x^2 + 1;$

gives the error

```
Error, unable to match delimiters
```

and we can click with the mouse on this same input line and correct it as

$(x^2 + 1) ;$

and the command will be executed after we press <Enter>. In earlier versions of Maple or in the Classic Worksheet, every line needed to be ended with a semicolon; this is not the case in the Maple Worksheet. However, it is still the case that ending a line with a ":" will suppress the output.

Maple is a powerful tool for symbolically solving equations and manipulating them. We can integrate, differentiate, solve differential equations exactly, apply various techniques from linear algebra, and much more. We note that Maple 10 allows for entering objects such as fractions and integrals from a palette but since earlier versions of Maple do not have this feature, all code here is entered only from the keyboard.

In Maple, it is a good idea to always start your sessions by typing restart on the first line. If funny things start happening, you can go back to the beginning and re-execute everything, the first line clearing the Maple memory. To assign a variable a value, we use a ":=" instead of just an "=". It is often useful to assign each line a name so that you can refer to it later. One common way is to label them eq1, eq2, etc. (even if you technically have an expression instead of an equation). The exception for labeling statements is for any command that plots. If you label one of these, it will return the calculated points and not the desired plot. Sometimes this is desirable but most of the times it is not.

Anything following a "#" is ignored by Maple. Maple is also case sensitive. That is, the variables t and T are two different variables. Also, typing $\wedge$ will automatically go to a superscript whereas typing _ will cause everything that follows to be a subscript. Variables also continue to be defined/hold a value until we overwrite them or clear them. In contrast to MATLAB, if you execute a command and want to go back and change it, you can go and correct the line where the error occurred or where the change is desired. Once this is re-executed, the new value is the one in Maple's memory.

While everything in Maple can be entered with keyboard input, students usually find the Expression palette easier to navigate. We assume that here. Finally, when multiplying two things, we can use the palette to generate the "·" or equivalently just type "*" and the "·" will appear.

If we want to plot the function $y = f(x)$ we need to know the x-values that we are going to plug into our function. Let's try to plot $f(x) = x^3 - x$ between the x-values of -2 and 2. We first click on $f := a \to y$ in the Expression palette and then replace the a with x and the y with our function. The result will be the first line below; the second line plots the function.

$f := x \to x^3 - x$
$plot(f(x), x = -2..2)$

If we want to label the axes, title the graph, and change the x-y values that we see in the window to, say, $x \in [-\pi/2, \pi/2]$ and $y \in [-2, 2]$, we could do so with the commands

$$plot\left(f(x), x = -\frac{\pi}{2}..\frac{\pi}{2}, y = -2..2, labels = [x, y], title = "\text{Plot of f(x)=x}^\wedge\text{3-x}"\right)$$

We note that the fraction $\frac{\pi}{2}$ was entered using the Expression palette (for the fraction symbol $\frac{a}{b}$) and the Common Symbols palette (for the symbol π). We could also do things like plot the graph in a different color or super-impose another graph on it. Suppose we wanted to also graph the function $f(x) = x^2 - 1$ in the color green and on the same graph. We could use the commands

$g := x \to x^2 - 1$

$plot\left([f(x), g(x)], x = -\dfrac{\pi}{2}..\dfrac{\pi}{2}, y = -2..2, labels = [x, y], color = [red, green],\right.$

$\qquad title = "\text{Plot of f(x)=x\textasciicircum3-x and x\textasciicircum2 -1}"\right)$

A.2.1 Some Helpful Maple Commands

Maple also has a help section that can be accessed with the mouse or the keyboard. If you know the command you want to use but forget the proper syntax, the keyboard version is probably the way to go. Suppose the command you have questions about is the `solve` command. In the Maple window, you could enter `?solve` to open the help page for this command, you could enter `??solve` to see the syntax for this command, or you could enter `???solve` to see some examples of this command in use.

Some commands that you might use are:
`simplify, expand, evalf, evalc, allvalues, eliminate, coeff, diff, Diff, exp, fsolve, lhs, rhs, plot, roots, solve, subs, sqrt, assume, combine, mtaylor.` There are also commands for linear algebra and vector calculus including `Eigenvectors, Eigenvalues, Jacobian, Transpose, Matrix, SubMatrix, Multiply, Trace, MatrixInverse, Row, Column.`

This list is not exhaustive but the commands frequently come up in computations. There are also numerous options for plotting and the reader is again referred to the help menu for instructions. *Some of the above commands, as well as others that are not on the list, require a specific* **Maple package** *to be loaded first.* Four of the common packages that we will take advantage of are `plots, DEtools, LinearAlgebra, VectorCalculus`. E.g., we load the linear algebra package by typing `with(LinearAlgebra):` and note that it's usually best to end the line with a colon. If we ended with a semicolon instead, we would see all the commands that are included in this package. If we are just beginning, this may be helpful but after a while it's not necessary to see these all the time. See the help menu for instructions on using the various commands and additional packages.

Example 1:
Some basic commands and examples of plotting in Maple.
$Pi + \exp(1) + \ln(4)$

$\pi + e^3 + \ln\left(\dfrac{1}{2}\right)$

#The Pi *symbol above is from the Common Symbols palette; for the* e^3 *first go to Expression palette and click on* a^b*, then highlight the a and click on* e *in the Common Symbols and then highlight the b and type 3; the* ln *is also entered from the palette*

$\left\{ evalf \left(\text{Pi} + \exp(1) + \ln(4) \right), evalf \left(\pi + e^3 + \ln \left(\frac{1}{2} \right) \right) \right\}$

$eq1 := x^2 - 1 + \sin(x)$

$plot \left(eq1, x = -2..\frac{\pi}{2} \right)$

$plot \left(eq1, x = -2..\frac{\pi}{2}, y = -1.5..0.5, thickness = 4, labels = [x, y], \right.$
  $labelfont = [12, 12], color = blue, legend = ["x^2-1+\sin(x)"], linestyle =$
  $3, title = "A \text{ basic plot }")$

$f := x \rightarrow \sqrt{x+2} + \ln(x+2)$ #*Entered from Expression palette,* $f := a \rightarrow y$

$plot(f(x), x = -2..1)$

$plot([eq1, f(x)], x = -2..1, y = -1.5..2, labels = [x, y], color =$
  $[blue, maroon], legend = ["x^2-1+\sin(x)", "\text{sqrt}(x+2)+\ln(x+2)"],$
  $linestyle = [1, 2], title = "\text{Superimposed plots}")$

$with(plots) :$

$eq2 := plot(\ln(x), x = 0..3, y = -2..2, linestyle = 1, color = blue)$

$eq3 := plot \left(e^{-x^2}, x = 0..3, y = -2..2, linestyle = 3, color = black \right)$

$display([eq2, eq3], title = "\ln(x) \text{ vs. } \exp(-x^2)")$

Example 2: Solving algebraic equations.

$restart$

$f := x \rightarrow x^4 + 2 \cdot x^3 - 8 x^2$ #*The $\cdot$ is from the Expression palette or typed as* $*$

$f(3)$

$subs(x = 3, f(x))$

$sol1 := solve(f(x), x)$

$sol1[4]$

$roots(f(x))$

$factor(f(x))$

$g := x \rightarrow 1 - x^2$

$f(x) \cdot g(x)$

$factor(f(x) \cdot g(x))$

$expand(f(x) \cdot g(x))$

$\dfrac{f(x)}{x}$

$simplify \left(\dfrac{f(x)}{x} \right)$

$h := x \rightarrow x^2 + 2 \cdot x + 4$

$sol2 := solve(h(x), x)$

$sol2[1]$

$Re(sol2[1])$

$Im(sol2[1])$

$2 \cdot i + sol2[1]$ #*Enter i from expression palette or type as* I

$evalc(2 \cdot i + sol2[2])$

$eq1 := x^4 + 3 \cdot x^2 - 7 \cdot x + 6$

$eq2 := solve(eq1)$

allvalues(*eq2*[1])
evalf(*allvalues*(*eq2*[1]))
[*evalf*(*allvalues*(*eq2*[1])), *evalf*(*allvalues*(*eq2*[2])), *evalf*(*allvalues*(*eq2*[3])),
  evalf(*allvalues*(*eq2*[4]))]

A.2.2 Programming in Maple

We can also write code in Maple. This is often done with a procedure. One thing to keep in mind when writing these, Maple keeps track of what is **global** and what is **local**. Any intermediate computations that are done are not saved nor can they be displayed. Thus if you need an intermediate computation for later, be sure to declare it as a **global** variable.

Example 3: Create a **procedure** in Maple that approximates $\sin x$ by its Taylor series about $x = 0$. The **procedure** takes N and $x1$ as inputs, where N is the number of terms in the expansion and the comparison between $\sin x$ and its Taylor series will be done over $[-x1,x1]$.

restart;
  #*when entering the code below(from the line containing* **proc** *until the line containing* **end proc***), you must use "Shift-Enter" to go to the next line!*
SineTaylorSeries := **proc**(*N*, *x1*)
local *SinePlot*, *k*;
description "this function compares the Taylor expansion plot of sine about
  x=0 to the actual function";
  #*This procedure takes* N, $x1$ *as inputs.* N*is number of terms in the expansion of sin(x) about x=0 with the comparison over [-x1, x1].*
SinePlot := 0;
for *k* **from** 0 **to** *N* **do**
$$SinePlot := SinePlot + \frac{(-1)^k \cdot x^{2 \cdot k + 1}}{(2 \cdot k + 1)!}$$
end do;
plot([*SinePlot*, sin(*x*)], *x* = -*x1*..*x1*, *color* = [*blue*, *red*], *style* = [*line*, *point*])
end proc : #*now you may finally just click "Enter"*
SineTaylorSeries(2, 3) #*this will compare the (2+1) term Taylor expansion
                              of sine over [-3,3]*
SineTaylorSeries(3, 2)
SineTaylorSeries(3, 6)

One final comment: the reader should also be aware of the difference between the various palettes and their typed versions. For example, if we type **pi** and subtract from it the symbol π from the Common Symbols palette, that is, **pi**$-\pi$, the answer does not simplify; however, **Pi**$-\pi$ simplifies to 0. A similar thing happens for Greek letters spelled out vs. entered from the Greek palette: Gamma$-\Gamma$ does not simplify but GAMMA$-\Gamma$ simplifies to 0.

Exercises

The remainder of this lab will make use of many Maple commands, including plot (using options such as title, color), solve, subs, evalf, simplify, rhs. Submit only the answers and output requested to the problems given here. (Remember that you can write over an incorrect Maple command and re-execute it, thus saving lots of space due to typos/errors.) Highlight or clearly mark the answers that you are asked to provide.

1. Enter the commands given in Maple Example 1 and submit both your input and output.

2. Enter the commands given in Maple Example 2 and submit both your input and output.

3. Enter the commands given in Maple Example 3 and submit both your input and output.

4. Determine which of e^π or π^{e^1} is the larger value. Use Digits:=15 to write your answers correct to 12 decimal places.

5. Determine which of $\sin(\ln(\sqrt{.7}))$ or $\ln(\sin(\sqrt{.7}))$ is the larger value. Use Digits:=15 to write your answers correct to 12 decimal places.

6. Plot $y = \sin 2x$, $-3\pi \le x \le 3\pi$ dotted and red. Label the axes.

7. Plot $y = (1-x)^2 - 2$, $-1 \le x \le 3$ dash-dot and cyan. Label the axes.

8. Plot $y = \tan x$, $-\frac{\pi}{2} < x < \frac{\pi}{2}$ with blue points. Label the axes.

9. Plot $y = e^{-x^2}$, $-5 \le x \le 5$ solid and magenta. Label the axes.

10. Plot $x = \ln t$, $.1 \le t \le 5$ and only show x-values from -3 to 2. Label the axes.

11. Plot $x = \sqrt{t}, 0 \le t \le 4$ and use title and legend. Label the axes.

12. Superimpose the plots $y = \sin x, \sin 2x, \cos x, \cos 2x$ on $-2\pi \le x \le 2\pi$ in the same figure.

13. Superimpose the plots of $y = \tan^{-1} x, y = \sin\left(\frac{1}{x}\right)$ on $0.1 \le x \le 10$ in the same figure.

14. Superimpose the plots of $y = \sin x, y = \cos x$, $0 \le x \le \pi$ in the same figure.

15. Use fsolve to find the intersection point of $y = \sin x, y = x^2 - 1$ on $0 \le x \le \pi/2$ to 8 decimal places.

16. Use `fsolve` to find the intersection point of $y = \tan x, y = x^2 + 1$ on $3\pi \leq x \leq 7\pi/2$ to 8 decimal places.

17. We know from Calculus that the Taylor expansion of $\cos x$ about $x = 0$ is given by

$$\cos x = 1 - \frac{x^2}{2!} + \frac{x^4}{4!} - \cdots = \sum_{k=0}^{\infty} \frac{(-1)^k x^{2k}}{(2k)!}, \quad \text{for } |x| < \infty.$$

Write a **procedure** in Maple that will take N, $x1$, and $x2$ (respectively, the number of terms in this expansion and the interval of comparison $[x1, \ x2]$) as inputs and will compare the actual function value with its approximation. Use the interval $[0, 6\pi]$ and compare the actual function with its expansion of 6 terms, 15 terms, and 30 terms (3 separate plots). Limit your viewing window on the vertical axis to $[-1.5 \ 1.5]$.

18. From Calculus, we know that the infinite series

$$\sum_{k=1}^{\infty} \frac{1}{k} = 1 + \frac{1}{2} + \frac{1}{3} + \frac{1}{4} + \frac{1}{5} + \frac{1}{6} + \cdots$$

diverges. Write a Maple **procedure** that will calculate the first N terms of this series. Run the code for $N = 100, 1000, 10,000$. Now use the Maple command **sum** and calculate the same 3 sums.

19. Find all roots of $x^2 + 3x + 2$.

20. Find all roots of $x^2 + 3x + 4$.

21. Find all roots of $2x^2 - 3x + 4$.

22. Find all roots of $x^2 + 4x + 4$.

23. Find all roots of $x^3 + 3x + 4$.

24. Find all roots of $x^4 + 2x^2 + 1$.

25. Find all roots of $x^4 + 2x^2 + 5$.

26. Find all roots of $x^4 + 2x^2 + x - 5$.

A.3   Mathematica

The code in this book was written using Mathematica 10. Certain code may not work in earlier versions.

In Mathematica, we execute commands by entering them and holding down <Shift> and then pressing <Enter>. Unlike MATLAB and the Classic Work-sheet version of Maple, there is no command prompt (so just start typing). Any errors that were made can be corrected on the same line that they were typed. For example, the input

```
y[x-]=x^2+1;
```

gives the error (click on the "+" in the box to the right)

Syntax::sntxf: "y[x-" cannot be followed by "]". More...

and we can click with the mouse on this same input line and correct it as

```
y[x_]=x^2+1;
```

and the command will be executed after pressing <Shift> <Enter>.

Mathematica is comparable to Maple in its ability to symbolically solve and manipulate equations (both are quite good). We can integrate, differentiate, solve differential equations exactly, apply various techniques from linear algebra, and much more.

In correcting input lines, Mathematica's kernel (i.e., computation engine) sometimes gets confused. When this happens, you should go up to **Kernel** $\longrightarrow$ **Quit Kernel** $\longrightarrow$ **Local** or simply type **Quit[]** and then re-execute the relevant commands. If your line ends with a semicolon ";" then the output is suppressed. If your line has nothing then the output is shown. To assign a variable a value, we use "=" unless it is a function. For functions we must type the variable name followed by an underscore as in the first example above. It is often useful to assign each line a name so that you can refer to it later. One common way is to label them **eq1, eq2**, etc. (even if you technically have an expression instead of an equation). Plots can also be labeled, too, which helps when you want to superimpose various plots.

Anything inside (* *) is ignored by Mathematica. Mathematica is also case sensitive. That is, the variables **t** and **T** are two different variables. Variables also continue to be defined/hold a value until we overwrite them or clear them.

If we want to plot the function $y = f(x)$ we need to know the x-values that we are going to plug into our function. It is often useful to enter the equation first to make sure there are no typos and then ask Mathematica to plot it. Let's try to plot $f(x) = x^3 - x$ between the x-values of -2 and 2. We could enter this equation with the keyboard only as

```
f[x_]=x^3-x
```

but it's often easier to enter it using the palette as

```
f[x_]=x³ − x
Plot[f[x],{x,-2,2}]
```

The first line defines the equation as a function of the variable x and can be read as "f is the function defined as $x^3 - x$," while the second line says to plot the function f[x] from $x = -2$ to $x = 2$. Note that the underscore was needed to define the function but not when it was referred to at a later step. If we want to label the axes, title the graph, and change the x-y values that we see in the window to, say, $x \in [-\pi/2, \pi/2]$ and $y \in [-2, 2]$, we could do so with the commands

$$\text{Plot}\left[f[x], \left\{x, -\frac{\pi}{2}, \frac{\pi}{2}\right\}, \text{PlotRange} \rightarrow \{-2, 2\}, \text{AxesLabel} \rightarrow \{"x", "y"\},\right.$$
$$\left.\text{PlotLabel} \rightarrow "\text{Plot of } f(x) = x^3 - x"\right]$$

We can also do things like plot the graph in a different color or superimpose another graph on it. Suppose we wanted to also graph the function $f(x) = x^2 - 1$ in the color green and on the same graph.

$$\text{g}[x\_]=x^2-1$$
$$\text{Plot}\left[\{f[x], g[x]\}, \left\{x, -\frac{\pi}{2}, \frac{\pi}{2}\right\}, \text{PlotRange} \rightarrow \{-2, 2\}, \text{AxesLabel} \rightarrow \{"x", "y"\},\right.$$
$$\left.\text{PlotStyle} \rightarrow \{\text{Blue}, \text{Green}\}, \text{PlotLabel} \rightarrow "\text{Plot of } x^3 - x \text{ vs. } x^2 - 1"\right]$$

A.3.1 Some Helpful Mathematica Commands

Mathematica has a help section that can be accessed with the mouse to find many commands and the proper syntax to use them. Some commands that you might use are: Simplify, FullSimplify, Expand, N, ReplaceAll, Coefficient, D, E, Solve, NSolve, NDSolve, Plot FindRoot, ExpandAll, Series, Eliminate, Integrate. There are also commands for linear algebra including Eigenvalues, Eigenvectors, Eigensystem, CharacteristicPolynomial, Row, Column MatrixForm, LinearSolve, Tr, Det, Transpose.

This list is not exhaustive but the commands frequently come up in computations. We can sometimes enter a command in two different ways—with the keyboard or with the palette. E.g., Integrate can be used or we could click on the integral sign. There are also numerous options for plotting and the reader is again referred to the help menu for instructions. *Additional commands may be available by loading a specific* **Mathematica package**. We refer the reader to the help menu for instructions on using the various commands and additional packages.

The reader is encouraged to go through the following example in detail, making sure that the syntax of each line is understood.

Example 1: Some sample commands from Mathematica.

$\pi + e^1 + \text{Log}[4]$ (*entered from palette*)
N[Pi E^1+Log[4]] (*entered from keyboard*)
N[$\pi + e^1 + \text{Log}[4]$] (*gives a decimal answer*)
eq1[x_]=$x^2 - 1 + \text{Sin}[x]$
Plot[eq1[x], $\{x, -2, \frac{\pi}{2}\}$]
Plot $\left[\text{eq1}[x], \left\{x, -2, \frac{\pi}{2}\right\}, \text{PlotRange} \rightarrow \{-1.5, 1.5\}, \text{AxesLabel} \rightarrow \{"x", "y"\},\right.$
   PlotStyle$\rightarrow$\{Blue, Dotted, Thick\}, PlotLegends$\rightarrow$"Expressions",
   PlotLabel$\rightarrow$"A basic plot"]
f[x_]= $\sqrt{x+2} + \text{Log}[x+2]$
Plot[f[x],{x,-2,1}]
Plot [\{eq1[x], f[x]\}, $\{x, -2, 1\}$, PlotRange$\rightarrow$\{-1.5, 2\}, AxesLabel$\rightarrow$\{"x", "y"\},
   PlotStyle$\rightarrow$\{\{Blue, Dashed, Thick\}, \{Magenta, Thick\}\},
   PlotLegends$\rightarrow$"Expressions", PlotLabel$\rightarrow$"Superimposed plots"]
eq2[x_] = Plot[Log[x], $\{x, 0, 3\}$, PlotRange$\rightarrow$\{-2, 2\}, PlotStyle$\rightarrow$\{Blue\}]
eq3[x_] = Plot $\left[e^{-x^2}, \{x, 0, 3\}, \text{PlotRange} \rightarrow \{-2, 2\},\right.$

   PlotStyle$\rightarrow$\{Black, Dashed, Thick\}$\Big]$
Show[eq2[x], eq3[x], PlotLabel$\rightarrow$ "ln(x) vs. exp($-x^2$)"]

Example 2: Solving algebraic equations with Mathematica.

Quit[ ](*open bracket, close bracket, no space*)
f[x_]= $x^4 + 2x^3 - 8x^2$
(*Note: multiplication is done with '*' or with a <Space>*)
f[3]
ReplaceAll[f[x], x$\rightarrow$3]
Sol1 = Solve[f[x] == 0, x] (*'double equal' is needed for solving*)
Sol1[[4]]
Sol1[[4]][[1]]
Roots[f[x] == 0, x]
Factor[f[x]]
g[x_]=$1 - x^2$
f[x] g[x]
Factor[f[x] g[x]]
Expand[f[x] g[x]]
$\frac{f[x]}{x}$

FullSimplify$\left[\frac{f[x]}{x}\right]$
h[x_]= $x^2 + 2x + 4$
Sol2 = Solve[h[x] == 0, x]
Sol2[[1]]
Sol2a = Sol2[[1]][[1]][[2]]

```
Re[Sol2a]
Im[Sol2a]
2 i + Sol2a  (*i entered from palette*)
eq1[x_] = x⁴ + 3 x² - 7 x + 6
eq2 = Solve[eq1[x] == 0, x]
N[eq2]
```

A.3.2 Programming in Mathematica

We can also write code in Mathematica. This is often done with functions and various combinations of `For`-statements, `if`-statements, etc. We will use ":=" to tell Mathematica that the rhs will be the delayed value assigned.

Example : Create functions using `Module` in Mathematica that approximates $\sin x$ by its Taylor series about $x = 0$. You should create a function that takes N and $x1$ as inputs, where $N+1$ is the number of terms in the expansion and the comparison between $\sin x$ and its Taylor series will be done over $[-x1, x1]$.

```
Quit[]
SineTaylorSeries[N_, x1_] := Module[{k, x},
  SinePlot[x_] = 0;
  For[k = 0, k ≤ N, k++,
                                   (-1)ᵏx²ᵏ⁺¹
    SinePlot[x_] = SinePlot[x] + ───────────];
                                    (2k + 1)!
  Plot[{Sin[x], SinePlot[x]}, {x, -x1, x1},
    PlotStyle→ {Blue, {Red, Dashing[Tiny]}}]]
SineTaylorSeries[3,6]
SineTaylorSeries[2,3]
SineTaylorSeries[8,10]
```

Exercises

The remainder of this lab will make use of many Mathematica commands, including `Plot`, `Solve`, `NSolve`, `FullSimplify`. Submit only the answers and output requested to the problems given here. (Remember that you can write over an incorrect Mathematica command and re-execute it, thus saving lots of space due to typos/errors.) Highlight or clearly mark the answers that you are asked to provide.

1. Enter the commands given in Mathematica Example 1 and submit both your input and output.

2. Enter the commands given in Mathematica Example 2 and submit both your input and output.

3. Enter the commands given in Mathematica Example 3 and submit both your input and output.

4. Determine which of e^π or π^{e^1} is the larger value. Use `Digits:=15` to write your answers correct to 12 decimal places.

5. Determine which of $\sin(\ln(\sqrt{.7}))$ or $\ln(\sin(\sqrt{.7}))$ is the larger value. Use `Digits:=15` to write your answers correct to 12 decimal places.

6. Plot $y = \sin 2x$, $-3\pi \le x \le 3\pi$ dotted and red. Label the axes.

7. Plot $y = (1-x)^2 - 2$, $-1 \le x \le 3$ dashed and magenta. Label the axes.

8. Plot $y = \tan x$, $-\frac{\pi}{2} < x < \frac{\pi}{2}$ with blue points. Label the axes.

9. Plot $y = e^{-x^2}$, $-5 \le x \le 5$ solid and green. Label the axes.

10. Plot $x = \ln t$, $.1 \le t \le 5$ and only show x-values from -3 to 2. Label the axes.

11. Plot $x = \sqrt{t}, 0 \le t \le 4$ and use title and legend. Label the axes.

12. Superimpose the plots $y = \sin x, \sin 2x, \cos x, \cos 2x$ on $-2\pi \le x \le 2\pi$ in the same figure.

13. Superimpose the plots of $y = \tan^{-1} x, y = \sin\left(\frac{1}{x}\right)$ on $0.1 \le x \le 10$ in the same figure.

14. Superimpose the plots of $y = \sin x, y = \cos x$, $0 \le x \le \pi$ in the same figure.

15. Use NSolve to find the intersection point of $y = \sin x, y = x^2 - 1$ on $0 \le x \le \pi/2$ to 8 decimal places.

16. Use NSolve to find the intersection point of $y = \tan x, y = x^2 + 1$ on $3\pi \le x \le 7\pi/2$ to 8 decimal places.

17. We know from Calculus that the Taylor expansion of $\cos x$ about $x = 0$ is given by

$$\cos x = 1 - \frac{x^2}{2!} + \frac{x^4}{4!} - \cdots = \sum_{k=0}^{\infty} \frac{(-1)^k x^{2k}}{(2k)!}, \quad \text{for } |x| < \infty.$$

Write a function(s) in Mathematica that will take N, $x1$, and $x2$ (respectively, the number of terms in this expansion and the interval of comparison $[x1, x2]$) as inputs and will compare the actual function value with its approximation. Use the interval $[0, 6\pi]$ and compare the actual function with its expansion of 6 terms, 15 terms, and 30 terms (3 separate plots). Limit your viewing window on the vertical axis to $[-1.5\ 1.5]$.

18. From Calculus, we know that the infinite series

$$\sum_{k=1}^{\infty} \frac{1}{k} = 1 + \frac{1}{2} + \frac{1}{3} + \frac{1}{4} + \frac{1}{5} + \frac{1}{6} + \cdots$$

diverges. Write a function(s) in Mathematica that will calculate the first N terms of this series. Run the function for $N = 100, 1000, 10,000$. Now use the Mathematica command Sum and calculate the same 3 sums.

19. Find all roots of $x^2 + 3x + 2$.

20. Find all roots of $x^2 + 3x + 4$.

21. Find all roots of $2x^2 - 3x + 4$.

22. Find all roots of $x^2 + 4x + 4$.

23. Find all roots of $x^3 + 3x + 4$.

24. Find all roots of $x^4 + 2x^2 + 1$.

25. Find all roots of $x^4 + 2x^2 + 5$.

26. Find all roots of $x^4 + 2x^2 + x - 5$.

Appendix B

B.1   A Primer on Matrix Algebra

This section is a concise introduction to matrix algebra. We assume no prior knowledge of matrices.

As a motivation, we consider a system of two first-order differential equations

$$\frac{dx}{dt} = -5x + 8y$$

$$\frac{dy}{dt} = -4x + 7y.$$

We want to find a compact way to write and ultimately solve this system. We can write the coefficients of the variables in a rectangular array. For reasons we will see shortly, the way we do this is

$$\begin{bmatrix} \dfrac{dx}{dt} \\ \dfrac{dy}{dt} \end{bmatrix} = \begin{bmatrix} -5 & 8 \\ -4 & 7 \end{bmatrix} \begin{bmatrix} x \\ y \end{bmatrix}.$$

We say that

$$\begin{bmatrix} \dfrac{dx}{dt} \\ \dfrac{dy}{dt} \end{bmatrix} \quad \text{and} \quad \begin{bmatrix} x \\ y \end{bmatrix}$$

are **vectors** and

$$\begin{bmatrix} -5 & 8 \\ -4 & 7 \end{bmatrix}$$

is the **coefficient matrix** of our original system of equations. If we are not dealing with differential equations, the left-hand side may be constant but we can still write our system in the form of vectors and matrices.

Simply put, a **matrix** is a rectangular array of numbers. The horizontal lists are called rows, and the vertical lists are called columns. For example, let

$$\mathbf{A} = \begin{bmatrix} 2 & 3 & 4 \\ 0 & 1 & -2 \end{bmatrix}.$$

This matrix has two rows: $\begin{bmatrix} 2 & 3 & 4 \end{bmatrix}$ and $\begin{bmatrix} 0 & 1 & -2 \end{bmatrix}$. It has three columns: $\begin{bmatrix} 2 \\ 0 \end{bmatrix}$, $\begin{bmatrix} 3 \\ 1 \end{bmatrix}$, and $\begin{bmatrix} 4 \\ -2 \end{bmatrix}$.

We refer to the size of a matrix as its number of rows by its number of columns. A 2×3 matrix (read "two by three matrix") is a matrix with two rows and three columns. Although we sometimes will talk abstractly about vectors, for now when we say **vector**, we mean the traditional understanding as a matrix with either one row or one column; a vector with only one row is called a **row vector**, and a vector with only one column is called a **column vector**. Our concepts of **vector spaces** (see Section 5.3) apply to these "vectors" as well.

We frequently need to refer to specific elements of a matrix, and we use the row and column information to provide an address for each element. For example, the ijth element is the element in the ith row and the jth column (rows are always listed before columns). Thus we denote a generic matrix by

$$\mathbf{A} = \begin{bmatrix} a_{11} & a_{12} & \cdots & a_{1n} \\ a_{21} & a_{22} & \cdots & a_{2n} \\ \vdots & \vdots & \ddots & \vdots \\ a_{m1} & a_{m2} & \cdots & a_{mn} \end{bmatrix}$$

or more succinctly by $\mathbf{A} = (a_{ij})_{m \times n}$. We call a matrix **square** if the number of rows is equal to the number of columns.

Matrices can be considered to be a number system much as the integers, the real numbers, or the complex numbers are number systems. As with other number systems we define ways to add, subtract, multiply, and divide matrices. Thus we construct an algebra for matrices which we call, appropriately enough, matrix algebra. We again note that a vector (in the traditional multivariable calculus context) is simply a matrix with only one column or one row and thus the statements about matrices apply to vectors as well. Two matrices can be added provided that they have the same size; thus two 2×3 matrices can be added together, but a 2×3 cannot be added to a 3×5 matrix. Addition and subtraction are defined elementwise. Thus $\mathbf{A} + \mathbf{B} = \mathbf{C}$ means that for all i and j, $c_{ij} = a_{ij} + b_{ij}$. Similarly $\mathbf{A} - \mathbf{B} = \mathbf{C}$ means $c_{ij} = a_{ij} - b_{ij}$. For example

$$\begin{bmatrix} a & b \\ c & d \end{bmatrix} + \begin{bmatrix} e & f \\ g & h \end{bmatrix} = \begin{bmatrix} a+e & b+f \\ c+g & d+h \end{bmatrix},$$

$$\begin{bmatrix} a & b \\ c & d \end{bmatrix} - \begin{bmatrix} e & f \\ g & h \end{bmatrix} = \begin{bmatrix} a-e & b-f \\ c-g & d-h \end{bmatrix}.$$

In algebra on the real numbers, 0 is a distinguished number called the additive identity. This means that for any real a, $a + 0 = a$ and $0 + a = a$. There is a matrix (for each size of matrix) playing this same role for matrix arithmetic. It is called the **zero matrix** and is the matrix consisting of all zeros, and denoted by $\mathbf{0}$. It is easy to verify that

$$\mathbf{A} + \mathbf{0} = \mathbf{0} + \mathbf{A} = \mathbf{A}.$$

There are actually two number systems involved in matrix algebra: the matrices themselves and the elements that make up the matrices. We distinguish these by calling the former matrices and the latter **scalars**. Technically the 1×1 matrix (a) is a matrix, while a is a scalar. Whenever feasible, matrices are denoted by capital letters and scalars are denoted by lower case letters. We occasionally need to multiply scalars and matrices together. If s is a scalar and $\mathbf{A}$ is a matrix, we define a new matrix $s\mathbf{A}$ to be the matrix that results from multiplying all the entries of $\mathbf{A}$ by s. Thus if $\mathbf{B} = s\mathbf{A}$, then $b_{ij} = sa_{ij}$. For example,

$$s \begin{bmatrix} a & b \\ c & d \end{bmatrix} = \begin{bmatrix} sa & sb \\ sc & sd \end{bmatrix}.$$

Multiplication in matrix algebra is, in general, much stranger than multiplication in the real numbers. While the multiplication by a scalar that we just defined is similar to addition and subtraction of matrices, there is a major conceptual difference. Multiplication of a matrix by a scalar is the multiplication of two different types of objects, namely, scalars and matrices. It is useful to define a multiplication between two matrices. But first we consider the special case of multiplying a matrix and a vector. If we consider an $m \times n$ matrix, the vector we consider must be of size $n \times 1$. It is perhaps most consistent to consider the multiplication of a matrix $\mathbf{A}$ and a vector $\mathbf{x}$ as a *linear combination of the columns of* $\mathbf{A}$. The idea of a linear combination is the same as that used in Section 3.2 in which we combined solutions of a linear equation and the result was still a solution of the same linear equation. Here, the equivalent statement is that we take a linear combination of the columns (of size $m \times 1$) of $\mathbf{A}$ and end up with a vector (still of size $m \times 1$) that is in the **column space** of $\mathbf{A}$ (see Section 5.2 for further discussion of this). For example,

$$\begin{bmatrix} 1 & 2 & 3 \\ 4 & 5 & 6 \\ 7 & 8 & 9 \end{bmatrix} \begin{bmatrix} 2 \\ 1 \\ -1 \end{bmatrix} = 2 \cdot \begin{bmatrix} 1 \\ 4 \\ 7 \end{bmatrix} + 1 \cdot \begin{bmatrix} 2 \\ 5 \\ 8 \end{bmatrix} - 1 \cdot \begin{bmatrix} 3 \\ 6 \\ 9 \end{bmatrix}$$

$$= \begin{bmatrix} 1 \\ 7 \\ 13 \end{bmatrix}.$$

In order to be able to use the rows of the vector $\mathbf{x}$ as coefficients used to take linear combinations of the columns of $\mathbf{A}$, it is essential that the number of columns of $\mathbf{A}$ be equal to the number of rows of $\mathbf{x}$. Thus, we have the rule

$$\mathbf{A}_{m \times n} \mathbf{x}_{n \times 1} = \mathbf{b}_{m \times 1}, \tag{B.1}$$

which governs the size of the matrices and vectors that we can multiply together. In general, we have

$$
\begin{bmatrix} a_{11} & a_{12} & \cdots & a_{1n} \\ a_{21} & a_{22} & \cdots & a_{2n} \\ \vdots & \vdots & \ddots & \vdots \\ a_{m1} & a_{m2} & \cdots & a_{mn} \end{bmatrix} \begin{bmatrix} x_1 \\ x_2 \\ \vdots \\ x_n \end{bmatrix} = x_1 \begin{bmatrix} a_{11} \\ a_{21} \\ \vdots \\ a_{m1} \end{bmatrix} + x_2 \begin{bmatrix} a_{12} \\ a_{22} \\ \vdots \\ a_{m2} \end{bmatrix} + \cdots + x_n \begin{bmatrix} a_{1n} \\ a_{2n} \\ \vdots \\ a_{mn} \end{bmatrix}
$$

$$
= \begin{bmatrix} x_1 a_{11} + x_2 a_{12} + \cdots + x_n a_{1n} \\ x_1 a_{21} + x_2 a_{22} + \cdots + x_n a_{2n} \\ \vdots \\ x_1 a_{m1} + x_2 a_{m2} + \cdots + x_n a_{mn} \end{bmatrix}
$$

$$
= \begin{bmatrix} b_1 \\ b_2 \\ \vdots \\ b_m \end{bmatrix}. \tag{B.2}
$$

This is simply an alternate way of writing this system of m equations in n unknowns and we refer to the matrix $\mathbf{A}$ in such situations as the coefficient matrix of the linear system.

When we consider multiplication of two matrices, $\mathbf{AB}$, we can easily think of doing matrix vector multiplication, with the vectors $\mathbf{x}$ of the previous matrix-vector multiplication now being the columns of $\mathbf{B}$. Let

$$
\mathbf{A} = (a_{ij})_{m \times n} \text{ and } \mathbf{B} = (b_{ij})_{n \times p}.
$$

We can think of the resulting columns of $\mathbf{C}_{m \times p} = \mathbf{AB}$ as

$$
c_j = b_{1j} \begin{bmatrix} a_{11} \\ a_{21} \\ \vdots \\ a_{m1} \end{bmatrix} + b_{2j} \begin{bmatrix} a_{12} \\ a_{22} \\ \vdots \\ a_{m2} \end{bmatrix} + \cdots + b_{nj} \begin{bmatrix} a_{1n} \\ a_{2n} \\ \vdots \\ a_{mn} \end{bmatrix}. \tag{B.3}
$$

In terms of multiplication by the elements of the matrices, each entry, c_{ij}, of the product $\mathbf{C} = \mathbf{AB}$ is defined by

$$
c_{ij} = a_{i1}b_{1j} + a_{i2}b_{2j} + \cdots + a_{ir}b_{rj}.
$$

In a sense what we have done is multiply the ith row of $\mathbf{A}$ by the jth column of $\mathbf{B}$. For the 2×2, we have

$$
\begin{bmatrix} a & b \\ c & d \end{bmatrix} \begin{bmatrix} e & f \\ g & h \end{bmatrix} = \begin{bmatrix} ae + bg & af + bh \\ ce + dg & cf + dh \end{bmatrix}.
$$

You can multiply matrices of different sizes, but the number of columns of the matrix on the left must equal the number of rows of the matrix on the right:

$$
\begin{bmatrix} 1 & 3 \\ 2 & 1 \end{bmatrix} \begin{bmatrix} -1 & 3 & 2 \\ 2 & -3 & 4 \end{bmatrix} = \begin{bmatrix} 5 & -6 & 14 \\ 0 & 3 & 8 \end{bmatrix}.
$$

Another way of viewing multiplication of matrices, $\mathbf{AB}$, is $\mathbf{A}$ acting on the columns of $\mathbf{B}$:

$$\mathbf{AB} = \mathbf{A} \begin{bmatrix} | & | & & | \\ b_1 & b_2 & \cdots & b_n \\ | & | & & | \end{bmatrix} = \begin{bmatrix} | & | & & | \\ \mathbf{A}b_1 & \mathbf{A}b_2 & \cdots & \mathbf{A}b_n \\ | & | & & | \end{bmatrix}. \quad (\text{B.4})$$

We could use this to show a few nice rules for matrix multiplication:

Matrix multiplication is associative: $\quad \mathbf{A(BC)} = \mathbf{(AB)C} = \mathbf{ABC}$,

Matrix operations are distributive: $\quad \begin{cases} \mathbf{A(B + C)} = \mathbf{AB} + \mathbf{AC}, \\ \mathbf{(B + C)D} = \mathbf{BD} + \mathbf{CD}. \end{cases}$

However, matrix multiplication does not commute in general: $\mathbf{AB} \neq \mathbf{BA}$. The specific example

$$\begin{bmatrix} 1 & 2 \\ 3 & 4 \end{bmatrix} \begin{bmatrix} 0 & 1 \\ 1 & 0 \end{bmatrix} = \begin{bmatrix} 2 & 1 \\ 4 & 3 \end{bmatrix},$$

$$\begin{bmatrix} 0 & 1 \\ 1 & 0 \end{bmatrix} \begin{bmatrix} 1 & 2 \\ 3 & 4 \end{bmatrix} = \begin{bmatrix} 3 & 4 \\ 1 & 2 \end{bmatrix},$$

shows that the order of the matrices when we multiply does make a difference.

After you have practiced multiplying some matrices, you should become convinced that, while more complicated than real number multiplication, matrix multiplication is not very hard to do. Despite not being very hard, it is cumbersome and not very pleasant if the matrices are large and many need to be multiplied together. Matrix multiplication for large matrices is ideal for a calculator or computer. Indeed many calculators and software packages are able to perform matrix multiplication, and we assume that you have access to one and have mastered its use from this point on.

In what follows, we frequently need to multiply a matrix by itself repeatedly. The analogous notation from real number multiplication carries through. The product $\mathbf{AA}$ is written $\mathbf{A}^2$, $\mathbf{AAA}$ is $\mathbf{A}^3$, and in general the product of n copies of $\mathbf{A}$ is $\mathbf{A}^n$.

While dealing with notation for powers of matrices, we should mention the transpose operation. The **transpose** of a matrix is a new matrix which is the same as the old except the rows and columns have been switched. The transpose of an $m \times n$ matrix is then an $n \times m$ matrix. The notation for the transpose of $\mathbf{A}$ is $\mathbf{A}^T$. We always use a capital T for transpose.

The transpose of the following two matrices is calculated here.

$$\begin{bmatrix} -1 & 3 & 2 \\ 2 & -3 & 4 \end{bmatrix}^T = \begin{bmatrix} -1 & 2 \\ 3 & -3 \\ 2 & 4 \end{bmatrix}$$

and

$$\begin{bmatrix} -2 & 3 & 1 \\ 5 & 1 & 4 \\ 2 & -3 & -4 \end{bmatrix}^T = \begin{bmatrix} -2 & 5 & 2 \\ 3 & 1 & -3 \\ 1 & 4 & -4 \end{bmatrix}.$$

We note that the transpose can also be useful to simplify the typesetting. When referring to a column vector, we usually write it in transposed form $[x_1 \ x_2 \ \ldots \ x_n]^T$. Since the transpose flips the rows and columns, repeating this process will give us our original matrix: $(\mathbf{A}^T)^T = \mathbf{A}$. Multiplying two matrices and taking the transpose of the result is equivalent to doing the transpose of each matrix but with the order switched:

$$(\mathbf{AB})^T = \mathbf{B}^T \mathbf{A}^T. \tag{B.5}$$

We can also define two numbers associated with a given square matrix. The **trace** of matrix

$$\mathrm{Tr}(\mathbf{A}) = \begin{bmatrix} a & b \\ c & d \end{bmatrix} = a + d.$$

That is, the trace is defined as the sum of the (upper left to lower right) diagonal elements of a matrix. In terms of notation for an $n \times n$ matrix $\mathbf{A}$, we can write

$$\mathrm{Tr}(\mathbf{A}) = \sum_{j=1}^{n} a_{jj}.$$

Further, there is a number associated with a matrix called its **determinant**. The determinant of a matrix is a very useful theoretical tool that comes up in many different places in this book. The determinant of a 2×2 matrix

$$\mathbf{A} = \begin{bmatrix} a & b \\ c & d \end{bmatrix}$$

is $ad - cb$. Often straight vertical lines are used to denote a determinant. Thus

$$\det(\mathbf{A}) = \begin{vmatrix} a & b \\ c & d \end{vmatrix} = ad - cb.$$

We present two methods of calculating the determinant of a 3×3 matrix. The first is easier to use but does not generalize to higher dimensions, while the latter is sometimes harder for the student to grasp but extends easily to larger square matrices.

For the first way, we write the first two columns of $\mathbf{A}$ as a 4th and 5th column:

$$\begin{matrix} a & b & c & a & b \\ d & e & f & d & e \\ g & h & i & g & h \end{matrix}.$$

We then proceed in an analogous way to the 2×2 case. That is, we begin at the top left, going down the diagonal and multiplying the entries. We repeat this for the next two entries in the first row. Then we subtract off the reverse diagonals beginning with the first row, 3rd column entry and moving to the next two:

Subtract these products $= -gec - hfa - idb$

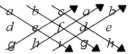

Add these products $= +aei + bfg + cdh$

Then the formula for the 3×3 case is

$$\det \mathbf{A} = aei + bfg + cdh - gec - hfa - idb.$$

The second approach for a 3×3 matrix also holds for larger square matrices: **expansion-by-minors**. To expand on the first row, we take the entry in the $(1,1)$ position, multiply by $(-1)^{1+1}$, and multiply it by the determinant of the matrix that remains when we cross out the first row and first column (called a **minor**); then add to it the entry in the $(1,2)$ position, multiplied by $(-1)^{1+2}$, and multiplied times the determinant of the matrix obtained by crossing out the first row and second column; then add the $(1,3)$ entry mulitplied by $(-1)^{1+3}$, and multiplied times the determinant of the matrix obtained by crossing out the first row and third column. For example, if

$$\mathbf{A} = \begin{bmatrix} a & b & c \\ d & e & f \\ g & h & i \end{bmatrix},$$

then

$$\det(\mathbf{A}) = a \begin{vmatrix} e & f \\ h & i \end{vmatrix} - b \begin{vmatrix} d & f \\ g & i \end{vmatrix} + c \begin{vmatrix} d & e \\ g & h \end{vmatrix}$$
$$= aei - ahf - bdi + bgf + cdh - cge,$$

which is the same formula obtained with the other method of calculation. Observe that this method is a recursive definition since the determinant of a 4×4 matrix involves computing four determinants of 3×3 matrices, each of which requires the computation of three 2×2 matrices. Here we defined expansion on the first row. We can expand on any row or column in the same way as described above with the sign changing as $(-1)^{i+j}$ where (i,j) is the position we are. We direct the reader to any introductory matrix algebra book for details or for other methods of computing a determinant.

Example 2 Find the determinant of each matrix. Use the first method to find $|B|$ and the second method to find $|C|$.

$$A = \begin{bmatrix} 1 & 3 \\ 0 & 2 \end{bmatrix}, \qquad B = \begin{bmatrix} 1 & 4 & 0 \\ -2 & 3 & 1 \\ 0 & 2 & 1 \end{bmatrix}, \qquad C = \begin{bmatrix} 1 & 2 & 3 \\ 2 & 3 & 1 \\ 3 & 5 & 4 \end{bmatrix}.$$

We apply the relevant definitions of the determinant from above:

$$|A| = 1 \cdot 2 - 0 \cdot 3 = 2,$$

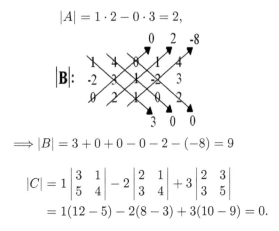

$$\implies |B| = 3 + 0 + 0 - 0 - 2 - (-8) = 9$$

$$|C| = 1 \begin{vmatrix} 3 & 1 \\ 5 & 4 \end{vmatrix} - 2 \begin{vmatrix} 2 & 1 \\ 3 & 4 \end{vmatrix} + 3 \begin{vmatrix} 2 & 3 \\ 3 & 5 \end{vmatrix}$$

$$= 1(12 - 5) - 2(8 - 3) + 3(10 - 9) = 0.$$

Before moving on, we state (without proof) a few useful properties of the determinant

---

Some Properties of det(A)

1. if $\mathbf{A}$ has 2 rows (or 2 columns) that are equal, $\det(\mathbf{A}) = 0$
2. $\det(\mathbf{A}^T) = \det(\mathbf{A})$
3. $\det(\mathbf{AB}) = \det(\mathbf{BA}) = \det(\mathbf{A})\det(\mathbf{B})$
4. $\det(c\mathbf{A}) = c^n \det(\mathbf{A})$ for an $n \times n$ matrix
5. exchanging two rows (or two columns) in $\mathbf{A}$ changes the sign of $\det(\mathbf{A})$
6. subtracting m times row i from row j has the same determinant as the original matrix

---

This last point will be useful for Gaussian elimination, covered in Section 5.2.

The last operation we consider at this time is the inverse. The motivation comes from high school (or earlier!) but is useful for putting the current discussion in its proper context. When working with the real numbers we have both 2 and $\frac{1}{2}$. We know these numbers are multiplicative inverses of each other because $2 \times \frac{1}{2} = 1$ and $\frac{1}{2} \times 2 = 1$. In an abstract algebra course, we express the fact that $\frac{1}{2}$ is the multiplicative inverse of 2 by writing it as 2^{-1}. Similarly $2 = \left(\frac{1}{2}\right)^{-1}$. Here the symbol "$-1$" can be interpreted either as physically inverting the number, or as the abstract inverse, i.e., 2^{-1} is the number which when multiplied by 2 gives 1. If you have not thought about it before, notice that all division can be expressed as multiplying by a multiplicative inverse. This is the idea that is carried over into matrix multiplication. The symbol $\mathbf{I}$ will denote any square matrix (same number of rows as columns) with ones on the diagonal and zeros elsewhere and is called an **identity matrix**. For example, the 3×3 identity matrix is the matrix

$$\mathbf{I} = \begin{bmatrix} 1 & 0 & 0 \\ 0 & 1 & 0 \\ 0 & 0 & 1 \end{bmatrix}.$$

This matrix plays the role of "1" in matrix multiplication and it's an easy calculation to see that $\det(\mathbf{I}) = 1$. You should verify that

$$\mathbf{AI} = \mathbf{A}$$

and

$$\mathbf{IA} = \mathbf{A}.$$

Because of this, the identity matrix is frequently called the multiplicative identity. The **inverse of a matrix** $\mathbf{A}$ is denoted by $\mathbf{A}^{-1}$. The matrix $\mathbf{A}^{-1}$ is defined to be the matrix such that $\mathbf{AA}^{-1} = \mathbf{I}$ and $\mathbf{A}^{-1}\mathbf{A} = \mathbf{I}$ if it exists. If you followed the discussion about 2 and $\frac{1}{2}$, then you should see this as exactly analogous. The existence of $\mathbf{A}^{-1}$ is an issue. By the way $\mathbf{A}^{-1}$ is defined, it needs to be multiplied by $\mathbf{A}$ on both sides, so $\mathbf{A}$ has to be a square matrix for $\mathbf{A}^{-1}$ to exist. A definition and theorem help us at this point.

Definition B.1.1

Let $\mathbf{A}$ be an $n \times n$ matrix.
a. If $\det(\mathbf{A}) \neq 0$, then $\mathbf{A}$ is said to be *nonsingular* (invertible).
b. If $\det(\mathbf{A}) = 0$, then $\mathbf{A}$ is said to be *singular* (noninvertible).

Although we mentioned determinants at an earlier point, they tie in nicely here.

THEOREM B.1.1

Suppose $\mathbf{A}$ is an $n \times n$ matrix. Then $\mathbf{A}^{-1}$ exists if and only if $\mathbf{A}$ is nonsingular.

Determining if the inverse of a matrix exists is a straightforward, tedious calculation. Finding the inverse of a matrix is a non-trivial, though also completely algorithmic, process. It is also computationally expensive and thus not usually calculated in practice unless necessary. We put the discussion of techniques for finding one by hand in Appendix B.2. As with matrix multiplication, computing inverses is not usually done by hand, and the packages of MATLAB®, Maple, and Mathematica are quite capable of doing these computations. The computer labs at the end of this appendix show how to compute matrix inverses with the computer.

Example 3: Determine whether each matrix is singular or nonsingular:

$$\mathbf{A} = \begin{bmatrix} 1 & 3 \\ 0 & 2 \end{bmatrix}, \qquad \mathbf{B} = \begin{bmatrix} 1 & 4 & 0 \\ -2 & 3 & 1 \\ 0 & 2 & 1 \end{bmatrix}, \qquad \mathbf{C} = \begin{bmatrix} 1 & 2 & 3 \\ 2 & 3 & 1 \\ 3 & 5 & 4 \end{bmatrix}.$$

Solution

From the previous example, we calculated

$$|\mathbf{A}| = 2, \quad |\mathbf{B}| = 9, \quad |\mathbf{C}| = 0$$

and thus Definition B.1.1 lets us conclude that $\mathbf{A}, \mathbf{B}$ are nonsingular whereas $\mathbf{C}$ is singular.

We will learn how to calculate inverses in Section B.2 but we give the inverses here to illustrate their use. In the previous example, the inverses of $\mathbf{A}$ and $\mathbf{B}$ are

$$\mathbf{A}^{-1} = \frac{1}{2} \begin{bmatrix} 2 & -3 \\ 0 & 1 \end{bmatrix}, \qquad \mathbf{B}^{-1} = \frac{1}{9} \begin{bmatrix} 1 & -4 & 4 \\ 2 & 1 & -1 \\ -4 & -2 & 11 \end{bmatrix}.$$

Multiplication shows that $\mathbf{A}^{-1}\mathbf{A} = \mathbf{I} = \mathbf{A}\mathbf{A}^{-1}$:

$$\mathbf{A}^{-1}\mathbf{A} = \frac{1}{2} \begin{bmatrix} 2 & -3 \\ 0 & 1 \end{bmatrix} \begin{bmatrix} 1 & 3 \\ 0 & 2 \end{bmatrix} = \frac{1}{2} \begin{bmatrix} 2 & 0 \\ 0 & 2 \end{bmatrix} = \begin{bmatrix} 1 & 0 \\ 0 & 1 \end{bmatrix}$$

and similarly

$$\mathbf{A}\mathbf{A}^{-1} = \begin{bmatrix} 1 & 3 \\ 0 & 2 \end{bmatrix} \frac{1}{2} \begin{bmatrix} 2 & -3 \\ 0 & 1 \end{bmatrix} = \frac{1}{2} \begin{bmatrix} 2 & 0 \\ 0 & 2 \end{bmatrix} = \begin{bmatrix} 1 & 0 \\ 0 & 1 \end{bmatrix}.$$

We can similarly show that $\mathbf{B}^{-1}$ is indeed the inverse of $\mathbf{B}$.

A few other miscellaneous properties for an $n \times n$ matrix (where we assume $\mathbf{A}, \mathbf{B}$ are both nonsingular):

---

1. $\det(\mathbf{I}) = 1$
2. $(\mathbf{A}^{-1})^{-1} = \mathbf{A}$
3. $\det(\mathbf{A}^{-1}) = \frac{1}{\det(\mathbf{A})} = \det(\mathbf{A})^{-1}$
4. $(\mathbf{AB})^{-1} = \mathbf{B}^{-1}\mathbf{A}^{-1}$
5. $(\mathbf{A}^T)^{-1} = (\mathbf{A}^{-1})^T$

Property (4) can be remembered with the socks-shoes analogy: If $\mathbf{A}$ is the action "put on your shoes" and $\mathbf{B}$ is the action "put on your socks," then $\mathbf{AB}$

applied to your foot, $\mathbf{x}$, can be thought of as put on your socks $\mathbf{Bx}$ and then your shoes $\mathbf{A}(\mathbf{Bx})$. The inverse of this is take them off, which you must do in reverse order. With $\mathbf{y}$ as your foot with shoe and sock, you first take off your shoes $\mathbf{A}^{-1}\mathbf{y}$ and then take off your socks $\mathbf{B}^{-1}(\mathbf{A}^{-1}\mathbf{y})$.

Matrices can also be used to succinctly write a system of linear equations. For example,

$$
\begin{aligned}
a_{11}x_1 + a_{12}x_2 + \cdots + a_{1n}x_n &= b_1 \\
a_{21}x_1 + a_{22}x_2 + \cdots + a_{2n}x_n &= b_2 \\
\vdots \qquad \vdots \qquad \ddots \qquad \vdots \qquad \vdots & \\
a_{n1}x_1 + a_{n2}x_2 + \cdots + a_{nn}x_n &= b_n
\end{aligned}
\tag{B.6}
$$

can be written
$$\mathbf{Ax} = \mathbf{b},$$

where

$$
\mathbf{A} = \begin{bmatrix} a_{11} & a_{12} & \cdots & a_{1n} \\ a_{21} & a_{22} & \cdots & a_{2n} \\ \vdots & \vdots & \ddots & \vdots \\ a_{n1} & a_{n2} & \cdots & a_{nn} \end{bmatrix}, \quad \mathbf{x} = \begin{bmatrix} x_1 \\ x_2 \\ \vdots \\ x_n \end{bmatrix}, \quad \text{and} \quad \mathbf{b} = \begin{bmatrix} b_1 \\ b_2 \\ \vdots \\ b_n \end{bmatrix}. \tag{B.7}
$$

We define $\mathbb{R}^n$ as the set of all vectors with n real entries and thus $\mathbf{x}$ is a vector in $\mathbb{R}^n$. It is of great interest to find a vector $\mathbf{x}$ that will solve the system given a matrix $\mathbf{A}$ and a vector $\mathbf{b}$. In terms of theory, the solution will exist and be unique when $\mathbf{A}^{-1}$ exists and will be given by

$$\mathbf{x} = \mathbf{A}^{-1}\mathbf{b}.$$

Example 4 Find the unique solution of the system

$$
\begin{aligned}
x + 3y &= -1 \\
2y &= 4
\end{aligned}
$$

by writing the system in matrix-vector form and computing $\mathbf{A}^{-1}\mathbf{b}$.

Solution

We can rewrite this system in its equivalent matrix-vector form:

$$
\begin{bmatrix} 1 & 3 \\ 0 & 2 \end{bmatrix} \begin{bmatrix} x \\ y \end{bmatrix} = \begin{bmatrix} -1 \\ 4 \end{bmatrix}.
$$

We saw previously that $\mathbf{A}^{-1}$ exists and is given by

$$
\mathbf{A}^{-1} = \frac{1}{2} \begin{bmatrix} 2 & -3 \\ 0 & 1 \end{bmatrix}.
$$

Thus the unique solution is given by

$$
\mathbf{x} = \mathbf{A}^{-1}\mathbf{b} = \frac{1}{2} \begin{bmatrix} 2 & -3 \\ 0 & 1 \end{bmatrix} \begin{bmatrix} -1 \\ 4 \end{bmatrix} = \begin{bmatrix} -7 \\ 2 \end{bmatrix}.
$$

The reader may question why we would solve the system this way when we could just solve the last equation for y and substitute it into the other. The reason lies in the insight that we will be able to gain by considering the system in matrix-vector form. We will be able to deal with significantly more difficult systems with this same method and solving one equation for a variable and substituting it into another becomes a clumsy way (at best) to solve the system.

We give an important theorem that summarizes some of the topics we have discussed:

THEOREM B.1.2

The following are equivalent characterizations of the $n \times n$ matrix $\mathbf{A}$:

a) $\mathbf{A}$ is invertible.

b) The system $\mathbf{Ax} = \mathbf{b}$ has a unique solution $\mathbf{x}$ for all $\mathbf{b}$ in $\mathbb{R}^n$.

c) The system $\mathbf{Ax} = \mathbf{0}$ has $\mathbf{x} = \mathbf{0}$ as its unique solution.

d) $\det(\mathbf{A}) \neq 0$.

e) The n columns of $\mathbf{A}$ are linearly independent.

Although we leave the mechanics of calculating an inverse to Appendix B.2, we know whether or not an inverse exists simply by calculating its determinant. Solutions may exist, however, even when $\mathbf{A}^{-1}$ does not exist. It is *very* computationally expensive to calculate the inverse of a matrix and this is never how a vector $\mathbf{x}$ is found. But writing the solution in this way is very useful for conceptually understanding what we are trying to accomplish.

Geometry of Solving Ax=b

In attempting to find an $\mathbf{x}$ that satisfies $\mathbf{Ax=b}$ for a given $\mathbf{A}$, $\mathbf{b}$, we could have three possible situations: **unique solution**, **infinite solutions**, or **no solution**. If we have either a unique solution or infinite solutions, we say the system of equations is **consistent**, whereas if there is no solution we say the system of equations is **inconsistent**. In terms of two variables, consider the following equations and their graphs in Figure B.1.

$$\begin{aligned} x + 2y &= 4 \\ -x + y &= 2 \end{aligned} \qquad \begin{aligned} 3x - 3y &= -6 \\ -x + y &= 2 \end{aligned} \qquad \begin{aligned} -x + y &= -1 \\ -x + y &= 2 \end{aligned}$$

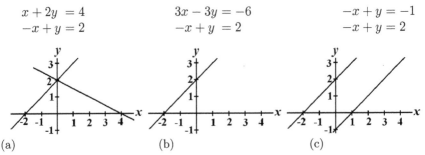

FIGURE B.1: (a) Consistent, unique solution; (b) consistent, infinite solutions; (c) inconsistent, no solution.

For two variables, the geometric interpretation of unique, infinite, or no solutions is easily seen. In higher dimensions where the number of equations is the same as the number of unknown variables, we are also concerned with finding the intersection of lines, but it is just harder to graph. Instead of the traditional graphing of the equations as given in Figure B.1, it will often be useful to rewrite the equations as $\mathbf{Ax}=\mathbf{b}$ and consider the columns of $\mathbf{A}$ as the vectors. The equations of three pictures, which are redrawn in Figure B.2, then give us

$$\begin{bmatrix} 1 & 2 \\ -1 & 1 \end{bmatrix} \begin{bmatrix} x_1 \\ x_2 \end{bmatrix} = \begin{bmatrix} 4 \\ 2 \end{bmatrix} \tag{B.8}$$

$$\begin{bmatrix} 3 & -3 \\ -1 & 1 \end{bmatrix} \begin{bmatrix} x_1 \\ x_2 \end{bmatrix} = \begin{bmatrix} -6 \\ 2 \end{bmatrix} \tag{B.9}$$

$$\begin{bmatrix} -1 & 1 \\ -1 & 1 \end{bmatrix} \begin{bmatrix} x_1 \\ x_2 \end{bmatrix} = \begin{bmatrix} -1 \\ 2 \end{bmatrix}. \tag{B.10}$$

In the first case, (B.8), we are then asking if there is a linear combination of $\begin{bmatrix} 1 \\ -1 \end{bmatrix}$ and $\begin{bmatrix} 2 \\ 1 \end{bmatrix}$ that will give the vector $\begin{bmatrix} 4 \\ 2 \end{bmatrix}$. These vectors are not multiples of each other and thus are linearly independent. Thus, any vector in the two-dimensional space can be written as a linear combination of them and, consequently, we can obtain a unique solution to our system of equations.

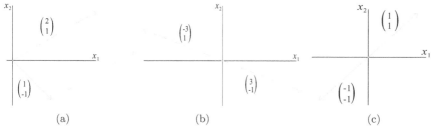

(a)   (b)   (c)

FIGURE B.2: (a) Consistent, unique solution because the vectors are linearly independent; (b) consistent, infinite solutions because although vectors are linearly dependent, the vector $[-6 \ \ 2]^T$ lies in the same line; (c) inconsistent, no solution because the vectors are linearly dependent and vector $[-1 \ \ 2]^T$ is not in the same line.

In the second system of equations, (B.9), we observe that $\begin{bmatrix} 3 \\ -1 \end{bmatrix}$ and $\begin{bmatrix} -3 \\ 1 \end{bmatrix}$ are multiples of each other. Thus, they can only give a solution that lies along one of the vectors. Since $\begin{bmatrix} -6 \\ 2 \end{bmatrix}$ is a multiple of either of these vectors, there is an infinite number of solutions.

In the third system of equations, (B.10), we have $\begin{bmatrix} -1 \\ -1 \end{bmatrix}$ and $\begin{bmatrix} 1 \\ 1 \end{bmatrix}$ and again observe that these are multiples of each other. However, because the vector $\begin{bmatrix} -1 \\ 2 \end{bmatrix}$ is *not* a multiple of these, we can never find a linear combination of these vectors that will work.

The interpretation of matrix-vector multiplication as taking linear combinations of the columns extends nicely to more equations as well. If we consider $\mathbf{Ax} = \mathbf{b}$ as

$$\begin{bmatrix} 1 & 2 \\ 4 & 3 \\ 3 & 1 \end{bmatrix} \begin{bmatrix} x_1 \\ x_2 \end{bmatrix} = \begin{bmatrix} b_1 \\ b_2 \\ b_3 \end{bmatrix}, \tag{B.11}$$

we would like to know if we can find a linear combination of the columns of $\mathbf{A}$, $\begin{bmatrix} 1 \\ 4 \\ 3 \end{bmatrix}$ and $\begin{bmatrix} 2 \\ 3 \\ 1 \end{bmatrix}$, that will give us $\mathbf{b}$; see Figure B.3. The geometrical answer is that $\mathbf{b}$ must lie in the plane that contains the two column vectors (and also includes the origin). We can equivalently say that it's the plane of all possible combinations of the two column vectors (i.e., its the column space of the matrix). We revisit these last two examples in Section 5.2 after covering Gaussian elimination.

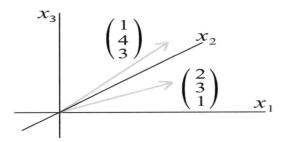

FIGURE B.3: In order for there to be a solution to (B.11), $\mathbf{b}$ must lie in the plane that contains the two column vectors $[1 \ 4 \ 3]^T$ and $[2 \ 3 \ 1]^T$.

● ● ● ● ● ● ● ● ● ● ●

Problems

In Problems **1–3**, write the following systems using vectors and matrices.

1. $\begin{cases} 3x + 5y = 7 \\ x + y \ \ = -1 \end{cases}$　**2.** $\begin{cases} 2x + 5y - z = 2 \\ -x + 2y \ \ = -1 \\ x + 4z \ \ = 0 \end{cases}$　**3.** $\begin{cases} x' = x - 5y \\ y' = 2x + y \end{cases}$.

In Problems **4–5**, calculate (by hand) $\mathbf{AB}$ and $\mathbf{BA}$ for the following matrices.

4. $\mathbf{A} = \begin{bmatrix} 2 & -3 \\ -1 & 4 \end{bmatrix}$, $\mathbf{B} = \begin{bmatrix} -2 & 2 \\ 1 & -5 \end{bmatrix}$

$$\textbf{5. A} = \begin{bmatrix} 1 & 2 & -3 \\ 0 & -1 & 4 \\ 3 & 0 & 2 \end{bmatrix}, \quad \textbf{B} = \begin{bmatrix} 3 & -2 & 2 \\ -1 & 1 & -5 \\ 0 & 2 & 2 \end{bmatrix}$$

In Problems **6–37**, consider the following matrices:

$$\textbf{A} = \begin{bmatrix} 3 & 0 \\ -1 & 2 \\ 1 & 1 \end{bmatrix}, \quad \textbf{B} = \begin{bmatrix} 4 & -1 \\ 0 & 2 \end{bmatrix}, \quad \textbf{C} = \begin{bmatrix} 1 & 4 & -2 \\ 6 & 3 & 1 \end{bmatrix}$$

$$\textbf{D} = \begin{bmatrix} 1 & 5 & 2 \\ -1 & 0 & 3 \\ 4 & 2 & 1 \end{bmatrix}, \quad \textbf{E} = \begin{bmatrix} 1 & -3 & 4 \\ 5 & 1 & 1 \\ 3 & 2 & 1 \end{bmatrix}$$

Compute the following expressions, when possible. If it is not possible, state why it is not possible.

6. D + E	**7. D − E**	**8. 2D + 3E**	**9. 2A**
10. B + C	**11. 4E − 2D**	**12. Tr(E)**	**13. Tr(B)**
14. Tr(3D)	**15. Tr(D + C)**	**16. A^T**	**17. D^T**
18. D^T + E^T	**19. A^2**	**20. AC**	**21. DE**
22. ED	**23. AB**	**24. BA**	**25. BA^T**
26. EA	**27. AE**	**28. A^{T}E**	**29. ABC**
30. det(B)	**31. det(D)**	**32. det(E)**	**33. det(B^2)**
34. det(DE)	**35. det(ED)**	**36. det(E)det(D)**	**37. det(D^T)**

B.2 Matrix Inverses, Cramer's Rule

This section of Appendix B is essential if the course is to be used as a one-quarter course that combines linear algebra and differential equations, a one-semester course in differential equations that stresses the key influences of linear algebra, or if the reader seeks a deeper understanding of the theory of differential equations as seen in Chapters 3 through 5. If none of these applies, the reader may skip this material unless a brief review of Cramer's rule (used for Section 4.5 on Variation of Parameters) is needed and this part of the current section is self-contained.

We focus on a system of linear equations given as (B.6) and written in matrix-vector notation as

$$\textbf{Ax} = \textbf{b}.$$

We will typically be given a vector **b** in $\mathbb{R}^m$ and a matrix **A** in $\mathbb{R}^{m \times n}$ (that is, an $m \times n$ matrix with real entries). Our goal will be to find a vector **x** in $\mathbb{R}^n$ that will make the equation $\textbf{Ax} = \textbf{b}$ a true statement.

We make note of two areas of active research at this point. Many times, a mathematical model gives rise to a system that has more variables than equations, thus leading to an $m \times n$ matrix where $m < n$ and both are large, say $50,000 \times 100,000$. Although the methods we learn in linear algebra apply, these kinds of matrices are usually best dealt with in the subject of *optimization*. Many other mathematical models of physical systems give rise to very large **square** systems, on the order of a matrix $\mathbf{A}$ that is $100,000 \times 100,000$. Although we will learn how to solve $\mathbf{Ax} = \mathbf{b}$ by hand in this section, we clearly would not want to solve that large of a system by hand. What do we do? Interested students should take a course in **numerical linear algebra** and explore the wealth of algorithms that efficiently give us our desired solution. There is a tremendous number of unsolved problems in these areas and researchers (maybe even your professors!) are actively seeking solutions today.

B.2.1   Calculating the Inverse of a Matrix

The reader should be sure to be familiar with Section 5.2 and elementary row operations in order to follow this section. With this knowledge, we can now find the inverse of a matrix. How? Let us think what Gaussian elimination accomplished for us. We began with

$$\mathbf{Ax} = \mathbf{b}$$

and reduced this to a system

$$\mathbf{Ux} = \mathbf{b_u}$$

where $\mathbf{U}$ is an upper triangular matrix and the $\mathbf{b_u}$ denotes the vector $\mathbf{b}$ after we have performed the same elementary row operations that gave us $\mathbf{U}$. Our answer is the vector $\mathbf{x}$ and we can also write the answer (if it exists, of course) as

$$\mathbf{x} = \mathbf{A}^{-1}\mathbf{b} \quad \text{or equivalently} \quad \mathbf{x} = \mathbf{U}^{-1}\mathbf{b_u}.$$

In a previous example, we considered $\mathbf{Ax} = \mathbf{b}$, where

$$\mathbf{A} = \begin{bmatrix} -2 & 0 & 4 \\ 4 & 2 & -4 \\ 2 & 4 & 2 \end{bmatrix}, \quad \mathbf{b} = \begin{bmatrix} 2 \\ -2 \\ -2 \end{bmatrix}.$$

We reduced the system to row-echelon form, which can equivalently be written as

$$\mathbf{Ux} = \mathbf{b_u} \text{ with } \mathbf{U} = \begin{bmatrix} 1 & 0 & -2 \\ 0 & 1 & 2 \\ 0 & 0 & 1 \end{bmatrix}, \quad \mathbf{b_u} = \begin{bmatrix} -1 \\ 1 \\ 2 \end{bmatrix}.$$

We illustrate below how Gauss-Jordan elimination can also be used to give us the matrix inverses. For now, assume that we are able to calculate the inverses and in doing so we obtain

$$\mathbf{A}^{-1} = \frac{1}{2} \begin{bmatrix} 5 & 4 & -2 \\ -4 & -3 & 2 \\ 3 & 2 & -1 \end{bmatrix}, \quad \text{or } \mathbf{U}^{-1} = \begin{bmatrix} 1 & 0 & 2 \\ 0 & 1 & -2 \\ 0 & 0 & 1 \end{bmatrix} \text{ with } \mathbf{b_u} = \begin{bmatrix} -1 \\ 1 \\ 2 \end{bmatrix}.$$

The reader should verify that in either case we end up with the desired solution.

In the above example, we were able to obtain $\mathbf{A}^{-1}$, but this will not always be the case as mentioned previously. There are many examples in which our system cannot be solved with the use of our three elementary row operations. In this case, we say the system (or matrix) is **singular** and the inverse of $\mathbf{A}$ does not exist. Assuming that $\mathbf{A}^{-1}$ exists, we were able to proceed with Gauss-Jordan elimination and end up with the system

$$\mathbf{x} = \mathbf{A}^{-1}\mathbf{b}.$$

Thus the inverse of our matrix is related to the elementary row operations of Gauss-Jordan elimination. If we instead wrote our original system as

$$\mathbf{Ax} = \mathbf{Ib} \tag{B.12}$$

and reduced this through Gauss-Jordan elimination to

$$\mathbf{Ix} = \mathbf{A}^{-1}\mathbf{b}, \tag{B.13}$$

we would have the inverse of the matrix on the right-hand side. Note that we did not need to keep track of $\mathbf{b}$ in this case. Let us again consider our previous example given as (5.8) with coefficient matrix

$$\mathbf{A} = \begin{bmatrix} -2 & 0 & 4 \\ 4 & 2 & -4 \\ 2 & 4 & 2 \end{bmatrix}. \tag{B.14}$$

Instead of constructing the augmented matrix with $\mathbf{b}$ as a fourth column, we consider the matrix

$$\mathbf{A} = \left[\begin{array}{ccc|ccc} -2 & 0 & 4 & 1 & 0 & 0 \\ 4 & 2 & -4 & 0 & 1 & 0 \\ 2 & 4 & 2 & 0 & 0 & 1 \end{array} \right]. \tag{B.15}$$

According to our previous discussion, if we do Gauss-Jordan elimination on the left half while keeping track of the resulting effect on the right half of the matrix, our final answer will be the inverse of $\mathbf{A}$. We will repeat the elementary row operations from above and will show all intermediate steps:

$$R_2 - (-2R_1) \to R_2 \text{ gives} \quad \left[\begin{array}{ccc|ccc} -2 & 0 & 4 & 1 & 0 & 0 \\ 0 & 2 & 4 & 2 & 1 & 0 \\ 2 & 4 & 2 & 0 & 0 & 1 \end{array} \right] \tag{B.16}$$

$$R_3 - (-R_1) \to R_3 \text{ gives} \quad \left[\begin{array}{ccc|ccc} -2 & 0 & 4 & 1 & 0 & 0 \\ 0 & 2 & 4 & 2 & 1 & 0 \\ 0 & 4 & 6 & 1 & 0 & 1 \end{array} \right] \tag{B.17}$$

$$R_3 - 2R_2 \to R_3 \quad \text{gives} \quad \left[\begin{array}{ccc|ccc} -2 & 0 & 4 & 1 & 0 & 0 \\ 0 & 2 & 4 & 2 & 1 & 0 \\ 0 & 0 & -2 & -3 & -2 & 1 \end{array}\right] \quad \text{(B.18)}$$

$$\begin{array}{l} \frac{-1}{2}R_1 \to R_1, \ \frac{1}{2}R_2 \to R_2, \\ \frac{-1}{2}R_3 \to R_3 \quad \text{gives} \end{array} \left[\begin{array}{ccc|ccc} 1 & 0 & -2 & \frac{-1}{2} & 0 & 0 \\ 0 & 1 & 2 & 1 & \frac{1}{2} & 0 \\ 0 & 0 & 1 & \frac{3}{2} & 1 & \frac{-1}{2} \end{array}\right] \quad \text{(B.19)}$$

$$R_2 - 2R_3 \to R_2 \quad \text{gives} \quad \left[\begin{array}{ccc|ccc} 1 & 0 & -2 & \frac{-1}{2} & 0 & 0 \\ 0 & 1 & 0 & -2 & \frac{-3}{2} & 1 \\ 0 & 0 & 1 & \frac{3}{2} & 1 & \frac{-1}{2} \end{array}\right] \quad \text{(B.20)}$$

$$R_1 - (-2R_3) \to R_1 \quad \text{gives} \quad \left[\begin{array}{ccc|ccc} 1 & 0 & 0 & \frac{5}{2} & 2 & -1 \\ 0 & 1 & 0 & -2 & \frac{-3}{2} & 1 \\ 0 & 0 & 1 & \frac{3}{2} & 1 & \frac{-1}{2} \end{array}\right]. \quad \text{(B.21)}$$

The right half of this matrix is exactly the inverse of $\mathbf{A}$ as the reader should check. But that was a lot of work! Now you will perhaps understand why it's not even desirable to calculate the inverse with a computer. Nevertheless, it IS instructive to understand the manipulations that are possible. In the computer lab, we give sample code that will do these elementary row operations and then the built-in commands that will do the same thing.

Example 1: Find the inverse of the general 2×2 matrix.

Solution

In order for $\mathbf{A}^{-1}$ to exist, we need $\det(\mathbf{A}) \neq 0$. If we assume the form

$$\begin{bmatrix} a & b \\ c & d \end{bmatrix},$$

this means that $ad - bc \neq 0$. Going through the steps of Gauss-Jordan elimination gives

$$\left[\begin{array}{cc|cc} a & b & 1 & 0 \\ c & d & 0 & 1 \end{array}\right] \quad \text{(B.22)}$$

$$R_2 - \frac{c}{a}R_1 \to R_2 \quad \text{gives} \quad \left[\begin{array}{cc|cc} a & b & 1 & 0 \\ 0 & d - \frac{bc}{a} & \frac{-c}{a} & 1 \end{array}\right] \quad \text{(B.23)}$$

$$\frac{a}{ad - bc}R_2 \to R_2 \quad \text{gives} \quad \left[\begin{array}{cc|cc} a & b & 1 & 0 \\ 0 & 1 & \frac{-c}{ad-bc} & \frac{a}{ad-bc} \end{array}\right] \quad \text{(B.24)}$$

$$R_1 - bR_2 \to R_1 \quad \text{gives} \quad \left[\begin{array}{cc|cc} a & 0 & 1 + \frac{bc}{ad-bc} & \frac{-ba}{ad-bc} \\ 0 & 1 & \frac{-c}{ad-bc} & \frac{a}{ad-bc} \end{array}\right] \quad \text{(B.25)}$$

$$\frac{1}{a}R_1 \to R_1 \quad \text{gives} \quad \left[\begin{array}{cc|cc} 1 & 0 & \frac{d}{ad-bc} & \frac{-b}{ad-bc} \\ 0 & 1 & \frac{-c}{ad-bc} & \frac{a}{ad-bc} \end{array}\right]. \quad \text{(B.26)}$$

Since each term of the inverse has the same denominator, we factor it out to give

$$\begin{bmatrix} a & b \\ c & d \end{bmatrix}^{-1} = \frac{1}{ad - bc} \begin{bmatrix} d & -b \\ -c & a \end{bmatrix}, \tag{B.27}$$

and we clearly see why we need $ad - bc \neq 0$.

Example 2 Find the inverse of a diagonal matrix, if it exists.

Solution

An $n \times n$ diagonal matrix has the form

$$\begin{bmatrix} d_1 & 0 & \cdots & 0 \\ 0 & d_2 & \cdots & 0 \\ \vdots & & \ddots & \vdots \\ 0 & 0 & \cdots & d_n \end{bmatrix}, \tag{B.28}$$

with $d_i \neq 0$, $i = 1, \cdots, n$ in order to have a non-zero determinant. We set up the augmented matrix

$$\left[\begin{array}{cccc|cccc} d_1 & 0 & \cdots & 0 & 1 & 0 & \cdots & 0 \\ 0 & d_2 & \cdots & 0 & 0 & 1 & \cdots & 0 \\ \vdots & & \ddots & \vdots & \vdots & & \ddots & \vdots \\ 0 & 0 & \cdots & d_n & 0 & 0 & \cdots & 1 \end{array} \right] \tag{B.29}$$

and then perform one step (on each row) of Gauss-Jordan elimination:

$$\frac{1}{d_1} R_1 \to R_1, \ \frac{1}{d_2} R_2 \to R_2, \\ \cdots, \frac{1}{d_n} R_n \to R_n \ \text{gives} \quad \left[\begin{array}{cccc|cccc} 1 & 0 & \cdots & 0 & \frac{1}{d_1} & 0 & \cdots & 0 \\ 0 & 1 & \cdots & 0 & 0 & \frac{1}{d_2} & \cdots & 0 \\ \vdots & & \ddots & \vdots & \vdots & & \ddots & \vdots \\ 0 & 0 & \cdots & 1 & 0 & 0 & \cdots & \frac{1}{d_n} \end{array} \right]. \tag{B.30}$$

Thus, for a diagonal matrix in which none of the diagonal entries are zero, its inverse is obtained by replacing each diagonal entry with its reciprocal.

B.2.2 Cramer's Rule

This section of the Appendix can stand alone from the rest, with the possible exception of Appendix B.1, which the reader may need to read parts of in order to take the determinant of a matrix. Cramer's rule is really just another way to solve a system of equations. However, *it is computationally inefficient to use Cramer's rule for even moderately large systems*, so we'll skip the theory and just give the formulas for the 2×2, 3×3, and general $n\times n$ cases. We stress that Cramer's rule only works with systems of n equations in n unknowns. In Section 4.5, we need to solve a system of equations and Cramer's rule is a straightforward way to do this. Any of the methods of this section will work, too, but would be a bit more cumbersome because we may be dealing with functions instead of scalars as the entries in our matrices.

Let us consider the 2×2 linear system

$$a_{11}x + a_{12}y = b_1 \tag{B.31}$$
$$a_{21}x + a_{22}y = b_2. \tag{B.32}$$

Cramer's rule states that the solution is

$$x = \frac{\begin{vmatrix} b_1 & a_{12} \\ b_2 & a_{22} \end{vmatrix}}{\begin{vmatrix} a_{11} & a_{12} \\ a_{21} & a_{22} \end{vmatrix}}, \quad y = \frac{\begin{vmatrix} a_{11} & b_1 \\ a_{21} & b_2 \end{vmatrix}}{\begin{vmatrix} a_{11} & a_{12} \\ a_{21} & a_{22} \end{vmatrix}}. \tag{B.33}$$

Mechanically, we obtained the solution to the first variable by replacing the first column of the coefficient matrix by the right-hand side in the numerator (and then took its determinant) and divided by the determinant of the coefficient matrix. We note that the denominator cannot be zero because otherwise our answer will be undefined or indeterminate. But the denominator is the determinant of the coefficient matrix, which is zero only when there is no inverse. In this situation, the system either has no solution or possibly infinitely many solutions. In the case of a non-zero determinant, we will always be able to solve the system.

For the 3×3 linear system

$$a_{11}x + a_{12}y + a_{13}z = b_1 \tag{B.34}$$
$$a_{21}x + a_{22}y + a_{23}z = b_2 \tag{B.35}$$
$$a_{31}x + a_{32}y + a_{33}z = b_3, \tag{B.36}$$

we again replace the first column of the coefficient matrix in the numerator by the right-hand side. For simplicity, let $\mathbf{A}$ denote the coefficient matrix and $|\mathbf{A}|$ its determinant. Then Cramer's rule states that the solution is

$$x = \frac{\begin{vmatrix} b_1 & a_{12} & a_{13} \\ b_2 & a_{22} & a_{23} \\ b_3 & a_{32} & a_{33} \end{vmatrix}}{|\mathbf{A}|}, \quad y = \frac{\begin{vmatrix} a_{11} & b_1 & a_{13} \\ a_{21} & b_2 & a_{23} \\ a_{31} & b_3 & a_{33} \end{vmatrix}}{|\mathbf{A}|}, \quad z = \frac{\begin{vmatrix} a_{11} & a_{12} & b_1 \\ a_{21} & a_{22} & b_2 \\ a_{31} & a_{32} & b_3 \end{vmatrix}}{|\mathbf{A}|}. \tag{B.37}$$

For the general linear system

$$\begin{aligned}
a_{11}x_1 + a_{12}x_2 + \cdots + a_{1n}x_n &= b_1 \\
a_{21}x_1 + a_{22}x_2 + \cdots + a_{2n}x_n &= b_2 \\
\vdots \quad \vdots \quad \ddots \quad \vdots \quad \vdots \\
a_{n1}x_1 + a_{n2}x_2 + \cdots + a_{nn}x_n &= b_n,
\end{aligned}$$

(B.38)

we will again let $\mathbf{A}$ denote the coefficient matrix and Cramer's rule can be stated as

$$x_1 = \frac{\begin{vmatrix} b_1 & a_{12} & \cdots & a_{1n} \\ b_2 & a_{22} & \cdots & a_{2n} \\ \vdots & \vdots & & \vdots \\ b_n & a_{n2} & \cdots & a_{nn} \end{vmatrix}}{|\mathbf{A}|},$$

where the first column in the numerator is replaced by $\mathbf{b}$,

(B.39)

$$x_i = \frac{\begin{vmatrix} a_{11} & \cdots & b_1 & \cdots & a_{1n} \\ a_{21} & \cdots & b_2 & \cdots & a_{2n} \\ \vdots & & \vdots & & \vdots \\ a_{n1} & \cdots & b_n & \cdots & a_{nn} \end{vmatrix}}{|\mathbf{A}|},$$

where the ith column in the numerator is replaced by $\mathbf{b}$ for $i = 2, \cdots, n-1$,

(B.40)

$$x_n = \frac{\begin{vmatrix} a_{11} & a_{12} & \cdots & b_1 \\ a_{21} & a_{22} & \cdots & b_2 \\ \vdots & \vdots & & \vdots \\ a_{n1} & a_{n2} & \cdots & b_n \end{vmatrix}}{|\mathbf{A}|},$$

where the nth column in the numerator is replaced by $\mathbf{b}$.

(B.41)

Example If possible, solve the following system using Cramer's rule:

$$\begin{aligned}
2x + 3y &= 4 \\
-x + 2y &= 5.
\end{aligned}$$

(B.42)

Solution
In order to see if the system can be solved by Cramer's rule, we need to calculate the determinant of the coefficient matrix:

$$|\mathbf{A}| = \begin{vmatrix} 2 & 3 \\ -1 & 2 \end{vmatrix} = 4 - (-3) = 7.$$

Because $|\mathbf{A}| \neq 0$, we can solve the system. Applying (B.33) gives

$$x = \frac{\begin{vmatrix} 4 & 3 \\ 5 & 2 \end{vmatrix}}{|\mathbf{A}|} = \frac{-7}{7} = -1, \quad y = \frac{\begin{vmatrix} 2 & 4 \\ -1 & 5 \end{vmatrix}}{|\mathbf{A}|} = \frac{14}{7} = 2.$$

Example 4: If possible, solve the following system using Cramer's rule:

$$\begin{aligned} 2x + 3y &= 4 \\ -x + 2y + z &= -5 \\ 7y + 2z &= -6. \end{aligned} \tag{B.43}$$

Solution

In order to see if the system can be solved by Cramer's rule, we need to calculate the determinant of the coefficient matrix:

$$|\mathbf{A}| = \begin{vmatrix} 2 & 3 & 0 \\ -1 & 2 & 1 \\ 0 & 7 & 2 \end{vmatrix} = 2(4 - 7) - 3(-2 - 0) + 0(-7 - 0) = 0.$$

Because $|\mathbf{A}| = 0$, we cannot use Cramer's rule to solve the system. If we instead tried to solve the system by Gaussian elimination, we would obtain, after two steps of elimination,

$$\begin{bmatrix} 2 & 3 & 0 & 4 \\ 0 & 7/2 & 1 & -3 \\ 0 & 0 & 0 & 0 \end{bmatrix}. \tag{B.44}$$

This can be solved to give

$$x = \frac{23 + 3z}{7}, \quad y = \frac{-6 - 2z}{7}, \quad z = \text{anything}.$$

Thus we have infinitely many solutions for this system.

Example 6 If possible, solve the following system using Cramer's rule:

$$2x + 3y = 4$$
$$-x + 2y + z = -5$$
$$3y - 2z = -4. \tag{B.45}$$

Solution
In order to see if the system can be solved by Cramer's rule, we need to calculate the determinant of the coefficient matrix:

$$|\mathbf{A}| = \begin{vmatrix} 2 & 3 & 0 \\ -1 & 2 & 1 \\ 0 & 3 & -2 \end{vmatrix} = 2(-4 - 3) - 3(2 - 0) + 0(-3 - 0) = -20.$$

Because $|\mathbf{A}| \neq 0$, we can solve the system. Applying (B.37) gives

$$x = \frac{\begin{vmatrix} 4 & 3 & 0 \\ -5 & 2 & 1 \\ -4 & 3 & -2 \end{vmatrix}}{|\mathbf{A}|}, \quad y = \frac{\begin{vmatrix} 2 & 4 & 0 \\ -1 & -5 & 1 \\ 0 & -4 & -2 \end{vmatrix}}{|\mathbf{A}|}, \quad z = \frac{\begin{vmatrix} 2 & 3 & 4 \\ -1 & 2 & -5 \\ 0 & 3 & -4 \end{vmatrix}}{|\mathbf{A}|}$$

$$= \frac{-70}{-20} = \frac{7}{2} \qquad = \frac{20}{-20} = -1 \qquad = \frac{-10}{-20} = \frac{1}{2}$$

as the solution to our system.

Problems

In Problems **1–6**, determine which of the following 3×3 matrices are in row-echelon form.

1. $\begin{bmatrix} 1 & 0 & 0 \\ 0 & 1 & 0 \\ 0 & 0 & 1 \end{bmatrix}$ 2. $\begin{bmatrix} 1 & 2 & 0 \\ 0 & 1 & 0 \\ 0 & 0 & 0 \end{bmatrix}$ 3. $\begin{bmatrix} 1 & 0 & 0 \\ 0 & 1 & 0 \\ 0 & 2 & 0 \end{bmatrix}$

4. $\begin{bmatrix} 1 & 3 & 4 \\ 0 & 0 & 1 \\ 0 & 0 & 0 \end{bmatrix}$ 5. $\begin{bmatrix} 1 & 5 & -3 \\ 0 & 1 & 1 \\ 0 & 0 & 0 \end{bmatrix}$ 6. $\begin{bmatrix} 1 & 2 & 3 \\ 0 & 0 & 0 \\ 0 & 0 & 1 \end{bmatrix}$

In Problems **7–12**, solve each of the following systems by Gauss-Jordan elimination.

7. $\begin{cases} 2x_1 - 3x_2 = -2 \\ 2x_1 + x_2 = 1 \end{cases}$

8. $\begin{cases} x_1 + x_2 + 2x_3 = 8 \\ -x_1 - 2x_2 + 3x_3 = 1 \\ 3x_1 - 7x_2 + 4x_3 = 10 \end{cases}$

9. $\begin{cases} 2x_1 + 2x_2 + 2x_3 = 0 \\ -2x_1 + 5x_2 + 2x_3 = 1 \\ 8x_1 + x_2 + 4x_3 = -1 \end{cases}$

10. $\begin{cases} -2b + 3c = 1 \\ 3a + 6b - 3c = -2 \\ 6a + 6b + 3c = 5 \end{cases}$

11. $\begin{cases} x - 2y + 4z = 2 \\ 2x - 3y + 5z = 3 \\ 3x - 4y + 7z = 7 \end{cases}$ 12. $\begin{cases} 2x + 3y - 2z = 2 \\ x - 2y + 3z = 2 \\ 4x - y + 5z = 1 \end{cases}$

In Problems **13–15**, use Cramer's rule to solve the following systems of equations. You must evaluate the determinants by hand. (You may, of course, check your answer with the computer.)

13. $\begin{cases} 3x + y = 2 \\ 2x - 3y = 5 \end{cases}$ 14. $\begin{cases} x + 4z = 2 \\ 2x - 3y = 3 \\ 3x - 4y + 6z = 0 \end{cases}$ 15. $\begin{cases} 2x + 3y - 2z = 2 \\ x - 2y + 3z = 2 \\ 4x - y + 4z = 1 \end{cases}$

In Problems **16–19**, solve using (i) Cramer's rule, (ii) Gauss-Jordan elimination, (iii) built-in commands of MATLAB, Maple, or Mathematica.

16. $\begin{cases} 7x_1 - 2x_2 = 3 \\ 3x_1 + x_2 = 5 \end{cases}$ 17. $\begin{cases} 4x + 5y = 2 \\ 11x + y + 2z = 3 \\ x + 5y + 2z = 1 \end{cases}$

18. $\begin{cases} x - 4y + z = 6 \\ 4x - y + 2z = -1 \\ 2x + 2y - 3z = -20 \end{cases}$ 19. $\begin{cases} 4x + y + z + w = 6 \\ 3x + 7y - z + w = 1 \\ 7x + 3y - 5z + 8w = -3 \\ x + y + z + 2w = 3 \end{cases}$

20. Use Cramer's rule to solve $\begin{cases} x = x' \cos\theta - y' \sin\theta \\ y = x' \sin\theta + y' \cos\theta \end{cases}$

for x', y' in terms of x and y.

In Problems **21–26**, find the inverse of the matrices using Gauss-Jordan elimination. If the matrix is 2×2, additionally find the inverse using the formula for a 2×2 matrix inverse.

21. $\begin{bmatrix} 1 & 2 \\ -1 & -1 \end{bmatrix}$ 22. $\begin{bmatrix} 3 & -4 \\ 1 & 5 \end{bmatrix}$ 23. $\begin{bmatrix} 1 & 2 \\ 2 & 7 \end{bmatrix}$,

24. $\begin{bmatrix} 1 & 5 \\ 3 & -6 \end{bmatrix}$ 25. $\begin{bmatrix} 1 & 2 & -4 \\ -1 & -1 & 5 \\ 2 & 7 & -3 \end{bmatrix}$ 26. $\begin{bmatrix} 1 & 3 & -4 \\ 2 & 5 & -1 \\ 3 & 13 & -6 \end{bmatrix}$

B.3   Linear Transformations

In Section 6.1, we considered the equations

$$\frac{dx}{dt} = y$$

$$\frac{dy}{dt} = -\frac{k}{m}x \qquad \text{(B.46)}$$

and

$$\frac{dx}{dt} = -5x + 8y$$

$$\frac{dy}{dt} = -4x + 7y, \qquad \text{(B.47)}$$

and discussed how the vector field can be used to describe the motion of a trajectory (projection of the solution onto the x-y plane) through a given point (x, y). In the first system, Equation (B.46), we can write

$$\begin{bmatrix} x' \\ y' \end{bmatrix} = \begin{bmatrix} 0 & 1 \\ -\frac{k}{m} & 0 \end{bmatrix} \begin{bmatrix} x \\ y \end{bmatrix}$$

and thus view the right-hand side as "the coefficient matrix $\mathbf{A}$ multiplied by the vector $\begin{bmatrix} x \\ y \end{bmatrix}$." The result gives the instantaneous rate of change of the trajectory through that point. For example, if we consider the trajectory that passes through $(1, 2)$ (for, say, $m = 2, k = 3$), we find its instantaneous rate of change is

$$\begin{bmatrix} 0 & 1 \\ -\frac{3}{2} & 0 \end{bmatrix} \begin{bmatrix} 1 \\ 2 \end{bmatrix} = \begin{bmatrix} 2 \\ -\frac{3}{2} \end{bmatrix}$$

and the motion of the trajectory is right and down as was shown in the picture of the vector field (see Figure 6.2).

For the second system, Equation (B.47), we can again consider the trajectory through the point $(1, 2)$ and calculate the instantaneous rate of change as

$$\begin{bmatrix} -5 & 8 \\ -4 & 7 \end{bmatrix} \begin{bmatrix} 1 \\ 2 \end{bmatrix} = \begin{bmatrix} 11 \\ 10 \end{bmatrix}.$$

Thus the motion of the trajectory is right and up and we again saw this in the vector field (see Figure 6.3).

In both of these systems, multiplication of a vector by a coefficient matrix describes the slope of the solution through the given (x, y) pair. So we may think of *a matrix doing something to or acting on a vector;* this gives us a useful interpretation of matrix-vector multiplication.

With this as our motivation, we will now consider some useful ideas from linear algebra. For the purposes of this section we adopt the viewpoint of $\mathbf{A}$ *acting* on a vector $\mathbf{x}$. In the case of a square $n \times n$ matrix $\mathbf{A}$, the vector $\mathbf{x}$ gets moved from one location in $\mathbb{R}^n$ to another; geometrically, the vectors in $\mathbb{R}^n$ have been moved according to the "action" of $\mathbf{A}$, with perhaps some vectors (e.g., $\mathbf{0}$) being left unchanged. In the case of a rectangular $m \times n$ matrix $\mathbf{A}$, we have a situation where $\mathbf{A}$ takes a vector from one vector space $(\mathbb{R}^n)$ to another vector space $(\mathbb{R}^m)$. Note that the origin in $\mathbb{R}^n$ is mapped to the origin in $\mathbb{R}^m$.

Example 1 The matrix-vector multiplication

$$\begin{bmatrix} -1 & 3 & 2 \\ 1 & 1 & 0 \\ 2 & -3 & 4 \end{bmatrix} \begin{bmatrix} -1 \\ 2 \\ 3 \end{bmatrix} = \begin{bmatrix} 13 \\ 1 \\ 4 \end{bmatrix} \quad \text{and} \quad \begin{bmatrix} -1 & 3 & 2 \\ 1 & 1 & 0 \\ 2 & -3 & 4 \end{bmatrix} \begin{bmatrix} 0 \\ 0 \\ 0 \end{bmatrix} = \begin{bmatrix} 0 \\ 0 \\ 0 \end{bmatrix}$$

shows that a 3×3 matrix takes a vector in $\mathbb{R}^3$ to a vector in $\mathbb{R}^3$, whereas

$$\begin{bmatrix} -1 & 3 & 2 \\ 1 & 1 & 0 \end{bmatrix} \begin{bmatrix} -1 \\ 2 \\ 3 \end{bmatrix} = \begin{bmatrix} 13 \\ 1 \end{bmatrix} \quad \text{and} \quad \begin{bmatrix} -1 & 3 & 2 \\ 1 & 1 & 0 \end{bmatrix} \begin{bmatrix} 0 \\ 0 \\ 0 \end{bmatrix} = \begin{bmatrix} 0 \\ 0 \end{bmatrix}$$

shows that a 2×3 matrix takes a vector in $\mathbb{R}^3$ to a vector in $\mathbb{R}^2$. Note that in both cases, the origin was mapped to the origin (in a different space in the case of a rectangular matrix).

Regardless of whether our matrix $\mathbf{A}$ is square or rectangular, our definition of matrix multiplication gives us the following:

> **PROPOSITION B.3.1**
>
> Let c_1, c_2 be any scalars, let $\mathbf{x}, \mathbf{y}$ be vectors in $\mathbb{R}^n$, and let $\mathbf{A}$ be an $m \times n$ matrix. Then
>
> $$\mathbf{A}(c_1\mathbf{x} + c_2\mathbf{y}) = c_1(\mathbf{A}\mathbf{x}) + c_2(\mathbf{A}\mathbf{y}) \tag{B.48}$$
>
> and we say $\mathbf{A}$ is a linear transformation.

In this case, we say that $\mathbf{A}$ takes $\mathbb{R}^n$ into $\mathbb{R}^m$. A matrix is one example of a linear transformation but there are many other examples. When the domain and range are the same space, as is the case when $\mathbf{A}$ is a square matrix, we say that $\mathbf{A}$ is a **linear operator**.

Example 2 Let D denote the differentiation operator, d/dt, from Section 3.1.1. D is a linear transformation that takes polynomials of degree n to polynomials of degree $n - 1$:

$$D(a_n t^n + \cdots + a_1 t + a_0) = n a_n t^{n-1} + \cdots + a_1. \tag{B.49}$$

If $\mathbf{x}$ and $\mathbf{y}$ are polynomials then we can easily verify that

$$D(c_1\mathbf{x} + c_2\mathbf{y}) = c_1(D\mathbf{x}) + c_2(D\mathbf{y}).$$

It is a natural question to ask, if all matrices are linear transformations, is it the case that all linear transformations can somehow be represented by a matrix? The answer is yes (for $n < \infty$) but we refer the reader to other texts for the details [47]. Here, instead, we focus on writing the above example in matrix notation and then examining linear transformations as matrix multiplications, as this gives us geometrical insight into our problem.

Example 3 In the last example it was shown that D is a linear transformation; can we represent this linear transformation by a matrix? For a fixed $n < \infty$, consider a polynomial $a_n t^n + \cdots + a_1 t + a_0$, writing the coefficients of the powers as the entries in a vector $\mathbf{x}$ of length $n + 1$

$$\mathbf{x} = \begin{bmatrix} a_n \\ a_{n-1} \\ \vdots \\ a_1 \\ a_0 \end{bmatrix}. \tag{B.50}$$

Differentiation can be represented with the matrix

$$D = \begin{bmatrix} 0 & 0 & 0 & \cdots & 0 & 0 & 0 \\ n & 0 & 0 & \cdots & 0 & 0 & 0 \\ 0 & n-1 & 0 & \cdots & 0 & 0 & 0 \\ 0 & 0 & n-2 & \cdots & 0 & 0 & 0 \\ & \vdots & & \ddots & & \vdots & \\ 0 & 0 & 0 & \cdots & 2 & 0 & 0 \\ 0 & 0 & 0 & \cdots & 0 & 1 & 0 \end{bmatrix}. \tag{B.51}$$

For instance, consider the polynomial $\mathbf{x}(t) = 2t^4 - t^3 + \sqrt{2}t^2 + t - 1$. Then

$$D\mathbf{x} = \begin{bmatrix} 0 & 0 & 0 & 0 & 0 \\ 4 & 0 & 0 & 0 & 0 \\ 0 & 3 & 0 & 0 & 0 \\ 0 & 0 & 2 & 0 & 0 \\ 0 & 0 & 0 & 1 & 0 \end{bmatrix} \begin{bmatrix} 2 \\ -1 \\ \sqrt{2} \\ 1 \\ -1 \end{bmatrix} = \begin{bmatrix} 0 \\ 8 \\ -3 \\ 2\sqrt{2} \\ 1 \end{bmatrix}, \tag{B.52}$$

and this last vector can then be rewritten as the polynomial $8t^3 - 3t^2 + 2\sqrt{2}t + 1$, which is the derivative of $2t^4 - t^3 + \sqrt{2}t^2 + t - 1$.

Some linear transformations with simple geometric interpretations

Interpreting a linear transformation as a matrix that takes vectors from one space to another is often useful. We present four examples that have a concrete geometrical interpretation. The vectors, as we will see, remain in the same space. For simplicity, we will consider vectors in the plane $\mathbb{R}^2$.

1. Stretch. In this situation, a vector is stretched (or shrunk) by a factor k. A matrix that will do this is one of the form

$$\mathbf{A} = \begin{bmatrix} k & 0 \\ 0 & k \end{bmatrix}. \tag{B.53}$$

This matrix can be written more compactly as $\mathbf{A} = k\mathbf{I}$, where $\mathbf{I}$ is the 2×2 identity matrix. We know that multiplication by the identity leaves things unchanged. Also, multiplication of a scalar k times a vector (or matrix) requires us to multiply *every entry* by this factor of k. Thus $k\mathbf{I}$ will lengthen any vector by a factor of k if $k > 1$ (and will reverse the orientation and lengthen by a factor of k if $k < -1$) and will shorten any vector by a factor of k if $0 < k < 1$ (and will reverse the orientation and shorten by a factor of k if $-1 < k < 0$). We have encountered this situation before with eigenvalues and eigenvectors.

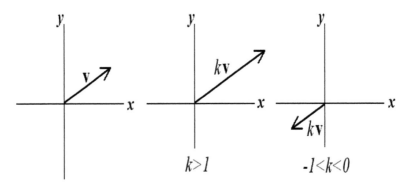

FIGURE B.4: Vector $\mathbf{v}$ stretched by a factor of k. Note that the direction may change depending on the sign of k.

Example 4 Recall that for a matrix $\mathbf{A}$, the eigenvalue λ and eigenvector $\mathbf{v}$ satisfy

$$\mathbf{Av} = \lambda\mathbf{v}.$$

In other words, $\mathbf{A}$ stretches $\mathbf{v}$ by a factor of λ.

2. Rotation. In this case, a vector is rotated counterclockwise about the origin by some angle ϕ. The matrix that will do this is of the form

$$\mathbf{A} = \begin{bmatrix} \cos\phi & -\sin\phi \\ \sin\phi & \cos\phi \end{bmatrix}. \tag{B.54}$$

Consider the vector $(1 \ 0)^T$, which lies along the x-axis. If this vector through the origin is rotated through an angle of $\phi = \pi/4$ rad, the matrix simplifies to

$$\mathbf{A} = \begin{bmatrix} 1/\sqrt{2} & -1/\sqrt{2} \\ 1/\sqrt{2} & 1/\sqrt{2} \end{bmatrix} \tag{B.55}$$

and thus

$$\begin{bmatrix} 1/\sqrt{2} & -1/\sqrt{2} \\ 1/\sqrt{2} & 1/\sqrt{2} \end{bmatrix} \begin{bmatrix} 1 \\ 0 \end{bmatrix} = \begin{bmatrix} 1/\sqrt{2} \\ 1/\sqrt{2} \end{bmatrix} = \frac{1}{\sqrt{2}} \begin{bmatrix} 1 \\ 1 \end{bmatrix} \approx \begin{bmatrix} .7071 \\ .7071 \end{bmatrix}. \qquad (B.56)$$

Recalling from multivariable calculus that the Euclidean length of a vector is the square root of the sum of the squares, we see that the resulting vector is again of length one. It should also be apparent that the vector is oriented at an angle of $\pi/4$ rad.

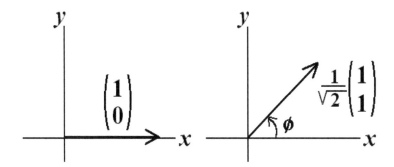

FIGURE B.5:   Vector $[1 \ 0]^T$ rotated through an angle of $\phi = \pi/4$.

Example 6 We consider the system

$$\frac{dx}{dt} = ax - by$$

$$\frac{dy}{dt} = bx + ay, \qquad (B.57)$$

where a, b are real and $b \neq 0$. In this situation the eigenvalues of the coefficient matrix $\mathbf{A}$ are $\lambda_1 = a + ib$, $\lambda_2 = a - ib$. We saw in Section 6.1 that solutions were rotated because of the complex eigenvalues. We will not solve the system now but simply want to examine the right-hand side $\mathbf{Ax}$, where $\mathbf{x} = (x \ y)^T$. If we use polar coordinates and let

$$r = \sqrt{a^2 + b^2}, \quad \tan\phi = \frac{b}{a}, \qquad (B.58)$$

then *multiplication of a vector* $\mathbf{v}$ *by* $\mathbf{A} = \begin{bmatrix} a & -b \\ b & a \end{bmatrix}$ *corresponds, in the case* $b > 0$, *to a counterclockwise rotation through* ϕ *rad, which is then followed by a stretching or shrinking of the length of the vector by a factor of* r; *see Hirsch and Smale [22] for a more in-depth discussion. If* $b < 0$, *the rotation is clockwise, whereas if* $b > 0$ *the rotation will be counterclockwise. Our rotation* matrix is

$$\mathbf{A}_\phi = \begin{bmatrix} \cos\phi & -\sin\phi \\ \sin\phi & \cos\phi \end{bmatrix}. \tag{B.59}$$

Because we have $a = r\cos\phi, b = r\sin\phi$, we can write

$$\begin{bmatrix} a & -b \\ b & a \end{bmatrix} = \begin{bmatrix} r & 0 \\ 0 & r \end{bmatrix} \begin{bmatrix} \cos\phi & -\sin\phi \\ \sin\phi & \cos\phi \end{bmatrix}.$$

This last equality gives us the interpretation of a rotation followed by a stretch.

It is probably natural for the reader to question the usefulness of the interpretation of Example 5 in the case when our matrix is not initially of the form

$$\begin{bmatrix} a & -b \\ b & a \end{bmatrix},$$

where the eigenvalues of the original matrix are $a \pm ib$. For a general matrix with complex (nonreal) eigenvalues, we can use a similarity transformation to convert the original matrix to the above form. The details of this can be found in Appendix B.4.1.

3. Projection. We can also write a matrix that will project any vector onto a given line. If we consider a vector and a line on which we wish to project, the projection is defined as the component of the vector in the direction of the line. In this situation, we can write a matrix in the form

$$\mathbf{A} = \begin{bmatrix} \cos^2\phi & \cos\phi\,\sin\phi \\ \cos\phi\,\sin\phi & \sin^2\phi \end{bmatrix}, \tag{B.60}$$

where ϕ (measured counterclockwise from the origin) is the angle of the line through the origin that we will project onto. Let us again look at a simple example by considering the vector $[1 \ \ 2]^T$ and projecting it onto the y-axis. The y-axis is $\pi/2$ rad counterclockwise from the positive x-axis and thus $\phi = \pi/2$ rad. Our matrix in this case becomes

$$\mathbf{A} = \begin{bmatrix} 0 & 0 \\ 0 & 1 \end{bmatrix}, \tag{B.61}$$

and applying this matrix to our vector gives

$$\begin{bmatrix} 0 & 0 \\ 0 & 1 \end{bmatrix} \begin{bmatrix} 1 \\ 2 \end{bmatrix} = \begin{bmatrix} 0 \\ 2 \end{bmatrix}. \tag{B.62}$$

This answer is what we expected given that the component in the direction of the y-axis is 2. The reader should again try a few additional examples so

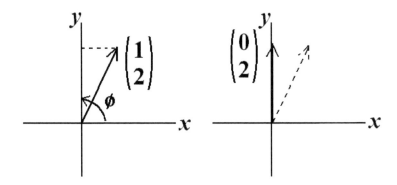

FIGURE B.6: Vector $[1\ \ 2]^T$ projected onto the y-axis.

as to be convinced that the above matrix projects any given vector passing through the origin onto a line passing through the origin with angle ϕ.

Example 6 Consider

$$A = \begin{bmatrix} 1 & 0 \\ 0 & 0 \end{bmatrix}.$$

For any $\mathbf{x} = [x\ \ y]^T$, we see that $\mathbf{Ax} = [x\ \ 0]^T$, a projection onto the x-axis ($\phi = 0$).

4. Reflection. Sometimes we wish to reflect a vector over a given line. We again consider a line through the origin that makes an angle ϕ with the x-axis, where ϕ is measured counterclockwise. The matrix that will do this reflection is

$$\mathbf{A} = \begin{bmatrix} 2\cos^2\phi - 1 & 2\cos\phi\ \sin\phi \\ 2\cos\phi\ \sin\phi & 2\sin^2\phi - 1 \end{bmatrix}. \tag{B.63}$$

As an example, let us reflect the vector $[1\ \ 1]^T$ across the $\pi/6$ rad line. Our linear transformation and vector multiply as

$$\mathbf{A} = \begin{bmatrix} 1/2 & \sqrt{3}/2 \\ \sqrt{3}/2 & -1/2 \end{bmatrix} \begin{bmatrix} 1 \\ 1 \end{bmatrix} = \begin{bmatrix} \sqrt{3}/2 + 1/2 \\ (\sqrt{3} - 1)/2 \end{bmatrix} \approx \begin{bmatrix} 1.366 \\ 0.366 \end{bmatrix}. \tag{B.64}$$

The angle that this vector makes with the positive x-axis is

$$\tan\theta = \frac{(\sqrt{3} - 1)/2}{(\sqrt{3} + 1)/2} \implies \theta = 15° \approx 0.2618 \text{ rad.}$$

The reader should again try a few additional examples so as to be convinced that the above matrix reflects any given vector passing through the origin across a line passing through the origin with angle ϕ.

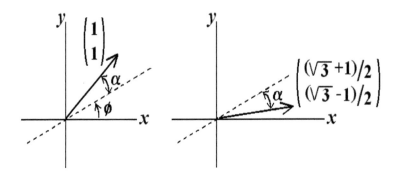

FIGURE B.7:   Vector $[1\ 1]^T$ reflected across the line $\pi/6$.

Other Linear Transformations

In dealing with differential equations, most of the linear transformations that we encounter will have a nice interpretation as a matrix. More generally, we consider the following:

Definition B.3.1

Let V and W be vector spaces. The function $T : V \to W$ is called a linear transformation of V into W if

(1) $T(\mathbf{u} + \mathbf{v}) = T(\mathbf{u}) + T(\mathbf{v})$

(2) $T(c\mathbf{u}) = cT(\mathbf{u})$

both hold true for all $\mathbf{u}$, $\mathbf{v}$ in V and for any scalar c.

We observe that matrices are clearly linear transformations by this definition, with $T(\mathbf{v}) = \mathbf{A}\mathbf{v}$.

Example 7   Consider the differential operator, $D = \frac{d}{dt}$, from Section 3.1.1. We define $C^1[a,b]$ to be the set of all functions with continuous first derivatives on $[a,b]$ and $C^0[a,b]$ to be the set of all continous functions on $[a,b]$. For any functions $x(t), y(t)$ in $C^1[a,b]$ and any constant c, we see that (1) $D(x+y) = Dx + Dy$, and (2) $D(cx) = cD(x)$. Thus, we say that D is a linear transformation from $C^1[a,b]$ into $C^0[a,b]$.

Example 8   Consider $f : \mathbb{R} \to \mathbb{R}$ defined by $f(x) = \cos(x)$. This is not a linear transformation because $\cos(x + y) \neq \cos x + \cos y$ for all x, y, which is required of a linear transformation. For example, take $x = y = \frac{\pi}{2}$ to see linearity does not always hold.

If we let $\mathbf{v}$ be a vector in V and $\mathbf{w}$ be a vector in W such that $T(\mathbf{v}) = \mathbf{w}$, we say that **w** is the **image** of **v** under T. The set of all images of the vectors in V is called the **range** of T. In Section 5.3.1, we discussed the column space and nullspace of a matrix. Analogous concepts hold for all linear tranformations.

Definition B.3.2

Let $T : V \to W$ be a linear transformation and consider $T(\mathbf{v}) = \mathbf{0}$. The set of all vectors **v** in V that satisfy this equation is called the kernel of T, often denoted ker(T). The dimension of the kernel of T is called the nullity of T, often denoted nullity(T). The dimension of the range of T is called the rank of T, often denoted by rank(T).

In our discussion of vector spaces, the set of vectors that satisfied $\mathbf{Av} = \mathbf{0}$ lived in the **nullspace** of **A**. We also discussed the **rank** of a matrix and it is the same here. We finish with what hopefully appears to be a familiar theorem (cf. Theorem 5.3.2).

THEOREM B.3.4

Let $T : V \to W$ be a linear transformation from an n-dimensional vector space V into a vector space W.

$$\text{rank}(T) + \text{nullity}(T) = n.$$

Example 9 Let's consider the projection of a vector in $\mathbb{R}^2$ to the x-axis (still in $\mathbb{R}^2$). Then (B.60) gives us the linear transformation

$$\mathbf{A} = \begin{bmatrix} 1 & 0 \\ 0 & 0 \end{bmatrix}$$

describing this projection. By inspection, we see that ker(**A**) consists of all vectors of the form $\begin{bmatrix} 0 \\ c \end{bmatrix}$ for any constant c. Because $\dim(\mathbb{R}^2) = 2$ and nullity(**A**) $= 1$, we have that rank(T) $= 2 - $ nullity(T) $= 1$.

Problems

In Problems **1–9**, consider the given vector **x**. Find the vectors that result from each of the following:
(a) stretch by a factor of c (sketch the original vector and the resulting vector)
(b) rotation by an angle of ϕ (sketch the original vector, the angle of rotation,

and the resulting vector)

(c) projection onto the line that makes an angle ϕ with the x-axis (sketch the original vector, the line of projection, and the resulting vector)

(d) reflection through the line that makes an angle ϕ with the x-axis (sketch the original vector, the line of reflection, and the resulting vector)

1. $\mathbf{x} = [1 \quad 3]^T$ for (a) stretch by $c = \frac{3}{2}$, (b) rotation by $\phi = \pi/2$, (c) projection onto the line at angle $\phi = \pi/2$, (d) reflection through the line at an angle $\phi = \pi/2$

2. $\mathbf{x} = [-1 \quad -2]^T$ for (a) stretch by $c = -2$, (b) rotation by $\phi = \pi$, (c) projection onto the line at angle $\phi = \pi$, (d) reflection through the line at an angle $\phi = \pi$

3. $\mathbf{x} = [2 \quad 1]^T$ for (a) stretch by $c = -1$, (b) rotation by $\phi = -\pi/4$, (c) projection onto the line at angle $\phi = 0$, (d) reflection through the line at an angle $\phi = 0$

4. $\mathbf{x} = [-2 \quad 3]^T$ for (a) stretch by $c = 3$, (b) rotation by $\phi = \pi/3$, (c) projection onto the line at angle $\phi = \pi/3$, (d) reflection through the line at an angle $\phi = \pi/3$

5. $\mathbf{x} = [1 \quad -3]^T$ for (a) stretch by $c = \frac{1}{2}$, (b) rotation by $\phi = \pi/3$, (c) projection onto the line at angle $\phi = \pi/3$, (d) reflection through the line at an angle $\phi = \pi/3$

6. $\mathbf{x} = [-3 \quad 1]^T$ for (a) stretch by $c = 2$, (b) rotation by $\phi = -\pi/2$, (c) projection onto the line at angle $\phi = \pi/3$, (d) reflection through the line at an angle $\phi = \pi/6$

7. $\mathbf{x} = [-1 \quad -1]^T$ for (a) stretch by $c = \frac{5}{2}$, (b) rotation by $\phi = \pi/3$, (c) projection onto the line at angle $\phi = \pi$, (d) reflection through the line at an angle $\phi = \pi$

8. $\mathbf{x} = [2 \quad 0]^T$ for (a) stretch by $c = 3$, (b) rotation by $\phi = \pi/4$, (c) projection onto the line at angle $\phi = \pi/3$, (d) reflection through the line at an angle $\phi = \pi/6$

9. $\mathbf{x} = [3 \quad 2]^T$ for (a) stretch by $c = 2$, (b) rotation by $\phi = \pi$, (c) projection onto the line at angle $\phi = \pi/2$, (d) reflection through the line at an angle $\phi = \pi/2$

10. Let $\mathbf{A}$ be a 2×2 reflection matrix. Show that $\mathbf{A}^2 = I$. This shows that if $\mathbf{y}$ is a reflection of $\mathbf{x}$, then applying the reflection matrix again gives us the original vector.

11. Consider the reflection matrix given in Equation (B.63). Let m be the slope of the line of reflection. By setting $m = \tan\phi$ and using basic trig identities, show that the reflection matrix can also be written

$$\frac{1}{m^2 + 1} \begin{bmatrix} 1 - m^2 & 2m \\ 2m & m^2 - 1 \end{bmatrix}. \tag{B.65}$$

12. Determine whether $T : \mathbb{R}^2 \to \mathbb{R}^2$, $T(x, y) = (x, 1)$ is a linear transformation.

13. Determine whether $T : \mathbb{R}^3 \to \mathbb{R}^3$, $T(x, y, z) = (x + y, x - y, z)$ is a linear transformation.

14. Let P be the vector space of polynomial functions. Show that the definite integral, $T : P \to \mathbb{R}$, $T(p) = \int_a^b p(x)dx$, is a linear transformation.

Let $T : V \to W$ be a linear transformation and $\mathbf{u}$, $\mathbf{v}$ be vectors in V. In Problems **15**–**17**, show that given properties hold.

15. $T(\mathbf{0}) = \mathbf{0}$ 16. $T(-\mathbf{v}) = -\mathbf{v}$ 17. $T(\mathbf{u} - \mathbf{v}) = T(\mathbf{u}) - T(\mathbf{v})$

B.4   Coordinates and Change of Basis

In this section, we examine how to represent a vector given a basis, and how changing from one basis to another affects this representation. It is best understood after Sections B.3 and 5.4. In terms of vectors in $\mathbb{R}^n$, we are used to thinking of the **standard basis**, $\left\{ \begin{bmatrix} 1 \\ 0 \\ 0 \end{bmatrix}, \begin{bmatrix} 0 \\ 1 \\ 0 \end{bmatrix}, \begin{bmatrix} 0 \\ 0 \\ 1 \end{bmatrix} \right\}$. For example, if we consider the vector

$$\mathbf{w} = \begin{bmatrix} 3 \\ -2 \\ 5 \end{bmatrix}, \tag{B.66}$$

we probably interpret it as "3 units in the x-direction, -2 units in the y-direction, and 5 units in the z-direction." This is correct if we are using the standard basis (often denoted $\{\mathbf{e}_1, \mathbf{e}_2, \mathbf{e}_3\}$) and we could write

$$\begin{bmatrix} 3 \\ -2 \\ 5 \end{bmatrix} = 3 \begin{bmatrix} 1 \\ 0 \\ 0 \end{bmatrix} - 2 \begin{bmatrix} 0 \\ 1 \\ 0 \end{bmatrix} + 5 \begin{bmatrix} 0 \\ 0 \\ 1 \end{bmatrix} = 3\mathbf{e}_1 - 2\mathbf{e}_2 + 5\mathbf{e}_3. \tag{B.67}$$

Writing $\mathbf{w}$ as we did in (B.66) is called the **coordinate vector of $\mathbf{w}$ relative to the standard basis**. The **coordinates of x relative to the basis** are exactly the coefficients that we used to write it as a linear combination of the standard basis vectors. We also need to specify that we have an **ordered basis**, which just means that we need to keep the same order of the basis vectors when we refer to the basis. The representation of $\mathbf{w}$ would probably change if we changed the basis.

For example, if our basis B is

$$\begin{bmatrix} 1 \\ 0 \\ 0 \end{bmatrix}, \begin{bmatrix} 1 \\ 1 \\ 0 \end{bmatrix}, \begin{bmatrix} 1 \\ 1 \\ 1 \end{bmatrix}, \tag{B.68}$$

then the vector $\mathbf{w}$ is written

$$[\mathbf{w}]_B = \begin{bmatrix} 5 \\ -7 \\ 5 \end{bmatrix} = 5 \begin{bmatrix} 1 \\ 0 \\ 0 \end{bmatrix} - 7 \begin{bmatrix} 1 \\ 1 \\ 0 \end{bmatrix} + 5 \begin{bmatrix} 1 \\ 1 \\ 1 \end{bmatrix}. \tag{B.69}$$

How did we find the coordinate vector of $\mathbf{w}$ relative to a given basis? We want to find a linear combination of the ordered basis B that gives us the coordinates of $\mathbf{w}$ in the standard basis:

$$c_1 \begin{bmatrix} 1 \\ 0 \\ 0 \end{bmatrix} + c_2 \begin{bmatrix} 1 \\ 1 \\ 0 \end{bmatrix} + c_3 \begin{bmatrix} 1 \\ 1 \\ 1 \end{bmatrix} = \begin{bmatrix} 3 \\ -2 \\ 5 \end{bmatrix}. \tag{B.70}$$

We could have put this as a problem from Appendix B.2, because we simply need to solve a system of three equations in three unknowns. In matrix notation we have

$$\underbrace{\begin{bmatrix} 1 & 1 & 1 \\ 0 & 1 & 1 \\ 0 & 0 & 1 \end{bmatrix}}_{P} \underbrace{\begin{bmatrix} c_1 \\ c_2 \\ c_3 \end{bmatrix}}_{[\mathbf{w}]_B} = \underbrace{\begin{bmatrix} 3 \\ -2 \\ 5 \end{bmatrix}}_{[\mathbf{w}]_S}, \tag{B.71}$$

and we can easily solve this to find the coefficients, which are the coordinates of $\mathbf{w}$ relative to B. Here, we let P denote the **transition matrix from B to S**. The formula

$$P[\mathbf{w}]_S = [\mathbf{w}]_B$$

is the change of basis from B to S. If we wanted the change of basis from S to B, we could use the following theorem:

THEOREM B.4.1

Let P be the transition matrix from S to B. Then P is invertible and

$$[\mathbf{w}]_S = P^{-1}[\mathbf{w}]_B$$

is the transition matrix from B to S.

In the above discussion, we could have used any two ordered bases in place of B and S.

It is often of interest to find the matrix that will take an ordered basis B to another ordered basis B', instead of being concerned only with how the coordinate representation of one vector changes.

THEOREM B.4.2

Let $\{\mathbf{v}_1, \mathbf{v}_2, \cdots, \mathbf{v}_n\}$ be the ordered basis B, and $\{\mathbf{u}_1, \mathbf{u}_2, \cdots, \mathbf{u}_n\}$ be the ordered basis B' for a given vector space V. Then the bases can be related by

$$\mathbf{v}_1 = c'_{11}\mathbf{u}_1 + c'_{21}\mathbf{u}_2 + \cdots c'_{n1}\mathbf{u}_n \qquad (B.72)$$
$$\mathbf{v}_2 = c'_{12}\mathbf{u}_1 + c'_{22}\mathbf{u}_2 + \cdots c'_{n2}\mathbf{u}_n \qquad (B.73)$$
$$\vdots = \vdots \qquad (B.74)$$
$$\mathbf{v}_n = c'_{1n}\mathbf{u}_1 + c'_{2n}\mathbf{u}_2 + \cdots c'_{n3}\mathbf{u}_n. \qquad (B.75)$$

Moreover, applying Gauss-Jordan elimination to the system

$$[B \mid B'] = \begin{bmatrix} v_{11} & v_{12} & \cdots & v_{1n} & u_{11} & u_{12} & \cdots & u_{1n} \\ v_{21} & v_{22} & \cdots & v_{2n} & u_{21} & u_{22} & \cdots & u_{2n} \\ \vdots & \vdots & \ddots & \vdots & \vdots & \vdots & \ddots & \vdots \\ v_{n1} & v_{n2} & \cdots & v_{nn} & u_{n1} & u_{n2} & \cdots & u_{nn} \end{bmatrix} \qquad (B.76)$$

gives us

$$[B \mid B'] \longrightarrow [I \mid P^{-1}], \qquad (B.77)$$

where P^{-1} is the transition matrix from B to B'.

Example Consider the following two bases in $\mathbb{R}^3$:

$$B = \left\{ \begin{bmatrix} 2 \\ 1 \\ 0 \end{bmatrix}, \begin{bmatrix} -4 \\ -1 \\ 3 \end{bmatrix}, \begin{bmatrix} 5 \\ 2 \\ -1 \end{bmatrix} \right\} \quad \text{and} \quad B' = \left\{ \begin{bmatrix} 1 \\ 0 \\ 0 \end{bmatrix}, \begin{bmatrix} 0 \\ 1 \\ 0 \end{bmatrix}, \begin{bmatrix} 0 \\ 0 \\ 1 \end{bmatrix} \right\}.$$

The transition matrix from B to B' is found by writing the augmented matrix

$$\begin{bmatrix} 2 & -4 & 5 & 1 & 0 & 0 \\ 1 & -1 & 2 & 0 & 1 & 0 \\ 0 & 3 & -1 & 0 & 0 & 1 \end{bmatrix}$$

and row-reducing it via Gauss-Jordan elimination to the form

$$\begin{bmatrix} 1 & 0 & 0 & -5 & 11 & -3 \\ 0 & 1 & 0 & 1 & -2 & 1 \\ 0 & 0 & 1 & 3 & -6 & 2 \end{bmatrix}.$$

The transition matrix from B to B' is thus given by

$$P^{-1} = \begin{bmatrix} -5 & 11 & -3 \\ 1 & -2 & 1 \\ 3 & -6 & 2 \end{bmatrix}.$$

B.4.1 Similarity Transformations

Many times it is convenient to use the standard basis. But when we are trying to understand the qualitative behavior of solutions of differential equations near an equilibrium point (in Sections 6.1 and 6.2), it is often useful to convert to **eigencoordinates** where the eigenvectors will be a basis for the space. This is because the resulting matrix will often be diagonal (and at least always in **Jordan form**, where the matrix is **block diagonal**); see Section 5.6.

In (B.3), we gave the interpretation of matrix-matrix multiplication being equivalent to multiplying the columns of the right matrix with the left matrix to obtain the columns of the result. That is, for $\mathbf{AB} = \mathbf{C}$, we let $\mathbf{b_j}$ and $\mathbf{c_j}$ denote the columns of $\mathbf{B}$ and $\mathbf{C}$, respectively, and we write

$$\mathbf{Ab_j} = \mathbf{c_j}.$$

Using this interpretation we can write our eigenvectors as columns of a matrix $\mathbf{V}$. Then our system becomes

$$\mathbf{AV} = \mathbf{V\Lambda}, \tag{B.78}$$

where

$$\mathbf{\Lambda} = \begin{bmatrix} \lambda_1 & 0 & \cdots & 0 \\ 0 & \lambda_2 & \cdots & 0 \\ \vdots & 0 & \ddots & 0 \\ 0 & 0 & \cdots & \lambda_n \end{bmatrix} \quad \text{and} \quad \mathbf{V} = \begin{bmatrix} v_{11} & v_{12} & \cdots & v_{1n} \\ v_{21} & v_{22} & \cdots & v_{2n} \\ \vdots & \vdots & \ddots & \vdots \\ v_{n1} & v_{n2} & \cdots & v_{nn} \end{bmatrix}. \tag{B.79}$$

(Note that the right-hand side is *not* $\mathbf{\Lambda V}$.) If we assume that our matrix $\mathbf{A}$ has n linearly independent eigenvectors, then our matrix $\mathbf{V}$ is invertible and we can write

$$\mathbf{\Lambda} = \mathbf{V}^{-1}\mathbf{AV}. \tag{B.80}$$

Note that $\mathbf{\Lambda}$ is a diagonal matrix. Thus this equation says that left multiplying our original matrix $\mathbf{A}$ with $\mathbf{V}^{-1}$ and right multiplying it with $\mathbf{V}$ yields a diagonal matrix. This process is called a **diagonal similarity transformation**. Our original matrix really did not matter other than we needed it to have n linearly independent eigenvectors, which may or may not happen for a given matrix.

In terms of coordinates and change of bases, we are given a matrix $\mathbf{A}$ relative to the standard basis, S, and we want to find the matrix relative to the basis of the eigenvectors, B'. The transition matrix from B' to S is exactly our matrix $\mathbf{V}$ that has the eigenvectors as the columns. The transition matrix from S to B' is thus $\mathbf{V}^{-1}$. We combine this together to obtain the matrix relative to the basis of eigenvectors as

$$\mathbf{V}^{-1}\mathbf{AV}.$$

Example 2 Diagonalize the matrix

$$\mathbf{A} = \begin{bmatrix} -1 & 4 \\ 3 & 3 \end{bmatrix}$$

by using a diagonal similarity transformation.

Solution

We need to find the two eigenvectors of the matrix first. If they are linearly independent, then we will be able to proceed. We could do this by hand or with the computer to find the eigenvalue-eigenvector pairs as

$$\left\{ -3, \begin{bmatrix} -2 \\ 1 \end{bmatrix} \right\} \left\{ 5, \begin{bmatrix} 1 \\ \frac{3}{2} \end{bmatrix} \right\}.$$

We can thus write the matrix of linearly independent eigenvectors and calculate its inverse:

$$\mathbf{V} = \begin{bmatrix} -2 & 1 \\ 1 & \frac{3}{2} \end{bmatrix}, \quad \mathbf{V}^{-1} = \begin{bmatrix} \frac{-3}{8} & \frac{1}{4} \\ \frac{1}{4} & \frac{1}{2} \end{bmatrix}.$$

The original matrix $\mathbf{A}$ can be diagonalized as

$$\mathbf{V}^{-1}\mathbf{A}\mathbf{V} = \begin{bmatrix} -3 & 0 \\ 0 & 5 \end{bmatrix}.$$

If we have complex eigenvalues, we usually do *not* diagonalize the matrix simply because this would introduce complex variables into a problem that originally had real numbers. In Section B.3, we considered the system

$$\frac{dx}{dt} = ax - by$$

$$\frac{dy}{dt} = bx + ay, \qquad \text{(B.81)}$$

where a, b are real and $b \neq 0$. We saw that the eigenvalues are $a \pm ib$ and **we interpreted the action of the matrix as a rotation through an angle θ followed by a stretch by a factor r, where $r = \sqrt{a^2 + b^2}$ and $\tan\theta = b/a$ are the standard polar coordinates.** Thus we can think of (B.81) as being the desired form of a system whose coefficient matrix has complex eigenvalues. Analogous to the case of real eigenvalues, we can use a similarity transformation to convert the matrix to this form.

Example 3 Consider

$$\mathbf{A} = \begin{bmatrix} 3 & 2 \\ -4 & -1 \end{bmatrix}.$$

The eigenvalues can be easily calculated as $1 + 2i, 1 - 2i$ with eigenvectors $(1 \quad -1+i)^T, (1 \quad -1-i)^T$, respectively. We previously used the eigenvectors as the columns of our matrix $\mathbf{V}$. Our eigenvectors are now complex but they still give us the insight that we need. We write one of the eigenvectors in the form $\mathbf{v}_1 + i\mathbf{v}_2$. If we take the first one, this would give

$$\begin{bmatrix} 1 \\ -1+i \end{bmatrix} = \begin{bmatrix} 1 \\ -1 \end{bmatrix} + i \begin{bmatrix} 0 \\ 1 \end{bmatrix}.$$

We then use $\mathbf{v}_1$ and $\mathbf{v}_2$ as the respective columns of our matrix $\mathbf{V}$:

$$\begin{aligned} \mathbf{V}^{-1}\mathbf{AV} &= \begin{bmatrix} 1 & 0 \\ -1 & 1 \end{bmatrix}^{-1} \begin{bmatrix} 3 & 2 \\ -4 & -1 \end{bmatrix} \begin{bmatrix} 1 & 0 \\ -1 & 1 \end{bmatrix} \\ &= \begin{bmatrix} 1 & 0 \\ 1 & 1 \end{bmatrix} \begin{bmatrix} 3 & 2 \\ -4 & -1 \end{bmatrix} \begin{bmatrix} 1 & 0 \\ -1 & 1 \end{bmatrix} \\ &= \begin{bmatrix} 1 & 2 \\ -2 & 1 \end{bmatrix}. \end{aligned} \tag{B.82}$$

If we had chosen the second eigenvector, we would have had $\begin{bmatrix} 1 & -2 \\ 2 & 1 \end{bmatrix}$. In either case we see the eigenvalues in the matrix as $1 \pm 2i$.

In general, two $n \times n$ matrices $\mathbf{A}$ and $\mathbf{B}$ are **similar** if there exists an invertible matrix $\mathbf{P}$ such that

$$\mathbf{B} = \mathbf{P}^{-1}\mathbf{AP}.$$

Based on the previous discussion, we can state the following theorem.

THEOREM B.4.3

Similar matrices have the same eigenvalues.

Problems

In Problems 1–4, consider the bases B and B' and vector $\mathbf{w}$.
(a) Find the transition matrix from B' to B.
(b) Find the transition matrix from B to B'.
(c) Compute the coordinate matrix $[\mathbf{w}]_B$, for the given $\mathbf{w}$.
(d) Use part (c) to compute $[\mathbf{w}]_{B'}$.

1. In $\mathbb{R}^2$, $\mathbf{w} = \begin{bmatrix} 3 \\ -5 \end{bmatrix}$ for bases

$$B = \left\{ \begin{bmatrix} 1 \\ 0 \end{bmatrix}, \begin{bmatrix} 0 \\ 1 \end{bmatrix} \right\}, \quad B' = \begin{bmatrix} 2 \\ 1 \end{bmatrix}, \begin{bmatrix} -3 \\ 4 \end{bmatrix}$$

2. In $\mathbb{R}^2$, $\mathbf{w} = \begin{bmatrix} 3 \\ -5 \end{bmatrix}$ for bases

$$B = \left\{ \begin{bmatrix} 1 \\ 2 \end{bmatrix}, \begin{bmatrix} 4 \\ -1 \end{bmatrix} \right\}, \quad B' = \begin{bmatrix} 1 \\ 3 \end{bmatrix}, \begin{bmatrix} -1 \\ -1 \end{bmatrix}$$

3. In $\mathbb{R}^3$, $\mathbf{w} = \begin{bmatrix} -5 \\ 8 \\ -5 \end{bmatrix}$ for bases

$$B = \left\{ \begin{bmatrix} -3 \\ 0 \\ -3 \end{bmatrix}, \begin{bmatrix} -3 \\ 2 \\ -1 \end{bmatrix}, \begin{bmatrix} 1 \\ 6 \\ 1 \end{bmatrix} \right\}, \quad B' = \left\{ \begin{bmatrix} -6 \\ -6 \\ 0 \end{bmatrix}, \begin{bmatrix} -2 \\ -6 \\ 4 \end{bmatrix}, \begin{bmatrix} -2 \\ -3 \\ 7 \end{bmatrix} \right\}.$$

4. In $\mathbb{R}^3$, $\mathbf{w} = \begin{bmatrix} -4 \\ 7 \\ -4 \end{bmatrix}$ for bases

$$B = \left\{ \begin{bmatrix} 2 \\ 1 \\ 2 \end{bmatrix}, \begin{bmatrix} 2 \\ -1 \\ 1 \end{bmatrix}, \begin{bmatrix} 1 \\ 1 \\ 2 \end{bmatrix} \right\}, B' = \left\{ \begin{bmatrix} 3 \\ 1 \\ -5 \end{bmatrix}, \begin{bmatrix} 1 \\ 1 \\ -3 \end{bmatrix}, \begin{bmatrix} -1 \\ 0 \\ 2 \end{bmatrix} \right\}.$$

In Problems **5–10**, if possible, find a matrix **P** that diagonalizes the matrix **A** and then determine $\mathbf{P}^{-1}\mathbf{AP}$. If it is not possible, explain why.

5. $\begin{bmatrix} 2 & 0 \\ 1 & 2 \end{bmatrix}$ **6.** $\begin{bmatrix} 1 & 0 \\ 7 & -1 \end{bmatrix}$ **7.** $\begin{bmatrix} 2 & -3 \\ 1 & -1 \end{bmatrix}$

8. $\begin{bmatrix} 3 & 0 & 0 \\ 0 & 2 & 0 \\ 0 & 1 & 2 \end{bmatrix}$ **9** $\begin{bmatrix} 2 & 0 & -2 \\ 0 & 3 & 0 \\ 0 & 0 & 5 \end{bmatrix}$ **10** $\begin{bmatrix} 0 & 0 & -2 \\ 1 & 2 & 1 \\ 1 & 0 & 3 \end{bmatrix}$

Appendix B Computer Labs

Appendix B Computer Lab: MATLAB

MATLAB Example 1: Enter the code below to use some of the basic linear algebra commands for manipulating matrices and vectors.

```
>> clear all
>> A=[1 3 5;1 7 9;1 2 3] %creates the matrix A
>> A+A
>> 2*A %same result as A+A
>> A*A
>> A^2 %same result as A*A
>> B=inv(A) %calculates the inverse of A
>> B*A %verifies that B is the inverse of A
>> A' %transpose of A
>> trace(A)
>> det(A)
>> x=[8 17 30] %a row vector
>> x=[8; 17; 30] %a column vector
>> A*x %matrix-vector multiplication, note the
   %dimensions must match
>> x'*A %vector-matrix multiplication, note the
   %dimensions must match
>> b=[0; 1; 3] %a column vector
>> y=A\b %solution of the system Ay=b
>> A=[A;2 4 6; 8 10 12] %appends two rows as the 4th
   %and 5th rows of A
>> A(:,4)=[1;2;3;4;5] %appends one column as the 4th
   %column of A
>> A(:,3) %selects the 3rd column of A
>> A(4,:) %selects the 4th row of A
>> A(2:4,3:4) %selects the submatrix of A consisting of
   %the elements in rows 2 through 4 and columns 3 through 4
```

MATLAB Example 2: Enter the code that uses elementary row operations for row reduction, followed by the built-in solver for $\mathbf{Ax} = \mathbf{b}$. Consider the augmented matrix

$$\begin{bmatrix} 2 & 1 & 4 & 1 \\ 2 & 1 & -1 & 0 \\ 4 & 3 & 2 & -1 \end{bmatrix}. \tag{B.83}$$

```
>> clear all
>> A=[2,1,4,1;2,1,-1,0;4,3,2,-1] %creates augmented matrix
```

```
>> A(2,:)=A(2,:)-A(1,:)%row 1 of A subtracted from row 2
   %and inserted as the new row 2; A is overwritten
>> A(3,:)=A(3,:)-2*A(1,:)  %2 times row 1 of A subtracted
   %from row 3 and inserted as the new row 3
>> A=A([1,3,2],:)  %swaps 2nd and 3rd rows of A
>> A(1,:)=A(1,:)/2 %multiplies row 1 by 1/2
>> A(3,:)=-A(3,:)/5 %multiplies row 3 by -1/5
>> A(1,:)=A(1,:)-A(2,:)/2 %row 2 of A, mult by 1/2, then
   %subtracted from row 1 and inserted as new row 1
>> A(2,:)=A(2,:)+6*A(3,:)  %row 3 of A, mult by -6, then
   %subtracted from row 2 and inserted as the new row 2
>> A(1,:)=A(1,:)-5*A(3,:)  %row 3 of A, mult by 5, then
   %subtracted from row 1 and inserted as the new row 1
>> %We are now in reduced-row echelon form; alternatively,
>> B=[2,1,4,1;2,1,-1,0;4,3,2,-1] %re-creates original matrix
>> rref(B)
```

MATLAB Example 3: Enter the code that finds the eigenvalues and eigenvectors (via the built-in commands) and then uses them to diagonalize the matrix **A**.

```
>> clear all
>> A=[-1, 4; 3, 3]
>> [v,d]=eig(A) %eigenvalues AND eigenvectors of A
>> v %shows the matrix v, which has the eigenvectors as its
   % columns; we should verify that we have lin ind columns
>> inv(v)*A*v %diagonalizes A
```

MATLAB Exercises
Turn in both the commands that you enter for the exercises below as well as the output/figures. These should all be in one document. Please highlight or clearly indicate all requested answers. Some of the questions will require you to modify the above MATLAB code to answer them.

1. Enter the commands given in MATLAB Example 1 and submit both your input and output.

2. Enter the commands given in MATLAB Example 2 and submit both your input and output.

3. Enter the commands given in MATLAB Example 3 and submit both your input and output.

In Exercises 4–17, consider: $\mathbf{A} = \begin{bmatrix} 1 & 3 & 2 \\ -1 & 2 & 1 \end{bmatrix}$, $\mathbf{B} = \begin{bmatrix} 4 & -1 \\ 0 & -2 \end{bmatrix}$,

$$\mathbf{C} = \begin{bmatrix} 3 & 3 \\ 1 & 0 \\ -1 & 1 \end{bmatrix}, \ \mathbf{C_1} = \begin{bmatrix} 1 & 8 & -2 \\ 1 & 0 & 3 \\ 3 & -2 & 1 \end{bmatrix}, \ \mathbf{C_2} = \begin{bmatrix} 1 & -3 & 3 \\ 3 & 1 & 1 \\ 2 & 4 & 2 \end{bmatrix}$$

4. Compute (a) $\mathbf{C}_1 + \mathbf{C}_2$; (b) $\mathbf{C}_1 - \mathbf{C}_2$
5. Compute (a) $2\mathbf{C}_1 + 3\mathbf{C}_2$; (b) $2\mathbf{C}$
6. Compute (a) $4\mathbf{C}_2 - 2\mathbf{C}_1$; (b) $\mathrm{Tr}(\mathbf{C}_2)$
7. Compute (a) $\mathrm{Tr}(\mathbf{B})$; (b) $\mathrm{Tr}(3\mathbf{C}_1)$
8. Compute (a) $\mathbf{C}^T$; (b) $\mathbf{A}^T$
9. Compute (a) $\mathbf{C} + \mathbf{C}_2{}^T$; (b) $\mathbf{CA}$
10. Compute (a) $\mathbf{C}_1\mathbf{C}_2$; (b) $\mathbf{C}_2\mathbf{C}_1$
11. Compute (a) $\mathbf{CB}$; (b) $\mathbf{BC}^T$
12. Compute (a) $\mathbf{C}_2\mathbf{C}$; (b) $\mathbf{C}^T\mathbf{C}_2$
13. Compute (a) $\mathbf{CBA}$; (b) $\det(\mathbf{B})$
14. Compute (a) $\det(\mathbf{B}^2)$; (b) $\det(\mathbf{C}_1\mathbf{C}_2)$
15. Compute (a) $\det(\mathbf{C}_2\mathbf{C}_1)$; (b) $\det(\mathbf{C}_2)\det(\mathbf{C}_1)$
16. Compute (a) $\det(\mathbf{C}_1{}^T)$; (b) $\mathbf{B}^{-1}$
17. Compute (a) $\mathbf{C}_1{}^{-1}$; (b) $\mathbf{C}_2{}^{-1}$
18. Use elementary row operations to find the reduced-row echelon form of
$$\begin{bmatrix} 2 & 6 & -2 & -1 \\ 0 & -3 & 1 & 0 \\ 0 & 0 & 2 & 2 \end{bmatrix}$$
19. Use elementary row operations to find the reduced-row echelon form of
$$\begin{bmatrix} 1 & 3 & -6 & 0 \\ 0 & -2 & 1 & 1 \end{bmatrix}$$
20. Use elementary row operations to find the reduced-row echelon form of
$$\begin{bmatrix} 2 & 3 & -3 & 0 \\ 0 & 1 & -1 & 2 \\ 3 & 1 & 0 & -2 \end{bmatrix}$$
21. Use elementary row operations to find the reduced-row echelon form of
$$\begin{bmatrix} -2 & 4 & -3 & 0 \\ 1 & 1 & -1 & 5 \\ 5 & 1 & 0 & -5 \end{bmatrix}$$
22. Use elementary row operations to find the reduced-row echelon form of
$$\begin{bmatrix} -2 & 4 & 0 & 2 \\ 1 & 1 & 3 & 2 \\ 1 & -1 & 1 & 0 \end{bmatrix}$$
23. Use elementary row operations to find the reduced-row echelon form of
$$\begin{bmatrix} 2 & 6 & 0 & 4 \\ -1 & 1 & 4 & 2 \\ 1 & -1 & -4 & -2 \end{bmatrix}$$
24. Use a similarity transformation to diagonalize $\begin{bmatrix} 4 & -8 & -10 \\ -1 & 6 & 5 \\ 1 & -8 & -7 \end{bmatrix}$.

25. Use a similarity transformation to diagonalize $\begin{bmatrix} 2 & -2 & 1 \\ -3 & 3 & 1 \\ 3 & -2 & 0 \end{bmatrix}$.

26. Use a similarity transformation to diagonalize $\begin{bmatrix} 1 & -2 & -1 \\ -1 & 0 & -1 \\ 1 & 1 & 2 \end{bmatrix}$.

27. Use a similarity transformation to diagonalize $\begin{bmatrix} -2 & 2 & 0 \\ 4 & 0 & 0 \\ \frac{-10}{3} & \frac{5}{3} & 1 \end{bmatrix}$.

28. Use a similarity transformation to diagonalize $\begin{bmatrix} \frac{-2}{3} & \frac{-2}{3} & 0 \\ \frac{-4}{3} & \frac{-4}{3} & 0 \\ \frac{3}{3} & \frac{-1}{3} & -1 \end{bmatrix}$.

Appendix B Computer Lab: Maple

Maple Example 1: Enter the code below to use some of the basic linear algebra commands for manipulating matrices and vectors.

```
restart
with(LinearAlgebra) #loads the package for linear algebra
A := Matrix([[a, b], [c, d]])
B := MatrixInverse(A) #inverse of A
A⁻¹ #also the inverse of A
MatrixMatrixMultiply(A, B) #matrix multiplication
A.B #also matrix multiplication
simplify(A.B)
subs(a = 3, b = 1, c = 4, d = 1, A)
C := Matrix([[1, 3, 5], [1, 7, 9], [1, 2, 3]])
C + C
2 · C
Transpose(C)
Trace(C)
Determinant(C)
x := Vector([8, 17, 30]) #a column vector
< 8, 17, 30 > #also a column vector
xT := Transpose(x) #its transpose gives row vector
x%T #also a tranpose operation
C.x
```

$xT.C$
$b := Vector([0, 1, 3])$
$y := LinearSolve(C, b)$ #solution of Cy=b
C
$Column(C, 3)$ #selects 3rd column of C
$Row(C, 2)$ #selects 2nd row of C
$C1 := < C, < 2, 4, 6 >^{\%T}, < 8, 10, 12 >^{\%T} >$ #appends two rows as the 4th
     and 5th rows of C
$C2 := < C1| < 1, 2, 3, 4, 5 >>$ #appends a column as the 5th column of A
$DeleteColumn(C2, 1..2)$ #deletes columns 1 through 2 in C
$DeleteRow(C2, 1..3)$ #deletes rows 1 through 3 in C
$SubMatrix(C2, 2..4, 3..4)$ #selects the submatrix of A consisting of the
     elements in rows 2 through 4 and columns 3 through 4

Maple Example 2: Enter the code that uses elementary row operations for
row reduction, followed by the built-in solver for $\mathbf{Ax} = \mathbf{b}$. Consider the
augmented matrix

$$\begin{bmatrix} 2 & 1 & 4 & 1 \\ 2 & 1 & -1 & 0 \\ 4 & 3 & 2 & -1 \end{bmatrix}. \tag{B.84}$$

$restart$
$with(Student[LinearAlgebra]) :$ #loads the package that allows ero
$A := Matrix([[2, 1, 4, 1], [2, 1, -1, 0], [4, 3, 2, -1]])$
$A1 := AddRow(A, 2, 1, -1)$ #row 1 of A, multiplied by −1,
     then added to row 2 and inserted as the new row 2
$A2 := AddRow(A1, 3, 1, -2)$ #row 1 of A1, multiplied by −2,
     then added to row 3 and inserted as the new row 3
$A3 := SwapRow(A2, 2, 3)$ #swaps the 2nd and 3rd rows of A2
$A4 := MultiplyRow(A3, 1, 1/2)$ #multiplies row 1 by 1/2
$A5 := MultiplyRow\left(A4, 3, -\dfrac{1}{5}\right)$
$A6 := AddRow\left(A5, 1, 2, -\dfrac{1}{2}\right)$
$A7 := AddRow(A6, 2, 3, 6)$
$A8 := AddRow(A7, 1, 3, -5)$
$A9 := GaussianElimination(A)$
$BackwardSubstitute(A9)$
$ReducedRowEchelonForm(A)$

Maple Example 3: Enter the code that finds the eigenvalues and eigenvectors (via the built-in commands) and then uses them to diagonalize the
matrix $\mathbf{A}$.

restart
with(LinearAlgebra) :
$A := Matrix([[-1, 4], [3, 3]])$
$Ev := Eigenvectors(A)$ #*eigenvalues AND eigenvectors of A*
$Ev1 := Column(Ev[2], 1)$ #*first eigenvector*
$Ev2 := Column(Ev[2], 2)$ #*second eigenvector*
$V := Ev[2]$ #*matrix with eigenvectors as columns*
$V^{-1}.A.V$ #*diagonalizes A*

Maple Exercises
Turn in both the commands that you enter for the exercises below as well as the output/figures. These should all be in one document. Please highlight or clearly indicate all requested answers. Some of the questions will require you to modify the above Maple code to answer them.

1. Enter the commands given in Maple Example 1 and submit both your input and output.

2. Enter the commands given in Maple Example 2 and submit both your input and output.

3. Enter the commands given in Maple Example 3 and submit both your input and output.

In Exercises 4–17, consider: $\mathbf{A} = \begin{bmatrix} 1 & 3 & 2 \\ -1 & 2 & 1 \end{bmatrix}$, $\mathbf{B} = \begin{bmatrix} 4 & -1 \\ 0 & -2 \end{bmatrix}$,

$$\mathbf{C} = \begin{bmatrix} 3 & 3 \\ 1 & 0 \\ -1 & 1 \end{bmatrix}, \mathbf{C_1} = \begin{bmatrix} 1 & 8 & -2 \\ 1 & 0 & 3 \\ 3 & -2 & 1 \end{bmatrix}, \mathbf{C_2} = \begin{bmatrix} 1 & -3 & 3 \\ 3 & 1 & 1 \\ 2 & 4 & 2 \end{bmatrix}$$

4. Compute (a) $\mathbf{C_1} + \mathbf{C_2}$; (b) $\mathbf{C_1} - \mathbf{C_2}$

5. Compute (a) $2\mathbf{C_1} + 3\mathbf{C_2}$; (b) $2\mathbf{C}$

6. Compute (a) $4\mathbf{C_2} - 2\mathbf{C_1}$; (b) $\text{Tr}(\mathbf{C_2})$

7. Compute (a) $\text{Tr}(\mathbf{B})$; (b) $\text{Tr}(3\mathbf{C_1})$

8. Compute (a) $\mathbf{C}^T$; (b) $\mathbf{A}^T$

9. Compute (a) $\mathbf{C} + \mathbf{C_2}^T$; (b) $\mathbf{CA}$

10. Compute (a) $\mathbf{C_1}\mathbf{C_2}$; (b) $\mathbf{C_2}\mathbf{C_1}$

11. Compute (a) $\mathbf{CB}$; (b) $\mathbf{BC}^T$

12. Compute (a) $\mathbf{C_2}\mathbf{C}$; (b) $\mathbf{C}^T\mathbf{C_2}$

13. Compute (a) $\mathbf{CBA}$; (b) $\det(\mathbf{B})$

14. Compute (a) $\det(\mathbf{B}^2)$; (b) $\det(\mathbf{C_1}\mathbf{C_2})$

15. Compute (a) $\det(\mathbf{C_2}\mathbf{C_1})$; (b) $\det(\mathbf{C_2})\det(\mathbf{C_1})$

16. Compute (a) $\det(\mathbf{C_1}^T)$; (b) $\mathbf{B}^{-1}$

17. Compute (a) $\mathbf{C_1}^{-1}$; (b) $\mathbf{C_2}^{-1}$

18. Use elementary row operations to find the reduced-row echelon form of
$$\begin{bmatrix} 2 & 6 & -2 & -1 \\ 0 & -3 & 1 & 0 \\ 0 & 0 & 2 & 2 \end{bmatrix}$$

19. Use elementary row operations to find the reduced-row echelon form of
$$\begin{bmatrix} 1 & 3 & -6 & 0 \\ 0 & -2 & 1 & 1 \end{bmatrix}$$

20. Use elementary row operations to find the reduced-row echelon form of
$$\begin{bmatrix} 2 & 3 & -3 & 0 \\ 0 & 1 & -1 & 2 \\ 3 & 1 & 0 & -2 \end{bmatrix}$$

21. Use elementary row operations to find the reduced-row echelon form of
$$\begin{bmatrix} -2 & 4 & -3 & 0 \\ 1 & 1 & -1 & 5 \\ 5 & 1 & 0 & -5 \end{bmatrix}$$

22. Use elementary row operations to find the reduced-row echelon form of
$$\begin{bmatrix} -2 & 4 & 0 & 2 \\ 1 & 1 & 3 & 2 \\ 1 & -1 & 1 & 0 \end{bmatrix}$$

23. Use elementary row operations to find the reduced-row echelon form of
$$\begin{bmatrix} 2 & 6 & 0 & 4 \\ -1 & 1 & 4 & 2 \\ 1 & -1 & -4 & -2 \end{bmatrix}$$

24. Use a similarity transformation to diagonalize $\begin{bmatrix} 4 & -8 & -10 \\ -1 & 6 & 5 \\ 1 & -8 & -7 \end{bmatrix}$.

25. Use a similarity transformation to diagonalize $\begin{bmatrix} 2 & -2 & 1 \\ -3 & 3 & 1 \\ 3 & -2 & 0 \end{bmatrix}$.

26. Use a similarity transformation to diagonalize $\begin{bmatrix} 1 & -2 & -1 \\ -1 & 0 & -1 \\ 1 & 1 & 2 \end{bmatrix}$.

27. Use a similarity transformation to diagonalize $\begin{bmatrix} -2 & 2 & 0 \\ 4 & 0 & 0 \\ \frac{-10}{3} & \frac{5}{3} & 1 \end{bmatrix}$.

28. Use a similarity transformation to diagonalize $\begin{bmatrix} \frac{-2}{3} & \frac{-2}{3} & 0 \\ \frac{-4}{3} & \frac{-4}{3} & 0 \\ \frac{2}{3} & \frac{-1}{3} & -1 \end{bmatrix}$.

Appendix B Computer Lab: Mathematica

Mathematica Example 1: Enter the code below to use some of the basic linear algebra commands for manipulating matrices and vectors.

```
Quit[]
A={{1,3,5},{1,7,9},{1,2,3}}
A//MatrixForm (*displays matrix in a nicer visual form*)
MatrixForm[A] (*alternate syntax*)
(A+A)//MatrixForm
   (*parentheses in above line allow calculations to be done
later*)
(2A)//MatrixForm (*same result as A+A*)
(A.A)//MatrixForm
(MatrixPower[A,2])//MatrixForm (*same result as A.A*)
(B=Inverse[A])//MatrixForm (*calculates the inverse of A*)
(B.A)//MatrixForm (*verifies that B is the inverse of A*)
(Transpose[A])//MatrixForm (*transpose of A*)
Tr[A]
Det[A]
(X = {{8, 17, 30}})//MatrixForm (*a row vector*)
(X = {{8}, {17}, {30}})//MatrixForm (*column vector*)
(A.X)//MatrixForm (*matrix-vector multiplication, note the*)
   (*dimensions must match*)
(Transpose[X].A)//MatrixForm (*vector-matrix multiplication,*)
   (*note the dimensions must match*)
(b={{0},{1},{3}})//MatrixForm (*a column vector*)
y=LinearSolve[A,b] (*solution of Ay=b*)
(A=Join[A,{{2,4,6}},{{8,10,12}}])//MatrixForm
   (*appends two rows as the 4th and 5th rows of A*)
(A=Join[A,{{1},{2},{3},{4},{5}},2])//MatrixForm
   (*appends column by placing one additional entry at end of
each row*)
(Take[A,All,{3,3}])//MatrixForm (*selects the 3rd column of A*)
(Take[A,{4,4}])//MatrixForm (*selects the 4th row of A*)
(Take[A,{2,4},{3,4}])//MatrixForm
   (*submatrix of A consisting of the elements in rows 2*)
   (*through 4 and columns 3 through 4*)
```

Mathematica Example 2: Enter the code that uses elementary row operations for row reduction, followed by the built-in solver for $\mathbf{Ax} = \mathbf{b}$. Consider the augmented matrix

$$\begin{bmatrix} 2 & 1 & 4 & 1 \\ 2 & 1 & -1 & 0 \\ 4 & 3 & 2 & -1 \end{bmatrix}. \tag{B.85}$$

```
Quit[]
(A={{2,1,4,1},{2,1,-1,0},{4,3,2,-1}})//MatrixForm (*the
augmented matrix*)
A[[2]] -= A[[1]]
    (*row 1 of A, subtracted from row 2 and *)
    (*inserted as the new row 2*)
A//MatrixForm (*visually nicer form*)
A[[3]] -=2 A[[1]]
    (*row 1 of A, mult by 2, then subtracted from row 3 and*)
    (*inserted as the new row 3*)
A//MatrixForm
A[[{2, 3}]] = A[[{3, 2}]]
    (*swaps 2nd and 3rd rows of A*)
A[[1]] = 1/2*A[[1]]
    (*multiplies row 1 of A by 1/2*)
A//MatrixForm
A[[3]] = -1/5*A[[3]]
    (*multiplies row 3 of A by -1/5*)
A//MatrixForm
A[[1]] -= 1/2*A[[2]]
    (*row 2 of A, mult by 1/2, then subtracted from row 1 and*)
    (*inserted as the new row 1*)
A[[2]] -= -6*A[[3]]
    (*row 3 of A, mult by -6, then subtracted from row 2 and*)
    (*inserted as the new row 2*)
A//MatrixForm
A[[1]] -= 5*A[[3]]
    (*row 3 of A, mult by 5, then subtracted from row 1 and*)
    (*inserted as the new row 1*)
A//MatrixForm
    (*We are now in reduced-row echelon form;*)
    (*alternatively, we could have used:*)
(B={{2,1,4,1},{2,1,-1,0},{4,3,2,-1}})//MatrixForm (*the
augmented matrix*)
(RowReduce[B])//MatrixForm
```

Mathematica Example 3: Enter the code that finds the eigenvalues and eigenvectors (via the built-in commands) and then uses them to diagonalize the matrix **A**.

```
Quit[]
(A = {{-1,4},{3,3}})//MatrixForm
eq1=Eigensystem[A] (*eigenvalues and eigenvectors of A*)
(eq2a=eq1[[2]][[1]])//MatrixForm (*first eigenvector*)
(eq2b=eq1[[2]][[2]])//MatrixForm (*second eigenvector*)
(V = Transpose[eq1[[2]]])//MatrixForm
   (*eigenvectors are the columns of V*)
(eq3 = Inverse[V].A.V)//MatrixForm (*diagonalizes A*)
```

Mathematica Exercises
Turn in both the commands that you enter for the exercises below as well as the output/figures. These should all be in one document. Please highlight or clearly indicate all requested answers. Some of the questions will require you to modify the above Mathematica code to answer them.

1. Enter the commands given in Mathematica Example 1 and submit both your input and output.
2. Enter the commands given in Mathematica Example 2 and submit both your input and output.
3. Enter the commands given in Mathematica Example 3 and submit both your input and output.

In Exercises 4–17, consider: $\mathbf{A} = \begin{bmatrix} 1 & 3 & 2 \\ -1 & 2 & 1 \end{bmatrix}$, $\mathbf{B} = \begin{bmatrix} 4 & -1 \\ 0 & -2 \end{bmatrix}$,

$$\mathbf{C} = \begin{bmatrix} 3 & 3 \\ 1 & 0 \\ -1 & 1 \end{bmatrix}, \quad \mathbf{C_1} = \begin{bmatrix} 1 & 8 & -2 \\ 1 & 0 & 3 \\ 3 & -2 & 1 \end{bmatrix}, \quad \mathbf{C_2} = \begin{bmatrix} 1 & -3 & 3 \\ 3 & 1 & 1 \\ 2 & 4 & 2 \end{bmatrix}$$

4. Compute (a) $\mathbf{C_1} + \mathbf{C_2}$; (b) $\mathbf{C_1} - \mathbf{C_2}$
5. Compute (a) $2\mathbf{C_1} + 3\mathbf{C_2}$; (b) $2\mathbf{C}$
6. Compute (a) $4\mathbf{C_2} - 2\mathbf{C_1}$; (b) $\text{Tr}(\mathbf{C_2})$
7. Compute (a) $\text{Tr}(\mathbf{B})$; (b) $\text{Tr}(3\mathbf{C_1})$
8. Compute (a) $\mathbf{C}^T$; (b) $\mathbf{A}^T$
9. Compute (a) $\mathbf{C} + \mathbf{C_2}^T$; (b) $\mathbf{CA}$
10. Compute (a) $\mathbf{C_1}\mathbf{C_2}$; (b) $\mathbf{C_2}\mathbf{C_1}$
11. Compute (a) $\mathbf{CB}$; (b) $\mathbf{BC}^T$
12. Compute (a) $\mathbf{C_2}\mathbf{C}$; (b) $\mathbf{C}^T\mathbf{C_2}$

13. Compute (a) $\mathbf{CBA}$; (b) $\det(\mathbf{B})$

14. Compute (a) $\det(\mathbf{B}^2)$; (b) $\det(\mathbf{C_1C_2})$

15. Compute (a) $\det(\mathbf{C_2C_1})$; (b) $\det(\mathbf{C_2})\det(\mathbf{C_1})$

16. Compute (a) $\det(\mathbf{C_1}^T)$; (b) $\mathbf{B}^{-1}$

17. Compute (a) $\mathbf{C_1}^{-1}$; (b) $\mathbf{C_2}^{-1}$

18. Use elementary row operations to find the reduced-row echelon form of
$$\begin{bmatrix} 2 & 6 & -2 & -1 \\ 0 & -3 & 1 & 0 \\ 0 & 0 & 2 & 2 \end{bmatrix}$$

19. Use elementary row operations to find the reduced-row echelon form of
$$\begin{bmatrix} 1 & 3 & -6 & 0 \\ 0 & -2 & 1 & 1 \end{bmatrix}$$

20. Use elementary row operations to find the reduced-row echelon form of
$$\begin{bmatrix} 2 & 3 & -3 & 0 \\ 0 & 1 & -1 & 2 \\ 3 & 1 & 0 & -2 \end{bmatrix}$$

21. Use elementary row operations to find the reduced-row echelon form of
$$\begin{bmatrix} -2 & 4 & -3 & 0 \\ 1 & 1 & -1 & 5 \\ 5 & 1 & 0 & -5 \end{bmatrix}$$

22. Use elementary row operations to find the reduced-row echelon form of
$$\begin{bmatrix} -2 & 4 & 0 & 2 \\ 1 & 1 & 3 & 2 \\ 1 & -1 & 1 & 0 \end{bmatrix}$$

23. Use elementary row operations to find the reduced-row echelon form of
$$\begin{bmatrix} 2 & 6 & 0 & 4 \\ -1 & 1 & 4 & 2 \\ 1 & -1 & -4 & -2 \end{bmatrix}$$

24. Use a similarity transformation to diagonalize $\begin{bmatrix} 4 & -8 & -10 \\ -1 & 6 & 5 \\ 1 & -8 & -7 \end{bmatrix}$.

25. Use a similarity transformation to diagonalize $\begin{bmatrix} 2 & -2 & 1 \\ -3 & 3 & 1 \\ 3 & -2 & 0 \end{bmatrix}$.

26. Use a similarity transformation to diagonalize $\begin{bmatrix} 1 & -2 & -1 \\ -1 & 0 & -1 \\ 1 & 1 & 2 \end{bmatrix}$.

27. Use a similarity transformation to diagonalize $\begin{bmatrix} -2 & 2 & 0 \\ 4 & 0 & 0 \\ \frac{-10}{3} & \frac{5}{3} & 1 \end{bmatrix}$.

28. Use a similarity transformation to diagonalize $\begin{bmatrix} \frac{-2}{3} & \frac{-2}{3} & 0 \\ \frac{-4}{3} & \frac{-4}{3} & 0 \\ \frac{2}{3} & \frac{-1}{3} & -1 \end{bmatrix}$.

It always seems to start from some lecture notes...it has been
one interesting and crazy path!

Thanks (again) for listening....

Section 1.1

1. (i) 2nd order, (ii) linear, (iii) N/A

3. (i) 2nd order, (ii) linear, (iii) N/A

5. (i) 3rd order, (ii) linear, (iii) IVP

7. (i) 1st order, (ii) nonlinear, (iii) IVP

9. (i) 2nd order, (ii) nonlinear, (iii) N/A

11. (i) 2nd order, (ii) nonlinear, (iii) BVP

13. $y'(x) = 6x^2 \Rightarrow x(6x^2) = 3(2x^3)$, which is true for all x.

15. $y'(x) = 0 \Rightarrow 0 = x^3(2-2)^2 = 0$, which is true for all x.

17. $y'(x) = e^x - 1 \Rightarrow (e^x - 1) + (e^x - x)^2 = e^{2x} + (1 - 2x)e^x + x^2 - 1$, which is true for all x.

19. $y'(x) = 2x - x^{-2}$ and $y''(x) = 2 + x^{-3}$. Substitution shows its equation holds for all $x \neq 0$

21. Note the solution does not exist for $x = 3$. $y'(x) = \frac{1}{(x-3)^2} \Rightarrow \frac{1}{(x-3)^2} = \left(\frac{-1}{x-3}\right)^2$ which is true when $x \neq 3$.

23. $y = e^x$, $y' = e^x$, $y'' = e^x \Rightarrow y'' - 3y' + 2y = e^x - 3e^x + 2e^x = 0$; $y = e^{2x}$, $y' = 2e^{2x}$, $y'' = 4e^{2x} \Rightarrow y'' - 3y' + 2y = 4e^{2x} - 6e^{2x} + 2e^{2x} = 0$.

25. (b) e^{-3x}, (c) xe^{-3x}, and (e) $2e^{-3x} + xe^{-3x}$ are all solutions; (a) e^x and (d) $4e^{3x}$ are not solutions.

27. (b) e^{3x}, (c) e^{4x}, and (e) $e^{3x} + 2e^{4x}$ are all solutions; (a) e^{2x} and (d) e^{5x} are not solutions.

29. $y'(x) = re^{rx}, y''(x) = r^2 e^{rx} \Rightarrow (r^2 e^{rx}) + 3(re^{rx}) + 2(e^{rx} = e^{rx}(r^2 + 3r + 2) = 0 \Rightarrow r = -2, -1$, which is true for all x.

31. $y'(x) = rxe^{rx} + e^{rx}$, $y''(x) = r^2 xe^{rx} + 2re^{rx}$. Substituting and equating coefficients gives $r = -3$, which is true for all x.

Section 1.2

1. $y(x) = \sin x + C$

3. $x(t) = \pm\sqrt{2t - t^2 + C}$

5. $y(x) = C \csc^2 x$

7. $y = \frac{C}{(x^2+1)^2}$, where $C = \pm e^{C_1}$

9. $y = \log_{10} |-10^x + C|$

11. $y = \pm e^{1/2} e^{x^2 - 1}$ and is even since we have only an x^2 term.

13. $\sin y + 1 = \frac{2 + 2\sin 3}{e^x + 1}$

15. $y(x) = \tan\left(\frac{\pi}{2} - \arctan(x^2)\right)$

17. $\frac{e^y}{(y+2)^2} = \frac{e^{-1}}{x+4}$

19. $y(x) = \int_0^x e^{t^2}\,dt$

21. (a) $y(x) = \arctan\left(1 - \frac{1}{x}\right)$

23. $y(x) = C(x+1)e^{-x}$

25. Substitute $u = 2y + x$. Solution is $2y - 2\ln|2 + x + 2y| + 4 + 2\ln 2 = 0$

27. Substitute $u = 4x + 2y - 1$. Solution is $x = -2\left(\ln\left(\sqrt{4x + 2y - 1} + 2\right)\right) - \sqrt{4x + 2y - 1}$

29. It represents the long-term population that the system can sustain.

Section 1.3

1. $y = e^{-x}\left(\frac{e^{2x}}{2} + C\right)$

3. $y = e^{-2x}\left(\frac{-3xe^{2x}}{2} + \frac{3}{4}e^{2x} + C\right)$

5. $y = e^{x^3}\left(-\frac{e^{-x^3}}{3} + C\right)$

7. Must consider both $x < 0$, $x > 0$. Each gives $y(x) = \frac{x^2}{3} + \frac{C}{x}$.

9. $y = (1 + x^2)(x - \arctan x + C)$

11. $y = e^{-x}\left(\frac{e^x}{2}(\cos x + \sin x) + C\right)$

13. $y = 4e^x(x + 1)$

15. $y = \sin x - \cos x$

17. $y = \frac{\sin x}{x} + x^{-1}$

19. $x = y^2 + Cy$

21. $x(y) = 2\sin^2 y + C\sin y$

Section 1.4

1. (a) $v(4) = 128$ ft/s; (b) Ave velocity=64 ft/s; (c) 256 ft tall

3. $v_0 = 3g = (3\text{ s})(32\text{ ft/s}^2) = 96$ ft/s

5. 1.82 sec, 17.1 m, 2.04 sec, 20.4 m

7. $v(t) = 60\left(\frac{1 + e^{-.326t}}{1 - e^{-.326t}}\right)$

9. 40 min

11. 64.5°F

13. 10 min

15. $s(t) = \frac{400}{3} - \frac{325}{3}e^{(-3t/50)} \rightarrow$ conc= $s(t)/25 - t$

17. 24 min

19. 200 days

21. $(c \pm x)y = 2a^2$

23. About 8am.

Section 1.5

1. Equation is not exact.

3. $x^2y - \frac{1}{3}y^3 = C$

5. $x^2y^3 + y = C$

7. $x^3 + x^3\ln y - y^2 + C$

9. $x^2y + 2y^2 + x = 2$. Only the top curve in this implicit solution passes through the IC.

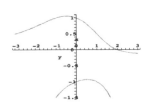

Graph for 1.5#9.

11. $x^2 - 3x^3y^2 + y^4 = -1$. Only the top curve in this implicit solution passes through the IC.

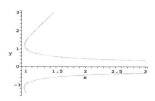

Graph for 1.5#11.

13. $-xe^{-y} - y^2 = -e^{-3} - 9$

15. $A = -2$; solution is $\frac{y}{x^2} - \frac{y}{x} = C$

17. $N(x,y) = x^2y + \phi(y)$

Section 1.6

1. $y = \frac{Ce^x}{e^x - 1}$

3. $\frac{1}{2}x^2 + xy + ln(y) = C$

5. $y = -2x\sin(x) + 2x^2\cos(x) + Cx$

7. $ye^{x^3} - \frac{e^{x^3}}{3} = C$

11. $y = \frac{1}{x+1+Ce^x}$

13. $\ln|y| = 4x - x^2 + C$

15. $y^2 = -2x + Cx^2$

17. $\sqrt{y} = x^2 \ln|x| + Cx^2$

19. $y = (-3x^2 + C|x|^3)^{1/3}$

21. $N(t) = \frac{1}{\frac{1}{K} + Ce^{-rt}}$

23. $y(x) = Cx^2 - x$

25. $y = \pm x/\sqrt{\ln x + C}$

27. $y(x) = \frac{Cx^2}{1 - Cx}$

29. $y = C(y^2 - x^2)$

31. $\arctan\left(\frac{y}{x}\right) + \frac{1}{2}\ln\left(1 + \left(\frac{y}{x}\right)^2\right) = -x$

33. $\frac{1}{\sqrt{y(y-2x)}} = C$

37. Substitution gives $\frac{dy}{dx} = \frac{r\cos t\,dt + \sin t\,dr}{r\sin t\,dt + \cos t\,dr}$ and simplification gives $\frac{dr}{dt} = r\left(\frac{-N\cos t + M\sin t}{N\sin t + M\cos t}\right)$.

39. $\frac{dr}{dt} = r(\sec^2 t + \tan t)$ separates and solves to $\sqrt{x^2 + y^2} = \exp\left[\frac{y}{x} - \ln(\cos(\tan^{-1}\left(\frac{y}{x}\right))) + C\right]$

41. Lines are parallel, so there is no point of intersection.

43. $\ln|x + 4| = -5\ln\left|1 - \frac{y-1}{x+4}\right| - 4(x + 4)$

45. $\tan^{-1}\left(\frac{x-3}{y+2}\right) = \frac{1}{2}\ln|x - 3| + C$

Chapter 1 Review

1. False. An IC is needed for an IVP.

3. False. We do not need to solve for $y(x)$.

5. False. Solutions need only be defined for x in the interval $(x_0 - h, x_0 + h)$.

7. True. This form can always be separated.

9. $y'(x) = Ce^x - 2x - 2 \Rightarrow x^2 + y(x) = x^2 + (Ce^x - x^2 - 2x - 2) = y'(x)$ for all x

11. $y' = \frac{1}{(x+C)^2}, \Rightarrow$ $(y - 3)^2 = \left(\frac{-1}{(x+C)}\right)^2 = y'$

13. $y = \pm\sqrt{Ce^{1/x} + 2}$

15. $y = \frac{1}{1+x}$

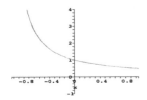

Graph for Chapter 1 Review#**15**.

17. $y = \ln \left| \frac{-\ln |x| + C}{e^{1/x}} \right|$

19. $\cot \left(\frac{1}{2} \ln \left(\frac{y}{x} \right) \right) = \ln x + C$

21. $\arcsin \left(\frac{y}{x} \right) = \ln x + C$

23. $y \ln x + \frac{1}{4} y^4 = C$

25. $\frac{x^3}{y^2} + x + \frac{5}{y} = C$

27. $\mu(x) = x^{-3}, \frac{-y^2}{2x^2} - y = C$

29. $y^{-3} = \frac{3x}{2} + C x^3$

31. $y = e^{-2x^2} \left[\frac{5}{8} e^{2x^2} (2x^2 - 1) + C \right]$

33. $y^{-2} = e^{x^2/2} (4 e^{-x^2/2} + C)$

35. The soup should be ready 34 minutes before closing.

37. The tank will fill in 15 minutes; $y(15) = 12.43$ lbs of salt.

39. $y = x^2$

Section 2.1

1. b

3. a

5. b

7. d

9. $y' = \cos y$, see figure

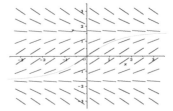

Graph for 2.1#**9**.

11. $y' = \sin y$, see figure

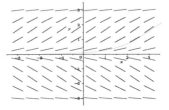

Graph for 2.1#**11**.

13. $y' = \cos x$, see figure

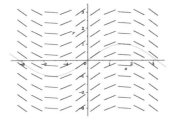

Graph for 2.1#**13**.

15. $y' = \sin x$, see figure

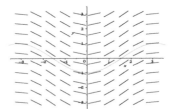

Graph for 2.1#**15**.

17. $y' = x + y$, see figure

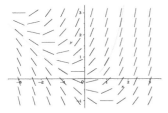

Graph for 2.1#**17**.

19. $y' = e^{x^2}$, see figure

Graph for 2.1#**19**.

21. $y' = \frac{x-1}{y-1}$, see figure

Graph for 2.1#**21**.

23. $y' = \frac{x^2-1}{y^2+1}$, see figure

Graph for 2.1#**23**.

25. $y' = xy(x+2)$, see figure

Graph for 2.1#**25**.

Section 2.2

1. Yes, the Theorem guarantees existence or uniqueness of a solution at $(0,0)$.

3. The theorem does *not* guarantee that a solution exists or is unique on some interval.

5. Theorem does not guarantee existence or uniqueness of a solution at $(1,0)$.

7. (i) Solutions exist for all (x,y); (ii) solutions are unique everywhere except possibly when $y = -2$.

9. (i) Solutions exist for all (x,y); (ii) solutions are unique everywhere except possibly when $y = x$.

11. (i) Solutions exist everywhere except possibly when $y = 0$; (ii) solutions are unique everywhere except possibly when $y = 0$.

13. (i) Solutions exist everywhere except possibly when $y = -x$; (ii) solutions are unique everywhere except possibly when $y = -x$.

15. Solutions will exist everywhere and will be unique everywhere except possibly along $y = 2$. Separation of variables $\Rightarrow y = 2 + (2x)^{5/2}$ passes through $(0,2)$. Since $y = 2$ also passes through $(0,2)$, the solution is not unique.

17. Solutions will exist everywhere and will be unique everywhere except possibly along $y = 0$. Separation of variables $\Rightarrow y = (x - 1 + \sqrt{3})^2$ passes through $(1, 3)$. There is no problem regarding uniqueness at the given point $(1, 3)$.

19. Solutions will exist everywhere and will be unique everywhere except possibly along $y = 1$. Separation of variables $\Rightarrow y = 1 + (3x)^{3/2}$ passes through $(0, 2)$. Since $y = 1$ also passes through $(0, 1)$, the solution is not unique.

Section 2.3

1. (ii) $y^* = -\frac{3}{2}$ is unstable; (iii) for $y_0 < -\frac{3}{2}$, $y \to -\infty$ as $x \to \infty$; for $y_0 > -\frac{3}{2}$, $y \to \infty$ as $x \to \infty$;

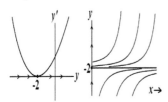

2.3#**1** (i), (ii), (iv).

3. (ii) $y^* = -2$ is stable; $y^* = 3$ is unstable; (iii) for $y_0 < -2$, $y \to -2$ as $x \to \infty$; for $-2 < x_0 < 3$, $y \to 3$ as $x \to \infty$; for $y_0 > 3$, $y \to \infty$ as $x \to \infty$;

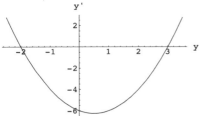

2.3#**3** (i), (ii), (iv).

5. (ii) $y^* = -2$ is half-stable; (iii) for $y_0 < -2$, $y \to -2$ as $x \to \infty$; for $y_0 > -2$, $y \to \infty$ as $x \to \infty$;

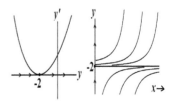

2.3#**5** (i), (ii), (iv).

7. (ii) $y^* = 0$ is half-stable, $y^* = 2$ is stable; (iii) for $y_0 < 0$, $y \to 0$ as $x \to \infty$; for $0 < y_0$, $y \to 2$ as $x \to \infty$;

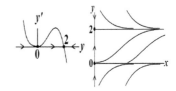

2.3#**7** (i), (ii), (iv).

9. (ii) $y^* = -3, 3$ are unstable, $y^* = 2$ is stable; (iii) for $y_0 < -3$, $y \to -\infty$ as $x \to \infty$; for $-3 < y_0 < 2$, $y \to 2$ as $x \to \infty$; for $y_0 > 3$, $y \to \infty$ as $x \to \infty$;

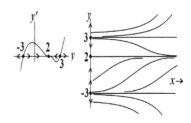

2.3#**9** (i), (ii), (iv).

11. (ii) $y^* = \pm\pi$ are stable, $y^* = 0$ is unstable; (iii) for $y_0 < -\pi$, $y \to -\pi$ as $x \to \infty$; for $-\pi < y_0 < 0$, $y \to -\pi$ as $x \to \infty$; for $0 < y_0 < \pi$, $y \to \pi$ as $x \to \infty$; for $y_0 > \pi$, $y \to \pi$ as $x \to \infty$;

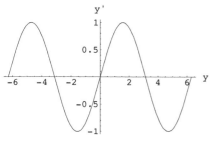

2.3#11 (i), (ii), (iv).

13. (ii) $v^* = \sqrt{\dfrac{gm}{k}}$ is stable; $v^* = -\sqrt{\dfrac{gm}{k}}$ is unstable but not physically meaningful (so ignore for rest of problem); (iii) for $v_0 > -\sqrt{\dfrac{gm}{k}}$, $v \to \sqrt{\dfrac{gm}{k}}$ as $t \to \infty$;

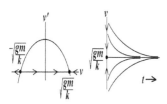

2.3#13 (i), (ii), (iv).

15. (ii) $x^* = 2$ is stable; (iii) for $x_0 \in (-\infty, \infty)$, $x \to 2$ as $t \to \infty$;

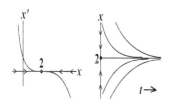

2.3#15 (i), (ii), (iv).

17. (ii) $x^* = 0$ is unstable; $x^* = 1$ is stable; (iii) for $x_0 < 0$, $x \to -\infty$ as $t \to \infty$; for $x_0 > 0$, $x \to 0$ as $t \to \infty$;

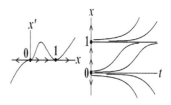

2.3#17 (i), (ii), (iv).

19. (ii) $x^* = 1, -1$ is stable; $x^* = \dfrac{1}{2}$ is unstable; (iii) If $x > 1$, $x \to 1$ as $t \to \infty$. If $\frac{1}{2} < x < 1$, $x \to 1$ as $t \to \infty$.
If $-1 < x < \frac{1}{2}$, $x \to -1$ as $t \to \infty$.
If $x < -1$, $x \to -1$ as $t \to \infty$.

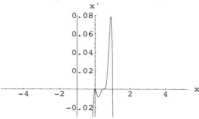

2.3#19 (i), (ii), (iv).

21. $f(y) = y^2 - 1 \Rightarrow f'(y) = 2y$. Equilibria are $y^* = \pm 1$;
$f'(-1) < 0 \Rightarrow y^* = -1$ is stable;
$f'(1) > 0 \Rightarrow y^* = 1$ is unstable.

23. $f(y) = y^3 + 1 \Rightarrow f'(y) = 3y^2$. Only real equilibrium is $y^* = -1$; $f'(-1) > 0 \Rightarrow y^* = -1$ is unstable.

25. $r < 0 \Rightarrow y^* = -\sqrt{-r}$ is stable and $y^* = \sqrt{-r}$ is unstable. $r > 0 \Rightarrow$ no equilibria $\Rightarrow$ saddlenode bifurcation.

27. $r < 0 \Rightarrow y^* = 0$ is stable, $y^* = -r > 0$ is unstable. $r > 0 \Rightarrow y^* = 0$ is unstable, $y^* = -r < 0$ is stable $\Rightarrow$ transcritical bifurcation.

29. It's stable.

Section 2.4

1. (a) $x = 0$ is half-stable, $x = 1$ is stable. For logistic equation, $x = 0$ is unstable and $x = k$ is stable. Yes, they are different. (b) For small x, logistic model growth is larger.

3. $x = 0$ - Stable, $x = 2$ - Unstable, $x = 10$ - Stable

5. (a) $x^* = 0, 1, 4$; $x^* = 0$ is stable; $x^* = 1$ is unstable; $x^* = 4$ is stable. The stability results do not differ from that of the Allee effect. (b) For small x, this model is approximately $x' = -4x^2$ whereas the comparable Allee model would be $x' = -4x$; the population in this model dies off at a much slower rate than in the Allee model.

7. (a) $x^* = 0, a, 5$; $x^* = 0$ is unstable; $x^* = a$ is stable; $x^* = 5$ is unstable.
(b) For $0 < x_0 < 5$, $x \Rightarrow a$ as $t \Rightarrow \infty$; for $x_0 > 5$, $x \Rightarrow \infty$ as $t \Rightarrow \infty$.
(c) The parameter a could be a measure of the health of the individual or the strength of their immune system. If a person had a compromised immune system, then the body will not be able to keep the bacteria in check at low levels but may still be able to keep it in check at higher levels.

9. $x' = x(x-1)(x-6)(x-10)$ is one possibility.

11. (a) $x' = x^2(2-x)^2(x-4)$;
(b) $x^* = 0, 2, 4$; $x^* = 0$ is half-stable; $x^* = 2$ is half-stable; $x^* = 4$ is unstable.

13. Let $x = \frac{N}{K}$, $\tau = rt$.

15. Let $x = \frac{g}{k_4}$, $\tau = \frac{tk_3}{k_4}$, $r = \frac{k_2 k_4}{k_3}$, $s = \frac{k_1 s_0}{k_3}$.

Section 2.5

1. $y' = x^3$, $y(1) = 1$.

(a)

x_i	Euler y_i	Explicit $y(x_i)$
1.0	1	1
1.1	1.1	1.1160
1.2	1.2331	1.2684
1.3	1.4059	1.4640
1.4	1.6256	1.7104

(b)

x_i	Runge-Kutta y_i	Explicit $y(x_i)$
1.0	1	1
1.1	1.1160	1.1160
1.2	1.2684	1.2684

3. $y' = x^4 y$, $y(1) = 1$.

(a)

x_i	y_i	$y(x_i)$
1.0000	1.0000	1.0000
1.1000	1.1000	1.1299
1.2000	1.2611	1.3467
1.3000	1.5225	1.7205
1.4000	1.9574	2.4004

(b)

x_i	y_i	$y(x_i)$
1.0000	1.0000	1.0000
1.1000	1.1299	1.1299
1.2000	1.3467	1.3467
1.3000	1.7204	1.7205
1.4000	2.4003	2.4004

5. $y' = \frac{\sin x}{y^3}$, $y(\pi) = 2$.

(a)

x_i	y_i	$y(x_i)$
3.1416	2.0000	2.0000
3.2416	2.0000	1.9994
3.3416	1.9988	1.9975
3.4416	1.9963	1.9944
3.5416	1.9925	1.9901

(b)

x_i	y_i	$y(x_i)$
3.1416	2.0000	2.0000
3.2416	1.9994	1.9994
3.3416	1.9975	1.9975
3.4416	1.9944	1.9944
3.5416	1.9901	1.9901

7. $y' = e^{-y}$, $y(0) = 2$.

(a)

x_i	y_i	$y(x_i)$
0.0	2	2
0.1	2.2	2.2155
0.2	2.4300	2.4641
0.3	2.6929	2.7492
0.4	2.9917	3.0743
0.5	3.3298	3.4433
0.6	3.7107	3.8603
0.7	4.1383	4.3299
0.8	4.6165	4.8568

(b)

x_i	y_i	$y(x_i)$
0.0	2	2
0.1	2.0134	2.0134
0.2	2.0267	2.0267
0.3	2.0398	2.0398
0.4	2.0527	2.0527
0.5	2.0655	2.0655
0.6	2.0781	2.0781
0.7	2.0905	2.0905
0.8	2.1028	2.1028

9. $y' = y + \cos x$, $y(0) = 0$.

(a)

x_i	y_i	$y(x_i)$
0	0	0
0.1000	0.1000	0.1050
0.2000	0.2095	0.2200
0.3000	0.3285	0.3450
0.4000	0.4568	0.4801
0.5000	0.5946	0.6253
0.6000	0.7418	0.7807
0.7000	0.8986	0.9466
0.8000	1.0649	1.1231

(b)

x_i	y_i	$y(x_i)$
0	0	0
0.1000	0.1050	0.1050
0.2000	0.2200	0.2200
0.3000	0.3450	0.3450
0.4000	0.4801	0.4801
0.5000	0.6253	0.6253
0.6000	0.7807	0.7807
0.7000	0.9466	0.9466
0.8000	1.1231	1.1231

11. $y' = x + y$, $y(0) = 0$.

(a)

x_i	y_i	$y(x_i)$
0	0	0
0.1000	0	0.0052
0.2000	0.0100	0.0214
0.3000	0.0310	0.0499
0.4000	0.0641	0.0918
0.5000	0.1105	0.1487
0.6000	0.1716	0.2221
0.7000	0.2487	0.3138
0.8000	0.3436	0.4255

(b)

x_i	y_i	$y(x_i)$
0	0	0
0.1000	0.0052	0.0052
0.2000	0.0214	0.0214
0.3000	0.0499	0.0499
0.4000	0.0918	0.0918
0.5000	0.1487	0.1487
0.6000	0.2221	0.2221
0.7000	0.3138	0.3138
0.8000	0.4255	0.4255

13. See graph.

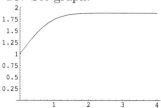

Graph for 2.5#**13**.

15. $y' = x^3 y - x^2 y^2$, $y(-1) = 1$. Take $h = .001$ viewing window to be $-1 < x < 10, -1 < y < 10$.

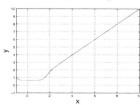

Graph for 2.5#**15**.

17. $y' = y\sqrt{x^2 + y^2 + 1} + \cos(xy)$, $y(0) = 1$. Take $h = .01$ viewing window to be $-.5 < x < 1, -1 < y < 20$.

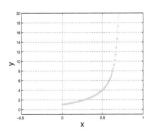

Graph for 2.5#17.

Section 2.6

1. $y = ce^{kx}$, $y' = kce^{kx}$, $y'' = k^2ce^{kx} \Rightarrow$
$yy'' = ce^{kx} * k^2ce^{kx} = c^2k^2e^{2kx} = (y')^2$

3. Substitution gives $v\frac{dv}{dx} = -g\frac{R^2}{x^2}$. Separating, solving, and applying the IC gives $v = \pm\sqrt{2gR(\frac{R}{x} - 1) + v_0^2}$. The "+" corresponds to the object going away from center of earth; the object "escapes" if the radicand is always nonnegative. Since $\frac{R}{x} \to 0$ as $x \to \infty \Rightarrow -2gR + v_0^2 \geq 0$.

Chapter 2 Review

1. False. Neither f nor $\frac{\partial f}{\partial y}$ is continuous everywhere.

3. False. RK use four function evaluations to calculate the next step.

5. True.

7. (i) Solutions exist everywhere; (ii) solutions are unique everywhere.

9. (i) Solutions exist everywhere; (ii) solutions are unique everywhere except possibly along $y = 0$.

11. (i) Solutions exist everywhere except possibly when $y = \frac{\pi}{2} \pm n\pi$ for $n = 0, 1, 2, \cdots$; (ii) solutions are unique everywhere except possibly when $y = \frac{\pi}{2} \pm n\pi$ for $n = 0, 1, 2, \cdots$.

13. (i) Solutions exist everywhere except possibly along $x = -1$; (ii) solutions are unique except possibly along $x = -1$.

15. See graph.

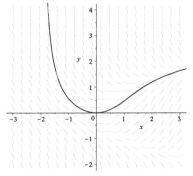

Graph for Chap 2 Review #15.

17. (a) $x^* = a$ is stable; $x^* = 0, 4$ are unstable. (b) $x_0 \in (0, a) \to a$ as $t \to \infty$; $x_0 \in (a, 4) \to a$ as $t \to \infty$; $x_0 > 4 \to \infty$ as $t \to \infty$.

19. Left picture matches (d); right picture matches (b).

21. $x^* = 0$ half-stable; $x^* = 2$ unstable; $x^* = 4/3$ stable. See graph.

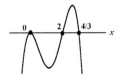

Graph for Chap 2 Review # **21.**

23. $x^* = -3/2, 5$ are both stable; $x^* = -1$ unstable. See graph.

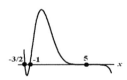

Graph for Chap 2 Review # **23.**

25. $x^* = -2, 4/\sqrt{3}$ are both stable; $x^* = -4/\sqrt{3}, 2$ are both unstable. See graph.

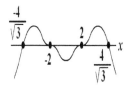

Graph for Chap 2 Review # **25**.

27. $r < 1 \Rightarrow y^* = \sqrt{1-r}$ is stable and $y^* = -\sqrt{1-r}$ is unstable. $r > 1 \Rightarrow$ no equilibria $\Rightarrow$ saddlenode bifurcation. See graph.

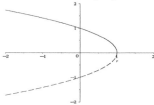

Graph for Chap 2 Review # **27**.

29. $r < 1 \Rightarrow y^* = 0$ is unstable, $y^* = 1 - r$ is stable. $r > 1 \Rightarrow y^* = 0$ is stable, $y^* = 1 - r$ is unstable $\Rightarrow$ transcritical bifurcation. See graph.

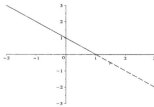

Graph for Chap 2 Review # **29**.

31. $y' = \frac{1-y^2}{2x}$, $y(1) = \pi$.

x_i	RK:y_i	Euler:y_i
1.0000	3.1416	3.1416
1.1000	2.7742	2.6981
1.2000	2.5144	2.4127
1.3000	2.3210	2.2118
1.4000	2.1713	2.0621
1.5000	2.0522	1.9459
1.6000	1.9550	1.8531
1.7000	1.8743	1.7770
1.8000	1.8061	1.7135

33. $y' = \frac{y-y^2}{x+1}$, $y(2) = 0$.
All answers are 0 because $y_0 = 0$ is an equilibrium point.

35. $y' = \frac{y(x+y)}{x^2}$, $y(\sqrt{2}) = 1$.

x_i	RK:y_i	Euler:y_i
1.51	1.1251	1.1207
1.61	1.2592	1.2495
1.71	1.4030	1.3868
1.81	1.5571	1.5332
1.91	1.7222	1.6891

37. $y' = \frac{-(2xy^2 - y)^2}{x}$, $y(\pi) = 1$.

x_i	RK:y_i	Euler:y_i
3.24	.5678	.11153
3.34	.5110	.11150
3.44	.4724	.11148
3.54	.4435	.11146
3.64	.4206	.11144

Section 3.1

1. A unique solution is guaranteed because coefficients are constant. Substitution gives $c_1 = c_2 = 2$.

3. A unique solution is not guaranteed because $a_2(x) = x$ is zero at the IC. Substitution gives $c_1 = c_2 = 0$.

5. $a_2(x) = x^2 = 0$ when $x = 0$, so Thm 3.1.1 does not apply.
Any constants c_1 and c_2 will work.

7. Solutions are guaranteed everywhere except when $x = 0$.

9. Solutions are guaranteed for all (x, y).

11. Solutions are guaranteed when $x > 0$.

13. Solutions are guaranteed everywhere except when $x = \frac{\pi}{2} + n\pi$, $n = 0, 1, 2, \cdots$.

15. $P(D)f_1 = 0$, $P(D)f_2 = \cos x + 4 + \sin x$, $P(D)f_3 = -e^{-2x}$

17. $P(D)f_1 = 8x + x^3$, $P(D)f_2 = 0$, $P(D)f_3 = -3\sin 2x$

19. $P(D)f_1 = 4e^x$, $P(D)f_2 = 0$, $P(D)f_3 = 2\cos x$

21. $Q(D)P(D)(y) = -3\sin x + 3x\cos x$; $P(D)Q(D)(y) = 3x\cos x$

23. $PQ(y) = 1 - x\sin x$; $QP(y) = -\cos x - x\sin x$

25. $Q(D)P(D)(y) = -5\cos x + 3\sin x - 8x + 3x^2$; $P(D)Q(D)(y) = -5\cos x + 3\sin x - 8x + 3x^2$

27. $Q(D)P(D)(y) = 5x^3 + 15x^2 + 21x + 21$; $P(D)Q(D)(y) = 5x^3 + 12x^2 + 21x + 16$

29. $(D - 3)(D + 1)$

31. $(D^2 + 1)^2$

33. $(D - 1)(D^2 + D + 1)$

35. $(D + 8)(D + 2)(y) = 0$

37. $D(D + 3)^2(y) = 0$

39. $(D - 2)(D^2 + 2D + 4)(y) = 0$

Section 3.2

1. By definition: Set $c_1 x + c_2(2x) = 0$ for all x. E.g., choose $c_1 = -2, c_2 = 1$.

3. By Wronskian: $W(x) = \det \begin{pmatrix} e^x & x + 1 \\ e^x & 1 \end{pmatrix} = e^x(x + 2) \neq 0$.

5. Any solution looks like $y = Ce^{-\int p(x)\,dx}$. If y_1 and y_2 are solutions, they differ by a multiplicative constant. Substituting then gives $W(x) = 0$ for all x, which shows linear dependence.

7. Lin. independent because $W(x) \neq 0$.

9. Linearly dependent. Choose, for example, $c_1 = 2$, $c_3 = -1$, and $c_2 = 8$.

11. Linearly dependent. Choose, e.g., $c_1 = 4$, $c_2 = -1.5$, $c_3 = -1$.

13. Lin. independent because $W(x) \neq 0$.

15. Lin. independent because $W(x) \neq 0$.

17. Linearly independent.

19. Set $c_1 x + c_2|x| = 0$ for all x.
(a) On $[0, 1]$, we have $|x| = x$ so choose $c_1 = -c_2$. Thus set is linearly dependent on $[0, 1]$.
(b) On $[-1, 0]$, we have $|x| = -x$ so choose $c_1 = c_2$. Thus set is linearly dependent on $[-1, 0]$.
(c) On $[-1, 1]$, we need results of both **a** and **b** to hold, which can only happen when $c_1 = c_2 = 0$. Thus set is linearly independent on $[-1, 1]$.
(d) $W(x) = \det \begin{pmatrix} x & -x \\ 1 & -1 \end{pmatrix} = 0$ on $[-1, 0]$. $W(x) = \det \begin{pmatrix} x & x \\ 1 & 1 \end{pmatrix} = 0$ on $[0, 1]$. Thus $W(x) = 0$ for all x even though $\{x, |x|\}$ is linearly independent.

21. $y = c_1 x + c_2 x \ln x$, $y' = c_1 + c_2(1 + \ln x)$, $y'' = \frac{c_2}{x}$
$x^2 y'' - xy' + y = c_2 x - c_1 x - c_2 x - c_2 x \ln x + c_1 x + c_2 x \ln x = 0$
Letting $c_1 = 3$, $c_2 = -4$ satisfies the IC. Theorem not violated since $a_0(x) \neq 0$ on the specified interval.

23. $y = c_1 x^2 + c_2 x$,
$y' = 2c_1 x + c_2$, $y'' = 2c_1$
$x^2 y'' - 2xy' + 2y = 2x^2(c_1) - 2x(2c_1 x^2 + c_2) + 2(c_1 x^2 + c_2 x) = 0$
Theorem does NOT guarantee existence of unique solution on $(0, \infty)$ since $a_0(x) = 0$ at $x = 0$. $y = c_1 x^2 + x$ is the general solution that satisfies IC (not unique).

25. There are two functions and the equation is 2nd order. Both functions are solutions and $W(x) = 4 \neq 0 \Rightarrow$ functions are linearly independent.

27. (i) There are the same number of functions as the order of the DE; (ii) each of the two functions are solutions; (iii) the two solutions are linearly independent, since $W(0) = 5 \neq 0$.

29. (i) There are the same number of functions as the order of the DE; (ii) each of the two functions are solutions; (iii) the two solutions are linearly independent, since $W(0) = \frac{1}{2} \neq 0$.

31. (i) There are the same number of functions as the order of the DE; (ii) each of the three functions are solutions; (iii) the three solutions are linearly independent, since $W(x) = 1 \neq 0$.

33. (i) There are the same number of functions as the order of the DE;

(ii) each of the five functions are solutions; (iii) the five solutions are linearly independent, since
$W(x) = 8 \cos^4 x + 8 \cos^2 x \sin^2 x \neq 0$.

35. No. $y = x^2 + 4$ is not a solution.

37. No, it's not a fundamental set of solutions because $\sin x$ is not a solution.

39. Yes, it's a fundamental set of solutions.

41. No, it's not a fundamental set of solutions because there are not enough functions to consider.

43. Yes, it's a fundamental set of solutions.

45. Yes, it's a fundamental set of solutions.

47. Yes, it's a fundamental set of solutions.

49. Yes, it's a fundamental set of solutions.

51. Yes, it's a fundamental set of solutions.

Section 3.3

1. $y = xe^x$

3. $y = xe^{3x}$

5. $y = e^{-3x}$

7. $y = \sin 5x$

9. $y = 1$

11. $y = x$

13. $y = x + 1$

15. $y'' - y = 0 \Rightarrow e^{-x}, e^x$ are solutions.

17. $y'' - 3y' + 2y = 0 \Rightarrow e^{2x}, xe^x$ are solutions.

Section 3.4

1. (i) $u_1' = u_2$, $u_2' = -\frac{4}{7}u_2 + \frac{3}{7}u_1$, IC: $u_1(0) = 0, u_2(0) = 1$; (ii) with $h = .5, y(5) \approx 5.9617$; with $h = .01$, $y(5) \approx 5.9619$.

3. (i) $u_1' = u_2$, $u_2' = -2u_2 - 10u_1 + \sin x$, IC: $u_1(\pi) = e, u_2(\pi) = 1$; (ii) with $h = .5$, $y(8.14) \approx 0.12947$; with $h = .01$, $y(8.14) \approx 0.09996$.

5. (i) $u_1' = u_2$, $u_2' = \frac{-3u_1 + x}{x+2}$, IC: $u_1(0) = 0, u_2(0) = 4$; (ii) with $h = .5$, $y(5) \approx -1.9947$; with $h = .01$, $y(5) \approx -1.9994$.

7. (i) $u_1' = u_2$, $u_2' = \frac{-3u_1^2 + e^x}{x^2+2}$, IC: $u_1(0) = 1, u_2(0) = 2$; (ii) with $h = .5, y(5) \approx 5.7432$; with $h = .01$, $y(5) \approx 5.7317$.

9. (i) $u_1' = u_2$, $u_2' = u_3, u_3' = -\frac{1}{8}u_3$, IC: $u_1(0) = 1, u_2(0) = 0, u_3(0) = 2$; (ii) with $h = .5$, $y(5) \approx 21.5134$; with $h = .01$, $y(5) \approx 21.5134$.

11. (i) $u_1' = u_2$, $u_2' = u_3, u_3' = \frac{-xu_3u_2 - \sin u_1 + e^{-x}}{x^2}$, IC: $u_1(1) = 1$, $u_2(1) = 3, u_3(1) = 1$; (ii) with $h = .5, y(6) \approx 17.8569$; with $h = .01$, $y(6) \approx 17.8326$.

13. (i) $u_1' = u_2$, $u_2' = u_3, u_3' = \frac{1}{8}u_3$, IC: $u_1(1) = 1, u_2(1) = 1, u_3(1) = 0$; (ii) with $h = .5, y(6) \approx -29.9164$; with $h = .01, y(6) \approx -29.9121$.

15. (i) $u_1' = u_2$, $u_2' = u_3, u_3' = u_4$, $u_4' = \frac{1}{x+2}(-3u_2 + 2u_1)$, IC: $u_1(-1) = 3, u_2(-1) = -1, u_3(-1) = 0, u_4(-1) = 1$; (ii) with $h = .5, y(4) \approx 80.0743$; with $h = .01, y(4) \approx 80.1260$.

Section 3.5

1. $4e^{i\pi/2}$

3. $\sqrt{2}e^{i3\pi/4}$

5. $2e^{i2\pi/3}$

7. $5e^{i\arctan(3/4)}$

9. $\sqrt{5}e^{-i\arctan(1/2)}$

11. $z = \pm i$

13. $z = i, \pm\frac{\sqrt{3}}{2} + i\frac{1}{2}$

15. $z = \sqrt{2} \pm i\sqrt{2}, -\sqrt{2} \pm i\sqrt{2}$

17. $p(1+i) = 3i; p(2i) = -7 + 2i$

19. $p(i) = i; p(2i) = -i$

21. $p(2i) = \frac{-e^z}{3}; p(2i) = e^z(\frac{1}{5} + i\frac{2}{5})$

23. $\text{Re}(e^{ix})$

25. $\text{Im}(e^{(1+2i)x})$

27. $\text{Re}(\sqrt{2}e^{i(x+\pi/4)})$

29. $\text{Re}(\sqrt{5}e^{-x+i(3x-\arctan(1/3))})$

31. $2i\cos 2x$

33. $\cos^2 x - \sin^2 x + i2\cos x \sin x$

35. $-\sin x + i\cos x$

37. $\frac{1}{5}e^x[\cos 2x + 2\sin 2x + i(\sin 2x - 2\cos 2x)]$

39. $A_1 = -i, A_2 = 1$

41. $A_1 = \frac{1+i}{3}, A_2 = \frac{1+2i}{3}$

43. $x' = ie^{ix}, x'' = -e^{ix} \Rightarrow x'' + x = 0$

Section 3.6

1. $y(x) = c_1 e^{-6x} + c_2 e^{-2x}$

3. $y(x) = c_1 e^{-x} + c_2 e^{4x}$

5. $y(x) = c_1 + c_2 e^{2x}$

7. $y(x) = c_1 e^{\frac{-x}{2}} + c_2 x e^{\frac{-x}{2}}$

9. $y(x) = c_1 e^{2x} \sin x + c_2 e^{2x} \cos x$

11. $y(x) = c_1 e^{2x} + c_2 e^{-2x} + c_3 \sin 2x + c_4 \cos 2x$

13. $y(x) = c_1 \cos 2x + c_2 \sin 2x + c_3 x \cos 2x + c_4 x \sin 2x$

15. $y(x) = c_1 e^x + c_2 x e^x + c_3 x^2 e^x$

17. $y(x) = c_1 e^{-2x} + c_2 e^{-x} + c_3 e^x + c_4 e^{2x}$

19. $y(x) = c_1 e^{-3x} + c_2 e^{-x} + c_3 e^x + c_4 e^{3x} + c_5$

21. $y(x) = c_1 + c_2 x + c_3 e^x + c_4 x e^x + c_5 \cos x + c_6 \sin x$

23. $y(x) = -e^{-4x} + e^{3x}$

25. $y(x) = x e^{-2x}$

27. $y = -\frac{e^{-\pi-x}}{3} \sin 3x$

29. $y = \cos 2x + \frac{1}{2} \sin 2x$

31. $D^2(D+2)(y) = 0$

33. $D^4((D-2-3i)(D-2+3i))^3(y) = 0 \Rightarrow$
$D^4(D^2 - 4D + 13)^3(y) = 0$

Section 3.7

1. $x(t) = \frac{1}{6} \cos 16t;$ $\frac{1}{6}$ ft; $\frac{\pi}{8}$ sec; $\frac{8}{\pi}$ oscillations/sec

3. $x(t) = \frac{1}{4} \sin 8t;$ $\frac{1}{4}$ ft; $\frac{\pi}{4}$ sec; $\frac{4}{\pi}$ oscillations/sec

5. $\frac{1}{2\pi} \sqrt{K(\frac{1}{I_1} + \frac{1}{I_2})}$

7. $I = \frac{V}{R}(1 - e^{\frac{-Rt}{L}})$

9. $Q = Q_0 e^{-t/(CR)}$

11. $Q = VC$

13. Underdamped.

15. Overdamped.

17. Critically damped.

19. Overdamped.

21. Critically damped.

23. $\sqrt{\frac{b^2}{2m} - \frac{k}{m}} \approx i\sqrt{\frac{k}{m}}\sqrt{1 - \frac{b^2}{2k}} \approx i\sqrt{\frac{k}{m}}(1 - \frac{b^2}{4k}) \approx i\sqrt{\frac{k}{m}}$ for small b.

25. $1 > k > 0$

27. $b = 2\sqrt{3}$

29. $\frac{1}{2} < m$

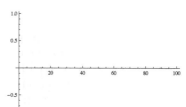

31.
Graph for 3.7#31.

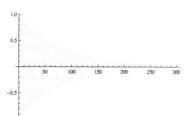

33.
Graph for 3.7#33 with $\omega = 10$.

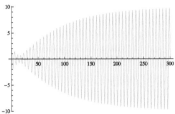

Graph for 3.7#33 with $\omega = 1$.
It seems $\omega = 1$ gives the largest oscillations. Amplitude ≈ 10.

Chapter 3 Review

1. False. We only can conclude that this is a possibility.

3. True. The method won't work without a solution.

5. False. Only constant-coefficient differential operators commute.

7. A unique solution is guaranteed because coefficients are constant. Substitution gives $c_1 = c_2 = 2$.

9. A unique solution is not guaranteed at $x = 0$. The IC requires $c_2 = 0$ with any c_1.

11. A unique solution is guaranteed for all x.

13. A unique solution is guaranteed for all $x > 0$.

15. Linearly independent.

17. Linearly independent.

19. Linearly independent.

21. Linearly dependent.

23. Yes, it forms a fundamental set of solutions.

25. No, it is not a fundamental set of solutions because e^{-x} is not a solution.

27. No, it is not a fundamental set of solutions because neither is a solution.

29. No, it is not a fundamental set of solutions because neither $e^{2x} \sin x$ or $e^{2x} \cos x$ are solutions.

31. $y_2 = e^{5x}$

33. $y_2 = e^{-5x}$

35. $(D - 4)(D + 3)(y) = 0$

37. $D(D - 4)(D + 4)(y) = 0$

39. $(D - 1)^3(y) = 0$

41. $y(5) \approx 1.63314$

43. $y(6) \approx 18.56345$

45. $y(4) \approx -.52660$

47. $\text{Im}(3e^{i2x})$

49. $\text{Re}(e^{-2x+ix})$

51. $p(i) = \frac{1-i}{2}; p(2i) = \frac{1-i}{6}$

53. $p(2i) = \frac{-e^z}{3}; p(2i) = e^z\left(\frac{1}{5} + i\frac{2}{5}\right)$

55. $y(x) = c_1 e^{2x} + c_2 e^{3x}$

57. $y(x) = c_1 e^{-x} + c_2 + c_3 x + c_4 e^{4x}$

59. $y(x) = c_1 e^{-x} + c_2 x e^{-x} + c_3 x^2 e^{-x} + c_4$

61. $y(x) = c_1 e^{-7x} + c_2 e^{-2x} + c_3 + c_4 e^{2x}$

63. $y(x) = (3x + 1)e^{-3x}$

65. $y(x) = \frac{1}{2}e^{-4x}(3e^{2x} - 1)$

67. $y(x) =$
$\frac{1}{3}e^{2(x-\pi/2)}(5\cos 3x - 6\sin 3x)$

69. $y(x) = 2 - \sin x$

71. Overdamped.

73. Overdamped.

75. Critically damped.

77. $b > 4$

79. $k = \frac{1}{8}$

81. $m = \frac{1}{4}$

Section 4.1

1. $y_p = \frac{e^x}{2}$

3. $y_p = \frac{-3}{50}\cos x - \frac{21}{50}\sin x$

5. $y_p = \frac{3}{2}\cos x$

7. $y_p = 2(\frac{1}{6}) - 12(\frac{x}{6} + \frac{5}{36}) + 6(\frac{e^x}{2})$,
now simplify.

9. Resonant frequency $\omega = 2$, amplitude 0.2.

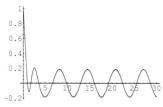

Graph for 4.1#**9**, $\omega = 1$.

11. Resonant frequency $\omega = \sqrt{1/2}$, amplitude 0.6.

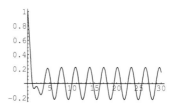

Graph for 4.1#**9**, $\omega = 2$.

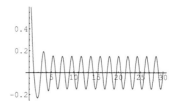

Graph for 4.1#**9**, $\omega = 3$.

13. Resonant frequency $\omega = \sqrt{3}$, amplitude 0.3.

Section 4.2

1. $y_p = A + Bx + Cx^2 + Ee^{-x}$

3. $y_p = Ax\sin x + Bx\cos x$
$+ Ce^x \sin x + Ee^x \cos x$

5. $y_p = Axe^{2x}\sin x + Bxe^{2x}\cos x$
$+ Ce^{2x}\sin x + Ee^{2x}\cos x$
$+ F\sin x + G\cos x$

7. $y_p = A\sin x + B\cos x$

9. $A = \frac{1}{9}$

11. $A = \frac{11}{17}, B = \frac{-7}{17}$

13. $A = 0, B = \frac{8}{441}, C = \frac{1}{21}, E = 0$

15. $A = \frac{-5}{36}, B = \frac{-1}{6}, C = \frac{1}{5}, E = \frac{3}{50}$

17. $y(x) = c_1 e^{-2x} + c_2 e^{4x} - 3e^{-3x} - \frac{1}{2}e^{2x}$

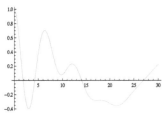

Graph for 4.1#11, $\omega = \sqrt{1/18}$.

Graph for 4.1#13, $\omega = \sqrt{3}$.

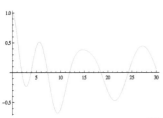

Graph for 4.1#11, $\omega = \sqrt{5/18}$.

19. $y(x) = c_1 e^x \sin x + c_2 e^x \cos x + \left(\frac{x}{5} + \frac{2}{25}\right) \cos x - \frac{2}{25}(5x + 7)\sin x$

21. $y(x) = c_1 e^x \sin 2x + c_2 e^x \cos 2x - \frac{1}{4} x e^x \cos 2x$

23. $y(x) = c_1 e^x \sin 2x + c_2 e^x \cos 2x + \frac{1}{2} x e^x + \frac{1}{16} e^x - \frac{1}{4} x e^x \cos 2x$

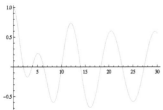

Graph for 4.1#11, $\omega = \sqrt{1/2}$.

25. $y(x) = c_1 e^{-3x} + c_2 e^{-x} + \frac{e^{-x}}{16}(4x - 2 + e^{2x})$

27. $y(x) = c_1 e^{-3x} \sin x + c_2 e^{-3x} \cos x + c_3 e^{-4x} + \frac{1}{4} x^2 e^{-4x} + \frac{1}{2} x e^{-4x} + \frac{1}{4} e^{-4x}$

29. $y = \frac{7}{2} e^{2x} - \frac{7}{2} e^{-2x} - 8x$

31. $y(x) = (1 - x) e^x$

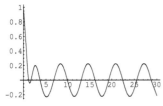

Graph 4.1#13, $\omega = 1$.

33. $y = \frac{21}{20} e^{-3x} \cos x - \frac{101}{60} e^{-3x} \sin x + 3x e^{-3x} - \frac{1}{20} e^{3x} \cos x - \frac{1}{60} e^{3x} \sin x$

35. a. $y(x) = c_1 e^{4x} - \frac{1}{4}\left(x^2 + \frac{x}{2} + \frac{1}{8}\right)$;
c. $y(x) = c_1 e^x + \frac{1}{3} e^{4x}$

Section 4.3

1. $y_p = A_1 + A_2 x + A_3 x^2 + B e^{-x}$

3. $y_p = A_1 x \sin x + A_2 x \cos x + B_1 e^x \sin x + b_2 e^x \cos x$

5. $y_p = A_1 x e^{2x} \sin x + A_2 x e^{2x} \cos x + A_3 e^{2x} \sin x + A_4 e^{2x} \cos x + B_1 \sin x + B_2 \cos x$

7. $y_p = A_1 \sin x + A_2 \cos x$

Graph for 4.1#13, $\omega = \sqrt{2}$.

9. $A = \frac{1}{9}$

11. $A_1 = \frac{11}{17}, A_2 = \frac{-7}{17}$

13. $A_1 = 0, A_2 = \frac{8}{441}, A_3 = \frac{1}{21}, A_4 = 0$

15. $A_1 = \frac{-5}{36}, A_2 = \frac{-1}{6}, B_1 = \frac{1}{5}, B_2 = \frac{3}{50}$

17. $y(x) = c_1 e^{-2x} + c_2 e^{4x} - 3e^{-3x} - \frac{1}{2}e^{2x}$

19. $y(x) = c_1 e^x \sin x + c_2 e^x \cos x + (\frac{x}{5} + \frac{2}{25}) \cos x - \frac{2}{25}(5x + 7) \sin x$

21. $y(x) = c_1 e^x \sin 2x + c_2 e^x \cos 2x - \frac{1}{4}xe^x \cos 2x$

23. $y(x) = c_1 e^x \sin 2x + c_2 e^x \cos 2x + \frac{1}{2}xe^x + \frac{1}{16}e^x - \frac{1}{4}xe^x \cos 2x$

25. $y(x) = c_1 e^{-3x} + c_2 e^{-x} + \frac{e^{-x}}{16}(4x - 2 + e^{2x})$

27. $y(x) = c_1 e^{-3x} \sin x + c_2 e^{-3x} \cos x + c_3 e^{-4x} + \frac{1}{4}x^2 e^{-4x} + \frac{1}{2}xe^{-4x} + \frac{1}{4}e^{-4x}$

29. $y = \frac{7}{2}e^{2x} - \frac{7}{2}e^{-2x} - 8x$

31. $y(x) = (1 - x)e^x$

33. $y = \frac{21}{20}e^{-3x} \cos x - \frac{101}{60}e^{-3x} \sin x + 3xe^{-3x} - \frac{1}{20}e^{3x} \cos x - \frac{1}{60}e^{3x} \sin x$

35. (a) $y(x) = c_1 e^{4x} - \frac{1}{4}(x^2 + \frac{x}{2} + \frac{1}{8})$;
(c) $y(x) = c_1 e^x + \frac{1}{3}e^{4x}$

Section 4.4

1. $y_p = \frac{e^{-x}}{P(-1)} = \frac{e^{-x}}{2}$

3. $y_p = \frac{4e^{3x}}{P(2)} = \frac{e^{-2x}}{2}$

5. $y_p = \frac{3}{P(0)} = \frac{3}{4}$

7. $y_p = \text{Re}(3\frac{e^{i2x}}{P(2i)})$
$= \frac{-3}{10}\cos 2x + \frac{3}{5}\sin 2x$

9. $y_p = \frac{-xe^{2x}}{P'(2)} = \frac{-xe^{2x}}{4}$

11. $y_p = \frac{xe^{-3x}}{P'(-3)} = \frac{-xe^{-3x}}{6}$

13. $y_p = \frac{2x^2 e^{-x}}{P''(-1)} = x^2 e^{-x}$

15. $y_p = \text{Re}(\frac{-2xe^{(-1+i)x}}{P'(-1+i)})$
$= -xe^{-x} \sin x$

17. $P(D + 1)v = x \Rightarrow y_p = ve^x = (\frac{1}{5}x - \frac{2}{25})e^x$

19. $P(D + 0)v = x^2 + 3x - 1 \Rightarrow y_p = \frac{1}{9}x^2 + \frac{1}{3}x - \frac{11}{81}$

21. $Q(D) = D - 3, \ Q(D + (-3))u = x \Rightarrow y_p = (\frac{-1}{36}x - \frac{1}{12}x^2)e^{-3x}$

23. $Q(D) = D + 1 + i, \ Q(D + (-1 + i))u = (D + 2i)u = x^2 \Rightarrow u = \frac{1}{2}x + (\frac{1}{4} - \frac{1}{2}x^2)i \Rightarrow y_p = \frac{1}{4}x^2 e^{-x} \cos x + \frac{1}{12}x(2x^2 - 3)e^{-x} \sin x$

25. $y(x) = c_1 e^{-2x} + c_2 e^{4x} - 3e^{-3x} - \frac{1}{2}e^{2x}$

27. $y(x) = c_1 e^x \sin x + c_2 e^x \cos x + (\frac{x}{5} + \frac{2}{25}) \cos x - \frac{2}{25}(5x + 7) \sin x$

29. $y(x) = c_1 e^{-3x} \sin x + c_2 e^{-3x} \cos x + c_3 e^{-4x} + \frac{1}{4}x^2 e^{-4x} + \frac{1}{2}xe^{-4x}$

31. $y = \frac{7}{2}e^{2x} - \frac{7}{2}e^{-2x} - 8x$

33. $y(x) = (1 - x)e^x$

35. (a) $y(x) = c_1 e^{4x} - \frac{1}{4}(x^2 + \frac{x}{2} + \frac{1}{8})$;
(c) $y(x) = c_1 e^x + \frac{1}{3}e^{4x}$

Section 4.5

1. $y(x) = c_1 + c_2 x + x \arctan x - \frac{1}{2}\ln(x^2 + 1)$

3. $y = C_1 e^x + C_2 xe^x + -e^x \ln x - e^x$

5. $y = C_1 \sin 2x + C_2 \cos 2x$
$+ \sin 2x \cos x + \sin x \cos x \ln\left(\tan \frac{x}{2}\right)$
$- \cos 2x \sin x$

7. $y = C_1 + C_2 e^{-x} + C_3 e^{2x} - \frac{x^4}{8} +$
$\frac{x^3}{4} - \frac{9x^2}{8} + \frac{15x}{8} - \frac{17}{16}$

9. $y(x) = \frac{e^{2x}}{6}(8 - 3e^x + e^{3x})$

11. $y(x) = \cos x + \ln(\cos x) \cos x$
$+ 2 \sin x + x \sin x$

13. $y(x) = \frac{1}{6}(3 + 2 \cos x + \cos 2x)$

15. $y(x) = -1 + \cos x + \sin x$
$+ \ln(\cos(\frac{x}{2}) + \sin(\frac{x}{2})) \sin x$
$- \ln(\cos(\frac{x}{2}) - \sin(\frac{x}{2})) \cos x$

17. $y(x) = \frac{1}{4}x(\ln x - 1) + \frac{c_1}{x}$
$+ c_2 x + c_3$

19. $y(x) = \frac{x}{2} + c_1 \sin(\ln x) + c_2 \cos(\ln x)$

Section 4.6

1. $y(x) = c_1 x^{-2} + c_2 x^{-1}$

3. $y(x) = c_1 \sin(2 \ln x) + c_2 \cos(2 \ln x)$

5. $y(x) = c_1 x^{-3} + c_2 x^{-7}$

7. $y(x) = c_1 x + c_2 x^{3/2}$

9. $y(x) = c_1 x^2 + c_2 x^2 \ln x$

10. $y(x) = c_1 x^{-1+\sqrt{3}} + c_2 x^{-1-\sqrt{3}}$

11. $y(x) = c_1 x^{-2+\sqrt{7}} + c_2 x^{-2-\sqrt{7}}$

13. $y(x) = c_1 x^{1/7} \sin\left(\frac{\sqrt{6}}{7} \ln x\right)$
$+ c_2 x^{1/7} \cos\left(\frac{\sqrt{6}}{7} \ln x\right)$

15. $y(x) = c_1 x^{-5/3} \sin\left(\frac{2\sqrt{2}}{3} \ln x\right)$
$+ c_2 x^{-5/3} \cos\left(\frac{2\sqrt{2}}{3} \ln x\right)$

17. $W(x) = \frac{1}{x}(y_1 y_2' - y_1' y_2) \neq 0$,
since y_1, y_2 are linearly independent.

19. $y = \frac{A}{x} + \frac{B}{x} \ln x + \ln x - 2$

Section 4.7

1. $g(\omega) = \frac{5}{\sqrt{81+\omega^4-18\omega^2}}$, $\omega_{res} = 3$

Gain for 4.7#**1**.

3. $g(\omega) = \frac{2}{\sqrt{81+\omega^4+18\omega^2}}$, $\omega_{res} = 0$

Gain for 4.7#**3**.

5. $I(t) = C_1 e^{-40t} + C_2 e^{-200t}$
$+ \frac{632}{6817} \cos(10t) + \frac{192}{6817} \sin(10t)$

11. $mx'' = -kx + b(y' - x')$; $y = \sin \omega t \Rightarrow mx'' + bx' + kx = b\omega \cos \omega t$

Chapter 4 Review

1. True.

3. True.

5. False.

7. $y_p = 2(-\frac{25}{3}) + \frac{1}{2}(6x + 1)$

9. $y_p = \frac{13}{25}e^{2x} - \frac{46}{289} \cos 2x - \frac{14}{289} \sin 2x$

11. $y_p = -xe^{-3x} \cos 4x + \frac{1}{13}e^{4x}$

13. $y(x) = c_1 e^{4x} + c_2 - \frac{1}{17}(4 \sin x + \cos x)$

15. $y(x) = c_1 e^{2x} + c_2 x e^{2x} + \frac{1}{16} e^{-2x}$

17. $y(x) = c_1 \sin(\frac{5x}{2})$
$+ c_2 \cos(\frac{5x}{2}) + \frac{2}{29} e^{-x}$

19. $y(x) = c_1 e^{-5x} + c_2 e^{2x} - \frac{1}{100}(10 + 3e^{3x} + 10xe^{3x})$.

21. $y(x) = \frac{c_1}{x} + \frac{c_2}{\sqrt{x}} + 3 + \frac{x}{6}$

23. $y(x) = c_1 \sin\left(\frac{1}{\sqrt{3}} \ln x\right)$
$+ c_2 \cos\left(\frac{1}{\sqrt{3}} \ln x\right) + \frac{x^2}{13}$

25. $g(w) = \frac{3}{\sqrt{w^4 - 30w^2 + 289}}$, $w_{res} = \sqrt{15}$

Section 5.1

1. $\begin{bmatrix} \cos t & e^t \\ 2t & 3 \end{bmatrix}$; $\begin{bmatrix} -\cos t & e^t \\ \frac{1}{3}t^3 & \frac{3}{2}t^2 \end{bmatrix} + \mathbf{C}$

3. $\begin{bmatrix} \cos t & e^t \\ 2t & 3 \end{bmatrix}$;
$\begin{bmatrix} -\cos t & e^t \\ \frac{t^3}{3} & \frac{3t^2}{2} \end{bmatrix} + \mathbf{C}$

5. $\begin{bmatrix} 0 & -e^{-t} & 3e^{3t} \\ 1 & 0 & 6e^{3t} \\ e^t & -e^{-t} & 9e^{3t} \end{bmatrix}$;
$\begin{bmatrix} t & -e^{-t} & \frac{1}{3}e^{3t} \\ \frac{1}{2}t^2 & 0 & \frac{2}{3}e^{3t} \\ e^t & -e^{-t} & e^{3t} \end{bmatrix} + \mathbf{C}$

7. $\begin{bmatrix} e^{-t}(\cos t - \sin t) & -3e^{3t} & -3e^{-3t} \\ \frac{1}{2\sqrt{t}} & 4t^3 & -\sin t \\ \frac{4}{3}t^{1/3} & 0 & 3te^{3t} + e^{3t} \end{bmatrix}$
$\begin{bmatrix} t & -e^{-t} & \frac{e^{3t}}{3} \\ \frac{t^2}{2} & 0 & \frac{2e^{3t}}{3} \\ e^t & -e^{-t} & e^{3t} \end{bmatrix} + \mathbf{C}$

9. $\frac{d\mathbf{x_1}}{dt} = \begin{pmatrix} -4e^{4t} \\ -8e^{4t} \end{pmatrix} = A\mathbf{x_1}$;

$\frac{d\mathbf{x_2}}{dt} = \begin{pmatrix} e^t \\ 3e^t \end{pmatrix} = A\mathbf{x_2}$

11. $\frac{d\mathbf{x_1}}{dt} = \begin{pmatrix} -4e^{-2t} \\ -6e^{-2t} \end{pmatrix} = A\mathbf{x_1}$;

$\frac{d\mathbf{x_2}}{dt} = \begin{pmatrix} -3e^{-3t} \\ -6e^{-3t} \end{pmatrix} = A\mathbf{x_2}$

13. $\frac{d\mathbf{x_1}}{dt} = \begin{pmatrix} e^t \\ 0 \\ 0 \end{pmatrix} = A\mathbf{x_1}$;

$\frac{d\mathbf{x_2}}{dt} = \begin{pmatrix} 0 \\ -e^{-t} \\ 0 \end{pmatrix} = A\mathbf{x_2}$;

$\frac{d\mathbf{x_3}}{dt} = \begin{pmatrix} -2e^{2t} \\ 0 \\ 2e^{2t} \end{pmatrix} = A\mathbf{x_3}$

15. $\frac{d\mathbf{x_1}}{dt} = \begin{pmatrix} -2e^{-2t} \\ 6e^{-2t} \\ -2e^{-2t} \end{pmatrix} = A\mathbf{x_1}$;

$\frac{d\mathbf{x_2}}{dt} = \begin{pmatrix} e^{-t} \\ -2e^{-t} \\ 0 \end{pmatrix} = A\mathbf{x_2}$;

$\frac{d\mathbf{x_3}}{dt} = \begin{pmatrix} e^t \\ 0 \\ -e^t \end{pmatrix} = A\mathbf{x_3}$

17. Linearly independent.

19. Linearly dependent.

21. Linearly independent.

23. Linearly independent.

25. Not a fundamental set of solutions.

27. Yes, it is a fundamental set of solutions.

29. Yes, it is a fundamental set of solutions.

Section 5.2

1. $\begin{bmatrix} 1 & 4 & 3 \\ 2 & -5 & -1 \end{bmatrix}$

3. $\begin{bmatrix} 5 & 8 & -2 & -1 \\ 1 & -4 & \frac{1}{2} & 0 \end{bmatrix}$

5. $\begin{bmatrix} 1 & 2 & -7 & 5 & 1 \\ 1 & 0 & -1 & 3 & -2 \\ 2 & 3 & -9 & 3 & 8 \end{bmatrix}$

7. $x_2 = -2, x_1 = 11$

9. $x_3 = 1, x_2 = \frac{1}{3}, x_1 = \frac{-1}{2}$

11. $x_2 = t, x_3 = 1 + 2t, x_1 = 6 + 9t$

13. $x_1 = -1, x_2 = 2$

15. Infinite solutions: $x_2 = t, x_1 = 3 - 2t$

17. $x_1 = 4, x_2 = -3, x_3 = 2$

19. Infinite solutions: $x_3 = t, x_2 = 2 + 2t, x_1 = -1 - 7t$

Section 5.3

1. (i) There are the same number of vectors as the dimension of the vector space; (ii) each of the two vectors is in the correct space, $\mathbb{R}^2$ in this case; (iii) the two vectors are linearly independent.

3. (i) There are the same number of vectors as the dimension of the vector space; (ii) each of the two vectors is in the correct space, $\mathbb{R}^2$ in this case; (iii) the two vectors are linearly independent.

5. (i) There are the same number of vectors as the dimension of the vector space; (ii) each of the three vectors is in the correct space, $\mathbb{R}^3$ in this case; (iii) the three vectors are linearly independent.

7. (i) There are the same number of vectors as the dimension of the vector space; (ii) each of the four vectors is in the correct space, $\mathbb{R}^4$ in this case; (iii) the four vectors are linearly independent.

9. Yes, it is a basis.

11. Yes, it is a basis.

13. No, it's not a basis because vectors aren't in $\mathbb{R}^3$.

15. No, it's not a basis because vectors aren't in $\mathbb{R}^3$.

17. Yes, it is a basis.

19. No, it's not a basis because there aren't enough vectors to span.

21. No, it is not a basis because vectors are linearly independent.

23. One needs to show that each of the 10 conditions in Definition 5.3.1 holds. For example, show that (i) a 3×3 matrix added to one of the same size is again a 3×3 matrix; (v) the 3×3 matrix of zeros added to any 3×3 matrix $\mathbf{A}$ is just $\mathbf{A}$.

25. Label the matrices $A_1 \cdots A_4$. It is clear that

$$c_1 A_1 + c_2 A_2 + c_3 A_3 + c_4 A_4 = \begin{bmatrix} c_1 & c_2 \\ c_3 & c_4 \end{bmatrix}$$

spans $\mathbb{R}^{2 \times 2}$ and it is also clear that

$$c_1 A_1 + c_2 A_2 + c_3 A_3 + c_4 A_4 = \begin{bmatrix} c_1 & c_2 \\ c_3 & c_4 \end{bmatrix}$$

$$= \begin{bmatrix} 0 & 0 \\ 0 & 0 \end{bmatrix} \text{ only if }$$

c_1, c_2, c_3, c_4 all are 0. Hence, they are linearly independent and so they form a basis.

27. One needs to show that each of the 3 conditions in Definition 5.3.5 holds. For example, show that (i) $(0,0,0)$ is in the plane $x = 0$. Similarly for (ii) and (iii).

29. One needs to show that each of the 10 conditions in Definition 5.3.1 holds. For example, show that (i) an $m \times n$ matrix added to one of the same size is again a $m \times n$ matrix; (v) the $m \times n$ matrix of zeros added to any $m \times n$ matrix $\mathbf{A}$ is just $\mathbf{A}$.

31. $R(A) = \left\{ \begin{bmatrix} 1 \\ 0 \end{bmatrix}, \begin{bmatrix} 0 \\ 1 \end{bmatrix} \right\}$,
$\text{null}(A) = \emptyset$

33. $R(A) = \left\{ \begin{bmatrix} 1 \\ 0 \end{bmatrix}, \begin{bmatrix} 0 \\ 1 \end{bmatrix} \right\}$,
$\text{null}(A) = \left\{ \begin{bmatrix} 4 \\ -3 \\ 5 \end{bmatrix} \right\}$

35. $R(A) = \left\{ \begin{bmatrix} 1 \\ 0 \end{bmatrix}, \begin{bmatrix} 0 \\ 1 \end{bmatrix} \right\}$,
$\text{null}(A) = \left\{ \begin{bmatrix} 3 \\ -5 \\ 0 \\ 4 \end{bmatrix}, \begin{bmatrix} -1 \\ 1 \\ 2 \\ 0 \end{bmatrix} \right\}$

37. $R(A) = \left\{ \begin{bmatrix} 1 \\ -2 \\ 1 \end{bmatrix} \right\}$,
$\text{null}(A) = \left\{ \begin{bmatrix} 3 \\ 1 \end{bmatrix} \right\}$

39. Any matrix with reduced echelon form that contains two non-zero pivots.

41. Answers will vary. For example,
$\begin{bmatrix} 1 & 0 & 0 \\ 0 & 1 & 0 \\ 0 & 0 & 1 \end{bmatrix}$

43. Answers will vary. For example,
$\begin{bmatrix} 1 & 0 & 0 \\ 0 & 0 & 0 \\ 0 & 0 & 0 \end{bmatrix}$

45. $\text{null}(\mathbf{A}) = \left\{ \begin{bmatrix} -1 \\ 0 \\ 1 \end{bmatrix}, \begin{bmatrix} 1 \\ 1 \\ 0 \end{bmatrix} \right\}$,
$R(\mathbf{A}) = \left\{ \begin{bmatrix} 1 \\ 1 \\ 1/2 \end{bmatrix} \right\}$.

47. $R(\mathbf{A}^T) = \left\{ \begin{bmatrix} 1 \\ -1 \\ 1 \end{bmatrix} \right\}$,
$\text{null}(\mathbf{A}^T) = \left\{ \begin{bmatrix} -1 \\ 1 \\ 0 \end{bmatrix}, \begin{bmatrix} -1/2 \\ 0 \\ 1 \end{bmatrix} \right\}$.

49. $R(\mathbf{A}^T) = \left\{ \begin{bmatrix} 0 \\ 0 \\ 0 \\ 1 \end{bmatrix}, \begin{bmatrix} 1 \\ -1 \\ 1 \\ 0 \end{bmatrix} \right\}$,
$\text{null}(\mathbf{A}^T) = \left\{ \begin{bmatrix} -1 \\ 1 \\ 0 \end{bmatrix} \right\}$

Section 5.4

1. $\lambda_1 = 2, \lambda_2 = -1$,
$\mathbf{v}_1 = \begin{bmatrix} 5 \\ 2 \end{bmatrix}, \mathbf{v}_2 = \begin{bmatrix} 1 \\ 1 \end{bmatrix}$

3. $\lambda_1 = -3, \lambda_2 = -1$,
$\mathbf{v}_1 = \begin{bmatrix} -1 \\ 1 \end{bmatrix}, \mathbf{v}_2 = \begin{bmatrix} -2 \\ 1 \end{bmatrix}$

5. $\lambda_1 = -4, \lambda_2 = -1$,
$\mathbf{v}_1 = \begin{bmatrix} 1 \\ 0 \end{bmatrix}, \mathbf{v}_2 = \begin{bmatrix} 7 \\ 3 \end{bmatrix}$

7. $\lambda_1 = 1, \lambda_2 = 3$,
$\mathbf{v}_1 = \begin{bmatrix} -1 \\ 1 \end{bmatrix}, \mathbf{v}_2 = \begin{bmatrix} 1 \\ 1 \end{bmatrix}$

9. $\lambda_1 = -2, \lambda_2 = 2$,
$\mathbf{v}_1 = \begin{bmatrix} 1 \\ -1 \end{bmatrix}, \mathbf{v}_2 = \begin{bmatrix} 1 \\ 1 \end{bmatrix}$

11. $\lambda_1 = 3, \lambda_2 = 2,$
$$\mathbf{v}_1 = \begin{bmatrix} 1 \\ 1 \end{bmatrix}, \mathbf{v}_2 = \begin{bmatrix} 2 \\ 1 \end{bmatrix}$$

13. $\lambda_1 = 5, \lambda_2 = 1,$
$$\mathbf{v}_1 = \begin{bmatrix} 1 \\ 1 \end{bmatrix}, \mathbf{v}_2 = \begin{bmatrix} 1 \\ -3 \end{bmatrix}$$

15. $\lambda_1 = 0, \lambda_2 = 4,$
$$\mathbf{v}_1 = \begin{bmatrix} -1 \\ 1 \end{bmatrix}, \mathbf{v}_2 = \begin{bmatrix} 1 \\ 1 \end{bmatrix}$$

17. $\lambda_1 = 8, \lambda_2 = -5,$
$$\mathbf{v}_1 = \begin{bmatrix} 6 \\ 1 \end{bmatrix}, \mathbf{v}_2 = \begin{bmatrix} -1 \\ 2 \end{bmatrix}$$

19. $\lambda_{1,2} = \pm i$
$$\mathbf{v}_1 = \begin{bmatrix} i \\ 1 \end{bmatrix}, \mathbf{v}_2 = \overline{\mathbf{v}_1}$$

21. $\lambda_{1,2} = \pm i$
$$\mathbf{v}_1 = \begin{bmatrix} 3 + i \\ 5 \end{bmatrix}, \mathbf{v}_2 = \overline{\mathbf{v}_1}$$

23. $\lambda_{1,2} = -2 \pm 2i$
$$\mathbf{v}_1 = \begin{bmatrix} 1 + 2i \\ 1 \end{bmatrix}, \mathbf{v}_2 = \overline{\mathbf{v}_1}$$

25. $\lambda_{1,2} = -5 \pm i$
$$\mathbf{v}_1 = \begin{bmatrix} 2 + 2i \\ 1 \end{bmatrix}, \mathbf{v}_2 = \overline{\mathbf{v}_1}$$

27. $\lambda_{1,2} = \pm i\sqrt{5}$
$$\mathbf{v}_1 = \begin{bmatrix} -2 \\ -1 + i\sqrt{5} \end{bmatrix}, \mathbf{v}_2 = \overline{\mathbf{v}_1}$$

29. $\lambda_1 = 2, \lambda_2 = 1, \lambda = -4;$
$$\mathbf{v}_1 = \begin{bmatrix} 1 \\ 2 \\ 0 \end{bmatrix}, \mathbf{v}_2 = \begin{bmatrix} 0 \\ 0 \\ 1 \end{bmatrix} \mathbf{v}_3 = \begin{bmatrix} 1 \\ -1 \\ 1 \end{bmatrix}$$

31. $\lambda_{1,2} = 2 \pm 2i, 1; \mathbf{v}_1 = \begin{bmatrix} -1 + 2i \\ 1 \\ 0 \end{bmatrix},$
$$\mathbf{v}_2 = \overline{\mathbf{v}_1}, \mathbf{v}_3 = \begin{bmatrix} -3 \\ -1 \\ 5 \end{bmatrix}$$

33. a. $\lambda_1 = -1, \lambda_2 = 3, \lambda = 3;$
$$\mathbf{v}_1 = \begin{bmatrix} 0 \\ 0 \\ 1 \end{bmatrix}, \mathbf{v}_2 = \begin{bmatrix} 1 \\ 2 \\ 0 \end{bmatrix}, \mathbf{v}_3 = \begin{bmatrix} 0 \\ -3 \\ 1 \end{bmatrix}$$

35. $u_1' = 2u_1, u_2' = 3u_2;$
$x = 2c_1 e^{2t} + c_2 e^{3t}, y = c_1 e^{2t} + c_2 e^{3t}$

37. $u_1' = 5u_1, u_2' = -3u_2;$
$x = 4c_1 e^{-t}, y = c_1 e^{-t} + c_2 e^{-3t}$

39. $u_1' = -u_1, u_2' = u_2;$
$x = c_1 e^{-t} + c_2 e^{t}, y = c_1 e^{-t} + 3c_2 e^{t}$

41. $\begin{bmatrix} 1 & -1 \\ 1 & 1 \end{bmatrix} \begin{bmatrix} 3^{30} & 0 \\ 0 & 1 \end{bmatrix} \begin{bmatrix} -1 & 1 \\ 1 & 1 \end{bmatrix}^{-1}$

43. $\lambda = \pm 3i$

45. $\lambda = 0, \pm i\sqrt{3}$

Section 5.5

1. $\mathbf{x}(t) = c_1 \begin{bmatrix} 2 \\ 1 \end{bmatrix} e^{-2t} + c_2 \begin{bmatrix} 1 \\ -1 \end{bmatrix} e^{t}$

3. $\mathbf{x}(t) = c_1 \begin{bmatrix} -1 \\ 2 \end{bmatrix} e^{-2t} + c_2 \begin{bmatrix} 1 \\ -3 \end{bmatrix} e^{-t}$

5. $\mathbf{x}(t) = c_1 \begin{bmatrix} 1 \\ 1 \end{bmatrix} e^{4t} + c_2 \begin{bmatrix} 1 \\ 3 \end{bmatrix} e^{2t}$

7. $\mathbf{x}(t) = c_1 \begin{bmatrix} 1 \\ 1 \end{bmatrix} e^{-2t} + c_2 \begin{bmatrix} 1 \\ 3 \end{bmatrix} e^{3t}$

9. $\mathbf{x}(t) = c_1 \begin{bmatrix} 1 \\ 1 \end{bmatrix} e^{5t} + c_2 \begin{bmatrix} -1 \\ 3 \end{bmatrix} e^{t}$

11. $\mathbf{x}(t) = c_1 \begin{bmatrix} 1 \\ 1 \end{bmatrix} e^{4t} + c_2 \begin{bmatrix} 1 \\ 2 \end{bmatrix} e^{-2t}$

13. $\mathbf{x}(t) = c_1 \begin{bmatrix} 2 \\ 1 \end{bmatrix} e^{-2t} + c_2 \begin{bmatrix} 1 \\ 1 \end{bmatrix} e^{t}$

15. $\mathbf{x}(t) = c_1 \begin{bmatrix} 3 \\ 1 \end{bmatrix} e^{-5t} + c_2 \begin{bmatrix} 2 \\ 1 \end{bmatrix} e^{t}$

17. $\mathbf{x}(t) = c_1 \begin{bmatrix} 1.732 \\ 1 \end{bmatrix} e^{3.732t}$

$+ c_2 \begin{bmatrix} -1.732 \\ 1 \end{bmatrix} e^{0.268t}$

19. $\mathbf{x}(t) =$

$c_1 \left(\begin{bmatrix} 0 \\ 1 \end{bmatrix} \cos 2t - \begin{bmatrix} 1 \\ 0 \end{bmatrix} \sin 2t \right)$

$+ c_2 \left(\begin{bmatrix} 0 \\ 1 \end{bmatrix} \sin 2t + \begin{bmatrix} 1 \\ 0 \end{bmatrix} \cos 2t \right)$

21. $\mathbf{x}(t) =$

$c_1 \left(\begin{bmatrix} 0 \\ 1 \end{bmatrix} \cos 2t - \begin{bmatrix} 1 \\ 0 \end{bmatrix} \sin 2t \right)$

$+ c_2 \left(\begin{bmatrix} 0 \\ 1 \end{bmatrix} \sin 2t + \begin{bmatrix} 1 \\ 0 \end{bmatrix} \cos 2t \right)$

23. $\mathbf{x}(t) =$

$e^{-t} c_1 \left(\begin{bmatrix} 2 \\ -3 \end{bmatrix} \cos 2t - \begin{bmatrix} 0 \\ -1 \end{bmatrix} \sin 2t \right)$

$+ c_2 \left(\begin{bmatrix} 2 \\ -3 \end{bmatrix} \sin 2t + \begin{bmatrix} 0 \\ -1 \end{bmatrix} \cos 2t \right)$

25. $\mathbf{x}(t) =$

$e^{-3t} c_1 \left(\begin{bmatrix} 0 \\ 1 \end{bmatrix} \cos 2t - \begin{bmatrix} 2 \\ 0 \end{bmatrix} \sin 2t \right)$

$+ e^{-3t} c_2 \left(\begin{bmatrix} 0 \\ 1 \end{bmatrix} \sin 2t + \begin{bmatrix} 2 \\ 0 \end{bmatrix} \cos 2t \right)$

27. $x(t) = c_2 e^{2t} + c_3 e^{5t}$,
$y(t) = -c_2 e^{2t} + 2c_3^{5t}$,
$z(t) = c_1 e^t + 4c_2 e^{2t} + \frac{1}{4}c_3 e^{5t}$

29. $x(t) = e^{-t}(c_1 \sin 2t + c_2 \cos 2t)$,
$y(t) = e^{-t}(-c_1 \cos 2t + c_2 \sin 2t) + \frac{1}{2}c_3 e^t$, $z(t) = c_3 e^t$

31. $x = c_1 e^{t/2} \sin(\frac{\sqrt{15}}{2}t)$
$+ c_2 e^{t/2} \cos(\frac{\sqrt{15}}{2}t))$,
$y = \frac{1}{4} e^{t/2}((3c_1 + \sqrt{15}c_2) \sin(\frac{\sqrt{15}}{2}t)$
$+(3c_2 - \sqrt{15}c_1) \cos(\frac{\sqrt{15}}{2}t))$, $z = c_3 e^{-t}$

33. $r = -1, -3, x = c_1 e^{-t} + c_2 e^{-3t}$
$x' = y; y' = -3x - 4y$,
Eigenvalues $\lambda_{1,2} = -1, -3$, Eigenvectors $v_{1,2} = [1 \quad -1]^T, [1 \quad -3]^T$.
Solution is again $x = c_1 e^{-t} + c_2 e^{-3t}$.

35. $x(t) = -896.64 e^{-.16597t}$
$- 1398.4 e^{-.29403t} + 12295.1$,
$y(t) = 66.898 e^{-.16597t}$
$+ 2342.9 e^{-.29403t} + 9590.2$,
$z(t) = 829.74 e^{-.16597t}$
$- 944.50 e^{-.29403t} + 8114.8$.

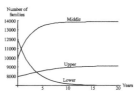

Graph for 5.5#35.

Section 5.6

1. $\mathbf{x}(t) = e^t \begin{bmatrix} c_1 + c_2 t \\ \frac{c_2}{2} \end{bmatrix}$

3. $\mathbf{x}(t) = e^{3t} \begin{bmatrix} c_1 + c_2 t \\ 2c_1 + c_2(2t - \frac{1}{2}) \end{bmatrix}$

5. $\mathbf{x}(t) = e^{-t} \begin{bmatrix} c_1 + c_2(t - 1) \\ c_1 + c_2 t \end{bmatrix}$

7. $\mathbf{x}(t) = e^{-t} \begin{bmatrix} c_1 + c_2 t \\ \frac{1}{4}c_2 \end{bmatrix}$

9. $\mathbf{x}(t) = e^{-t} \begin{bmatrix} c_1 + c_2(t + \frac{1}{2}) \\ c_1 + c_2 t \end{bmatrix}$

11. $\mathbf{x}(t) = e^t \begin{bmatrix} -c_1 + c_2(-t + \frac{1}{2}) \\ c_1 + c_2 t \end{bmatrix}$

13. $\lambda_1 = 2, \lambda_2 = 2, \lambda_3 = 1$;

$\mathbf{v}_1 = \begin{bmatrix} -1 \\ 1 \\ 0 \end{bmatrix}, \mathbf{v}_2 = \begin{bmatrix} 1 \\ -1 \\ 1 \end{bmatrix}$

$\mathbf{u}_1 = \begin{bmatrix} -2 \\ 1 \\ 0 \end{bmatrix}$

$\mathbf{x}(t) = c_1 t e^{2t} + c_2 e^{2t} - c_3 e^t$
$\mathbf{y}(t) = c_1(e^{2t} - t e^{2t}) - c_2 e^{2t} + c_3 e^t$
$\mathbf{z}(t) = c_3 e^t$

15. $\lambda_1 = -1$, $\lambda_2 = 1$, $\lambda_3 = 1$;

$$\mathbf{v}_1 = \begin{bmatrix} -1 \\ 0 \\ 0 \end{bmatrix}, \quad \mathbf{v}_2 = \begin{bmatrix} -1 \\ 2 \\ -4 \end{bmatrix}$$

$$\mathbf{u}_1 = \begin{bmatrix} 1 \\ 1 \\ 0 \end{bmatrix}$$

$x(t) = c_1 e^{-t} - \dfrac{1}{2} c_2 e^t$
$+ c_3 (\frac{3}{4} e^t - \frac{1}{2} t e^t)$, $y(t) = c_2 e^t + c_3 t e^t$,
$z(t) = -2 c_2 e^t + c_3 (e^t - 2t e^t)$

17. $\lambda_1 = 2$, $\lambda_2 = 2$, $\lambda_3 = 2$;

$$\mathbf{v}_1 = \begin{bmatrix} 0 \\ 0 \\ 1 \end{bmatrix}, \quad \mathbf{v}_2 = \begin{bmatrix} 0 \\ 1 \\ 0 \end{bmatrix}$$

$$\mathbf{u}_1 = \begin{bmatrix} -1 \\ 1 \\ 0 \end{bmatrix}$$

$x(t) = c_3 e^{2t}$
$y(t) = c_2 e^{2t}$
$z(t) = (c_3 t + c_2 t + c_1) e^{2t}$

19. $\lambda_1 = 2$, $\lambda_2 = 2$, $\lambda_3 = 2$;

$$\mathbf{v} = \begin{bmatrix} 1 \\ -1 \\ 1 \end{bmatrix}, \quad \mathbf{u}_1 = \begin{bmatrix} 1 \\ 0 \\ 1 \end{bmatrix}$$

$$\mathbf{u}_2 = \begin{bmatrix} 1 \\ 0 \\ 0 \end{bmatrix}$$

$x(t) = -e^{-t}(c_1 + t c_2 + c_2 + t^2 c_3 + 2t c_3 + 2 c_3)$, $y(t) = e^{-t}(c_1 + t c_2 + t^2 c_3)$,
$z(t) = -e^{-t}(c_1 + t c_2 + c_2 + t^2 c_3 + 2t c_3)$

21. $\lambda_1 = 1$, $\lambda_2 = 1$, $\lambda_3 = 1$;

$$\mathbf{v} = \begin{bmatrix} -1 \\ -1 \\ 0 \end{bmatrix}, \quad \mathbf{u}_1 = \begin{bmatrix} 0 \\ 2 \\ -1 \end{bmatrix}$$

$$\mathbf{u}_2 = \begin{bmatrix} 1 \\ 0 \\ 0 \end{bmatrix}$$

$x(t) = e^t(c_1 + t c_2 + t^2 c_3)$, $y(t) = e^t(c_1 + t c_2 - 2 c_2 + t^2 c_3 - 4t c_3 + 2 c_3)$,
$z(t) = e^t(c_2 + 2t c_3)$

23. Matrix already in Jordan normal form with only 3 eigenvectors,

one for each of $\lambda = 1, 2, 3$. Generalized eigenvectors are the basis vectors, $\mathbf{e}_1, \ldots, \mathbf{e}_6$.
$x_1(t) = C_1 e^t$, $x_2(t) = (C_2 t + C_4) e^{2t}$,
$x_3(t) = C_2 e^{2t}$,
$x_4(t) = (C_6 + C_5 t + \frac{1}{2} C_3 t^2) e^{3t}$,
$x_5(t) = (C_3 t + C_5) e^{3t}$, $x_6(t) = C_3 e^{3t}$.

Section 5.7

1. $\begin{bmatrix} e^{3t} & 0 \\ 0 & e^{-2t} \end{bmatrix}$

3. $\begin{bmatrix} e^{3t} & \frac{2}{9}(e^{-6t} - e^{3t}) \\ 0 & e^{-6t} \end{bmatrix}$

5. $\begin{bmatrix} 2e^{-t} - e^{-3t} & e^{-3t} - e^{-t} \\ 2e^{-t} - 2e^{-3t} & -e^{-t} + 2e^{-3t} \end{bmatrix}$

7. $\begin{bmatrix} \frac{e^t + e^{-t}}{2} & \frac{e^t - e^{-t}}{2} \\ \frac{e^t - e^{-t}}{2} & \frac{e^t + e^{-t}}{2} \end{bmatrix}$

9. $\begin{bmatrix} e^{3t} \cos 2t & e^{3t} \sin 2t \\ -e^{3t} \sin 2t & e^{3t} \cos 2t \end{bmatrix}$

11. $\begin{bmatrix} e^{-t} \cos 3t & -e^{-t} \sin 3t \\ e^{-t} \sin 3t & e^{-t} \cos 3t \end{bmatrix}$

13. $\begin{bmatrix} a_1 & a_2 & a_3 \\ 0 & e^t & 0 \\ b_1 & b_2 & b_3 \end{bmatrix}$, where

$a_1 = \dfrac{2e^{-2t} + e^t}{3}$,
$a_2 = \dfrac{-2e^{-2t} + 2e^t + 3te^t}{9}$,
$a_3 = \dfrac{-2e^{-2t} + 2e^t}{3}$, $b_1 = \dfrac{-e^{-2t} + e^t}{3}$,
$b_2 = \dfrac{e^{-2t} - e^t + 3te^t}{9}$, $b_3 = \dfrac{e^{-2t} + 2e^t}{3}$.

15. $\begin{bmatrix} e^{2t} & a_2 & a_3 & a_4 \\ b_1 & b_2 & b_3 & b_4 \\ 0 & 0 & e^{-3t} & 0 \\ c_1 & c_2 & c_3 & c_4 \end{bmatrix}$,

where $a_2 = -e^{-t} + e^{2t}$,
$a_3 = \dfrac{-e^{-3t} - 5e^{-t} + 6e^{2t}}{10}$,
$a_4 = e^{-t} - e^{2t}$,
$b_1 = \dfrac{e^{2t} - e^{4t}}{2}$,

$b_2 = \frac{2e^{-t}+5e^{2t}+3e^{4t}}{10}$,

$b_3 = \frac{-3e^{-3t}+e^{-t}+3e^{2t}-e^{4t}}{10}$,

$b_4 = \frac{-2e^{-t}-5e^{2t}+7e^{4t}}{10}$,

$c_1 = \frac{e^{2t}-e^{4t}}{2}$,

$c_2 = \frac{-8e^{-t}+5e^{2t}+3e^{4t}}{10}$,

$c_3 = \frac{2e^{-3t}-4e^{-t}+3e^{2t}-e^{4t}}{10}$,

$c_4 = \frac{8e^{-t}-5e^{2t}+7e^{4t}}{10}$.

17. $\mathbf{J}^2 = \begin{bmatrix} 1 & -2 \\ 0 & 1 \end{bmatrix}$,

$\mathbf{J}^3 = \begin{bmatrix} -1 & 3 \\ 0 & -1 \end{bmatrix}$,

$\mathbf{J}^4 = \begin{bmatrix} 1 & -4 \\ 0 & 1 \end{bmatrix}$

19. $\begin{bmatrix} a^n & na^{n-1} \\ 0 & a^n \end{bmatrix}$

21. $\begin{bmatrix} -e^{-3t}+2e^{-t} & -e^{-t}+e^{-t} \\ 2e^{-t}-2e^{-3t} & 2e^{-3t}-e^{-t} \end{bmatrix}$

23. $\frac{1}{2}\begin{bmatrix} e^t+e^{-t} & e^t-e^{-t} \\ e^t-e^{-t} & e^t+e^{-t} \end{bmatrix}$

25. $\begin{bmatrix} e^{3t}\cos(2t) & e^{3t}\sin(2t) \\ -e^{3t}\sin(2t) & e^{3t}\cos(2t) \end{bmatrix}$

27. $\begin{bmatrix} e^{5t} & 0 \\ \frac{1}{4}e^{5t}-\frac{1}{4}e^{-3t} & e^{-3t} \end{bmatrix}$

29. $p(\mathbf{A}) = \mathbf{A}^2 - \mathbf{A} - 2\mathbf{I} = 0$

31. $p(\mathbf{A}) = \mathbf{A}^2 - 3\mathbf{A} + 2\mathbf{I} = 0$

33. $p(\mathbf{A}) = \mathbf{A}^2 - 4\mathbf{A} + 3\mathbf{I} = 0$

35. $p(\mathbf{A}) = \mathbf{A}^2 - 2\mathbf{A} + 2\mathbf{I} = 0$

37. $p(\mathbf{A}) = \mathbf{A}^2 - 2\mathbf{A} + 5\mathbf{I} = 0$

39. $p(\mathbf{A}) = \mathbf{A}^2 + 4\mathbf{A} + 13\mathbf{I} = 0$

Section 5.8

1. $x(t) = \frac{-c_1}{5}e^{-2t}(-6+e^{5t})$
$- \frac{2c_2}{5}e^{-2t}(-1+e^{5t})$,
$y(t) = \frac{3c_1}{5}e^{-2t}(-1+e^{5t})$
$+ \frac{c_2}{5}e^{-2t}(-1+6e^{5t})$

3. $x(t) = c_1e^{4t}$,
$y(t) = \frac{-c_1}{2}e^{2t}(-1+e^{2t})$
$+ c_2e^{2t}$

5. $x(t) = c_1e^t - c_3e^{-t}(-1+e^{2t})$,
$y(t) = c_1e^t(-1+e^t) + c_2e^{2t}$
$+ c_3e^{-t}(-2+e^{2t}+e^{3t})$,
$z(t) = c_3e^{-t}$

7. $x(t) = c_1e^{3t} + c_2e^{-2t} + \frac{5}{3}$,
$y(t) = -3e^{3t}c_1 - \frac{1}{2}c_2e^{-2t} - \frac{3}{2}$

9. $x(t) = c_2e^{4t} - \frac{1}{17}\cos t - \frac{4}{17}\sin t$,
$y(t) = c_1e^{2t} - \frac{1}{2}c_2e^{4t} - \frac{6}{85}\cos t - \frac{7}{85}\sin t$

11. $x(t) = (t+c_2)e^t + c_3e^{-t}$,
$y(t) = c_1e^{2t} - (1+t+c_2)e^t - \frac{1}{5}\cos t - \frac{2}{5}\sin t$, $z(t) = c_3e^{-t}$

13. $x(t) = e^{2t}c_1 + c_2e^{-2t} - e^t - \frac{1}{5}\cos t + \frac{1}{5}\sin t$,
$y(t) = c_3e^t$, $z(t) = c_1e^{2t} - \frac{1}{3}e^{-2t}c_2 - \frac{1}{5}\sin t - \frac{2}{3}e^t$

Chapter 5 Review

1. False. A full set of eigenvectors may not exist.

3. False.

5. False.

7. True.

9. Yes, it is a fundamental set.

11. Yes, it is a fundamental set.

13. No, it is not a fundamental set.

15. Yes, it is a fundamental set.

17. Yes, it forms a basis.

19. Yes, it forms a basis.

21. No, it is not a basis because vectors are linearly dependent.

23. No, it is not a basis because vectors are linearly dependent.

25. Yes, it forms a basis.

27. $R(A) = \left\{ \begin{bmatrix} -2 \\ 2 \\ -1 \end{bmatrix} \right\}$,

$\text{null}(A) = \left\{ \begin{bmatrix} 2 \\ 1 \end{bmatrix} \right\}$

29. $R(A) = \left\{ \begin{bmatrix} -1 \\ -1 \end{bmatrix}, \begin{bmatrix} 2 \\ 1 \end{bmatrix} \right\}$,

$\text{null}(A) = \left\{ \begin{bmatrix} 2 \\ 0 \\ 1 \end{bmatrix} \right\}$

31. $R(A) = \left\{ \begin{bmatrix} -1 \\ -1 \\ 2 \end{bmatrix}, \begin{bmatrix} 2 \\ 7 \\ 1 \end{bmatrix} \right\}$,

$\text{null}(A) = \left\{ \begin{bmatrix} -2 \\ -6 \\ 5 \end{bmatrix} \right\}$

33. $\lambda_1 = 2, \lambda_2 = -2$,
$\mathbf{v}_1 = \begin{bmatrix} 1 \\ 1 \end{bmatrix}, \mathbf{v}_2 = \begin{bmatrix} 1 \\ -3 \end{bmatrix}$;
(ii) $x(t) = c_1 e^{2t} + c_2 e^{-2t}$,
$y(t) = c_1 e^{2t} - 3 c_2 e^{-2t}$

35. $\lambda_1 = 2, \lambda_2 = 3$,
$\mathbf{v}_1 = \begin{bmatrix} -1 \\ 1 \end{bmatrix}, \mathbf{v}_2 = \begin{bmatrix} 1 \\ -2 \end{bmatrix}$,

37. $\lambda_1 = -2, \lambda_2 = -1$,
$\mathbf{v}_1 = \begin{bmatrix} 1 \\ -2 \end{bmatrix}, \mathbf{v}_2 = \begin{bmatrix} -1 \\ 1 \end{bmatrix}$;
$x(t) = -c_1 e^{-t} + c_2 e^{-2t}, y(t) = c_1 e^{-t} - 2 c_2 e^{-2t}$

39. $\lambda_1 = -5, \lambda_2 = -1$,
$\mathbf{v}_1 = \begin{bmatrix} 1 \\ 1 \end{bmatrix}, \mathbf{v}_2 = \begin{bmatrix} -1 \\ 5 \end{bmatrix}$;
$x(t) = c_1 e^{-5t} + c_2 e^{-t}$,
$y(t) = c_1 e^{-5t} + 5 c_2 e^{-t}$

41. (iv) $x(t) = 2 c_1 \cos 4t + 2 c_2 \sin 4t$,
$y(t) = -c_1 \sin 4t + c_2 \cos 4t$

43. $\lambda_1 = 4i, \lambda_2 = -4i$,
$\mathbf{v}_1 = \begin{bmatrix} 1 \\ -4i \end{bmatrix}, \mathbf{v}_2 = \begin{bmatrix} 1 \\ 4i \end{bmatrix}$

45. $\lambda_1 = 2 + i, \lambda_2 = 2 - i$,
$\mathbf{v}_1 = \begin{bmatrix} -3 - i \\ 1 \end{bmatrix}, \mathbf{v}_2 = \begin{bmatrix} -3 + i \\ 1 \end{bmatrix}$

47. $\lambda_1 = 4 + i\sqrt{3}, \lambda_2 = 4 - i\sqrt{3}$,
$\mathbf{v}_1 = \begin{bmatrix} 2 \\ 1 + i\sqrt{3} \end{bmatrix}, \mathbf{v}_2 = \begin{bmatrix} 2 \\ 1 - i\sqrt{3} \end{bmatrix}$

49. $u_1' = -2u_1, u_2' = 2u_2$;
$x = c_1 e^{2t} + c_2 e^{-2t}$,
$y = c_1 e^{2t} + \frac{1}{2} c_2 e^{-2t}$

51. $\lambda_{1,2,3} = -2, 2, 3$, $v_1 = \begin{bmatrix} 2 \\ -1 \\ 2 \end{bmatrix}$,

$v_2 = \begin{bmatrix} 1 \\ -1 \\ 1 \end{bmatrix}, v_3 = \begin{bmatrix} 2 \\ -1 \\ 1 \end{bmatrix}$

53. $\lambda_{1,2,3} = -1, 1, 5$, $v_1 = \begin{bmatrix} -1 \\ -1 \\ 1 \end{bmatrix}$,

$v_2 = \begin{bmatrix} 1 \\ 1 \\ 1 \end{bmatrix}, v_3 = \begin{bmatrix} 1 \\ -1 \\ 1 \end{bmatrix}$

55. $\lambda_{1,2,3} = 2, 1, 0$, $v_1 = \begin{bmatrix} 1 \\ -1 \\ 1 \end{bmatrix}, v_2 = \begin{bmatrix} -1 \\ -1 \\ 2 \end{bmatrix}, v_3 = \begin{bmatrix} -1 \\ -1 \\ 1 \end{bmatrix}$

57. $\lambda_1 = -1, \mathbf{v}_1 = \begin{bmatrix} 1 \\ 0 \end{bmatrix} \mathbf{u}_1 = \begin{bmatrix} 0 \\ 1 \end{bmatrix}$
$\mathbf{x} = \begin{bmatrix} c_1 e^{-t} + c_2 t e^{-t} \\ c_2 e^{-t} \end{bmatrix}$

59. $\lambda_1 = 1, \mathbf{v}_1 = \begin{bmatrix} 1 \\ \frac{3}{2} \end{bmatrix} \mathbf{u}_1 = \begin{bmatrix} 0 \\ -1/4 \end{bmatrix}$
$\mathbf{x} = \begin{bmatrix} c_1 e^t + c_2 t e^t \\ \frac{3}{2} c_1 e^t + \frac{3}{2} c_2 t e^t - \frac{1}{4} c_2 e^t \end{bmatrix}$

61. $\lambda_1 = 3$, $\mathbf{v}_1 = \begin{bmatrix} 1 \\ 0 \end{bmatrix}$ $\mathbf{u}_1 = \begin{bmatrix} 0 \\ 1/2 \end{bmatrix}$

$\mathbf{x} = \begin{bmatrix} c_1 e^{3t} + c_2 t e^{3t} \\ \frac{1}{2} c_2 e^{3t} \end{bmatrix}$

63. $\lambda_{1,2,3} = 1, -1 \pm 2i$, $\mathbf{v}_1 = \begin{bmatrix} 0 \\ 1 \\ 2 \end{bmatrix}$,

$\mathbf{v}_2 = \begin{bmatrix} i \\ 1 \\ 0 \end{bmatrix}$ $\mathbf{v}_3 = \begin{bmatrix} -i \\ 1 \\ 0 \end{bmatrix}$;

65. $\lambda_{1,2,3} = -1, -1 \pm i$, $\mathbf{v}_1 = \begin{bmatrix} 0 \\ 1 \\ 1 \end{bmatrix}$,

$\mathbf{v}_2 = \begin{bmatrix} i \\ 0 \\ 1 \end{bmatrix}$ $\mathbf{v}_3 = \begin{bmatrix} -i \\ 0 \\ 1 \end{bmatrix}$;

67. (i) $\lambda_{1,2,3} = 2, -2 \pm 4i$, $\mathbf{v}_1 = \begin{bmatrix} 3 \\ 0 \\ 5 \end{bmatrix}$,

$\mathbf{v}_2 = \begin{bmatrix} -i \\ 1 - i \\ 1 \end{bmatrix}$ $\mathbf{v}_3 = \begin{bmatrix} i \\ 1 + i \\ 1 \end{bmatrix}$;

69. (i) $\lambda_1 = 1, \lambda_2 = -1, \lambda_3 = 5$,

$\mathbf{v}_1 = \begin{bmatrix} 1 \\ 1 \\ 1 \end{bmatrix}$, $\mathbf{v}_2 = \begin{bmatrix} 1 \\ 1 \\ -1 \end{bmatrix}$, $\mathbf{v}_2 = \begin{bmatrix} 1 \\ -1 \\ 1 \end{bmatrix}$;

71. (i) $\lambda_1 = -3, \lambda_2 = 5, \lambda_3 = 1$,

$\mathbf{v}_1 = \begin{bmatrix} -1 \\ -1 \\ 1 \end{bmatrix}$, $\mathbf{v}_2 = \begin{bmatrix} 1 \\ -1 \\ 1 \end{bmatrix}$ $\mathbf{v}_3 = \begin{bmatrix} 1 \\ 1 \\ 1 \end{bmatrix}$

73. (i) $\lambda_1 = 1, \lambda_2 = 1, \lambda_3 = 2$,

$\mathbf{v}_1 = \begin{bmatrix} -2 \\ 1 \\ 0 \end{bmatrix}$, $\mathbf{v}_2 = \begin{bmatrix} 0 \\ 0 \\ 1 \end{bmatrix}$ $\mathbf{v}_3 = \begin{bmatrix} 1 \\ -1 \\ 1 \end{bmatrix}$

75. (i) $\lambda_1 = -2, \lambda_2 = -2, \lambda_3 = -1$,

$\mathbf{v}_1 = \begin{bmatrix} -1 \\ 1 \\ 1 \end{bmatrix}$, $\mathbf{u}_1 = \begin{bmatrix} -2 \\ 0 \\ 1 \end{bmatrix}$ $\mathbf{v}_3 = \begin{bmatrix} 12 \\ 0 \\ 0 \end{bmatrix}$

77. (i) $\lambda_1 = 9, \lambda_2 = -12, \lambda_3 = -12$,

$\mathbf{v}_1 = \begin{bmatrix} 1/3 \\ 0 \\ 1/3 \end{bmatrix}$, $\mathbf{v}_2 = \begin{bmatrix} 0 \\ 4 \\ -2 \end{bmatrix}$,

$\mathbf{u}_1 = \begin{bmatrix} 2/3 \\ 0 \\ -1/3 \end{bmatrix}$

79. (i) $\lambda_{1,2,3} = 6$,

$\mathbf{v}_1 = \begin{bmatrix} 0 \\ -18 \\ 9 \end{bmatrix}$, $\mathbf{u}_1 = \begin{bmatrix} -3 \\ 15 \\ -6 \end{bmatrix}$, $\mathbf{u}_2 = \begin{bmatrix} 1 \\ 0 \\ 0 \end{bmatrix}$

81. $\begin{bmatrix} e^{2t} & 0 \\ 0 & e^t \end{bmatrix}$

83. $\begin{bmatrix} e^{3t} & -e^t + e^{3t} \\ 0 & e^t \end{bmatrix}$

85. $\begin{bmatrix} \frac{1}{3}(e^t + 2e^{4t}) & \frac{1}{3}(e^{4t} - e^t) \\ \frac{1}{3}(2e^{4t} - 2e^t) & \frac{1}{3}(2e^t + e^{4t}) \end{bmatrix}$

87. $\begin{bmatrix} e^{3t} \cos t & -e^{3t} \sin t \\ e^{3t} \cos t & -e^{3t} \sin t \end{bmatrix}$

89. $x(t) = c_1 e^{2t} + c_2 e^{-2t}$, $y(t) = c_1 e^{2t} - 3c_2 e^{-2t}$

91. $x(t) = c_1 e^t + c_2 e^{5t}$, $y(t) = 5c_1 e^t + c_2 e^{5t}$

93. $x(t) = c_1 e^{-2t} + c_2 e^{2t} + c_3 e^{3t} + \frac{13}{3} e^t$, $y(t) = \frac{-1}{2} c_1 e^{-2t} - c_2 e^{2t} - \frac{1}{2} c_3 e^{3t} - \frac{11}{3} e^t$, $z(t) = c_1 e^{-2t} + c_2 e^{2t} + \frac{1}{2} c_3 e^{3t} + \frac{11}{3} e^t$

Section 6.1

1. $\gamma = -3 \Rightarrow$ Saddle

Graph for 6.1#1.

3. $\gamma = 6$, $\beta = 5$, $\beta^2 - 4\gamma = 1 \Rightarrow$ Unstable Spiral

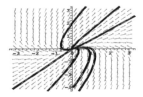

Graph for 6.1#**3**.

5. $\gamma = -5 \Rightarrow$ Saddle

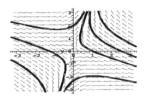

Graph for 6.1#**5**.

7. $\gamma = 5, \beta = 6, \beta^2 - 4\gamma = 16 \Rightarrow$ Unstable Node

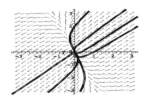

Graph for 6.1#**7**.

9. $\gamma = 11, \beta = 4, \beta^2 - 4\gamma = -28 \Rightarrow$ Unstable Spiral

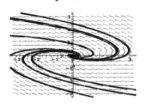

Graph for 6.1#**9**.

11. $\gamma = 8, \beta = -4, \beta^2 - 4\gamma = -16 \Rightarrow$ Stable Spiral

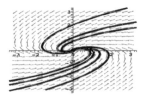

Graph for 6.1#**11**.

13. $\gamma = 7, \beta = 4, \beta^2 - 4\gamma = -12 \Rightarrow$ Unstable Spiral

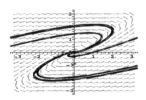

Graph for 6.1#**13**.

15. $\gamma = -40 \Rightarrow$ Saddle

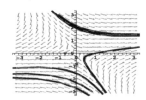

Graph for 6.1#**15**.

17. (a) $d > b$, $1 + d > 0$ $4(d - b) < 1 + d$;
(c) $d > b$, $1 + d > 0$, $4(d - b) > 1 + d$;
(e) $d < b$

19. $\beta = 2$, $\gamma = -3$. Then $x'' - 2x' - 3x = 0$, with roots $\lambda_{1,2} = -1, 3$. Thus, $x(t) = c_1 e^{-t} + c_2 e^{3t}$ and substituting into $x'(t) = -5x(t) + 8y(t)$ and solving gives $y(t) = \frac{1}{2}c_1 e^{-t} + c_2 e^{3t}$.

21. (i) $\lambda_1 = 3, \lambda_2 = 3 + 6i, \lambda_3 = 3 - 6i$; (ii) unstable;

23. (i) $\lambda_1 = 1, \lambda_2 = -1, \lambda_3 = 5$; (ii) saddle;

25. (i) $\lambda_1 = -3, \lambda_2 = 5, \lambda_3 = 1$; (ii) saddle

27. $(i)\lambda_{1,2} = 2 \pm 2i, 1$; (ii) unstable

Section 6.2

1. $(x^*, y^*) = (-2, 0)$ is a saddle; $(2, 0)$ is a linear center.

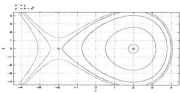

6.2#1

3. $(x^*, y^*) = (0, 0)$ is a saddle; $(-2, 2)$ is an unstable spiral.

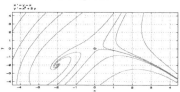

6.2#3

5. $(x^*, y^*) = (0, 0)$ is a linear center; $(1, 0)$ is a saddle; $(-1, 0)$ is a saddle;

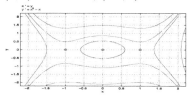

6.2#5

7. $(x^*, y^*) = (1, 1)$ is a stable spiral; $(1, 1)$ is a saddle.

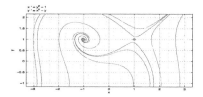

6.2#7

9. $(x^*, y^*) = (0, 0)$ is an unstable node; $(3, 0)$ is a stable node; $(0, 2)$ is a stable node; $(1, 1)$ is a saddle.

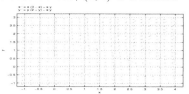

6.2#9

11. (a) $x' = \theta, \theta' = \frac{-g}{m} \sin \theta$; (b) Equilibria (θ, x): stable at $(\pm 2\pi, 0), (0, 0)$ and saddle at $(\pm \pi, 0)$; (e) see Figure 2.27; pendulum at stable equilibria is hanging straight down, at saddle equilibria is inverted straight up, center orbits is back and forth motion, orbits above saddle are whirling over the top motion.

13. Solutions do not cross in t-x-x' space, which is guaranteed by the Existence and Uniqueness Theorem. However, when you project these solutions onto the x-x' phase plane, the trajectories may cross.

15. (a) $(0, 0, 0)$, $\lambda_{1,2,3} = -10.525$, $-.475, -2.667 \Rightarrow$ stable node;
(b) $(0, 0, 0)$, $\lambda_{1,2,3} = -11.844$, $0.844, -2.667 \Rightarrow$ saddle;
$(-1.633, -1.633, 1), (1.633, 1.633, 1)$,
$\lambda_{1,2,3} = -11.242, -1.212 \pm i1.809$
$\Rightarrow$ both stable spiral; (c) $(0, 0, 0)$,
$\lambda_{1,2,3} = -21.939, 10.939, -2.667 \Rightarrow$

saddle; $(-8, -8, 24)$ and $(8, 8, 24)$, $\lambda_{1,2,3} = -13.682, 0.008 \pm i9.672 \Rightarrow$ both saddles (unstable spiral but stable in 3rd direction).

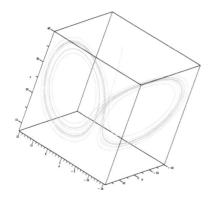

6.2#15; IC (1,0,40) with $r = 25$.

Section 6.3

1. For $r > 0$, $(x^*, y^*) = (\sqrt{r}, 0)$, $(-\sqrt{r}, 0)$; for $r > 0$, $(\sqrt{r}, 0)$ is a stable node, $(-\sqrt{r}, 0)$ is a saddle; for $r = 0$, we have $(x^*, y^*) = (0, 0)$ is the only equilibrium and its stability cannot be determined by linearization; for $r < 0$ there are no equilibria. The system undergoes a saddlenode bifurcation.

3. $(x^*, y^*) = (0, 0), (r, r)$. For $r < 0$, $(0, 0)$ is stable and for $r > 0$ it is a saddle. For $r < 0$, (r, r) is a saddle and for $r > 0$ it is a stable. The system undergoes a transcritical bifurcation.

5. For $r > 0$, $(x^*, y^*) = (\sqrt{r}, 0)$, $(-\sqrt{r}, 0), (0, 0)$; for $r > 0$, $(\pm\sqrt{r}, 0)$ are both stable nodes, $(0, 0)$ is a saddle; for $r = 0$, we have $(x^*, y^*) = (0, 0)$ is the only equilibrium and it is again a stable node; the system undergoes a supercritical pitchfork bifurcation.

7. See figures for $r = -.25, .25$.

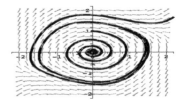

6.3#7 with $r = -.25$; an unstable limit cycle is present because one trajectory spirals into the origin while the other spirals away from the origin.

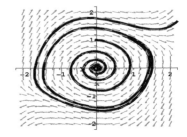

6.3#7 with $r = .25$; there is no unstable limit cycle and both trajectories spiral out of the origin.

9. Rewrite as $x' = y$, $y' = -y(x^2 - \epsilon) - x$. Equil. at $(0, 0)$. Characteristic polynomial is $\lambda^2 - \epsilon\lambda + 1$; imaginary part requires $\epsilon = 0$. There is a stable limit cycle for $\epsilon > 0$ and no limit cycle for $\epsilon < 0$.

11. Substitute coefficients to obtain $\mu = \frac{-1}{8} < 0 \Rightarrow$ supercritical Hopf bifurcation.

Section 6.4

1. We're given $\frac{1}{\alpha} = 3.5$ days. Thus $\beta \approx \frac{1.41}{3.5} \approx .403$.

3. (a) $i - \frac{\alpha}{\beta} \ln s + s = C$
(b) $i^* - \frac{\alpha}{\beta} \ln s^* + s^* = i(0) - \frac{\alpha}{\beta} \ln s(0) +$

$s(0)$

(c) $\frac{\beta}{\alpha} = \frac{\ln\left(\frac{s(0)}{s^*}\right)}{1-s^*}$

5. (a) $s' = \mu \Rightarrow s(t)$ always becomes and remains positive gives; (b) $i' = 0 \Rightarrow i_0 = 0$ gives $i(t) = 0$.

9. (a) $S'+I'+R' = 0 \Rightarrow N =$constant $\Rightarrow$ ignore R' equation; (b) $(S,I) = (S(0),0)$; (c) requiring Tr< 0 gives the condition.

11. $R_0 = \frac{\beta}{\mu+\alpha}\frac{\mu}{\mu+\phi}$ and this reduces to the previous expression $R_0 = \frac{\beta}{\mu+\alpha}$ when $\nu = 0$ (as it should).

13. Consider S, E, I for the variables in the reduced system. $R_0 = \frac{\beta}{\mu+\alpha}\frac{\delta}{\delta+\mu}$. Average life span for the infectious class is $\frac{1}{\mu+\alpha}$, so $\frac{\beta}{\mu+\alpha}$ represents the total adequate contact number of a typical infection. The quantity $\frac{\delta}{\delta+\mu}$ is the probability that an individual in the exposed class becomes infected. Thus R_0 represents the total number of new cases generated by a typical infection during its lifetime.

Section 6.5

1. (a) x, y are two species competing for the same food supply, growing logistically in the absence of the other; (b),(c) $(0,0)$ is unstable, $(0,2)$ is stable, $(3,0)$ is stable, $(1,1)$ is a saddle; (d) none; (e) see figure; (f) one of the populations will die off and this result depends only on the initial conditions.

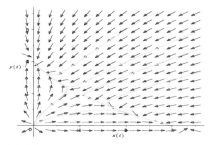

6.5#1

3. (a) x is prey with logistic growth and y predator; (b),(c) $(0,0)$ is a saddle, $(3,0)$ is a saddle, $(2,1)$ is a stable spiral; (d) none; (e) see figure; (f) for non-zero initial populations, all solutions approach a stable point; this is likely a result of the coefficients of xy in the two equations being the same (not realistic).

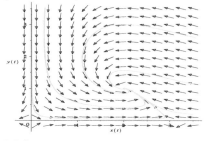

6.5#3

5. (a) x is prey with logistic growth, y predator, d measures quickness of predator death; (b),(c) $(0,0)$ is a saddle, $(1,0)$ is a saddle for $0 < d < 1$ and stable for $d > 1$, $(d, 1 - d)$ relevant only when $0 < d < 1$ and stable for this condition; (d) $(1,0)$ and $(d, 1 - d)$ undergo a transcritical bifurcation with each other when $d = 1$; (e) see figure; (f) for non-zero initial populations, all solutions approach a stable point for $d < 1$ and approach $(1,0)$ for $d > 1$.

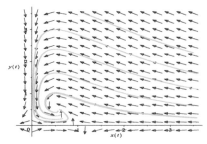

6.5#**5**, $0 < d < 1$

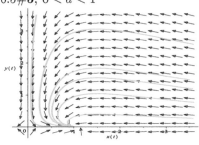

6.5#**5** $d > 1$

Chapter 6 Review

1. True.

3. True because the trajectories are projections of the solutions.

5. False. There additionally needs to be a limit cycle created or destroyed when the sign switches.

7. Stable node.

9. Unstable spiral.

11. Unstable spiral.

13. Saddle.

15. Unstable spiral.

17. Unstable spiral.

19. Unstable spiral.

21. $\lambda_1 = -27, \lambda_{2,3} = -9, -9$; Stable node.

23. $\lambda_1 = 3, \lambda_2 = -2, \lambda_3 = 2$; Saddle.

25. $\lambda_1 = -27, \lambda_2 = -18, \lambda_3 = -9$; Stable node.

27. (i) $(x^*, y^*) = (0, 1), (0, -1)$; (ii) $(0, 1)$ is a saddle, $(0, -1)$ is a linear center; the phase portrait suggests it's a nonlinear center.

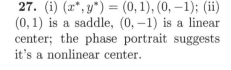

Chapter 6 Review #**27** (iii)-(iv).

29. (i) $(x^*, y^*) = (0, 0), (0, 2), (-1, 1)$; (ii) $(0, 0)$ is a saddle, $(0, 2)$ is a saddle, $(-1, 1)$ is linear center; the phase portrait suggests it's a nonlinear center.

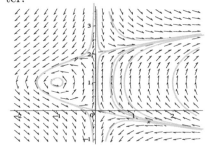

Chapter 6 Review #**29** (iii)-(iv).

31. Saddlenode bifurcation at $r = 0$.

33. Subcritical pitchfork bifurcation.

35. (a) $N^* = \Lambda/\mu$; (b) the limiting system is constant so we can substitute $S = N^* - I = \frac{\Lambda}{\mu} - I$; (c) $I^* = 0$ is stable when $\beta < \alpha + \mu$ and unstable for $\beta > \alpha + \mu$, $I^* = \frac{\Lambda(\beta - \alpha - \mu)}{\beta\mu}$ is not

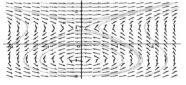

biologically relevant for $\beta < \alpha + \mu$ and is biologically relevant and stable for $\beta > \alpha + \mu$; (d) $R_0 = \frac{\beta}{\alpha+\mu}$.

Section 7.1

1. $2(1 - U_4(t))$

3. $5U_6(t)$

5. $2(U_5(t) - U_7(t))$

7. $t(1 - U_3(t)) + U_3(t)$

9. $t(1 - U_1(t)) + U_1(t) - U_2(t)$

11. $1 + U_2(t) + U_3(t) - 3U_5(t)$

13. $\cos 0 = 1$

15. e^2

17. $\frac{4}{5}$

19. $1 - e^{-2}$

21. 1

Section 7.2

1. $F(s) = \frac{2}{s^3}$

3. $F(s) = \frac{1}{s^2} - \frac{3}{s}$

5. $F(s) = \frac{2e^{-2s}}{s^2} - \frac{1-2s}{s^2}$

7. $F(s) = \frac{1}{s^2-1}$

9. $\frac{2}{s} - \frac{e^{-4s}}{s}$

11. $F(s) = \frac{e^{-2s}(e^{2s}s+e^s(2+s)-2(1+2s(1+s)))}{s^3}$

13. $F(s) = \frac{1}{s^2} - \frac{5}{s}$

15. $F(s) = \frac{s^2+2}{s^3+4s}$

17. $F(s) = \frac{4s}{s^4+10s^2+9}$

19. $F(s) = \frac{6}{s^4+10s^2+9}$

21. $F(s) = \frac{e^{-4s}}{s}$

23. $F(s) = \frac{4}{s}(1 - e^{-10s})$

25. $F(s) = \frac{6}{s}(e^{-9s} - e^{-3s})$

27. $F(s) = \frac{1}{s}(1 - e^{-2s} + e^{-6s})$

29. $F(s) = \frac{2}{s^2}(1 - (5s + 1)e^{-5s}) + \frac{10e^{-5s}}{s}$

31. $F(s) = \frac{e^{-7s}(-1+e^{3s}-3s)}{s^2} + \frac{3e^{-7s}}{s}$

33. $F(s) = \frac{\alpha}{s^2+\alpha^2}$

35. (c) $\frac{\pi}{2}$ − arctan s

37. (a) $\sqrt{\frac{\pi}{s}}$ (b) $\frac{\sqrt{\pi}}{2s^{3/2}}$ (c) $\frac{2\sqrt{\pi}}{4s^{5/2}}$

39. Use the linearity property of the Laplace transform.

Section 7.3

1. $f(t) = 3\sin t \cos t$

3. $f(t) = 5te^{3t}$

5. $f(t) = e^{-2t}\cos\sqrt{3}t$

7. $f(t) = \frac{1}{2}(1-\cos\sqrt{2}t+\sqrt{2}\sin\sqrt{2}t)$

9. $f(t) = \frac{1}{2}t(t+4)e^{-3t}$

11. $f(t) = e^{-3t}(5 - 4e^t)$

13. $f(t) = (2\sin 3(t-\pi)+5\cos 3(t-\pi))U_\pi(t)$

15. $f(t) = (t-4)U_4(t) - (t-7)U_7(t)$

17. $y(t) = 2e^t$

19. $y(t) = \frac{3}{2}\sin 2t + 2\cos 2t$

21. $y(t) = 2\sin(3t)$

23. $y(t) = e^{3t}$

25. $y(t) = 2e^{-t}(\cos 2t + 2\sin 2t)$

27. $y(t) = \frac{-1}{2}(12te^t + 5e^{2t} - 15)e^{3t}$

29. $y(t) = \frac{1}{7}e^{-5t}(49e^{7t}\sin t$
$- 7e^{7t}\cos t + 4e^{7t} - 4)$

31. $y(t) = \frac{1}{36}e^t(114te^{3t} - 209e^{3t} + 198e^{2t} + 153e^t - 178)$

Section 7.4

1. $F(s) = \frac{1}{(s-2)^2}$

3. $F(s) = \frac{6}{(s+2)^4}$

5. $F(s) = \frac{s-4}{(s-4)^2+4}$

7. $F(s) = \frac{e^{-s}(3s+2)}{s^2}$

9. $F(s) = e^{-s\pi}$

11. $F(s) = e^{-s(4-\pi)}$

13. $e^{3t}\sin t$

15. $\frac{1}{2}e^{-t}(4\cos(2t) + \sin(2t))$

17. $(1 - 2t)e^{-2t}$

19. $(t - 3)U_3(t)$

21. $\frac{1}{3}(1 - 3e^{-3t+6})U_2(t)$

23. $t_0 = \pm\frac{1}{6}$

25. The graph is that of U_0 because $\int_{0^-}^{0^+} \delta(v)dv = 1$ so that $f(t) = \int_{0^-}^{t} \delta(v)dv = 1$ for $t > 0$.

27. $x(t) = e^{3t} - e^{2t} + \frac{1}{6}U_1(t)(1 + 2e^{3t-3} - 3e^{2t-2})$

29. $x(t) = \cos 2t + \frac{1}{2}U_{4\pi}(t)\sin 2t$

31. $x(t) = \frac{2}{5}e^{-t}\sin t - \frac{1}{5}e^{-t}\cos t + \frac{2}{5}\sin t + \frac{1}{5}\cos t - U_{\frac{\pi}{2}}(t)e^{-t+\pi/2}\cos 2t$

33. (a) $f(t)$ represents a forcing of the support with 1 for $t < 1$, 2 for $1 < t < 2$, and 0 otherwise.

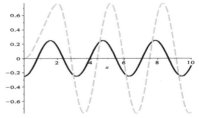

7.4#33(c)
(c) The solid black curve is the unforced solution while the dashed blue curve is the forced solution.

Section 7.5

1. pole: $s = 4$

3. poles: $s = \pm 2i$; zero: $s = -2$

5. poles: $s = -1, -2$; zeros: $s = -2$

7. poles: $s = -1 \pm i$; zero: $s = -3$

9. $H(s) = \frac{2s^2+s+1}{s^2+3s+4}$

11. $H(s) = \frac{s^2+1}{s^2+4s+13}$

13. $H(s) = \frac{1}{RCs+Q}$

15. $H(s) = \frac{1}{s^2+9} \Rightarrow$ poles $s = \pm 3i \Rightarrow$ solutions oscillate.

17. $H(s) = \frac{2s+1}{s^2+3s+2} \Rightarrow$ poles $s = -1, -2 \Rightarrow$ solutions decay.

19. $H(s) = \frac{s^2-4}{s^2+4} \Rightarrow$ poles $s = \pm 2i \Rightarrow$ solutions oscillate.

21. $H(s) = \frac{s^2-1}{s^2+2s+2} \Rightarrow$ poles $s = -1 \pm i \Rightarrow$ solutions decay with oscillations.

23. $H(s) = \frac{s^3-2s^2}{(s^2+2s+2)(s^2-1)} \Rightarrow$ poles $s = \pm 1, -1 \pm i \Rightarrow$ solutions grow without bound.

25. $s = 0, \pm 2i \Rightarrow$ oscillatory behavior

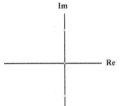

Problem **25**(a).

27. $s = -2 \pm 3i \Rightarrow$ damped oscillatory behavior

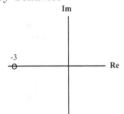

Problem **27**(a).

29. $s = \pm i \Rightarrow$ oscillatory behavior

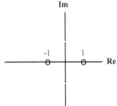

Problem **29**(a).

31. $s = -1 \pm i\sqrt{2} \Rightarrow$ damped oscillatory behavior

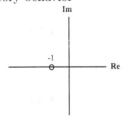

Problem **31**(a).

33. (a) $H(s) = \frac{F_0/m}{s^2 + 2\zeta\omega_n + \omega_n^2}$;
(b) $H(i\omega) = \frac{F_0/m}{-\omega^2 + i2\zeta\omega_n + \omega_n^2}$.

35. (a) $H(s) = \frac{bs}{ms^2 + bs + k}$;
$H(\omega) = \frac{b\omega}{\sqrt{b^2\omega^2 + m^2(\omega^2 - \omega_n^2)^2}}$

Section 7.6

1. $x(t) = 2e^t - e^{2t} + 2, y(t) = e^t(1 - e^t)$

3. $x(t) = -e^{-t}(3te^t + 1),$
$y(t) = e^{-t}(1 - e^t + 2te^t)$

5. $x(t) = e^t(5e^{3t} - 2), y(t) = 4e^t(e^{3t} - 1)$

7. $x(t) = 2\cos t + 8\sin t, y(t) = \cos t - 13\sin t + \sinh t$

9. $x(t) = 2e^{-2t}(e^t + 1), y(t) = 3 + 2e^{-2t}(e^t - 4)$

11. $x(t) = -3e^t + 2, y(t) = -2e^t - 8$

13. (c) $x(t) = -4 + \frac{t^3}{3} + 5t - 2t^2 + 5e^{-t}$

Section 7.7

1. $\frac{t^3}{3}$

3. $\frac{1}{128}((4t8t^2 - 3) + 3\sin 4t)$

5. $t + tU_0(t)$

7. $f(t) = e^{-3t}(e^t - 1)$

9. $f(t) = \frac{2}{9}\sin^2(\frac{3t}{2})$

11. $f(t) = \frac{1}{54}(9t^2 - 6t - 2e^{-3t} + 2)$

17. (a) $\sin kt(\cos kt \sin t + \sin kt - \cos t \sin kt)$, (b) $\sin kt(-\cos kt + \cos t \cos kt + \sin t \sin kt)$,
(c) $\cos kt(\cos kt \sin t + \sin kt - \cos t \sin kt)$,
(d) $\frac{1}{2}t\sin kt$

Chapter 7 Review

1. False. See Theorem 7.1.3.

3. True. See Theorem 7.3.1.

5. True. Uniqueness of Laplace transforms.

7. $\frac{1}{s}$

9. $F(s) = \frac{48}{s^5}$

11. $F(s) = \frac{2}{(1+s)^3}$

13. $\frac{s}{s^2-b^2}$

15. $\frac{2bs}{(s^2+b^2)^2}$

17. $\frac{s}{(s^2+b^2)^2}$

19. $\frac{n!}{s^{n+1}}$

21. $1 + e^{-3}$

23. 1

25. $s = \pm 2 \Rightarrow$ solutions grow without bound; $\frac{7}{2}e^{-2t}(e^{4t}+1)$

27. $s = -3, -3 \Rightarrow$ solutions decay without oscillating; $5(1-3t)e^{-3t}$

29. $s = -4 \pm i \Rightarrow$ damped oscillations; $3e^{-4t}\sin t$

31. $H(s) = \frac{1}{s^2+5s+6} \Rightarrow$ poles $s = -2, -3 \Rightarrow$ solutions decay; no zeros.

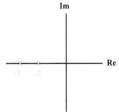

Problem **31**(b).

33. $H(s) = \frac{4s+8}{(s^2+4s+8)(s^2+1)} \Rightarrow$ poles $s = -2 \pm 2i, \pm i \Rightarrow$ solutions oscillate; e^{-2t} will have no effect.

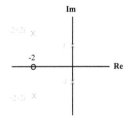

Problem **33**(b).

35. $x(t) = \frac{1}{3}e^t(e^{3t}+17)$

37. $x(t) = \frac{2}{9}\sin 9t + \cos 9t$

39. $x(t) = 3e^{-t}(\sin 2t + 2\cos 2t)$

41. $x(t) = \frac{-1}{9}e^{2t}(47e^{3t}-72e^t-15t-20)$

43. $x(t) = \frac{1}{180}e^{-2t}(-390te^{2t}+67e^{5t}+455e^{2t}-135e^t-207)$

45. $N'(t) = rN - (U_0 - U_1)\cdot 30 \cdot (1-t), N(0) = 20$. No, because the solution never reaches 0.

47. Follow steps in 7.2 Example 3.

49. (a) $I(t) = \frac{E_0}{R}(U_1 - U_2)(1 - e^{-Rt/L})$, (b) $I(t) = \frac{U_1-U_0}{L^2+R^2}e^{-Rt/L} \cdot (Le^{Rt/L}\cos t - Re^{Rt/L}\sin t - L$

53. $x(t) = \frac{-1}{63}e^{-10t}(77e^{9t}-144e^{7t}-59), y(t) = \frac{1}{315}e^{-10t}(756e^{10t}-385e^{9t}-1080e^{7t}-236)$

55. $x(t) = \frac{1}{2}(3-5e^{2t}), y(t) = -2(4+e^{2t})$

57. $\frac{3t^4}{4}$

59. $\frac{1}{64}(4t(8t^2-3)+3\sin 4t)$

61. $f(t) = e^{-4t}(e^t-1)$

63. $f(t) = \frac{1}{2}\sin^2 t$

65. $f(t) = \frac{1}{128}(e^{-8t}+\sin 8t-\cos 8t)$

Section 8.1

1. $R = 10, I = (-10, 10)$

3. $R = \infty, I = (-\infty, \infty)$

9. $|x| < 1$

Section 8.2

1. $y = 1 + x + \frac{x^2}{2} + \frac{2x^3}{3} + \frac{7x^4}{12} + \cdots$

3. $y = \frac{x^2}{2} + \frac{x^3}{6} + \frac{x^4}{6} + \cdots$

5. $y = 1 + x + x^2 + \cdots$

7. $y = x + x^2 + \frac{x^3}{2} + \cdots$

9. $y = 1 + x + \frac{x^3}{3} - \frac{x^4}{3} + \cdots$

11. $y = 1 + 2(x-1) + 4(x-1)^2 - \frac{25}{3}(x-1)^3 + \cdots$

13. $y = 1 - x - \frac{x^3}{3} + \frac{x^4}{12} - \cdots$

15. $y = c_0\left(1 - \frac{x^3}{6} + \frac{3x^5}{40} + \cdots\right)$
$+ c_1\left(x - \frac{x^3}{6} - \frac{x^4}{12} + \frac{3x^5}{40} + \cdots\right)$

17. $y = \frac{1}{4}(4e^{x^2} - x)$

Section 8.3

1. $x = 0$ and $x = 3$ are regular singular points

3. $x = -2$ is a regular singular point; $x = 0$ is an irregular singular point.

5. $x = \frac{-1}{2}$ is a regular singular point; $x = 0$ is an irregular singular point.

7. $T_5(x) = 5x - 20x^3 + 16x^5$
$T_6(x) = -1 + 18x^2 - 48x^4 + 32x^6$

Section 8.4

1. $y = c_1 \sum_{n=0}^{\infty} \frac{x^n}{(n+1)!}$

3. $y = 3c_1 x^{\frac{1}{2}} \sum_{n=0}^{\infty} \frac{2^{n+1}(n+1)}{(2n+3)!} x^n$
$+ c_2 x^{-1}\left(1 - \sum_{n=1}^{\infty} \frac{2^{n-1}(n-1)!}{(2n-2)!} x^n\right)$

5. $y_1(x) = c_0\left(-1 + \frac{1}{x}\right)$
$y_2(x) = c_0\left(x - \frac{x^2}{3} + \frac{x^3}{12} - \frac{x^4}{60}\right)$;
$y = c_1 y_1 + c_2 y_2$

7. $y_1(x) = c_0\left(\frac{1}{\sqrt{x}} - \frac{x^{3/2}}{2}\right)$,
$y_2(x) = c_0\left(x - \frac{x^3}{14}\right)$;
$y = c_1 y_1 + c_2 y_2$

9. $y_1(x) = c_1 \sum_{n=0}^{\infty} (-1)^n \frac{x^n}{(n!)^2}$,
$y = c_1 y_1(x) + c_2\left(y_1(x)\ln x - 2\sum_{n=2}^{\infty}(-1)^n \frac{1+\frac{1}{2}+\cdots+\frac{1}{n}}{(n!)^2} x^n\right)$

11. $y_1(x) = x + x^2$,
$y = c_1 y_1(x) + c_2\left(y_1(x)\ln x - 2x^2 - \sum_{n=2}^{\infty}(-1)^n \frac{x^{n+1}}{n(n-1)}\right)$

Section 8.5

1. Apply Ratio Test.

5. $y = \frac{u(x)}{\sqrt{x}} \Rightarrow y' = \frac{\sqrt{x}u' - (1/2)ux^{-1/2}}{x}$,
$y'' = \frac{x^{3/2}u'' - x^{1/2}u' + (3/4)ux^{-1/2}}{x^2}$;
substitute and simplify.
$y_1 = \sqrt{\cos x}\sqrt{x} \Rightarrow u = \cos x$; again,
substitute and simplify. Similarly,
$y_1 = \sqrt{\sin x}\sqrt{x} \Rightarrow u = \sin x$.

7. $x^2 \frac{d^2 y}{dx^2} + x\frac{dy}{dx} + x^2 y = 0$

Chapter 8 Review

1. False. See Theorem 8.1.1.

3. If you think this is true, then you have missed the point of this chapter.

5. False, check out Taylor's theorem.

9. $y = 3 + 12x + 12x^2 + 8x^3$
$+ 4x^4 + \cdots$

11. $y = 1 + 2x + x^2 + \frac{x^3}{6}$
$+ \frac{x^4}{24} + \cdots$

13. $y = x + x^2 + x^3 + \frac{7x^4}{12}$
$+ \frac{3x^5}{10} + \cdots$

15. $y = 2 + 16x + 256x^2 + \frac{14338}{3}x^3$
$+ \frac{286736}{3}x^4 + \cdots$

17. $y = 1 + 4(x - 1) + 7(x - 1)^2 +$
$\frac{20}{3}(x - 1)^3 + \frac{23}{6}(x - 1)^4 + \cdots$

19. $y = c_1 x^{\frac{4}{3}}\left(1 - \frac{3x^2}{16} + \frac{9x^4}{896} - \cdots\right)$
$+ c_2 x^{\frac{2}{3}}\left(1 - \frac{3x^2}{8} + \frac{9x^4}{320} - \cdots\right)$

20. $y = c_1 x^{\frac{1}{3}}\left(1 - \frac{3x^2}{16} + \frac{9x^4}{896} - \cdots\right)$
$+ c_2 x^{\frac{-1}{3}}\left(1 - \frac{3x^2}{8} + \frac{9x^4}{320} - \cdots\right)$

21. $y = c_1\left(1 + x + \frac{3x^2}{10} + \cdots\right)$
$+ c_2 x^{\frac{1}{3}}\left(1 + \frac{7x}{12} + \frac{5x^2}{36} - \cdots\right)$

Appendix B.1

1. $\begin{pmatrix} 3 & 5 \\ 1 & 1 \end{pmatrix}\begin{bmatrix} x \\ y \end{bmatrix} = \begin{bmatrix} 7 \\ -1 \end{bmatrix}$

3. $\begin{bmatrix} x' \\ y' \end{bmatrix} = \begin{bmatrix} 1 & -5 \\ 2 & 1 \end{bmatrix}\begin{pmatrix} x \\ y \end{pmatrix}$

4. $\mathbf{AB} = \begin{bmatrix} -7 & 19 \\ 6 & -22 \end{bmatrix}$

5. $\mathbf{AB} = \begin{bmatrix} 1 & -6 & -14 \\ 1 & 7 & 13 \\ 9 & -2 & 10 \end{bmatrix}$

$\mathbf{BA} = \begin{bmatrix} 9 & 8 & -13 \\ -16 & -3 & -3 \\ 6 & -2 & 12 \end{bmatrix}$

7. $\mathbf{D} - \mathbf{E} = \begin{bmatrix} 0 & 8 & -2 \\ -6 & -1 & 2 \\ 1 & 0 & 0 \end{bmatrix}$

9. $\begin{pmatrix} 6 & 0 \\ -2 & 4 \\ 2 & 2 \end{pmatrix}$

11. $\begin{pmatrix} 2 & -22 & 12 \\ 22 & 4 & -2 \\ 4 & 4 & 2 \end{pmatrix}$

13. $\mathrm{Tr}\mathbf{E} = 3$

15. Not possible because the matrices are different sizes.

17. $\begin{pmatrix} 1 & -1 & 4 \\ 5 & 0 & 2 \\ 2 & 3 & 1 \end{pmatrix}$

19. Not possible: Number of columns of A (2) does not equal the number of rows of A (3).

21. $\begin{bmatrix} 32 & 6 & 11 \\ 8 & 9 & -1 \\ 17 & -8 & 19 \end{bmatrix}$

23. $\mathbf{AB} = \begin{bmatrix} 12 & -3 \\ -4 & 5 \\ 4 & 1 \end{bmatrix}$

25. $\begin{bmatrix} 12 & -6 & 3 \\ 0 & 4 & 2 \end{bmatrix}$

27. Not possible: Number of columns of A (2) does not equal the number of rows of E (3).

29. $\begin{pmatrix} -6 & 39 & -27 \\ 26 & -1 & 13 \\ 10 & 19 & -7 \end{pmatrix}$

31. 55

33. 64

35. 1815

37. 55

Appendix B.2

1. Yes, in row-echelon form.

3. No, not in row-echelon form.

5. Yes, in row-echelon form.

7. $\begin{bmatrix} 2 & -3 & -2 \\ 2 & 1 & 1 \end{bmatrix}$

$\Rightarrow \begin{bmatrix} 1 & 0 & \frac{1}{8} \\ 0 & 1 & \frac{3}{4} \end{bmatrix}$

$\Rightarrow x_1 = \frac{1}{8}, x_2 = \frac{3}{4}$

9. $\begin{bmatrix} 2 & 2 & 2 & 0 \\ -2 & 5 & 2 & 1 \\ 8 & 1 & 4 & -1 \end{bmatrix}$

$\Rightarrow \begin{bmatrix} 1 & 1 & 1 & 0 \\ 0 & 7 & 4 & 1 \\ 0 & 0 & 0 & 0 \end{bmatrix}$

$\Rightarrow$ infinite solutions, $x_1 = -\frac{1}{7}(1 + 3x_3), x_2 = \frac{1}{7}(1 - 4x_3)$

11. $\begin{bmatrix} 1 & -2 & 4 & 2 \\ 2 & -3 & 5 & 3 \\ 3 & -4 & 7 & 7 \end{bmatrix}$

$\Rightarrow \begin{bmatrix} 1 & 0 & 0 & 6 \\ 0 & 1 & 0 & 8 \\ 0 & 0 & 1 & 3 \end{bmatrix}$

$\Rightarrow x = 6, y = 8, z = 3$

13. $x = \frac{1}{8}, y = \frac{3}{4}$

15. Determinant is zero, so no solution.

17. $x = \frac{3}{11}, y = \frac{2}{11}, z = -\frac{1}{11}$

19. $x = 1, y = 0, z = 2, w = 0$

21. $\begin{bmatrix} -1 & -2 \\ 1 & 1 \end{bmatrix}$

23. $\frac{1}{3}\begin{bmatrix} 7 & -2 \\ -2 & 1 \end{bmatrix}$

25. $\frac{1}{2}\begin{bmatrix} -32 & -22 & 6 \\ 7 & 5 & -1 \\ -5 & -3 & 1 \end{bmatrix}$

Section B.3

1. (a) $\begin{bmatrix} 3/2 \\ 9/2 \end{bmatrix}$; (b) $\begin{bmatrix} -3 \\ 1 \end{bmatrix}$;

(c) $\begin{bmatrix} 0 \\ 3 \end{bmatrix}$; (d) $\begin{bmatrix} -1 \\ 3 \end{bmatrix}$.

3. (a) $\begin{bmatrix} -2 \\ -1 \end{bmatrix}$; (b) $\begin{bmatrix} \sqrt{2} + \sqrt{2}/2 \\ -\sqrt{2} + \sqrt{2}/2 \end{bmatrix}$;

(c) $\begin{bmatrix} 2 \\ 1 \end{bmatrix}$; (d) $\begin{bmatrix} 2 \\ -1 \end{bmatrix}$.

5. (a) $\begin{bmatrix} 1/2 \\ -3/2 \end{bmatrix}$; (b) $\begin{bmatrix} 1/2 + 3\sqrt{2}/2 \\ \sqrt{3}/2 - 3/2 \end{bmatrix}$;

(c) $\begin{bmatrix} \frac{1-3\sqrt{3}}{4} \\ \frac{\sqrt{3}-9}{4} \end{bmatrix}$; (d) $\begin{bmatrix} \frac{-1-3\sqrt{3}}{2} \\ \frac{\sqrt{3}-3}{2} \end{bmatrix}$.

7. (a) $\begin{bmatrix} -5/2 \\ -5/2 \end{bmatrix}$; (b) $\begin{bmatrix} (1-\sqrt{3})/2 \\ (-3-\sqrt{3})/2 \end{bmatrix}$;

(c) $\begin{bmatrix} -1 \\ 0 \end{bmatrix}$; (d) $\begin{bmatrix} -1 \\ 1 \end{bmatrix}$.

9. (a) $\begin{bmatrix} 6 \\ 4 \end{bmatrix}$; (b) $\begin{bmatrix} -3 \\ -2 \end{bmatrix}$;

(c) $\begin{bmatrix} 0 \\ 2 \end{bmatrix}$; (d) $\begin{bmatrix} -3 \\ 2 \end{bmatrix}$.

13. Yes.

15. $T(\mathbf{0}) = T(0 \cdot \mathbf{u}) = 0 \cdot T(\mathbf{u}) = \mathbf{0}$

Section B.4

1. (a) $\frac{1}{11}\begin{bmatrix} 4 & 3 \\ -1 & 2 \end{bmatrix}$, (b) $\begin{bmatrix} 2 & -3 \\ 1 & 4 \end{bmatrix}$,

(c) $[w]_B = \begin{bmatrix} 3 \\ -5 \end{bmatrix}$,

(d) $[w]_{B'} = \begin{bmatrix} -\frac{3}{11} \\ \frac{13}{11} \end{bmatrix}$

3. (a) $\begin{bmatrix} \frac{3}{4} & \frac{3}{4} & 0 \\ -\frac{3}{4} & -\frac{17}{12} & -\frac{3}{2} \\ 0 & \frac{2}{3} & 1 \end{bmatrix}$,

(b) $[w]_B = \begin{bmatrix} -\frac{31}{9} \\ 5 \\ -\frac{1}{3} \end{bmatrix}$

5. Matrix of eigenvectors is $P = \begin{bmatrix} 0 & 1 \\ 0 & 0 \end{bmatrix}$, which is not invertible. Hence P^{-1} does not exist.

7. $P = \begin{bmatrix} \frac{3+i\sqrt{3}}{2} & \frac{3-i\sqrt{3}}{2} \\ 1 & 1 \end{bmatrix}$;

$P^{-1}AP = \begin{bmatrix} \frac{1+i\sqrt{3}}{2} & 0 \\ 0 & \frac{1-i\sqrt{3}}{2} \end{bmatrix}$

9. $P = \begin{bmatrix} -2 & 0 & 1 \\ 0 & 1 & 0 \\ 3 & 0 & 0 \end{bmatrix}$; $P^{-1}AP = \begin{bmatrix} 5 & 0 & 0 \\ 0 & 3 & 0 \\ 0 & 0 & 2 \end{bmatrix}$

References

[1] M.L. Abell and J.P. Braselton. *Modern Differential Equations: Theory, Applications, Technology*, Saunders College Publishing, Fort Worth, Texas, 1996.

[2] L. Almada, R. Rodriguez, M. Thompson, L. Voss, L. Smith, and E.T. Camacho. *Deterministic and Small-World Network Models of College Drinking Patterns*. Technical Report, Department of Mathematics & Statistics, California State Polytechnic University, Pomona, 2006 (<http://www.public.asu.edu/~etcamach/AMSSI/>).

[3] William A. Adkins and Steven H. Weintraub. *Algebra*, Springer Graduate Texts in Mathematics Vol. 136, corrected second printing 1999.

[4] B.W. Banks. *Differential Equations with Graphical and Numerical Methods*, Prentice Hall, Englewood Cliffs, New Jersey, 2001.

[5] E. Beltrami. *Mathematical Models in the Social and Biological Sciences*, Jones and Bartlett Publishers, Boston, 1993.

[6] C.M. Bender and S.A. Orszag. *Advanced Mathematical Methods for Scientists and Engineers*, McGraw-Hill, New York, 1978.

[7] R.L. Bewernick, J.D. Dewar, E. Gray, N.Y. Rodriguez, and R.J. Swift. *Population Processes*, Technical Report, Department of Mathematics & Statistics, California State Polytechnic University, Pomona, 2005 (<www.amssi.org>).

[8] W.E. Boyce and R.C. DiPrima. *Elementary Differential Equations*, John Wiley & Sons, New York, 4th ed., 1986.

[9] F. Brauer and C. Castillo-Chavez. *Mathematical Models in Population Biology and Epidemiology*, Springer, New York, 2001.

[10] R.L. Burden and J.D. Faires. *Numerical Analysis*, Brooks/Cole, Pacific Grove, 2001.

[11] E.T. Camacho. *Mathematical Models of Retinal Dynamics*, PhD thesis, Center for Applied Mathematics, Cornell University, 2003.

[12] E. Camacho, R. Rand, and H. Howland. Dynamics of two van der Pol oscillators coupled via a bath. *International Journal of Solids and Structures*, 41: 2133-2143, 2004.

[13] C. Corduneanu. *Principles of Differential and Integral Equations*, Chelsea Publishing Co., New York, 1977.

[14] J. Cannarella and J. Spechler. Epidemiological modeling of online social network dynamics, `http://arxiv.org/pdf/1401.4208v1.pdf`, 2014.

[15] D.J. Daley and J. Gani. *Epidemic Modelling*, Cambridge University Press, U.K., 1999.

[16] L. Edelstein-Keshet. *Mathematical Models in Biology*, Birkhäuser Mathematics Series. McGraw-Hill, New York, 1988.

[17] M.E. Gilpin. Do hares eat lynx? *Amer. Nat.*, 107:727–730, 1973.

[18] M.E. Gilpin and Francisco J. Ayala. Global models of growth and competition. *Proc. Natl. Acad. Sci. USA*, 70(12, Part I):3590–3593, 1973.

[19] C.A.S. Hall. An assessment of several of the historically most influential theoretical models used in ecology and of the data provided in their support. *Ecol. Model.*, 43:5–31, 1988.

[20] R. Haubrich. Frobenius, Schur, and the Berlin algebraic tradition. *Mathematics in Berlin*. Berlin, 83–96, 1998.

[21] H. Hethcote. The mathematics of infectious diseases. *SIAM Rev*, 42: 599–653, 2000.

[22] M. Hirsch and S. Smale. *Differential Equations, Dynamical Systems, and Linear Algebra*, Academic Press, London, 1974.

[23] K. Hoffman and R. Kunze. *Linear Algebra*, 2nd edition, Prentice Hall, 1971.

[24] J.H. Hubbard and B.H. West. *Differential Equations: A Dynamical Systems Approach, Part I, One Dimensional Equations*, Springer-Verlag, New York, 1990.

[25] W. Kermack and A. McKendrick. Contributions to the mathematical theory of epidemics—I. *Proc. R. Soc.*, 115A:700, 1927.

[26] I. Kyprianidis. Dynamics of a Nonlinear Electrical Oscillator Described by Duffing's Equation, http://www.math.upatras.gr/~crans.

[27] J. Guckenheimer and P. Holmes. *Nonlinear Oscillations, Dynamical Systems, and Bifurcations of Vector Fields*, Springer, New York, 1983.

[28] A.C. Lazer and P.J. McKenna. Large amplitude periodic oscillations in suspension bridges: some new connections with nonlinear analysis. *SIAM Rev.*, 32 (December):537–578, 1990.

[29] N.N. Lebedev. *Special Functions & Their Applications*, Dover, New York, 1972.

[30] H.E. Leipholz. *Stability Theory*, Academic Press, New York, 1970.

[31] G.N. Lewis. The Collapse of the Tacoma Narrows Suspension Bridge, in *Differential Equations with Boundary-Value Problems*, 5th ed, by D.G. Zill and M.R. Cullen, 2000.

[32] D.O. Lomen and D.Lovelock. *Differential Equations: Graphics, Models, Data*, Wiley, 1998.

[33] D. Ludwig, D.D. Jones, and C.S. Holling. Qualitative analysis of insect outbreak systems: the spruce budworm and forest. *J. Animal Ecol.*, 47:315–332, 1978.

[34] M. Martelli. *Introduction to Discrete Dynamical Systems and Chaos*, John Wiley & Sons, Inc., New York, 1999.

[35] R.E. Mickens. *Applications of Nonstandard Finite Difference Schemes*, World Scientific, Singapore, 2000.

[36] K.H. Lundberg, H.R. Miller, and D.L. Trumper. Initial Conditions, Generalized Functions, and the Laplace Transform: Troubles at the Origin. *Control Systems, IEEE*, 27: 22-35, 2007.

[37] D.D. Mooney and R.J. Swift. *A Course in Mathematical Modeling*, The Mathematical Association of America, Washington, D.C., 1999.

[38] C.R. Nave. *HyperPhysics website at Georgia State University*, http://hyperphysics.phy-astr.gsu.edu/hbase/hframe.html, 2000.

[39] G.M. Odell. Qualitative Theory of Systems of Ordinary Differential Equations, Including Phase Plane Analysis and the Use of the Hopf Bifurcation Theorem in *Mathematical Models in Molecular and Cellular Biology*, Cambridge University Press, Cambridge, 1980.

[40] M. Olinick. *An Introduction to Mathematical Models in the Social and Life Sciences*, Addison-Wesley Publishing Company, Reading, Massachusetts, 1978.

[41] Alan S. Perelson and Patrick W. Nelson. Mathematical Analysis of HIV-I: Dynamics in Vivo. *SIAM Review*, 41(1): 3-44, 1999.

[42] L.F. Richardson. Generalized foreign policy, *Br. J. Psychol. Monogr. Suppl.*, 23:130–148, 1939.

[43] C.C. Ross. *Differential Equations: An Introduction with Mathematica*, Springer-Verlag, New York, 1995.

[44] J.T. Sandefur. *Discrete Dynamical Systems: Theory and Applications*, Clarendon Press, Oxford, 1990.

[45] H.M. James, N.B. Nichols, and R.S. Phillips. *Theory of Servomechanisms*, New York, McGraw-Hill, 1947.

[46] T.Kanamaru. van der Pol Oscillator, Scholarpedia, 2(1):2202, 2007. http://www.scholarpedia.org/article/Van_der_Pol_oscillator

[47] G. Strang. *Linear Algebra and Its Applications*, Harcourt Brace Jovanovich Publishers, San Diego, 1988.

[48] S.H. Strogatz. *Nonlinear Dynamics and Chaos, with Applications to Physics, Biology, Chemistry, and Engineering*, Addison-Wesley Publishing Company, Reading, Massachusetts, 1994.

[49] K.R. Symon. *Mechanics*, Addison-Wesley Publishing Company, Reading, Massachusetts, 1971.

[50] S. Utida. Cyclic fluctuations of population density intrinsic to the host-parasite system. *Ecology*, 38:442–449, 1957.

[51] G.N. Watson. *A Treatise on the Theory of Bessel Functions*, Cambridge at the University Press, London, second edition, 1966.

[52] S.H. Weintraub. *Jordan Canonical Form: Application to Differential Equations*, Morgan & Claypool Publishers, 2008.

Index